Lecture Notes in Networks and Systems

Volume 140

The series "Lecture Notes in Networks and Systems" publishes the latest developments in Networks and Systems—quickly, informally and with high quality. Original research reported in proceedings and post-proceedings represents the core of LNNS.

Volumes published in LNNS embrace all aspects and subfields of, as well as new challenges in, Networks and Systems.

The series contains proceedings and edited volumes in systems and networks, spanning the areas of Cyber-Physical Systems, Autonomous Systems, Sensor Networks, Control Systems, Energy Systems, Automotive Systems, Biological Systems, Vehicular Networking and Connected Vehicles, Aerospace Systems, Automation, Manufacturing, Smart Grids, Nonlinear Systems, Power Systems, Robotics, Social Systems, Economic Systems and other. Of particular value to both the contributors and the readership are the short publication timeframe and the world-wide distribution and exposure which enable both a wide and rapid dissemination of research output.

The series covers the theory, applications, and perspectives on the state of the art and future developments relevant to systems and networks, decision making, control, complex processes and related areas, as embedded in the fields of interdisciplinary and applied sciences, engineering, computer science, physics, economics, social, and life sciences, as well as the paradigms and methodologies behind them.

**** Indexing: The books of this series are submitted to ISI Proceedings, SCOPUS, Google Scholar and Springerlink ****

More information about this series at http://www.springer.com/series/15179

Nikhil Marriwala · C. C. Tripathi ·
Dinesh Kumar · Shruti Jain
Editors

Mobile Radio Communications and 5G Networks

Proceedings of MRCN 2020

Springer

Editors
Nikhil Marriwala
Department of Electronics and Communication Engineering
University Institute of Engineering and Technology (UIET)
Kurukshetra University
Kurukshetra, Haryana, India

C. C. Tripathi
University Institute of Engineering and Technology (UIET)
Kurukshetra University
Kurukshetra, Haryana, India

Dinesh Kumar
Department of Electrical and Computer System Engineering
RMIT University
Melbourne, VIC, Australia

Shruti Jain
Department of Electronics and Communication Engineering
Jaypee University of Information Technology
Waknaghat, Himachal Pradesh, India

ISSN 2367-3370 ISSN 2367-3389 (electronic)
Lecture Notes in Networks and Systems
ISBN 978-981-15-7129-9 ISBN 978-981-15-7130-5 (eBook)
https://doi.org/10.1007/978-981-15-7130-5

This Springer imprint is published by the registered company Springer Nature Singapore Pte Ltd.
The registered company address is: 152 Beach Road, #21-01/04 Gateway East, Singapore 189721, Singapore

Preface

This conference provides a platform and aid to the researches involved in designing systems that permit the societal acceptance of ambient intelligence. The overall goal of this conference is to present the latest snapshot of the ongoing research as well as to shed further light on future directions in this space. This conference aims to serve for industry research professionals who are currently working in the field of academia research and research industry to improve the lifespan of the general public in the area of recent advances and upcoming technologies utilizing cellular systems, 2G/2.5G/3G/4G/5G and beyond, LTE, WiMAX, WMAN, and other emerging broadband wireless networks, WLAN, WPAN, and other homes/personal networking technologies, pervasive and wearable computing and networking, small cells and femtocell networks, wireless mesh networks, vehicular wireless networks, cognitive radio networks and their applications, wireless multimedia networks, green wireless networks, standardization activities of emerging wireless technologies power management, signal processing, and energy conservation techniques. This conference will provide support to the researchers involved in designing decision support systems that will permit the societal acceptance of ambient intelligence. It presents the latest research being conducted on diverse topics in intelligent technologies to advance knowledge and applications in this rapidly evolving field. The conference is seen as a turning point in developing the quality of human life and performance in the future; therefore, it has been identified as the theme of the conference. Authors were invited to submit papers presenting novel technical studies as well as position and vision papers comprising hypothetical/speculative scenarios.

5G technology is a truly revolutionary paradigm shift, enabling multimedia communications between people and devices from any location. It also underpins exciting applications such as sensor networks, smart homes, telemedicine, and automated highways. This book will provide a comprehensive introduction to the underlying theory, design techniques, and analytical tools of 5G and wireless communications, focusing primarily on the core principles of wireless system design. The book will begin with an overview of wireless systems and standards. The characteristics of the wireless channel are then described, including their fundamental capacity limits. Various modulation, coding, and signal processing schemes are then discussed in detail, including the state-of-the-art adaptive modulation, multicarrier, spread spectrum, and multiple antenna techniques. The book will be a valuable reference for engineers in the wireless industry. This book will be extremely valuable not only to graduate students pursuing research in wireless systems but also to engineering professionals who have the task of designing and developing future 5G wireless technologies.

For the proper review of each manuscript, every received manuscript was first checked for plagiarism and then the manuscript was sent to three reviewers. In this process, the committee members were involved, and the whole process was monitored and coordinated by the general chair. The technical program committee involved senior academicians and researchers from various reputed institutes. The members were from India as well as abroad. The technical program mainly involves the review of the paper. A total of **273** research papers were received, out of which 64 papers were accepted and registered and presented during the three-day conference, and the acceptance ratio is **22.3%**.

An overwhelming response was received from the researchers, academicians, and industry from all over the globe. The papers were from **Malaysia**, **Nigeria**, **Iran**, making it truly International and Pan India from **Guwahati**, **Kerala**, **Tamil Nadu**, **Raipur**, **Pune**, **Hyderabad**, **Rajasthan**, **Uttarakhand**, **Raipur**, **Ranchi** and not to mention our neighbouring states. The authors are from premium institutes, IITs, NITs, Central Universities, NSIT, PU, and many other reputed institutes.

Organizers of MRCN 2020 are thankful to the University Institute of Engineering and Technology (UIET) which was established by Kurukshetra University in 2004 to develop as a "Centre of Excellence" and offer quality technical education and to undertake research in engineering and technology. The ultimate aim of the UIET is to become a role model for engineering and technology education not only for the State of Haryana but for the world over to meet the challenges of the twenty-first century.

The editors would like to express their sincere gratitude to general chairs, plenary speakers, invited speakers, reviewers, technical programme committee members, international advisory committee members, and local organizing committee members of MRCN-2020; without whose support, the quality and standards of the conference could not be maintained. Special thanks to Springer and its team for this valuable publication. Over and above, we would like to express our deepest sense of gratitude to UIET, Kurukshetra University, Kurukshetra, for hosting this conference. **We are thankful to the Technical Education Quality Improvement Programme (TEQP-III) for sponsoring the International Conference MRCN-2020 event**.

Kurukshetra, India
Kurukshetra, India
Melbourne, Australia
Waknaghat, India

Nikhil Marriwala
C. C. Tripathi
Dinesh Kumar
Shruti Jain

Contents

Editors and Contributors

About the Editors

Dr. Nikhil Marriwala is working as Assistant Professor and Head of the Department Electronics and Communication Engineering Department at University Institute of Engineering and Technology, Kurukshetra University, Kurukshetra. He did his PhD. from NIT, Kurukshetra in the department of Electronics and Communication Engineering. He did his post-graduation in Electronics and Communication Engineering from IASE University, Sardarshahar, Rajasthan and did his B-Tech in Electronics and Instrumentation from MMEC, Mullana affiliated to Kurukshetra University, Kurukshetra. He has more than 18 years of experience in teaching graduate and postgraduate students. More than 31 students have completed their M-Tech dissertation under his guidance. His areas of interests are Software Defied Radios, Cognitive Radios, Soft Computing, Wireless Communications, Wireless Sensor Networks, Fuzzy system design, and Advanced Microprocessors. He has published more than 5 book chapters in different International books, has authored more than 10-books with Pearson, Wiley, etc. and has more than 30 publications to his credit in reputed International Journals in SCI/Scopus/Web of Science Journals and 20 papers in International/National conferences. He also has 02 patents published to his credit. He has been Chairman of Special Sessions in more than 5 International/National Conferences and has delivered a keynote address at more than 2 International conferences. He is having additional charge of Training and Placement Office, UIET, Kurukshetra University, Kurukshetra and heading the T&P cell for more than 10 years now. He has conducted more than 10 Faculty Development Programs in the emerging areas of IoT, AI, AR/VR, Robotics and Blockchain and has been organizing secretary of more than two International Conferences. He is a reviewer for many reputed journals such as the IEEE Access, International Journal of Communication Systems-Wiley, IEEE Signal Processing Letters, Journal of Organizational and End User Computing (JOEUC), IJMTIE, IJICST, Egyptian Informatics Journal – Elsevier etc. He was awarded by 'Career Guru of the Month' award by Aspiring Minds.

Prof. C. C. Tripathi did his Ph.D. in Electronics from Kurukshetra University, Kurukshetra followed by Master of Engineering in Microelectronics from BITS Pilani. Since 2016, he has been working as a Director, University Institute of Engineering Technology (an autonomous institute), Kurukshetra University, Kurukshetra. He is heading the institute comprising of more than 70 faculty members, above 100 non-teaching technical, laboratory & administrative staff and above 1600 students. As a Director, he is also heading institute academic bodies like board of studies, academic council with four UG, eight PG programs in engineering and spearheading research in various engineering and applied sciences departments in the institute. His specialization is in Microelectronics, RF MEMS for Communication, Entrepreneurship and Industrial Consultancy etc. He has developed Micro-fabrication R&D Lab and RF MEMS R&D lab at the institute. He is a member of more than 14 Professional Bodies and Senior Member IEEE Society. He has published more than 70 papers in Scopus / Web of Science Journals and 33 in National / International conferences. He has guided 07 numbers of Ph.D. students and 44 number M.Tech/ M.E. student's thesis. He has also filled 4 Indian patents and developed more than 20 electro-optical products as an import substitute. He has been pivotal in implementing technical education quality improvement program (TEQIP-II) grants amounting Rs.10.00 Crores by preparing Institution Development Plan (IDP) during 2011-2015. Presently, He is also project Director TEQIP-III amounting grant of Rs.7.00 Crores (2016-2020). He has conducted more than 20 Faculty Development Programs in the area of Microelectronics, Flexible Electronics, Microwave & RF MEMS devices, Entrepreneurship development etc. He has played a key role in successfully organizing two National Conferences on the topic such as Converging Technology Beyond 2020.

Prof. Dinesh Kumar B.Tech from IIT Madras, and PhD from IIT Delhi, is a Professor at RMIT University, Melbourne, Australia. He has published over 400 papers, authored 5 books and is on a range of Australian and international committees for Biomedical Engineering. His passion is for affordable- diagnostics and making a difference for his students. His work has been cited over 5600 times and he has also had multiple successes with technology translation. He is the member of Therapeutics Goods Administration (TGA), Ministry of Health (Australia) for medical devices. He is also on the editorial boards for IEEE Transactions of Neural Systems and Rehabilitation Engineering and Biomedical Signals and Controls. He has been the chair of large number of conferences and given over 50 key-notes speeches.

Dr. Shruti Jain is Associate Professor in the Department of Electronics and Communication Engineering at Jaypee University of Information Technology, Waknaghat, H.P, India and has received her Ph.D. in Biomedical Image Processing. She has a teaching experience of around 15 years. Her research interests are Image and Signal Processing, Soft Computing, Bio-inspired Computing and Computer-Aided Design of FPGA and VLSI circuits. She has published more than 10 book chapters, 60 papers in reputed journals, and 40 papers in International conferences. She has also published five books. She is a senior member of IEEE, life member and Editor in Chief of Biomedical Engineering Society of India and a member of IAENG. She has completed one externally funded project and one in the pipeline. She has guided 01 Ph.D. student and now has 06 registered students. She is a member of the Editorial Board of many reputed journals. She is also a reviewer of many journals and a member of TPC of different conferences. She was awarded by Nation Builder Award in 2018-19.

Contributors

Abhay Agarwal Panipat Institute of Engineering & Technology, Samalkha, Panipat, India

Ankita Aggarwal Department of Electronics and Communication Engineering, Punjabi University, Patiala, India;
CEC, Mohali, India

Nivedita Agnihotri Department of Chemistry, Maharishi Markandeshwar (Deemed to be University), Mullana, Ambala, Haryana, India

Rajesh Agnihotri Department of Applied Science, UIET, Kurukshetra University, Kurukshetra, Haryana, India

Darpan Anand Computer Science and Engineering, Chandigarh University, S.A. S. Nagar, Punjab, India

Mamta Arora Department of Computer Science and Engineering, University Institute of Engineering and Technology (U.I.E.T), Kurukshetra University, Kurukshetra, Haryana, India;
Department of Computer Science and Technology, Manav Rachna University, Faridabad, Haryana, India

Shakti Arora Panipat Institute of Engineering & Technology, Samalkha, Panipat, India

Muzhgan Athaee Department of Computer Application, National Institute of Technology Kurukshetra, Kurukshetra, India

Vijay Anant Athavale Panipat Institute of Engineering & Technology, Samalkha, Panipat, Haryana, India

Aashi Bansal Department of Computer Engineering & Applications, GLA University, Mathura, UP, India

Achint Bansal Department of Computer Engineering & Applications, GLA University, Mathura, UP, India

Ankit Bansal Gulzar Group of Institutes, Ludhiana, Punjab, India

Harsh Bansal Department of Electronics and Communication Engineering, Amity School of Engineering and Technology (ASET), Guru Gobind Singh Indraprastha University (GGSIPU), New Delhi, India

Amit Batra Department of CSE, University Institute of Engineering and Technology, Kurukshetra University, Kurukshetra, India

Urmila Berar Department of Applied Science, UIET, Kurukshetra University, Kurukshetra, Haryana, India

Shikha Bhardwaj ECE Department, DCRUST, Sonepat, India;
ECE Department, UIET, Kurukshetra University, Kurukshetra, Haryana, India

Lakshya Bhardwaj School of Computer Science, University of Petroleum and Energy Studies, Dehradun, India

Meena Chaudhary Manav Rachna University, Faridabad, Haryana, India

Poonam Chaudhary Department of Computer Science and Engineering, The NorthCap University, Gurgaon, India

Deepika Chauhan Shivajirao Kadam Institute of Technology and Management, Indore, MP, India

Vineeta Choudhary Ujjain Engineering College, Ujjain, Madhya Pradesh, India

Pawan Kumar Dahiya ECE Department, DCRUST, Sonepat, Haryana, India

Twinkle Dalal J.C. Bose University of Science and Technology, Faridabad, India

Mayank Dave Department of Computer Engineering, National Institute of Technology, Kurukshetra, India

Reeta Devi Kurukshetra University, Kurukshetra, Haryana, India

Himani Dhakla Department of Electrical Engineering U.I.E.T, Kurukshetra University, Kurukshetra, India

Sanjeev Dhawan Faculty of Computer Science and Engineering, Department of Computer Science and Engineering, University Institute of Engineering and Technology (U.I.E.T), Kurukshetra University, Kurukshetra, Haryana, India

Chetna Dhonchak Department of Chemistry, Maharishi Markandeshwar (Deemed to be University), Mullana, Ambala, Haryana, India

Divya University Institute of Engineering and Technology, Kurukshetra University, Kurukshetra, Haryana, India

Akshay Kumar Dogra Department of Electronics and Communication Engineering, Dr. B.R. Ambedkar National Institute of Technology, Jalandhar, India

Vidya Kant Dwivedi Bansal College of Engineering, Mandideep, India

Irraivan Elamvazuthi Department of Electrical and Electronics Engineering, Universiti Teknologi PETRONAS, Perak, Malaysia

Sarika Gambhir University Institute of Engineering and Technology, Kurukshetra University, Kurukshetra, India

Jatinder Garg Baba Hira Singh Bhattal Institute of Engineering and Technology, Lehragaga, Punjab, India

Sonu Bala Garg IK Gujral Punjab Technical University, Jalandhar, Punjab, India

Vijay Kumar Garg Department of Electrical Engineering U.I.E.T, Kurukshetra University, Kurukshetra, India

Vimal Gaur Maharaja Surajmal Institute of Technology, New Delhi, India

Vinay Gautam Chitkara University Institute of Engineering and Technology, Chitkara University, Rajpura, Punjab, India

Shilpy Ghai Department of Computer Science and Engineering, MMEC, Maharishi Markandeshwar Deemed to be University, Mullana, Ambala, India

Savita Gill Department of Applied Science, University Institute of Engineering and Technology, Kurukshetra, India

Krishan Gopal Department of Electronics and Communication Engineering, University Institute of Engineering and Technology, Kurukshetra University, Kurukshetra, Haryana, India

Sumeet Goyal Chandigarh Group of Colleges, Landran, Mohali, Punjab, India

Jyoti Grover Malaviya National Institute of Technology, Jaipur, India

Anuj Kumar Gupta Department of Computer Science and Engineering, Chandigarh Group of Colleges, Landran, Mohali, Punjab, India

Lakshay Gupta Department of Electronics and Communication Engineering, Amity School of Engineering and Technology (ASET), Guru Gobind Singh Indraprastha University (GGSIPU), New Delhi, India

Monish Gupta Department of Electronics and Communication Engineering, University Institute of Engineering and Technology, Kurukshetra University, Kurukshetra, Haryana, India

Muskan Gupta MSM Institute of Ayurveda, BPSMV, Sonipat, India

Vishal Gupta University Institute of Engineering & Technology, Kurukshetra University, Kurukshetra, India

Himanshu Maggu Panipat Institute of Engineering & Technology, Samalkha, Panipat, India

Anurag Jain Department of Virtualization, School of Computer Science, University of Petroleum and Energy Studies, Dehradun, India

Deepika Jain ECE, IKGPTU Jalandhar, Punjab, India

Shruti Jain JUIT, Solan, Himachal Pradesh, India

Ajay Jangra Maharishi Markandeshwar (Deemed To Be University) MMDU, Mullana, Ambala, India;
University Institute of Engineering and Technology, Kurukshetra University, Kurukshetra, India

Sahil Jindal University Institute of Engineering and Technology, Kurukshetra University, Kurukshetra, India

Akanksha Joshi Department of Applied Sciences, Limerick Institute of Technology, Limerick, Ireland

Nidhi Kabta Computer Science and Engineering, Chandigarh University, S.A.S. Nagar, Punjab, India

Anshul Kalia CT University, Ludhiana, India

Vaishali Kalra Department of Computer Science and Engineering, The NorthCap University, Gurgaon, India

Nikhil Kalwit Swami Vivekanand College of Engineering, Indore, MP, India

Anjali Karhana Computer Science and Engineering, Chandigarh University, S.A. S. Nagar, Punjab, India

Amandeep Kaur Department of CSE, M.M. University, Sadopur, Ambala, Haryana, India

Bikram Pal Kaur IT Department, Chandigarh Engineering College, Mohali, Punjab, India

Gurmeet Kaur Department of Electronics and Communication Engineering, Punjabi University, Patiala, India

Gurpreet Kaur CT University, Ludhiana, India

Harleen Kaur Department of Electronics, Punjab Technical University, Jalandhar, India

Iqbal Kaur Department of Basic and Applied Sciences, Punjabi University, Patiala, Punjab, India

Navneet Kaur Department of Chemistry, Maharishi Markandeshwar (Deemed to be University), Mullana, Ambala, Haryana, India

Pravleen Kaur School of Computer Science, University of Petroleum and Energy Studies, Dehradun, India

Vishal Kharbanda Department of Electronics and Communication Engineering, Amity School of Engineering and Technology (ASET), Guru Gobind Singh Indraprastha University (GGSIPU), New Delhi, India

Ankita Khot Department of Computer Engineering, National Institute of Technology, Kurukshetra, India

Dinesh Kumar Kurukshetra University, Kurukshetra, Haryana, India;
J C Bose University of Science and Technology, YMCA, Faridabad, Haryana, India

Rajiv Kumar Department of Chemistry, DCRUST, Murthal, Sonepat, Haryana, India

Rajneesh Kumar Department of Computer Science and Engineering, MMEC, Maharishi Markandeshwar Deemed to be University, Mullana, Ambala, India

Vijay Kumar Department of Computer Science and Engineering, MMEC, Maharishi Markandeshwar Deemed to be University, Mullana, Ambala, India

Meet Kumari Punjabi University, Patiala, Punjab, India

Parveen Lata Department of Basic and Applied Sciences, Punjabi University, Patiala, Punjab, India

Shilpa Mahajan The NorthCap University, Gurugram, India

Ritvik Maheshwari Amity School of Engineering and Technology, Amity University, Lucknow, India

Kamal Malik CT University, Ludhiana, India

Manisha Malik CT University, Ludhiana, India

Leeladhar Malviya Shri G. S. Institute of Technology and Science, Indore, M.P, India

Neeraj Mangla CSE Department, M. M. Engineering College, Maharishi Markandeshwar (Deemed to be University) (MMDU), Mullana, Ambala, India

Nikhil Marriwala Department of Electronics and Communication Engineering, University Institute of Engineering and Technology, Kurukshetra University, Kurukshetra, India

Abhishek Mishra Department of Computer Engineering & Applications, GLA University, Mathura, UP, India

Sumita Mishra Amity School of Engineering and Technology, Amity University, Lucknow, India

Piyush Moghe Swami Vivekanand College of Engineering, Indore, MP, India

Sonia Nain Department of Chemistry, DCRUST, Murthal, Sonepat, Haryana, India

Rajit Nair Jagran Lakecity University, Bhopal, India

Perumal Nallagownden Faculty of Engineering, Multimedia University, Cyberjaya, Selangor, Malaysia

Dhiraj Nitnaware Institute of Engineering & Technology, Devi Ahilya University, Indore, M.P, India

Gitanjali Pandove ECE Department, DCRUST, Sonepat, Haryana, India

Mohit Pant Ujjain Engineering College, Ujjain, India

Rashmi Pant Ujjain Engineering College, Ujjain, Madhya Pradesh, India

Ruchi Pasricha Chandigarh Group of Colleges, Mohali, Punjab, India

Piyush Pawar Swami Vivekanand College of Engineering, Indore, MP, India

Chandra Prakash School of Computing, DIT University, Dehradun, Uttarakhand, India

T. Ganga Prasad ECE, VFSTR, Vignan University, Guntur, Andhra Pradesh, India

Priyanka Department of Applied Science, University Institute of Engineering and Technology, Kurukshetra, India

Jyoti Pruthi Manav Rachna University, Faridabad, Haryana, India

Gobbi Ramasamy Smart Assistive and Rehabilitative Technology (SMART) Research Group, Seri Iskandar, Malaysia

Arun Kumar Rana Maharishi Markandeshwar (Deemed to be University), Mullana, India

Virender Ranga Department of Computer Engineering, National Institute of Technology Kurukshetra, Kurukshetra, Haryana, India

Priyanka Rani IK Gujral Punjab Technical University, Jalandhar, Punjab, India

Rekha Rani Department of Computer Science, Guru Nanak College, Budhlada, Punjab, India

MSS. Rukmini ECE, VFSTR, Vignan University, Guntur, Andhra Pradesh, India

Rupali M. M. Engineering College, Maharishi Markandeshwar (Deemed to be University) (MMDU), Mullana, Ambala, India

Abdulrashid Sabo Sule Lamido University Jigawa, Kano, Nigeria

B. S. Saini Department of Electronics and Communication Engineering, Dr. B.R. Ambedkar National Institute of Technology, Jalandhar, India

Indu Saini Department of Electronics and Communication Engineering, Dr. B.R. Ambedkar National Institute of Technology, Jalandhar, India

Rakesh Kumar Saini School of Computing, DIT University, Dehradun, Uttarakhand, India

Ayodeji Olalekan Salau Department of Electrical/Electronics and Computer Engineering, Afe Babalola University, Ado-Ekiti, Nigeria

Archit Sharma University Institute of Engineering and Technology, Kurukshetra University, Kurukshetra, India

Ashish Sharma Department of Computer Engineering and Applications, GLA University, Mathura, UP, India

Deepti Sharma Department of Computer Science and Engineering, Maharishi Markandeshwar (Deemed to be University), Mullana, Ambala, India

Manvinder Sharma Jaipur National University, Jaipur, India;
Chandigarh Group of Colleges, Landran, Mohali, Punjab, India

Nisha Sharma The NorthCap University, Gurugram, India

Preeti Sharma Bansal College of Engineering, Mandideep, India

Reecha Sharma Punjabi University, Patiala, Punjab, India

Sharad Sharma Maharishi Markandeshwar (Deemed to be University), Mullana, India

Srishti Sharma Department of Computer Science and Engineering, The NorthCap University, Gurgaon, India

Sudhir Sharma Department of Electrical Engineering DAVIET, Punjab Technical University, Jalandhar, India

Anu Sheetal GNDU Regional Campus Gurdaspur, Punjab, India

Chaitanya Singh Chameli Devi Group of Institutions, Indore, MP, India

Harmeet Singh CT University, Ludhiana, India

Kulvinder Singh Faculty of Computer Science and Engineering, Department of Computer Science and Engineering, University Institute of Engineering and Technology (U.I.E.T), Kurukshetra University, Kurukshetra, Haryana, India

Kulwant Singh Sant Longowal Institute of Engineering and Technology, Longowal, Punjab, India

Manjeet Singh J.C. Bose University of Science and Technology, Faridabad, India

Narinder Singh Department of Computer Science, Guru Nanak College, Budhlada, Punjab, India

Niharika Singh Smart Assistive and Rehabilitative Technology (SMART) Research Group, Seri Iskandar, Malaysia

Parul Singh Department of Computer Engineering, National Institute of Technology Kurukshetra, Kurukshetra, Haryana, India

Sohni Singh Chandigarh Group of Colleges, Landran, Mohali, Punjab, India

Sakshi Singhal Department of Computer Engineering, National Institute of Technology Kurukshetra, Kurukshetra, Haryana, India

Bhim Sain Singla College of Engineering and Management, Punjabi University, Patiala, Punjab, India

Deepak Sood Department of Electronics and Communication Engineering, University Institute of Engineering and Technology, Kurukshetra University, Kurukshetra, Haryana, India

Soujanya Computer Science and Engineering, Chandigarh University, S.A.S. Nagar, Punjab, India

Shailu Srivastava JUIT, Solan, Himachal Pradesh, India

Tanima Department of Electronics and Communication Engineering, Dr. B.R. Ambedkar National Institute of Technology, Jalandhar, India

Rohit Tanwar School of Computer Science, University of Petroleum and Energy Studies, Dehradun, India

Neeraj Thakur Computer Science and Engineering, Chandigarh University, S.A. S. Nagar, Punjab, India

Kshitij Tripathi Department of Computer Engineering and Applications, GLA University, Mathura, UP, India

Hitender Kumar Tyagi Kurukshetra University, Kurukshetra, Haryana, India

Rohit Vaid Department of Computer Science and Engineering, MMEC, Maharishi Markandeshwar Deemed to be University, Mullana, Ambala, India

Rashmi Vashisth Department of Electronics and Communication Engineering, Amity School of Engineering and Technology (ASET), Guru Gobind Singh Indraprastha University (GGSIPU), New Delhi, India

Rahul Verma Department of Electronics and Communication Engineering, Amity School of Engineering and Technology (ASET), Guru Gobind Singh Indraprastha University (GGSIPU), New Delhi, India

Rohit Yadav Swami Vivekanad College of Engineering, Indore, M.P, India

Analysis and Application of Vehicular Ad hoc Network as Intelligent Transportation System

Vinay Gautam

Abstract Intelligent transportation systems (ITSs) play a major role to manage traffic in cities. This is used to keep control and manage traffic and re-route traffic based on different parameters which has been discussed in this paper in a detailed manner. Vehicular ad hoc network is one way of implementation of ITS and mostly used setting to manage traffic and progressing quickly with time. Individuals are doing research these days for the most part in the field of media transmission. VANET is the most developing exploration region in remote correspondence. Most VANET applications are based upon the information push correspondence model, where data is scattered to a lot of vehicles. The decent variety of the VANET applications and their potential correspondence conventions needs a precise writing review. In perspective on previously mentioned, in this paper, we have contemplated and examined the attributes and difficulties of different research works identified with the applications, conventions and security in VANET. In addition to the subsequent current works, this paper is concerned about to explore different issues related to VANET. The conceivable work found the advantages and disadvantages for the future research. At last, an unthinkable examination of the considerable number of conventions is given.

Keywords Intelligent transportation system · VANET · Convolutional neural network (CNN)

1 Introduction

An intelligent transportation system (ITS) is an innovative system which plays a major role to manage transportation using innovative services and facilitate safe, smarter and better coordination among users. Vehicular ad hoc network (VANET) is a mostly used and successful ITS setting which is adopted by most of the country to

V. Gautam (✉)
Chitkara University Institute of Engineering and Technology, Chitkara University, Rajpura, Punjab, India
e-mail: vinay.gautam@chitkara.edu.in

N. Marriwala et al. (eds.), *Mobile Radio Communications and 5G Networks*, Lecture Notes in Networks and Systems 140,
https://doi.org/10.1007/978-981-15-7130-5_1

manage traffic. VANET uses different settings to communicate such as the vehicle to vehicle and road-side unit (RSU) to automobile in a short range. The main aim of deploying VANET is to overcome the accident issues. It has a large variety of application for drivers to drive well on the roads in the urban region. The rate of accidents is increased day by day by increasing the population of vehicles as well; therefore, it is vital for the vehicles to impart the VANET system. Vehicular specially appointed systems have invigorated enthusiasm for both scholastic and industry settings in light of the fact that, once sent, they would carry another driving experience to drivers. The security and protection is a genuine test where one node communicates with another node in open environment and influence the message passing in VANETs. This paper starts with providing summaries of the existing research works and later on classifying gaps identified in the pre-existing works. On the other hand, privacy safeguarding strategies are checked on, and the exchange off among security and protection is talked about. This paper concerned a point of view toward how to recognize and repudiate malevolent hubs all the more productively and difficulties that have yet been unrevealed. The benefits of such a system extend on operations, management, planning and monitoring.

Management strategies for traffic incidents have become a promising area of research because it has pronounced prospective to mend automobile and road security and transportation efficiency. Recent research works have put great emphasis on VANETs. Also, we have seen different technologies like fog computing and crowdsensing in social Internet of vehicles to provide a timely response for device-to-device-enabled real-time traffic management. At long last, we present a portion of the exploration challenges that still should be tended to for across the board selection of adaptable, dependable, hearty and secure car crash the executive's models, conventions, advancements and services.

The complete paper is further divided into different sections. Section 2 covers a detailed review of intelligent Transport Systems. The comparative study is laid down in Sect. 3. The research challenges are explained in Sect. 4. Section 5 is conclusive section.

2 Related Works

Al-Mayouf et al. [1] presented an accident management system utilizing specially appointed vehicle systems combined with open transport cell innovation. All ambulances and vehicles were accepted to have an installed route instrument and the capacity to utilize remote correspondence. In this framework, a multi-hop ideal sending calculation was utilized by the continuous course intending to diminish traffic blockage. The outcome presumed that this framework gave the best course among source and goal.

Wu et al. [2] proposed a vehicle-to-roadside (V2R) arrangement based distributed clustering which is accomplished by conditional game. Here, the author uses matrices-based fuzzy algorithm to generate a cluster of vehicles. A virtualization plot for multi-bounce information conveyance was incorporated with the steering convention to guarantee high throughput and low postponement in multi-jump misfortune vehicular conditions.

Bichiou and Rakha [3] propose a optimal solution for the possible situation faced by moving Vehicle. The proposed calculation beats the other convergence control techniques by creating lower the fuel consumption and decrease in carbon dioxide emission.

Ke et al. [4] proposed a financially savvy strategy for the discovery of traffic state status and accomplished exact identification of moving articles in a dynamic video with convolutional neural system classifier. The test results demonstrated that the proposed strategy beats other cutting-edge strategies regarding exactness.

Liu et al. [5] proposed two intelligent transportation regulator using fog computing which conspires by expecting the traffic lights as haze gadgets that can oppose the assaults from malevolent vehicles and can keep away from the issue of single-point disappointment in vehicular system. They utilized cryptography-based security algorithm to control the vehicle fraudulent messages and to protect traffic light. The exploratory outcomes demonstrated that the proposed technique beats other cutting-edge strategies regarding accuracy.

Rausch et al. [6] proposed a self-composed traffic management technique to regulate traffic inflow in already congested road. It manages the inflow of vehicles by changing traffic lights to bypass blocked regions by updating the driver's course decision. In this way, they inspected four unmistakable episode situations in a lattice and genuine size street organize and thought about automated traffic light controls.

Wang et al. [7] give an outline of a few promising exploration territories for a crowdsensing-based structure to give an auspicious reaction to gadget to gadget empowered ongoing traffic board in varied SIoV. The member vehicles dependent on D2D interchanges coordinate direction and topology data to powerfully manage their social practices as indicated by system conditions. A genuine taxi direction examination-based execution assessment was given to show the viability of the planned structure.

An et al. [8] proposed a vigorous and productive system parceling calculation which can instinctively settle on the quantity of segments dependent on the system availability and traffic blockage designs. The proposed methodology is tried on the local arranging system of the USA. Numerical investigation on lambda decision and calculation affectability with respect to various information missing proportions was likewise performed and expounded.

Hussain et al. [9] foreseen dependable, effective, strong and safe shrewd transportation frameworks (ITS) by the coordination of VANET-based mists with Named Data Networking (NDN) as a result of the constraints and weaknesses of the current IP-based systems' administration and the requirement for productive substance conveyance. They went for the design and succinct naming instrument for NDN-VC.

Huang et al. [10] proposed a plan named meet-cloud dependent on meet-table (records the vehicles it experiences) and distributed computing to safely and precisely disperse negative messages (snippets of data that characterize the negative characteristics of vehicles) in VANET. The calculation for disseminating and redistributing negative messages was structured.

Manuar et al. [11] presents the idea of UAVs that can be helpful to implement transportation rules and also support traffic police, which can be enabled with DSRC interface. There were also some limitations like low battery life. For future purpose, fuel cells can be used to increase the efficiency of the UAVs.

Allan de Souza et al. [12] have introduced their vision on improving the traffic productivity. Here, traffic management system (TMS) accumulates data from heterogeneous sources which enhance the effectiveness of TMS. The TMS control traffic is based on three activities: data gathering, data procedure and administration conveyance. This helped in improving the efficiency of TMS and achieving the desired level of accuracy in TMS.

Gnoni and Saleh [13] give key thoughts and refreshed survey of NMS. It proposes fundamental well-being standards received in chance administration. The principle point of NMS is to distinguish the best method to beat the close to miss creating component. The aftereffect of NMS is that close to miss information can be translated considering well-being standards violated. At last, it was reasoned that NMS is one of the mainstays of execution of top to bottom perceptibility, and in the future, it will use various sources that will integrate data to improve the efficiency.

Osman et al. [14] described a smart transportation regulatory system which takes 12 min to ample one cycle and also reduces waiting time of a vehicle.

Daely et al. [15] designed a central Web server interface which receives data from LED streetlights and easy to use for all the users. This streetlight creates a psychological effect which increases the alertness in the user.

Raiyn [16] introduced a new scheme of traffic management (SUA) which has three unique stages: find the most suitable way, refresh it to automobile and allocate it to vehicle.

Thomas and Vidal [17] focused on machine learning technique to detect traffic accidents by collecting available data from sources. The result concluded that the amount of traffic affects the precision of machine learning. The future will concentrate on improving the data-collecting ability while using real-world data.

Uchida et.al [18]. introduced Markov chain algorithm based-traffic accident prevention system which identifies the dangerous conditions based on radio signals and wireless signals from multiple sensors of mobile devices. The future study of this paper includes additional decision progress based on the Markov chain model.

Maaloul et.al [19]. proposed a novel vision-based street mishap calculation in thruways and turnpikes, and it depends on versatile traffic movement stream demonstrating procedure utilizing measurement heuristic technique for mishap discovery. The methodology depended on OF estimation and heuristic calculation for versatile limits. After experimentation, the productivity and common sense of the proposed calculation was by utilizing just 240 casings for traffic movement demonstrating.

Christian et al. [20]. proposed intelligent driving diagnosis using accident risk map analysis. Experimental tests have also been done for intelligent driving systems. The first test confirms the associate execution, while the subsequent test depends on approving the client execution when the right hand is utilized and not utilized; both the tests are performed to decide the productivity of driving collaborator. The outcome presumed that the proposed framework demonstrated better execution than 90% of right alarms on every street, and also, the accidents along the road decreased by an average of 10%.

Qi et al. [21] presented emergency warning light strategies. A traffic light control system is used (deterministic and stochastic PNs) for the intersection dealing with accidents. Petri nets-based dynamic model is used to indicate the blockage of traffic flows. The traffic cannot just be controlled yet additionally thusly control different offices including loop detectors and cautioning lights. Their work will help in management of real-time traffic accidents at urban roads.

Wang et al. [22] presented radio frequency identification (RFID) sensor, Neverstop, which uses fuzzy control strategies and hereditary calculations in enormous information insightful transportation framework. Neverstop is built with devices that will automatically control traffic lights at intersections. It helps to reduce the waiting time.

de Souza et al. [23] introduced ICARUS an alarming and re-steering framework which aides in diminishing the blockage of vehicles utilizing a vehicular system. It receives notification about congestion of vehicles; then, it calculates new routes by using inter-vehicle communication. The result demonstrated that the proposed solution will reduce CO emissions, fuel consumption and travel time.

Liu et al. [24] presented many technologies that have potential to alleviate traffic congestion like 5G wireless networks, SDN VANETs and MEC cloud server. Software Defining Network (SDN) technologies can give high-data transmission and low inactivity correspondence benefits alongside programmability. Mobile Edge Computing (MEC) cloud server gives a constant or close to ongoing reaction speed for basic missions. Deep learning-based forecast calculations with elite figuring power and ultra-huge extra room are provided by RCCS technology.The result demonstrated that these technologies will help in decreasing traffic congestion and will help in improving the ability to manage urban traffic.

Chaolong et al. [25] presented real-time graphic visualization technique that helps to enhance the usefulness and precision of graphical analysis of huge traffic data, improving the human–computer interaction. It will collect real-time data and process through traffic monitoring system and provide graphs on road congestion to traffic control department for dealing with accidents.

de Souza and Villas [26] proposed a traffic management system (TMS) to enhance efficacy of transportation and reduce traffic congestion problems. This system will gather traffic-related data, recognize jamming and will recommend substitute paths. They also introduce FASTER, a fully distributed TMS that does not overload the communication channel and improve overall vehicle traffic efficiency.

Li and Song [27] proposed ART scheme for vehicular ad hoc network (VANET) which is capable to discover and handle the malicious attack. It evaluates trustworthiness of both mobile nodes and data in VANETs.

Sundar et al. [28] presents an intelligent traffic management system to smoothen the traffic for emergency services. In this system, the vehicles are deployed with

RFID tags and these tags are used to route the vehicle. The same tag information is used to identify the vehicles. Based on these RFID, the traffic light goes to green or red, so the traffic officer can move.

El Mouna Zhioua et al. [29] presented a unique calculation dependent on fluffy rationale for portal determination. This new approach uses standardized methodology which uses the CH as a default passage. Two bunching and CH choice calculations are considered, and re-enactment outcomes indicate that convention achieves superior and brings about terms of postponement and bundle misfortune than the deterministic methodology for the two calculations. Also, recreations show that an efficient CH political race calculation is essential to guarantee great exhibitions; with C-DRIVE, there is higher parcel misfortune midpoints saw than with center calculation.

Wang et al. [30] built up a half breed VANET-upgraded constant way making arrangements for vehicles to maintain a strategic distance from blockage in an ITS. We first propose a crossover VANET-upgraded ITS casing work with functionalities of constant traffic data assortment, including both V2V and V2R correspondence sin VANETs and cell interchanges in open transportation framework. At that point, an all-around ideal constant way arranging calculation is intended to improve generally spatial usage and diminish normal vehicle travel cost by methods for Lyapunov streamlining.

Lv et al. [31] proposed a novel approach with a SAE prototypical for traffic flow. The anticipated strategy can effectively find the dormant traffic flow highlight portrayal, for example, the nonlinear spatial and worldly relationships from the traffic information and applied the avaricious layer savvy solo learning calculation to retrain the profound system, and afterward, the fine-tuning procedure is used to refresh the prototype constraints to improve the forecast exhibition. The given technique is better than the contemporary strategies.

Pillay et al. [32] exhibited a hypothetical structure that deals with mishap causation and security. The twenty articles looked into recommend that, in the high risk ventures, for example, development, mining and human services, some level of advances is being made in understanding the causation of mishaps and how they could be avoided. There is additionally some proof that a portion of the methodologies being used is intelligent of the advancing idea of securing the board over various periods. It is recognized the hypothetical structure has been applied to an exceptionally modest number of papers, and further work is required to test its adequacy.

Cheng et al. [33] proposed and actualized a PSO approach for the two-path position issue for half and half VANET-sensor systems. An ILP model for the two-path issue is set up first. At that point, a Center PSO approach for the issue is proposed, and a hypothetical examination for the Center PSO is inferred. Vehicular specially appointed systems (VANETs) help improve traffic well-being and reduce traffic clog.

Yu et al. [34] describes direct conventions of VANET for storing, sharing and searching data. Three classifications such as unicast, geo-cast and communicate have been used here, acknowledged both innovative and financial difficulties in acquiring genuine world testbeds. This incorporates the customary VANET as well as a much sizable and heterogeneous system structure.

Feteiha et al. [35] proposed advanced vehicular relaying technique to enhance connectivity in crowded areas. This paper contributes by reducing error rate expression, diversity gain and outage expressions, and to benchmark to assess our analysis.

3 Comparative Analysis

Table 1 shows the detailed comparative analysis of different techniques with pros and cons.

4 Research Challenges

- Accumulation of data for preparing a dataset to support VANETs [1].
- Optimize performance of the cluster-based algorithm for dynamic VANETs [2].
- Lower delays when contrasted with other convergence control methodologies [3].
- CNN for legitimately start to finish discovery of the diverse traffic objects from the street can be improved.
- Quantitative assessment of the performance of the self-healing strategy is required.
- Challenges.
- Deep analysis of network model.
- Specific metrics for data trust evaluation.
- Flexible network infrastructure.
- Future explore is expected to align such an edge by thinking about the system homogeneity and usage contemplations concerning various leveled traffic organize control.
- Content replication and scalability.
- To decrease data transfer capacity and capacity necessities of the cloud administration and to move the processing prerequisites from cloud to the edge.
- Has very low efficiency which causes reduction in flight time.
- Complex integration of data, big data management issues and also hard to show alternative route guidance.
- The system does not know how to avoid overloading of data.
- It uses predefined thresholds and also works on single server.
- Only the big data issues are there.
- The proposed system requires complex and wide information.
- Data refining is not there.
- The decision process is not very good.
- This system can investigate the motor vehicles up to 3.5 tons only.
- The intelligent driving assistant requires more improvement.
- System has very low coverage which led to high delay.
- To reduce high cost and increase the coverage of system.

Table 1 Detailed comparative analysis

Citation	System/method	Technique	Merits	Demerits
Al-Mayouf et al. [1]	Vehicular ad hoc network	Multi-hop reliable transmission algorithm based on optimal selection and forwarding in wireless sensor network (WSN)	• Manage accidents makes real-time communication between vehicles • Enables ambulances to avoid congested road segments	• Server overload can happen due to the chaotic behavior of traffic • Lack of data availability
Wu et al. [2]	Coalitional game theory-based protocol	Multi-hop data delivery, fuzzy logic algorithm and Q-learning algorithm (a reinforcement learning approach)	• Improved the MAC layer contention efficiency for V2R communications • Better performance evaluation of multi-hop routes by employing reinforcement learning	• Performance issues due to vehicle movement, route span, wireless link quality for route selection and cluster maintenance overhead
Bichiou and Rakha [3]	Intersection management solution founded from optimal control theory	Intersection control strategy	• Lower delays • Reduced carbon dioxide emission	• The high computational cost associated with finding the optimum solution • Impractical for real-time implementation
Ke et al. [4]	Traffic congestion detection utilizing fusion of visual highlights and CNN	Convolutional neural network	• Foreground objects detection • Multidimensional congestion detection	• Learning rate of the model is slow • Average computational time is high resulting in the processing of low frames per second
Liu et al. [5]	Canny traffic light control in VANET utilizing mist figuring	VANET and fog computing	• VANET and system architecture • Secure and fog device friendly	–

(continued)

Table 1 (continued)

Citation	System/method	Technique	Merits	Demerits
Rausch et al. [6]	Self-organized traffic management strategy	Event-oriented route choice (orienteering)	• Purely decentralized • Exhibits self-organized nature • Reduces the impact of incidents • Scales immediately and tackles the disruption at exact time and location	• It does unequivocally not target streamlining traffic stream conveyance in the upset system • Queue length estimation on every road segment is a critical issue
Wang et al. [7]	Mobile crowdsensing in SloV	Intelligent transportation system (ITS), Social Internet of Vehicles (SloV)	• Provides opportune reaction in heterogeneous SIoV for traffic management • Makes a reasonable situational choice between traffic stacks in cell systems and transmission delay brought about by roadside units transferring	• Transmission delay brought about by the separation in a geological directing plan should be additionally researched • Building a precise model by extricating system parameters from genuine world datasets is important
An et al. [8]	Network partitioning algorithm	Macroscopic fundamental diagram	• It does not require a pre-characterized number of sub-networks as a displaying input	• Future look into is expected to adjust such a limit by thinking about the system homogeneity and usage contemplations with respect to progressive traffic organize control

(continued)

Table 1 (continued)

Citation	System/method	Technique	Merits	Demerits
Hussain et al. [9]	Named data networking	VANET-based clouds with NDN	• Greatly improves the entire philosophy of future associated vehicles • Concise naming mechanism for communicating nodes	• Scalability • Unpredictable availability of content for replication • Lack of multiple interfaces for connectivity • Federated mobility
Huang et al. [10]	Negative message	VANET and cloud computing	• Accomplishes great inclusion rate and great precise inclusion rate • Secure and works appropriately beneath a few assaults	• High cloud bandwidth cost • Large storage requirements
Manuar et al. [11]	UAV-enabled systems using ITS	ITS technologies used with latest research and also used DSRC interface	• Enforces traffic rules • Supports traffic police	• Low efficiency • Limited flight time
de Souza et al. [12]	Traffic management systems	Gathers information from heterogeneous sources	• More accuracy • Good efficiency	• Heterogeneous data integration • Big data issues • Alternative route guidance
Gnoni and Saleh [13]	Near miss management systems	Near miss data can be interpreted by safety principles	• Works on many safety principles • NMS is one of the pillars of implementation	• Does not know how to avoid information overload
Osman et al. [14]	Computer vision	By the help of computer vision can detect which road needs to be cleared	• Low processing cost • Reduces waiting time	• Runs on a single server • Uses predefined thresholds

(continued)

Table 1 (continued)

Citation	System/method	Technique	Merits	Demerits
Daely et al. [15]	LED streetlight using Web-based system	Zigbee-based system	• Provides flexible interface • LEDs decrease rate of accident	• Big data management issues could be there
Raiyn [16]	Using the SUA approach	Has three different phases for allocating the road to the vehicle	• Reduces processing time • Have features like autonomy and negotiation	• Requires system-wide information which is very complex
Thomas and Vidal [17]	Using GPS location	Machine learning can detect traffic accidents	• Collects data via machine learning	• Does not refine the data
Uchida et.al [18]	Using Markov chain algorithm	Mobile chain algorithm senses the dangerous conditions	• Efficiently identifies dangerous conditions	• Poor decision process
Maaloul et.al [19]	Adaptive video-based algorithm	Vision-based approach for adaptive threshold	• Higher efficiency	• It can only investigate motor up to 3.5 tons
Christian et.al [20]	Risk map analysis	Intelligent driving systems for safe driving	• Showed correct alerts on each road	• There could be improvement in intelligent driving assistant
Qi et al. [21]	Petri nets-based traffic light control system	• Deadlock recovery and live-lock prevention and conflict resolution strategies are developed • ITS technology	• Designed a crisis traffic light control framework for crossing points furnishing crisis reaction to manage mishaps • Decreases fuel utilization, mishap term, optional mishaps and roads turned parking lots	• High cost

(continued)

Table 1 (continued)

Citation	System/method	Technique	Merits	Demerits
Wang et al. [22]	Uses genetic algorithms and fuzzy control method	• RFID sensor helps in big data intelligent transportation system	• Controls the traffic lights at intersection automatically • Helps in reducing the average waiting time for vehicles	• The impediment depends on a functioning system between focal servers and every terminal gadget • If association is lost, Neverstop could lose a few capacities that rely upon organize running condition
de souza et al. [23]	Improvement of traffic condition through an alerting and re-routing system	• It will use inter-vehicle communications	• Reduces travel time, fuel consumption, CO emissions of vehicles • Reduces broadcast storm problem	• Low coverage • High delay
Liu et al. [24]	SDN-based heterogeneous network and 5G network technology	• Data collection • High definition cameras and high-performance sensors • Deep learning-based prediction algorithms	• Decreases traffic congestion • Improves ability to manage urban traffic • Shortens rescue time	• High cloud bandwidth cost
Chaolong et al. [25]	Real-time graphic visualization technology	• Networking front-end data acquisition • Video surveillance	• Reduces the deaths of malignant traffic accident • Reduces the accident rate • Decrease traffic congestion	• Low coverage • High cost
de Souza and Villas [26]	Faster	• Gather traffic-related data in central entity	• Improves traffic efficiency • Improves traffic congestion • Does not overload communication channel	• High cost • Network issues

(continued)

Table 1 (continued)

Citation	System/method	Technique	Merits	Demerits
Li and Song [27]	Attack-resistant trust management scheme	• VANETs	• Detects malicious attack • Evaluates the trustworthiness of data as well as nodes in VANETs	• Some RFID systems will only work within the range inches or centimeters
Sundar et al. [28]	Chip mechanism at safe place	• Chip makes it available to traffic controller to know about the traffic and ambulance location	• Designed to make way for ambulance to go quickly • Helps in finding the stolen vehicles	• The CH probably will not be the ideal portal to the framework
El Mouna Zhioua et al. [29]	Ad hoc networks	• Uses algorithms and fuzzy control method and intelligent transportation	• Provides information to drivers and prevents accidents	• Individual way arranging may prompt new clog whenever performed clumsily • The start to finish transmission delay cannot be disregarded in certain situations
Wang et al. [30]	A hybrid-VANET-upgraded ITS	• Utilizes both vehicular specially appointed systems (VANETs) and cell frameworks of the open transportation system	• Collected ongoing traffic data can be used for turnpike traffic flow management	• Shallow traffic models are still to some degree uninspiring
Lv et al. [31]	A stacked autoencoder model	• The proposed technique for traffic flow forecast has predominant performance	• Deep learning calculations can speak to traffic highlights without earlier knowledge	• With socio-specialized systems being used as the fundamental methodology

(continued)

Table 1 (continued)

Citation	System/method	Technique	Merits	Demerits
Pillay et al. [32]	Rules and guidelines suggestive of the mechanical age	• Safety management initiatives were dependent on behavior-based safety and human mistakes management controls	• Strategies being used are intelligent of the developing idea of well-being executives over the distinctive eras	• Need to consider moving vehicles and a general complex management system
Cheng et al. [33]	Vehicular ad hoc networks	• Vehicular ad hoc network (VANET) originated from mobile ad hoc network (MANET)	• Developed applications, steering conventions and recreation instruments • This gives a point by point depiction of the existing conventions in VANET	• The individual lumps are not expected to remain reserved in the system for exceptionally long on the grounds that the in-organize capacity size might be restricted
Yu et al. [34]	Hierarchical bloom-filter routing (HBFR), to handle mobility	• VANET will give different administrations, for example, correspondence, stockpiling, and figuring for a scope of utilizations from safe route	• The proactive substance steering may decrease the measure of traffic if the notice overhead is minimized • This causes execution debasement as far as expanded mistake likelihood	• This causes performance degradation in terms of increased error probability
Feteiha et al. [35]	Cooperative RVC-Net	• Cooperative RVC-Net as fundamental part of the Long-Term Evolution-Advanced (LTE-A); LTE	• Devising a powerful precoding transmission conspire and a hand-off choice strategy that significantly increments decent variety gains and decreases blunder rates	–

- To reduce network issues and the cost of system.
- Need to develop error-prone transmission media and system to increase security of VANETs.

5 Conclusion

Intelligent transportation systems (ITS) are innovative idea to enhance safe communication and coordination among users using ICT smart services. The vehicular ad hoc networks (VANETs) are the most upcoming remote arrange condition under intelligent transportation systems (ITSs) and are progressing quickly with time. The decent variety of the VANET applications and their potential correspondence conventions needs a precise writing review. In perspective on previously mentioned, in this paper, we have contemplated and examined the attributes and difficulties of different research works identified with the applications, conventions and security in VANET. In addition, subsequent to checking on the current works, we have discussed them, feature multilayer difficulties concerning the exhibition of the VANETs, the as of now proposed arrangements, and the conceivable future work found the advantages and disadvantages for the future research. At last, an unthinkable examination of the considerable number of conventions is given. In this paper, we have laid down summaries of latest research works in recent years and later on classifying gaps identified in the pre-existing works that challenge the existing technologies. In the future, we will use this literature to identify research problem in same domain.

References

1. Al-Mayouf YRB, Mahdi OA, Taha NA, Abdullah NF, Khan S, Alam M (2018) Accident Management system based on vehicular network for an intelligent transportation system in urban environments. J Adv Transp (2018)
2. Wu C, Yoshinaga T, Ji Y, Zhang Y (2018) Computational intelligence inspired data delivery for vehicle-to-roadside communications. IEEE Trans Veh Technol 67(12):12038–12048
3. Bichiou Y, Rakha HA (2018) Developing an optimal intersection control system for automated connected vehicles. IEEE Trans Intelligent Transportation Systems (2018).
4. Ke X, Shi L, Guo W, Chen D (2018) Multi-Dimensional Traffic Congestion Detection Based on Fusion of Visual Features and Convolutional Neural Network. IEEE Trans Intell Transp Syst 99:1–14
5. Liu J, Li J, Zhang L, Dai F, Zhang Y, Meng X, Shen J (2018) Secure intelligent traffic light control using fog computing. Future Generation Computer Systems 78:817–824 (in press)
6. Markus R, Lämmer S, Treiber M (2018) Self-healing road networks: a self-organized management strategy for traffic incidents in urban road networks. arXiv preprint arXiv:1811.11300
7. Wang X, Ning Z, Hu X, Ngai ECH, Wang L, Hu B, Kwok RYK (2018) A city-wide real-time traffic management system: Enabling crowdsensing in social Internet of vehicles. IEEE Commun Mag 56(9):19–25
8. An K, Chiu Y-C, Xianbiao Hu, Chen X (2018) A network partitioning algorithmic approach for macroscopic fundamental diagram-based hierarchical traffic network management. IEEE Trans Intell Transp Syst 19(4):1130–1139

9. Hussain R, Bouk SH, Javaid N, Khan AM, Lee J (2018) Realization of VANET-Based Cloud Services through Named Data Networking. IEEE Commun Mag 56(8):168–175
10. Huang B, Cheng X, Huang C, Cheng W (2018) Meet-cloud for secure and accurate distribution of negative messages in vehicular Ad hoc network. Tsinghua Sci Technol 23(4):377–388
11. Menouar H et al (2017) UAV-enabled intelligent transportation systems for the smart city: Applications and challenges. IEEE Commun Mag 55.3:22–28
12. Allan de Souza AM et al (2017) Traffic management systems: a classification, review, challenges, and future perspectives. Int J Distrib Sens Netw 13.41550147716683612
13. Gnoni MG, Saleh JH (2017) Nearmiss management systems and observability-in-depth: handling safety incidents and accident precursors in light of safety principles. Safetyscience 91:154–167
14. Osman T et al (2017) Intelligent traffic management system for cross section of roads using computer vision. In: 2017 IEEE 7th annual computing and communication workshop and conference (CCWC). IEEE
15. Daely PT et al (2017) Design of smart LED streetlight system for smart city with web-based management system. IEEE Sens J 17.18:6100–6110
16. Raiyn J (2017) Road traffic congestion management based on a search-allocation approach. Trans Telecommun J 18(1):25–33
17. Thomas RW, Vidal JM (2017) Toward detectingaccidentswithalreadyavailablepassivetrafficinformation. In: 2017 IEEE 7th annual computing and communication workshopandconference (CCWC). IEEE
18. Uchida N et al. (2017) Mobile traffic accident prevention system based on chronological changes of wireless signals and sensors.JoWUA 8.3:57–66
19. Maaloul B et al (2017) "Adaptive video-based algorithm for accident detection on highways. In: 2017 12th IEEE international symposium on industrial embedded systems(SIES). IEEE
20. Quintero MCG, Cuervo PAC (2017) Intelligent driving assistant based on accident risk maps analysis and intelligent driving diagnosis. In: 2017 IEEE intelligent vehicles symposium (IV). IEEE
21. Meng C, Wen J (2016) Emergency traffic-light control system design for intersections subject to accidents. IEEE Trans Intell Trans Syst 17.1:170–183
22. Wang et al (2016) Soft computing in big data intelligent transportation systems. Appl Soft Comput 38:1099–1108
23. de Souza AM et al (2016) ICARUS: improvement of traffic condition through an alerting and re-routing system. Comput Netw 110:118–132
24. Liu J et al (2017) High-efficiency urban-traffic management in context-aware computing and 5G communication. IEEE Commun Mag 55.1:34–40
25. Chaolong J et al (2016) Research on visualization of multi-dimensional real-time traffic data stream based on cloud computing. Engineering 137:709–718
26. de Souza AM, Villas L (2016) A fully-distributed traffic management system to improve the overall traffic efficiency. In: Proceedings of the 19th ACM international conference on modeling, analysis and simulation of wireless and mobile systems. ACM
27. Li W, Song H (2016) ART: an attack-resistant trust management scheme for securing vehicular ad hoc networks. IEEE Trans Intell Trans Syst 17.4:960–969
28. Sundar R, Hebbar S, Golla V (2014) Implementing intelligent traffic control system for congestion control, ambulance clearance, and stolen vehicle detection. IEEE Sens J 15(2):1109–1113
29. El Mouna Zhioua G et al (2014) A fuzzy multi-metric QoS-balancing gateway selection algorithm in a clustered VANET to LTE advanced hybrid cellular network. IEEE Trans Veh Technol 64.2:804–817
30. Wang M et al (2014) Real-time path planning based on hybrid-VANET-enhanced transportation system. IEEE Trans Veh Technol 64.5:1664–1678
31. Lv Y et al (2014) Traffic flow prediction with big data: a deep learning approach. IEEE Trans Intel Trans Syst 16.2:865–873
32. Pillay M (2015) Accident causation, prevention and safety management: a review of the state-of-the-art. Procedia Manufac 3:1838–1845

33. Cheng JJ et al (2015) Routing in internet of vehicles: a review. IEEE Trans Intell Trans Syst 16.5:2339–2352
34. Yu Y-T, Mario G, Sanadidi MY (2015) Scalable VANET content routing using hierarchical bloom filters. Wireless Commun Mobile Comput 15.6:1001–1014
35. Feteiha MF, Hassanein HS (2014) Enabling cooperative relaying VANET clouds over LTE-A networks. IEEE Trans Veh Technol 64(4):1468–1479

Comprehensive Study on Internet of Things (IoT) and Design Considerations of Various Microstrip Patch Antennas for IoT Applications

Sohni Singh, Bhim Sain Singla, Manvinder Sharma, Sumeet Goyal, and Abdulrashid Sabo

Abstract IoT has become very imperative and significant since it can be incorporated to almost everything nowadays such as smart cities, smart agriculture, smart homes, and so on. Microstrip patch antenna (MPA) is very easy to build and has low production cost which makes it a very good choice in large number of applications. These applications include wireless LAN, mobile satellite communication, and global system for mobile communication, missile, and so on. All these applications owe to certain advantages of microstrip patch antenna such as low profile, low cost, low mass, and very easy to integrate. The resonance of these antennas can be achieved at any frequency by varying various antenna parameters and shape of the patch. The antennas having different shapes of patches have been discussed in this paper. As the technology is growing at a very fast rate, the application of antenna in various fields plays a very crucial role. The antenna has widespread types, and each type has a particular type of application. Power consumption is reduced with the use of microstrip patch antenna. The specific IoT application defines the design and choice of antenna which depends upon various frequency bands and the transmission strength. The paper highlights various antennas which are used very frequently in the applications of IoT along with their frequency bands.

Keywords Microstrip patch antenna · IoT · Wireless applications · Bandwidth · Radiation pattern

S. Singh (✉) · M. Sharma · S. Goyal
Chandigarh Group of Colleges, Landran, Mohali, Punjab, India
e-mail: sohnibagga5@gmail.com

B. S. Singla
College of Engineering and Management, Punjabi University, Patiala, Punjab, India

A. Sabo
Sule Lamido University Jigawa, Kano, Nigeria

N. Marriwala et al. (eds.), *Mobile Radio Communications and 5G Networks*, Lecture Notes in Networks and Systems 140,
https://doi.org/10.1007/978-981-15-7130-5_2

1 Introduction

Internet of Things (IoT) basically allows the objects to directly communicate with the Internet without any human intervention. This means that the Internet receives information through the objects and the environment surrounding them. This information is collected with the help of sensors integrated in the object. These sensors help in collecting the information. The object must be able to reply to any interrogation done by the Internet. IoT has become very imperative and significant since it can be incorporated to almost everything nowadays such as smart cities, smart agriculture, smart homes, and so on [1, 2]. Besides the choice of communication protocols, choosing the perfect antenna system for a specified application is the critical component in the design of smart devices. A design challenge is imposed while making the choice of a right antenna for the particular application. The performance characteristics of the antenna such as noise, fading effects, radiation pattern, and efficiency need to be maintained while designing an IoT module [3]. The characteristics of the antenna deployed for application in IoT are sensing and communication capability and eco-friendliness. An efficient IoT-based antenna system can be realized if the antenna design works alongside the enhanced techniques required for interference, multiplexing, mitigation, and allocation of various radio resources. The various applications of IoT in antenna systems include the design of antenna for RFID tags and UWB, MIMO antenna system, and their techniques of transmission and antenna arrays in the design method of position tolerance [4]. Various technologies such as wireless communication and WiMAX are moving at a very fast pace. Wireless communication means communication through electromagnetic waves. Common examples of wireless communications are cordless computer peripherals, pagers and cellular phones, and Global Positioning System (GPS). Types of technology of wireless communication include wireless fidelity (Wi-Fi), wireless local area network (WLAN), and WiMAX technology. Wi-Fi is the IEEE standard 802.11 communication standard for WLAN. Computers are connected with the help of Wi-Fi network. Very broad area can be covered while roaming and maintaining a connection to the network. All this is provided with the latest technology of wireless communication. In this way, the freedom and flexibility of the user are enhanced. The installation is very easy without creating any panic for selecting a proper location. An antenna is required for all these applications. Antenna reduces power consumption and helps in improving transmission and reception and is therefore the foremost requirement in the devices used in cellular and mobile technologies. Antennas are designed to radiate and receive electromagnetic energy. The sinusoidal voltage in the transmission line creates an electric field which leads to the flow of current and creates a magnetic field. The electromagnetic waves are created due to time-varying magnetic field and electric field. In this way, the radiation occurs in the antenna. Antenna acts as transiting device between the waveguide and the free space [5, 6]. There are many advantages of microstrip antenna which include low cost, profile, and mass. It is very easy to fabricate as well. It is capable of operating at dual and triple frequency bands. Antennas play a crucial role in WLAN and WiMAX applications. There are also some drawbacks of microstrip patch antenna which include low gain, poor efficiency, and bandwidth. The wireless phones operating in dual band can operate in two networks having different frequencies and that is why the wireless phones are gaining much popularity. The topology of IoT is shown in Fig. 1. There are three different layers: device layer, connectivity layer, and application layer. The device layer contains the devices which contain sensors and actuators such as wearables, homes, and in industries. The device layer is connected to the connectivity layer through industrial gateways, LTE, consumer gateways, and Bluetooth. All these are connected to the service provider or Internet. The application layer contains the input–output devices, networking layer, security, and memory [7]. Table 1 shows the applications of IoT at different frequency bands.

The specific IoT application defines the design and choice of antenna which depends upon various frequency bands and the transmission strength. Various wireless networks such as 5G and IEEE utilize a wide range of spectrum unlike in the IoT network such as smart transmission and reception of data. The IoT platform includes antenna for various frequency bands which can be licensed or unlicensed [8]. The table highlights various antennas which are used very frequently in the applications of IoT along with their frequency bands.

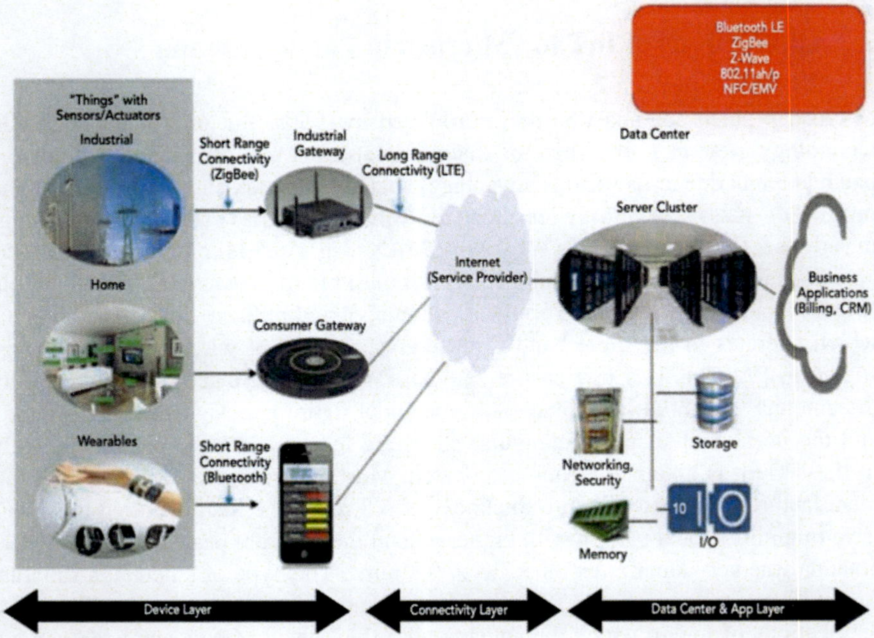

Fig. 1 Topology of IoT

Table 1 IoT applications for different frequency bands

Application	Technology	Frequency band
Smart homes/smart buildings	Wi-Fi	2.4 GHz
GPS	1575.42, 1227.6, 1176.45 MHz	
Zigbee	915 MHz, 2.4 GHz	
Z-wave	2.4 GHz	
Bluetooth	2.4 GHz	
HART	2.4 GHz	
ISA 100.11a	2.4 GHz	
Smart agriculture, smart cities	LoRa	433, 868, 915 MHz
Sigfox	868, 902 MHz	
Medical	MBAN/WBAN IEEE 802.15.6	400, 800, 900, 2.36, 2.4 GHz
Avionics (intra-com)	WAIC	4.2, 4.4 GHz

2 Applications of IoT for Microstrip Patch Antenna

Microstrip patch antenna was first introduced in 1950s. But only when the PCB technology grew in 1970s, then the microstrip antenna was considered. Microstrip patch antenna due to its various advantages such as low cost, simple structure, light profile, low mass is the most widespread used antenna and has extensive applications in various technologies such as WLAN, WiMAX, and wireless communications. The antenna can be very easily combined with monilithic microwave integrated circuit (MMIC) and have widespread applications in civil and military fields. The antenna which operates in the triple band in different networks of wireless technology is shown in [9]. This is a low-profile antenna. The microstrip feed line is given to the antenna with 50Ω impedance. A trapezoidal figure is designed in the ground, and the horizontal-shaped rectangular strips are inserted into the patch. The path of the current is obtained through the patch. Most antennas use FR4 substrate for manufacturing the patch with a thickness of 0.8 mm. The manufactured antennas have omnidirectional pattern with higher gain at the resonant peaks. The size of the antenna was very small, i.e., $38 \times 30 \times 0.8\ \text{mm}^3$. This type of antenna is suitable for multi-frequency processes. The design of the antenna is shown in Fig. 2. Various applications of IoT in which the antenna with this specific frequency band can be incorporated are smart homes and smart buildings where the technology used is Wi-Fi, GPS, Z-wave, Zigbee, and Bluetooth.

The inverted antenna in L-shaped which has the capability to operate in multiband is shown in [10]. Inverted L-strips are used in antenna for frequencies in WiMAX applications. Bandwidth of 270 MHz was attained at resonant peak of 2.6, 1000 MHz

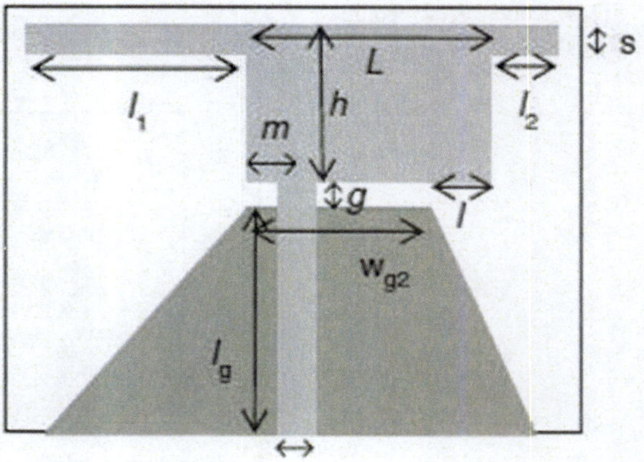

Fig. 2 Microstrip feed line antenna

bandwidth at resonant peak of 3.5 GHz, and bandwidth of 1250 MHz was attained at resonant peak of 5.5 GHz. Omnidirectional radiation pattern was obtained having different gains. The gain of 3.7 dBi was obtained with efficiency of 68% at 2.6 GHz and another gain of 3.1 dBi having efficiency 80% at 3.5 GHz resonant frequency. At 5.5 GHz resonant frequency, gain obtained was 4.8 dBi with 84% efficiency. The antenna design is shown in Fig. 3. Various IoT applications at this particular frequency bands are scientific, industrial, and medical applications.

A coplanar waveguide (CPW) fed antenna is shown in [11] which is much smaller in size. Dual band operation was achieved by inserting the slots into the patch. These slots gave rise to the excitation of multi-resonant modes. 230 MHz of bandwidth was achieved at the resonant frequency of 2.42 GHz. WLAN and C-band frequency bands were attained. Omnidirectional radiation pattern was obtained with a gain of

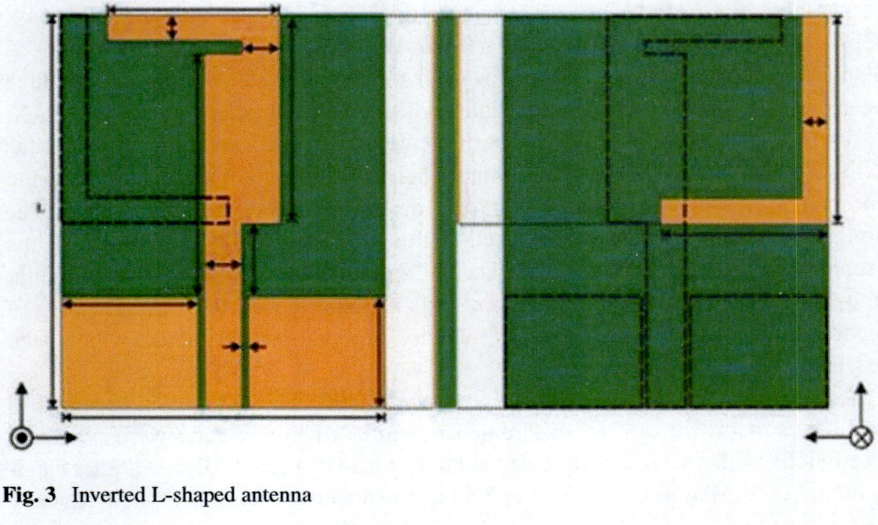

Fig. 3 Inverted L-shaped antenna

Fig. 4 Coplanar waveguide fed patch antenna

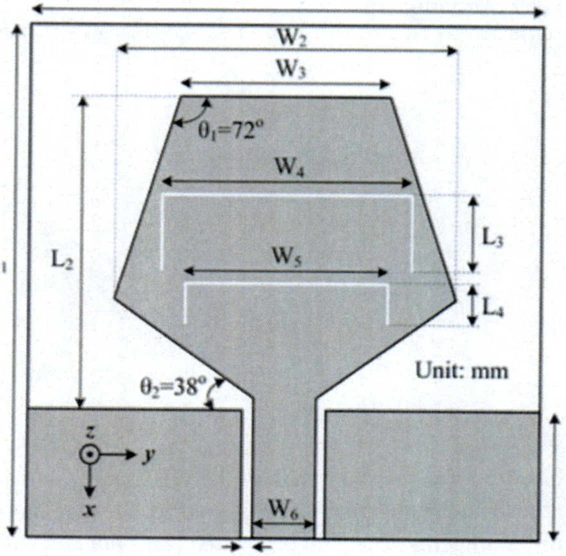

1.4 dBi for the given resonant frequency. The coplanar waveguide fed into a planar antenna is shown in [6]. The antenna is operated in triple band. Dimensions were $25 \times 25 \times 0.8\ mm^3$. The shape of the patch is hexagonal having two slots containing bent shapes in them. The antenna was capable of operating at three frequency bands. These frequency bands are 2.14–2.85, 3.29–4.08, and 5.02–6.09 GHz. Omnidirectional radiation pattern was obtained by the antenna at the operating frequencies with desired gain. The installation of this antenna into the wireless devices is very easy. The geometry of antenna is shown in Fig. 4. This low-profile antenna is suitable for 5G IoT applications and uses the technology of Bluetooth and Wi-Fi.

An antenna is the foremost requirement in the devices used in WLAN and WiMAX applications. A microstrip fed antenna for WLAN and WiMAX applications is planned in [7]. Fork-shaped band was used to cover the bandwidths for the three bands of WLAN (lower, medium, and higher) and the higher band of WiMAX. A double-shaped design is present at the lower side of substrate which covers the WiMAX band of 3.5 GHz. The omnidirectional radiation pattern was achieved having acceptable gains. The antenna was having the magnitude of $34.5 \times 18\ mm^2$. The multi-frequency processes can be used in this type of antenna. The design of the proposed antenna is shown in Fig. 5. The IoT applications employed at the particular frequency bans are smart traffic system. This includes the intelligent monitoring of the traffic, automatic identification of vehicles, and other traffic factors which need IoT technologies.

An antenna with CPW feed is represented in [8]. The antenna is operated in two bands at the slots. Two different bandwidths of the antenna were obtained. Bandwidth of 124 MHz was obtained at the resonant peak 2.45 GHz. Another bandwidth of 1124 MHz was obtained at 5.5 GHz resonant frequency. These frequency

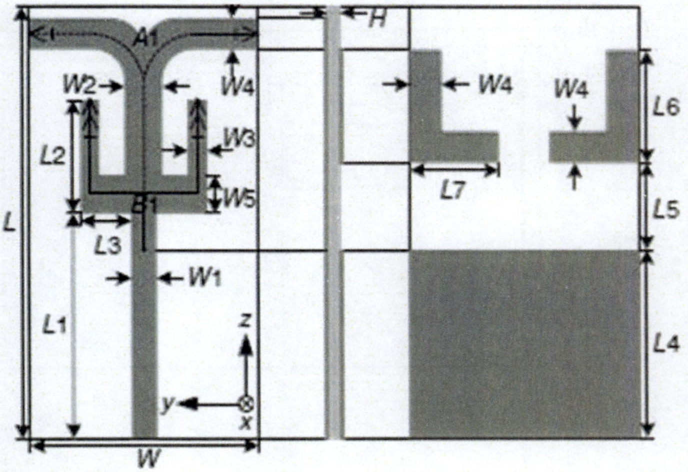

Fig. 5 Microstrip fed antenna

bands are operated at WLAN. The size of the antenna is very small, and therefore, it can be very easily installed in wireless devices operating in the latest technology of wireless communication. Triple frequency bands can be operated in this antenna. Higher bandwidth is achieved at the desired resonant frequencies. The antenna size is $20 \times 30\ mm^2$. WiMAX and WLAN applications in IoT were covered under three operating frequency bands. These three frequency bands are 2.14–2.52, 2.82–3.74, and 5.15–6.02 GHz. The results of hardware and software were almost the same. Very high gain was achieved at the resonant peaks. Multiple frequencies could also be performed by the antenna, and this type of antenna is shown in [10]. It is very compact in size. The antenna fits perfect for WLAN applications. The antenna consisted of three main components: a spherical ring, deserted ground structure, and a Y-shaped strip. The compact-sized antenna was capable of exciting the resonant modes due to the insertion of rings in antenna. Excellent gain and radiation pattern were achieved. This made the antenna perfect for WLAN applications. The design of the antenna is represented in Fig. 6.

An antenna which contained L-shaped strips and square slots is shown in [11]. The antenna has a very simple design and is compact. It is capable of operating at triple band making it suitable for both WiMAX and WLAN applications in IoT. A monopole radiator is contained in this antenna. The simulation of the antenna is done in the software. Then the fabrication is done on PCB followed by the testing on the video network analyzer (VNA) tool. Three bandwidths were provided by the antenna at the tri-band. These bandwidths are 480, 900, and 680 MHz. The design of the antenna fabricated is shown in Fig. 7.

An antenna capable of operating at tri-band for WLAN and WiMAX is shown in [12]. Inverted dual L-shaped strips are with Y-shaped strolled split ring which comprises the antenna. Three current paths were obtained with the designed structure, and the dimensions are adjusted according to that. The three bandwidths 430, 730, and

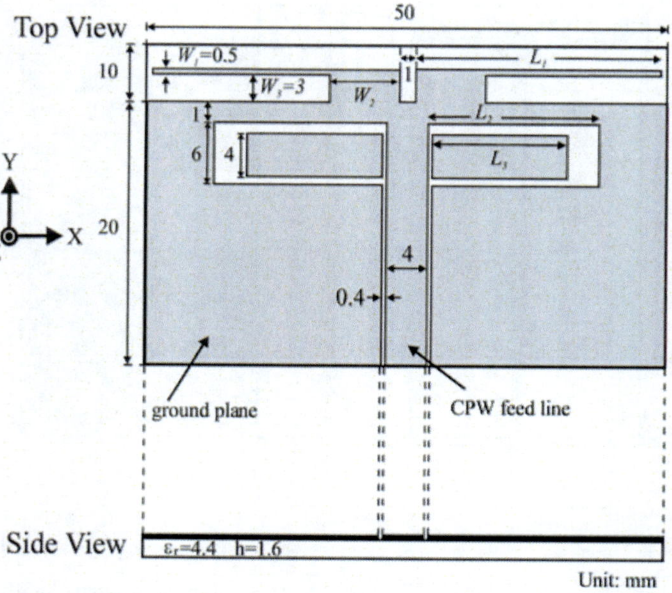

Fig. 6 CPW feed antenna design

Fig. 7 L-shaped strip antenna

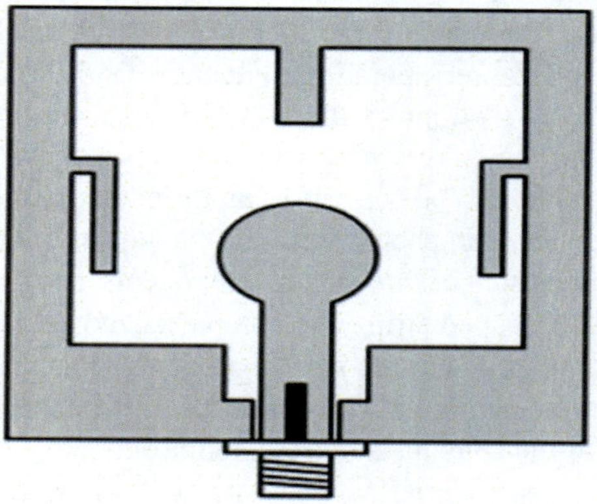

310 MHz are achieved. The frequency bands attained were 3.05–3.88 GHz and 5.57–5.88 GHz, respectively. All the operational bands were followed by antenna suitable for wireless applications. A coplanar waveguide (CPW) fed antenna is shown in [13] which operates at tri-band and covers almost all the bands for wireless applications. Three bandwidths were achieved: 2.34–2.50, 3.07–3.82, and 5.13–5.89 GHz. The antenna consists of following components: strips of S-shape and U-shape and a rectangular circle. The lower side of antenna contained three straight strips. This antenna is widely used for wireless applications because of its excellent gain and radiation pattern. The designed antenna geometry is shown in Fig. 8.

Fig. 8 Coplanar waveguide fed antenna

A compact antenna with multi-resonator structures is proposed [14]. T-shaped stub resonators and edges are contained in these structures. The operative bands at which the resonant frequencies were obtained cover almost the entire band of wireless communication. These three bands are 2.40–2.51, 3.35–3.94, and 5.02–6.634 GHz. The designed antenna geometry is represented in Fig. 9. Various IoT applications at the particular frequency bands are smart agriculture and in smart cities.

A compact antenna capable of providing tri-band operation for wireless applications is proposed [15]. The slots of L-shape and U-shape are etched into the patch of antenna. The ground plane is the lower side of the patch. The size of the antenna is

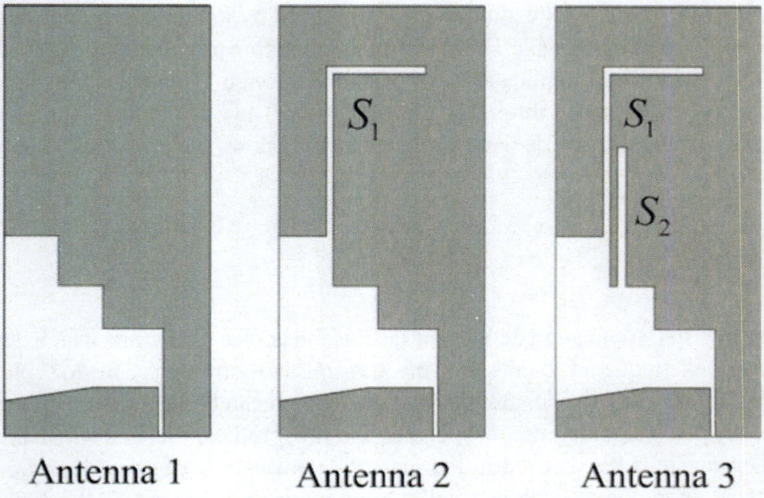

Fig. 9 Multi-resonator structure antenna

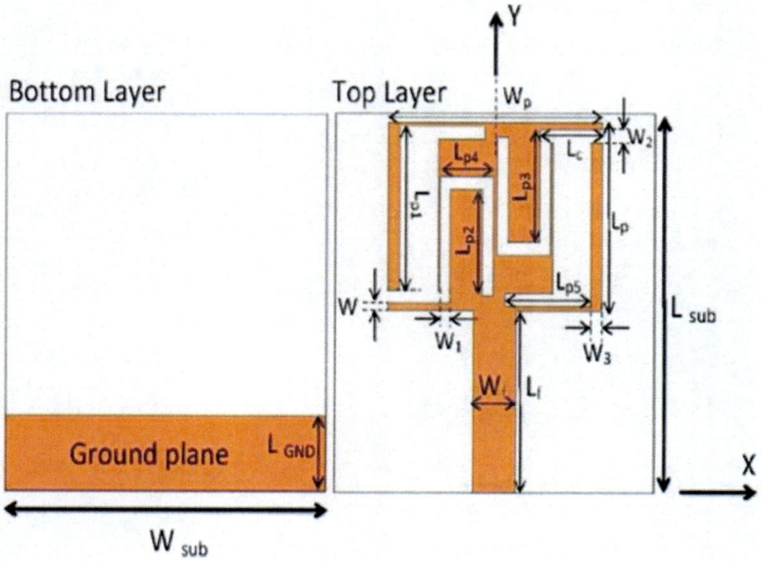

Fig. 10 Tri-band operation antenna design

15 × 15 × 1.6 mm^3 which is very small compared to the antennas designed earlier. Three operating bands were attained: 2.25–2.85, 3.4–4.15, and 4.45–8 GHz. Radiation patterns obtained were omnidirectional in nature, and gains of acceptable value were achieved at resonant peaks. The proposed antenna geometry is shown in Fig. 10.

A triple band antenna is presented [16]. The antenna patch consists of a C-shaped slot. For WiMAX applications, the antenna operates at 3.8 and 5.8 GHz. For Wi-Fi applications, the antenna operates at 9.3 GHz. The antenna is having the radiation efficiency of 92%. The reflection losses are reduced by itching hexagonal shapes at the ground surface of antenna. The Wi-Fi applications are provided through C-shaped slots, and the WiMAX applications are provided through T-shaped patch. The radar applications are provided through L-shaped patch. The polarization of the antenna is circular having ultra-wide bandwidth. Figure 11 shows the proposed antenna.

3 Conclusion

The specific IoT application defines the design and choice of antenna which depends upon various frequency bands and the transmission strength. The IoT platform includes antenna for various frequency bands which can be licensed or unlicensed. Microstrip patch antennas are inexpensive and low-profile structures which are very compact in size. MPA is used in the market extensively because of its properties. The latest technologies such as satellite communication, WiMAX, WLAN, wireless communications, missile technology, and radar are growing at such a rapid pace

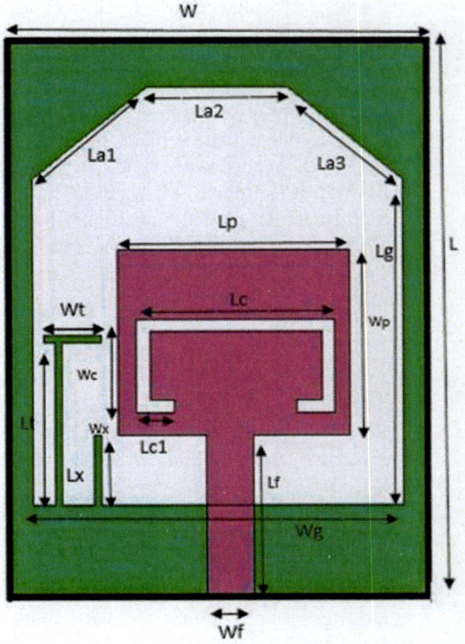

Fig. 11 C-shaped slot antenna

such that they need such type of antenna which can be easily attached with the device without making it heavier. So, the compact size and lightweight of microstrip patch antenna makes it the best choice for the devices employed in the latest technology of wireless communication. Different types of antenna along with their design and geometry along with their IoT applications are studied and discussed in the paper. Different antennas provide different IoT applications.

References

1. Lizzi L, Fabien F, Monin P, Danchesi C, Boudaud S (2016) Design of miniature antennas for IoT applications. In: 2016 IEEE sixth international conference on communications and electronics (ICCE), IEEE, pp 234–237
2. Jha KR, Bukhari B, Singh C, Mishra G, Sharma SK (2018) Compact planar multistandard MIMO antenna for IoT applications. IEEE Trans Antennas Propag 66(7):3327–3336
3. Pham C, Fabien F, Diop M, Lizzi L, Dieng O, Thiaré O (2017) Low-cost antenna technology for LPWAN IoT in rural applications. In: 2017 7th IEEE international workshop on advances in sensors and interfaces (IWASI), IEEE, pp 121–126
4. Bekasiewicz A, Koziel S (2016) Compact UWB monopole antenna for internet of things applications. Electron Lett 52(7):492–494
5. Sharma M, Singh H (2018) A review on substarte intergrated waveguide for mmW. Circ Comput Sci ICIC 2018, (June 2018), 137–138
6. Sharma M, Singh S, Khosla D, Goyal S, Gupta A (2018) Waveguide diplexer: design and analysis for 5G communication. In: 2018 fifth international conference on parallel, distributed

and grid computing (PDGC). IEEE, pp 586–590
7. Karagiannis V, Chatzimisios P, Vazquez-Gallego F, Alonso-Zarate J (2015) A survey on application layer protocols for the internet of things. Trans IoT Cloud Comput 3(1):11–17
8. Blaauw D, Sylvester D, Dutta P, Lee Y, Lee I, Bang S, Kim Y et al (2014) IoT design space challenges: circuits and systems. In: 2014 symposium on VLSI technology (VLSI-Technology): digest of technical papers. IEEE, pp 1–2
9. Thomas KG, Sreenivasan M (2009) Compact triple band antenna WLAN/WiMAX applications. IEEE Trans Electronics Lett 45(16):813
10. Lu JH, Huang BJ (2010) Planar multi-band monopole antenna with L-shaped parasitic strip for WiMAX application. IEEE Trans Electron Lett 46(10):671–672
11. Liu W-C, Chao-Ming Wu, Chu N-C (2010) A compact CPW-Fed slotted patch antenna for dual-band operation. IEEE Trans Antennas Wireless Propag Lett 9:110–113
12. Xu Y, Jiao YC, Luan YC (2012) Compact CPW-fed printed monopole antenna with triple- band characteristics for WLAN/WiMAX applications. IEEE Trans Electron Lett 48(24):1519–1520
13. Hu W, Yin Y-Z, Yang X, Fei P (2013) Compact multiresonator-loaded planar antenna for multiband operation. IEEE Trans Antennas Propag 61(5):2838–2841
14. Li X, Shi X-W, Hu W, Fei P, Yu JF (2013) Compact triband ACS- Fed monopole antenna employing open-ended slots for wireless communication. IEEE Trans Antennas Wirel Propag Lett 12:388–391
15. Moosazadeh M, Kharkovsky S (2014) Compact and small planar monopole antenna with symmetrical L- and U-shaped slots for WLAN/WiMAX application . Trans Antennas Wirel Propag Lett 13:388–391
16. Manshouri N, Najafpour M, Yazgan A, Maleki M, Kaya H (2014) A novel rectangular microstrip antenna for ultra-wideband applications with dual band-notched characteristic. In: Proceedings 22nd signal processing and communications applications conference (SIU), pp 1035–1038

Nonclassicality Used as Quantum Information Processing in Nonlinear Optical Systems

Priyanka and Savita Gill

Abstract We have studied nonclassical effects in pump mode in seven-wave mixing nonlinear process involving pump photons of different frequencies. The nonclassicality of a system increases with increase in the number of photons present in the system prior to interaction in nonlinear medium. We have found squeezing and sub-Poissonian photon statistics that can be used in obtaining noise reduction in signal processing. Further, we have also obtained antibunching of photons that can be used as a criteria of obtaining single-photon source that is primary requirement for quantum information processing.

Keywords Squeezing · Antibunching · Quantum fluctuation · Nonlinear optics

1 Introduction

A state can be defined as a nonclassical state which does not include any classical analogue. Squeezing, antibunching and sub-Poissonian photon statistics of light are the nonclassical states of light. Coherent state of electromagnetic field has equal uncertainty in both the quadrature components. Squeezed state of electromagnetic field is the minimum uncertainty state having less fluctuation in one quadrature component than the other quadrature. A review on squeezing was given by Dodonov [1] and Loudon et al. [2]. Squeezing has been based on theoretical investigation and experimental observation in nonlinear optical processes like harmonic generation [3], multiwave mixing [4–6], Raman and hyper-Raman [7, 8], nonlinear polarization rotation [9] and Faraday rotation of nonlinearity in atomic systems [10]. Mandal

Priyanka · S. Gill (✉)
Department of Applied Science, University Institute of Engineering and Technology, Kurukshetra 136119, India
e-mail: savita2015@kuk.ac.in

Priyanka
e-mail: chauhan7101@gmail.com

N. Marriwala et al. (eds.), *Mobile Radio Communications and 5G Networks*, Lecture Notes in Networks and Systems 140,
https://doi.org/10.1007/978-981-15-7130-5_3

[11] and Hillery [12] have given the concept of amplitude squeezing in nonlinear medium.

The concept of higher order antibunching was introduced by Lee [13], and it has been predicted in various processes [14–16]. Higher order antibunching and the use of antibunching in the detection of squeezing have also been reported [17–22]. Nonclassical states of electromagnetic field have interesting applications in quantum information processing like cryptography which is helpful in developing single-photon source [23], teleportation [24], dense coding [25] and quantum key distribution [26]. Strong optical nonlinearities can be studied for making optical modulators and laser second and third harmonic generators [27, 28]. Theoretical prediction of the present study of HOA can be experimentally verified with the help of homodyne experiment, since criteria of higher order antibunching appears in terms of factorial moment which can be measured by using homodyne photon counting experiments [29, 30].

Earlier, these nonclassical effects have been studied in recent past in multiwave mixing process involving incident pump photons having same frequencies [31–33], but in the present work, we have studied nonclassical effects, i.e., squeezing, sub-Poissonian photon statistics and antibunching of photons in seven-wave mixing process involving pump photons of different frequencies. In this paper, Sect. 2 gives the condition of nonclassicality of a nonlinear optical system. In Sects. 2 and 4, we derive an analytic expression of nonclassicality along with discussion of result. Finally, Sect. 5 is dedicated to conclusion.

2 Condition of Nonclassicality of a Nonlinear Optical System

2.1 Condition of Normal Squeezing

Normal squeezing or the first-order squeezing is defined as [12]

$$X_1 = \frac{1}{2}\left(A + A^+\right) \quad \text{and} \quad X_2 = \frac{1}{2i}\left(A - A^+\right) \tag{1}$$

where X_1 and X_2 are real and complex conjugate components of field amplitude, respectively.

where

$$\begin{aligned} A &= a\mathrm{e}^{i\omega t} \\ A^+ &= a^+\mathrm{e}^{-i\omega t} \end{aligned} \tag{2}$$

Operators X_1 and X_2 obey commutation relation as

$$[X_1, X_2] = (X_1 X_2 - X_2 X_1)$$
$$[X_1, X_2] = \frac{1}{4i}\left(A^2 + A^\dagger A - AA^\dagger - A^{\dagger 2} - A^2 - AA^\dagger + A^\dagger A + A^{\dagger 2}\right)$$
$$[X_1, X_2] = \frac{1}{2i}\left(A^\dagger A - AA^\dagger\right)$$

Using $AA^\dagger = A^\dagger A + 1$ in the above commutation expression, we get

$$[X_1, X_2] = \frac{i}{2} \tag{3}$$

The uncertainty in X_1 quadrature is

$$\begin{aligned}(\Delta X_1)^2 &= \langle\psi|X_1^2|\psi\rangle - \langle\psi|X_1|\psi\rangle^2 \\ &= \frac{1}{4}\langle\psi|a^2 + a^{\dagger 2} + aa^\dagger + a^\dagger a|\psi\rangle \\ &\quad - \frac{1}{4}\langle\psi|a + a^\dagger|\psi\rangle^2\end{aligned}$$

where α is expectation value of operator$A$.

$$(\Delta X_1)^2 = \frac{1}{4}(\alpha^2 + \alpha^{*2} + 2|\alpha|^2 + 1) - \frac{1}{4}(\alpha + \alpha^*)^2$$
$$(\Delta X_1)^2 = \frac{1}{4}(\alpha^2 + \alpha^{*2} + 2|\alpha|^2 + 1 - \alpha^2 - \alpha^{*2} - 2|\alpha|^2)$$
$$(\Delta X_i)^2 = \frac{1}{4} \quad \text{for i = 1 and 2} \tag{4}$$

The above relation holds for the coherent state.
Uncertainty criteria for squeezed light is given as

$$\Delta X_1 \Delta X_2 \geq \frac{1}{4} \tag{5}$$

X_i variable fulfills the condition of squeezing if

$$(\Delta X_i)^2 < \frac{1}{4}(i = 1 \text{ or } 2) \tag{6}$$

2.2 Condition of Photon Statistics

In sub-Poissonion photon statistics, deviation of photon number is less than the mean value of photon number present in the system and can be defined as

$$[\Delta N(t)]^2 < \langle N(t) \rangle \tag{7}$$

where N is number of photons present in the system.

2.3 Condition of Higher Order Antibunching (HOA) of Photons

HOA is expressed in terms factorial moment of number operator. HOA criteria was introduced by Lee [13] is given as

$$R(m, l) = \frac{\langle N_x^{m-1} \rangle \langle N_x^{l+1} \rangle}{\langle N_x^m \rangle \langle N_x^l \rangle} - 1 < 0 \tag{8}$$

where N is number operator. $\langle N^{(k)} \rangle = \langle N(N-1)(N-2)....(N-k+1) \rangle$ is kth factorial moment of number operator. Integers l and m satisfying condition $l \leq m \leq 1$ and x subscript denotes particular mode. $m = 1$ is chosen by Ba An [16] and criteria of lth-order antibunching is reduced to

$$A_{x,l} = \frac{\langle N_x^{l+1} \rangle}{\langle N_x^l \rangle \langle N_x \rangle} - 1 < 0 \tag{9}$$

and

$$\langle N_x^{l+1} \rangle < \langle N_x^l \rangle \langle N_x \rangle \tag{10}$$

By simplifying Eq. (8), we obtain condition of lth-order antibunching as

$$d(l) = \langle N_x^{l+1} \rangle - \langle N_x \rangle^{l+1} < 0 \tag{11}$$

From Eq. (11), we can see that for coherent state (Poissonian state) $d(l) = 0$ and for super-Poissonian state (photons are bunched) $d(l) > 0$. Thus, we can say that single-photon source used in quantum cryptography should satisfy the criteria given in Eq. (11) of HOA.

3 Seven-Wave Mixing Process in Pump Mode

3.1 *Squeezing in Seven-Wave Mixing Process*

In seven-wave mixing process, the interaction takes place in such a way that one photon of frequency ω_1 and two photons of frequency ω_2 each are absorbed and four photons of frequency ω_3 each are emitted, such that

$$\omega_1 + 2\omega_2 = 4\omega_3$$

Hamiltonian represents the total energy for the above process which is sum of kinetic energy and potential energy of system, and Hamiltonian for the above system is given as

$$\begin{aligned} H &= \omega_1 a^+ a + \omega_2 b^+ b + \omega_3 c^+ c \\ &\quad + g\big(ab^2 c^{+4} + a^+ b^{\dagger 2} c^4\big)\text{taking } (\hbar = 1) \end{aligned} \tag{12}$$

where g is coupling constant, $a^\dagger(a)$, $b^\dagger(b)$, $c^\dagger(c)$ are the creation (annihilation) operators, respectively. $A = a\exp i\omega_1 t$, $B = b\exp i\omega_2 t$, $C = c\exp i\omega_3 t$ are the slowly varying operators at frequencies ω_1, ω_2 and ω_3.

Time evolution of operator in B mode is given by Heisenberg equation of motion as

$$\frac{\mathrm{d}B}{\mathrm{d}t} = \frac{\partial B}{\partial t} + i[H, B] \tag{13}$$

Commutation relation of H and B operators is given as

$$[H, B] = (HB - BH)$$

Using H and B operator in the above expression, we get,

$$\begin{aligned} [H, B] &= (\omega_1 a^\dagger ab + \omega_2 b^\dagger bb + \omega_3 c^\dagger cb + gab^2 c^{\dagger 4} b \\ &\quad + ga^\dagger b^{\dagger 2} c^4 b - b\omega_1 a^\dagger a - b\omega_2 b^\dagger b \\ &\quad - b\omega_3 c^\dagger c - bgab^2 c^{\dagger 4} - bga^\dagger b^{\dagger 2} c^4)\exp(i\omega_2 t) \\ &= (\omega_2 b^\dagger bb - \omega_2 b^\dagger bb - \omega_2 b + ga^\dagger b^{\dagger 2} bc^4 \\ &\quad - ga^\dagger b^{\dagger 2} bc^4 - 2ga^\dagger b^\dagger c^4)\exp(i\omega_2 t) \\ [H, B] &= (-\omega_2 b - 2ga^\dagger b^\dagger c^4)\exp(i\omega_2 t) \end{aligned} \tag{14}$$

and

$$\frac{\partial B}{\partial t} = i\omega_2 b\exp(i\omega_2 t) \tag{15}$$

Using Eqs. (14) and (15) in Heisenberg equation of motion (13), we get

$$\begin{aligned}\dot{B} &= i\omega_2 b \exp(i\omega_2 t) \\ &\quad - i\omega_2 b \exp(i\omega_2 t) - 2iga^\dagger b^\dagger c^4 \exp(i\omega_2 t) \\ \dot{B} &= -2igA^\dagger B^\dagger C^4 \end{aligned} \tag{16}$$

Similarly,

$$\begin{aligned}\frac{\mathrm{d}A}{\mathrm{d}t} &= \frac{\partial A}{\partial t} + i[H, A] \\ \dot{A} &= -igB^{\dagger 2}C^4 \end{aligned} \tag{17}$$

and

$$\begin{aligned}\frac{\mathrm{d}C}{\mathrm{d}t} &= \frac{\partial C}{\partial t} + i[H, C] \\ \dot{C} &= -4igAB^2C^{\dagger 3} \end{aligned} \tag{18}$$

Now, second derivative of B [from Eq. (16)] is

$$\begin{aligned}\frac{\mathrm{d}^2 B}{\mathrm{d}t^2} &= -2ig[\dot{A}^\dagger B^\dagger C^4 + A^\dagger \dot{B}^\dagger C^4 + A^\dagger B^\dagger \dot{C} C^3 \\ &\quad + A^\dagger B^\dagger C\dot{C}C^2 + A^\dagger B^\dagger C^2\dot{C}C + A^\dagger B^\dagger C^3\dot{C}] \\ &= -2ig[(igB^2C^{\dagger 4})B^\dagger C^4 + A^\dagger(2igABC^{\dagger 4})C^4 \\ &\quad + A^\dagger B^\dagger(-4igAB^2C^{\dagger 3})C^3 \\ &\quad + A^\dagger B^\dagger C(-4igAB^2C^{\dagger 3})C^2 + A^\dagger B^\dagger C^2(-4igAB^2C^{\dagger 3})C \\ &\quad + A^\dagger B^\dagger C^3(-4igAB^2C^{\dagger 3})]\end{aligned}$$

So

$$\begin{aligned}\frac{\mathrm{d}^2 B}{\mathrm{d}t^2} &= -2ig[igB^2B^\dagger C^{\dagger 4}C^4 + 2igA^\dagger ABC^{\dagger 4}C^4 \\ &\quad - 4igA^\dagger AB^\dagger B^2C^{\dagger 3}C^3 - 4igA^\dagger AB^\dagger B^2CC^{\dagger 3}C^2 \\ &\quad - 4igA^\dagger AB^\dagger B^2C^2C^{\dagger 3}C - 4igA^\dagger AB^\dagger B^2C^3C^{\dagger 3}] \\ \frac{\mathrm{d}^2 B}{\mathrm{d}t^2} &= 2g^2[B^\dagger B^2C^{\dagger 4}C^4 + 2A^\dagger ABC^{\dagger 4}C^4 + 2BC^{\dagger 4}C^4 \\ &\quad - 16A^\dagger AB^\dagger B^2C^{\dagger 3}C^3 - 72A^\dagger AB^\dagger B^2C^{\dagger 2}C^2 \\ &\quad - 96A^\dagger AB^\dagger B^2C^\dagger C - 24A^\dagger AB^\dagger B^2]\end{aligned} \tag{19}$$

In this process, we assume interaction time is small ($\approx 10^{-12}s$). Using short-time approximation technique, expanding $B(t)$ according to the Taylor's expansion and taking terms up to second order in g^2t^2 as

$$B(t) = B(0) + \frac{t}{1!}\frac{\mathrm{d}B(0)}{\mathrm{d}t} + \frac{t^2}{2!}\frac{d^2B(0)}{\mathrm{d}t^2}$$

Using Eqs. (16) and (19) in the above expression, we get

$$\begin{aligned} B(t) = B - 2igtA^{\dagger}B^{\dagger}C^4 + g^2t^2(&B^{\dagger}B^2C^{\dagger 4}C^4 \\ &+ 2A^{\dagger}ABC^{\dagger 4}C^4 + 2BC^{\dagger 4}C^4 - 16A^{\dagger}AB^{\dagger}B^2C^{\dagger 3}C^3 \\ &- 72A^{\dagger}AB^{\dagger}B^2C^{\dagger 2}C^2 - 96A^{\dagger}AB^{\dagger}B^2C^{\dagger}C - 24A^{\dagger}AB^{\dagger}B^2) \end{aligned} \tag{20}$$

And complex conjugate of $B(t)$ is

$$\begin{aligned} B^{\dagger}(t) = B^{\dagger} &+ 2igtABC^{\dagger 4} \\ &+ g^2t^2(B^{\dagger 2}BC^{\dagger 4}C^4 + 2A^{\dagger}AB^{\dagger}C^{\dagger 4}C^4 \\ &+ 2B^{\dagger}C^{\dagger 4}C^4 - 16A^{\dagger}AB^{\dagger 2}BC^{\dagger 3}C^3 \\ &- 72A^{\dagger}AB^{\dagger 2}BC^{\dagger 2}C^2 - 96A^{\dagger}AB^{\dagger 2}BC^{\dagger}C \\ &- 24A^{\dagger}AB^{\dagger 2}B) \end{aligned} \tag{21}$$

Field amplitude quadrature component is defined as

$$X(t) = \frac{1}{2}\left[B(t) + B^{+}(t)\right] \tag{22}$$

To study the squeezing, initially we assume a quantum state which is the product of coherent state $|\alpha\rangle$ and $|\beta\rangle$ for pump mode A and B, respectively, and vacuum state $|0\rangle$ for mode C, i.e.,

$$|\psi\rangle = |\alpha\rangle_A|\beta\rangle_B|0\rangle_C \tag{23}$$

Using Eqs. (20), (21) and (23) in Eq. (22), we get

$$X(t) = \frac{1}{2}[B + B^{\dagger} - 24g^2t^2(A^{\dagger}AB^{\dagger}B^2 + A^{\dagger}AB^{\dagger 2}B)] \tag{24}$$

Expectation value of $X(t)$ is given as

$$\langle\psi|X(t)|\psi\rangle = \frac{1}{2}[\beta + \beta^* - 24g^2t^2(|\alpha|^2|\beta|^2\beta + |\alpha|^2|\beta|^2\beta^*] \tag{25}$$

where α and β are the expectation value of operators A and B, respectively, with $A^{\dagger}A = |\alpha|^2$ and $B^{\dagger}B = |\beta|^2$.

Squaring of Eq. (25), we have

$$\langle\psi|X(t)|\psi\rangle^2 = \frac{1}{4}[\beta^2 + \beta^{*2} + 2|\beta|^2 \\ - 24g^2t^2|\alpha|^2(2\beta^2|\beta|^2 + 2\beta^{*2}|\beta|^2 + 4|\beta|^4)] \tag{26}$$

now

$$X^2(t) = X(t)X(t)$$
$$X^2(t) = \frac{1}{4}[B^2 + B^{\dagger 2} + 2B^\dagger B + 1 \\ - 24g^2t^2(2A^\dagger AB^\dagger B^3 + 2A^\dagger AB^{\dagger 3}B \\ + A^\dagger AB^2 + A^\dagger AB^{\dagger 2} + 4A^\dagger AB^{\dagger 2}B^2 + 4A^\dagger AB^\dagger B)] \tag{27}$$

And expectation value of $X^2(t)$ is given as

$$\langle\psi|X^2(t)|\psi\rangle = \frac{1}{4}[\beta^2 + \beta^{*2} + 2|\beta|^2 + 1 \\ - 24g^2t^2|\alpha|^2(2\beta^2|\beta|^2 + 2\beta^{*2}|\beta|^2 \\ + \beta^2 + \beta^{*2} + 4|\beta|^4 + 4|\beta|^2)] \tag{28}$$

Subtracting Eq. (26) from Eq. (28), we get

$$[\Delta X(t)]^2 = \langle\psi|X^2(t)|\psi\rangle - \langle\psi|X(t)|\psi\rangle^2 \\ = \frac{1}{4}[1 - 24g^2t^2|\alpha|^2(\beta^2 + \beta^{*2} + 4|\beta|^2)] \tag{29}$$

So

$$[\Delta X(t)]^2 - \frac{1}{4} = -12g^2t^2|\alpha|^2|\beta|^2(\cos 2\theta + 2) \tag{30}$$

where θ is phase angle of field amplitude with $\beta = |\beta|\exp(i\theta)$. The right-hand side of Eq. (30) is negative, which shows the existence of squeezing for all values of θ for which $\cos 2\theta > 0$ in pump mode. Squeezing will be maximum, when $\theta = 0$.

3.2 *Sub-Poissonian Photon Statistics in Seven-Wave Mixing Process*

Condition of sub-Poissonian photon statistics is given as

$$[\Delta N_B(t)]^2 - \langle N_B(t)\rangle < 0 \tag{31}$$

where

$$[\Delta N_B(t)]^2 = \langle N_B^2(t)\rangle - \langle N_B(t)\rangle^2 \tag{32}$$

Using Eq. (20, 21), the expectation value of $N_B(t) = B^\dagger(t)B(t)$ is given as

$$< N_B(t) >= |\beta|^2 - 48g^2t^2|\alpha|^2|\beta|^4 \tag{33}$$

After squaring of the above Eq. (33), we get

$$\langle N_B(t)\rangle^2 = |\beta|^4 - 96g^2t^2(|\alpha|^2|\beta|^6) \tag{34}$$

as

$$N_B^2(t) = N_B(t)N_B(t)$$

Using $N_B(t)$ in the above expression, we get $N_B^2(t)$ as

$$N_B^2(t) = B^{\dagger 2}B^2 + B^\dagger B - 96g^2t^2(A^\dagger AB^{\dagger 3}B^3 + 2A^\dagger AB^{\dagger 2}B^2)$$

Expectation value of $N_B^2(t)$ is given as

$$\langle N_B^2(t)\rangle = |\beta|^4 + |\beta|^2 - 96g^2t^2(|\alpha|^2|\beta|^6 + 2|\alpha|^2|\beta|^4) \tag{35}$$

Using Eqs. (34, 35) in Eq. (32), we get

$$[\Delta N_B(t)]^2 = |\beta|^2 - 96g^2t^2(|\alpha|^2|\beta|^4) \tag{36}$$

Using Eqs. (33) and (36) in Eq. (31), we get

$$[\Delta N_B(t)]^2 - \langle N_B(t)\rangle = -144g^2t^2(|\alpha|^2|\beta|^4) \tag{37}$$

We are getting negative value and obey the sub-Poissonian photon statistics of light.

3.3 *Antibunching of Photons in Seven-Wave Mixing Process*

The expectation value of $N_B(t)$ using Eqs. (20, 21) is given as

$$< N_B(t) >= |\beta|^2 - 48g^2t^2|\alpha|^2|\beta|^4 \tag{38}$$

Now the expectation value of $N_B^2(t) = B^{\dagger 2}(t)B^2(t)$ is given as

$$\langle N_B^2(t)\rangle = \langle B^{\dagger 2}(t)B^2(t)\rangle = |\beta|^4 - 48g^2t^2(2|\alpha|^2|\beta|^6 + |\alpha|^2|\beta|^4) \tag{39}$$

Now using Eqs. (38, 39) in Eq. (11), we get

$$d_B(1) = \langle N_B^2(t) \rangle - \langle N_B(t) \rangle^2$$
$$d_B(1) = -48g^2t^2|\alpha|^2|\beta|^4 \quad (40)$$

In the above Eq. (40), we are getting negative value which shows the existence of antibunching and depends on number of pump photons present in the system.

4 Result

The presence of squeezing and antibunching in seven-wave mixing process involving pump photon of different frequencies is shown in Eqs. (30) and (40). Figs 1 and 2 represent a graph between field squeezing say S_x and antibunching say $d(1)$ with photon number in mode B, i.e., $|\beta|^2$ respectively. From Figs. 1 and 2, it is clear that for a fixed value of photon number in mode A, i.e., $|\alpha|^2$, squeezing and antibunching increases nonlinearly with increase in photon number in mode B, i.e., $|\beta|^2$. Further, it is also shown that S_x and antibunching increases with increase in photon number in coherent state A, i.e., $|\alpha|^2$.

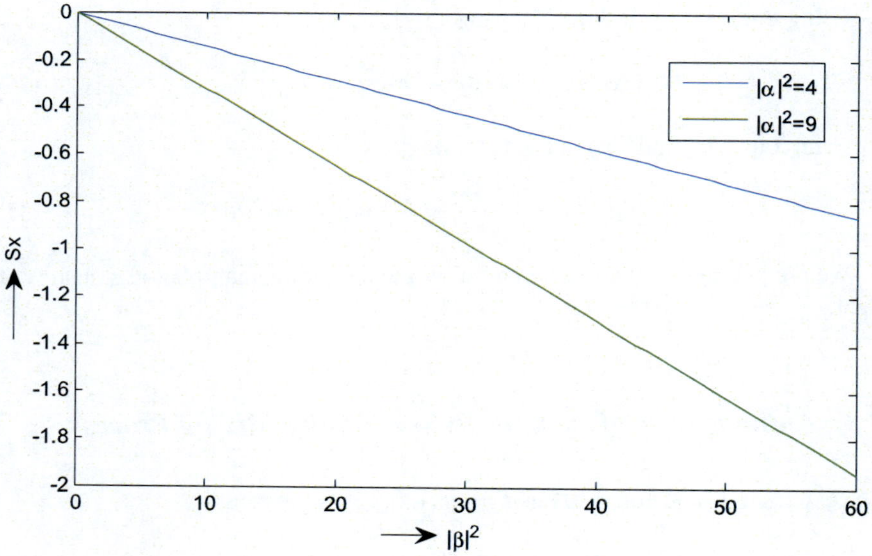

Fig. 1 Variation of field amplitude squeezing S_x with $|\beta|^2$ (taking $g^2t^2 = 10^{-4}$, $\theta = 0$ for maximum squeezing)

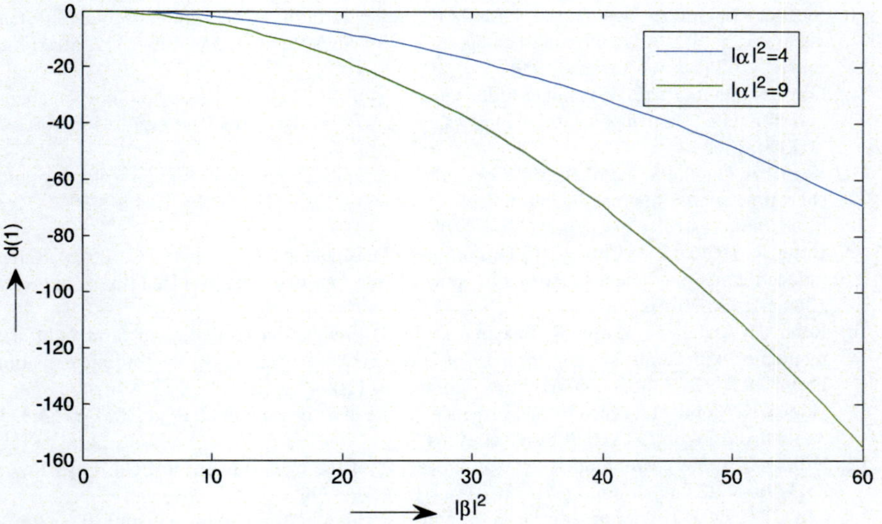

Fig. 2 Variation of antibunching with $|\beta|^2$ (taking $g^2t^2 = 10^{-4}$)

5 Conclusion

The result shows that seven-wave mixing nonlinear optical process can be used to create nonclassical effects of light, i.e., squeezing, sub-Poissonian photon statistics and antibunching. Further nonclassicality is directly related to pump photons of optical field thus nonclassicality of a system increases with increase in number of photons present in the system prior to interaction in nonlinear medium. Thus, we can conclude that the process of generation of squeezing is more suitable for noise reduction in various nonlinear processes and antibunching of photons can be used as a single-photon source that is the primary requirement for quantum information processing [34].

References

1. Dodonov VV (2002) Nonclassical states in quantum optics: a 'squeezed' review of the first 75 years. J Opt B Quantum Semiclassical Opt 4(1):R1–R33. https://doi.org/10.1088/1464-4266/4/1/201
2. Loudon R, Knight PL (2007) Squeezed light. J Mod Opt 34:709–759. https://doi.org/10.1080/09500348714550721
3. Gill S, Rani S, Singh N (2012a) Higher order amplitude squeezing in fourth and fifth harm—onic generation. Indian J Phys 86:371–375. https://doi.org/10.1007/s12648-012-0060-z
4. Gill S (2017) Non classical effects of light in higher order five wave mixing. Int J Adv Sci Eng Technol 5(3):16–20. IJASEAT-IRAJ-DOI-8716
5. Rani S, Lal J, Singh N (2011) Higher order amplitude squeezing in six wave mixing process. Int J Opt. 9:10.1155/2011/629605. Article ID 629605

6. Slusher RE, Hollberg LW, Yurke B, Mertz JC, Valley JF (1985) Observation of squeezed states generated by four-wave mixing in an optical cavity. Phys Rev Lett 55:2409–2412. https://doi.org/10.1103/PhysRevLett.55.2409
7. Giri DK, Gupta PS (2005) Higher order squeezing of the electromagnetic field in spontaneous and stimulated Raman processes. J modern opt 52:1769–1781. https://doi.org/10.1080/09500340500073065
8. Kumar A, Gupta PS (1996) Higher order amplitude squeezing in hyper Raman scattering under short time approximation. Quantum Semiclassical Opt J Eur Optical Soc Part B 8:1053–1060. https://doi.org/10.1088/1355-5111/8/5/010
9. Dong R, Heersink J, Corney JF, Drummond PD, Anderson UL, Leuchs G (2008) Raman induced limits to efficient squeezing in optical fibres. Opt Lett 33:116–118. https://doi.org/10.1364/OL.33.000116
10. Rebhi R, Anders K, Kaiser R, Lezama A (2015) Fluctuation properties of laser light after interaction with atomic system: comparison between two level and multi level atomic transition. Phys Rev A 92:033853–033861. https://doi.org/10.1103/PhysRevA.92.033853
11. Hong CK, Mandel L (1985) Higher order squeezing of a quantum field. Phys Rev Lett 54:323–325. https://doi.org/10.1103/PhysRevLett.54.323
12. Hillery M (1987) Amplitude- squared squeezing of the electromagnetic field. Phys Rev A 36:3796–3802. https://doi.org/10.1103/PhysRevA.36.3796
13. Lee CT (1990a) Higher order criteria for nonclassical effects in photon statistics. Phys Rev A 41:1721–1723. https://doi.org/10.1103/physreva.41.1721
14. Lee CT (1990b) Many—photon antibunching in generalized pair coherent state. Phys Rev A 41:1569–1575. https://doi.org/10.1103/Physreva.41.1569
15. Agarwal GS, Tara K (1991) Nonclassical properties of states generated by the excitations on a Coherent state. Phys Rev A 43:49495. 10.1103/PhysRevA.43.492
16. An NB (2002) Multimode higher-order antibunching and squeezing in trio cohrent states. J Opt B: Quantum Semiclass Opt 4:222–227. 10.1088/1464-4266/4/3/310
17. Kim Y, Yoon TH (2002) Higher order sub-poissonian photon statistics of light. Opt Commumn 212:107–114. https://doi.org/10.1016/S0030-4018(02)01981-8
18. Mishra DK (2010) Study of higher order non classical properties of squeezed kerr state. Opt Commun 283:3284–3290. https://doi.org/10.1016/j.optcom.2010.04.007
19. Perina J, Perinova V, Kodousek J (1984) On the relations of antibunching, sub-poissonian statistics and squeezing. Opt Commun 49:210–214. https://doi.org/10.1016/003-4018(84)90266-9
20. Prakash H, Mishra DK (2006) Higher order sub poissonian photon statistics and their use in detection of Hong and Mandel squeezing and amplitude squared squeezing. J Phys B At Mol Opt Phys 39:2291–2297. https://doi.org/10.1088/0953-4075/39/9/014
21. Thapliyal K, Pathak A, Sen B, Peřina J (2014) Higher-order nonclassicalities in a codirectional nonlinear optical coupler Quantum entanglement, squeezing and antibunching. Phys Rev A 90:013808–013817. https://doi.org/10.1103/PhysRevA.90.013808
22. Giri SK, Sen B, Raymond O, Pathak A (2014) Single-mode and intermodal higher order nonclassicalities in two—mode Bose—Einstein condensates. Phys Rev A 89:033628. 10.1103/PhysRevA.89.033628
23. Thapliyal K, Pathak A (2015) Applications of quantum cryptographic switch: various tasks related to controlled quantum communication can be performed using Bell states and permutation of particles. Quantum Inf Proces 14:2599–2616. https://doi.org/10.1007/s11128-015-0987-z
24. Thapliyal K, Verma A, Pathak A (2015) A general method for selecting quantum channel for bidirectional controlled state teleportation and other schemes of controlled quantum communication. Quantum Inf Proces 14:4601–4614. https://doi.org/10.1007/s11128-015-1124-8
25. Hsu MTL, Delaubert V, Bowen WP, Fabre C, Bachor HA, Lam PK (2006) A quantum study of multibit phase coding for optical storage. IEEE J Quantum Electron 42:1001–1007. https://doi.org/10.1109/JQE.2006.881634

26. Shukla C, Pathak A (2014) Orthogonal State based deterministic secure communication without actual transmission of the message qubits. Quantum Inf Proces 13:2099–2113. https://doi.org/10.1007/s11128-014-0792-0
27. Xie RH (2000) Handbook of advanced electronic and photonic materials and devices. In: Nalwa HS, Edition. 9:267–307
28. Xie RH, Rao Q, Jensen L (2003) Encyclopedia of nanoscience and nanotechnology. HS Nalwa, Edition
29. Erenso D, Vyas R, Singh S (2002) Higher order sub-poissonian photon statistics in terms of factorial moment. J Opt Soc America B 19:1471–1475. https://doi.org/10.1364/JOSAB.19.001471
30. Vyas V, Singh S (1989) Photon-counting statistics of the degenerate optical parametric oscillator. Phys Rev A 40:5147–5159. https://doi.org/10.1103/PhysRevA.40.5147
31. Gill S, Rani S, Singh N (2012b) Minimum total noise in wave mixing process. Int J Opt 6:1–4. https://doi.org/10.1155/2012/431826
32. Gill S (2016) Squeezing in stimulated seven wave mixing process. Int J Res Sci Innov 3:41–43
33. Pratap R, Giri DK, Prasad A (2014) Effects of squeezing and sub-poissonian of light in fourth harmonic generation upto first order Hamiltonian interaction. Opt Int J Light Electron Opt 125(3):1065–1070. https://doi.org/10.1016/j.ijleo.2-013.07.143
34. Beveratos A, Brouri R, Gacoin T, Villing A, Poizat JP, Grangier P (2002) Single pho- on quantum cryptography. Phys Rev Lett 89:187901–187904. https://doi.org/10.1103/PhysRevLett.89.187901

Investigation of SNR in VLC-Based Intelligent Transportation System Under Environmental Disturbances

Ritvik Maheshwari, Jyoti Grover, and Sumita Mishra

Abstract Various forms of wireless communications technologies have been proposed for intelligent transportation systems (ITSs). Recent events have illustrated that visible light communication (VLC) can play a significant role in achieving Vehicle-to-Vehicle (V2V) and Vehicle-to-Infrastructure (V2I) communication. Since it is energy efficient and available in abundance, it is in huge demand. Our prime objective of this experiment is to study the challenges that come during channel modulation and find optimal solutions to curb them. In this paper, we have done a comparative study of field of view (FOV) angles in different environmental conditions to evaluate the change in the signal-to-noise ratio (SNR). We performed a stress test on a previously laid model using NS 3.25 network simulator. This analytical approach helped us to successfully simulate SNR for different topological schemes and infer a hybrid model from it. Each topology scheme depicts a specific traffic scenario on the road.

Keywords Intelligent transportation system · VLC · SNR · Simulation · Topology

1 Introduction

Visible light communication (VLC) is a new standard that could transform the future of wireless communication. It is concerned with the transmission of data with help of visible light. This field is in huge demand due to following reasons:

R. Maheshwari (✉) · S. Mishra
Amity School of Engineering and Technology, Amity University, Lucknow, India
e-mail: ritvik.maheshwari@student.amity.edu

S. Mishra
e-mail: smishra3@lko.amity.edu

J. Grover
Malaviya National Institute of Technology, Jaipur, India
e-mail: jgrover.cse@mnit.ac.in

N. Marriwala et al. (eds.), *Mobile Radio Communications and 5G Networks*, Lecture Notes in Networks and Systems 140,
https://doi.org/10.1007/978-981-15-7130-5_4

- The excessive use of radio frequency (RF) has made it a scarce resource for communication.
- With the recent popularity of 5G [1], they are indeed the major causes of health-related problems to humans, and more importantly to animals.
  Since VLC fulfills demands of 5G like high efficiency, low latency, free available spectrum, low battery consumption, and its integration with 5G is now a hot topic [2].
- VLC is less prone to cyber attacks because it endures in a confined environment, whereas RF covers a wide area and hence more prone to security attacks.
- Data can be transferred with high speed using VLC, making it a fast and more secure mode of communication.
- VLC is economical also. So, there is no need to install equipment specifically for data transfer. We can illuminate the surroundings as well as transfer data using this technology.
- As mentioned in the paper [3], RF is classified as carcinogenic to humans which adds one more reason to not use it as a mode of communication.

There is a specific term light ad hoc network (LANET) [4] which deals in data transmission using light. Several researches have been conducted to make this happen, but only few of them take real-world challenges into consideration [5].

Real-world challenges can be environmental factor, system factors, physical factors which can hamper the VLC data transfer. Environmental factors can include fog, sunlight, artificial lights on roads, rain, etc., whereas system factors may be wrong channel modeling, and similarly physical factors may be huge crowded streets, obstacles blocking way of light, etc. The Numbers of cars on road are increasing due to perpetual and expeditious development of modernization and motorization, thus leading the problems such as increased traffic congestion and frequent accidents. There are several daily basis challenges [6–8] which we experience on road while driving and which are also responsible for an inefficient transmission of data from V2V like:

- **Non-Line of Sight Transmission (NLOS)**—There are several researches which talk vividly about the design of line-of-sight transmission (LOS) of data [9], which means V2V transmission of data without reflection of light, or the light from one vehicle would go straight to other one like peer-to-peer connection without any intermediate reflecting object. But, this situation is not always the same on road. There may be situation where we may need help of reflecting surfaces like wet roads or nearby reflecting surface to make the transmission happen. Secondly, the time taken by NLOS transmission of data to distant cars than car-to-car transmission can be far more or less efficient but also difficult to apply, so we must find ways to make it a useful tool in communication [10].
- **Visibility**—We often hear news about increasing car accidents on highways due to lack of visibility. Especially in winters, car gets into multiple collisions due to lack of visibility. Once a car gets rammed into another, there is no secure mechanism to transmit that data to upcoming fast cars leading to further collision. Visibility issues can be due to pollution, artificial light of advertisement boards,

glow sign boards or environmental factors like fog, etc. If we talk about visible light, the channel of data transmission must be crystal clear to prevent any delay. In this case, fog may pose a hindrance to it. If we find a way to transmit data in bad weather condition, we can prevent first collision and hence prevent upcoming ones. The above given challenges can we solved only if we switch to more advanced transportation scheme.

Transportation is a very important aspect in development and urbanization of a nation. It holds great importance in industrial sector [11]. Recent advancements in transportation have urged investors to invest in latest technological researches to make it more intelligent in nature. The term ITS basically means to make our transportation more autonomous in nature. Like installing LIDAR GUN on highways to penalize the over speeding vehicles, installing geographical positioning systems (GPS) in cars to detect current location of vehicle by police or ambulance in case of emergency, finding best route in case of traffic. ITS also includes ways to manage traffic more and more efficiently without much involvement of man and thus making it error free and cost effective. Nowadays, data transmission through visible light has also become a hot topic in ITS [12], and car companies are really interested in installing feasible technology to make it happen. IEEE standard 802.15.7 [13] discusses about ITS and several anomalies and challenges which we come across while dealing with day-to-day traffic. There is a research work [14, 15] which explains the recent trends in ITS and the following related terms which add more light to the topic and explains us how to deal with challenges incurred in ITS.

- Advanced Traveler Information System (ATIS)
- Advanced Traffic Management System (ATMS)
- Advanced Public Transportation System (APTS)
- Emergency Management System (EMS).

In this paper, we are able to graphically analyze which FOV angles at receiver end are suitable for different environmental conditions. This analysis helped us to contribute a hybrid model which can work in clear as well as foggy weather conditions.

The rest of this paper is organized as follows. In Sect. 2, we moved on to topological orientation of vehicles on road to depict real-time traffic scenarios. Then, in Sect. 3, we discussed the methodology of our research work from installation of software till execution of code, the parameters and formulations used to compute SNR. In Sect. 4, results are analyzed graphically along with suggestion of an efficient model. Finally, we draw conclusions in Sect. 5.

2 Topology

In network aspect, topology is the layout of nodes in a network [15]. In terms of V2V communication, topology can be understood as the arrangement of vehicles on road in a fashion to ensure efficient transmission of data [16]. Few researches have

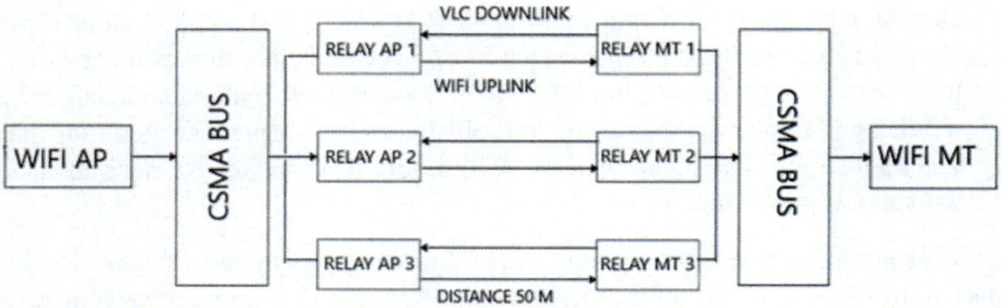

Fig. 1 Bus topology with three relay access points and three relay mount points

specifically taken topologies to showcase VLC networks [17]. We have simulated our calculation of SNR considering disturbances in the channel, and with the help of the following topology, we were able to depict the required traffic scenario like:

- **Bus Topology**—Fig. 1 describes bus topology; in vehicular scenario, we can take example of a 6-lane highways with infrastructure (storehouse and a center to transmit data) installed every 100 m; this infrastructure can be used for V2I transmission of data, where we can take it as carrier sense multiple access (CSMA) which transmits data (to vehicles on 3 lanes) of upcoming vehicles or accident so that cars in 3 lanes can be alert and further transmit to trailing vehicles. We have considered a hybrid environment where RF is used for uplink, whereas VLC is used for downlink to ensure better efficiency.

- **Point-to-Point Topology**—A simple transmission of data which involves single vehicle on both sides of the channel can be treated as peer to peer as shown in Fig. 2.

For example, suppose data transmitted to vehicle is connected to mobile phone, which means if a data is send to a car by means of light, it will ping to the mobile device connected to car. If a person is stuck with his car in parking lot because of wrongly parked car in front of him, he can use peer-to-peer communication and communicate with that person to remove his car, even if he is not near his car.

- **Wireless Topology**—If we talk about previous topologies, technically each one of them involves wired connection, but in real life, we need wireless connection, and seeing this, we were successful to construct a topology which is a mixture of wireless as well as peer to peer as shown in Fig. 3. This topology has several complexities during its implementation.

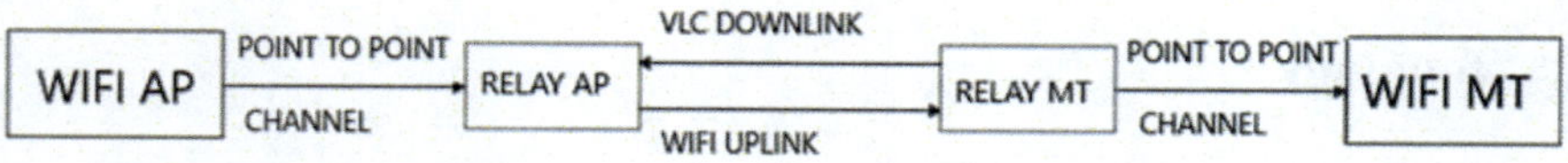

Fig. 2 Point-to-point topology with one relay access points and one relay mount points

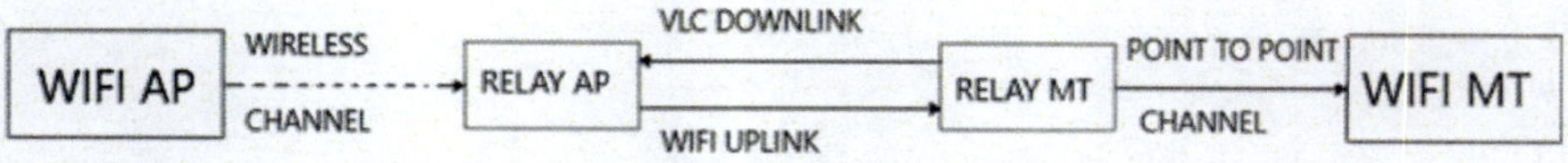

Fig. 3 Combination of wireless and peer-to-peer topology

3 Methodology

This section shows our whole process scenario to carry out experiment as shown in Fig. 4. First of all, we equipped our virtual box with operating system (OS) Ubuntu 16.04 because our code works on NS 3.25 simulator [18, 19] which works on this OS only. Then, after installing simulator on OS, we made our code feasible for different topologies and to work effectively in background noises like fog.

After building the module, we changed the field of view (FOV), i.e., receiver's angle, and performed a stress test by taking different angles into consideration. We considered two different environmental cases, where one was foggy and other was clear environment. This was taken into account because in real time weather is not always ideal for data transmission. In subsequent parts of the paper, we have shown the parameters and formulation used to do the above tests, as well as we have described our inferences and suggestions as perceived from the graphical comparative study.

3.1 Simulation and Formulations

In this section, a vivid description of parameters and formulas used in the computation is shown. In the following, Table 1 shows the parameters we have used to find the best optimal value of SNR.

The formulas given below have been incorporated from research work by Ketprom et al. In [20].

- **Background power** is calculated in the presence of fog, where FOV plays an important role.

$$P_{\mathrm{BG}} = H_{\mathrm{BKG}} \pi (\mathrm{FOV}) 2 \mathrm{Ar} \delta \lambda T F \tag{1}$$

H_{BKG} denotes background radiance, FOV is the field of view which is the angle of photo detector at receiver end to capture maximum light, Ar is the area of the photo detector, $\delta\lambda$ is the fiber optic bandwidth, and T_{F} denotes filter transitivity.

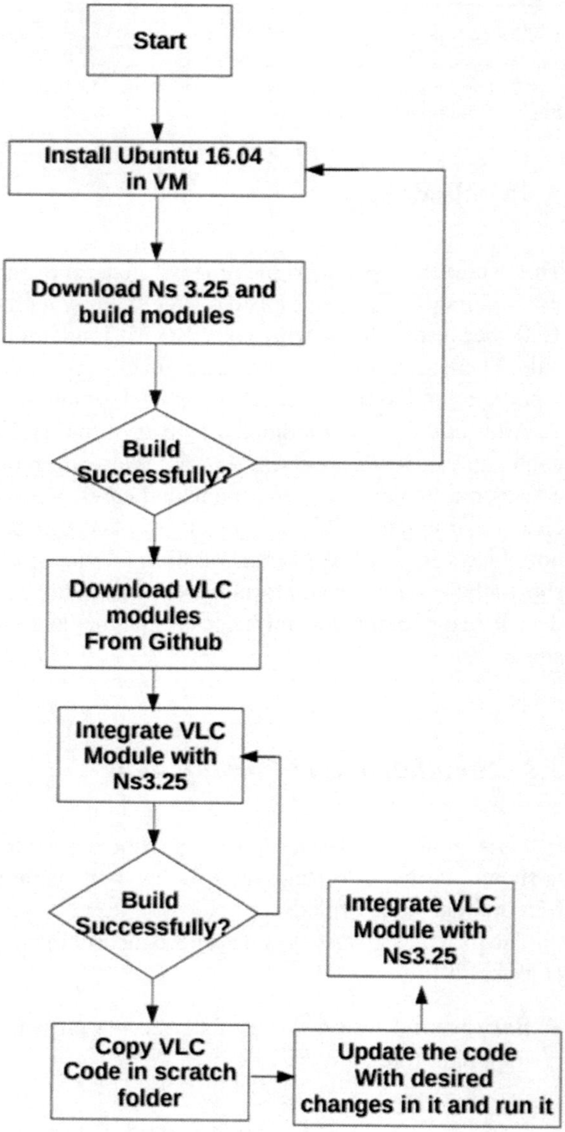

Fig. 4 Thorough flowchart of working of our experiment

- **Thermal noise** occurred due to heat generated among the circuits at receiver end.

$$s_{\mathrm{TH}}^2 = (4KT_{\mathrm{e}}FB)/R_{\mathrm{L}} \tag{2}$$

Table 1 Simulation parameters

Parameters	Standard values
Topology used	Bus, peer to peer, wireless
Field of view	70, 120, 170°
Lambertian angle	70°
Filter gain	1
Data rate	5 Mbps
Distance	1–25 m
Refractive index	1.5
Biasing at transmitter	0.5e−3 V
Duty cycle	0.85
Delay	6560 ns
Electric filter bandwidth	5 * 1e6
Band factor noise signal	10
Temperature	290 K
Photo detector area	1.0e−4
Boltzmann constant	1.381 * 1e−23
Background radiance	0.2 Wm^{-2} $nm^{-1}sr^{-1}$
Fiber optic bandwidth	10 NM
Filter transitivity	0.5
Resistance	100 Ω
Electric charge	1.6 * 1e−16 C
Speed of light	3 * 1e8 m/s
Planks constant	6.6 * 1e−34 J/s
Circuit noise figure	4

where K means Boltzmann constant, T_e means equivalent temperature, F denotes circuit noise figure, B tells about detector end electronic bandwidth, and finally R_L denotes load resistance.

- **Shot noise**—occurred due to junction of circuits present at receiver end.

$$s_{SS}^2 = 2qRP_SB \tag{3}$$

where q stands for electric charge, similarly P_S stands for reduced power, and finally R is

$$R = (nq\lambda)/(hc) \tag{4}$$

where q stands for electric charge, c is speed of light, n is the photo detector's quantum efficiency, and h is Planck's constant.

- **Background noise**—generated because of the disturbances in the channel, and here, it is due to fog.

$$s_{\mathrm{BG}}^2 = 2qRP_{\mathrm{BG}}B \tag{5}$$

- **Signal-to-noise ratio**—is the main formula of our computation which tells up to what extent signal is transmitted as compared to noise from transmitter to receiver end.

$$\mathrm{SNR} = R(P_{S1} - P_{S0}) / \left(\sqrt{\left(s_{SS}^2 + s_{\mathrm{TH}}^2 + s_{\mathrm{BG}}^2\right)} + \sqrt{\left(s_{TH}^2 + s_{\mathrm{BG}}^2\right)} \right) \tag{6}$$

Since we are using OOK modulation scheme, it consist of two states, 'on' and 'off.' Here in the formula, $P_{S1} - P_{S0}$ denotes change in power from 'on' state to 'off' state; in 'off' state, we only consider background power, and in 'on' state, we consider total power; so, here we will consider received power for our computation as it is the only change due to the involvement of fog.

In upcoming graphs, we considered all parameters for foggy channel whereas we ignored the parameter $s^2{}_{\mathrm{BG}}$ for clear channel.

4 Results and Analysis

This section presents the result of computations from above formulas and parameters, which are depicted in graphical form (see Figs. 5, 6 and 7).

If we observe the variations in graphs, we find few interesting observations. For example, in Fig. 5, SNR value in foggy channel is more than half of SNR in clear channel, but as we can see in Fig. 6, SNR change is 2.5 times between both channels, whereas for FOV 170° as shown in Fig. 7, SNR change is nearly 4–5 times.

This may be due to the fact that as we increase the angle, receiver is more prone to the disturbances in channel, which may be the reason that for higher values of FOV the effect of fog on SNR is much larger as compared for lower angle of FOV.

By observing the following graphs, we draw few inferences and few suggestions. These results were obtained when distance between transmitter and receiver was set to 25 m.

- We observe a consistent rise in SNR in clear environment at various angles, because as the angle increases, the capacity of receiver to capture light increases.
- We observe an inconsistent rise and fall in SNR in foggy channel at various FOV angles due to channel disturbances.
- Since SNR change is dependent on channel, so we propose a hybrid model.

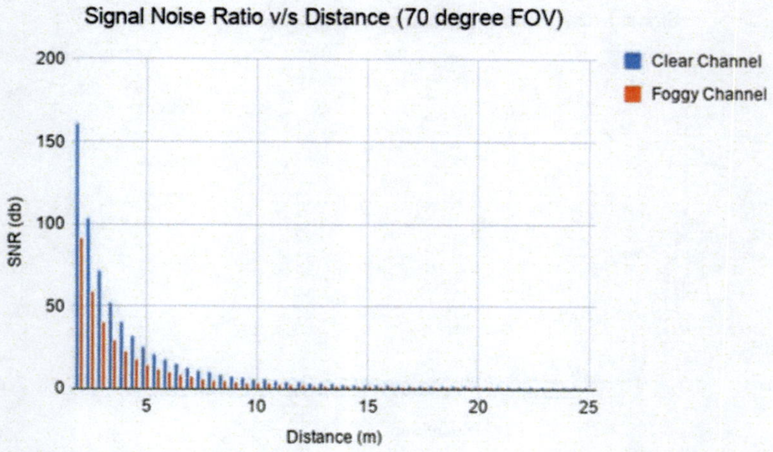

Fig. 5 Describing SNR in the presence and absence of fog when FOV is 70°

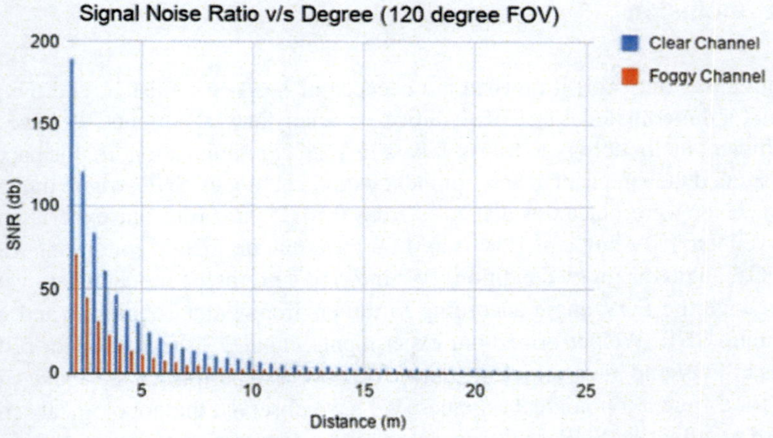

Fig. 6 Describing SNR in the presence and absence of fog when FOV is 120°

- In hybrid model, the vehicle must detect the environmental conditions, so that it can switch to the most suitable receiver angle for maximum efficiency.
- If it is clear conditions on road, the vehicle must increase the FOV angle to a value where we obtain maximum SNR as shown in Fig. 6 where FOV angle is 120°.
- But if environmental condition is not favorable, i.e., foggy, the vehicle must detect it and switch to that FOV angle where SNR for foggy channel is maximum as depicted in Fig. 7 where FOV angle is 70°.

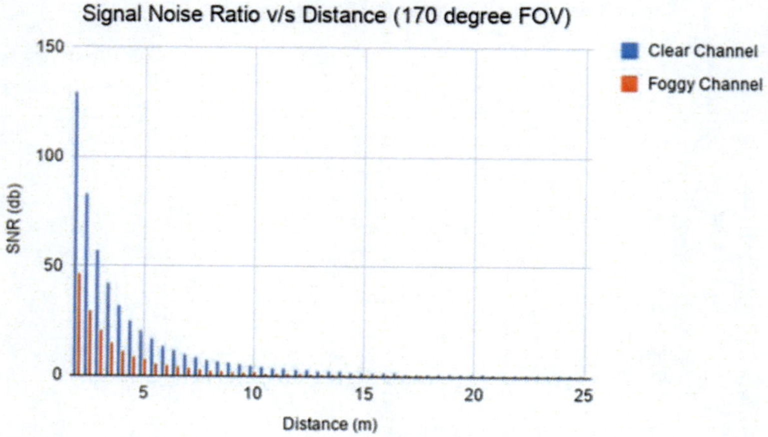

Fig. 7 Describing SNR in the presence and absence of fog when FOV is 170°

5 Conclusion

Major causes of accident on road are because of head-on collision, and the prime reason for this collision is lack of visibility, especially in winter nights. We have found the efficient methodology to transmit data in V2V communication. In this paper, we have studied the effects of an environment factor, i.e., fog on SNR, where transmitter and receiver were placed at distances from 0 to 25 m. From our experiment, we observed that FOV angle of 120° is best for clear environmental conditions whereas 70° FOV angle for foggy conditions. We inferred a prototype model where vehicles can change the FOV angle according to the environmental conditions and obtain maximum SNR. We performed our experiments in NS 3.25 simulator for different values of FOV and inter-vehicle distance. We have also performed the above simulations for different topological schemes. We have observed that topological schemes impose no effect on SNR, and their sole purpose is to define arrangement of nodes in a given network.

References

1. Demestichas P, Georgakopoulos A, Karvounas D, Tsagkaris K, Stavroulaki V, Jianmin L, Chunshan X, Jing Y (2013) 5G on the horizon: key challenges for the radio- access network. Veh Technol Mag IEEE 8(3):47–53
2. Feng L, Hu RQ, Wang J, Xu P, Qian Y (2016) Applying VLC in 5G networks: architectures and key technologies. IEEE Netw 30(6):77–83
3. International Agency for Research on Cancer, World Health Organization, IARC (2011) Classifies radio frequency electromagnetic fields as possible carcinogenic to humans. In: Press Release No. 208, 31 May 2011

4. Cen N, Jagannath J, Moretti S, Guan Z, Melodia T (2019) LANET: visible-light ad hoc networks. Ad Hoc Netw 84:107–123
5. Rehman S, Ullah S, Chong P, Yongchareon S, Komosny D (2019) Visible light communication: a system perspective- overview and challenges. Sensors 19:1153
6. Mohan D, Tsimhoni O, Sivak, Flannagan MJ (2009) Road safety in India: challenges and opportunities. In: UMTRi-2009–1
7. O'Brien DC et al (2008) Visible light communication: challenges and possibilities. In: IEEE 19th International Symposium PIMRC, pp 1–5, Sept 2008
8. Ayyash M, Elgala HKA, Jungnickel Vlkerand Little T, Shao S, Rahaim M, Schulz D, Hilt J, Freund R (2016) Coexistence of WiFi and LiFi towards 5G: concepts, opportunities, and challenges. IEEE Commun Mag 54:64–71
9. Cui K, Chen G, Xu Z, Roberts RD (2010) Line-of-sight visible light communication system design and demonstration. In: 7th international symposium on communication systems, networks digital signal processing, pp 621–625
10. Yarkan S, Arslan H (2006) Identification of LOS and NLOS for wireless transmission. In: CROWNCOM, pp 1–5
11. Yang Y, Bagrodia R (2009) Evaluation of VANET-based advanced intelligent transportation systems. In: Proceedings of the sixth ACM international work-shop on Vehicular InterNETworking (VANET '09). ACM, New York, pp 3–12
12. Ling Q, Baoshan L, Yongxing D (2019) Progress report on visible light communication in intelligent transportation environment. J Phys Confer Ser 1168:022050
13. Rajagopal S, Roberts RD, Lim S (2012) IEEE 802.15.7 visible light communication: modulation schemes and dimming support. IEEE Commun Mag 5:72–82
14. Singh B, Gupta A (2015) Recent trends in intelligent transportation systems: a review. J Trans Lit 9(2):30–34
15. Liu Q (2014) A study on topology in computer network. In: Proceedings 7th International Conference on Intelligent Computation Technology and Automation, ICICTA 2014, pp 45–48
16. Adnan-Quidan A, Morales-Cespedes M, Garcia-Armada A (2018) Aligning the light based on the network topology for visible light communications. In: IEEE international conference on communications workshops (ICC Workshops), pp 1–6
17. Pablo A (2017) Visible light communication networks under ring and tree topology constraints. J Comput Stan Interfaces 52:10–24
18. Network simulator 3 Homepage, https://www.nsnam.org. Last accessed 2019/11/30.
19. Aldalbahi A, Rahaim M, Khreishah A, Ayyash M, Ackerman R, Basuino J, Berreta W, Little T (2016) Extending ns3 to simulate visible light communication at network level. In: 23rd international conference on telecommunications (ICT), pp 1–6
20. Urachada K, Sermsak J, Yasuo K, Akira I, James R (2005) Channel modeling for optical wireless communication through dense fog. J Opt Netw 4:291–299

An Algorithm for Target Detection, Identification, Tracking and Estimation of Motion for Passive Homing Missile Autopilot Guidance

Manvinder Sharma and Anuj Kumar Gupta

Abstract The autonomous weapons which can identify the correct targets without any human intervention are in demand with the development of defense and war scenarios. The dynamics of the flight path is decided by the missile guidance system to achieve different types of mission objectives. Image processing equipped with intelligent sensors can identify any type of target other than traditional method of detection of only fire (aircrafts) or other signatures. The guidance system through image processing can differentiate between targets and can provide the latest error correction in the flight path. For detecting a particular object within an image, detection using point feature method is much effective technique. The point feature matching is done by comparing various correspondence points of object and analyzing the points between cluttered scene images to find a required object of interest in image. An algorithm which works on finding correspondence points between a target and reference images and detecting a particular object (target) is proposed in this paper. Tracking of object and estimation of motion model is also proposed by taking constant velocity and constant turn rate model. The real performance can be achieved by identifying the target in image using this detection approach and estimation of its motion.

Keywords SURF · Object recognition · Objects capture · Tracking · Motion estimation

M. Sharma
Jaipur National University, Jaipur, India
e-mail: manvinder.sharma@gmail.com

A. K. Gupta (✉)
Department of Computer Science and Engineering, Chandigarh Group of Colleges, Landran, Mohali, Punjab, India
e-mail: anuj.coecse@cgc.edu.in

N. Marriwala et al. (eds.), *Mobile Radio Communications and 5G Networks*, Lecture Notes in Networks and Systems 140,
https://doi.org/10.1007/978-981-15-7130-5_5

1 Introduction

A missile guidance system can be defined as the system which works in accordance with guidance law and has a group of variables of components that measures the position/location of object (missile) and changes the flight path with respect to its target [1]. A missile guidance works on an algorithm to detect the target and control its acceleration. The missile guidance system typically uses sensing (heat etc.)or detection of target, computing distance and miss distance, and control component [2].

Figure 1 shows the block diagram of the guidance methods. The missile seeker section involves the input or command to move a certain path reaching toward target. These may have some sort of sensors to seek the target or some kind of signal processing/image processing techniques. It is basically a component which generates the data to be fed as input to missile computer.

Guided missile is the term used for missile post World War II era. This was because the weapons were replaced by the technology of guided missile. Due to the advancement in the technology of guided missile, the accuracy in the weapons used in the military was increased significantly, though along with it the threat complexity was also increased. The applications of missile technology include guided projectiles, air-to-air and surface-to-air guided missiles, surface-to-surface aerodynamic guided missiles, etc. [3, 4].

Guidance system with control system functions like an autopilot in aircraft. A guidance and control system is a flying servomechanism which positions an object in accordance with variable signal. It operates on the principle of reducing difference between two quantities. The guidance system is used to detect the presence of target, tracking the target, determining the correct path toward target and generating electrical steering signals in terms of actual position of missile with respect to required path [5]. Figure 2 shows the guidance system which develops the input signal (variable signal). The variable input signal represents the path toward the target. The control system responds to guidance system signal and changes the position of wings, ignition and control servo unit. Figure 3 shows the control system

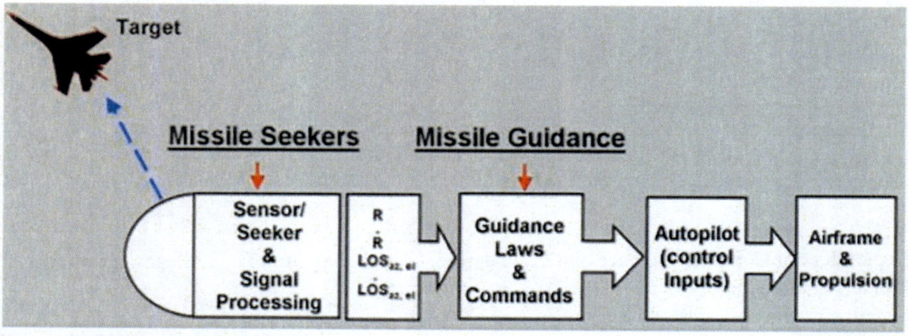

Fig. 1 Block diagram for guidance

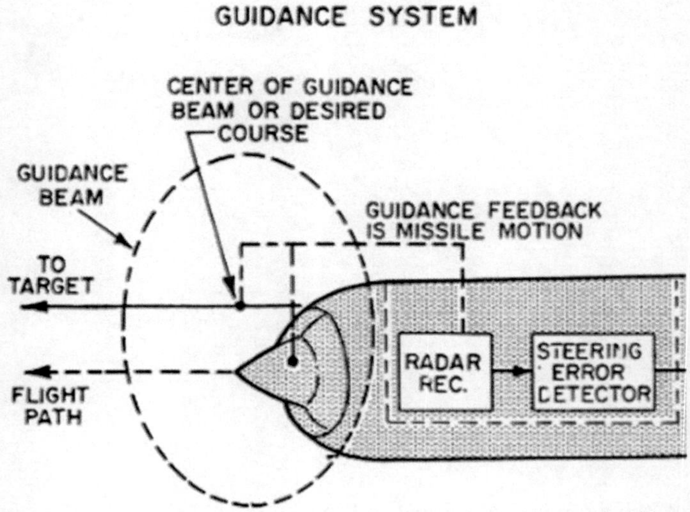

Fig. 2 Missile guidance system

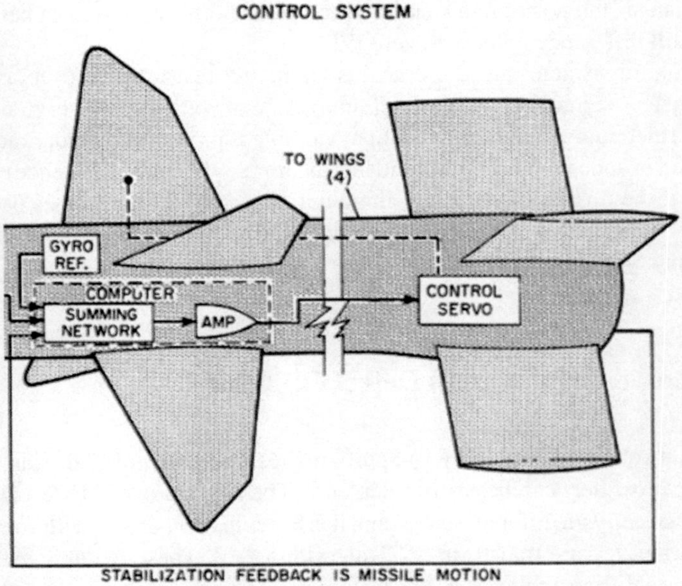

Fig. 3 Missile control system

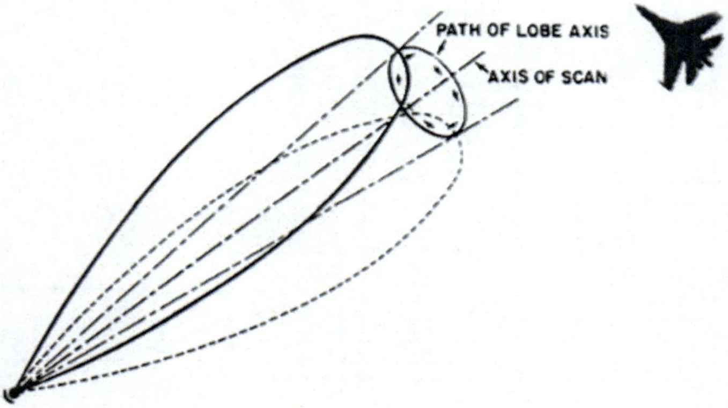

Fig. 4 Nutation axis and guidance

which takes the input variable signal and steers the missile toward the path [6]. The control system compares the path of target (input variable) and path of missile going on (output variable), if there is difference between the two variables, and then control system changes the position of missile to reduce the difference (error) between two variables till difference reduces to zero [7].

The control system mostly contains damping instruments (rate gyros and accelerometer), summing (addition and subtraction of voltages) and servo-amplifiers. For beam rider guidance system, the input variable is produced by radar receiver and comparison is done with center of guidance beam [8, 9]. Error (difference) voltage is produced if the missile is not flying along nutation axis (which defines path toward target), and control system makes corrective adjustments till the missile approaches nutation axis as shown in Fig. 4.

2 Proposed Algorithm to Detect Object

Speeded up robust feature (SURF) approximates Laplacian of Gaussian with box filter these box filters can be easily calculated. The advantage of SURF is that it can work concurrently on different scales, and it is faster in comparison with conventional scale-invariant feature transform SIFT algorithm [10]. These features make SURF to work on real-time application. For feature detection, local maximas within image are resolved by determinant of hessian which are then further used to select feature points in image [11].

Filtering images with a square and sum of image using integral image give faster result, and the formula is given as

Fig. 5 Proposed algorithm steps

$$S(x, y) = \left(\sum_{i=0}^{x} \sum_{j=0}^{y} I(i, j) \right) \tag{1}$$

The hessian matrix blob detector is used in SURF algorithm for the detection of the object in interest [12]. If the point is given as $p = (x, y)$ in an image I, the hessian matrix $H(p, \sigma)$ is rewritten as

$$H(p, \sigma) = \begin{pmatrix} L_x(p, \sigma) & L_{xy}(p, \sigma) \\ L_{yx}(p, \sigma) & L_{yy}(p, \sigma) \end{pmatrix} \tag{2}$$

where p is point in image and σ is scale.

$L_{xx}(x, \sigma)$ is convolution of second-order derivative of the Gaussian with the image. First, the determinant of the hessian matrix is calculated and then its non-maximal suppression which is used for SURF detection [13, 14].

Figure 5 shows the steps implemented in the proposed algorithm. The target data or the sample is provided in the first step. In the next step, the algorithm reads the image taken from the camera of the missile seeker. Then, the SURF algorithm extracts the features of input target image which has the following three steps: detection, description and matching [15]. Also, the features of image taken by seeker camera are extracted. The unique features are automatically identified in detection. The Gaussian kernel is approximated and the points of interest are selected by the use of spatial derivatives of Gaussian kernel and with local maxima in hessian distribution. Then, the description of each interest point is done. In the third step, matching of the convolved second-order derivative is done. 150 strongest feature points of target which is the big missile carrier truck were taken as reference image, and 350 strongest feature points of the image by seeker camera on missile were taken. Matching of the strongest features point is done, and the matching points are paired for removing the

Fig. 6 Image of target

outlines [16]. Finally, with the help of matched points, the target is located within scene of image by seeker camera is done. The satisfactory and fast results for colored images are obtained by implementing this approach.

3 Results and Discussion

The algorithm is implemented on MATLAB 2018a tool. The target was taken as aircraft which is shown in Fig. 6. The frame converted from camera video is shown in Fig. 7. Figure 8 shows feature point extraction of reference image (150) and image from camera (350) shown in Fig. 9. Figure 10 shows putatively matched points between target and input image. Figure 11 shows the detected target between the input images. Another scenario was taken in which the dummies are hanged with thread showing flying objects and from the seeker. The target reference image is taken as same. Algorithm again detected the target aircraft although the shape of blue aircraft (target) and yellow aircraft is same which is shown in Fig. 12. Figures 13 and 14 shows detection using different angles.

4 Motion Estimation and Tracking

For tracking of object, the image taken by seeker part of camera is preprocessed to extract he target. Using pcsegdist function, the point cloud segmentation belonging to target is classified into clusters. Using bounding box detection, each cluster is

Fig. 7 Scene taken from camera video

Fig. 8. 150 strongest points of target

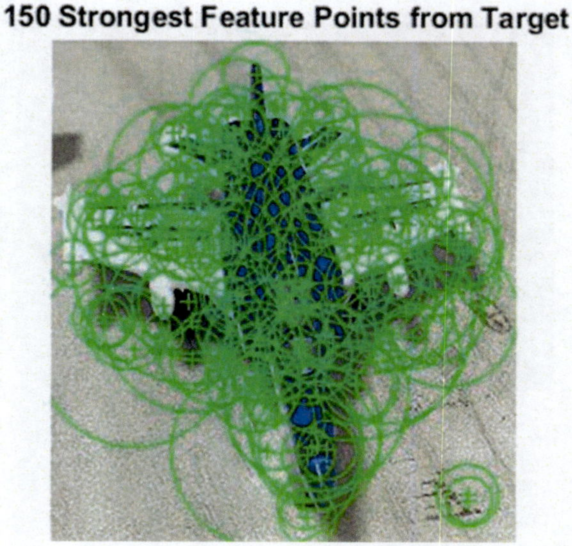

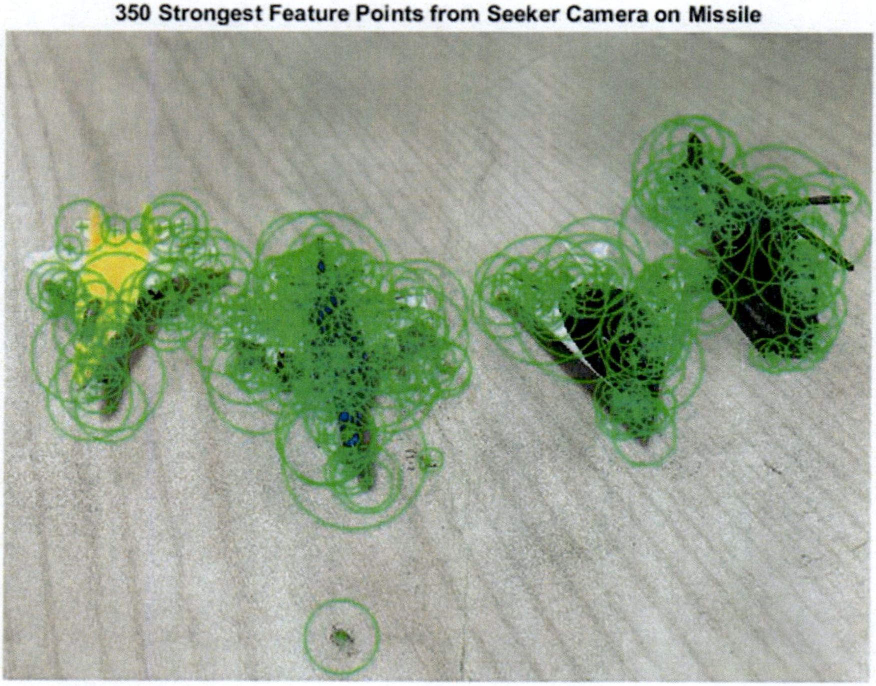

Fig. 9. 350 strongest feature points of seeker image

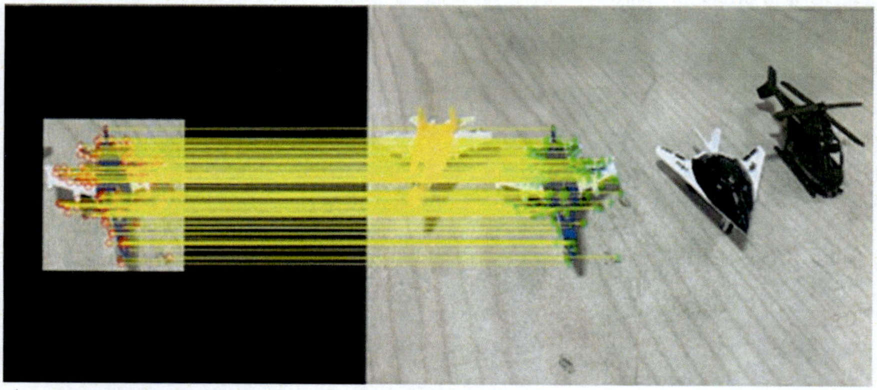

Fig. 10 Putatively matched points between target and seeker

converted as the following format [*x y z l w h*] where *l*, *w* and *h* refer to length, width and height of each cluster and *x*, *y* and *z* are axis positions of bounding box. Figure 15 shows the detected target using bird's eye view and tracking of detected target is estimated with two state-space models. Top view of detected target is shown in Fig. 16.

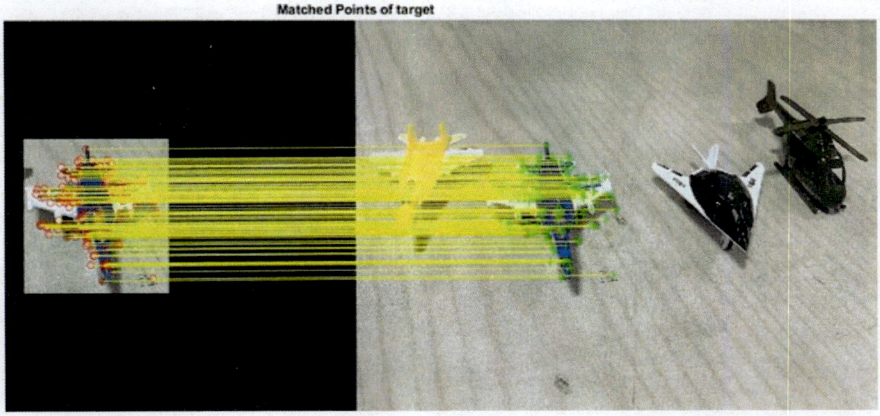

Fig. 11 Matching of feature points

Fig. 12 Detected target from seeker frame

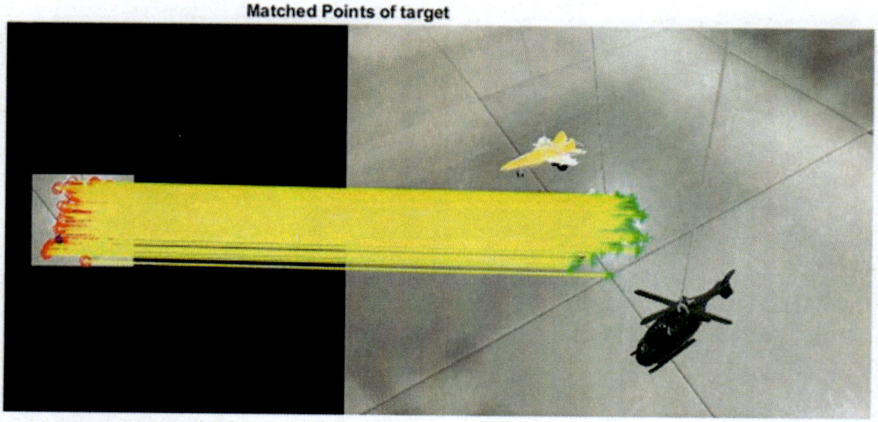

Fig. 13 Matched feature points of target

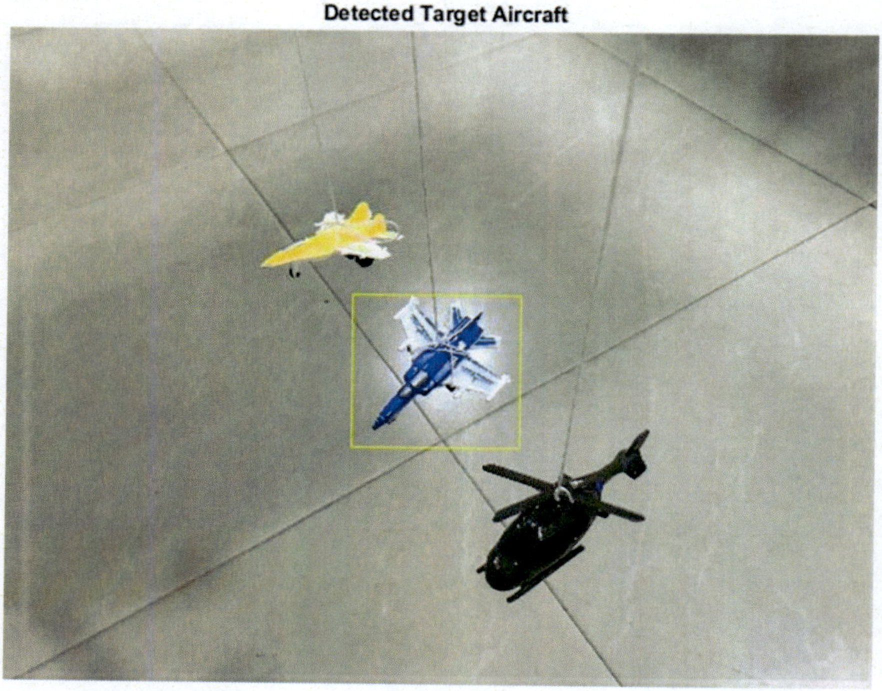

Fig. 14 Detected target aircraft at another angle

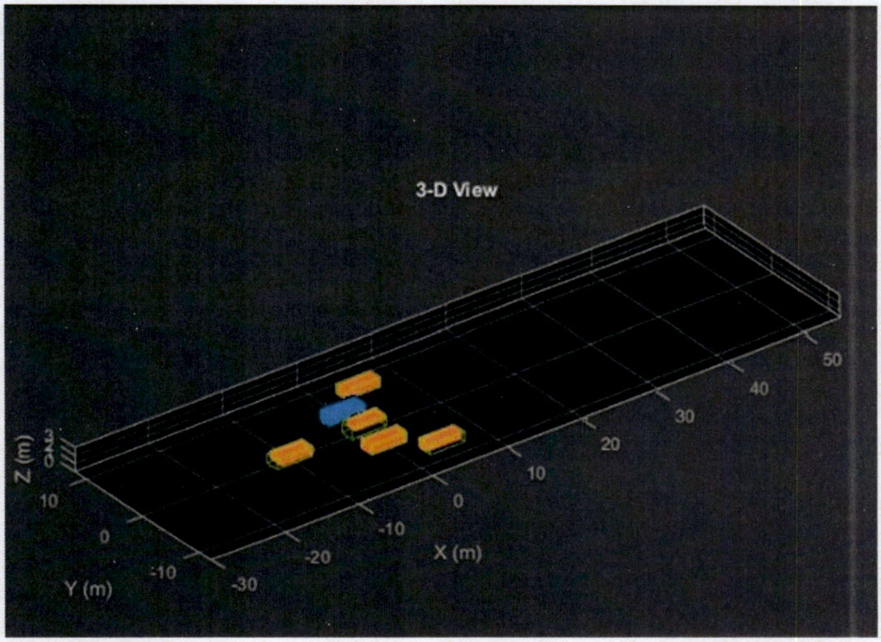

Fig. 15 Bird eye view of detected target

Fig. 16 Top view of detected target

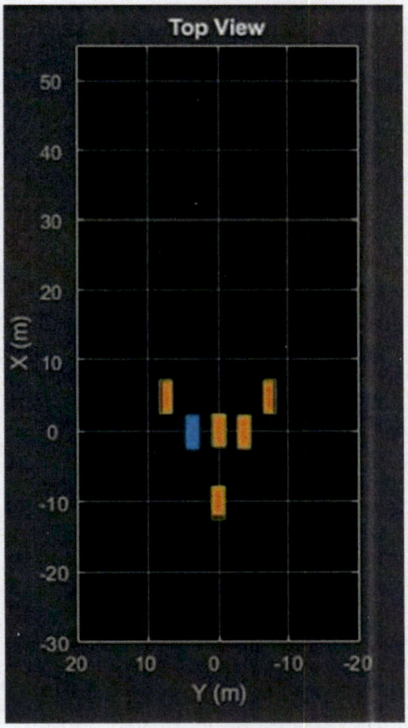

For tracking an object, the first step is defining its state as well as models which define transition of state and corresponding measurements which collectively known as state-space model of target [17]. Using a cuboid model, convention can be written as

$$x = [x_{ki}\theta lhw] \tag{3}$$

where x_{ki} is portion of state which controls kinematics.

θ is yaw angle and l, h and w is length, height and width of cuboid model.

For tracking, two models can be considered as two state-space models. The first state-space model is considered as constant velocity cuboid model, and the second state-space model is considered as constant turn rate model [18]. The kinematic part of state model for constant velocity can be mathematically described as

$$x_{cv} = x\dot{x}y\dot{y}z\dot{z}\theta lhw \tag{4}$$

And kinematic part of state model for constant turn rate can be written as

$$x_{ct} = x\dot{x}y\dot{y}\ddot{\theta}z\dot{z}\theta lhw \tag{5}$$

The probabilistic data is coupled with IMM filter [19, 20]. Using the supporting function, the IMM filter uses constant turn rate model (C_t) and constant velocity model (C_v) which helps to track switching between motion models. For events like lane changing or direction changing by target, these models achieve good estimation accuracy [21]. Figure 17 shows the estimation of maneuvering tracking of object when constant velocity (C_v) is taken more as compared to constant turn rate (C_t).In the first estimation model, the C_v is taken as 0.90 and C_t is taken as 0.10. The target direction can be tracked with C_v, C_t and $C_v - C_t$ mixed model.

In the second estimation, the values of C_v and C_t are taken as same. Figure 18 shows the target movement using the C_v, C_t and mixed velocities.

For the third estimation of tracking of target, the constant velocity is taken less as compared to constant turn rate. The value of C_v is taken as 0.10 and C_t is taken as 0.90. The model estimates the tracking of target which is shown in Fig. 19 with C_v, C_t and $C_v - C_t$ mixed. The probability of each model which is corrected with detection of target is updated by IMM filter, and error covariance is estimated to track the motion model probabilities.

5 Conclusion

The autonomous intelligent system and smart systems are required for the modern warfare so that they are self-initializing without the intervention of humans. Each target should be identified and hit very accurately with passive homing missiles as

Fig. 17 $C_v > C_t$

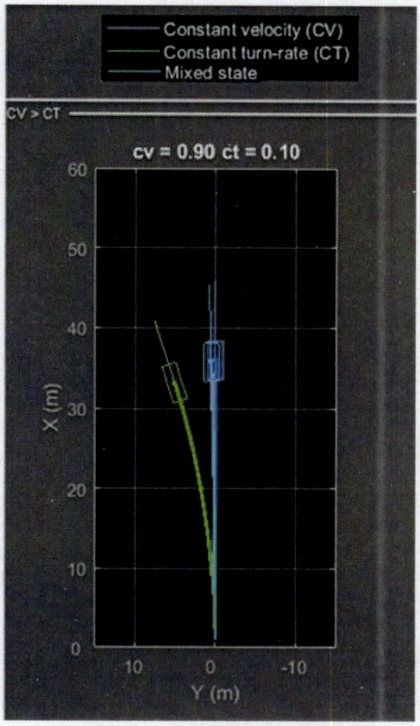

Fig. 18 $C_v = C_t$

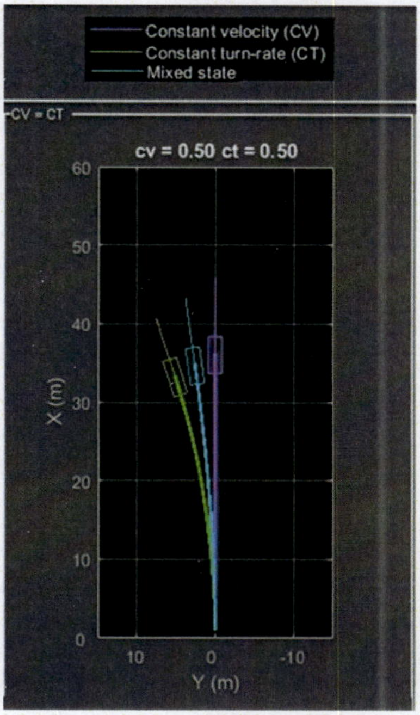

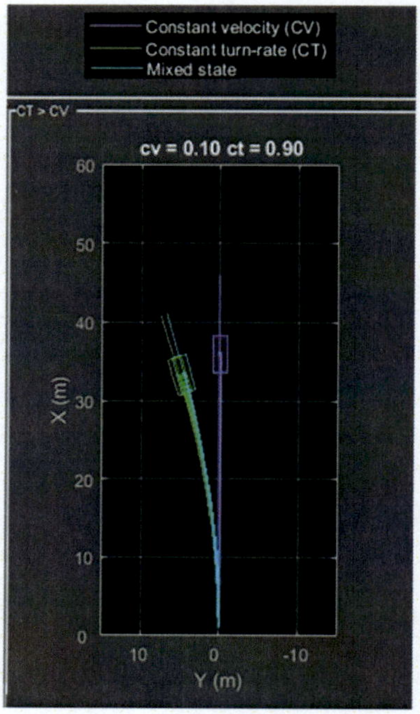

Fig. 19 $C_v < C_t$

they are the important elements of warfare. In this paper, work has been carried out for identifying the particular object (target) from a input seeker camera using modified SURF algorithm. The work presents mathematical modeling and basic notation for object characterization and identification. After simulation, the algorithm works faster and accurately identified the target out of two almost similar shape of aircrafts in experimental result. Also, the estimation of motion in direction of moving target is predicted. This model can control the path of the flight by correcting the error between the actual direction of missile and the identified object.

References

1. Yanushevsky R (2018) Modern missile guidance. CRC Press
2. Rao GA, Shripad PM (2005) New criterion for aircraft susceptibility to infrared guided missiles. Aerosp Sci Technol 9(8):701–712.
3. Peckham DH (1992) Guided missiles. U.S. Patent 5,139,215, issued August 18 1992
4. Sonawane HR, Mahulikar SP (2011) Tactical air warfare: Generic model for aircraft susceptibility to infrared guided missiles. Aerosp Sci Technol 15(4):249–260
5. Blakelock JH (1991) Automatic control of aircraft and missiles. Wiley, Hoboken
6. Jackson PB (2010) Overview of missile flight control systems. Johns Hopkins APL Tech Dig 29(1):9–24
7. Schumacher C, Khargonekar PP (1998) Stability analysis of a missile control system with a dynamic inversion controller. J Guidance Control Dyn 21(3):508–515

8. Guthrie JM, Casimir JB, Clark EA, Glen WA Jr, George EB, Donald IS (1961) Missile control system. U.S. Patent 3,010,677, issued 28 Nov 1961
9. Williams D, Jack R, Bernard F (1983) Design of an integrated strapdown guidance and control system for a tactical missile. In: Guidance and control conference, p 2169
10. Huijuan Z, Hu Q (2011) Fast image matching based-on improved SURF algorithm. In: 2011 international conference on electronics, communications and control (ICECC). IEEE, pp 1460–1463
11. Bouris D, Nikitakis A, Papaefstathiou I (2010) Fast and efficient FPGA-based feature detection employing the SURF algorithm. In: 2010 18th IEEE annual international symposium on field-programmable custom computing machines. IEEE, pp 3–10
12. Heo H, Lee WO, Lee JW, Park KR, Lee EC, Whang M (2011) Object recognition and selection method by gaze tracking and SURF algorithm. In: 2011 international conference on multimedia and signal processing, vol 1. IEEE, pp 261–265
13. He W, Yamashita T, Lu H, Lao S (2009) Surf tracking. In: 2009 IEEE 12th international conference on computer vision. IEEE, pp 1586–1592
14. Zhou D, Hu D (2013) A robust object tracking algorithm based on SURF. In: 2013 international conference on wireless communications and signal processing. IEEE, pp 1–5
15. Sakai Y, Oda T, Ikeda M, Barolli L (2015) An object tracking system based on sift and surf feature extraction methods. In: 2015 18th international conference on network-based information systems. IEEE, pp 561–565
16. Du M, Wang J, Li J, Cao H, Cui G, Fang J, Lv J, Chen X (2013) Robot robust object recognition based on fast surf feature matching. In: 2013 Chinese automation congress. IEEE, pp 581–586
17. Wang H, Suter D, Schindler K, Shen C (2007) Adaptive object tracking based on an effective appearance filter. IEEE Trans Pattern Anal Mach Intell 29(9):1661–1667
18. Morris DD, James MR (1998) Singularity analysis for articulated object tracking. In: Proceedings 1998 IEEE computer society conference on computer vision and pattern recognition (Cat. No. 98CB36231). IEEE, pp 289–296
19. Lopez R, Danès P, Royer F (2010) Extending the IMM filter to heterogeneous-order state space models. In: 49th IEEE conference on decision and control (CDC). IEEE, pp. 7369–7374
20. Punithakumar K, Kirubarajan T, Sinha A (2008) Multiple-model probability hypothesis density filter for tracking maneuvering targets. IEEE Trans Aerosp Electron Syst 44(1):87–98
21. McGinnity S, Irwin GW (2000) Multiple model bootstrap filter for maneuvering target tracking. IEEE Trans Aerosp Electron Syst 36(3):1006–1012

Community Discovery and Behavior Prediction in Online Social Networks Employing Node Centrality

Sanjeev Dhawan, Kulvinder Singh, and Amit Batra

Abstract The identification of community arrangement has taken a massive attention amid the investigators in the last few years who are concentrating on the characteristics of big complex graphs, e.g., biochemical systems, social networks, e-mail systems, the Internet, and food networks. In this paper, an attempt is made to recognize community structure employing community discovery algorithm having different purposes that identify different clusters employing node centrality. Several different clustering techniques can identify clusters of vertices termed as communities in real graphs that uncover the characteristics and hence structure of different communities. The suggested algorithm uncovers the big communities with high quality. The concept of node centrality is easier to employ to uncover best partitions and hence results in the best convergence of the aforesaid algorithm. Different investigations on actual world social networks exhibit the efficacy of the suggested technique.

Keywords Community discovery · Social nets · Optimization function · Modularity · Cluster · Label propagation · Objective function

1 Introduction

Community discovery is uncovering of sub-clusters in a network [1], which are termed as groups or communities. There are numerous applications of graph clustering in several fields of technology and informatics. Community structure, which is

S. Dhawan · K. Singh · A. Batra (✉)
Department of CSE, University Institute of Engineering and Technology, Kurukshetra University, Kurukshetra, India
e-mail: amitbatra2011@gmail.com

S. Dhawan
e-mail: sdhawan2015@kuk.ac.in

K. Singh
e-mail: ksingh2015@kuk.ac.in

N. Marriwala et al. (eds.), *Mobile Radio Communications and 5G Networks*, Lecture Notes in Networks and Systems 140,
https://doi.org/10.1007/978-981-15-7130-5_6

a crucial and significant characteristic of big graphs, can be characterized as the clustering of vertices into clusters; for example, there is a large number of links within clusters as compared to amid groups. A large real-world network may be depicted by networks comprising of vertices and links. The identification of community structure is a crucial and significant task of research among researchers concentrating on the functional characteristics of existing graphs, e.g., email systems, food graphs, social networks, the Internet, and biochemical networks. Nodes can be individuals or organizations, whereas the edges may define the relationships among these individuals. For instance, in a friendship net like Facebook, a vertex represents an individual and a link depicts the friendship amid two friends. Different communities may have distinct number of vertices, distinct number of links. The nodes of these clusters are having large number of connections to the nodes of their own clusters, while they are having limited number of connections to the nodes of another clusters. The cluster discovery algorithms which need the acquaintance of arrangement of the whole network are known as universal techniques as compared to local algorithms that require knowledge of only local structure. Community discovery algorithms can be used to discover several partitions using the implementation of several definitions of communities. No single demarcation of cluster exists. There are a number of distinct definitions, based on which different methodologies can be adopted to discover the structure of different communities in a large complex network. Community discovery in graphs or networks, also popularly known as clustering, is not a straightforward problem. As such the community detection problem can be solved by employing several different approaches. There are several guidelines based on which different algorithms can be compared and evaluated with each other. Aggarwal [2] noticed that the count of nodes and links in different parts of the social nets may be different; i.e., distinct components of the social nets may have distinct denseness and unreal communities may be detected in sub-parts of the network by applying global algorithms. Most popular and used definition of community needs that apiece individual vertex should have huge links in the interior of the cluster or group than across [3]. Gargi et al. [4] suggested that universal techniques are not scalable for complex and changing nets. There are some hybrid techniques which efficiently combines both global and local search on a network. The most often used quality metric is termed as modularity which was suggested by Newman along with Girvan [5]:

$$Q = \sum_{i \in C} \left[\frac{E_i^{\text{in}}}{E} - \left(\frac{2E_i^{\text{in}} + E_i^{\text{out}}}{2E} \right) \right] \tag{1}$$

where set of all communities is denoted by C, one of the cluster in C is i, the count of internal links in community i is denoted by E_i^{in}, the count of external links to vertices outside of community i is denoted by E_i^{out}, and the count of links is depicted by E. Alternatively, the modularity can also be defined as under:

$$Q = \frac{1}{2E} \sum_{ij} \left[A_{ij} - \left(\frac{k_i k_j}{2E} \right) \right] \delta_{c_i, c_j} \tag{2}$$

where degree of node i is depicted by k_i, the element of sociometric matrix is depicted by A_{ij} (where $A_{ij} = 1$, when link amid vertices i and j exist, and $A_{ij} = 0$ else). Kronecker delta symbol is denoted by δ_{c_i,c_j}, and the cluster to which vertex i is allotted is having label *ci*. As for most complex networks, there does not exist a normal division; as a ground reality, it becomes more cumbersome to evaluate different community discovery algorithms. The modularity metric may have a resolution limit, and it is not possible to discover the communities, when these communities are small. Furthermore, complex networks with heterogeneous degree distributions have to be dealt with efficiently by employing several methods along with heterogeneous sizes of different communities. In the last few years, many novel techniques centered on progression calculation for cluster discovery challenges are suggested. This is because of the fact that the evolutionary computation can be employed to solve the optimization problems efficiently, which requires an appropriate illustration for the challenge and the task to enhance its capabilities. Natural techniques like particle swarm [6] and ant colony enhancement [7] and bat techniques [8] along with genetic algorithms and all evolutionary approaches have contributed a great toward community discovery. Another technique known as modularity optimization for discovering communities results in a challenge which is called as NP-Hard optimization challenge [9]. As a matter of fact, several approximation techniques have been employed like spectral techniques [10] or genetic ones [11, 12], which pertain to evolution algorithms. Label propagation methods [13] have been explored by which information which can be utilized in community discovery task is obtained from observing the neighborhood of each of the nodes [14] or the connections between nodes and can also be employed in evolution algorithms to decrease the search space and to increase the convergence. The community discovery task can be spread up by utilizing the information about the problem in the initialization of operators (problem-specific operators in [15]) and hence by decreasing the solution space. There are several merits of evolutionary computation approaches for community discovery. During the task of community discovery, the number of communities is calculated automatically. Evolution techniques are population-centric techniques which provide efficient implementations appropriate for big networks. Genetic techniques employ the guidelines of choice and progress to obtain various answers to a given challenge. In the field of artificial intelligence, the genetic techniques [14] are employed as the optimization strategies. Genetic algorithms are the very useful methods to find an answer to a complex challenge about which very few facts are identified and a technique of rapidly discovering a valid answer to a subtle task or challenge. Individuals in genetic algorithm are termed as chromosomes, which form a solution. The genetic operations like mutation, choice and crossover amid chromosomes are carried out to produce chromosomes for new population in every iteration. A possible solution is represented by each chromosome, and the genetic algorithm gives the best chromosome for the selected objective function. At the start, all the chromosomes are initialized and more number of chromosomes evolve in further iterations. A fitness value is allotted to every chromosome by fitness function, and it is predicted how best a chromosome is to get a answer to a problem. Several optimization functions are suggested for multiobjective community discovery like Community Score and modularity Q utilized by BOCD multiobjective optimization technique [16], or Community Fitness and Community Score employed in the MOGA-Net [17], which utilizes an optimization framework called NGSA-II [18]. The suggested community discovery algorithms mostly optimize only single objective function, whereas multipurpose enhancement is employed for cluster discovery. In multipurpose genetic technique called MOCD [12], which employs the enhancement framework PESA-II [19], both parts of modularity are utilized as enhancement tasks. Centrality-based algorithms have been found as common in clustering of individuals and have been found as very useful for clustering big databases. As a matter of fact, centrality centric techniques are not observed as better options for recognizing groups of random structures. Clusters observed through these centrality-based functions have convex structures, and every group or module is presented by a core or midpoint. In this paper, efficacy of centrality centric cluster discovery in nets has been discussed and studied. In this paper, a multiobjective solution called Net-degree has been suggested employing standard principle centered on the vertex centrality. As a matter of fact, we employ centrality, whereas the notion of significance of organizations and individuals in social nets was suggested by researchers in the early days [20] using the adjacency notion of middle individual who is significant in cluster and who has the largest impact on another. An attempt is made to discover partitions in social networks in which the metric, say, modularity or cluster score is not optimized; however, actual clusters in social nets are identified.

2 The Proposed Algorithm

The proposed algorithm is multipurpose enhancement which employs the vertex significance for community discovery and the technique used is NGSA-II [18] as multiobjective optimization framework. Firstly, a fitness value is assigned to every chromosome in the answer space. The population in NGSA-II is sorted into an order of sub-populace centered on the Pareto dominance ordering. NGSA-II determines fitness metric employing all optimized objectives. NGSA-II is a multiobjective optimization and genetic technique that gives additional answers, whereas a single-purpose optimization technique gives single optimum answer that optimizes single-purpose function. Solutions can be poor than other solutions as per single objective; however, solutions should be superior than other answer as per at least one other purpose. The resemblance amid individuals of every sub-cluster is evaluated on the Pareto front that provides inverse front of answers. Several Pareto optimal answers constitute a Pareto optimal front [21] in a multiobjective optimization problem. Whereas in each generation, the best chromosome remains, there is no loss of good chromosome. As per one objective function, every Pareto optimal solution is considered as good; however, there is only single answer that optimizes the entire fitness task. As per at least one optimization criteria, every Pareto optimal solution is considered as good. The chromosomes are represented by locus-centric adjacent representation [22], and for every chromosome, an array of integers is employed. We employed the fitness task on the basis of vertex significance and changed the stages of the genetic technique to fulfill the requirements of genetic technique for cluster discovery. The operators like initialization, crossover, and mutation are also changed. In size *n* network, every chromosome comprises of n genes $< G_1, G_2, \ldots, G_n >$. In the algorithm, firstly, the initial population is created. Every individual pertains to the alike cluster as single neighbor. The neighbor label is assigned as an individual community number. If ith gene has value *j*, it is meant that the vertices *i* and *j* are linked to a connection in the network. The aforesaid statement confirms that the vertices *i* and *j* pertain to the alike cluster. Two neighbors are linked by an edge if two individuals pertain to the same community. All the components which are connected in network '*G*' pertain to the alike cluster in a locus-centric adjacent representation. The aforesaid decreases the invalid search and restricts the possible solution space. Firstly, primary populace comprising of '*N*' chromosomes is created. Afterward, in further repetitions, better and new populace are created. The evolution begins with a set of chromosomes as set of solutions which represent the initial population. The term generation is used for population in every iteration. In this algorithm, crossover between individuals and mutations is carried out during every iteration. The above-stated task is occurred in an iteration. The new population is created by taking the solution from the earlier iteration. The optimization function values are determined for every chromosome, and the ranking of chromosomes is performed. The offspring and current population together build a combined population. The clone of the current population is generated proportionately, and offspring population is generated by carrying out modularity crossover and mutation operator on the replica populace. *M* child chromosomes are generated by carrying out crossover and mutation. The two optimization functions are minimized. The determination of distance values of all nodes is done, and the new population is comprised of first *N* nodes. A fitness rank is given to every solution employing NGSA-II algorithm (first level is the better, second level is next-better, and so forth). Afterward, sorting of combined population is done by NGSA-II. The updation of major populace is done. All the stages are carried out repeatedly leaving only the initialization till a threshold is reached.

2.1 Objective Tasks

In social networks analysis, the concepts of centrality were first formed. Importance of central individuals is higher than others in a network. In a cluster or a network, the central individual is nearer to everyone. Communication can be carried out directly with others by the central individual and knowledge can be spread by the central node easily in the network. An attempt is made to use two functions in the suggested multiobjective community discovery, where the first function is centered on the vertex significance while the subsequent function is based on the proportion of external links. The community is viewed as a central node where this node is having high degree surrounded by low-density neighbors. In order to obtain the best partition, the aforesaid two functions need to be minimized. The nodes which are not connected to the significant vertex of cluster to which individual pertain are counted, and the proportion of total of isolated vertices and count of individual vertices in a net is determined by using the first function. By employing the concept of centrality indices, the individual nodes can be ranked and most significant nodes can be recognized. An attempt is made to use the vertex degree significance, that is described by the degree, where degree is the count of neighbors of a vertex. The value of the above ratio must be little for a better quality cluster arrangement, i.e., the clusters which are largely separated are associated very densely within; however, these communities are associated with every other sparsely.

$$F_1 = \frac{\sum_{i=1}^{k} n_i - \text{Com}_i}{n};$$
$$\text{Com}_i = \sum_{i,j \in C_i; j \in N_i} \text{Com}(i, j);$$
$$\text{Com}(i, j) = \{1\, \text{comm}(i) = \text{comm}(j) 0\, \text{comm}(i)\text{comm}(j)\} \tag{3}$$

wherever, count of vertices in cluster is denoted by n, the community C_i comprises of n_i number of nodes and the number of communities are depicted by k, and the neighbors of the node i are depicted by N_i. In order to obtain a community that should comprise of huge interior connections amid individual vertices within the cluster than external connections to other clusters, the neighbors of every vertex must be within the alike cluster. An attempt is made to employ the principle of vertex superiority which is determined from the proportion of neighbors of the node which do not pertain to the alike cluster as the vertex and the count of all neighbors of the vertex. The quality of community structure is evaluated by the first function, and an attempt is made to suggest the subsequent task that evaluates the measure of quality of every vertex in a cluster. The second function (F_2) employs the average proportion of external links of every vertex and the value of F_2 must be small for a better quality cluster arrangement, i.e., the clusters which are highly separable should be connected densely internally; however, these communities should be connected sparsely to every other.

$$F_2 = \sum_{i=1}^{n} F_2(i);$$
$$F_2(i) = \frac{\sum_{i,j \in E; j \in N_i} \text{Com}\,(i, j)}{\deg(i)};$$
$$\text{Com}\,(i, j) = \{0\, \text{comm}(i) = \text{comm}(j) 1\, \text{comm}(i) \neq \text{comm}(j)\} \tag{4}$$

wherever, node i is having degree deg (i); i.e., count of neighbors and the set of links are denoted by E. The minimum of these two optimization functions is searched by the NSGA-II algorithm. For a better quality cluster arrangement, the value of

F_0 should be high comprising of clusters that are connected with denseness in the interior; however, these communities are linked sparsely to every another.

$$F_0 = 1 - F_1 - F_2 \tag{5}$$

2.2 Objective Tasks Employing Count of Detected Clusters

In order to get the best partition, the following two functions need to be minimized. An attempt is made to use other set of tasks that comprises also the count of discovered clusters in the determination of fitness value. The ratio of nodes which are not connected to significant middle individual vertex of cluster to which the vertex pertains and degree of vertex is used by the first task of the set (F_3). For good quality clusters that are connected with denseness in the interior; however, connected sparsely to every another, the value of F_3 should be low.

$$F_3 = \frac{\sum_{i=1}^{k}(n_i - \mathrm{Com}_i \big/ n_i)}{k};$$

$$\mathrm{Com_i} = \sum_{i,j \in C_i} \mathrm{Com_i}(i, j);$$

$$\mathrm{Com}\ (i, j) = \{1\, \mathrm{comm}(i) = \mathrm{comm}(j) 0\, \mathrm{comm}(i) \neq \mathrm{comm}(j)\} \tag{6}$$

wherever, count of vertices of a cluster is denoted by n, the count of vertices of the cluster C_i is denoted by $n_{i,}$ and the number of communities is denoted by k. An attempt is made to employ the same principle of superiority as we did in task F_2 employing the ratio of neighbors of the node that does not pertain to the alike cluster as the individual vertex and count of all neighbors. The quality structure of every community is evaluated by the first function, and an attempt is made to suggest the subsequent task which determines the superiority of every individual vertex in a cluster. The second function (F_4) employs the average proportion of external links of every vertex. In order to obtain a community that should comprise of huge connections in the interior amid individual vertices within the cluster than external connections to other clusters, the neighbors of every vertex must be generally within the alike cluster. For a better quality cluster arrangement, the value of F4 should be low, with the communities that are connected densely internally and connected with other communities sparsely.

$$F_4 = \frac{\sum_{i=1}^{n} F_2(i)}{k};$$

$$F_2(i) = \frac{\sum_{i,j \in E} \mathrm{Com}(i, j)}{\deg(i)};$$

$$\text{Com}\ (i, j) = \{0\,\text{comm}(i) = \text{comm}(j)1\,\text{comm}(i) \neq \text{comm}(j)\} \quad (7)$$

wherever, the count of vertices of a cluster is denoted by n, count of clusters is denoted by k, the set of edges is depicted by 'E', and the neighbors of node i is depicted by 'N_i'. The minimum of these both optimization functions is searched by NSGA-II algorithm, and for a better quality cluster arrangement, the value of obtained function Fk should be high, when the communities are connected densely internally, but connected sparsely with other communities.

$$Fk = 1 - F_3 - F_4 \quad (8)$$

2.3 *Mutation*

The chromosome of a populace is pushed out of a local minimum through the process of mutation and hence gives to detect a good minimum. The mutation process centered on neighbor is employed which is used to mutate genes by taking into consideration only the effective connections. The chromosomes with large number of altered genes often do not persist. Random modifications are made in chromosomes through the process of mutation, and a anew genetic material is put to the populace. Portion of prevailing individuals is arbitrarily modified to generate the offspring. The one of node neighbors label is used to replace the community label of the node. Mutation should not be carried out on large number of genes of chromosomes and should not be carried out often as it results into a random search. If mutation probability is greater than the random value generated, then every node of individual is mutated. Usually, the mutation probability used is 0.95.

2.4 *Crossover*

The parts from two parent chromosomes are combined to generate two new child chromosomes. Hence, the characteristics of two parents are found in the offspring created. Firstly, the random selection of two chromosomes is performed, and in the first chromosome, one individual node is chosen in random fashion. Crossover is carried out on two chromosomes. If the crossover probability is greater than the random value, crossover is executed. Usually, the crossover probability of 0.8 is used. An attempt is made to use crossover operator carrying out two-way crossing used in [23]. The determination of cluster of selected vertex is done, and all vertices of the first chromosome that pertains to the community of selected vertex are also allotted to the alike cluster label in the subsequent chromosome. Afterward, swapping of second and first chromosome is carried out, and repetition of the alike task is done.

2.5 Crossover on the Basis of Objective Task

In order to carry out the crossover operation, an attempt is made to employ the tournament choice of an individual from a populace of individuals for parent's selection. Firstly, a community is created that comprises of the alike vertices as the largest fitness value community in the descendants. Crossover is carried out to inherit good quality communities from one generation to the next with the largest optimization function fitness values to quickly achieve a local minimum. As a matter of fact, the crossover is carried out only on the higher fitness value nodes as predefined threshold 0.95. Using the optimization fitness values, sorting of all communities from two parents is done. Only vertices which do not pertain to this cluster can be allotted to the other cluster. Afterward, second-highest value cluster is created in the offspring; however, only from nodes not assigned earlier. Afterward, on the offspring, some mutations are carried out, such that every vertex pertains to the cluster to which the maximum neighbors pertain. As soon as all the vertices in the descendants are allotted, the task of creating communities is stopped.

3 Experimentation Results

Zachary karate club net is a popular social net [24]. The network is divided into two components as of conflict among the instructor and the administrator. Two clusters are recognized in the resultant segregate as shown in Fig. 1 along with the value of modularity as 0.36 and the optimization function value F_0 0.74. The clusters detected by the technique presented in this paper match with the result provided by Zachary. The modularity value of 0.42 is the best in the partition with 4 communities; however, the optimization function F_0 (0.51) is smaller in this partition than the 2 communities partition. Figure 2 depicts three communities where the modularity value is 0.38 and with maximum value of function Fk (0.74). In Fig. 3, four communities are recognized where the modularity value of 0.42 is the highest value; however, the achieved value of Fk 0.65 is lower than for three communities partition in Fig. 2. By visualizing the networks, the correctness of above results can be confirmed. If a comparison is done between partitions with three and four communities, one can observe that in partition with three communities, and the community with center 1 is partitioned into 2 communities. However, the center node 1 of other community is linked to four nodes, namely (5, 6, 7, 11, 17) of a new created community.

4 Conclusion

The network community discovery in this paper is designed as an multiobjective.

Optimization problem motivated by node centrality. An attempt is made to change initialization, mutation, and crossover for good community discovery. Furthermore, in this paper, crossover on the basis of objective function is used to increase the goodness and convergence of the algorithm. The investigations have been performed on the actual existing online social nets and showed the efficacy of the suggested technique in discovering center-centric clusters. More work has been done recently in the past few years. Emotional community detection in social networks has been performed by Kanavos et al. [25]. An incremental method to detect communities in dynamic evolving social networks has been proposed by Zhao et al. [26]. A novel trust-based community detection algorithm in social networks has been proposed by Chen et al. [27]. Such huge information can be used to carry out further research on analysis of social nets.

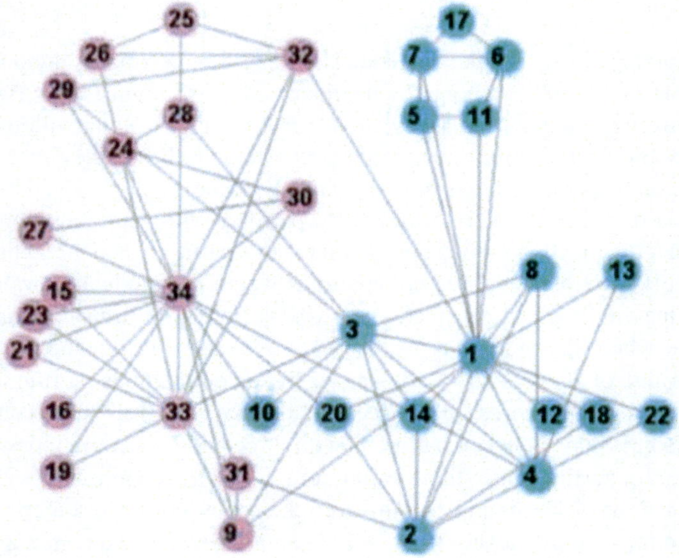

Fig. 1 Segregation of Zachary karate net into two clusters employing optimization function F_0 (0.74)

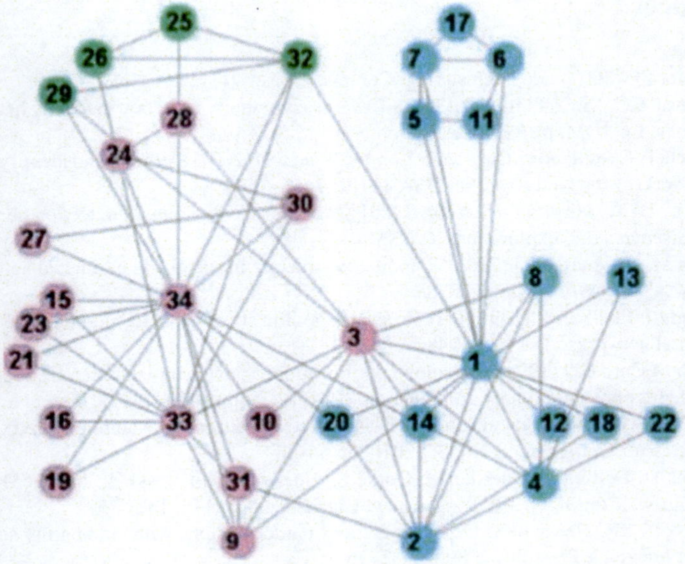

Fig. 2 Segregation of Zachary karate net into three clusters employing optimization function *Fk* (0.735)

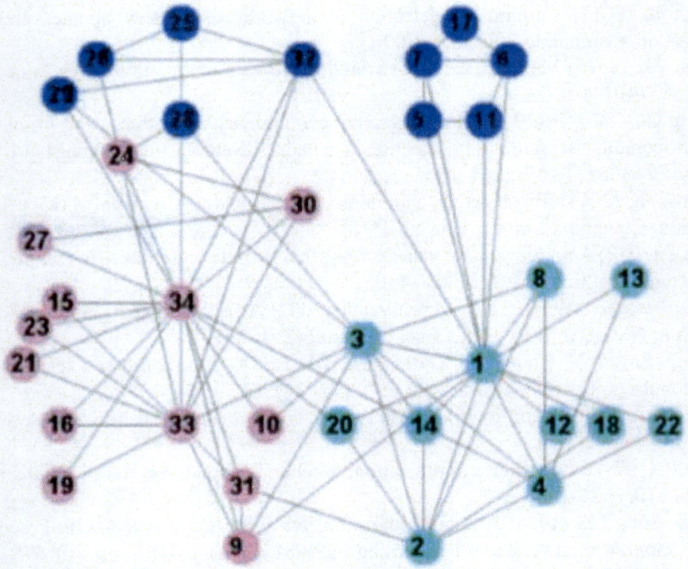

Fig. 3 Segregation of Zachary karate net into four clusters

References

1. Schaefer SE (2007) Graph clustering. Comput Sci Rev 1(1):27–64
2. Aggarwal CC, Xie Y, Philip SY (2011) Towards community detection in locally heterogeneous networks. In: SDM, pp 391–402
3. Radicchi F, Castellano C, Cecconi F, Loreto V, Parisi D (2004) Defining and identifying clusters in networks. Proc Natl Acad Sci USA 101(9):2658–2663
4. Gargi U, Lu W, Mirrokni VS, Yoon S (2011) Large-scale community detection on youtube for topic discovery and exploration. ICWSM
5. Girvan M, Newman MEJ (2002) Community structure in social and biological networks. Proc Natl Acad Sci USA 99:7821–7826
6. Kennedy J, Eberhart R (1995) Particle swarm optimization. In: IEEE international conference on neural networks, pp 1942–1948
7. Dorigo M,Caro GD (1999) Ant colony optimization: a new meta-heuristic. In: Proceedings of the congress on evolutionary computation. IEEE Press. pp 1470–1477
8. Yang XS (2010) A new meta heuristic bat-inspired algorithm. In: Nature Inspired Cooperative Strategies for Optimization-NICSO 2010, pp 65–74
9. Brandes U, Delling D, Gaertler M, Gorke R, Hoefer M, Nikoloski Z, Wagner D (2008) On modularity clustering. IEEE Trans Knowl Data Eng 20(2):172–188
10. Nadakuditi XR, Newman M (2014) Spectra of random graphs with community structure and arbitrary degrees. Phys Rev E 89(4):042816
11. Pizzuti C (2008) Ga-net: a genetic algorithm for community detection in social networks. PPSN, 1081–1090
12. Shi C, Yan Z, Cai Y, Wu B (2012) Multi-objective community detection in complex networks. Appl Soft Comput 12:850–859
13. Rizman K (2017) Community detection in networks using new update rules for label propagation. Computing 7(99):679–700
14. Rizman ŽK (2015) Maximal neighbor similarity reveals real communities in networks. Sci Rep 5(18374):1–10
15. Rizman ŽK, Žalik B (2017) Multi-objective evolutionary algorithm using problem specific genetic operators for community detection in networks. Neural Comput Appl 1–14. https://doi.org/10.1007/s00521-017-2884-0
16. Aggrawal R (2011) Bi-objective community detection (bocd) in networks using genetic algorithm. Contemp Comput 168(1):5–15
17. Pizzuti C (2012) A multiobjective genetic algorithm to find communities in complex networks. IEEE Trans Evol Comput 16(3):418–430
18. Deb K, Pratap A, Agarwal SA, Meyarivan T (2002) A fast and elitist multiobjective genetic algorithm: NSGA-II. IEEE Trans Evol Comput 6(2):182–197
19. Corne D, Jerram N, Knowles J, Oates M (2001) PESA-II:Region-based selection in evolutionary multi objective optimization. GECCO 283–290
20. Freeman LC (1979) Centrality in social networks: conceptual clarification. Soc Netw 1(3):215–239
21. Soland R (1979) Multicriteria optimization: a general characterization of efficient solutions. Decis Sci 10(1):26–38
22. Park YJ, Song MS (1998) A genetic algorithm for clustering problems. In: Proceedings 3rd annual conference on genetic programming (GP'98), Madison, USA, pp 568–575
23. Gong MG, Fu B, Jiao LC, Du HF (2011) Memetic algorithm for community detection in networks. Phys Rev E 006100
24. Zachary W (1977) An information flow model for conflict and fission in smallgroups. J Anthropol Res 33:452–473
25. Kanavos A, Perikos I, Hatzilygeroudis I, Tsakalidis A (2018) Emotional community detection in social networks. Comput Electr Eng 65:449–460

26. Zhao Z, Li C, Zhang X, Chiclana F, Viedma EH (2019) An incremental method to detect communities in dynamic evolving social networks. Knowledge-Based Syst 163(Chao Li):404–415
27. Chen X, Xia C, Wang J (2018) A novel trust-based community detection algorithm used in social networks. Chaos, Solitons Fractals 108:57–65

IoT–Blockchain Integration-Based Applications Challenges and Opportunities

Chaitanya Singh and Deepika Chauhan

Abstract In the era of the revolutionized world with a huge enhancement in the technology, various interconnecting devices are able to communicate in order to automate the today tasks for us. These IoT nodes require security, reliability, robustness and efficient management. This abundant information that shares in the network requires security measures. Blockchain technology has revolutionary impact on the world by transforming the way for sharing the information and provides a distributed environment with no centralized authority support in various industries. This paper emphasizes on the challenges faced in the integration of IoT and blockchain, surveys the various applications to analyse the potential of blockchain in the upgradation of IoT and addresses the development of some privacy preservation techniques for IoT system operating over blockchain infrastructure. This paper reflects the comparative analysis of various smart contract mechanisms used for the IoT platforms like Iota, Iotex and Iotchain. In this, we proposed the methodologies of integration used for the integration of both the technologies for the performance enhancement with respect to various parameters.

Keywords Blockchain · IoT · Consensus · Paxos · Smart contractor · Iota · Iotchain · Iotex

1 Introduction

The rapid evolution in the technology integrates the electronic devices and wireless networks to automate and provide advancement in the society. This advancement leads to the production of these miniature electronic devices in various areas and

C. Singh (✉)
Chameli Devi Group of Institutions, Indore, MP, India
e-mail: er.chaitanyasingh@gmail.com

D. Chauhan
Shivajirao Kadam Institute of Technology and Management, Indore, MP, India
e-mail: chauhandeepika522@gmail.com

N. Marriwala et al. (eds.), *Mobile Radio Communications and 5G Networks*, Lecture Notes in Networks and Systems 140,
https://doi.org/10.1007/978-981-15-7130-5_7

provides a new interface to the real world. The real world is now seen as a digital world. The Internet of thing has (IoT) is introduced as a collection of technologies integrated with each other from wireless sensor networks to radio frequency identification. IoT is the networking of physical electronic devices which communicate or share information between them with the help of sensors over the Internet [1]. IoT is the technology which helps to connect multiple devices along with the wireless network and done the communication between the devices using information sharing. IoT technology has played a prominent role in the upliftment and upgradation of society. It converts cities into smart cities, electrical grids into smart grids, houses into smart home, tourisms into smart tourism and health into smart health system. According to the Gartner Inc forecasts, the report on the automotive Internet of things (IoT) market which states that the IoT market will grow to 5.8 billion endpoints by 2020, which shows the 21% increase from 2019 [2].

IoT is the amalgam of hardware, software, data and services. The versatile nature of IoT has become very much popular in the technological trend as it covers all the industries. There are many advantages having a device based on IoT. The digital representation will generate by having various applications in various industries. The various sectors which undergo this automation are health care, smart cities, automobile automation, environment, smart grids, smart water management, smart transportation and large-scale deployments (Table 1).

IoT applications inculcate with specific features which make all the interconnected devices to generate enormous amount of data, long-duration connectivity support and high support for backups are required for the efficient working of devices connected. The devices uses are miniature category IoT applications that carry specific characteristics as they generate enormous amount of data, long-duration Internet connectivity support and high power support for backups and efficient working. As the

Table 1 IoT endpoint market by segment, 2018–2020, worldwide (Installed Base, Billions of Units)

Segment	2018	2019	2020
Utilities	0.98	1.17	1.37
Government	0.40	0.53	0.70
Building automation	0.23	0.31	0.44
Physical security	0.83	0.95	1.09
Manufacturing and natural resources	0.33	0.40	0.49
Automotive	0.27	0.36	0.47
Healthcare providers	0.21	0.28	0.36
Retail and wholesale trade	0.29	0.36	0.44
Information	0.37	0.37	0.37
Transportation	0.06	0.07	0.08
Total	3.96	4.81	5.81

devices used are miniature category; thus, they have some limitation in their functionality which includes power supply, storage memory, compute capacity, security, reliability, etc.

The heterogenic nature and integration in IoT pose various challenges for the technology. Another issues pose security and reliability of generated data as data needs to travel within the various networks for the efficient communication. Likewise in the centralized architecture, there exist issues related to fault tolerance, backup, recovery. Unauthorized entities can alter the information as per there convince or requirement; thus, the information shared in the network becomes unreliable. This brings us a need to verify the information which has been travelling in the network for the authenticity of the information.

One way to achieve the authenticity in IoT device by using the distributed services in which all the participants are authenticated users and also ensure mechanism to encrypt the data so that the data remains unalterable. For achieving authenticity we need to verifies all the users so that every user will able to verifies complete data from the initiation to the termination process that data should neither be altered nor tampered. To make the system reliable and fault tolerant the distributed services maintain the successful delivery of data to the destination.

Data immutability becomes a key challenge in the sectors in which traceability of asset is required. Likewise in the food industry if we need to determine the raw material used in the formation of the food products along with the destination where the food product has to be delivered finally. For example, if any food product firm has many clients and thousands of manufacturers through which the information has passed on for the process automation and regulate with some laws. Like if anybody wants to export or import any food product in different countries. In this scenario tracing of raw material used for the packaging, finishing, polishing, etc., and transportation of product between the intermediate destination would undergo with some laws which involve may participants who rely on some unautomated methodologies for the verification purpose which introduces the problem of tracking, delays in the delivery, unauthorized architecture or access, loss of product and unsuccessful delivery which has enormous impact on the economy of industry. Thus, IoT has the potential to transform and revolutionize the industry by digitizing and capturing the knowledge to control the process in real time to maintain the data privacy and reliability.

Thus, the reliability is the serious concern for the growing technology to overcome this new technology was introduced as the decentralized cryptocurrency which poses the potential to offer a better solution for the data reliability. The concept used in the bitcoin which revolutionized the mechanism of money transfer is based on the protocol which is called blockchain. This theory of blockchain is applicable for various use cases of digital world which includes voting system, smart contract, digital identity, supply chain management, fundraising, healthcare, food safety, etc. [3].

The work in the paper will project the below-mentioned information.

- Per lustration on blockchain technology with the illustration of unique features and challenges faced in the real-time application.
- Methodologies for performing the integration of IoT and blockchain.
- Analyse various challenges and advantages of IoT and blockchain integration for the real-time applications.

Chronicle organization of the paper is as follows. Section 2 introduces the theory of blockchain technology along with the challenges faced. Section 3 addresses the challenges faced during the integration of blockchain with IoT. Section 4 represents various blockchain–IoT-based applications. Section 5 shows the comparative analysis of various smart contract for IoT applications. Section 6 shows the conclusion and future work.

2 Blockchain

The issue of tampering of data can lead to serious issues in the economic world and thus needed a better solution to secure economic transaction. In reference to this problem, Satoshi Nakamoto in 2008 [3] presented two concepts that have a huge impact to resolve the issue of data privacy. The concept was introduced bitcoin, a virtual currency that maintains its value without the control of any centralized authority or any financial entity support rather than the coins and notes which are supervised by any centralized authority in the world. The concept was introduced the decentralized peer-to-peer communication system for the floating of information between the nodes or connected devices to make auditable and verifiable network.

The second concept gets the popularity which was blockchain itself. Blockchain is the mechanism which allows all the transactions to be verified by all the users in the network. This is an irreversible approach in which once you perform the transaction you cannot go back to it. This technology provides a distributed, secure, transparent and unalterable ledger. This is a public platform which allows all the users in the network to verify and access all the transactions since from the beginning to the last block added in the network. The blockchain protocol structured information in the form of set of transaction in the block. All the blocks in the network are interconnected to the previous block by the reference thus forms the chain of the block. To operate blockchain successfully, all the peers connected in the network require performing some basic functionality like routing, storage, wallet service, mining and recovery [4].

Blockchain is the distributed technology which helps to verify the economic transaction with the multiple users. The functioning of blockchain solely depends on three properties.

1. **Trustless**: There exist no needs to own digitally certified identity for the user. The entities involved in the transaction do not know each other but still able to exchange data without knowing each other.

2. **Permissionless**: There is no existence of any controlling authority that will give permission or take permission for performing the transactions.
3. **Irreversible**: They trust only on the quality of cryptographic algorithm used for the transactions to be performed successfully. A transaction once sent and accepted cannot be stopped.

Blockchain can be categorized into two types based on their functioning.

1. **Permissioned Blockchain**: This is the blockchain in which we limit the users who can participate in consensus algorithm. In this, limited users are able to validate the transaction and restrict users for smart contract generation.
2. **Permissionless Blockchain**: This is the blockchain technique in which any user can participate in the validation and verification of the transactions and has access to the consensus for generating the smart contract.

Functioning of blockchain undergoes the following steps which include routing, storage, wallet service and mining. Routing is the mechanism which is used to provide the efficient communication between the peers to peers in the interconnected network. Storage is the component which is required to save multiple copies of the transactions for the verifications, updations, backups and recovery purpose. Wallet services are required to authenticate the users by comparing their key pairs for all the users and for all the transactions. Mining algorithms are used to generate the block by using some algorithms or principles like proof of work. The nodes that perform proof of work are called miners, and they receive rewards and fees for decoding the transaction. Theory of proof of work is based on the election of leader after electing the leader, is responsible for validating the blocks and propagates the chain which means the addition of blocks in the blockchain.

Proof of work based on the mathematical computation, this is the widely used methodology in the blockchain but due to high criticization for the PoW as it is more expensive because it involves enormous computations (Fig. 1).

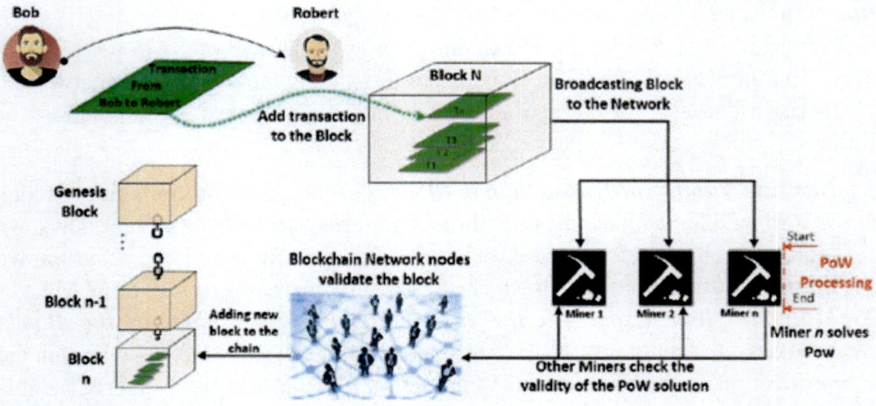

Fig. 1 Transaction block validation

Problem associated with the PoW is that multiple miners are involved in the process which leads to wastage of electricity and computational efforts.

2.1 Challenges

Blockchain provides highly secure architecture for the network still, and it undergoes some of the limitations.

2.1.1 Storage Capacity and Scalability

We know that blockchain is the database which is slow, immutable and highly redundant thus required huge scalability and maintenance. Due to the enormous growth in the transactions, the blockchain technology undergoes some challenges related to storage and scalability. The impact of growth in transactions is as follows.

- Average confirmation time for the transaction increases.
- The network transaction fees increases.
- The block difficulty is increased which is directly proportional to increase in computation power which in turn increases the electricity usage of the blockchain technology.
- The size of blockchain increase enormously leads the technology towards the difficulty in setting up for new full nodes which is required for the transaction complete processing and verification purpose.

Consider the example of bitcoin and ethereum algorithms, bitcoin processes an average seven (07) transaction per second and ethereum process an average twenty (20) transactions per second while the other payment gateways like PayPal manage two hundred transactions (200) per second and Visa manages average on 1667 transactions per second; thus, there is a need of scalability.

In blockchain technologies, the chains grow with the rate of 1 Mb per block in every 10 min in bitcoin and also store the copies among the nodes in the network.

To solve the issue of storage and scalability, we proposed some solutions as follows:

1. ***Increase Number of Transaction in Block***: The bitcoin algorithm's size of block is 1 Mb and liable to handle only three to four transactions per second. Thus, we require mechanism to increase the size of the block. In May 2018, a system we proposed called bitcoin cash which successfully increases its size to 32 MB.
2. ***Minimum Byte Usage for the Information Representation in the Block***: Blockchain technology stores information in the block which is used for the verification and validation of the transaction in the block. If we can reduce this information storage by any means than we can achieve better throughput. For this, we need to implement two things.

- Used hashing algorithm which generate a small signature for authentications. SCHNNOR signature is the way for implementing this.
- Use alternate data structure than Merkle tree for organizing the transaction inside a block. Merkelized abstract syntax tree is the alternative approach for handling in a better way.

3. ***Reduction of Time for Adding Block***: In proof-of-work algorithm of blockchain technology, time is the function of blockchain difficulty level. In bitcoin average time for creating block is 10 min, and in case of ethereum, the average time for creating block is 7 min. If we use any approach which helps to reduce this time, then the blockchain becomes efficient as there is increase in number of block addition in average time.
4. ***Increase the Connectivity Between Nodes***: In bitcoin, the transaction is communicated towards node twice.

- First time in the transaction broadcasting phase for the block addition.
- Secondly after the mining.

Thus, this complete process will require some time and also utilizing the network bandwidth for the processing which in turn generates some propagation delay. Thus for improving the performance, we need to reduce this bandwidth consumption and design mechanism to provide highly connected networks for the nodes to reduce the propagation delay in the network.

5. ***Speed of Consensus and Verification***: Speed of consensus plays vital role in the increase in throughput of the blockchain system, as in bitcoin proof of work leads towards high computation time and block difficulty totally depend on the scalability of the chain; thus to achieve high difficulty, we need to extend the chain which in turn decreases the speed of consensus. Thus to design a mechanism which helps to increase the speed and decrease the verification time for the block helps to increase the throughput.
   Alternative Options for this are:

- Litecoin is the alternative based on the concept of "SCRYPT" which helps to reduce the time from 10 min in bitcoin to 2.5 min in litecoin which work on the principle of serial execution of transaction rather than the parallelization of transactions [5].
- GHOST [6] is intended to improve the scalability of bitcoin by changing its chain selection rule. Off-chain solutions [7] are intended to perform transactions off the chain, increasing the bandwidth; at the same time, as it increases the probability of losing data.
- Bitcoin-NG is the alternative for a better performance in blockchain [8, 9] which overcome the limit of 10 min of bitcoin to as soon as possible.
- BigChainDB is the alternative work on the concept of big data distributed database and addition of blockchain characteristics which support high throughput and low latency [9].

6. ***Alternative Proofs***: This is an alternative which will help to increase the frequency of block addition as miners will not require spending their time in solving cryptographical puzzles.

7. ***Change in Storage Architecture***: In bitcoin, all the blocks are stored at all the nodes in the network which in turn needs enormous storage as the size of chain increases and the frequency of adding block is also increased which cause miners having minimum storage to be out of the market because miner node needs all the states of the transaction for the mining and validation purpose.
8. ***Off-Chaining***: This is the mechanism which promotes solving of puzzle off the chain which means do some processing out of the chain which helps to reduce the overhead over chain and reduce transaction processing cost [7, 8].

2.1.2 Security

Blockchain technology undergoes through various vulnerabilities and security threats. The most common threats occur are double-spending attach, 51% attack, man-in-the-middle attack and denial-of-service attacks.

- **51% Attack or Majority Attack**: This is a very common attack known as 51% attack and sometimes also known as majority attack. This attack can be done by increasing the computation power to the 51% of mining computation power. Many of the consensus algorithm proposed for the blockchain are subjected to the 51% attack. By getting 51% computation power, the attacker will get the value of nonce quickly and now he is authorized for permitting the next block for the addition in blockchain. After getting the authority, he/she would impact in the following way.
  - Apply restriction on block for verifying the transactions.
  - Restrict miners from mining any block.
  - Modify the transaction data which might leads towards the double-spending attack.
- **Double-Spending Attack**: This is the attack which uses the same coin twice in the transaction. This generally occurs when digital currency is disrupted or currency is stolen. Problem of double-spending can be solved by using the mechanism like proof of work for consensus. They generally occur due to the huge time variations and for fast payment scenarios and also can be achieved by the following ways.
  - By copying the digital currency and still having the original currency.
  - By simultaneously sending the coins to different peoples.
  - By reversing the transaction.
- Bitcoin manages the problem of double-spending by maintaining universal ledger for the verification of the transaction.
- **Race Attack**: This is the attack which occurs when two conflicting transactions are created by the attacker. This occurs due to the race of conformity of transaction first. In this, attacker sends first transaction to the victim directly who accept the transaction too quickly and then attacker sends another conflicting transaction returning the same amount to the actor is broadcasted to the network which makes

the first transaction invalid and probability of second transaction confirmation is high and the victim is cheated on.

- **Denial-of-Service Attack**: This is the attack in which possessed all the resources of the network becomes unavailable for the use by flooding of the transactions in a distributed way. This can be achieved by disconnecting mining pool, e-wallets, crypto exchange and other financial services of the network [10].
- **Consensus Delay**: This can be achieved by eclipse attack. Once the eclipse attack is done successfully, then victim node becomes unable to accept block from the network as timestamp of block does not match with the timestamp of the chain. Sometimes, it is also known as time jack attack.
- **Sybil Attack**: In this category attacker take control of multiple nodes in the network by creating fake id's and surrounded the victim node with all the controlled nodes of the network; thus, the node is subjected to Sybil attack. To overcome, this attack blockchain increases the cost of new id in blockchain and implement the consensus based on high computations.
- **Eclipse Attack**: These attacks generally focus on single node rather than the whole network node like Sybil attacks. These attacks make the node isolated from the other node by communication and get control over the chain by this victim node.
- **Mining Pool Attack**: Mining pools are required to mine large number of blocks by miners; they pool all the computation power and resources for better mining of the coins as this helps to share their rewards and obtains the highest computing power. The attackers are also become the part of these miners by exploiting the consensus algorithm internally and externally and attack over the network.
- **Finney Attack**: A Finney attack is possible when one transaction is pre-mined into a block and an identical transaction is created before that pre-mined block is released to the network, thereby invalidating the second identical transaction.
- **Forks**: Fork is the situation arises in which ideal single chain is breaking down into two or more valid chains which generally occur due to the software updations in the blockchain for the security concern. As all the nodes need to agree on the agreement update in consensus will lead the division of nodes into two types new node and old node which in turn create four scenarios.

  - The new nodes agree with the transactions of block which are sending by the old nodes.
  - The new nodes disagree with the transactions of block which are sending by the old nodes.
  - The old nodes agree with the transactions of block which are sending by the new nodes.
  - The old nodes disagree with the transactions of block which are sending by the new nodes.

Due do these four situations, fork can be categorized into three hard fork, soft fork and temporary fork.

- **Hard Fork**: When system comes with new agreement and not compatible with the older version and old nodes disagree with the new agreement because computing power of new nodes is high comparative to old nodes and break the chain into two parts. In this software validate according to older rule and block generated according to new rules. In this situation, we need to upgrade all the old nodes to the new version.

- **Soft Fork**: When system comes with new agreement and not compatible with the older version and new nodes disagree with the old agreement because computing power of new no Dark web is most comparative to old nodes but they work on same chain, and gradually, they upgrade themselves. In this block, generated are validated by older rules.
- **Temporary Fork**: When multiple miners done the mining of block, at the same time then entire network disagree with the choice of newly created block.

2.1.3 Data Privacy

In blockchain technology, key feature is transparency between the transactions and users. In this, all transactions are transparent from the initiation to the end of the transaction while some use cases require maintaining the privacy. In consensus algorithms, privacy cannot be achieved thus arise the problem of anonymity. To overcome this, blockchain provides two mechanisms zero cash and zero coin implementation which helps to generate anonymous transactions for the bitcoins by hiding the sender, receiver and interval details. Also, we can propose encrypted transaction to maintain the privacy.

2.1.4 Smart Contract

Smart contract is auto-executed line of code which is stored in the blockchain and work on the basis of some predetermined rules. They help to exchange property, shares, money or anything valuable. The execution of smart contract is based on the blockchain consensus algorithms and has several advantages over the traditional approach like cost reduction, efficient, transparent and fast. After having lot of advantages, it undergoes some limitations which are hacking, viruses, bugs, denial-of-service attack, poorly coded contract, loopholes and lost communication.

2.1.5 Legal Issues

Blockchain supports decentralized architecture; therefore, there exists no central control authority in case of fraudulent activity. Dark web is the most common fraudulent network which uses blockchain technology; thus, this is subjected to the legal conducts.

2.1.6 Consensus

These are the algorithm which helps to maintain the integrity of the information contained in the blocks and also guard transaction against the double-spending attack; thus, this is the key component of the blockchain as the value of consensus will decide the authenticity of transaction of the blocks.

Proof of Work (PoW)

Proofs of work are the algorithm used in the bitcoin; but nowadays, many other algorithms are available. Proof of work depends on the generation of coin on the basis of energy consumption.

Proof of Stack (PoS)

Proof of work required huge amount of computation for which high power consumption more electricity is required which makes the system costlier to overcome this high cost; an alternative was proposed by ethereum community called proof of stack which is cheaper and greener form of consensus. In this, mining depends on number of coins hold by the miners. It creates strong mechanism to defend against 51% attack. This undergoes with limitation as richer becomes richer which leads to the lowest number of participants which turns the system in long run to become centralized.

The Leased Proof of Stake (LPoS)

This is the enhanced version of the proof of stack which is used to overcome the limitation of proof of stack so that more participants should involve for the process of block addition. These algorithms allow participants to lease coins to the other participant so they become able to participate in the process of block selection and reduce the chance of controlling network by particular group or individual. Sharing of reward is done proportional to the coin held with the miners.

The Proof of Burn (PoB).

This is an alternative to proof of work which works on burning of coin means destruction of coin that is sending coin to the verifiable unspendable address known as an eater address for creating the new block in the chain. Eater address does not contain any private key; thus, nobody will gain access to that fund. **CounterParty** is based on PoB. However, this would lead to the increase in the price of bitcoin.

Proof of Importance

This concept was introduced by NEM which helps in the determination of the node which is eligible for adding the block in the blockchain. This process is named "harvesting" by the NEM. In exchange of these harvesting blocks, participated node gets transaction fees as a reward in the block. This protocol requires minimum of 10,000 vested XEM to become eligible for harvesting and harvesting can be done on the highest importance score.

3 IoT and Blockchain Integration

The biggest problem associated with IoT is the centralization control as all the communication between the nodes, validations, IoT device identification and verification should be controlled by the central authority generally cloud servers. But the problem exists with the present solutions that are high maintenance and infrastructure cost. But cloud server is vulnerable to single-point failure which impacts the complete ecosystem working which arises the need of peer-to-peer communication rather than client–server model [11].

Blockchain is tamperproof and decentralized technology and thus overcomes the limitations of IoT by resolving the issues of scalability, reliability and data privacy [12] (Fig. 2).

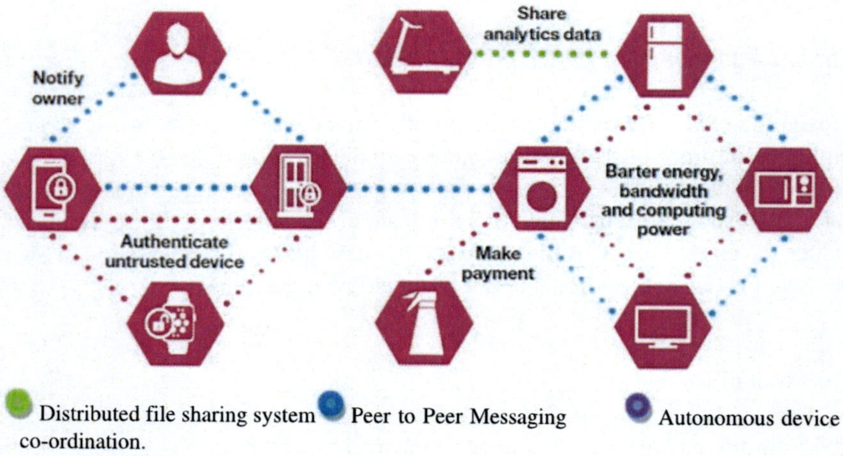

Fig. 2 IoT and blockchain integration

3.1 Advantages of IoT and Blockchain Integration

Integration of both the technology must require some enhancement in the new era to make result more reluctant.

- **Scalability and Decentralization**: The decentralized architecture makes the system to scale and helps to recover from single point of failure and bottleneck which in turns help to build fault-tolerant system and also provide extended storage facility which helps in the reduction of storage cost.
- **Unique Identity Management**: Blockchain integration will help to identify the object as each object is having a unique identity which in turn helps to identify the actual data provided by the device. It provides authorization and authentication of devices for the IoT available in the network.
- **Autonomy**: Blockchain technology empowers next-generation application features, making possible the development of smart autonomous assets and hardware as a service [13, 14]. With blockchain, devices are capable of interacting with each other without the involvement of any servers. IoT applications could benefit from this functionality to provide device-agnostic and decoupled-applications.
- **Reliability**: Integration helps to make IoT system more reliable. The trustless network of blockchain architecture helps IoT to secure the data and services by securing the transactions. Device connected in the network can communicate with each other by verifying and validating the service and data successfully. Blockchain enables accountability and traceability of service and data in IoT-based applications.
- **Security**: All the information related to the identity of devices and sensor has stored in the encrypted form in the form of transaction which make the system secure. The issue of hacking the sensor control information and data accessing can be reduced [15]. All the messages can be validated and verified by smart contract system; in this way, communication system has been improved.
- **Secure Code Deployment**: Immutable storage and tamper-proofing of transaction will help IoT to secure code used for deploying the sensors and devices in IoT.
- **Maintenance Cost Reduction**: Blockchain-based storage platform like SIA helps to reduce the storage cost required by IoT for storing sensor data and communication data which is of the enormous amount. Instead of using dedicated server, it works on the principle of rent out the storage for the system that also provides the monetizing facility to the data.

3.2 Integration Schemes for IoT and Blockchain

3.2.1 IoT–IoT Mechanism

This is the fastest and low latency-based approach in which IoT device can communicate with each other through routing protocol mechanism. In this approach, the interaction between devices can be performed without the use of blockchain. In this, some part of data is stored in the blockchain which required less storage space and the decentralization achieve is of lower quality.

3.2.2 IoT–Blockchain

In this approach, all the data related to the communication between the devices and the device associated data is store in the blockchain. In this approach, all the interaction goes through blockchain for enabling immutable records of interactions. In this, blockchain plays the role of controlling and monetizing the system and helps to increase the autonomy of the users. This is generally used for the use of cases based on rent such as Slock. Due to the storage of all, the details relevant to interaction in blockchain promotes an increase in bandwidth and enormous storage requirement.

3.2.3 Hybrid Approach

In this approach, combination of two approaches is used in which interaction performed using IoT–blockchain approach and data storage would use IoT–IoT approach; thus, the problem arises the identification of interaction which goes through blockchain. To overcome the interaction selection issue, we integrate some other technology too like cloud computing for storage and networking support, fog computing for mathematical computations [16, 17].

3.3 Challenges in Blockchain IoT Integration

3.3.1 Storage Capacity and Scalability

Storage and scalability are the major concern in the blockchain-based IoT integration system as we know IoT device process huge amount of data, and blockchain has some limitation over the number of transactions process per second as blockchain is not designed for processing huge amount of data or storing huge amount of data; thus, we can overcome this by using cloud computing in which cloud is used for storage and blockchain is used for securing the interaction between the IoT devices.

3.3.2 Security

IoT networks are subjected to various forms of vulnerabilities which make the security major concern; but due to integration with the blockchain, IoT undergoes some improvement but still their exist some issues related to security. Blockchain integration with IoT ensures that the data in the chain are immutable can able identify the change made in the transaction but if the data arrives to the blockchain transaction is corrupted data, in that case, the data remains corrupted in the blockchain.

There exist various for the corrupt data like malicious software's, viruses, bugs, environmental impact, failure of device, addition of noise. Along with these situations, there exist various types of threat which impact the data like denial-of-service attack and eavesdropping [18].

3.3.3 Data Privacy

The problem of data privacy in transparent and public blockchain has already been discussed, together with some of the existing solutions. However, the problem of data privacy in IoT devices entails more difficulty, since it starts at data collection and extends to the communications and application levels. Securing the device so that data is stored securely and not accessed by people without permission is a challenge since it requires the integration of security cryptographic software into the device. These improvements should take into account the limitation of resources of the devices and the restrictions related to economic viability. Many technologies have been used to secure communications using encryption (IPsec, SSL/TLS and DTLS).

3.3.4 Legal Issues

Some of the applications of IoT are unable to understand whether their network has to be managed by manufactures or make that network public to all the users. Because blockchain is an irreversible technology, it requires some legal regulation in a certain application which is not purely distributed or transparent. Thus, these regulation issues would impact the future of blockchain technology and IoT integration because it promotes the centralization or controlling characteristics which make the blockchain–IoT integration inefficient.

3.3.5 Consensus

There are various issues associated with the consensus algorithms like proof of work (PoW) as it consumes most of the computational power and a number of transaction processes per second are generally 3–7. PoS and DPoS are the algorithm which overcomes the limitation of power consumptions; but these algorithms suffer from the issue of liquidity of coins where poor becomes poorer and richer becomes richer. On the other hand, PBFT is the highest performer algorithm for the network but the identity of each node is known and thus ensures the issues related to anonymity.

4 IoT–Blockchain-Based Platforms and Applications

4.1 Smart Contract

Smart contract is the client management tool which is used for agreement between the nodes of the network in blockchain technology without the need of traditional contract agreement system. This helps to reduce the transactional cost to some extent like reaching agreement, formalization and enforcement. This can be used in various use cases like banking industry, insurance, telecommunication, education, real state, music industry. This is basically set of rules written in some programming language like solidity. This still undergoes with some limitations like.

- Complicated to implement.
- Practically still immutable

Ethereum and bitcoin are some examples for implementing smart contract system.

4.2 Hyperledger

Hyperledger was implemented by linux foundation, and this is an open-source platform which is used for the blockchain-based project deployment and generally used to implement permissioned blockchain. There are various advantages of hyperledger as it offers the platform for implementing high level of trust and helps to exchange asset on any network, optimized network performance and scalability. It supports multilateral transaction system. This platform helps developers to create plugging for individual workflow. This promotes modular architecture to the developers along with highly immutable distributed ledger support. Limitations of hyper ledger generally faced are complex architecture design, network fault-tolerant, insufficient skilled programmer and minimum use case support.

4.3 Lisk Framework

This is an open-source framework used for blockchain accessibility generally designed for Java script developers. This is the platform which works on modular architecture approach and provides developers with the toolkit for the implementation of applications based on blockchain along with the customization ease for the applications [19, 20]. It will help to maintain and implement connectivity between the modules of lisk. Limitation of lisk framework is extended block time which creates delay in the confirmation process of blocks. There are various use cases for the lisk like Java applications, decentralized application directories and blockchain.

4.4 *Quorum*

Quorum is a blockchain platform based on ethereum with high support for private, public transactions and smart contract system. This supports high transactional support with the support of 100 transactions per second. This is implemented for permissioned blockchain which is simple and based on major voting support system [20, 21]. This is very fast mechanism generally used for financial services. It allows multiple consensus mechanisms and achieves data privacy through cryptography and segmentation.

5 Experimental Analysis of Various Smart Contracts Used in IoT

In this, we are using the three platforms Iota, Iotex and Iotchain smart contracts and generate the results as follows (Figs. 3, 4, 5, 6, 7, 8, 9, 10, 11, 12, 13, 14, 15, 16, 17, 18, 19, 20, 21, 22, 23, 24 and 26; Table 2).

6 Conclusion and Future Work

Integration of IoT and blockchain undergoes some of the challenges which are mentioned in this paper. This integration model of two major technologies is having positive impact on the interaction between the citizens, government bodies and organizations. This integration promotes security of data and identity authentication but still it faces some challenge in case of scalability and storage. This paper projected

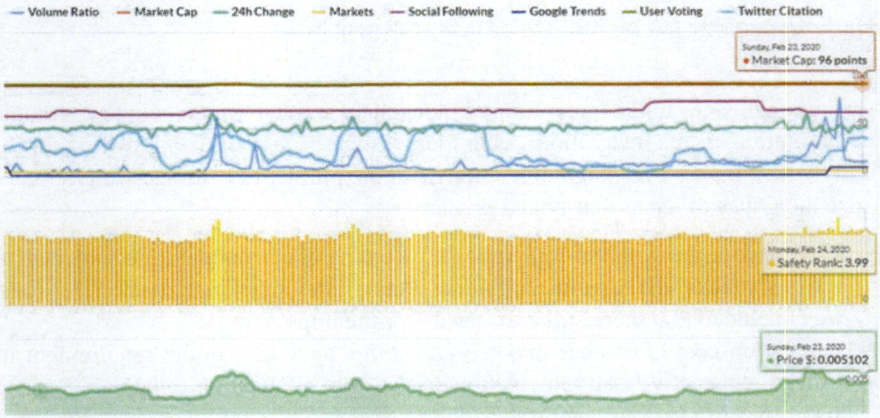

Fig. 3 Graph shows Feb 24, 2020, Market cap, safety rank and price of Iotex

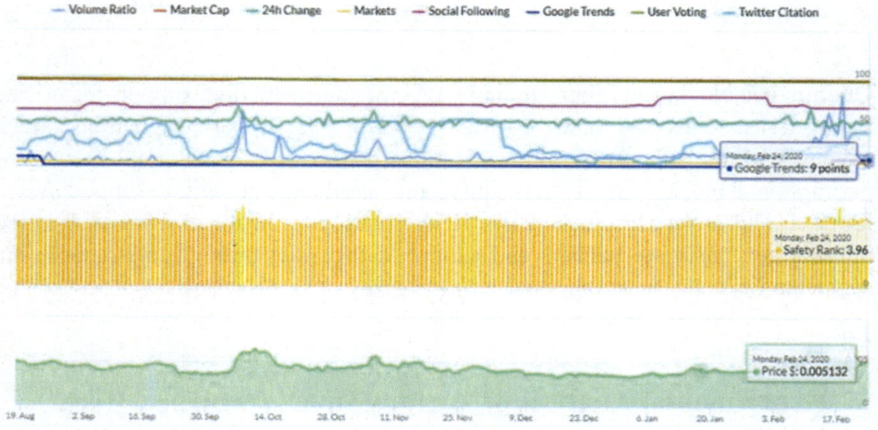

Fig. 4 Graph shows Feb 24, 2020, Google trend analysis of Iotex

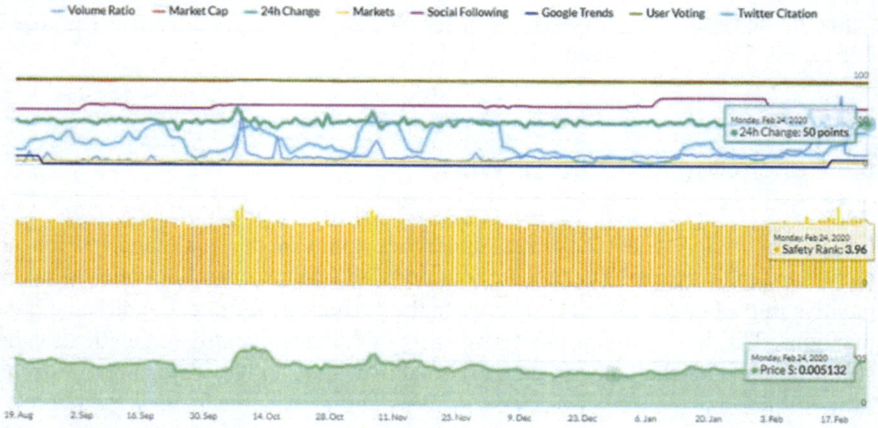

Fig. 5 Graph shows Feb 24, 2020. change in 24 analysis of Iotex

various challenges in the integration of blockchain and IoT which help to overcome the limitation of this integration. As in today's era, the size of data grows exponentially which makes the integration suffer with the problem of storage which in turn arise the issues of data security and privacy.

- The major issues with this integration are floating of corrupt data and subjected to various attacks like man-in-the-middle attack, time hijacking, attack 51%; thus, we concluded that there still exist needs for the improvement.
- Integration cost is another major issue in this approach as the requirement of storage, connectivity and interaction varies from application to application and also subject to various legal issues.

Fig. 6 Graph shows Feb 24, 2020, Social Following analysis of Iotex

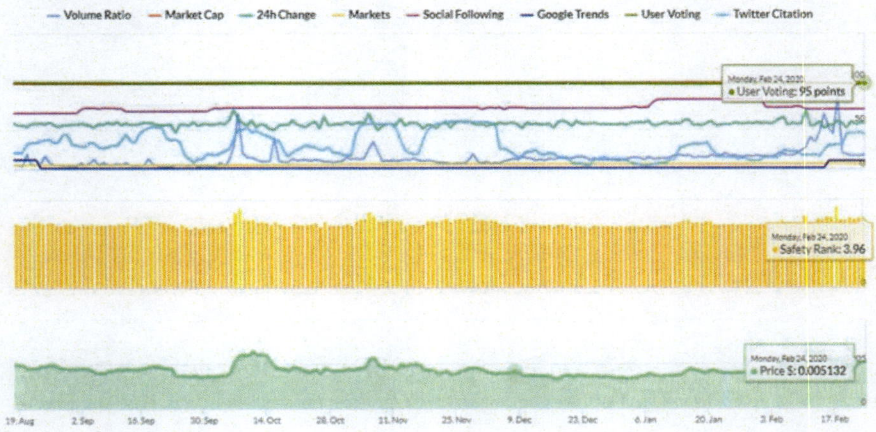

Fig. 7 Graph shows Feb 24, 2020, User Voting analysis of Iotex

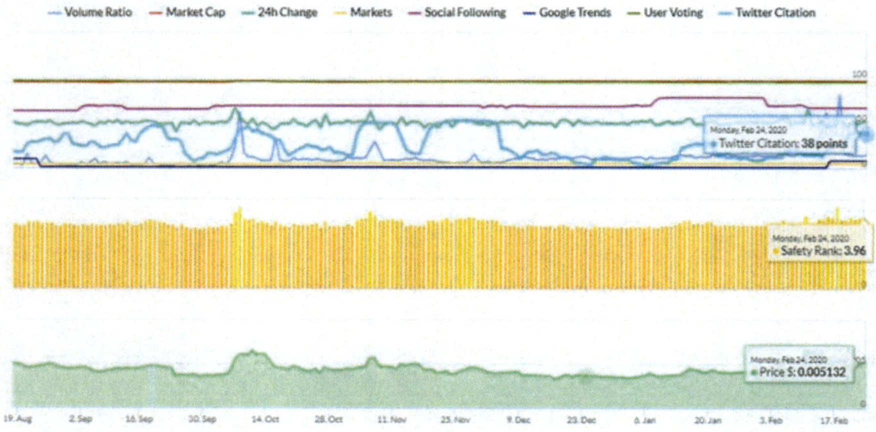

Fig. 8 Graph shows Feb 24, 2020, Twitter Citation analysis of Iotex

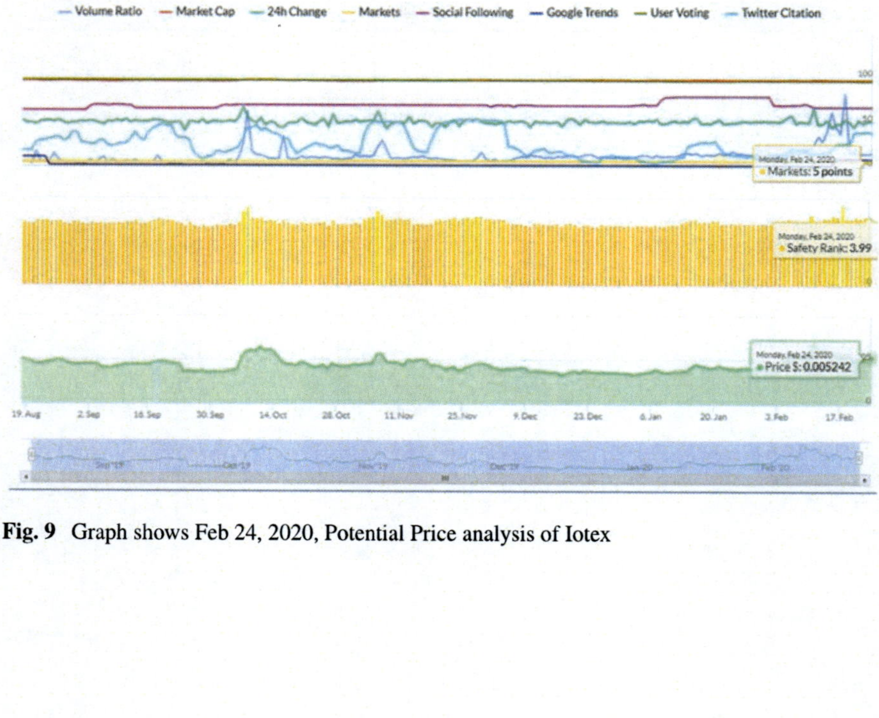

Fig. 9 Graph shows Feb 24, 2020, Potential Price analysis of Iotex

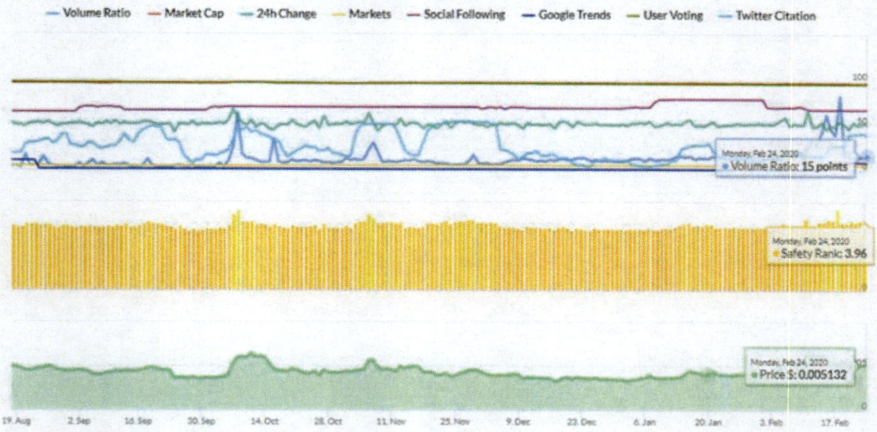

Fig. 10 Graph shows Feb 24, 2020, Volume Point analysis of Iotex

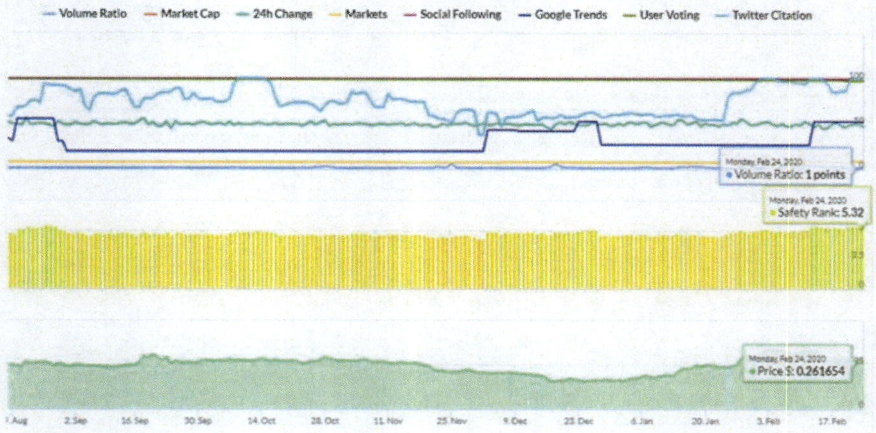

Fig. 11 Graph shows Feb 24, 2020, Safety Rank and Potential Price analysis of Iota

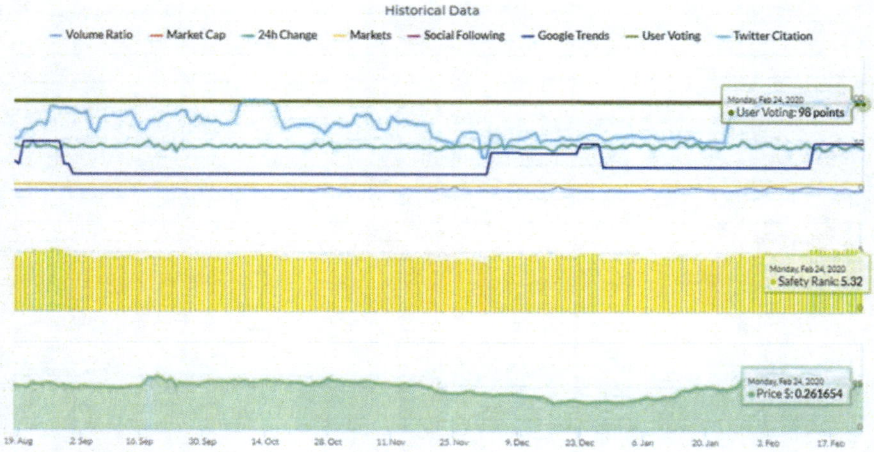

Fig. 12 Graph shows Feb 24, 2020, User Voting analysis of Iota

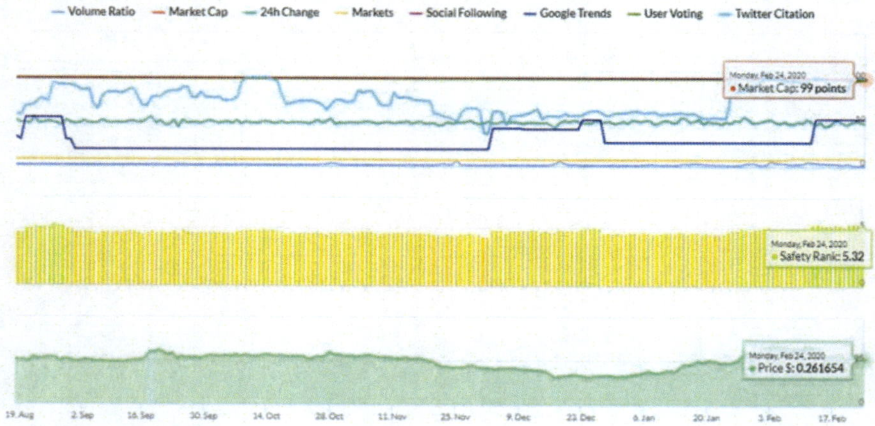

Fig. 13 Graph shows Feb 24, 2020, Market Cap analysis of Iota

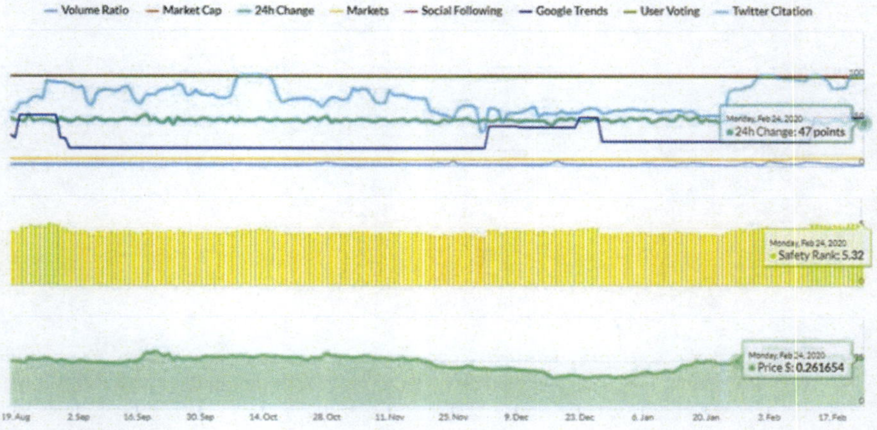

Fig. 14 Graph shows Feb 24, 2020, change in 24 analysis of Iota

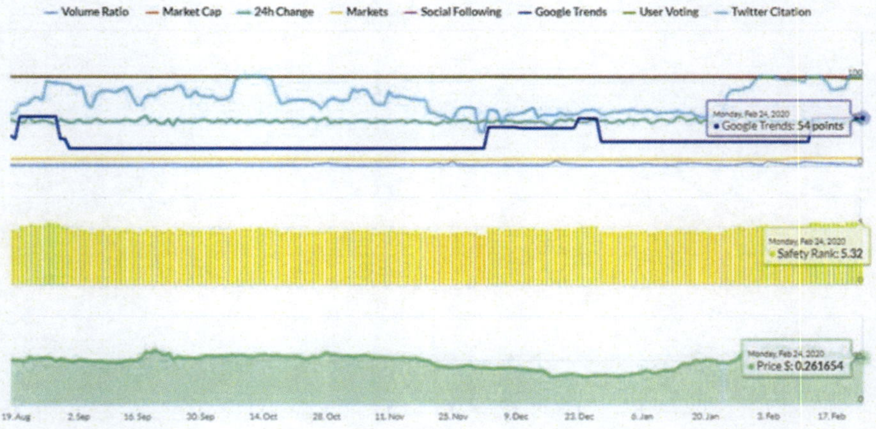

Fig. 15 Graph shows Feb 24, 2020, Google Trend analysis of Iota

Fig. 16 Graph shows Feb 24, 2020, Twitter Citation analysis of Iota

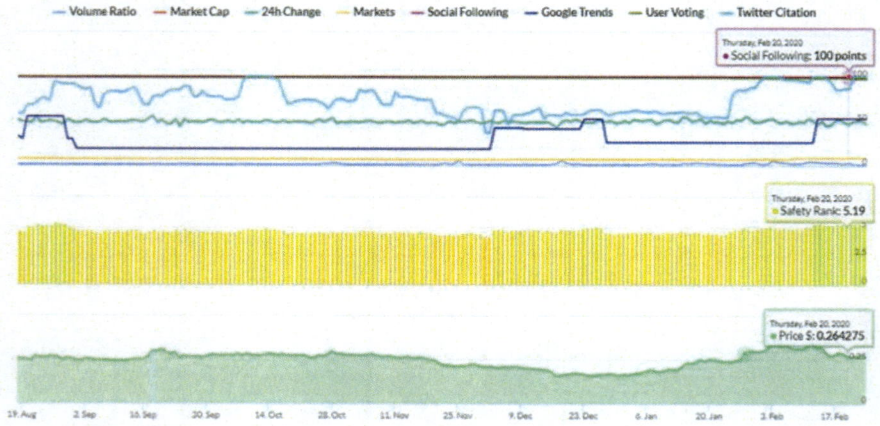

Fig. 17 Graph shows Feb 24, 2020, Social Following analysis of Iota

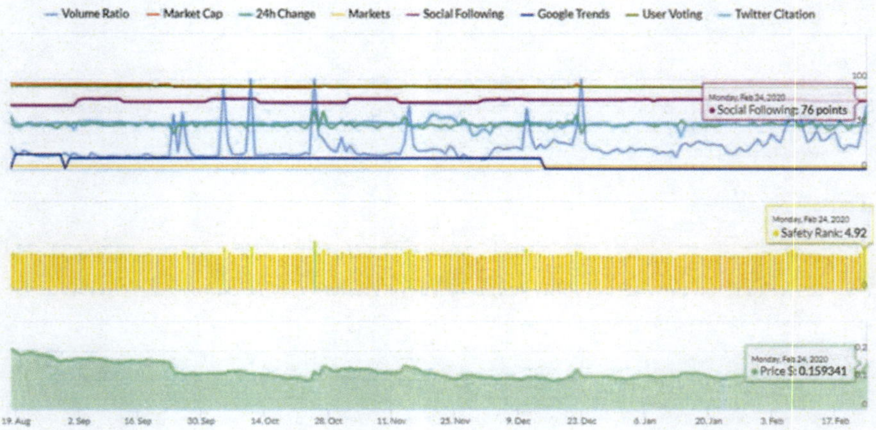

Fig. 18 Graph shows Feb 24, 2020, Social Following, Safety Rank and Price analysis of Iotchain

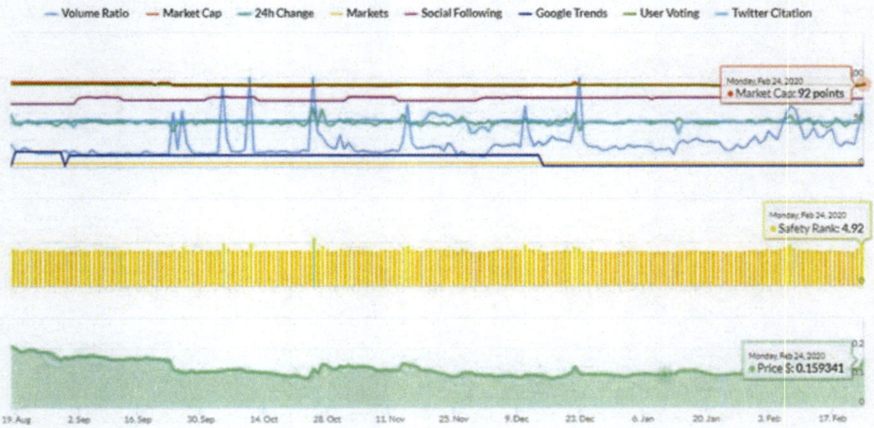

Fig. 19 Graph shows Feb 24, 2020, Market Cap analysis of Iotchain

Fig. 20 Graph shows Feb 24, 2020, Market Cap analysis of Iotchain

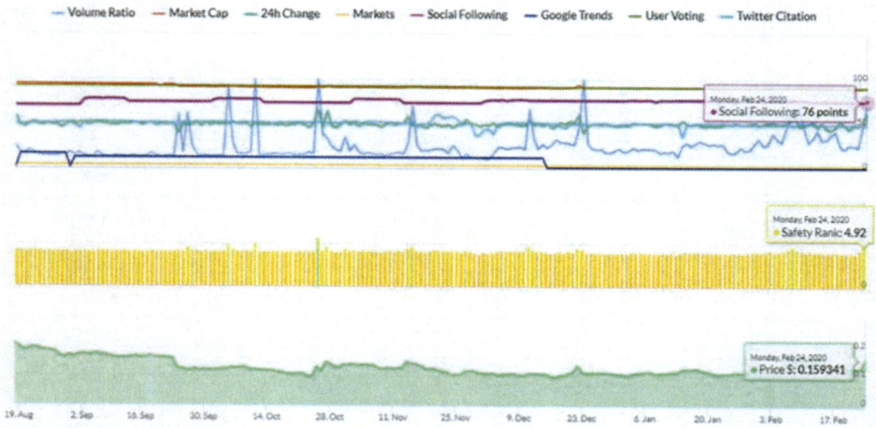

Fig. 21 Graph shows Feb 24, 2020, Social Following analysis of Iotchain

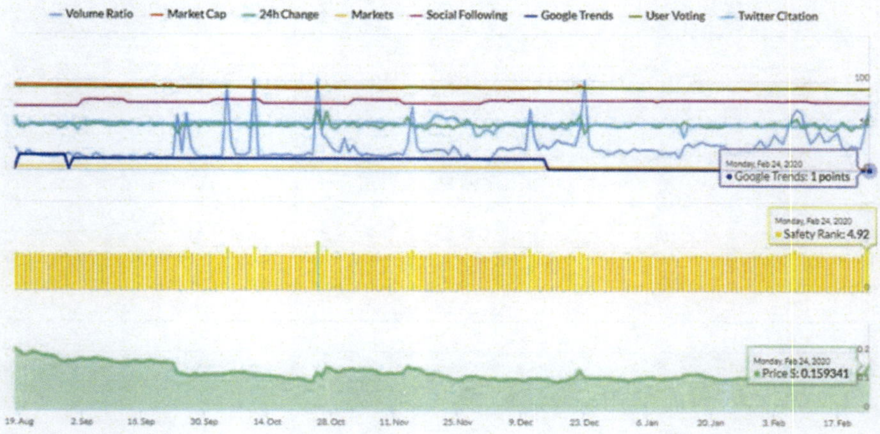

Fig. 22 Graph shows Feb 24, 2020, Google trends analysis of Iotchain

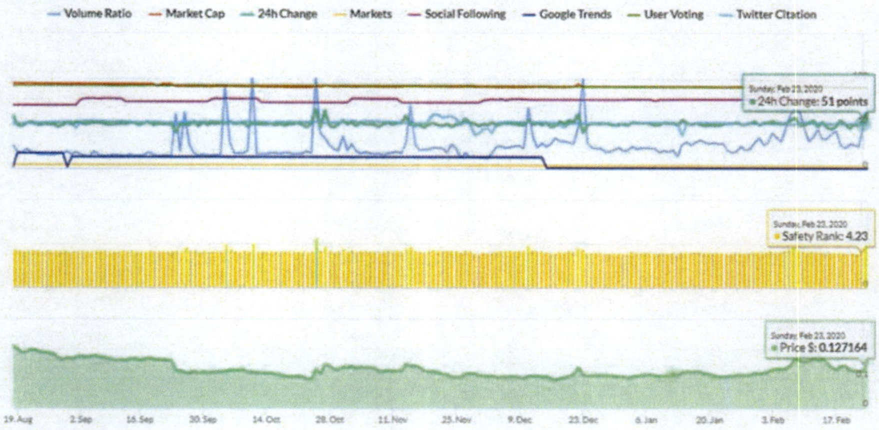

Fig. 23 Graph shows Feb 24, 2020, change in 24 analysis of Iotchain

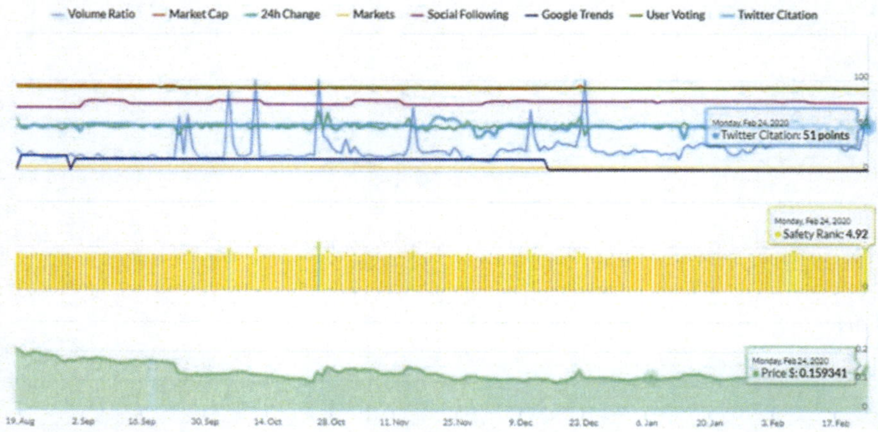

Fig. 24 Graph shows Feb 24, 2020, Twitter Citation analysis of Iotchain

Fig. 25 Graph shows Feb 24, 2020, User Voting analysis of Iotchain

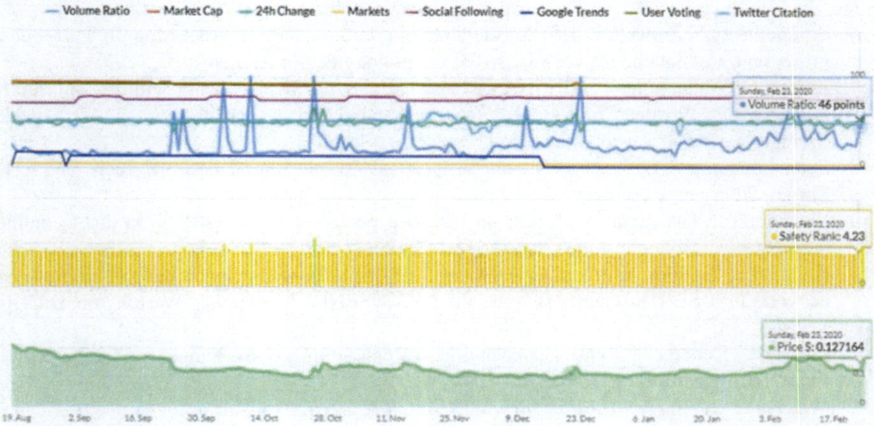

Fig. 26 Graph shows Feb 24, 2020, Volume Ratio analysis of Iotchain

Table 2 Comparative analysis of Iotex, Iota and Iotchain smart contract used for IoT

Sn	Parameter	Iotex	Iota	Iotchain
1	Market Cap	96 Points	99 Points	92 Point
2	Safety Rank	3.99	5.32	4.92
3	Potential Prize	$0.005242	$0.322677	$0.22751
4	24 h changes	51 Points	47 Points	51 Points
5	Social Following	66 Points	100 Points	76 Point
6	Google trends	9 Points	54 Points	1 Point
7	Twitter Citations	38 Points	100 Points	51 Point
8	User Voting	95 Points	98 Points	91 Points
9	Potential Profit	+35.2%	+23.3%	+42.8%
10	Market Cap Prize	$28,308,936 USD	$727,274,738 USD	$13,896,869 USD
11	Volume used Price	$4,227,073 USD	$9,415,959 USD	$10,142,777 USD
12	Volume Point	15 Point	1 Point	46 Points
13	Current Price	$0.005132	$0.261654	$0.159341

References

1. Díaz M, Martín C, Rubio B (2016) State-of-the-art, challenges, and open issues in the integration of internet of things and cloud computing. J Netw Comput Appl 67:99–117
2. Aayush, Internet of Things (IoT) (2017) Introduction, applications and future scope. https://www.gkmit.co/blog/internet-of-things-iot-introduction-applications-and-future-scope,2017
3. Toshendrakumar Sharma (2019) Block chain counsil.org Available online: https://www.blockchain-council.org/author/toshendra/. Accessed 29 Nov 2019
4. Antonopoulos AM (2014) Mastering bitcoin: unlocking digital cryptocurrencies, O'Reilly Media, Inc.
5. Litecoin (2011) https://litecoin.org/. Accessed 12 Sept 2019

6. Sompolinsky Y, Zohar A (2013) Accelerating bitcoin's transaction processing. In: Fast money grows on trees, not chains, vol 881. IACR Cryptology EPrint Archive
7. Decker C, Wattenhofer R (2015) A fast and scalable payment network with bitcoin duplex micropayment channels. In: Symposium on self-stabilizing systems, Edmonton, AB, Canada, Springer, pp 3–18
8. Stathakopoulou C, Decker C, Wattenhofer R (2015) A faster Bitcoin network, Tech. rep., ETH, Zurich, Semester Thesis
9. BigchainDB: The scalable blockchain database powering IPDB (2017). Available online: https://www.bigchaindb.com/. Accessed 12 Oct 2019.
10. Heilman E, Kendler A, Zohar A, Goldberg S (2015) Eclipse attacks on bitcoin'speerto- peer network. In: USENIX security symposium, USA,USENIX Association, Washington, D.C., pp 129–144
11. Buzby JC, Roberts T (2009) The economics of enteric infections: human foodborne disease costs. Gastroenterology 136(6):1851–1862
12. Malviya H (2019) How Blockchain will Defend IOT, 2016. Available online: https://ssrn.com/abstract=2883711. Accessed 14 Oct 2019
13. Chain of Things (2017) Available online: https://www.blockchainofthings.com/. Accessed 04 Oct 2019
14. Filament (2017) Available online: https://filament.com/. Accessed 04 Oct 2019
15. Aazam M, HuhE-N (2014) Fog computing and smart gateway based communicationfor cloud of things. In: Proceedings of the 2nd international conference on future internet of things and cloud, FiCloud-2014, Barcelona, Spain, pp 27–29
16. Ethembedded (2017) Available online: https://ethembedded.com/. Accessed 10 Sept 2019
17. Raspnode (2017) Available online: https://raspnode.com/. Accessed 10 Sept 2019
18. Roman R, Zhou J, Lopez J (2013) on the features and challenges of security andprivacy in distributed internet of things. Comput Netw 57(10):2266–2279
19. The Lisk Protocol (2017) Available online: https://docs.lisk.io/docs/the-liskprotocol. Accessed 1 Aug 2019
20. Quorum Whitepaper (2016) Available online: https://github.com/jpmorganchase/quorum-docs/blob/master/Quorum%20Whitepaper%20v0.1.pdf. Accessed 1 Aug 2019
21. Chronicled (2017) Available online: https://chronicled.com/. Accessed 13 Aug 2019

Analysis on Detection of Shilling Attack in Recommendation System

Sanjeev Dhawan, Kulvinder Singh, and Sarika Gambhir

Abstract Recommendation system is a system that attempts to predict the outcome of user according to his/her interest. Recommendation system mostly uses collaborative filtering algorithms. Although this recommendation system is successful in e-commerce sites, these collaborative filtering-based recommendation systems are exposed to shilling attacks. In this, attackers insert false profile information to have an impact on prediction or recommendation of the recommendation system. The consequence of shilling attack on recommendation system, categorization of shilling attacks, detection algorithms and evaluation metrics are provided by this paper.

Keywords Collaborative filtering · Detection algorithm · Recommendation system · Shilling attack

1 Introduction

There is large amount of information stored on Internet, and number of users are using Internet which causes the problem of information overload. Due to this, the users do not get the relevant information as they desire. Google and AltaVista have solved this problem to some extent, but prioritization and personalization were absent. These problems are solved by recommendation system. Recommendation system is an information filtering system that predicts the outcome according to user interest. It has solved the information overload problem [1] by producing the relevant information from large amount of data available on Internet. It helps in decision-making process

S. Dhawan · K. Singh · S. Gambhir (✉)
University Institute of Engineering and Technology, Kurukshetra University, Kurukshetra, India
e-mail: sarikagambhir@gmail.com

S. Dhawan
e-mail: sdhawan2015@kuk.ac.in

K. Singh
e-mail: ksingh2015@kuk.ac.in

N. Marriwala et al. (eds.), *Mobile Radio Communications and 5G Networks*, Lecture Notes in Networks and Systems 140,
https://doi.org/10.1007/978-981-15-7130-5_8

[2] and increases the business revenue in e-commerce setting. It also reduces the cost of transaction by buying the items from online shopping sites.

Recommender system is of three types, content based, collaborative filtering and hybrid filtering. In content-based recommender system, recommendation is based on the content rather on the other user opinion. Collaborative filtering is of two types—user based and item based. In user-based filtering, recommendation is produced by finding a subset of users that are similar to the target user. İn item-based filtering, recommendation is produced by finding a subset of similar items rated by people. Hybrid recommendation system is the combination of both content-based and collaborative filtering algorithm. Collaborative filtering (CF)-based recommender system is affected by shilling attacks. Collaborative filtering-based system is more prone to shilling attacks than the content-based recommender system. CF-based recommendation system attainment depends on how well it handles and discovers shilling attacks.

This design of paper is like this: Sect. 2 describes shilling attack, Sect. 3 consists of categorization of shilling attack, Sect. 4 explains related work, Sect. 5 contains detection attributes, Sect. 6 consists of evaluation metrics, and finally Sect. 7 describes conclusion and future work.

2 Shilling Attack

Shilling attacks are those that affect the recommendation system. In these attacks, malicious users are inserted into existing dataset in order to affect the result of recommendation systems. Product sellers generate these kinds of attacks. Strategy of attack is to.

- Firstly create fake profiles in e-commerce sites;
- Give target items to high or low ratings.

E-commerce sites are successful by the recommendation system. However, the collaborative filtering algorithm used in recommendation system is prone to shilling attacks. For instance, there are six users and six items shown in Table 1. We want to predict rating of user 1 that is the target user on item 6, incorporating the applicable criteria that follow.

Table 1 Without attack

Users	i_1	i_2	i_3	i_4	i_5	i_6	User 1 similarity with others
U_1	5	3	3	2		?	1.00
U_2	4		2		1	4	−1.00
U_3	3	3	1		2	1	0.76
U_4	4	1	1	2		3	0.72
U_5	3	3		3	4	2	0.94

Table 2 With attack

Users	i_1	i_2	i_3	i_4	i_5	i_6	User 1 similarity with others
U_1	5	3	3	2		?	1.00
U_2	4		2		1	4	−1.00
U_3	3	3	1		2	1	0.76
U_4	4	1	1	2		3	0.72
U_5	3	3		3	4	2	0.94
U_6	4	2		3	3	5	0.98

Here we assume $k = 1$; the similarity of user 1 with other users is shown in Table 1. Pearson's correlation coefficient (PCC) is used for calculating this similarity. According to this similarity that is shown in Table 1, user 1 and user 5 are most comparable, so the correct rating is 2 by user 1 on item 6.

Table 2 shows the user 6 is an attacker. Now according to the similarity calculated in Table 2, it shows that user 6 is an attacker more similar to user 1. The value of rating for user 1 on item 6 is 5; therefore, the accuracy of the recommendation system is reduced by the shilling attack.

3 Categorization of Shilling Attack

Shilling attacks are categorized as intent based and knowledge required by the attacker profiles.

3.1 Intent Based

Intent-based shilling attacks are categorized as push and nuke. Push attacks are the attacks in which attacker gives higher rating to target item to increase its popularity. Nuke attacks are those in which attacker gives lowermost rating to target item to reduce its popularity [3].

3.2 Based on Knowledge

Shillings attacks are classified as average, random and bandwagon attacks based on knowledge required for creation of shilling attack profiles.

True user and fake user profiles are shown in Figs. 1 and 2, respectively.

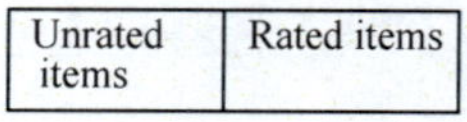

Fig. 1 Genuine user profile

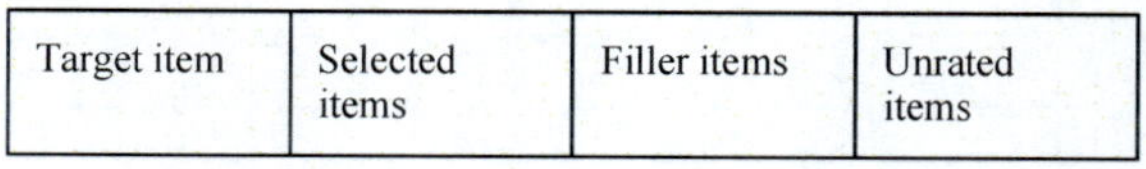

Fig. 2 Attacker profile

True user profile is divided into two parts, unrated items and rated items, as shown in Fig. 1.

Fake user profile.

An attacker profile is divided into four partitions—target item, selected items set, filler items set and unrated items set as shown in Figs. 2 and 3.

In different types of attack, different ratings are provided to select, filler and target items for creating the attacker profile as shown in Table 3.

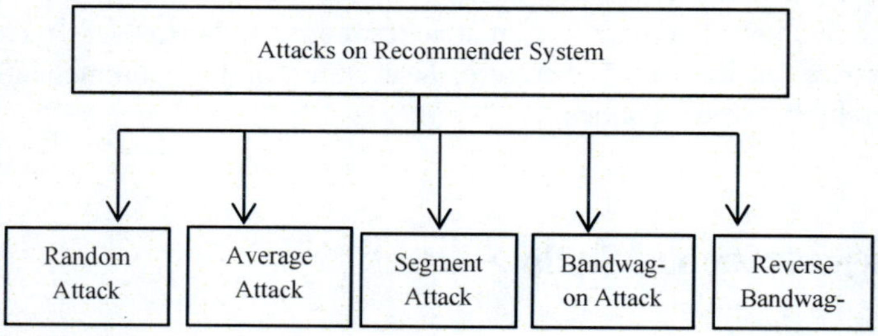

Fig. 3 Types of attack

Table 3 Types of attacks on recommender system

Attacks	I_s		I_f		I	I_t
	Items	Rating	Items	Rating		Push/nuke
Random	¢	¢	Random	System mean	¢	max/min
Average	¢	¢	Random	Item mean	¢	max/min
Segment	Segmented items	max/min	Random	$r_{\max}/r_{\min}$	¢	max/min
Bandwagon	Popular	max/min	Random	System mean/item mean	¢	max/min
Reverse bandwagon	Unpopular	max/min	Random	System mean	¢	max/min

4 Related Work

In this section, a number of detection algorithms have been proposed against recommender system which is based on supervised and unsupervised learning. Initially, the term shilling along with two attack models average and random was introduced by Riedl and Lam [4]. They analyzed that item-based CF approach is less prone to the shilling attacks than the user-based CF. Sarwar et al. [5] also worked on item-based CF. It has been verified that item-based algorithm is beneficial against user-based algorithms as the effect of attack is not much that it can change the outcome of system in item-based algorithm. Item-based approach also suffers from shilling attacks as proved by Mobasher et al. [6]. They also analyzed that segment attacks are effective on item-based CF. Burke et al. [7] discussed the segment attacks. Segment attack focused on subset of users with specific interest. They also analyzed that segment attacks are low-cost attacks. Chirita et al. [8] discussed many metrics for examining rating patterns which are used to identify the malicious users. Rating deviation mean agreement is given by these authors for detecting the behavior of attackers. Supervised, unsupervised and semi-supervised techniques aid in shilling attack detection. To identify shilling profiles, many classifications and clustering algorithm have been used. Bruke et al. [3] explained various detection attributes and used KNN classifier to classify the profiles. The result of this method shows that the classifier with model specific attributes performed better. In the same way, Kumar et al. [9] also used supervised machine leaning techniques for detection of shilling attacks. They compared the performance of six classifiers. They also analyzed that NN, SVM and random forest have performed better than the other classifiers. They collaborate these three classifiers and developed a new one which outdoes in many cases. Zhang and Zhou [10] offered an ensemble model for detection of shilling attack profiles based on BP neural network. It gave higher precision than the previous approach. Patel et al. [11] also proposed a classification technique for detection of shilling attack based on the attributes RDMA, DegSim and length variance. The classification method, decision tree, is used. It gives better accuracy in any of filler size and attack size. Davoudi and Chatterjee [12] proposed the classification technique for the identification of shilling attack profiles that is based on social interaction between the attackers and authentic users. Some researchers used unsupervised techniques for the detection of attack. Brayan et al. [13] proposed UnRAP algorithm. Hv score metric is considered for finding the attack profiles. The profile which is having high Hv score value is an attacker. Wang et al. [14] proposed AP-UnRAP algorithm for detection of group users. It gave the better performance than UnRAP algorithm. This algorithm is effective for popular and segment attack. But this algorithm performed significantly for segment attack when the larger attack size is to be used. Zhou et al. [15] discussed a new unsupervised algorithm for the detection of shilling attack based on target item analysis. This algorithm used RDMA and dig similarity attributes for the detection of fake profiles, and these have been refined by using target item analysis.

5 Detection Attributes

Detection of attributes is divided into two generic attributes and model specific attributes.

5.1 Generic Attributes

5.1.1 Rating Deviation Mean Agreement

It identifies user rating disagreement with genuine users. It is defined as

$$\text{RDMAu} = \frac{1}{Nu} \sum_{i=0} \frac{|ru, i - ri|}{|RU, i|}$$

where

N_u Total items rated user u,
$r_{u,i}$ u user rating to item i,
r_i Item i average rating,
$R_{U,i}$ Total ratings delivered by all users for item i.

5.1.2 Weighting Degree of Agreement:

$$\text{WDAu} = \sum_{i=0} \frac{|ru, i - ri|}{|RU, i|}$$

where

N_u Total items rated by user u,
$r_{u,i}$ User rating to item i,
r_i i item average rating,
$R_{U,i}$ Total ratings delivered by all users for item i.

5.1.3 Weighting Deviation from Mean Agreement

It uses too much weight on rating deviations for smaller items that can help detect anomalies. It is defined as

$$\text{WDMAu} = \frac{1}{Nu} \sum_{i=0} \frac{|ru, i - ri|}{|RU, i|^2}$$

where

N_u Total items rated by user *u*,
$r_{u,i}$ Rating of user u to item *i*,
r_i Item *i* average rating,
$R_{U,i}$ total ratings delivered by all users for item *i*.

5.1.4 Length Variance

It is used to determine how long the length variation of a given profile from the average length of the dataset. It is defined as shown below [7]

$$\text{lengthVaru} = \frac{nu - nu}{\sum_{k \in U}(nu - nu)^2}$$

nu Profile average length in a system.

5.2 *Model Specific Attributes*

These attributes are used for detection of specific type of attacks. For example, some attributes will be used for detection of average attack, some for random attack, etc. These kinds of attributes are defined below.

5.2.1 Filler Mean Variance

It is defined as

$$\text{FMVu} = \sum_{i \in Lf} \frac{(rx, i - ri)^2}{|Lf|}$$

It is used to detect the average attack.
where

L_f The filler item set,
$r_{x,i}$ User *u* rating to item *i*,
r_i The average of ratings assigned to item *i*.

5.2.2 Filler Mean Target Difference

It is calculated using formula given below

$$\mathrm{FMTD} = \left| \frac{\sum_{i \in L_s} rx, i}{|Ls|} - \frac{\sum_{i \in Lf} rx, i}{|Lf|} \right|$$

where

L_s The selected item set,
L_f Filler item set,
$r_{x,i}$ Rating produced by user u to item *i*.

It is used to detect the segment and bandwagon attacks.

5.2.3 Mean Variance

$$\mathrm{MeanVar}(r\mathrm{target}, j) = \frac{\sum_{i \in (Pj - \mathrm{target})} (ri, j - ri)^2}{|k|}$$

where

Pj User *j* profile,
r_{target} Hypothesized rating to target item,
$r_{i,j}$ Rating given by user *j* rating to an item *i*
r_i Item *i* mean across all users.

It is used to detect the average attack (Table 4).

6 Evaluation Metrics

6.1 *Prediction Shift*

It is the difference between prediction of an item before and after an attack is applied. It is calculated by the formula as described below

$$\mathrm{PredShift}_{u,i} = p_{u,i} - p'_{u,i}$$

$p_{u,i}$ Prediction of an item *i* before an attack
p'_s Prediction of an item *i* after an attack

Table 4 Detection attributes [15]

Attribute type	Attribute	Equation	Significance
Generic	RDMA	$\text{RDMAu} = \frac{1}{Nu}\sum_{i=0} \frac{\lvert ru,i-ri\rvert}{\lvert RU,i\rvert}$	Rating deviation from mean agreement
Generic	WDA	$\text{WDAu} = \sum_{i=0} \frac{\lvert ru,i-ri\rvert}{\lvert RU,i\rvert}$	Weighted degree of agreement
Generic	WDMA	$\text{WDMAu} = \frac{1}{Nu}\sum_{i=0} \frac{\lvert ru,i-ri\rvert}{\lvert RU,i\rvert^2}$	Weighted deviation from mean agreement
Generic	LengthVsar	$\text{lengthVaru} = \frac{nu-nu}{\sum_{k\in U}(nu-nu)^2}$	Length variance
Generic	DegSim	$\text{DegSim} = \frac{\sum_{i=1}^{x} Zi,j}{x}$	Degree of similarity with top N neighbors
Generic	TWDMA	$\text{TWDMAu}P = \frac{\text{RDMAu}}{\text{Trustu}}$	Trust into RDMA
Generic	UnRAP	$\text{Hv}(u) = \frac{\sum_{i\in I}(rui-rUi-ruI+rUI)^2}{\sum_{j\in J}(rui-ruI)^{?2}}$	Unsupervised retrieval of attack profile
Model specific	FMV	$\text{FMVu} = \sum_{i\in Lf} \frac{(rx,i-ri)^2}{\lvert Lf\rvert}$	Frequency mean variance
Model specific	FMTD	$\text{FMTD} = \left\lvert \frac{\sum_{i\in Ls} rx,i}{\lvert Ls\rvert} - \frac{\sum_{i\in Lf} rx,i}{\lvert Lf\rvert} \right\rvert$	Frequency mean target difference
Model specific	MeanVar	$\text{MeanVar}(\text{rtarget}, j) = \frac{\sum_{i\in(Pj-\text{target})}(ri,j-ri)^2}{\lvert k\rvert}$	Mean variance
Model specific	FMD	$\text{FMDu} = \frac{1}{\lvert Uu\rvert}\sum_{i=1}^{Uu} \lvert ru,i-ri\rvert$	Filler mean difference
Model specific	FAC	$\text{FACu} = \frac{\sum_i (ru,i-r)}{\sqrt{\sum_i}}$	Filler average correlation

6.2 Recall

It is defined as fraction of relevant data that area unit retrieved as attacker

$$\text{Recall} = \frac{\text{true positive}}{\text{true positive} + \text{false negative}}$$

6.3 Precision

It is defined as the fraction of retrieved value is really an attacker

$$\text{Precision} = \frac{\text{true positive}}{\text{true positive} + \text{false positive}}$$

6.4 F1 Measure

It is the combination of precision and recall

$$\text{F1 measure} = \frac{2 * \text{precison} * \text{recall}}{\text{precision} + \text{recall}}$$

6.5 Mean Absolute Error

It measures how close the probable prediction to the actual values

$$\text{MAE} = \frac{1}{N}\sum_{t=1}^{N}\frac{\text{At} - \text{Ft}}{\text{At}}$$

A_t Actual value
F_t Predicted value.

See Table 5.

7 Conclusion and Future Work

In this paper, the effect of shilling attack on recommender system, their types, related work and detection attributes and evaluation metrics are discussed. From the above analysis discussed, it is observed that supervised learning performs better than unsupervised learning. It means that supervised algorithms enhance the accuracy of recommendation system. A method is required to be developed that works with different filler size and attack size. In future, we will work on supervised learning method for detection of shilling attack that enhances the accuracy of recommendation system.

Table 5 Evaluation metrics [15]

Metric	Significance	
Precision	Fraction of retrieved value is really an attacker	$\frac{TP}{TP+FP}$
Recall	Fraction of relevant data that area unit retrieved as an attacker	$\frac{TP}{TP+FN}$
F1 measure	Combination of precision and recall(measure the accuracy of detection algorithm)	$F1\text{measure} = \frac{2*\text{precison}*\text{recall}}{\text{precision}+\text{recall}}$
Prediction shift	Measure the effect of shilling attack	PredShift $_{u,i} = p_{u,i} - p_{u,i}$ $p_{u,i}$ = prediction of an item i before an attack $p'_{u,i}$ = prediction of an item i after an attack
Mean absolute error	Measures how close the probable predictions to the actual value	$\text{MAE} = \frac{1}{N}\sum_{t=1}^{N}\frac{\text{At}-\text{Ft}}{\text{At}}$ A_t = actual value F_t = predicted value

References

1. Konstan JA, Riedl J (2012) Recommender systems: from algorithms touser experience. User Model User-Adapt Interact 22:101–123
2. Pathak B, Garfinkel R, Gopal R, Venkatesan R, Yin F (2010) Empirical analysis of the impact of recommender systems on sales. J Manage Inf Syst 27(2):159–188
3. Burke R, Mobasher B, Williams C, Bhaumik R (2006) Classification features for attack detection in collaborative recommender systems. In: ACM. USA
4. Lam SK, Riedl j (2004) Shillinsg recommender systems for fun and profit. In: Proceedings of the 13th international conference on World Wide Web. ACM, pp 393–402
5. Sarwar B, Karpis G, Konstan J, Riedl J (2001) Item based collaborative filtering recommendation. ACM, HongKong, pp 285–295
6. Mobasher B, Burke R, Bhaumik R, Williams C (2005) Effective attack models for shilling item-based collaborative filtering systems. In: Proceeding Of the 2005 WebKDD workshop. Chicago
7. Bruke R, Mobasher B, Bhaumik R, Williams C (2005) Segment based injection attacks against collaborative filtering recommender systems. In: Data mining fifth IEEE international conference 2005
8. Chirita P-A, Zamfir WN, Cristian (2005) Preventing shilling attacks in online recommender systems. In: Proceedings of the 7th annual ACM international workshop on Web information and data management, Germany, 4 Nov 2005
9. Kumar A, Rana DGAPS (2015) Ensemble approach to detect profile injection attack in recommender system. In: International conference on advances in computing, communications and informatics (ICACCI), Kochi
10. Zhang F, Zhou Q (2015) Ensemble detection model for profile injection attacks in collaborative recommender systems based on BP neural network. IET Inf Secur 9:24–31
11. Patel K, Thakar A, Shah C, Makvana K (2016) A novel supervised approach to detection of shilling attack in collaborative filtering based recommendation system. Int J Comput Sci Inf
12. Davoudi A, Chatterjee M (2017) Detection of profile injection attack in social recommender systems using outlie analysis. In: IEEE international conference on big data

13. Bryan K, O'Mahony M, Cunningham P (2008) Unsupervised retrieval of attack profiles in collaborative recommender systems. In: Recsys'08 Proceedings 2008 Acm conference on recommender systems. pp 155–162
14. Lu G, Engineering S (2014) A group attack detector for collaborative filtering recommandation. IEEE Internet Comput 2–5
15. Zhou W, Wen J, Koh YS, Xiong Q, Gao M, Dobbie G, Alam S (2015) Shilling attacks detection in recommender systems based on target item analysis. PLoS One 10(7)
16. Burke RD, Mobasher B, Williams CA, Bhaumik R (2006) Detecting profile injection attacks in collaborative recommender systems. In: IEEE conference on e-commerce technology. pp 23–30
17. Gunes I, Kaleli C, Bilge A, Polat H (2012) Shilling attacks against recommender systems: a comprehensive survey. Artif Intell Rev, pp 1–33
18. Bilge ZO, Polat H (2014) A novel shilling attack detection method. Procedia Comput Sci 31:165–174
19. Wang JZ, Wang JJ, Zhang ZG, Guo SP, Xu H, Yu B (2011) Forecasting stock indices with back propagation neural network. Expert Syst Appl 38(11):14346–14355
20. Alostad JM (2019) Improving the shilling attack detection in recommender systems using an SVM gaussian mixture model. World Scientific Publishing Co. Pte. Ltd. 18(01):1–18

A Comparative Analysis of Improvements in Leach Protocol: A Survey

Amandeep Kaur and Rajneesh Kumar

Abstract LEACH is the hierarchal and distributed clustering protocol of wireless sensor networks (WSNs) used to divide the network into clusters. There has been extensive research done in LEACH protocol over the past few years to overcome its limitations. This paper presents the survey of improvements in LEACH that has been done over the years and also discusses the advantages and disadvantages of all improvements.

Keywords Wireless sensor network · Clustering · LEACH · Cluster head · Base station

1 Introduction

WSNs consist of large number of sensor nodes (SNs) with limited sensing, computing and computation capabilities. SNs deployed randomly in sensing area with one or more base station (BS) depending on application requirement. WSN used variety of applications including pressure, temperature, humidity monitoring, disaster management, forest fire detection, security surveillance and so on [1–3]. As these SNs have limited energy and communication in WSN takes more energy as compared to computation so as to save overall energy of network, clustering technique is designed. LEACH protocol is the basic clustering protocol in which the network is divided into clusters. In LEACH, cluster head (CH) is selected based on certain criteria for each cluster that is responsible for collecting data from its members and transmitting aggregating data to BS.

A. Kaur (✉)
Department of CSE, M.M. University, Sadopur, Ambala, Haryana, India
e-mail: amandeepkaur@mmumullana.org

R. Kumar
Computer Science and Engineering Department, Maharishi Markandeshwar (Deemed to be University), Mullana, Ambala, Haryana, India
e-mail: drrajneeshgujral@mmumullana.org

N. Marriwala et al. (eds.), *Mobile Radio Communications and 5G Networks*, Lecture Notes in Networks and Systems 140,
https://doi.org/10.1007/978-981-15-7130-5_9

Low Energy-Efficient Adaptive Clustering Hierarchy (LEACH): LEACH [4] is an energy-efficient clustering protocol. It works in several rounds, and each round consists of two phases: set-up phase and steady phase.

Set-up Phase: Network is partitioned into clusters, and CH is elected for each cluster in set-up phase. For selection of CH, every SN produces a number between 0 and 1 randomly and then number is compared with threshold value $T(n)$ that is calculated by the following Eq. 1.

$$T(n) = \begin{cases} \dfrac{P}{1 - P * \left(r \bmod \frac{1}{P}\right)} & : \text{if } n \in G \\ 0: \text{ otherwise} \end{cases} \quad (1)$$

where P is the desired percentage of SNs to become CHs, r is current round, and G is the set of SNs that have not participated in CH selection process in previous $1/P$ rounds.

SN is elected as CH if its random number is less than threshold value $T(n)$, otherwise it behaves like a normal node. Once CHs are selected, each CH broadcasts a message in network and SNs join its closest CH to form cluster. Each CH generates a TDMA schedule to avoid collision and transmit these schedules to their members.

Steady-state phase: In this phase, awake SNs send the sense data of whole cluster to their CH and rest nodes remain in sleep state as per TDMA schedule to conserve the overall energy consumption of network. After receiving data from all nodes, CH transmits the aggregated data to sink or BS.

Advantages of LEACH:

1. LEACH is a distributed clustering protocol which reduces the communication between SNs and BS, thereby improving energy consumption of network.
2. LEACH protocol gives all SN equal chance to be CH which causes uniform energy distribution throughout network.
3. In LEACH, SN follows wake up as per their TDMA schedule which allows rest nodes to go into sleep state and increases network lifetime.

Disadvantages of LEACH:

1. LEACH protocol CH selection criteria are purely based on randomness which causes non-uniformity in network.
2. In LEACH, CH transmits data to BS that causes energy loss in CH very quickly. As a result, network lifetime degrades.
3. LEACH follows random CH selection criteria so the less energy node may be elected as CH. Hence, CH dies quickly by affecting robustness of network.

Improvements in LEACH: To overcome the limitation of LEACH protocol, various improvements have been done during last few years. This section describes all LEACH improvements as per chronological order.

2 LEACH Protocols with Single-Hop Communication

LEACH protocol is originally meant for single-hop communication as shown in Fig. 1 for enhancing network lifetime. Single-hop communication works better for small size network due to less routing overhead. There has been many improvements done in LEACH for single-hop communication which includes CH selection process, cluster form (square, rectangle, hexagonal and so on), inter- and intra-cluster formation, etc., during last few years. The following section discusses the improvements in LEACH.

a. **K-LEACH (K-medoids-LEACH)** [6] **(2013)**: Authors use K-medoids clustering for cluster formation in network. K-LEACH improves LEACH protocol by uniformly distributed CH throughout the network results in balanced energy consumption. But it requires precision and is complex enough.
b. **LEACH-TLCH (LEACH with two level CHs)** [7] **(2013)**: Uses same CH selection method as of LEACH. But the difference is that if current energy of selected CH is below the average energy of all SNs and distance between CH and BS is more than average distance of all nodes to BS, then the node having the highest energy is selected as secondary CH. This secondary CH is responsible for collecting data from its member nodes and sends aggregated data to BS. This protocol outperforms of average energy consumption in rounds. But due to selection of secondary CH, this protocol incurs extra overhead.
c. **An Extended Vice Cluster Selection Approach** [8] **(2013)**: This approach is the improvement of V-LEACH protocol. Improved V-LEACH protocol selects the CH basis on three parameters: maximum residual energy, minimum distance and minimum energy. In addition to CH, each cluster has vice CH that behaves as CH only when CH dies. This protocol increases the network lifetime by using vice CH as compared to LEACH but incurs extra burden because of selection of one more CH for each cluster.

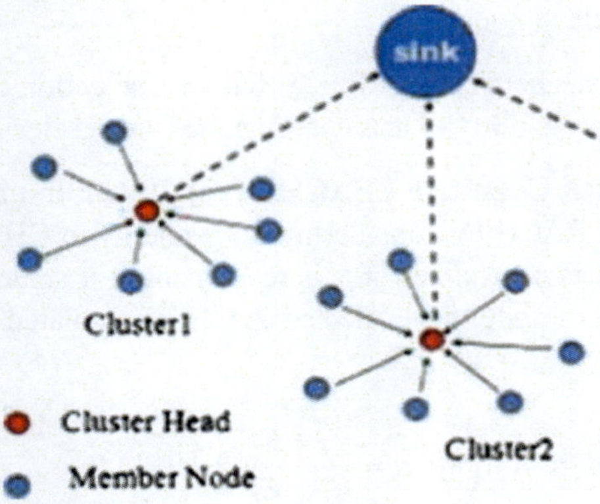

Fig. 1 Single-hop communication [5]

d. **LS-LEACH (Lightweight Secure LEACH)** [9] **(2013)**: LS-LEACH adds the security feature into LEACH protocol. This protocol uses energy-efficient and lightweight authentication scheme. In LS-LEACH, every node uses one secret key for sharing with BS and second shared with other nodes of network. Private key is used for sharing data between CH and BS, and group key is for joining CHs. The main feature of LS-LEACH is that CH and nodes both authenticate each other to prevent network from attackers. But drawback is lot of energy is wasted in transferring of keys between nodes.

e. **IB-LEACH (Intra-balanced LEACH)** [10] **(2014)**: IB-LEACH protocol reduces the burden of CH by distributing load to aggregator nodes. IB-LEACH protocol set-up phase is same as of LEACH but in addition to CH each cluster has aggregator node which transmits sensed data of its members to BS. The role of CH is only to manage cluster activities including creation of TDMA schedule, selection of aggregators and so on. IB-LEACH protocol reduces the energy consumption of network by reducing the load of CH but incurs control message overhead in selection of aggregators.

f. **CogLEACH (Cognitive LEACH)** [11] **(2014)**: CogLEACH is spectrum-aware LEACH protocol that considers number of idle channels in the selection process of CH. The probability of CH selection is calculated using the following Equation

$$P_i = \min\left(k\frac{C_i}{C_t}, 1\right) \tag{2}$$

$$C_t = \sum_{i=1}^{n} C_i \tag{3}$$

where

k Total number of CHs in each round.
n Total number of nodes in the network.
C_i Number of idle channels in node i.

CogLEACH protocol performs better than LEACH but CHs selection criteria do not consider energy so there is probability of less energy node can be selected as CH.

g. **CogLEACH-C (Centralized Cognitive LEACH)** [12] **(2015)**: In this, author solves the problem in CogLEACH by considering node energy in CH selection process. As CogLEACH-*C* is centralized, BS is responsible for selecting CHs in network. The probability of node to be selected as CH is calculated using the following Equation

$$P(t)_i = \min\left(k * n * \frac{C_i * E_i}{C_t * E_t}, 1\right) \tag{4}$$

where E_i and E_t are residual and average energy of network.

CogLEACH-C protocol enhances the network lifetime but due to its centralized nature it suffers from scalability problem and cannot be used for large WSN.

h. **EE-LEACH (Energy-Efficient LEACH)** [13] **(2015)**: EE-LEACH overcomes the non-uniform CH distribution limitation of LEACH. CH selection is based on spatial density of network and residual energy of node. This protocol uses conditional probability for data aggregation and Gaussian distribution for better network coverage. The main advantage of EE-LEACH protocol is uniform CH distribution throughout network and less energy consumption of node.
i. **MaximuM-LEACH** [14] **(2016)**: The drawback of LEACH is that some CH has more SNs and others have less due to its random nature. So to overcome these, authors proposed MaximuM-LEACH centralized clustering protocol. In this protocol, node joins its nearest CH only if CH members are less than particular number, otherwise node joins second nearest CH and so on. BS station selects the node as CH whose energy is greater than average energy of all SNs. As equal number of SNs are assigned to all CHs so MaximuM-LEACH ensures the load balancing and uniform energy consumption of network, therefore improving network lifetime. But the problem with this protocol is how particular number is calculated is not clearly explained by the author.
j. **Improved LEACH** [15] **(2016)**: Network performance metrics such as energy consumption and network lifetime directly depends on the CH selection process. Improved LEACH introduced new score function in place of probability function used to select CH in the network. Score function of node is calculated using the following Eqs. 5, 6 and 7:

$$\text{SCRFN1} = \frac{E_{\text{current}}}{E_{\max}} \tag{5}$$

$$\text{SCRFN2} = \frac{d_{\text{bs}}}{d_{\text{far}}} \tag{6}$$

$$\begin{aligned}\text{SCRFN} &= \sum_{i=1}^{n} w_i * \text{SCRFN}_i \\ &= \left[w_1 \left[\frac{E_{\text{current}}}{E_{\max}} \right] n + w_2 \left[\frac{d_{\text{bs}}}{d_{\text{far}}} \right] * \left[\begin{matrix} \dfrac{P}{1 - P * \left(r \bmod \frac{1}{P} \right)} : \text{if } n \in G \\ 0: \text{ otherwise} \end{matrix} \right] \right]\end{aligned} \tag{7}$$

E_{current} is residual energy, and $E_{\max}$ is maximum energy of node. d_{bs} is distance of node to BS, and d_{far} is distance of BS to farthest node of cluster. w_1 and w_2 are weight given to energy and distance function. Data transmission from nearest node preserves its energy. Therefore, parameters such as distance of node to BS and distance of BS to farthest node in cluster and energy select the optimal CH. Optimal CH automatically increases the network lifetime and performance of network. The limitation of improved LEACH is that weight function is not explained by authors.

k. **LEACH-CC (LEACH-Central Constrained)** [16] **(2016)**: LEACH-C [17] selects the node as CH if its residual energy is more than average energy of SNs. However, the limitation of this method is that if average energy of network is low and node whose energy even more than average value can die after the current round. LEACH-CC is centralized clustering protocol which improves the LEACH-C protocol by redefining CH selection approach. According to LEACH-CC protocol, BS selects the k. CHs if the difference between SN energy and average of all nodes energy is greater or equal to average energy of node for current round. This protocol ensures that low energy nodes do not participate in CH selection process. But as LEACH-CC is centralized, every node requires to transmit its location information to BS which is energy consuming and costly in terms of GPS.

l. **NR-LEACH (Node-Ranked LEACH)** [18] **(2017)**: Authors have introduced a clustering technique based on node rank. Each node uses residual energy, link connections with its neighbours and received signal strength to calculate its rank. Based on these parameters, the importance of node in networks calculated. Node with higher rank is selected as CH. These protocols increase the energy efficiency of network but it have more overhead of rank calculation.

m. **CH-LEACH** [19] **(2017)**: In CH-LEACH protocol, CH selection is based on centralized k-means clustering approach. Also authors consider the two position of BS: one is at centre and second is at edge of network in order to improve overall energy consumption of network. The main drawback of CH-LEACH is to predict the accurate value of k to find optimal clusters

n. **EiP-LEACH** [20] **(2017)**: One of the shortcomings of LEACH protocol is that less energy nodes may be selected as CH. To resolve this limitation in EiP-LEACH, authors have proposed modification in threshold calculation of basic LEACH protocol. The threshold value is calculated using the following equation:

$$T(n) = \begin{cases} \dfrac{P}{1 - P * \left(r \mod \frac{1}{P}\right)} * \dfrac{E_n}{E_{\text{intial}_{\text{average}}}} : \text{if}\, n \in G \\ 0\text{:} \quad \text{otherwise} \end{cases} \tag{8}$$

where

E_n Current energy of SN.
$E_{\text{intial}_{\text{average}}}$ Average initial energy of whole network.

EiP-LEACH protocol ensures that the low energy node is not selected as CH. It suffers from scalability problem.

o. **ECO-LEACH (Energy Harvesting and Cooperative LEACH)** [21] **(2018)**: This protocol improves the LEACH by replacing its random CH selection with dutycycling and cooperative transmission. CHs are selected as per their duty cycle that determines how often node behaves as CH in given time horizon. For example, if time horizon = 4, then node behaves as CH after every four rounds. Moreover, cooperative transmission strategy enables nodes to relay undelivered packets from CHs to the BS and also from members to CHs. ECO-LEACH protocol improves lifetime but bears control message overhead.

p. **ET-LEACH** [22] **(2019)**: The first step in ET-LEACH is to partition the network field into equal size squares. CH selection is on the basis of threshold value which is set half of the initial energy. The node is selected as CH if its energy is greater than threshold value. ET-LEACH protocol outperforms as compared to LEACH by reducing the overall total number of nodes participated in CH selection process. ET-LEACH also distributes CH uniformly throughout the network. Soft threshold value is chosen to reduce performance overhead. This protocol results in extra overhead as due to square partitioning of network.

q. **Modified HeteroLEACH** [23] **(2018)**: Conventional LEACH protocol has been proposed for homogeneous network. HETEROLEACH is an extension of LEACH which supports heterogeneous network. In this protocol, only nodes whose energy is more than threshold value participate in CH selection, and nodes are associated with CH based on predefined range. CH transmits the aggregated data of their members to BS to minimize redundancy and improving the network lifetime. But the problem is control message overhead for inter-cluster communication.

r. **TH-LEACH** [24] **(2018)**: TH-LEACH protocol reduces the overhead for CH selection of LEACH in every round. Based on the value of p (percentage of node to be CH), few nodes are selected as heterogeneous whose energy level is increased depending on initial energy level. Heterogeneous nodes are selected as CH for first round and distributed throughout the whole network. In further rounds, CHs are selected based on energy level, and SN serves as CH if its energy is greater than average energy of all other nodes in network. CH will be re-elected only if its energy is less than $T(n)$, i.e. total energy required to serve one round and calculated using following equation.

$$T(n) = (n-1) * L * E_{\text{elec}} + n * L * E_{\text{DA}} + \left(L * E_{\text{elec}} + L * E_{\text{mp}} * d_{\text{BS}}^4\right) \quad (9)$$

n Total Number of Nodes.
L Frame Bits.
E_{elec} Energy expenditure for transmitting or receiving one bit.
E_{DA} Energy required for data aggregation.
E_{mp} 0.0013PJ//bit/m4.
d Distance between CH and BS.

TH-LEACH protocol avoids the CH selection in every round and saves the network energy to enhance network lifetime. But the main demerit is to select coordinates of CH for the first round which is not possible for large WSN.

s. **Re-LEACH (Reappointment LEACH)** [25] **(2019)**: TH-LEACH protocol reduces the overhead for CH selection of LEACH in every round. In Re-LEACH, protocol consists of three phases: set-up phase, steady phase and reappointment phase. Set-up and steady phase work similar to LEACH protocol. According to this phase, if SN is elected as CH, it continues to behave as CH for next several rounds till it lost its energy completely. In this protocol, authors also modify threshold value $T(x)'$ for selecting CH given in the following equation:

$$T(x)' = \frac{K}{\left[N - K\left(\left(\text{floor}\left(\frac{r1}{T}\right) \bmod \frac{N}{K}\right)\right)\right]} \tag{10}$$

where

- N Total number of SNs in network,
- T Time for reappointment,
- K Total number of CHs,
- $r1$ 0 in reappointment of every new round.

Re-LEACH protocol reduces the overhead of CH selection in every round and improves the energy efficiency of network. But the disadvantage is that selecting new CH after the current CH loses its energy can create holes in network.

3 LEACH Protocols with Multi-Hop Communication

Traditional LEACH is not suitable for large WSN because more energy of CH is wasted for direct communication with BS. To reduce energy wastage, multi-hop communication between CHs and BS has been proposed in literature as shown in Fig. 2.

t. **E2ACM** [26] **(2014)**: Authors proposed E2ACM multi-hop clustering protocol in which CHs are selected based on LEACH protocol. The E2ACM protocol selects next CH to participate in routing based on strength factor S_f which is calculated as Eq. 11:

$$S_f = \frac{w_1 * b}{w_2 * n * w_3 * d} \tag{11}$$

where

b	Is battery life of CH.
n	Is the number of nodes registered with the CH.
d	is distance between neighbours CHs.
w_1, w_2 and w_3	Are the weights assigned battery life, cluster density and distance.

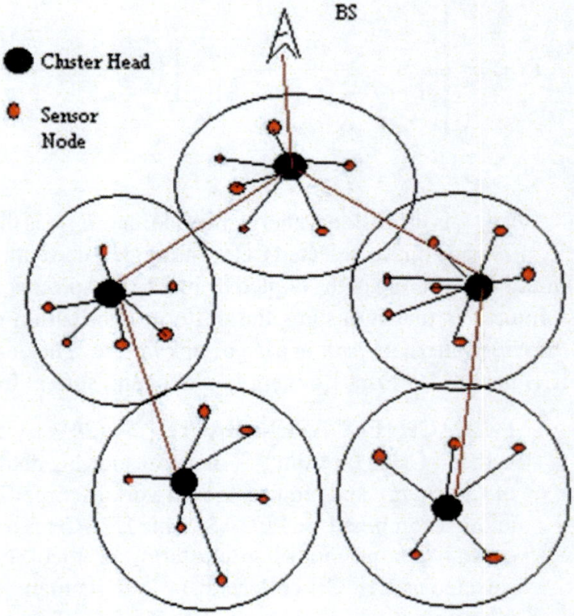

Fig. 2 Multi-hop communication [26]

The E2ACM protocol improves the network lifetime by reducing the energy consumption of CH required for transmission of packet to CH. The second advantage of protocol is that it provides alternate path in case of link failure. But E2ACM protocol suffers from hotspot problem as only CH near to BS is responsible for sending aggregated packets to BS.

u. **DL-LEACH** [27] **(2015)**: In LEACH protocol, node sends data to CH which is responsible for transmitting it to BS. It consumes a lot of energy if CH is far from BS. DL-LEACH protocol resolves this issue with two layers of multi-hop routing. DL-LEACH follows same CH selection process as of LEACH. Before sending data node, compare its distance with BS and CH. SN transmits data to BS directly if distance of node to BS is less than that of CH.
v. **DCDA** [28] **(2015)**: In DCDA protocol, authors modify conventional LEACH in order to minimize its two limitations. First is that some portion of network is left without CHs and some has more than one CHs, and second is less energy node may be selected as CH. DCDA protocol overcomes these weaknesses by dividing network into equal size grids and also selects CHs based on residual energy [28] as given in the following Eq. 12:

$$T(n) = \begin{cases} \frac{P}{1 - P * \left(r \mod \frac{1}{P}\right)} * \left[\left(\frac{E_{\text{current}}}{E_{\max}}\right) + (rs * p)\left(1 - \frac{E_{\text{current}}}{E_{\max}}\right) \text{if } i \epsilon G\right] \\ 0: \text{ otherwise} \end{cases} \tag{12}$$

E_{current} is the residual energy of node, and $E_{\max}$ is the maximum energy of network. In steady phase, selected CHs send aggregated data of their members to BS through multi-hop routing path adopted from E2ACM protocol [26]. The advantage of DCDA protocol is that it ensures the uniform distribution of CH in network, and it also provides alternate path in case of link failure. The demerit of DCDA protocol is that CH near to BS loses its energy quickly and suffers from hotspot problem.

w. **P-LEACH (PEGASIS-LEACH)** [29] **(2016)**: One of the drawbacks of LEACH is that if CH is far from BS, then transmitting data from CH to BS consumes a lot of CH energy and reduces the network lifetime. To overcome this in P-LEACH, authors combined the PEGASIS and LEACH protocol. P-LEACH is chain-based centralized and multi-hop clustering protocol. Node with maximum energy is selected as CH. CH collects data from its members and transmits it to next CH in chain until data reaches to BS. P-LEACH protocols reduce the overall energy consumption of network but incur extra overhead in chain formation.
x. **CL-LEACH (Cross-Layer LEACH)** [30] **(2016)**: In CL-LEACH protocol, authors proposed cross-layer techniques to improve energy efficiency of network. CHs are selected based on residual energy and distance of node to BS. CL-LEACH uses multi-hop communication to transmit data from node to BS. Node whose energy is greater than threshold is selected as intermediate node in multi-hop communication. Major strength of CL-LEACH is caching in which source node checks its route to destination if there is path in cache source, node follows that path, thereby reducing overall routing time. But CL-LEACH involves more overhead in maintaining multi-hop routing path.

4 Conclusion and Future Directions

Table 1 shows the comparative analysis of improvements of single and multi-hop LEACH in chronological order. This survey concludes that major LEACH modification is proposed in literature to overcome its following limitations: First, CH selection criteria are purely based on randomness which causes less energy node may be elected as CH. To overcome this drawback, different LEACH variants use distance and energy parameters for choosing CH. Second: CH is responsible for transmitting data to BS so the energy of CHs depletes very quickly, and as a result, network lifetime degrades that can be enhanced using multi-hop LEACH, and third is some portion of network left without CH as LEACH selects CH randomly. Authors solve this demerit by dividing network into grids, square and so on. However, the

Table 1 Comparative analysis of improvements in LEACH protocols with single and multi-hop communication

Protocol name	Year	Communication pattern	Centralized or distributed	Selection parameters	Advantages	Disadvantages
K-LEACH	2013	Single-hop	Distributed	Energy	Uniform CH distribution	Complex
LEACH-TLCH	2013	Single-hop	Distributed	Energy and distance	Average energy consumption is less	Incurs extra overhead in selection of secondary CH
Improved V-LEACH	2013	Single-hop	Distributed	Minimum distance, maximum residual energy, minimum energy	Improves network lifetime	Incurs extra overhead in selection of vice CH
LS-LEACH	2013	Single-hop	Distributed	Same as LEACH	Add security feature	Energy wastage due to transfer of keys
IB-LEACH	2014	Single-hop	Distributed	Same as LEACH	Minimize the load of CH	Incurs extra overhead in selection of aggregator
CogLEACH	2014	Single-hop	Distributed	Idle channels	Improves throughput and network lifetime	CHs are not selected based on energy so less energy node can be selected as CH
CogLEACH-C	2015	Single-hop	Centralized	Idle channels and energy	Enhance network lifetime	Suffers from scalability problem and cannot be used for large WSN
EE-LEACH	2015	Single-hop	Distributed	Spatial density and residual energy	Less energy consumption due to aggregation	More complexity
E2ACM	2015	Multi-hop	Distributed	Same as LEACH	Reduce overall energy consumption	Hotspot problem

(continued)

Table 1 (continued)

Protocol name	Year	Communication pattern	Centralized or distributed	Selection parameters	Advantages	Disadvantages
DCDA	2015	Multi-hop	Distributed	Same as LEACH	Ensure overall network coverage	Extra overhead for partitioning the network into grids
DL-LEACH	2015	Multi-hop	Distributed	Same as LEACH	SNs transmit data to BS if distance between node, and BS is less than distance between SN and CH	Suffers from short node lifespan
MaximuM-LEACH	2016	Single-hop	Centralized	Energy	Ensures load balancing	Number of nodes that will join CH is not clearly defined
Improved LEACH	2016	Single-hop	Distributed	Energy and distance	Minimize the average energy consumption per SN	Weight function is not clearly defined
LEACH-CC	2016	Single-hop	Centralized	Residual energy	Low energy node is not selected as CH	Energy wastage due to location transfer to BS
P-LEACH	2016	Multi-hop	Centralized	Energy	Increase energy efficiency	Packet delivery delay
CL-LEACH	2016	Multi-hop	Distributed	Residual energy and distance	Reduce packet delivery delay by using caching	Incurs extra overhead in chain formation
NR-LEACH	2017	Single-hop	Distributed	Residual energy, link connections and signal strength	Increases energy efficiency	More overhead of rank calculation
CH-LEACH	2017	Single-hop	Centralized	k-mean clustering	Use two BS to improve energy efficiency	Value of k is not explained

(continued)

Table 1 (continued)

Protocol name	Year	Communication pattern	Centralized or distributed	Selection parameters	Advantages	Disadvantages
EiP-LEACH	2017	Single-hop	Distributed	Energy	Low energy node is not selected as CH	Scalability problem
ECO-LEACH	2018	Single-hop	Distributed	CHs are selected based on duty cycle, i.e. time horizon	Improve network lifetime with duty cycling and cooperative transmission	Suffers from control message overhead
ET-LEACH	2018	Single-hop	Distributed	Initial energy	Uniform CH distribution	Extra overhead for partitioning the network into squares
Modified Hetero LEACH	2018	Single-hop	Distributed	Energy	Minimize data redundancy	Control message overhead for inter-cluster communication
TH-LEACH	2018	Single-hop	Distributed	Energy	Avoids the CH selection in every round	Select the coordinates of CH for first round which is not appropriate for large network
Re-LEACH	2019	Single-hop	Distributed	Total number of SNs and CHs	Avoids the selection of CH for several rounds	Hotspot problem

LEACH has been broad area of research for authors in the last years but there is scope for this protocol for IoT and machine learning techniques in future.

References

1. Akyildiz IF, Su W, Sankarasubramaniam Y, Cayirci E (2002) Wireless sensor networks: a survey. Comput Netw 38(4):393–422
2. Al-Karaki JN, Kamal AE (2004) Routing techniques in wireless sensor networks: a survey. IEEE Wirel Commun 11(6):6–28
3. Arampatzis T, Lygeros J, Manesis S (2005) A survey of applications of wireless sensors and wireless sensor networks. InIntelligent Control, 2005. In: Proceedings of the 2005 IEEE international symposium on, mediterrean conference on control and automation , pp 719–724
4. Heinzelman W, Chandrakasan A, Balakrishnan H (2000) Energy-eficient communication protocol for wireless sensor networks. In: Proceeding of the Hawaii international conference on system sciences, Hawaii, Jan 2000
5. Usman MJ, Xing Z, Chiroma H, Gital AY, Abubakar AI, Usman AM, Herawan T (2014) Modified low energy adaptive clustering hierarchy protocol for efficient energy consumption in wireless sensor networks. Int Rev Comput Soft (IRECOS), 9(11):pp. 1904–1915
6. Bakaraniya P, Mehta S (2013) K-LEACH: an improved LEACH protocol for lifetime improvement in WSN. Int J Eng Trends Technol 4(5):1521–1526
7. Fu C, Jiang Z, Wei WE, Wei A (2013) An energy balanced algorithm of LEACH protocol in WSN. Int J Comput Sci Issues (IJCSI) 10(1):354
8. Ahlawat A, Malik V (2013) An extended vice-cluster selection approach to improve v leach protocol in WSN. In: Advanced computing and communication technologies (ACCT). pp 236–240
9. Alshowkan M, Elleithy K, AlHassan H (2013) LS-LEACH: a new secure and energy efficient routing protocol for wireless sensor networks. In: Proceedings of the 2013 IEEE/ACM 17th international symposium on distributed simulation and real time applications 2013 Oct 30, pp 215–220
10. Salim A, Osamy WL, Khedr AMM (2014) IBLEACH: intra-balanced LEA CH protocolfor wireless sensor networks. Wirel netw 20(6):1515–1525
11. Eletreby RM, Elsayed HM, Khairy MM (2014) CogLEACH: a spectrum aware clustering protocol for cognitive radio sensor networks. In: 2014 9th international conference on cognitive radio oriented wireless networks and communications (CROWNCOM), 2014 Jun 2. pp 179–184
12. Latiwesh A, Qiu D (2015) Energy efficient spectrum aware clustering for cognitive sensor networks: CogLEACH-C. In: Communications and networking in China (ChinaCom).pp 515–520
13. Arumugam GS, Ponnuchamy T (2015) EE-LEACH: development of energy-efficient LEACH Protocol for data gathering in WSN. EURASIP J Wirel Commun Netw
14. Kumar SA, Ilango P, Dinesh GH (2016) A modified LEACH protocol for increasing lifetime of the wireless sensor network. Cybern Inf Technol 16(3):154–164
15. Amirthalingam K (2016) Improved leach: a modified leach for wireless sensor network. In: IEEE international conference on advances in computer applications (ICACA), 24 Oct 2016. pp 255–258
16. Heinzelman WB, Chandrakasan AP, Balakrishnan H (2002) An application-specific protocol architecture for wireless microsensor networks. IEEE Trans Wirel Commun 1(4):660–670
17. Ma Z, Li G, Gong Q (2002) Improvement on LEACH-C protocol of wireless sensor network (LEACH-CC)[J]. Int J Future Gener Commun Netw 9(2):183–192
18. Al-Baz A, El-Sayed A (2018) A new algorithm for cluster head selection in LEACH protocol for wireless sensor networks. Int J Commun Syst 31(1)

19. Abushiba W, Johnson P, Alharthi S, Wright C (2017) An energy efficient and adaptive clustering for wireless sensor network (CH-leach) using leach protocol. In: 2017 13th International computer engineering conference (ICENCO). pp 50–54
20. Bongale AM, Swarup A, Shivam S (2017) EiP-LEACH: Energy influenced probability based LEACH protocol for wireless sensor network. In: 2017 International Conference on emerging trends & innovation in ICT (ICEI) 3 Feb 2017, pp 77–81
21. Bahbahani MS, Alsusa E (2018) A cooperative clustering protocol with duty cycling for energy harvesting enabled wireless sensor networks. IEEE Trans Wireless Commun 17(1):101–111
22. Prabha PN, Ali A (2018) Energy efficient threshold based cluster head selection and optimized routing in LEACH. In: International conference on intelligent data communication technologies and internet of things, pp 1475–1486
23. Sharma YK, Kumar S (2018) A clusterhead selection technique for a heterogeneous WSN and Its lifetime enhancement using hetero leach protocol. In: Proceedings of the international conference on microelectronics, computing & communication systems. pp 247–257
24. Sarkar P, Kar C (2018) TH-LEACH: threshold value and heterogeneous nodes-based energy-efficient LEACH protocol. In: Algorithms and applications, pp 41–49
25. Pandey S, Kumar R (2019) Re-LEACH: an energy-efficient secure routing protocol for wireless sensor networks. In: International conference on computer networks and communication technologies, Springer, Singapore, pp 777–787
26. Kaur A, Gujral RK (2014) E2ACM: energy efficient adaptive cluster based multilevel routing protocol for wireless sensor networks. Int J Comput Appl 90(12)
27. Lee CH, Lee JY (2015) DL-LEACH: hierarchical dual-hop routing protocol for wireless sensor network. J Inst Internet, Broadcast Commun 15(5):139–145
28. Kaur A, Gujral RK (2015) Distributed clustering based data aggregation algorithm for grid based WSN. Int J Comput Sci Eng 12(3):61–66
29. Razaque A, Abdulgader M, Joshi C, Amsaad F, Chauhan M (2016) P-LEACH: energy efficient routing protocol for wireless sensor networks. In: Systems, applications and technology conference (LISAT), pp 1–5
30. Marappan P, Rodrigues P (2016) An energy efficient routing protocol for correlated data using CL-LEACH in WSN. Wireless Netw 22(4):1415–1423

Deep Learning in Health Care: Automatic Cervix Image Classification Using Convolutional Neural Network

Mamta Arora, Sanjeev Dhawan, and Kulvinder Singh

Abstract Cervical malignancy can be viably counteracted if identified in the precancerous stage. In order to appropriately treat cervical cancer, making an accurate determination of a patient's cervical type is critical. However, doing so can be difficult, even for trained healthcare providers, because of the thin-line difference among the various cervix types. Kaggle and Mobile ODT have distributed a gathering of a few thousand commented on photographs of cervices. In this paper, we utilize profound learning approaches in computer vision, for example, convolutional neural networks and transfer learning. We try different things to experiment the models, for example, batch normalization, image augmentation, and dataset methodologies, for example, cropping the images. The initiations are utilized for the preprepared model Inception v3 which was prepared on the ImageNet dataset of 1.2 million pictures.

Keywords Deep learning · Convolutional neural network · Transfer learning · Cervical cancer

1 Introduction

Cervical cancer is the cancer of cervix which is reported as the second most common cancer in women in age between 15 and 44 years worldwide [1]. The cause of the cervix cancer is a virus called human papillomavirus (HPV). The virus can damage

M. Arora (✉)
Department of Computer Science and Engineering, University Institute of Engineering and Technology (U.I.E.T), Kurukshetra University, Kurukshetra, Haryana, India
e-mail: imamta.arora@gmail.com

Department of Computer Science and Technology, Manav Rachna University, Faridabad, Haryana, India

S. Dhawan · K. Singh
Faculty of Computer Science and Engineering, Department of Computer Science and Engineering, University Institute of Engineering and Technology (U.I.E.T), Kurukshetra University, Kurukshetra, Haryana, India

N. Marriwala et al. (eds.), *Mobile Radio Communications and 5G Networks*, Lecture Notes in Networks and Systems 140,
https://doi.org/10.1007/978-981-15-7130-5_10

the cell in the cervix. The dysplasia cells look like cancer cells, but they are not malignant. These cells are known as cervical intraepithelial neoplasia (CIN). Based on the severity, the CIN is broadly classified into three grades (CIN1/CIN2/CIN3). Over period of time, CIN1 will relapse by body immune system. Thus, the major goal is to diagnose CIN2 and CIN3.

The treatment will be proven effective if right type of cervix is identified at early stage. According to cervical cancer statistic [2], the early detection of cervical cancer can increase the five-year survival rate to 91%. The commonly used cervical cancer screening methods are Pap test, HPV testing, colposcopy, and digital cervicography. These are effective methods of screening but suffers from low sensitivity in detecting CIN2/3+.

Moreover, these tests require the expert personnel and laboratory setup for conducting the test. The entire above-mentioned screening test requires human interventions, thus are more prone to human error. Deep learning and computer vision have proven effective in healthcare domain for classification of medical imaging.

The experimental work carried out in this paper is based on the image dataset which is available on Kaggle. The Kaggle proposed a challenge in June 2017 by making thousands of cervix images available to the public dataset through Intel and Mobile ODT challenge. The challenge was based on differentiating the cervix image into one of the three possible types. We proposed a novel algorithm for giving the solution to the aforesaid problem.

2 Cervix Taxonomy

The cervical cancer is the cancer of cervix which is lower part of the uteri. It is associated between the uteri and vaginal divider. The cervix can be think of alike little donut with a modest opening from the center. The cervix position gets changed all through the life cycle of women. The cervix is envisioned using instrument called speculum. The expert practitioners look at the cervix and give the prospective determination by inserting the speculum into vaginal cannel. It is brought about by an infection called human papillomavirus (HPV). This HPV is present in every person. Some of HPVs are not risky, while others can cause growth of abnormal cells in the body. The diseases caused due to HPV grew gradually. It approximately takes between 10 and 20 years to reach from precancerous to cancerous stage. In the event that cervical disease is identified and treated in premalignant growth organize, at that point cervical malignant growth can be averted.

The rate of survival can be increased with the early identification. The cervix is broadly classified into three types, namely Types I/II/III as shown in Fig. 1. There is a thin-line difference between three types of cervix which is difficult to diagnose for healthcare providers. With the advancement of technology, the task can become easier for the healthcare providers.

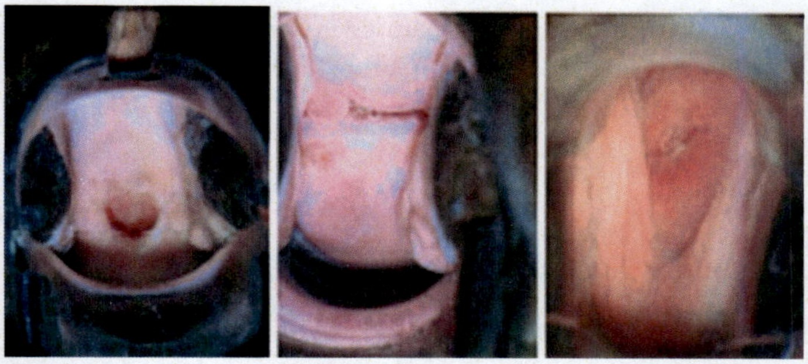

Fig. 1 Cervix images [3]

3 Literature Survey

As the cervical cancer takes decades in developing to cancerous stage, the mortality rate can be increased if it is diagnosed early. Due to the less awareness of cervical cancer screening methods, the patient comes to the doctor when it starts showing symptoms [4]. To serve the mass population, there exist the requirements of computer-assisted algorithms. The recently published studies show the state of the art in medical imaging achieved using artificial intelligence techniques, viz. machine learning and deep learning [5–8]. Some of the recent use cases of medical domain are elaborated in detail below.

The authors Wang et al. in the study [9] used two machine learning algorithms, viz. extreme learning machine (ELM) and support vector machine (SVM) for classification and detection of breast tumor. They worked on mammography images. These images belong to the women aged between 32 and 70. In the first phase, images are preprocessed for removing the noise using the median filter. The preprocessed images are then segmented using region growth, morphological operations, and modified wavelet transformation technique. The result proves that SVM performs better than EVM.

The published study [10] dealt with profound neural system for identifying and foreseeing tumor of the cyst. They connected different picture enlargement systems for expanding the extent of the harmful mass and generous lone sores pictures. Along these lines, the model had the option to report the area under the bend up to 87%.

Instances of successful convolutional designs are Inception v3 [11] and ResNet [12], which we look at in detail in later segments. Past work has likewise discovered methods like dropout [13], batch normalization [14], and image augmentation [15] to help with preparing and summing up profound neural systems. Transfer learning proved more success in this domain. It uses the pretrained weight of earlier models and applies to the problems having fewer amounts of data [15].

In the published study [16], author Lei Lei et al. use cervigram dataset provided by Kaggle for cervix-type classification. They have experimented four deep neural networks, namely Inception v4, AlexNet, ResNet, and VGG16. Results show that Inception v4 outperforms other networks with the accuracy of 74%. The author

Chaitanya et al. [17] utilized transfer learning in diagnosis of cervix image in Types I/II/III. They additionally utilized the optimization techniques like dropout, batch normalization, and image augmentation.

4 Dataset

The images used in experiment are provided by the Kaggle community [11]. The provided dataset contains hundreds of real cervical images. The multiple cervix images of same women are obtained from EVA system and sent to the physician. The best image is kept in the dataset. The additional set consists of annotated images of the same patient from different camera settings. The initial dataset the competition provides consists of the following distribution of classes (Table 1).

The amount and distribution of the additional data are as follows (Table 2).

As the efficiency of deep learning algorithms is directly proportionate to the amount of data, additional data significantly boosts the amount of data we have.

5 Proposed Methodology

The proposed architecture based on transfer learning is shown in Fig. 2. It accepts the set of cervix images for training stage. As images in training set are of different sizes, these images are cropped randomly into size of 224 × 224. To increase the size of available data, the technique of image augmentation is applied. It increases the data by flipping, rotating, and mirroring the images. In the diagram, a sample of horizontal flip is shown. Then, the images are normalized channel-wise for red, green, and blue. The normalized images act as input to convolutional neural network which gives the probability of each type of class as an output using softmax function.

Table 1 Image distribution in original dataset

Type	No. of images	Percentage (%)
Type 1	250	17
Type 2	781	53
Type 3	450	30

Table 2 Image distribution in additional dataset

Type	No. of Images	Percentage (%)
Type 1	1191	18
Type 2	781	53
Type 3	450	29

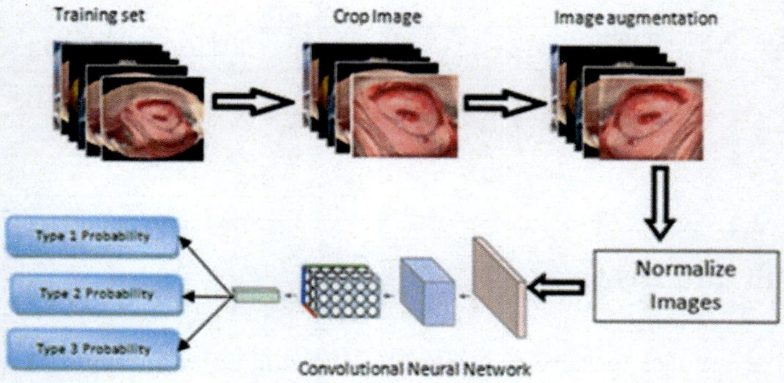

Fig. 2 Proposed architecture

The class which is having the maximum probability will be selected as predicted class for that instance.

Our proposed approach for cervix transformation zone classification relies principally on the transfer learning. It uses pretrained Inception v3 model which is trained on ImageNet. Only one fully connected and softmax layer is added on the pretrained model, which was trained on training set. All the methods are implemented in Python. The model is trained and tested on Intel I5 dual-core processor.

5.1 Convolutional Neural Network(ConvNet)

The convolutional neural network or ConvNet is a type of neural network with thousands of hidden layers. It is comprised of different back-to-back layers. The contribution to the convolutional system takes crude pixel estimations of a picture. The output or final layer consists of number of neurons corresponds to number of target classes [18]. In the proposed architecture, the output layer of the convolutional neural system will contain three neurons.

5.2 Pooling Layer

The function of the pooling layer is dimensionality reduction which in turn reduces the feature map. There are various kinds of pooling activation function which exists like max pooling, min pooling, average pooling, etc. The example of max pooling function is described with the help of example. The pixels of image are arranged in 8 by 8 matrices. The max pooling function reduces it to 4 by 4 matrices by selecting maximum pixel value from each quadrant.

Table 3 Exprimental results

Type	Precision	Recall	F1 score
Type 1	0.34	0.22	0.27
Type 2	0.61	0.71	0.65
Type 3	0.41	0.42	0.44

6 Result and Discussion

The proposed model for cervix images classification either in Type 1/Type 2/Type 3 is evaluated using accuracy matrices. The evaluation metrics used for the system is F1 score, recall, and precision. The classification report for the proposed model where 249 bottom layers are freeze and only top two layers trained is shown as below (Table 3).

7 Conclusion and Future Work

The work done in this paper portrayed the advancement and execution of convolutional-based algorithm for cervix image classification. The work demonstrates the usage of transfer learning approach to deal with this issue. We found that using the approach of transfer learning respect great results, as this was a fine segregation task absent much information. We found that including a custom, complex design over the preprepared model on ImageNet gave considerable lift to our outcomes. In future, investigation will be done by making three unique models for each class that will play out the paired characterization. At that point by utilizing the group system, we can show signs of improvement forecast of the image to the class which it belongs to.

References

1. Organization WH (2019) World cancer report, WHO Press
2. Statistics (2019) [Online]. Available: https://www.cancer.net/cancer-types/cervical-cancer/statistics. Last accessed on 10 Dec 2019
3. "intel-mobileodt-cervical-cancer-screening," Kaggle, [Online]. Available: https://www.kaggle.com/c/intel-mobileodt-cervical-cancer-screening. Last accessed 10 Dec 2019
4. LeCun YB, Hinton G (2015) Deep learning. Nature 521:436–444
5. Cireşan DC, Giusti A, Gambardella LM, Schmidhuber J (2013) Mitosis detection in breast cancer histology images with deep neural networks. In: Medical image computing and computer-assisted intervention—MICCAI, Berlin, Heidelberg
6. Holger RR, Lu L, Seff A, Cherry KM, Hoffman J, Wang S, Liu J, Turkbey E, Summers RM (2014) A new 2.5D representation for lymph node detection using random sets of deep convolutional neural network observations. In: International conference on medical image computing and computer-assisted intervention

7. Shin H-C, Orton MR, Collins DJ, Doran SJ, Leach MO (2013) Stacked auto encoders for unsupervised feature learning and multiple organ detection in a pilot study using 4D patient data. IEEE Trans pattern Anal Mach Intell 35(8):1930–1943
8. Song D, Kim E, Huang X, Patruno J, Muñoz-Avila H, Heflin J, Long LR, . Antani S (2015) Multimodal entity conference for cervical dysplasia diagnosis. IEEE Trans Med Imaging 34(1):229–245
9. Kamnitsas K, Ledig C, Newcombe VFJ, Simpson JP, Kane AD, Menon DK, Rueckert D, Glocker B (2017) Efficient multi-scale 3d cnn with fully connected crf for accurate brain lesion segmentation. Med Image Anal 36:61–78
10. Wolterink JM, Leiner T, de Vos BD, Van Hamersvelt RW, Viergever MA, Išgum I (2016) Automatic coronary calcium scoring in cardiac CT angiography using paired convolutional neural networks. Med Image Anal 34:123–136
11. Kaggle (2017) Cervical cancer screening. Intel and mobile ODT, 2017. [Online]
12. He K, Zhang X, Ren S, Sun J (2016) Deep residual learning for image recognition. In: IEEE conference on computer vision and pattern recognition (CVPR), Las Vegas, NV, USA
13. Srivastava N, Hinton G, Krizhevsky A, Sutskever I, Salakhutdinov R (2014) Dropout: a simple way to prevent neural from over fitting. J Mach Learn Res 15(1):1929–1958
14. Ioffe S, Szegedy C (2015) Batch normalization: accelerating deep network training by reducing internal covariate shift. Comput Res Repository (CoRR). abs/1502.03167, 2015.
15. Krizhevsky A, Ilya S, Hinton GE (2012) ImageNet classification with deep convolutional neural networks. Adv Neural Inf Process Syst 25:1097–1105
16. Lei L, Xiong R, Zhong H (2017) Identifying cervix types using deep convolutional networks. Springer, California
17. Asawa C, Homma Y, Stuart S (2017) Deep learning approaches for determining optimal cervical cancer treatment. Stanford, USA
18. Arora M, Dhawan S, Singh K (2018) Deep neural network for transformation zone classification. In: Proceedings of first international conference on secure cyber computing and communication (ICSCCC), Jalandhar, pp 213–216

Review Paper on Leaf Diseases Detection and Classification Using Various CNN Techniques

Twinkle Dalal and Manjeet Singh

Abstract Majority of Indian population depends on agribusiness for its survival, and it plays a vital role in every nation's economy. The disease is spread to other plants. Early detection of disease is a significant thing. Detection models to detect the disease are built by direct observation of every plant. This is essential as we can take parameters to restrict. Hence, healthy cropping is necessary for the growing agricultural economy. A better yield of crop is dependent on many factors including disease detection. The on time disease detection helps the farmers to save their crop yield as the remedies can be given on time. In order to solve the problem, various convolution neural network architectures have been designed and tested on labelled data to obtain high accuracy in classification and detection of disease. This work deals with the brief and detailed study of various techniques used for classification and detection of disease in plants based on feature extraction and different training methods.

Keywords Convolutional neural network · Image processing · Plant diseases detection

1 Introduction

India is a country of wide diversity of crops in which majority of population relies on the agricultural sector. The sensitivity of crops and the climatic condition proves to be a main reason behind which leads to diseases in plants which further affects the crop yield [1]. Out of the total diseases affected, plants make upto 10–30% of overall yield loss. So timely and robust detection and classification of diseases play an important role in increasing the quantity of the yield, and also it helps us to avoid

T. Dalal (✉) · M. Singh
J.C. Bose University of Science and Technology, Faridabad, India
e-mail: twinkledalal003@gmail.com

M. Singh
e-mail: mstomer2000@yahoo.com

N. Marriwala et al. (eds.), *Mobile Radio Communications and 5G Networks*, Lecture Notes in Networks and Systems 140,
https://doi.org/10.1007/978-981-15-7130-5_11

the diseases which can somehow reduce the total loss incurred. For ready reference, the conventional model of convolutional neural network framework is explained in brief manner herein below:

CNN is an abbreviated form of convolutional neural network composed of four main components

1. Convolution layer
   It basically extracts the features of an image while maintaining the spatial relation among the pixels by using a matrix that moves over the picture.
2. Nonlinearity (ReLU layer)
   The ReLU function is an abbreviated form of rectified linear unit function denoted by $f(x) = \max(x, 0)$ with input x, i.e. input given in the form of matrix. ReLU considers all negative values in the matrix to zero, and the rest are kept constant. ReLU speeds up the training process which can further improve the neural network. ReLU gives unchanged value when multiplied by itself so we can say that it is idempotent.
3. Pooling layer
   Pooling can be defined as the sampling after reducing the dimensionality with the help of ReLU. Pooling is of different types: Average, Max, sum, etc., depending on the type of pooling the element is selected from the feature map.
4. Fully connected layer (classification layer)
   It allows the network to learn about the nonlinear combination of the features [2] obtained from CNN after pooling.

This paper presents an absolute comparison of different systems for automatic image-based classification of diseases in plants [3].

The detailed study of these systems reveals that the different proposed systems are tested on different leaves of plants individually to detect the disease in plants. While all these plants have fruit that show the disease as well, our aim is to detect the disease at an earlier stage, before it affects the fruit, so as to decrease losses [3]. For this reason, the proposed system classifies plant species and diseases based only on the leaf [3].

After doing the detailed study of the various systems used for the work defined above, it is clear that the proposed system is based on convolution neural network [3], which is very well-known deep learning technique usually used in the field of computer vision for image detection.

Few CNN architecture, models are experimented with a purpose to present a complete comparison of the system in terms of accuracy, validation loss, speed of training and size of model.

The organization of this paper is as follows. Section 2 comprises a literature survey, whereas Sect. 3 consists of a brief study of various models based on CNN techniques, and Sects. 4 and 5 consist of comparisons table based on various factors some of which are mentioned above. And finally, Sect. 6 consists of the conclusion.

2 Literature Survey

An intensive survey in the field of CNN is performed and organized as follows:

Support vector machine (SVM): SVM classifier uses an unsupervised learning method. The SVM method aims to find out the linear combination of features in pattern recognition in order to separate different classes. The learning data found adjacent to the hyper-plane are called support vectors. SVM can separate sets of data as both linearly indistinguishable and distinguishable [4, 5].

K-nearest neighbour (KNN): The KNN [6, 7] classifier uses a supervised learning method. This method is a nonparametric and intense-based learning algorithm that makes a classification of the existing learning data when new data enter. The principle of this method is to assign new data input to the adjacent cluster in a previously established sample set. The distance between these two data points is calculated by using various distance functions. The most widely functions used are Euclidean distance and Manhattan distance.

Extreme learning machine: ELM is proposed by Huang et al. [8] based on single layer feed forward networks. In ELM, the hidden weights of the layer are generated randomly, and the output weights are calculated using least square technique.

Recent study in convolutional works in convolutional neural networks is presented as follows:

- AlexNet [9]
  Krizhevsky et al. [10] proved for the very first time that the convolution neural network can be a possible solution for recognition of images when working with a huge amount of data, and it is even a better approach than usual traditional method.
  In 2012, Krizhevsky et al. stepped inside the visualization or can be termed as image classification, known as ImageNet, with their proposed CNN network architecture popularly known as AlexNet, and they won by a great margin. AlexNet can be termed as the first step towards the recent trend convolutional neural networks. In AlexNet, ReLU has been used along with other components such as local response normalization and pooling. The network is trained on several numbers of GPUs. In the coming years, champions of the ImageNet entered the deep learning. One of the most important rebellion point of the network is its training on multiple GPU which in turn can increase the training speed in huge data. AlexNet is made up of five convolutional layers each preceded by ReLU layer. Another layer known as normalization layers is added to help generalization [10]. Output of the last convolution layer, i.e. fifth convolutional layer is provided as an input to the fully connected layer connected network after performing pooling. As stated earlier, fully connected layers are helpful in calculating the probability of class. Fully connected layer consists of dropout layers which are needed to avoid over fitting. The final or end of fully connected layer (FC8) denotes the class probabilities of image which is supplied as an input.
- SqueezeNet

After the release of AlexNet, several architectures of CNN were proposed with a primary objective of increasing the accuracy. SqueezNet is proposed by Iandola et al. [11]. To achieve the objective of lightweight model, the SqueezNet is made up of model three design strategy. That are reducing filter size, decreasing the input channels and down sampling late in the network [11]. SqueezeNet appoints new fire model. Fire module is made up of squeeze layer with 1 × 1 filters (which in turn decreases input channel of 3 × 3 filters) and expand layer with combination of 1 × 1 and 3 × 3 filters (reducing the filter size). Squeeze and expand layers are followed by the ReLU layers [11].

- LeNet [12]
  The LeNet model mainly comprises two parts: the first one is the self-explanatory feature extraction model, and the second one is the classification model.

  Feature extraction model
  The feature extraction model trains the network to identify unique high-level features extracted from the input. The LeNet comprises a sequence of convolution and pooling layers.

  - Convolution layer.
  - Max-pooling layer.

  Classification model
  The classification step consists of fully connected layers in which every neuron supplies a full connection to each and every learned feature maps obtained from the preceding layer. These fully connected layers rely on the Softmax activation function to calculate the score of classes.

- VGG19
  Simonyan et al. [13] demonstrated that CNN must include more depth for representing visual data in the hierarchical form to improve the performance of the network. The team obtained 7.3% of error rate in image classification which is less than as compared to preceding results obtained. They worked on network which has 19 layers composed of 3 × 3 filters which has stride 1 and pooling layers (max) with stride 2 to obtain more rigid features and their multiple combinations. The two blocks of 3 by 3 layers (convolution) are same as the field (receptive) of 5 by 5 in which an individual block of three layers has an efficient field(receptive) of 7 × 7. This method is made up of 3 fully connected layers (FC) after the convolutional layer. More depth and fully connected layers give result in a huge number of learning power as well as arguments. VGG-19 contains 1.44 × 108 parameters.
- Inception-v3
  This method is made of inception modules to build a deeper model while aiming increment of width [14]. The traditional filters are used to gather information about linear functions of the inputs, whereas with the introduction of inception module helps in obtaining higher learning abilities and selection power by introducing complicated filters which exploit the cross-channel as individually.

By evaluating the 1 by 1 convolution, the inception hinders the correlation of cross-channels followed by correlation and cross-spatial via 5 by 5 and 3 by 3 filters. By performing computational in parallel manner and then clubbing the abstracted feature maps of each and every convolution filters, this technique offers the privilege of abstraction at multiple levels from an individual input.

- ResNet15 [2]
  Residual neural network is abbreviated as ResNet developed by Kaiming [15]. He [15] had 3.57% error in classification which gives the much better results as compared to all other state of networks (convolutional neural network). Beyond a particular limit, increment in number of layers in a network can improve the training and validation loss to a great extent. The problem of disappearing gradient in network can affect the learning and make it slow and inappropriate. This problem can be solved by creating a shortcut of the input to the result of a convo (convolutional) layer in order to hinder the wrapping representation of the images. The block popularly known as residual block can lead to the addition to the depth of the network, whereas there is no change in the number of parameters. There are two blocks of residual block one as an identity function and the other as bundle of two convo (convolution) layers.
- Xception [2]
  Xception can be termed as an enhanced version of inception model in which replacement of modules is done with respect to the depth [14]. This architecture of convolutional neural network comprises 36 separable convolutions with respect to depth, and mapping of spatial relationship and cross-channels is completely decoupled.
- GoogleNet
  GoogleNet [16] can be termed as deep neural network a structure that helps to reduce thee computational time as well as complexity. It uses the modules of inception which uses more than one convolutional layer parallel to obtain multiple feature points. GoogleNet consists of batch normalization in place of dropout. In batch normalization, mean and variance are obtained at each layer to normalize.
- CaffeNet [17]
  CaffeNet is a form of deep learning or a form of a deep CNN composed of more than one layer gradually that extract or compute features from the image supplied as input. To be more specific, it contains five convolutional and eight learning layers along with three fully connected layers. This architecture of CNN can be termed as the initial point, but can be altered to work with five categories (class). The final layer is modified, and the result obtained from Softmax was adjusted according to the requirement (Tables 1, 2 and 3).

Neural networks served as the initial step towards AI by producing a model which behaves similar to a human learning process. The above table presents a comparative study based on the different parameters. The different model trained the dataset either using transfer learning or started training the dataset from the scratch or from both

Table 1 Summary of performance (merits and demerits) measures based on classification models

Classifier	Description	Merits	Demerits
K-nearest neighbour [17, 18]	(a) It is nonparametric and statistical classifier (b) Weight is assigned to the adjacent neighbour (c) Classification is done based on distance metric (d) It calculates distance using Euclidean distance	(a) Do not require classes to be linearly separable	(a) More time-consuming testing process because it requires calculation of distance to all known instances
Support vector machine [19, 20]	(a) This objective of this algorithm is to define decision boundaries in N-dimensional plane where *N* denotes the number of features	(a) Accuracy achieved in classification is very high as compared to other techniques	(a) Requires high training time with huge dataset (b) Selection of kernel function and kernel parameters is difficult for mapping original data into high-dimensional data
Extreme learning machine [21]	(a) An efficient learning algorithm for the single hidden layer feed forward neural networks	(a) It is robust and can handle noisy data (b) Well suited to analyse complex numbers	(a) Quite huge training time and processing time are required

the techniques. Transfer learning refers to the problem that is concerned on collecting and storing knowledge obtained while solving a particular problem and using it for a different but similar problem. The accuracy of various models has been recorded and presented above.

3 Conclusion

Various CNN architectures like VGG-19, Inception-v3, Xception, ResNet-50 are deeply studied with respect to different parameters according to their suitability and compared the performance. Execution of all models has been estimated with the help of changing the number of pictures (images), changing in weight and bias learning rate. Change in the number of images visibly affected the performance of the models. The main objective or the primary aim is to find an optimized model which can help the farmers to classify the problem before it becomes uncontrollable. In this work, we

Table 2 Comparison of different CNN models based on different parameters

Authors	Species	Data set	Number of classes	Total number of images	Best accuracy achieved till date
Atabay (2017)	Tomato	Plant village subset	10	19,742	97.53
Brahimi et al. (2017)	Tomato	Plant village subset	9	14,828	99.18
Barbedo (2018b)	12 crop plants	Barbedo (2016) (15% controlled, 85% in field)	56	1385	87
Brahimi et al. (2018)	14 crop	Plant village	39	54,323	99.76
Fuentes et al. (2017)	Tomato	Own dataset	10	5000	85.98
Fuentes et al. (2018)	Tomato	Own dataset	12	8927	96.25
Cruz et al. (2017)	Olive	Own dataset (controlled)	3	299	96.60
Liu et al. [13]	Apple	Own dataset (controlled and field)	4	1053	97.62
Ferentinos (2018)	25 crop	Plant village dataset	58	87,848	95
DeChant et al. (2017)	Maize	Own dataset (field)	2	1796	96.70
Picon et al. (2018)	Wheat	Johannes et al. (2017) extended (field)	4	8178	97
Oppenheim and Shani (2017)	Potato	Own dataset (controlled)	5	400	96
Ramcharan et al. (2017)	Cassava	Own dataset	6	2756	93
Too et al. (2018)	14 crop plants	Plant village	38	54,306	99.75
Sladojevic et al. [17]	Apple, pear, cherry, peach, grapevine	Own dataset (Internet)	15	4483	96.3

Table 3 Comparison table based on training strategy and use of CNN model either for detection or classification

Authors	Detection or classification	Deep convolutional neural network	Training strategy
Atabay (2017)	Classification	VGG16, 19, custom architecture	From scratch and transfer learning
Brahimi et al. (2017)	Classification	AlexNet, GoogleNet	From scratch and transfer learning
Barbedo (2018b)	Classification	GoogleNet	Transfer learning
Brahimi et al. (2018)	Classification	AlexNet, DenseNet169, Inception v3, ResNet34, SqueezeNet1-1.1, VGG13	From scratch and transfer learning
Fuentes et al. (2017)	Classification	AlexNet, ZFNet, GoogleNet, VGG16, ResNet50, 101, ResNetXt-101	Transfer learning
Fuentes et al. (2018)	Detection	AlexNet, ZFNet, GoogleNet, VGG16, ResNet50, 101, ResNetXt-101	Transfer learning
Cruz et al. (2017)	Classification	LeNet	Transfer learning
Liu et al. [13]	Classification	AlexNet, GoogleNet, ResNet20, VGG16 and custom architecture	From scratch and transfer learning
Ferentinos (2018)	Classification	AlexNet, AlexNetOWTBn, GoogleNet, Overfeat, VGG	Unspecified
DeChant et al. (2017)	Detection	Custom three stages architecture	From Scratch
Picon et al. (2018)	Classification	Custom ResNet50, ResNet50	Transfer learning
Oppenheim and Shani (2017)	Classification	VGG	Unspecified
Ramcharan et al. (2017)	Classification	Inception V3	Transfer learning
Too et al. (2018)	Classification	Plant village	Transfer learning
Sladojevic et al. [17]	Classification	CaffeNet	Transfer learning

presented the review of the use of various neural network models in the agricultural field to detect and classify plant diseases detection and classification. The literature study shows that the texture, colour and morphological characteristics are best suited for the purpose of identification and classification of plant diseases. In the nutshell, we would like to conclude that automatic detection and classification of disease in plants would reduce the cost incurred because of an expensive domain expert. And also, it will prove to be a boomb to farmers which would help them to improve the yield of crop which will further add to enhance the quality of Indian gross domestic product.

References

1. Tm P, Pranathi A, SaiAshritha K, Chittaragi NB, Koolagudi SG (2018) Tomato leaf disease detection using convolutional neural networks. In: 2018 eleventh international conference on contemporary computing (IC3). IEEE, pp 1–5
2. Bhatt P, Sarangi S, Pappula S (2017) Comparison of CNN models for application in crop health assessment with participatory sensing. In: 2017 IEEE global humanitarian technology conference (GHTC). IEEE, pp 1–7
3. Gandhi R, Nimbalkar S, Yelamanchili N, Ponkshe S (2018) Plant disease detection using CNNs and GANs as an augmentative approach. In: 2018 IEEE international conference on innovative research and development (ICIRD). IEsEE, pp 1–5
4. Behmann J et al (2015) A review of advanced machine learning methods for the detection of biotic stress in precision crop protection. Precis Agric 16.3:239–260
5. Prince G, Clarkson JP, Rajpoot NM (2015) Automatic detection of diseased tomato plants using thermal and stereo visible light images. PLoS ONE 10(4):e0123262
6. Arti S (2016) Machine learning for high-throughput stress phenotyping in plants. Trends Plant Sci 21(2):110–124
7. Mohanty Sharada P, Hughes DP, Salathé M (2016) Using deep learning for image-based plant disease detection. Front Plant Sci 7:1419
8. Khirade SD, Patil AB (2015) Plant disease detection using image processing. In: 2015 international conference on computing communication control and automation (ICCUBEA). IEEE
9. Durmuş H, Güneş EO, Kırcı M (2017) Disease detection on the leaves of the tomato plants by using deep learning. In: 2017 6th international conference on agro-geoinformatics. IEEE, pp 1–5
10. Krizhevsky A, Sutskever I, Hinton GHE (2012) ImageNet classification with deep convolutional neural networks. In: Advances in neural information processing systems
11. Mohanty SP, Hughes DP, Salathe M (2016) Using deep learning for image-based plant disease detection. Front Plant Sci. arXiv: 1604.03169, 22 September 2016
12. Amara J, Bouaziz B, Algergawy A (2017) A deep learning-based approach for banana leaf diseases classification. Datenbanksysteme für Business, Technologie und Web (BTW 2017)-Workshopband
13. Liu B, Zhang Y, He D, Li Y (2017) Identification of apple leaf diseases based on deep convolutional neural networks. Symmetry 10(1):11
14. Szegedy C, Liu W, Jia YQ, Sermanet P, Reed S, Anguelov D, Erhan D, Vanhoucke V, Rabinovich R (2014) Going deeper with convolutions. In: Proceedings of the 2014 IEEE conference on computer vision and pattern recognition. Columbus, OH, USA, 24–27 June 2014, pp 1–9
15. He K, Zhang X, Ren S, Sun J (2016) Deep residual learning for image recognition. In: IEEE conference on CVPR, pp 770–778

16. Jeon WS, Rhee SY (2017) Plant leaf recognition using a convolution neural network. Int J Fuzzy Logic Intell Syst 17(1):26–34
17. Sladojevic S, Arsenovic M, Anderla A, Culibrk D, Stefanovic D (2016) Deep neural networks based recognition of plant diseases by leaf image classification. Computational intelligence and neuroscience
18. Ahmed F, AI-Mamun HA, Bari ASMH, Hossain E (2012) Classification of crops and weeds from digital images: a SVM approach. Elsevier
19. Owomugisha G, Quinn JA, Mwebaze E, Lwasa J (2014) Automated vision-based diagnosis of banana bacterial wilt disease and black Sigatoka disease. In: Proceedings of the international conference on the use of mobile ICT in Africa 2014. ISBN: 978-0-79721533-7
20. Mrunmayee D, Ingole AB (2015) Diagnosis of pomegranate plant diseases using neural network. In: 2015 fifth national conference on computer vision, pattern recognition, image processing and graphics (NCVPRIPG). IEEE
21. Mahlein A-K (2016) Plant disease detection by imaging sensors–parallels and specific demands for precision agriculture and plant phenotyping. Plant Dis 100(2):241–251
22. Kutty SB, Abdullah NE, Hashim H, Rahim AAA, Kusim AS, Yaakub TNT, Yunus PNAM, Rahman MFA (2013) Classification of watermelon leaf diseases using neural network analysis. In: 2013 IEEE business engineering and industrial applications [BEIAC]

Artificial Intelligence and Virtual Assistant—Working Model

Shakti Arora, Vijay Anant Athavale, Himanshu Maggu, and Abhay Agarwal

Abstract In twenty-first-century virtual assistant is playing a very crucial role in day to day activities of human. According to the survey report of Clutch in 2019, 27% of the people are using the AI-powered virtual assistant such as: Google Assistant, Amazon Alexa, Cortana, Apple Siri, etc., for performing a simple task, people are using virtual assistant designed with natural language processing. In this research paper, we have studied and analyzed the working model and the efficiency of different virtual assistants available in the market. We also designed an intelligent virtual assistant that could be integrated with Google virtual services and work with the Google virtual assistant interface. A comparative analysis of the traffic and message communication with length of conversation for approximately three days is taken as input to calculate the efficiency of the designed virtual assistant.

Keywords Virtual assistant · Artificial intelligence · NLP · IoT · NLG

1 Introduction

The virtual assistant is an independent entity which provides administrative services to the users. It also provides technical and social assistance to the users. The virtual assistant can work in various modes like secretarial work, customer service support, marketing on social media, as well as can do the web editing task for a particular site or app. Virtual assistant services can be used in any of the organization, and with the training of the virtual assistant, the efficiency of the work can be improved

S. Arora (✉) · V. A. Athavale · Himanshu Maggu · A. Agarwal
Panipat Institute of Engineering & Technology, Samalkha, Panipat 132102, India
e-mail: shakti.nagpal@gmail.com

V. A. Athavale
e-mail: vijay.athavale@gmail.com

N. Marriwala et al. (eds.), *Mobile Radio Communications and 5G Networks*, Lecture Notes in Networks and Systems 140,
https://doi.org/10.1007/978-981-15-7130-5_12

continuously. Several supporting technologies help the virtual assistant to work as a real time assistant. For example, Natural Language Processing (NLP), Artificial Intelligence (AI) and IoT supported tools and applications are working as a backbone for virtual assistant.

1.1 Difference Between Virtual Assistant and Chatbots

Chatbots and virtual assistants are the applications of artificial intelligence and both can be used and served as the business applications, and considered as the intelligent applications of AI. The chatbots are used as an interface for extracting the information and work as a medium for providing the information to the user and afterward users can verify their queries or ask any product-related information, can fix an appointments and meetings with the managers and or with any registered entity, while virtual assistants are used for making business plans, setting reminders, planning events, attending calls and so on.

Virtual assistant and chatbots can apply artificial intelligence. Intelligence worked on the trained model support high level language processing and support high language processing skills; wider scope and can work with a range of functions.

Simple rule-based tasks can be performed; Artificial Neural Networks, and Ranking models are used to design chatbots various API are available to design VA, C#, JavaScript, Python can be used to design chatbots.

The comparison between the chatbot and virtual assistant is given in Table 1 and it is more evident that virtual assistant uses Artificial Intelligence, Natural Language Processing and the artificial neural network which makes the virtual assistant more dynamic and productive.

In Fig. 1, a survey report published by Clutch [1] in 2019 is presented, which visualizes the adoption of virtual assistant integration with different devices.

Table 1 Comparison between virtual assistant and chatbots

Virtual assistant	Chatbots
Apply the artificial intelligence	Worked on the trained model
Support high level language processing	Does not support high language processing skills
Wider scope and can work with a range of functions	Simple rule-based tasks can be performed
Designed using artificial neural networks	Ranking models are used to design chatbots
Various APIs are available to develop	C#, JavaScript, Python can be used to develop chatbots

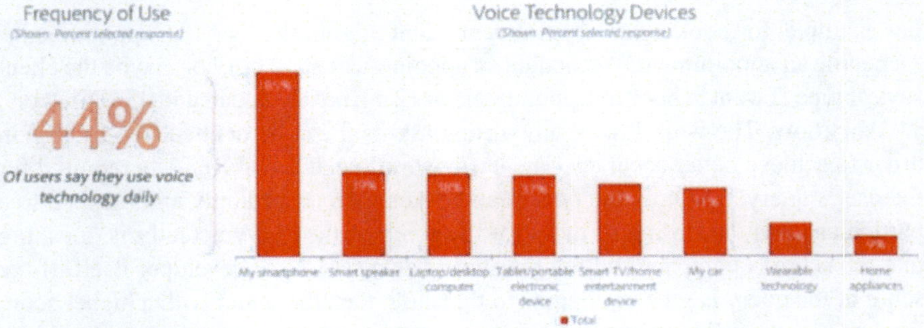

Fig. 1 Voice communication integration in different devices

Figure 1 shows the actual existence and working of virtual assistant in different domains. 85% smartphones are using virtual assistants, 19% smart speakers and 18% laptops in the market are using the services of the virtual assistant. Total 44% of the users are using the virtual assistant on different devices. With the advancement in IoT, such types of applications are embedded in the larger to smaller devices and are working efficiently.

2 Training Automation

Intent Identification: Intents are a special type of object that describes how to do something [2]. Intents are the actions, used to map the queries (asked by the users), to the most appropriate responses. It is the goal of the users in making a particular request. Context plays a vital role while taking these actions as they store the previous data and status of what has already been done in the conversation. Some events also occur while mapping a particular query to the response. These events are nothing but a special piece of code that performs a particular operation and allows us to call an intent-based non-verbal signal, like starting a new conversation or a button click. The designed intents activate the training phrases and train the machine based on inputs keywords and phrases provided by users.

Entity Identification: Entities tell the system or virtual assistant how the data from the user's expressions have been extracted [2]. When a user asks any query entity from that expression is identified. Entity can be a place or some sort of keyword. For example, show me any flight for *New York*, book an *appointment* with the director. Here, *New York* and *Appointment* are identified by the code and then the data is extracted from a particular user expression and helps to map the query to a particular response.

Training Phrase Organization: Training phrases are some example phrases for what a client or a user might ask. A lot of training phrases may be created for a particular intent [2]. When a user's utterance resembles one of these phrases, the intent has been matched and the user query has been mapped to a particular response.

For example, for booking an appointment client might ask, "book an appointment," "schedule an appointment," "schedule an appointment at 12 pm," or maybe the client says or type "I want to book an appointment at 12 pm next Monday with the Director."

Workflow: The workflow of any virtual assistant can be defined step by step in order to achieve better accuracy. The basic workflow, if there's no data required for the user's query, is simply the query can be stemmed accordingly and mapped to a particular intent according to the what the score of the query actually is, meaning that the score is compared with a threshold value set by the developer itself. If the score of the query is greater or equal to threshold then the intent with a higher score is taken and according to that, the VA responds to a particular user's query. But if the score is less than the threshold, then the query is forwarded to a "fallback" intent. Workflow can be as follow [1]:

- The user asks a query to a chat client.
- VA sends the phrase to NLP engine.
- NLP engine extracts user intents and entities from given phrases and sends them back to VA.
- Proper data service is called by intent and hence structured data by a given entity.
- Data is returned to VA.
- VA responds back to the client the same way as the client asks the query (Fig. 2).

Layers of VA: There are four layers in a chatbot. The layers define the workflow in a better and more understandable way. These are (a) UI layer, (b) integration layer, (c) machine learning layer and (d) data layer.

1. **UI Layer**: This is the closest layer to the end-users as shown in the Fig. 3. It is the only layer to which a user actually interacts with [4]. This layer is usually visual to the end-users.

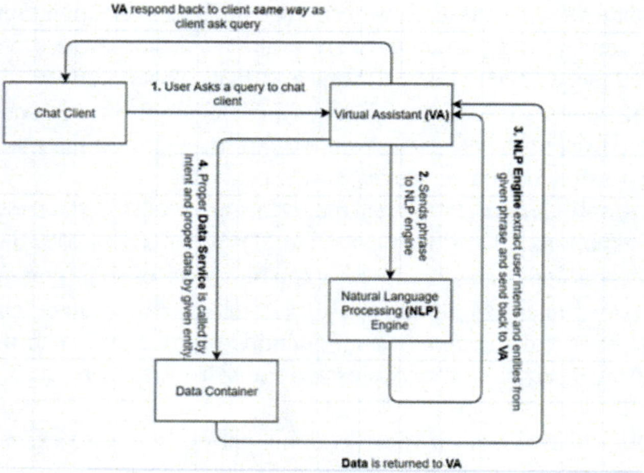

Fig. 2 Workflow of a VA [3]

2. **Middleware/Integration Layer**: This layer connects the UI layer to the conversational or ML layer. This layer consists of a special piece of code that calls a particular method to detect or map the user utterance to the particular action which could help in integration. Hence, this layer is called the integration one. This is the layer which enables the backend code to different platforms. Virtual assistant can work with google, webapp, twitter, facebook and many more.
3. **Machine Learning Layer**: This layer is also called the conversational layer. Conversational layer consists of NLP/NLU Natural l Language Undersatnding and a decision tree. It consists of your special piece of code including your intents, entities, context, etc. [4]. NLP is the component of an AI that makes it able to recognize, process and perform human language. NLP is based on deep learning and neural network to recognize patterns, categorize the patterns, contextualize them, translate them into text and perform a voice-based interaction [5]. NLU is the subgroup of NLP that deal with the best handle unorganized inputs and change them into a structured manner that software can understand and now the VA throws these utterances into the decision engine, and VA has the definite standard to encounter to exit the conversational loop and act upon [1, 4, 5]. That is how the user's queries are mapped to the most accurate and specific responses.
4. **Data Layer**: It is the farthest layer to the end-users. There is a data connector in this layer that is responsible for connecting and collecting the data to/from the cloud or any data source. This layer directly passes the structured data to Natural Language Generation (NLG) which converts structured data into specific text response and passes the response to the middleware/integration layer (Fig. 3).

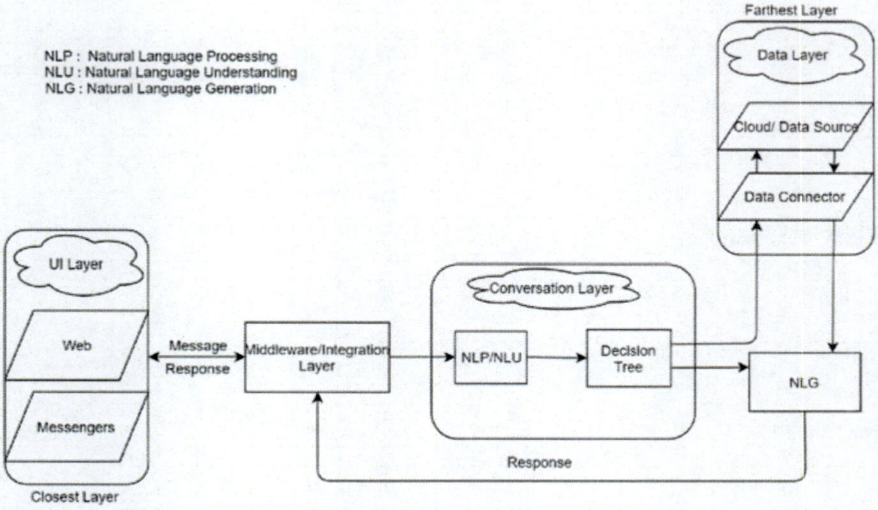

Fig. 3 Layers of VA [6]

3 Analytics

Precisely, potential test methods are used to retain data and to modify it into conversations with the user and to provoke and keep a conversation continuous. This covers the survey of input, way of generating output and "fallback" plan for when a direct/mapped response is not available. Abreast this, there is no method of data extraction for the VA to focus on. This, in turn, means that the VA's main focus is to assist the user through conversation and continuously redeveloping its unique library of data and metadata [7].

Analysis of the length of conversation and messages sent by users is done. The conversational length goes on increasing with the increase in the number of the user. The users love the interaction and behavior of virtual assistant which results in more engagement of users with the VA(s) for a longer duration. Hence, it is clear that the VA(s) is behaving more like a human than a machine. This shows the natural behavior of VA(s) (Figs. 4 and 5).

Fig. 4 Conversation length

Conversation length

Fig. 5 Text message

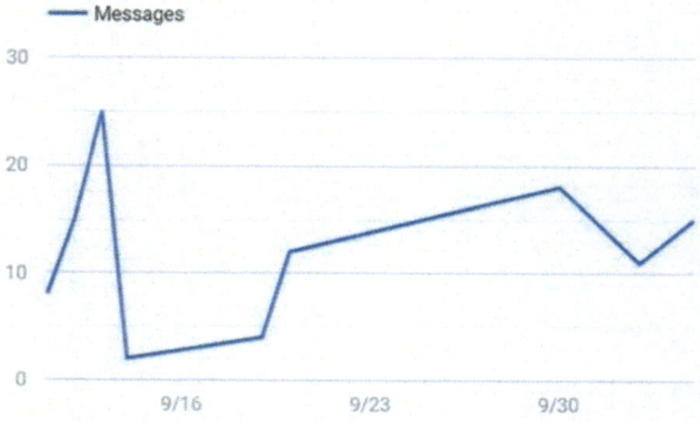

Text Message

4 Related Work

This paper introduces the usability of four voice-based and text-based VA (Google assistant, Cortana, Alexa and Siri). Google assistant answered 88% of questions correctly, while Siri scored 75%, Alexa scored 72.5% and Cortana came in with 63%. However, this test was on smart speakers and not smartphones [3, 8, 9]. To perform an action on virtual assistant via voice communication interface on different objects, some of the approaches work upon computers, tablets while some upon smart devices like Smartphones, Google Home, Alexa Echo dot, etc.

Alexa works on several platforms, like Amazon echo line speakers and smart devices, there is also an Alexa application for smartphones. Also, the application is an additive to other Alexa gadgets and not as a solitary AI-based VA. Alexa answered 72.5% of questions correctly and is improving day by day. The initial release of Alexa is in November 2014 and supports Fire OS 5.0 or later, iOS 8.0 or later Android 4.4 or later. In January 2019, Amazon's devices team announced that it had sold over 100 million Alexa-enabled devices [6, 10, 11] and is available in English, French, German, Punjabi, Hindi and many more languages. Alexa is of two types: Intelligent VA and cloud-based voice service.

Google Assistant works on several platforms like Android, Chrome OS, iOS, Linux, etc. There is even a Google Assistant application for smartphones. Also, the application is an additive to other Alexa gadgets and not as a solitary AI-based VA. Google Assistant answered 88% of questions correctly and is improving day by day. Google Assistant was unveiled on May 18, 2016, and has support for Android, Google Home, Wear OS by Google, Android TV, smart speakers, headphones, smart displays, iOS, Ubuntu, Raspberry Pi, etc. Google Assistant is available in English, Arabic, Bengali, Chinese, Dutch, French, German, Gujarati, Hindi, Indonesian, Italian, Japanese, Korean, Tamil and many more languages. It is basically a virtual assistant.

Siri works on several platforms like iOS gadgets, macOS and Apple TV too. However, Apple was the beginning head to set a voice-controlled VA on a phone, and Siri has expanded from being a fundamental attachment to your iPhone's capacities to something significantly smarter [12, 13]. Siri answered 75% of questions correctly and is improving day by day. The initial release of Siri is on October 12, 2011, and supports iOS 5 onward, macOS, TV OS, watch OS, audio OS, iPad OS and is available in English, French, German, Chinese, Dutch and many more languages. These are intelligent virtual assistants.

Cortana works on several platforms like Windows, iOS, Android, Xbox OS, PlayStation, etc. Cortana answered 63% of the questions correctly. Cortana was presented in 2014. Cortana also works with Apple iOS gadgets and Google Android and is available in English, French, German, Dutch and many more languages. Since launch, the total number of questions asked is 18 billion and the number of people with access to Cortana is around 800 million in 2019. These are personal digital assistant [14–16].

5 Future Work

Recognizing speech and natural language to make the conversation more natural is still the biggest task to work upon. Speech recognition and machine getting to know have persevered to be refined.

6 Conclusion

In this paper, a comparative analysis of chatbot and virtual assistant as well as analysis is done on the number of virtual assistants available in the market. The next generation of virtual assistants is also discussed. We designed a virtual assistant to converse with humans, with a consistent design. The designed VA(s) accepts both speech and contextual text for communication in both the input and output case. Also, in order to make the VA's conversation more interactive, we use a few technologies like speech recognition and the knowledge base. A sample of a few days of voice message and text messages are taken as testing to analyze the performance of designed virtual assistant. To make the conversation between human and software more natural, we use some techniques like natural language processing and machine learning in the machine learning layer.

References

1. https://www.prnewswire.com/news-releases/
2. https://clutch.co/developers/internetof-things/resources/iot-technology-smart-devices-home
3. https://en.wikipedia.org/wiki/Cortana
4. https://www.slideshare.net/AbedMatini/chatbot-presentation-iitpsa-22-Feb-2018
5. Google Cloud: Dialog Flow Documentation. https://cloud.google.com/dialogflow/docs/console
6. Proceedings of 10th conference of the Italian Chapter of AIS, 'Empowering society through digital innovations'. Università Commerciale Luigi Bocconi in Milan, Italy, 14 Dec 2013. ISBN: 978-88-6685-007-6
7. Mining Business Data. https://miningbusinessdata.com
8. https://en.wikipedia.org/wiki/Siri
9. https://en.wikipedia.org/wiki/Amazon_Alexa
10. Imrie P, Bednar P (2013) Virtual personal assistant. In: Martinez M, Pennarolaecilia F (eds) ItAIS 2013
11. Alexa vs Siri vs Google Assistant vs Cortana. https://www.newgenapps.com/blog/alexa-vs-Siri-vs-Cortana-vs-google-which-ai-assistant-wins
12. Google Assistant. https://en.wikipedia.org/wiki/Google_Assistant. Tulshan, Amrita & Dhage, Sudhir (2019)
13. Survey on Virtual Assistant: Google Assistant, Siri, Cortana, Alexa. In: 4th International symposium SIRS 2018, Bangalore, India, 9–22 Sept 2018, Revised Selected Papers. https://doi.org/10.1007/978-981-13-5758-9_17
14. Russel S, Norvig P (2009) Artificial intelligence: a modern approach, 3rd edn. Prentice Hall

15. https://www.google.com/search/about/learn-more. Accessed on 03 Nov 2016
16. Apple, ios—siri. https://www.apple.com/ios/siri. Accessed on 03 Nov 2016
17. A Glossary of Term of Humans. https://medium.com/@Wondr/ai-explained-for-humans-your-artificial-intelligence-glossary-is-is-right-here-6920279ff88f

Data Security in Wireless Sensor Networks: Attacks and Countermeasures

Ayodeji Olalekan Salau, Nikhil Marriwala, and Muzhgan Athaee

Abstract Secure routing of information is one of the major concerns of specialist working in the field of wireless sensor networks (WSNs) as sensor nodes gather information from the physical world and transmit to base stations for authorized users to access via the internet. In a WSN, sensor nodes are deployed most of the times in unattended locations where there is the possibility of authorized or unauthorized access. This and others have brought the need to secure transmitted information from intruders. Though there are many ways to provide security to a network, cryptography has been identified as the best way for these main three security requirements which include: data confidentiality, authentication, and integrity (CIA). In this paper, we present an overview of the present challenges in a WSN and outline a number of countermeasures to solve these challenges. To tackle the challenge of security encountered in existing WSN systems, we propose an efficient and robust encryption and decryption algorithms for secure communication. The major key feature of this algorithm is that it combines both substitution and transposition cipher techniques in order to achieve the encrypted text. The algorithm is the outcome of the inspection of various existing algorithms in this field.

Keywords Wireless sensor networks · Network attacks · Security · Data encryption and decryption

A. O. Salau (✉)
Department of Electrical/Electronics and Computer Engineering, Afe Babalola University, Ado-Ekiti, Nigeria
e-mail: ayodejisalau98@gmail.com

N. Marriwala
Department of Electronics and Communication Engineering, University Institute of Engineering and Technology, Kurukshetra University, Kurukshetra, India

M. Athaee
Department of Computer Application, National Institute of Technology Kurukshetra, Kurukshetra, India

N. Marriwala et al. (eds.), *Mobile Radio Communications and 5G Networks*, Lecture Notes in Networks and Systems 140,
https://doi.org/10.1007/978-981-15-7130-5_13

1 Introduction

In recent times, wireless sensor network (WSN) technology has experienced rapid growth in modern science and engineering applications. A WSN consists of small nodes which are powered by electronic devices that are non-rechargeable and have limited battery. The sensor nodes are made up of four components, namely: sensing unit, microcontroller, power supply, and transceiver. They consume energy while receiving, processing, and transmitting data. These networks are widely used in many applications such as military surveillance, battlefield, medical field, target tracking, precision agriculture, home applications, and many other commercial applications.

Nowadays, secure transmission is a challenge in WSN as unauthorized users try to gain access to transmitted information [1, 2]. This brings about the need to secure transmitted information from unauthorized users. Though there are many ways of providing security in a WSN, cryptography still remains the most widely used approach for tackling this challenge. Some of the widely used cryptographic techniques include: symmetric key ciphers and asymmetric key ciphers.

The major challenges faced when employing any efficient security scheme in a WSN are created by the memory, size of sensors, processing power, and type of tasks expected from the sensor nodes as well as its limited communication capacity [3, 4].

The rest of this paper is organized as follows. Section 2 presents the types of attacks which occur in a WSN. In Sect. 3, we discuss the security requirements in a WSN. Cryptographic techniques are discussed in Sect. 4, while in Sect. 5 we discuss the various attacks experienced by cryptosystems, and in Sect. 6, we present a review of related works. In Sect. 7, we present our proposed approach, and Sect. 8 concludes the paper.

2 Attacks in WSN

In recent years, there has been an increasing demand for communication devices, for instance, cordless and mobile phones, laptops, and teleconferencing equipment which use some form of wireless connection to communicate. Communication with any of these devices requires a secure means of transmission [5]. WSNs experience a wide range of attacks, therefore, in this section, we categorize the attacks into the major types.

1. **Outsider Versus Insider Attacks**: Attacks coming from outside the WSN are known as outsider attacks, while those coming from inside are known as insider attacks.
2. **Passive Versus Active Attacks**: Passive attacks occur often and thus can be detected easily. Typical examples are traffic monitoring, traffic analysis, and eavesdropping. These attacks usually lead to node outages, node malfunctioning, node subversion, message corruption, and denial of service, while in an active attack, information is being modified and they are difficult to get rid of [6]. Active

attacks occur when an unauthorized attacker monitors and modifies information within the network. These attacks include: routing attacks, spoofed and altered or replayed routing information, selective forwarding, sybil attacks, hello flood attacks, wormhole attacks, sinkhole attacks, Internet smurf attacks, homing, and misdirection.

3. **Jamming Attack**: Jamming attack could be viewed as a special case of denial of service. Radio frequency signals are emitted by the intruder to intentionally disturb signal transmission to sensing equipment. This attack could occur in the physical layer or the data link layer.
4. **Hello Flood Attack**: According to some protocols, sensor nodes are required to send messages to other senor nodes to show proximity; spiteful nodes could send messages to other nodes making them believe that the particular node is in close range. This attack usually occurs in the network and routing layer.
5. **Black Hole Attack**: This attack generates false messages. It basically changes the transmitted information, thus leading to two sets of information (real and false packets of information). Eavesdropper reprograms nodes with false information about route received from the base station and generates or reprograms nodes with fake information about routes; this is known as black hole. This attack occurs in the network and routing protocol layers.
6. **Sybil Attack**: In this attack, a single harmful node behaves like other nodes and sends several fake messages.

3 Security Requirements in WSN

The major security requirements for a WSN are as follows: firstly, data confidentiality: This deals with the privacy of data. It basically saves data from third parties by assuring that data should be received by authorized users only. Secondly, data integrity: This confirms that data during the sending process does not change, i.e., data should be transmitted accurately. Thirdly, data authentication: Confirmation of received information guarantees the unwavering quality of messages by distinguishing their roots. Information verification is accomplished through symmetric or deviated components while sending and accepting hubs offer mystery keys. Fourthly, data availability: This ensures that authorized users are not denied communication services or transmitted information. The fifth is access control: This is the process of granting entry to authorized users to ensure legitimacy. Access control determines who can access the system, what system resources can be accessed, and how the resources are to be used.

3.1 Security Service

The main objective of network security is to provide four basic security services. These services are:

1. **Confidentiality**: Only the intended persons should be able to access the information transmitted. No one can read the substance of data with the exception of the approved client or beneficiary.
2. **Authentication**: This service is concerned with the assurance that the communication is authentic, i.e., data received on the network has been sent only by a verified sender. This means the communication is between authorized users.
3. **Data Integrity**: This deals with recognizing any modification to the information. The information may be altered by an unintended entity intentionally or accidentally. Data integrity services check for whether transmitted information is changed or not.
4. **Non-repudiation**: This ensures that any user cannot refute the ownership of a message or activity. It provides assurance that the original creator of a message cannot refuse its transmission.

4 Cryptographic Techniques

A great part of the data imparted every day must be kept private. Data, for example, monetary reports, worker information, and therapeutic records should be conveyed in a way that guarantees secrecy and honesty. The issue of unsecured correspondence is exacerbated by the way information is transmitted over the internet in form of email or text.

The most common cryptographic techniques used in recent times to withstand the various cyber-attacks are categorized in Sect. 4.1.

4.1 Secret Key Cryptography

Secret key cryptography is a cryptographic scheme that uses symmetric key encryption for the encryption of the data. With secret key cryptography, both conveying parties utilize a similar key to encode and decode messages. Before any encrypted information can be sent over the network, both parties must have the key and should concur on the cryptographic calculation that they will use for encryption and decoding. An illustration of this scheme is presented in Fig. 1.

One of the real issues with secret key cryptography is the strategic issue of how to get the key from one party to the next without enabling access to an assailant. In the event that the two parties secure their information with secret key cryptography, and if an unauthorized user accesses their key, the user can see any hidden message sent

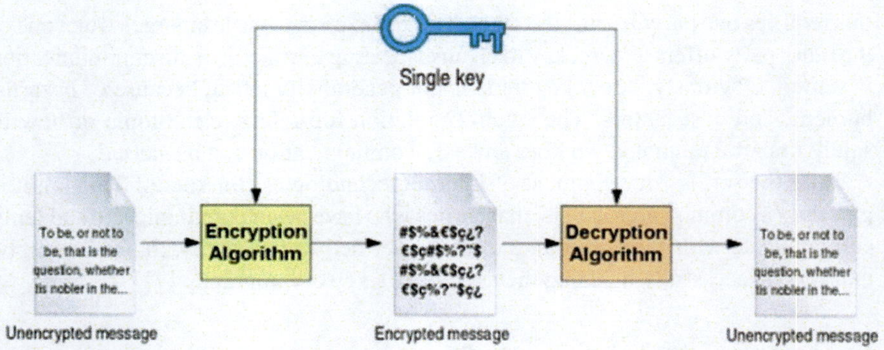

Fig. 1 Symmetric key encryption

between the two parties. Not exclusively can the unauthorized user decode the two authorized users messages but can likewise imagine that he is one of the authorized users and send encoded information to the other. In this situation, the receiving authorized user will not realize that the message originated from the unauthorized user rather than his counterpart.

At the point when the issue of mystery key dispersion is tackled, mystery key cryptography can be a profitable instrument. The calculations give fantastic security and encode information generally rapidly. Most of the delicate information sent in a secure socket layer (SSL) session is sent utilizing mystery key cryptography. Secret key cryptography is likewise called symmetric cryptography in light of the fact that a similar key is utilized to encode and decode both pieces of information. It is important to note that secret key cryptographic calculations incorporate the Data Encryption Standard (DES), triple-quality DES (3DES), Rivest Cipher 2 (RC2), and Rivest Cipher 4 (RC4). Symmetric key cryptography utilizes a similar key to scramble and decode information. Some normal symmetric key calculations are the DES, triple DES, Blowfish, and the Advanced Encryption Standard (AES). DES is insufficient on the grounds that it utilizes a 64-bit key and has been hacked severally. A special consideration should be given to those crypto-securities similar to Microsoft's Windows XP Encrypted File System (EFS) which are default to DES and must be changed to give great security. The principal advantage of the symmetric key cryptography is speed. The guideline issues with this framework are key dispersion and versatility. Keys should be appropriated safely, and each protected channel should be allocated a different key. Symmetric key frameworks give privacy, however, this does not guarantee authenticity of the message because the sender can deny having sent the message.

One major issue with using symmetric algorithm is the key exchange issue, which can exhibit a great conundrum. The other fundamental issue is that of trust between two parties that use a secret symmetric key. Issues of trust might be experienced when encryption is utilized for validation and trustworthiness checking. The asymmetric key can be utilized to check the personality of the other participating party, although this requires one party to trust the other. The key exchange issue emerges when one or the other party offers a secret key to an unsuspecting user before safe communication is started. Obviously, direct key trade is not generally practical because of hazards, burdens, and cost factors. The catch-22 relationship refers to the topic of how to safely impart a mutual key before any safe communication can be started.

With the serious development of Internet technologies (Internet of Things), it is presently a common prerequisite that parties who have never communicated suddenly communicate with each other in a safe and verified way. Luckily, this issue can be managed successfully by using the asymmetric key cryptography [7].

5 Attacks on Cryptosystem

In this present age, businesses as well as every part of human activities are driven by data. Thus, it is required that information be shielded from noxious exercises, for example, assaults. Assaults are normally sorted in light of the activity performed by the attacker. An attack can be categorized as passive or active [8].

1. **Passive Attack**: A passive attack is a system attack in which a network is observed, and now and then checked for open ports and vulnerabilities. The design is exclusively to pick up data about the objective, but no information is changed about the objective. Passive attacks incorporate dynamic observation and uninvolved surveillance. In a passive attack, an attacker scans the network for vulnerabilities without collaboration through strategies like session catch. In dynamic surveillance, the attacker draws in to the objective framework through strategies like port sweeps.
2. **Active Attack**: A dynamic assault is a system misuse in which a programmer endeavors to roll out improvements to information on the objective or information in transit to the objective [9].

5.1 Types of Active Attacks

- **Masquerade**: In a masquerade attack, the intruder tries to impersonate a specific client in the system to get entrance. A masquerade is achieved using stolen login IDs and passwords and by discovering security holes in programs or through bypassing the authentication mechanism [10]. A typical example of this is shown in Fig. 2.
- **Session Replay**: In a session replay attack, a programmer takes an approved client's sign-in data by taking the approved client's session ID. The attacker gains entrance and has the capacity to do anything the approved client can do on the site [11]. Figure 3 shows an illustration of session replay.

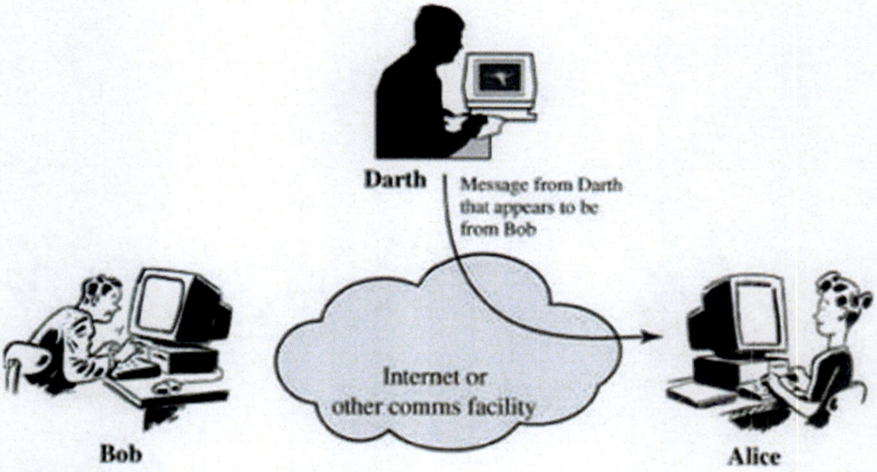

Fig. 2 Masquerade (active attack)

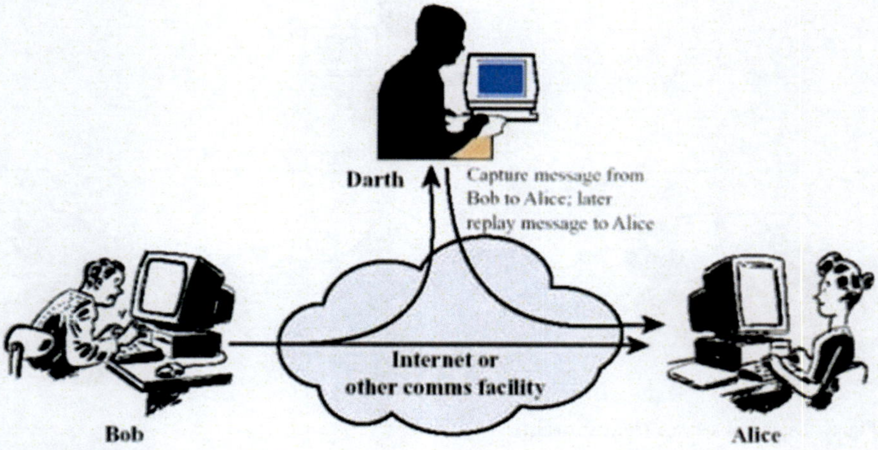

Fig. 3 Session replay (active attack)

- **Message Modification**: In message modification attack, the attacker adjusts bundle headers to guide a message to an alternate goal or changes the information on an objective machine. This is shown in Fig. 4.
- **Denial of Service**: In denial-of-service (DoS) attack, clients are denied access to a system or web asset [12]. This is largely achieved by overpowering the objective with more activity than it can deal with. This is shown in Fig. 5.

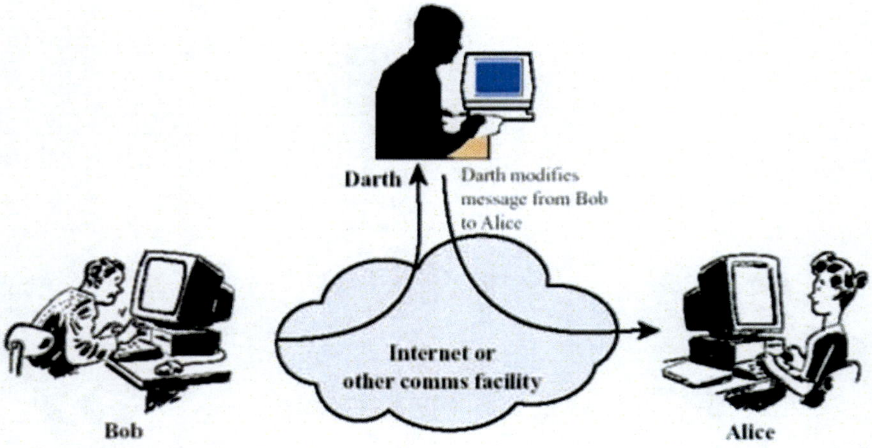

Fig. 4 Message modification (active attack)

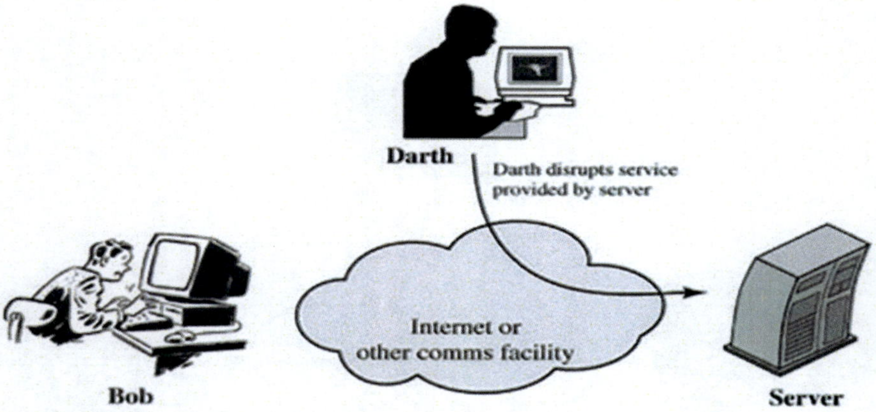

Fig. 5 Denial of service (active attack)

6 Existing Methods

6.1 Application Based Methods

Singh [13] proposed a new algorithm based on the ASCII value transformation with the help of a randomly generated secret key. The algorithm follows the symmetric key encryption as the process performs encryption and decryption using the same key. The algorithm goes with extracting minimum ASCII values for the input plaintext and the secret key that has been used. The idea is to perform a modulus operation on the input text value with the minimum ASCII value of the input and similarly perform same for the key values and minimum ASCII value of the key and store it in a separate array. The values of each index of both the arrays are added. The final sum is added to each of the ASCII values of the key in order to get the final ciphertext.

Saraswat [14] proposed a new methodology for encrypting a plaintext message using the extended version of the Vigenere table that has more characters than the simple Vigenere table. The numeric values are also included in this table for creating the extended version of the Vigenere table. The cross section of the table values gives the intermediate text, and after performing the Caesar cipher process, the final ciphertext is found.

Sukhraliya [15] proposed a methodology for the encryption of plaintext. The idea is influenced by the set theory of mathematics, i.e., to have a subset of the content of the plaintext, the plaintext is taken, and a random number is selected. The quotient and remainder are generated by dividing the plaintext values with a random number. The starting and ending values of the plaintext are used to make a subset value within that range and are separated with the mode of these values. The final values are converted to characters to get the encoded text.

Singh [16] has proposed a new symmetric key encryption algorithm. It is based on the mathematical principle of modulus and shifting of the values using the right shift. The input text is converted to corresponding ASCII values character by character, and the minimum value is extracted from them. The modulus operation is performed on the original ASCII value with the minimum one with the restriction that the values should not be greater than a particular value. Then, a random key is generated of four characters, and its minimum ASCII value is again used for performing the modulus operation on the key. The modulated key array is then right shifted, and the values are added with the mode content of the plaintext. The result is then converted to characters to get the encrypted text.

Chatterjee [17] proposed a new cryptographic algorithm called MSA algorithm based on the symmetric key and polyalphabetic substitution of the characters of the input text; arandom key was generated. The relative position of characters is important in this encryption scheme. A 4×4 key matrix is created which is a modified version of play fair encryption. A table is processed using cyclic shifting, right shift, and left shift. The combined pair of characters were exchanged or replaced with characters from the table. The text formed by the replacement of characters was used as the final ciphertext for the algorithm.

Bisht [18] provided a comparative study of various encryption algorithms. They compared both the symmetric and asymmetric key cryptography on different factors. The utilization of a key is important as far as encryption and decoding are the same or extraordinary. The calculation utilization is characterized by its sort whether symmetric or deviated. The length of the key is utilized by its deem, and the speed is portrayed as quick or moderate. The security is defined as great but not secure or minimally secure. The cost is more economic or unreasonable, and its execution is directed by its calculation utilization which is either basic or complex.

Rani [19] proposed a new system for detecting suspicious emails using triple DES algorithm. The system detects any type of security attack or suspicious email. The authors employed the use of cryptographic techniques for suspicious email detection.

The cryptographic technique used was the triple DES algorithm which is a private key cryptographic system. The system detects suspicious messages sent from the senders that are registered on a Web site. The Web site provides sign-up facility for new users to send and receive emails. Cryptographic systems were connected to identify suspicious messages. Triple DES calculations were connected with the first text given as a plaintext message, where the main key was utilized to DES-encode the message. The second key was utilized to DE decode the scrambled message. Since the second key is not the correct key, this decoding just scrambles the information further. The twice mixed message is then encoded again with the primary key to yield the last figure content. This three advance technique is called triple-triple-DES and is simply DES done three times with two keys utilized as a part of a specific arrange. Triple DES should likewise be possible with three separate keys rather than just two. In either case, the resultant key space is around 2^{112}. Triple DES algorithm is utilized by administrators to scramble the messages sent to clients or to send a few notices about alternate client's suspicious movement.

6.2 Technique Based Methods

1. **DES**: This data encryption standard algorithm works on 64-bit block of data. It uses a 64-bit key. The key is then converted to 56 bit. The plaintext is divided into two equal halves/blocks, and a key is used for encryption. Initial permutation is applied on plaintext, and then, each half goes through 16 rounds of encryption. After encryption, each half is combined and a final ciphertext is formed [20].
2. **AES**: It is a widely adopted symmetric encryption technique. It is six times faster than triple DES. It works on 128 bits of data with key size of 128 bit/192 bit/256 bit of length. The length of the key AES uses ten rounds for 128-bit keys, 12 rounds for 192-bit keys, and 14 rounds for 256-bit keys. The encryption process consists of byte substitution, rows shifting, and mixing of columns with XOR operation. The last text obtained after the encryption is called ciphertext. The reverse process is applied for decryption [21].
3. **HASH**: Hashing technique is applied to enumerate a complex illustration of a fix size message or file. It is called a 'fingerprint' or a 'message digest.'
4. **SHA-1**: This is a hashing technique similar to MD5 (a 128-bit message digest), but generates a digest of 20 bytes. Since it is of a huge digest size, it is less possible that two distinct messages will have the same SHA-1 message digest. As a result of this, it is approved over MD5.
5. **RSA**: RSA is a key generation and asymmetric encryption algorithm. For key generation, two large prime numbers are selected. Both the prime numbers are multiplied and stored. Encryption key is produced by applying formula, and the decryption key is then calculated. The encryption key is called public key, and decryption key is called private key.

6.2.1 Comparative Study

All the approaches discussed in this paper have been developed for secure communication. The parameters for the comparison among them are the complexity, algorithm type, i.e., symmetric or asymmetric, length of the key, and time of execution. It is common knowledge that the AES algorithm is more secure and robust than other cryptographic techniques. The time of execution of encryption is much less using this algorithm than the other algorithms. The existing work of authors in [14] takes in consideration alphabet encryption only, while the proposed algorithm works for alphanumeric and some special characters also. Authors in [16] solved the problem of the length of a key. The proposed algorithm takes the length of plaintext to be random according to the plaintext length making it tough for the cryptanalyst to get the key. The challenge with the work in [22] is that the execution time increases exponentially with a corresponding increase in the plaintext length.

7 Proposed Solution

Performing a cryptographic function in an efficient way has always been a challenge in computer network security. This research paper presents an efficient algorithm for performing encryption and decryption. The algorithm is a combination of an extended Vigenere table and transformational functions. The aim is to perform substitution and transposition cipher methods on plaintext in order to get a robust ciphertext. In the substitution cipher, the characters of the plaintext are replaced with other characters. Transposition cipher deals with the shifting of characters or changing the position of characters. The proposed algorithm was implemented with Java programming language. It blends both techniques together to produce a more effective secret text.

A simple Vigenere table just has few alphabets, but the extended Vigenere table used for this work contains all alphabets (A–Z) and digits (0–9), therefore making the encryption of the alphanumeric characters possible. This can help in the encryption of the secret messages in the defense services and even in any other organization. The algorithm takes in plaintext as input which contains all alphabets and digits. The key that must be entered must be of a fixed length that would be predefined.

The final ciphertext is obtained using the proposed algorithm. The proposed algorithm is presented sequentially with the following steps:

Step 1: Input the Plaintext and Key
The process for the algorithm starts with the input of the plaintext. The plaintext can be any of the alphabets (A–Z), digits (0–9), or some special characters (: ; < = > ? @). The plaintext can be of any length.
Key 1: This key is a randomly generated key which is of the same length as the plaintext.
Key 2: This is a user-defined key which contains a single numeric value having a value less than the size of the plaintext.

Step 2: Substitution Text Determination from Plaintext and Key 1
The following step introduces the modified Vigenere cipher approach which includes digits and special characters. The modified Vigenere cipher approach is used to get substitutions for alphabets [A–Z] only. The proposed algorithm applies substitution by combining the alphabets [A–Z], digits [0–9], and some special characters [: ; < = > ? @]. The substitution text can be determined by using the following mathematical formula: ST = (PT + K − 2 * 43) % 43 + 48, where ST = substitution text, PT = plain text, and K = key. This formula is applied to each character of the plaintext and key by taking their ASCII value. This actually keeps the value of the substitution text between the ASCII values' range of 48–90.
Step 3: Applying Mathematical Function on the Substitution Text Characters
The characters of the substitution text are taken, and a fixed value is added to each ASCII value of the characters. The fixed value added is the secret key 2. The value of the key 2 must be in the range of 2–5. The user will define the value for the key 2.
Step 4: Applying Transformational Function to the New Value of the Substitution Text
In this step, the substituted text is transposed making it more reliable, secure, and safe. The characters of the substituted text are written in a diagonal manner for the number of rows defined by key 2, and then read in the sequence of the rows formed.
Step 5: Final Ciphertext Determination
Encryption:

- **Input**: Plain Text, Key 1, Key 2. **Output**: Cipher Text: Take input of the Plain Text from the user.
- Generate random Key 1 equal to the length of Plain Text.
- Take input of another key i.e. Key 2 from user lying in the range of 2–5.
- Find the Substitution Text (ST) – each character. Calculate using the formula: ST = (Plain Text + Key 1–2 * 48) % 43 + 48
  If (ST > = 90), then ST = ST − 43
  If (ST <= 48), then ST = ST + 43
  ST = Substitution Text for each character.
- Add value of key 2 to ASCII value of each character to get Intermediate text.
  IT = ST + key 2
  IT = Intermediate Text for each character. Obtain the Transposition of the intermediate text as discussed in step 4 using the key 2.
- Final CIPHER TEXT = Transposed Text.
  End

Decryption:

- Begin: Take the cipher Text.
- Transpose the cipher text using the reverse transformational function used for encrypting the intermediate text (I.T.).
- Find the Substitution text (ST) from the intermediate text by using the formula: ST = IT – Key 2.
- Find the Final Substitution Text (FST) by using the formula: FST = (ST – Key 1 + 43) % 43 + 48.
    - If (FST > = 90), then FST = FST – 43.
    - If (FST < = 48), then FST = FST + 43.
- Convert the Final Substitution Text to Plain Text.
- End.

8 Conclusion

In this paper, a new cryptographic algorithm was presented as a solution to the limitations of existing cryptographic algorithms. The encryption and decryption of plaintext was carried out using the designed algorithm. The improved multilevel substitution technique as well as the transposition technique of encryption and decryption was integrated in this algorithm to obtain the final ciphertext. The algorithm uses two keys for encryption to make it very difficult for attackers to obtain the keys. The plaintext remains encrypted until the valid key is provided. Vigenere cipher is regard as a simple and weak method which is easy to detect by an intruder. To overcome the limitations of the Vigenere cipher table, the proposed multilevel substitution and transposition encryption scheme was developed. The encryption key is generated randomly, hence, the designed algorithm ensures non-repudiation; it also makes it difficult for cryptanalyst to guess the encryption key. At the same time, the computational complexity is much lesser than most modern ciphers, making it an appropriate choice for lightweight applications where resources are limited. As there are always certain limitations for any work, the designed algorithm has the limitation of not being able to encrypt small letter alphabets and some special characters. Because of this limitation, we converted all lowercase letters to uppercase letters at the beginning of the encryption process.

Therefore, it can be concluded that the designed algorithm is effective and more secure than existing algorithms. Our future work will be targeted at improving the cryptographic algorithm to make it more robust to tackle all types of printable characters.

References

1. Alotaibi M (2019) Security to wireless sensor networks against malicious attacks using Hamming residue method. EURASIP J Wirel Commun Netw, 1–7. https://doi.org/10.1186/s13638-018-1337-5
2. Li P, Sun L, Fu X, Ning L (2013) Security in wireless sensor networks. In: Wireless network security. Springer, Berlin, Heidelberg. https://doi.org/10.1007/978-3-642-36511-9_8
3. Liwandouw V, Wowor A (2017) The existence of cryptography: a study on instant messaging. Procedia Comput Sci 124:721–727
4. Chelli K (2015) Security issues in wireless sensor networks: attacks and countermeasures. In: Proceedings of the world congress on engineering. London, U.K.
5. Sanchez J, Correa R, Buena H, Arias S, Gomez H (2016) Encryption techniques: a theoretical overview and future proposals. In: Third international conference on eDemocracy & eGovernment, pp 60–64
6. Biswas S, Adhikari S (2015) A survey of security attacks, defenses and security mechanisms in wireless sensor network. Int J Comput Appl 131(17):28–35. https://doi.org/10.5120/ijca2015907654
7. Kuppuswamy P, Al-Khalidi SQY (2012) Implementation of security through simple symmetric key algorithm based on modulo. Int J Comput Technol 3:335–338
8. Joshi A, Wazid M, Goudar RH (2015) An efficient cryptographic scheme for text message protection against brute force & cryptanalytic attacks. Procedia Comput Sci 48:360–366
9. Ismail ES, Baharudin S (2012) Secure hybrid mode-based cryptosystem. Am J Appl Sci 9(3):289–292
10. Abd Elminaam DS, Abdual Kader HM, Hadhoud MM (2010) Evaluating the performance of symmetric encryption algorithms. Int J Netw Secur 10(3):213–219
11. Singh G, Singla AK, Sandha KS (2011) Throughput analysis of various encryption algorithms. Int J Comput Sci Technol 2(3):527–529
12. Adamu AA, Wang D, Salau AO, Ajayi O (2020) An integrated IoT system pathway for smart cities. Int J Emerg Technol 11(1):1–9
13. Shinghe SR, Patil R (2014) An encryption algorithm based on ASCII value. Int J Comput Sci Inf Technol 5(6):7232–7234
14. Saraswat CK, Sudhakar PT, Biswas P (2016) An extended hybridization of Vigenere and Caeser cipher techniques for secure communication. Procedia Comput Sci 92:355–360
15. Sukhraliya V, Chaudhary S, Solanki S (2013) Encryption and decryption algorithm using ASCII values with the substitution array approach. Int J Adv Res Comput Commun Eng 2(8):3094–3097
16. Singh U, Garg U (2013) An ASCII value based text data encryption system. Int J Sci Res Publ 3:1–5
17. Chatterjee J, Nath SD, Nath A (2011) A new symmetric key cryptography algorithm using extended MSA method: DJSA symmetric key algorithm. In: International conference on communication systems and network technologies, pp 89–94. https://doi.org/10.1109/CSNT.2011.25
18. Bisht N (2015) A comparative study of some symmetric and asymmetric key cryptography algorithms. Int J Innov Res Sci Eng Technol 4(3):552–565
19. Rani N (2015) Suspicious email detection system via triple des algorithm: cryptography approach. IJCSMC 4(5)
20. Messai M (2014) Classification of attacks in wireless sensor networks. International congress on telecommunication and application, Bejaia, Algeria. Available online https://arxiv.org/ftp/arxiv/papers/1406/1406.4516.pdf
21. Mathur A (2012) An ASCII value based data encryption algorithm and its comparison with other symmetric data encryption algorithms. Int J Comput Sci Eng (IJCSE) 4(9):1650–1657
22. Addagarla SK, Babji Y (2013) A comparative security study review on symmetric key cryptosystem based algorithms. Int J Comput Sci Mob Comput 2(7):146–151

Emerging Next-Generation TWDM-PON with Suitable Modulation Format

Meet Kumari, Reecha Sharma, and Anu Sheetal

Abstract The exponential growth of high-speed broadband services, an increase in the number of end-users, high bandwidth, and massive data rate demand innovative and emerging points to a multipoint network that satisfies the need for next-generation passive optical access network (NG-PON). Next-generation passive optical access network stage 2 (NG-PON2)-based time and wavelength division multiplexing passive optical network (TWDM-PON) are mostly considered as the most promising optical access candidate due to its high-speed services, low power consumption, back-to-back compatibility, and cost-effectiveness. In this paper, the eight wavelengths-based bidirectional 80/20 Gbps TWDM-PON system using return to zero (RZ), non-return to zero (NRZ), and carrier suppressed return to zero (CSRZ) data modulation formats have been proposed and analyzed by varying input power, transmission distance, and bit rate. Further, the system is evaluated in terms of minimum bit error rate (BER), quality factor (Q factor), received optical power (dBm), eye diagrams, and eye height. It has been observed that the most appropriate modulation format for downstream transmission is NRZ while for upstream transmission, the most suitable modulation formats are RZ and NRZ.

Keywords CSRZ · NG-PON2 · NRZ · RZ · TWDM-PON

1 Introduction

The present optical access networks (OANs) insist on the high speed, long reach, high end user, and a large number of users' handling applications. To support and satisfy the estimated demand for the OANs, various networks such as active and passive optical networks may be implemented, as they support the extensive capacity

M. Kumari (✉) · R. Sharma
Punjabi University, Patiala, Punjab, India
e-mail: meetkumari08@yahoo.in

A. Sheetal
GNDU Regional Campus Gurdaspur, Punjab, India

N. Marriwala et al. (eds.), *Mobile Radio Communications and 5G Networks*, Lecture Notes in Networks and Systems 140,
https://doi.org/10.1007/978-981-15-7130-5_14

backbone network, core network, and metro network [1, 2]. Presently, the passive optical networks (PONs) being deployed and considered as an attractive solution for future OANs. There are different PONs that are based on time-division multiplexing (TDM) such as gigabit passive optical network (GPON) standardized by international telecommunication union-telecommunication (ITU-T) and Ethernet passive optical network (EPON) standardized by institute of electrical and electronics engineers (IEEE). Also, the first generations PONs are next-generation passive optical network stage 1 (XG-PON1), ten gigabit Ethernet PON (10G-EPON) and next-generation passive optical network stage 2 (XG-PON2) standardized by ITU-T and IEEE. This is called next-generation passive optical network stage 1 (NG-PON1). But the NG-PON1-based systems do not meet the future network demands such as high capacity, high split ratio, high bit rate, bandwidth upgradability, maximum reuse of ODN, flexibility, minimum cost, and high performance. Therefore, full service access network (FSAN) group members recognized that hybrid time and wavelength division multiplex (TWDM) was the best solution for new generation PON system [3, 4]. Thus, TWDM-PON has been chosen as the innovative and primary candidate for next-generation passive optical network stage 2 (NG-PON2) by FSAN due to its attractive advantages of backward compatibility, high bit rate (40/10 Gbps for downstream/upstream), component availability, mature and simple technology. Moreover, it also provides a large number of users' handling capacity, significant cost-effectiveness, reliability, easy upgradability, and 'pay as you grow' features. Thus, TWDM-PON can be used for fiber to the home (FTTH) and business applications at a high data rate [2]. In TWDM-PON, at high data rate, the modulation formats become an important issue for optimum system performance. There are different types of modulation formats that have been proposed to enhance the performance of TWDM-PON [5, 6].

Amandeep Kaur et al. analyzed and investigate the performance of non-return to return (NRZ) and return to zero (RZ) modulation formats in co-existence 2.5 Gbps GPON/10 Gbps gigabit capable GPON (G-GPON) system at variable bit rate and transmission distance. It results show that RZ modulation format is better as compared to NRZ modulation format. It provides the faithful transmission of an input signal up to 90 and 140 km for downstream (at 1577 nm and 10 Gbps bit rate) and upstream (1270 nm and 2.5 Gbps) transmission in the presence of fiber nonlinearities [2].

Vivek Kachhatiya et al. optimize and compared the downstream transmission of wavelength routing PON (WR-PON) and power splitter-PON (PS-PON) at 10 Gbps over optical fibers like ITU-T G.652.A fiber, ITU-T G.652.B fiber, Alcatel 6912 tealight ultra 1625 fiber, corning leaf 1625 fiber, and lucent true-wave rs 1625 fiber. The result shows that the performance of the system is restricted due to fiber attenuation, dispersion, and nonlinearities for variable input power, non-return to zero external modulation (NRZ-EM), and return to zero external modulation (RZ-EM) [7].

The comprehensive literature review shows that modulation formats, viz. NRZ, RZ, etc., have a considerable impact on enhancing the performance of various PON and TWDM-PON systems. In this paper, the performance comparison of different modulation formats, viz. NRZ, RZ, carrier suppressed return to zero (CSRZ) in bidirectional 80/20 Gbs TWDM-PON for varying transmission distance, bit rate, input power under fiber attenuation, dispersion, and nonlinearities have been presented. The organization of the paper is as follows: Sect. 2 describes the architecture design of TWDM-PON at different modulation formats and system parameters used in work. Section 3 presents the results and discussions, and finally, conclusion is drawn in Sect. 4.

2 TWDM-PON Architecture for Different Modulation Formats

2.1 Downstream System Design

Figure 1 shows the eight wavelength () downstream architecture of TWDM-PON system at 10 Gbps using different modulation formats such as NRZ, RZ, and CSRZ over 10–80 km single mode fiber (SMF) distance. An optical line terminal (OLT) consists of eight downstream transmitters where each transmitter consists of continuous wave (CW) laser, pseudorandom bit sequence (PRBS) generator to generate the random bit sequence, data format generator (NRZ, RZ, or CSRZ), and mach-zehnder modulator (MZM) modulator to modulate the incoming signals at 10 Gbps. Here, the aggregate 80 Gbps downstream signals are transmitted using 8×1 Ideal Mux. Then, in optical distribution network (ODN), received multiplexed signals are passed through bidirectional SMF with reference wavelength = 1,550 nm (attenuation = 0.2 dB/km, dispersion = 16.75 ps/nm/km, and dispersion slope = 0.075 ps/nm^2/km) is utilized and split into eight optical network units (ONUs) using 1×2 bidirectional splitter. Each ONU consists of 1×8 splitter for a single wavelength. For the reception at ONU side, PIN photodiode, low pass Bessel filter, 3R regenerator, and bit error rate (BER) analyzer are used [1, 8].

2.2 Upstream System Design

The eight-wavelength upstream architecture of TWDM-PON system consists of eight ONUs. Each ONU modulated the upstream signal at 2.5 Gbps using different modulation formats such as NRZ, RZ, and CSRZ. An ONU section consists of eight upstream transmitters where each transmitter consists of CW laser, PRBS generator, data format generator (NRZ, RZ, or CSRZ), and MZM modulator at 2.5 Gbps [1, 10].

All ONUs transmit 2.5 Gbps signal at 1528 nm with two cascaded dynamic Y selects. These selectors allow the signal to be transmitted only at a definite time. The switching time for first selector uses 0–4.48×10^{-08} s range, whereas the second

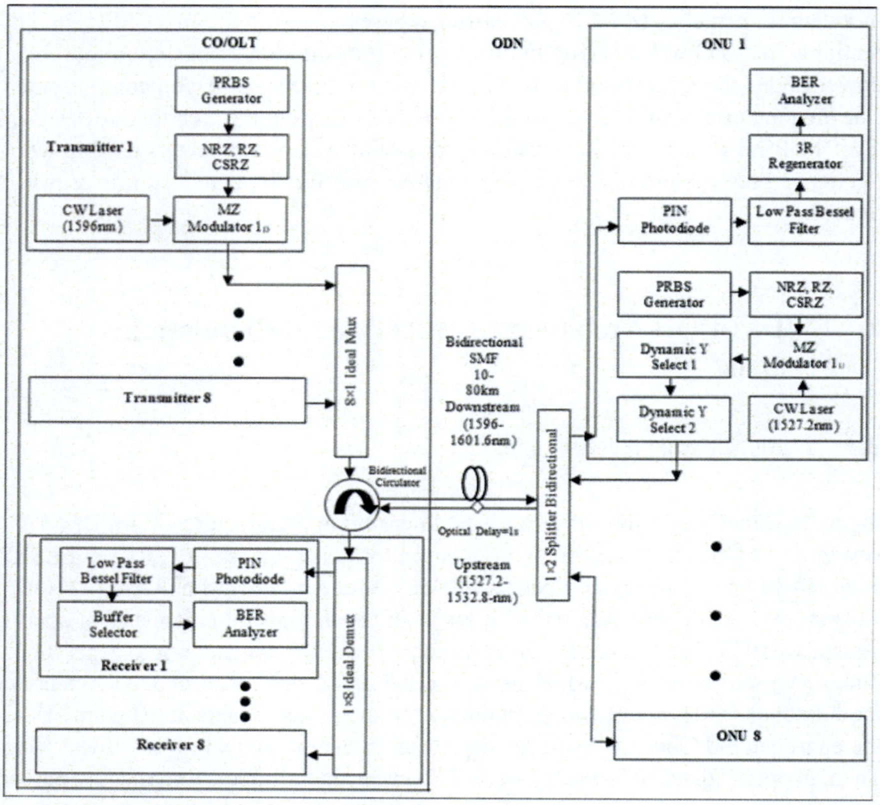

Fig. 1 Schematic diagram of bidirectional 40/10 Gbps OTDM-PON system up to 10–80 km at different modulation formats [9]

selector uses 6.4×10^{-09}–5.12×10^{-08} s range at 2.5 Gbps. For the reception at OLT side, PIN photodiode, low pass Bessel filter, 3R regenerator, and BER analyzer are used.

2.3 *Different Modulation Formats*

NRZ modulation technique minimizes the dispersion and inter-symbol interference (ISI) between transmission pulses. It has narrow optical spectrum due to lower on–off transactions [5].

RZ modulation technique has twice power peak as compared to NRZ modulation. It has property of reduced spectral efficiency and reduced dispersion [5].

CSRZ modulation technique is the special form of RZ modulation technique. It has pi phase shift as compared to RZ modulation between adjacent bits. It minimizes

Table 1 System parameters and their values [9]

S. No.	Component name	Component parameters	Parameter value	Unit
1	CW laser source	Output power	0–18	dBm
		linewidth	10	MHz
2	MZM modulator	Wavelength range	1596–1601.6 (downstream) 1527.2–1532.8 (upstream)	nm
		Extinction ratio	30	dB
3	PRBS sequence generator	Bit rate	10–20	Gbps
		Order	7	
4	8 × 1 ideal Mux	Input ports	8	
		Wavelength range	1596–1601.6 (downstream) 1527.2–1532.8 (upstream)	nm
		Channel spacing	0.8	nm
5	PIN diode	Responsivity	1	A/W
		Dark current	10	nA
		Thermal noise	1×10^{-22}	W/Hz
6	Low pass Bessel filter	Cut-off frequency	7.5–15	GHz
		Order	4	
7	Optical fiber	Reference wavelength	1550	nm
		Length	10–80	km
		Attenuation	0.2	dB/km
		Dispersion	16.75	ps/nm/km
		Dispersion slope	0.075	$ps/nm^2/k$
		Effective area	80	μm^2
		Nonlinear index of refraction	26×10^{-21}	m^2/W
		Temperature	300	K

the linear impairments, dispersion, and maximize the spectral efficiency to Tbps [5]. The various system parameters along with their values are given in Table 1.

3 Results and Discussion

The three modulation techniques, viz. NRZ, RZ, and CSRZ have been compared for variable input power, transmission distance as well as bit rate for asymmetrical and bidirectional eight wavelengths 80/20 Gbps TWDM-PON in terms of Q factor, BER value, received optical power, and eye diagrams.

3.1 Comparative Analysis for Downstream Transmission

The downstream transmission of TWDM-PON at 1596 nm channel using three modulation techniques, viz. NRZ, RZ, and CSRZ have been compared for variable fiber length (10–80 km at $P_{in} = 10$ dBm and 10 Gbps), input power (6–18 dBm over 60 km and 10 Gbps), and bit rate (2.5–20 Gbps at $P_{in} = 10$ dBm and 10 km).

Figure 2 shows a graphical representation of Q factor versus fiber length. It is clear that NRZ achieves stable performance and the other side CSRZ modulation format shows poor performance compares with NRZ and RZ formats. It also describes as the fiber length increases the performance of TWDM-PON system degrades for all three modulation formats due the presence of fiber dispersion and nonlinearities such as self-phase modulation (SPM), cross-phase modulation (XPM), and four-wave mixing (FWM) [7].

Also, Fig. 2 demonstrates the eye diagrams of NRZ, RZ, and CSRZ modulation formats over 60 km, 30 km, and 20 km, respectively, under fiber dispersion and nonlinearities. The results show that NRZ modulation has maximum eye opening and eye height as compared to other modulation formats. Hence, NRZ is a suitable choice form downstream transmission in TWDM-PON over long reach.

Figure 3 shows a graph of Q factor versus input power at channel 1596.8 nm for NRZ, RZ, and CSRZ modulation formats over 60 km at 10 Gbps. It indicates that NRZ modulation format provides better error-free performance at the input power from 8 to 16 dBm as compared to RZ and CSRZ modulation formats. Also, it is observed that optimum input power for NRZ modulation is 14 dBm. Thus, it shows that the input power of transmitter CW laser varies with the different modulation formats

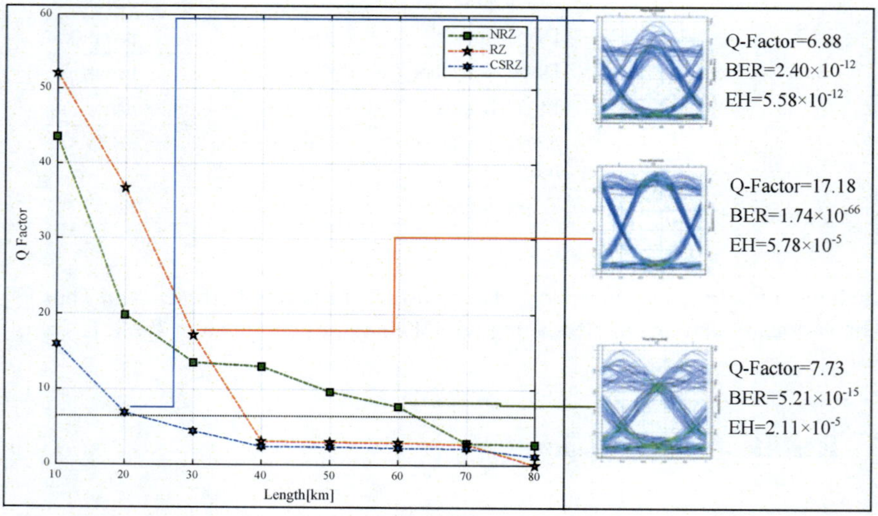

Fig. 2 Q factor versus length (km) for downstream TWDM-PON system for NRZ, RZ, and CSRZ modulation formats along with eye diagrams

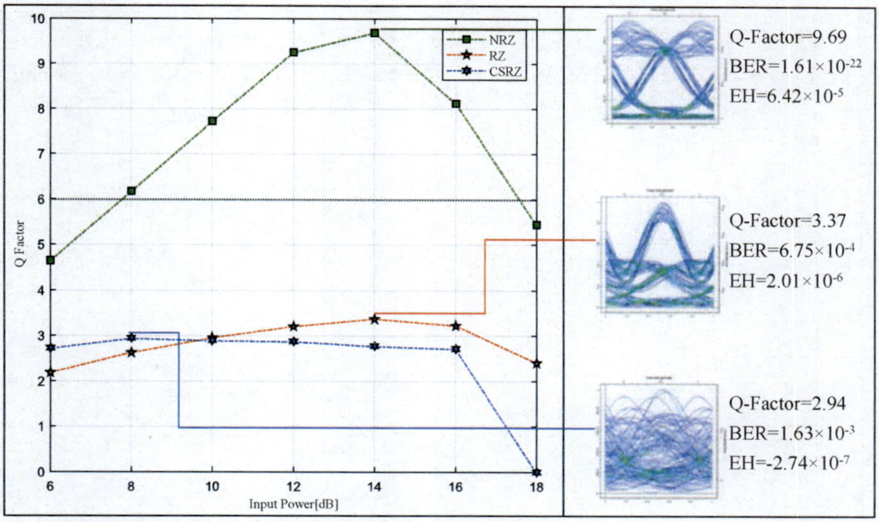

Fig. 3 Q factor versus input power (dBm) for TWDM-PON with NRZ, RZ, and CSRZ modulation along with eye diagrams at 14 dBm, 14 dBm, and 8 dBm optimum input power, respectively

and for fiber specifications. Again, it is observed that when input power is more significant than that of optimum input power, the performance of network degrades due to increasing fiber dispersion and nonlinearities. Hence, NRZ outperforms RZ and CSRZ modulation in terms of the performance of TWDM-PON for downstream transmission.

Further, Fig. 3 demonstrates the eye diagrams of NRZ, RZ, and CSRZ modulation formats at 14 dBm, 14 dBm, and 8 dBm input power, respectively. The results show that NRZ modulation has maximum eye opening and eye height as compared to other modulation formats. Hence, NRZ is a suitable choice form downstream transmission in TWDM-PON at high input power.

Figure 4 shows a graph of Q factor versus bit rate at channel 1596.8 nm for NRZ, RZ, and CSRZ modulation formats over 10 km at 10 Gbps. It shows that NRZ modulation format provides better error-free performance at the bit rate from 10 to 20 Gbps as compared to RZ and CSRZ modulation formats. It is also observed that the maximum bit rate for NRZ modulation is 20 Gbps. Moreover, the bit rate increases, the performance of the network degrades due to increasing fiber dispersion and nonlinearities. Hence, NRZ outperforms RZ and CSRZ modulation in terms of performance of TWDM-PON for downstream transmission at a high bit rate.

Figure 4 demonstrates the eye diagrams of NRZ, RZ, and CSRZ modulation formats at 20 Gbps, 18 Gbps, and 14 Gbps bit rates, respectively. The results show that NRZ modulation has maximum eye opening and eye height as compared to others modulation formats. Hence, NRZ is a suitable choice form downstream transmission in TWDM-PON at a high bit rate.

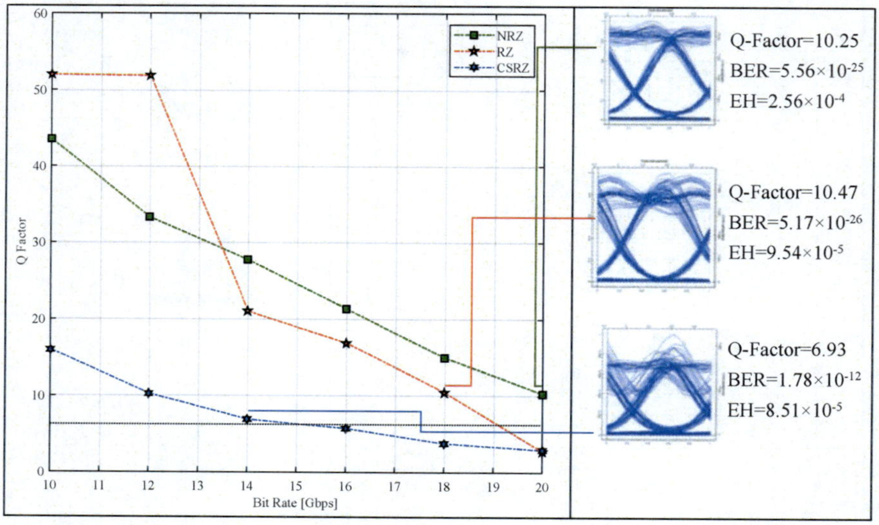

Fig. 4 Schematic diagram of bidirectional 40/10 Gbps OTDM-PON system up to 10–80 km at different modulation formats

3.2 *Comparative Analysis for Upstream Transmission*

The upstream TWDM-PON system at 1528 nm channel using three modulation techniques, viz. NRZ, RZ, and CSRZ have been compared for variable fiber length (10–80 km at $P_{in} = 0$ dBm and 2.5 Gbps), input power (6–18 dBm over 60 km and 2.5 Gbps), and bit rate (2.5–20 Gbps at $P_{in} = 0$ dBm and 10 km).

In Fig. 5, the Q factor has been observed versus the fiber length for upstream TWDM-PON over 60 km at 2.5 Gbps using different modulation formats to establish the maximum transmission distance with high reliability. For RZ modulation format, the fiber distance increases, the Q factor value decreases from 205.46 at 10 km to 35.30 at 80 km (for RZ). Also, for NRZ modulation format, the Q factor value decreases from 92.06 at 10 km to 6.06 at 80 km. In the same manner, for CSRZ modulation format, the Q factor value decreases from 41.87 at 10 km to 5.45 at 80 km. Thus, based on varying Q factor value, the RZ modulation format performs better than other modulation formats.

The eye diagrams for three modulation formats are shown in figure at 80 km. It is clear that the transmitted information is maintained and reconstructed effectively for short range distance for all modulation formats. However, as the transmission distance increases, the eye height (EH) decreases. At a distance of 80 km, the EH is minimum for all modulation formats, but that for NRZ and CSRZ modulation format is less than for RZ format, indicating the better performance in upstream TWDM-PON system.

The performance characteristics of TWDM-PON for 1528 nm channel at optimum input power form NRZ, RZ, and CSRZ modulation is shown in Fig. 6 as Q Factor

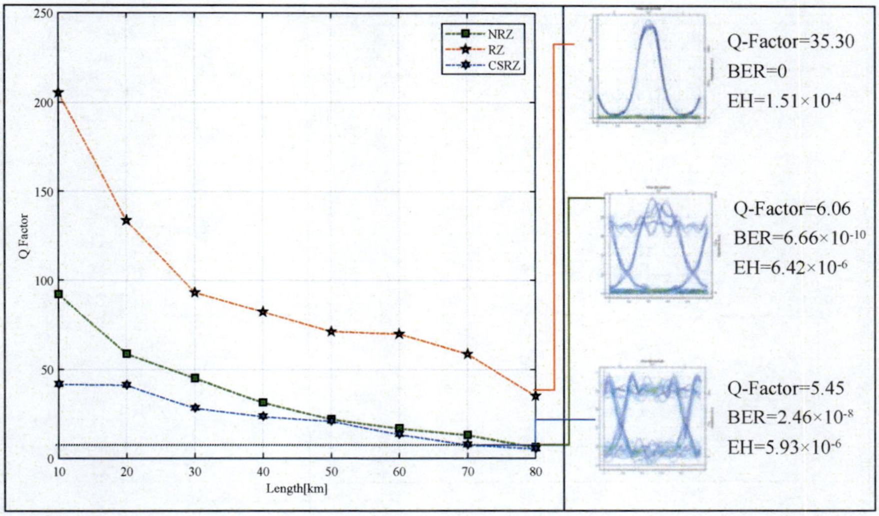

Fig. 5 Q factor versus length (km) for upstream TWDM-PON system for NRZ, RZ, and CSRZ modulation formats along with eye diagrams

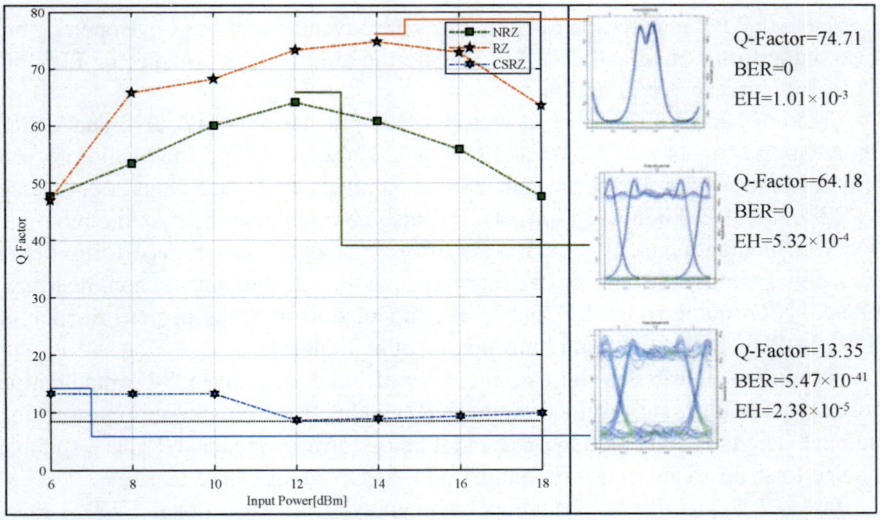

Fig. 6 Schematic diagram of bidirectional 40/10 Gbps OTDM-PON system up to 10–80 km at different modulation formats

versus optical input power over 60 km at 2.5 Gbps. It is observed from figure that RZ, NRZ, and CSRZ provide error-free performance at the optimum input power of 14 dbm, 12 dBm, and 6 dBm, respectively. It is also noticed that the optimization of optical input power for multi-wavelength TWDM-PON network with any modulation format is an essential step to minimize the fiber dispersion and nonlinearities. Hence,

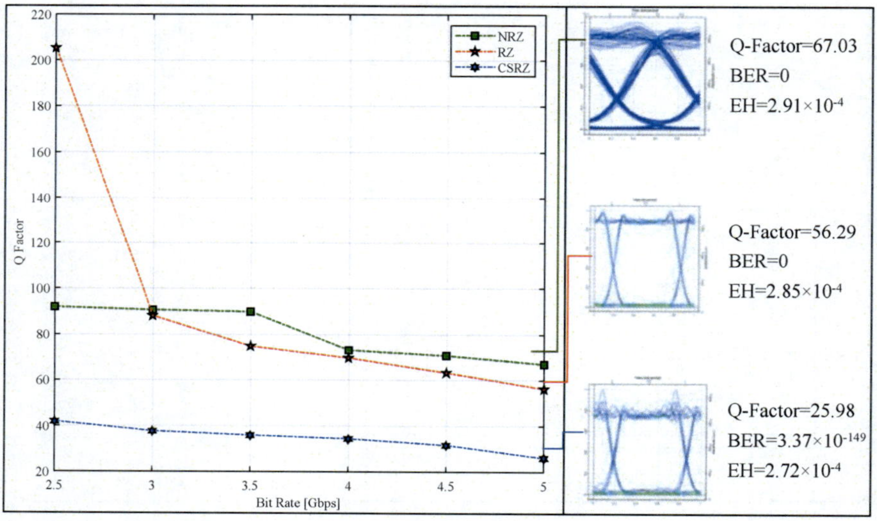

Fig. 7 Schematic diagram of bidirectional 40/10 Gbps OTDM-PON system up to 10–80 km at different modulation formats

it is clear that RZ modulation not only gives the advantages of large eye opening but also outperforms NRZ and CSRZ modulation in terms of performance of TWDM-PON for upstream transmission.

Figure 7 shows a graph of Q factor versus bit rate for NRZ, RZ, and CSRZ modulation formats over 10 km at 10 Gbps. It shows that NRZ modulation format provides better error-free performance at the bit rate from 2.5 to 5 Gbps as compared to RZ and CSRZ modulation formats. It is also observed that the maximum bit rate for NRZ modulation is 20 Gbps. It is also observed that the upstream performs better than downstream transmission in the presence of fiber dispersion and nonlinearities. Hence, NRZ outperforms RZ and CSRZ modulation in terms of performance of TWDM-PON for downstream transmission at high bit rate.

Figure 7 demonstrates the eye diagrams of NRZ, RZ, and CSRZ modulation formats at 20 Gbps. The results show that NRZ modulation has maximum eye opening and eye height as compared to other modulation formats. Hence, NRZ is a suitable choice form upstream transmission in TWDM-PON at high bit rate.

Tables 2, 3, and 4 show the comparative analysis of bidirectional TWDM-PON system using NRZ, RZ, and CSRZ modulation formats for varying transmission distance, input power, and bit rate. It is observed that for downstream transmission NRZ shows better performance than RZ and CSRZ modulation formats in terms of minimum BER and received optical power. While, for upstream transmission, RZ (both for varying fiber length and input power) and NRZ (for varying bit rate) perform better than CSRZ modulation format. Moreover, upstream transmission shows better performance as compared to downstream transmission for all three modulation formats.

Table 2 Comparative analysis of NRZ, RZ, and CSRZ modulation formats at $P_{in} = 10/0$ dBm (downstream/upstream) and 10/2.5 Gbps for varying transmission distance

Modulation format	Distance (km)	BER		Received optical power (dB)	
		Downstream	Upstream	Downstream	Upstream
NRZ	10	0	0	1.82	−31.13
	50	1.61×10^{-22}	5.24×10^{-106}	−6.17	−35.81
	80	2.93×10^{-3}	6.66×10^{-10}	−12.17	−41.67
RZ	10	0	0	−1.70	−37.18
	50	1.36×10^{-3}	0	−9.68	−38.28
	80	1	0	−13.68	−39.68
CSRZ	10	1.27×10^{-57}	0	−2.38	−37.18
	50	8.28×10^{-3}	3.71×10^{-96}	−10.38	−38.18
	80	1.11×10^{-2}	2.46×10^{-8}	−16.38	−39.37

Table 3 Comparative analysis of NRZ, RZ, and CSRZ over distance = 60 km at 10/2.5 Gbps for varying input power

Modulation format	Input power (dBm)	BER		Received optical power (dB)	
		Downstream	Upstream	Downstream	Upstream
NRZ	6	1.66×10^{-06}	0	−12.18	−43.09
	12	1.09×10^{-20}	0	−6.18	−37.09
	18	1.94×10^{-8}	0	−0.18	−31.09
RZ	6	1.38×10^{-2}	0	−15.70	−40.08
	12	6.75×10^{-4}	0	−13.02	−39.32
	18	7.30×10^{-3}	0	−12.64	−37.18
CSRZ	6	3.19×10^{-3}	5.47×10^{-41}	−16.39	−43.37
	12	1.99×10^{-3}	1.65×10^{-4}	−10.38	−37.37
	18	1	7.06×10^{-4}	−4.37	−31.37

4 Conclusion

In this paper, the bidirectional and asymmetrical 80/20 Gbps TWDM-PON for NRZ, RZ, and CSRZ modulation formats have been analyzed and compared by varying input power, transmission distance, and bit rate under the influence of fiber attenuation, dispersion, and nonlinearities. From the results, it is concluded that the fiber distance and bit rate increase, the performance of the proposed system in terms of Q factor, minimum BER, received optical power, eye height, and eye diagrams degrade. Also, the effect of increasing input power on the performance of the proposed system is investigated, and the results show that at optimum input power, the system performance increases. Further, the comparative performance of different modulation formats, i.e. NRZ, RZ, and CSRZ in the proposed system have been observed and

Table 4 Comparative analysis of NRZ, RZ, and CSRZ at $P_{in} = 10/0$ dBm (downstream/upstream) over distance = 10 km for varying bit rate

Modulation format	Bit rate [DW/UP] (Gbps)	BER		Received optical power (dB)	
		Downstream	Upstream	Downstream	Upstream
NRZ	10/2.5	0	0	1.82	−31.13
	14/3.5	1.26×10^{-169}	0	−7.28	−39.09
	20/5	5.56×10^{-25}	0	−7.39	−41.09
RZ	10/2.5	0	0	−1.70	−37.18
	14/3.5	2.41×10^{-98}	0	−11.09	−39.67
	20/5	2.69×10^{-3}	0	−11.62	−39.68
CSRZ	10/2.5	1.27×10^{-57}	0	−2.38	−39.37
	14/3.5	1.78×10^{-12}	1.22×10^{-281}	−2.49	−39.38
	20/5	1.23×10^{-3}	3.37×10^{-149}	−2.62	−39.39

results show that for downstream transmission NRZ shows better performance than RZ and CSRZ modulation formats. While for upstream transmission, RZ (both for varying fiber length and input power) and NRZ (for varying bit rate) perform better than CSRZ modulation format. Thus, the most suitable downstream fiber distance, input power, and bit rate for NRZ modulation format is 60 km, 14 dBm, and 20 Gbps per channel. Whereas, for upstream, most appropriate fiber distance and input power for RZ modulation are 80 km and 14 dBm. Moreover, for upstream, the maximum achievable bit rate is 5 Gbps for NRZ modulation format. Finally, the results also concluded that upstream transmission shows better performance as compared to the downstream transmission for all three modulation formats.

References

1. Kachhatiya V, Prince S (2016) Four-fold increase in users of time-wavelength division multiplexing (TWDM) passive optical network (PON) by delayed optical amplitude modulation (AM) upstream. Opt Fiber Technol 32:71–81
2. Kaur A, Singh ML, Sheetal A (2014) Simulative analysis of co-existing 2.5 G/10 G asymmetric XG-PON system using RZ and NRZ data formats. Opt Int J Light Electron Opt 125:3637–3640
3. Bindhaiq S, Zulki N, Supa AM, Idrus SM, Salleh MS (2017) 128 Gb/s TWDM PON system using dispersion-supported transmission method. Opt Fiber Technol 38:87–97
4. Nabih A, Rashed Z (2019) Comparison between NRZ/RZ modulation techniques for upgrading long haul optical wireless communication systems. J Opt Commun, 1–6
5. Ali F, Khan Y, Qureshi SS (2019) Transmission performance comparison of 16 * 100 Gbps dense wavelength division multiplexed long haul optical networks at different advance modulation formats under the influence of nonlinear impairments. J Opt Commun, 1–10
6. Alipour A, Mir A, Sheikhi A (2016) Ultra high capacity inter-satellite optical wireless communication system using different optimized modulation formats. Opt Int J Light Electron Opt 127:8135–8143

7. Kachhatiya V, Prince S (2018) Downstream performance analysis and optimization of the next generation passive optical network stage 2 (NG-PON2). Opt Laser Technol 104:90–102
8. Yao S, Fu S, Wang H, Tang M, Shum P, Liu D (2014) Performance comparison for NRZ, RZ, and CSRZ modulation formats in RS-DBS Nyquist WDM system. J Opt Commun Netw 6:355–361
9. Kumari M, Sharma R, Sheetal A (2019) Comparative analysis of high speed 20/20 Gbps for long-reach NG-PON2. J Opt Commun, 1–14
10. Sasikala V, Chitra K, Dwdm ÁCÁ (2018) Effects of cross-phase modulation and four-wave mixing in DWDM optical systems using RZ and NRZ signal. In: Lecture notes in electrical engineering, pp 53–63

Big Data and Analytics in Higher Educational Institutions

Ankit Bansal and Vijay Anant Athavale

Abstract The higher educational institutions are working in a rapidly increasingly complicated and competitive scenario. The contemporary challenges faced by the higher educational institutions in India are discussed in this paper. Big Data is a trending area of research that is using data analysis to notify decisions. Research on Big Data is mainly focused on evaluating real-time, compiled data and corresponds to heavy datasets to find repeating behavioral patterns instead of creating metadata of the status quo. This paper outlined the effect of Big Data analytics to improve faculty members' understanding of learners' perspectives and behavior toward the programs as well as opportunities and challenges associated with its knowledge and implementation.

Keywords Analytics · Big Data

1 Introduction

The higher educational institutions have been working in a rapidly increasingly complicated and competitive scenario for a decade. The pressure to revert to the economic, political and social change is day by day increasing like the growing need to increase the ratio of learners in some streams, implanting in-house program outcomes.

The huge amount of data is being collected and stored in various educational institutional databases. Most of them are giving a knowledge repository online. Data is

A. Bansal (✉)
Gulzar Group of Institutes, Ludhiana, Punjab, India
e-mail: erankitbansal@gmail.com

V. A. Athavale
Panipat Institute of Engineering & Technology, Panipat, Haryana, India
e-mail: vijay.athavale@gmail.com

N. Marriwala et al. (eds.), *Mobile Radio Communications and 5G Networks*, Lecture Notes in Networks and Systems 140,
https://doi.org/10.1007/978-981-15-7130-5_15

scattered in various systems like student information systems, social media repositories, learning management systems, personal computers and administrative systems, etc. In addition to availability in abundance of data, it comes in various formats.

The paper checks the various aspects of Big Data and analytics in outlining the challenges faced by higher educational institutions (see Fig. 1). The paper finds the main trends affecting higher educational institutions and explores the strengths of Big Data and analytics in addressing these issues. Then, it highlights opportunities and issues linked with the implementation of Big Data in higher educational institutions.

Despite rapidly increasing changes happening in the era of higher education, the role of Big Data in solving these issues is often unnoticed. As learning technologies are day by day increasing from all sides in higher educational institutions, huge datasets are generated. These datasets can be utilized to inform higher educational institutions to accept better in response to issues arising within and outside their scenarios [1].

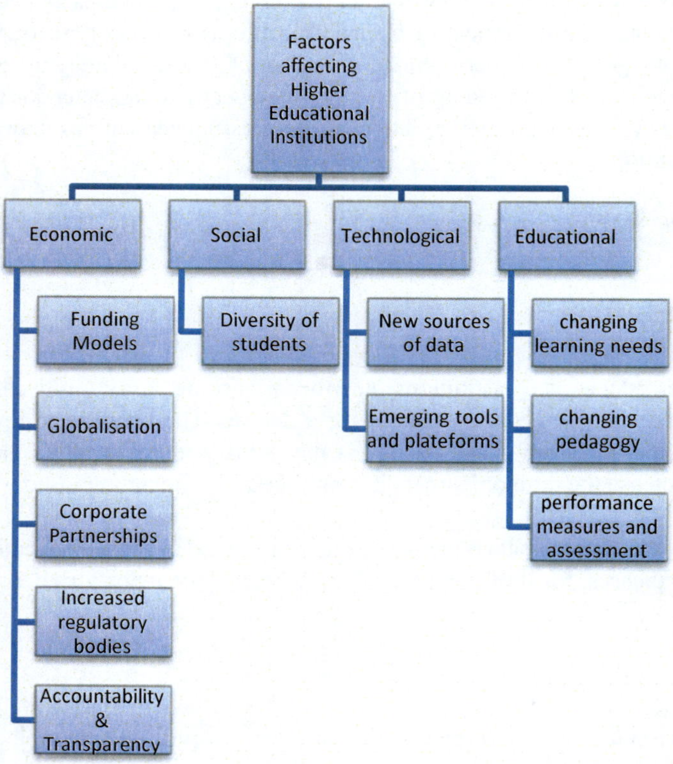

Fig. 1 Factors affecting higher educational institutions

2 Evolution of Big Data

Data is used by many organizations currently to take appropriate decisions about their plans. It is not new to use data to make good decisions; institutions have been saving and analyzing huge data since the occurrence of data warehouse systems. Generally, higher educational institutions are running on structured data [2].

Big Data elaborates data that is huge and runs too fast, thus, exceeding the processing speed of old database systems [3]. Big Data also includes innovative techniques to identify, save, distribute, manage and analyze huge data sets with disparate structures.

Douglas, L. in Gartner's report proposed a threefold definition encompassing the "three Vs" (volume, velocity and variety) [4].

Some other properties are also there such as data validity, which refers to the accuracy of data and volatility. Apart from it, there are three stages data collection, data analysis, visualization (see Fig. 2).

2.1 Data Collection

It is unbolting the value accumulated from Big Data. This requires finding data that can acknowledge useful information. Data must be filtered for relevance and saved in a useful form.

2.2 Data Analysis

After collection, it needs to be analyzed to generate the required information. However, with the increasing heterogeneity in the nature of data, it is a very complex process to manage and analyze heterogeneous data set. We can call the situation as the 'complexity' of Big Data.

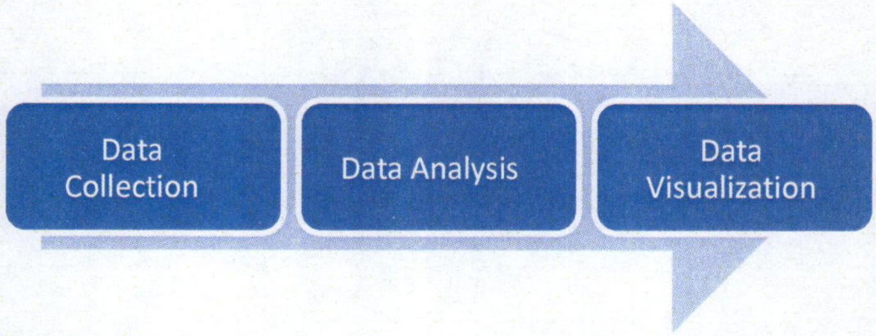

Fig. 2 Important stages of Big Data

2.3 Data Visualization

Here, users find the analyzed data that is understandable and integrated into existing processes, and hence it can be used in decision making.

3 Big Data and Analytics in Higher Education

Big Data analytics in higher educational institutions comprises four types of analytics which are mentioned in given Fig. 3.

3.1 Institutional Analytics

If we talk about higher educational institutions in India like any state technical university always want to find out where it stands. What is its reputation, where it stands

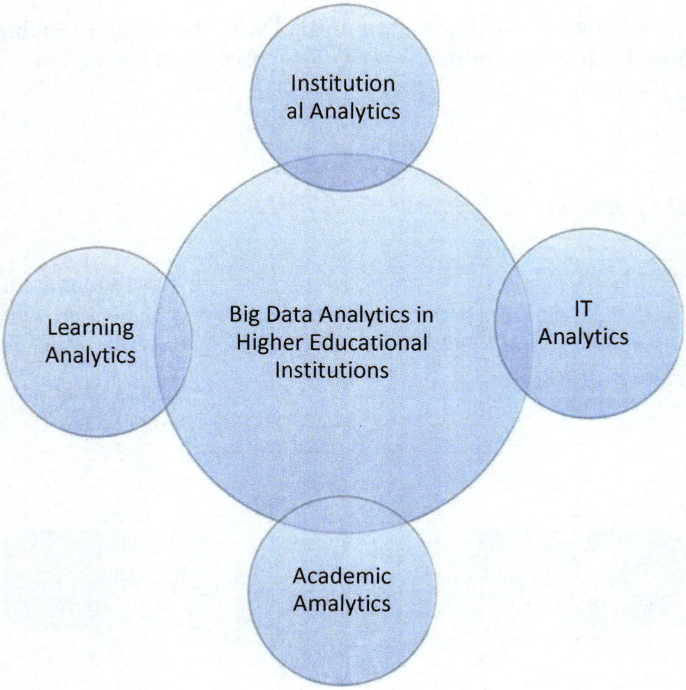

Fig. 3 Analytics in higher educational institutions

based on facts and figures or data? The data can be text, numbers, pictures, videos, audio and feelings.

Institutional analytics is used to quantify policy acceptance, remedies and instructions acceptance, and to find structural analytics. It empowers institutions to take effective decisions to improve their system.

3.2 *Information Technology (IT) Analytics*

IT analytics helps to develop and deploy technology and tools, processes, integrate data from a variety of systems.

IT analytics include how many times the particular data is being searched so that it can find out the importance of particular data.

3.3 *Academic Analytics*

Academics analytics play an important role in improving academics. As per NBA guidelines, the outcomes-based education model relies on continuous evaluation, continuous learning and continuous improvement. Academic analytics identify various shortcomings in various programs.

3.4 *Learning Analytics*

As higher educational institutions are using hybrid approaches to teaching in online scenarios and platforms. Learning analytics is to quantify student behaviors toward online content, number of visitors, bounce rate, the number of clicks or click-through rate, conversion rate and time spent on a page, acceptability of content with the positive comments of learners, cross-questions of learners and no. of assignments solved for that content.

This can be better understood by the concept of learning provided by MOOCs like NPTEL, Swayam, Udemy, etc. Learning analytics provide various facts and figures which are very helpful in understanding learners' behaviors, time spent on particular content and effectiveness of that particular data for comments and assignments. Also, it provides various gaps which can be taken as an opportunity to improve it.

4 Opportunities

Due to having huge student databases, higher educational institutions can analyze various datasets. Existing processes of teaching, learning, academic work and administration can be exceptionally improved with the help of Big Data and analytics in higher educational institutions [5], which can contribute to improving policies and practices.

Higher educational institutions can have the predictive analytic tools to improve learning outcomes for various learners as well as ways they can ensure high-quality standards academic programs with the help of Big Data.

As Big Data can be in the form of text, numeric, audio, video, facts and figures, images, feelings, emotions, etc., Big Data analytics can provide various ways to retrieve data-based content retrieval for the same. It will provide an opportunity to provide the higher education industry to find out various shortcomings, opportunities, gaps, etc.

Big Data analytics strengthen faculty to measure, watch and answer, in real-time to a learner's understanding. Faculty can make courses interesting for learners with different levels of knowledge. Using data analytics to understand, where each learner is struggling or improving can allow faculty to offer a different learning material for each learner within the same course. This will improve the interest of learners. Big Data analytics is very useful in understanding employment trends and to prepare introductory courses and fundamental learning principles (Fig. 4).

5 Challenges of Implementation

Generally, higher educational institutions collected data in their best capacities from the best possible ways. Also, they have their alumni as well as present students' database. By the time, this data keeps on increasing with the traces of their information which is continuously captured by the institutions from time to time, and not only by the institution, every stakeholder is playing his/her role in contributing to increasing data. But this data is scattered across the whole institution. Now there are so many challenges linked with this data like consolidation of data, quality of this data, acceptance of this data, difficulties in retrieving or accessing that data, skilled people to manage, access, compare, retrieve that data and to develop new algorithms for better analytics, patience due to time-consuming process, etc.

Another major issue is also to be mentioned here about ethics of data collection, institutional policies and blockades due to privacy concerns, security and ownership of that data.

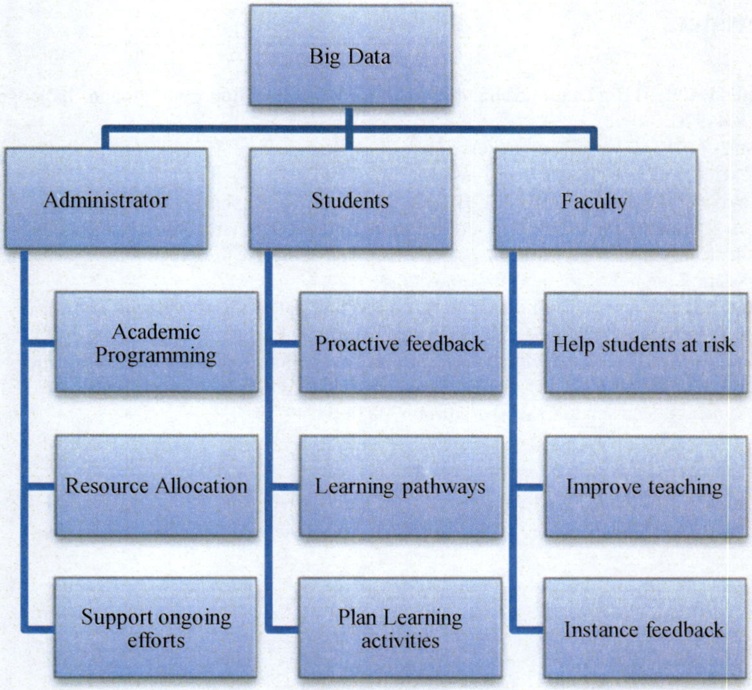

Fig. 4 Main Big Data opportunities of stakeholders

6 Conclusion

In this era of abundant data, higher educational institutions like other institutions share the same reasons for using Big Data analytics.

In higher educational institutions, though the data is day by day increasing but is scattered across the whole institution and in institution domain like stakeholders' personal computers and laptops, faculties, cloud and available with different extensions and formats, making it difficult to access, retrieve and consolidate. For efficient compilation and analyze, different datasets are needed, regardless of where it originated and aggregating data within the institution domain.

Big Data analytics is capable to figure out innovations, gaps, facts for improvement of results in growing businesses and services. Its institutional analytics can help any university or higher educational institutes to find their position across the globe. Its academic analytics provides various information related to the effectiveness and shortcomings of the programs. It is IT analysis focuses on the popularity and citations of data. Rest learning analytics is a hot and trendy aspect of Big Data analysis which can create a good knowledge repository.

In the future, Big Data analytics can help program developers or faculty members to organize new courses as per individual learners' understanding.

References

1. Daniel B (2015) Big Data and analytics in higher education: opportunities and challenges. BJET 46:904–920
2. Basu A (2013) Five pillars of prescriptive analytics success. Analytics. https://www.analytics-magazine.org. 8–12 March/April Issue
3. Manyika J, Chui M, Brown B, Bughin J, Dobbs R, Roxburgh C et al (2011) Big Data: the next frontier for innovation, competition, and productivity. Retrieved October 7 and 30, 2014, from https://www.mckinsey.com/Insights/MGI/Research/Technology_and_Innovation/Big_data_The_next_frontier_for_innovation
4. Douglas L (2011) 3D data management: controlling data volume, velocity and variety. Gartner Report. Retrieved December 30, 2013, from https://blogs.gartner.com/doug-laney/files/2012/01/ad949-3D-Data-Management-Controlling-Data-Volume-Velocity-and-Variety.pdf
5. Baer L, Campbell J (2011) Game changers. EDUCAUSE. Retrieved March 24, 2014, from https://net.educause.edu/ir/library/pdf/pub72034.pdf

A Hybrid Channel Access Method to Optimize Congested Switched Network

Shilpa Mahajan and Nisha Sharma

Abstract In this paper, a congestion control TDMA and FDMA-based MAC mechanism for optimizing data flow in a tree switched network is suggested. A method has been devised for minimizing the communication loss and delay for carrying heavy network traffic from data centers to fixed data points. A two-stage approach is adopted, where traffic analysis is done first and then node prioritization is made considering different quality parameters. Node's individual contributions are calculated and based on their contribution's frequency allocation using FDMA will be applied. Later, load on critical cross-points will be observed and TDMA scheduling method is devised. Simulations are done in NS2 environment. This method not only reduces communication losses but also improves throughput.

Keywords Switched · Network · MAC · TDMA · FDMA · Congestion

1 Introduction

A switched network [1] is a reliable dedicated network that aims to provide high speed communication link between servers' and sub-areas. The content level communication [2] analysis is done by observing node or network level communication. To provide high speed communication, these networks require high bandwidth. Common applications where such networks are used are in biometric application, big data processing and server-to-server communications.

Some basic properties of switched network are discussed below.

High Bandwidths

Switched network requires higher bandwidth for fast data transmission. The communication can be performed up to 1014 Hz.

S. Mahajan (✉) · N. Sharma
The NorthCap University, Sec-23 A, Gurugram, India
e-mail: shilpa@ncuindia.edu

N. Marriwala et al. (eds.), *Mobile Radio Communications and 5G Networks*, Lecture Notes in Networks and Systems 140,
https://doi.org/10.1007/978-981-15-7130-5_16

Low Transmission Loss

For checking data loss, amplifiers and repeaters can be used. The error checkers can also be applied for data validity.

Cross-Communication

Switched network works on topological network, where shared media is used, and some methods are required to be applied to avoid cross-communication problems.

Communication Security

Security is always a main concern in communicating network. Network is vulnerable to active and passive attacks. Measures should be ensured to avoid these attacks. Thus, low level security mechanisms are of utmost important.

Rigidness and Flexibility

Since Wired channels are dedicated channels, they are more relaible then wireless media. The technology is less complex and requires easy handling.

2 Related Work

Mukhopadhyay et al. [1] have proposed a method for congested switched wireless network. This paper improves network capabilities by suggesting method for mobile handling and uses capacity-driven estimation technique. A distributed component-driven architecture is also proposed for avoiding congestion. It improves the placement of switches/nodes in order to improve communication efforts.

Khoo et al. have worked on rerouting technique using disjoint analysis approach for congested networks [3]. Author defined a traffic-driven analysis on optical network to identify the traffic situation in the network. It works by applying traffic request on different network parts with cross-switch estimation under various parameters. An effective evaluation method is also suggested in switched network.

Fenglin et al. have provided a work on load balanced communication method in MPLS networks [4]. Author defined a traffic engineering method for route choice generation and resource utilization. A load balanced communication is provided by the model for generating alternative paths in the network. The communication behavior shows optimized throughput and reduces communication losses.

Nishijima et al. have devised work on synchronous TDMA for reducing communication delay [5]. A delay-driven communication forwarding is provided for reducing the delay and providing the optimized network communication.

Cheung et al. [6] have proposed a data offloading method for cellular congestion aware communication in switched network. It also identifies the problems relative to mobility, location dependency and user availability. The dynamic network connection and cost modeling are also defined in this work. Author provided a balanced network formation technique for optimizing static and dynamic behavior of the network [7, 3].

Girry et al. [8] have devised an approach on circuit switching to observe congestion problem for high and low level network traffic. A TCP connection-based method of path formation is defined for identification of interactive communication. A queue-driven approach for congestion analysis is proposed and reformation technique is also suggested during bottleneck situations.

Najah et al. [9] have suggested a capacity analysis-based communication method for behavior control in global environment. It also makes use of optical buffering method to provide the delay specific observation so that the controlled communication is performed in the network. The results are validated through extensive communication in switched network.

Gholami and Akbari [10] defined a rerouting approach for reducing congestion in data center. A fat tree topology is implemented in mininet emulator and open flow control mechanism is used for improving throughput and delay in a network. During congestion, packets will choose least loaded path computed from all the available paths in a network.

In a proposed work hybrid MAC congestion control (HMCC), an improvement to the MAC protocol is provided for optimizing the communication in congested switched network. The work has combined the TDMA and FDMA to achieve the optimization in the dedicated tree switched network. Analysis of nodes has been done on load factors and balancing techniques have been devised for optimizing the congestion problem. The protocol has been implemented in NS2 environment and results have been compared with other existing approach.

3 Contributions

Following goals have been considered to develop HMCC.

- To define a MAC [11] improved communication model for optimization of switched network.
- To apply frequency division based on node requirement prioritization for reducing communication delay.
- To apply TDMA to distribute the load for critical points for data loss reduction.
- To improve communication throughput and reduce communication delay.

4 Design Philosophy

In HMCC, an optimization to the switched network is defined on modifying the functional capabilities of MAC layer. The work is defined to combine the TDMA and FDMA [12, 13] modeling in MAC layer to achieve the load equalization and frequency distribution. Analysis of the network under load and communication rate parameters is done. Once the parametric derivations are obtained, the node requirement-based prioritization will be assigned. The prioritization will be used to

Fig. 1 HMCC layout

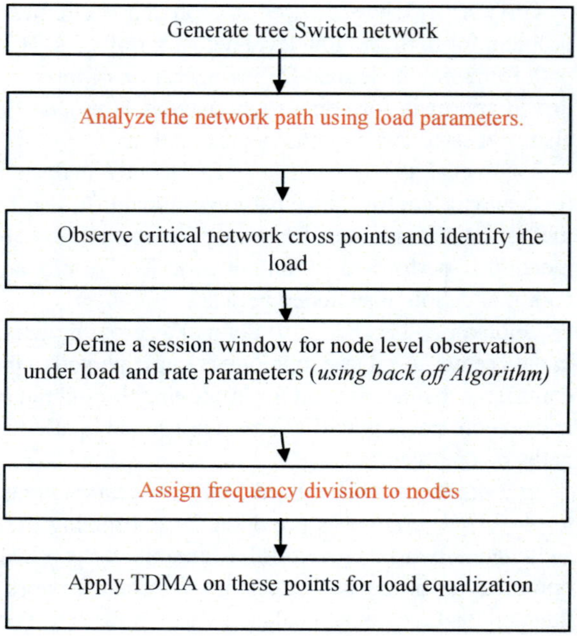

assign the frequency to different nodes based on the requirements. Later, the node load on various switched nodes will be analyzed. The TDMA will be applied to assign the communication session so that the load equalization will be achieved. Diagrammatic layout of HMCC is shown in Fig. 1.

4.1 Network Parametric Analysis

In order to analyze load condition of the path, it is necessary to find metric for congested node based on load rate and packet loss factors. Load rate can be defined as current link utilization over maximum load capacity. Link load utilization (lutilization) can be defined as

$$\text{lutilization} = \left(\frac{\text{Throughput}}{\text{data rate}}\right) * 100 \tag{1}$$

Here, data rate is the maximum data that can be sent through a link, whereas throughput is the actual data transmitted through a link per time unit. Thus,

$$T_{a\rightarrow b} = \frac{\sum_{a\rightarrow b} B}{\text{Time}} \tag{2}$$

Here, T is the notation used for data rate. $a \rightarrow b$ represents a link between a and b. B is the amount of data send from $a \rightarrow b$.

Other factors affecting the path load condition are packet loss. It is the number of packets received by a node to the total number of packets transmitted. It is calculated using packet loss indicator.

$$p_i = \frac{p_{Ti} - p_{Ri}}{p_{Ti}} \tag{3}$$

If the value of $p_i > 20$, then there may be a link failure due to node failure, collision or congestion.

4.2 HMCC Procedure

Algorithm 1

```
Notations
S          Switches
Snode           Source Node
Dnode          Destination Node
cur                Current Node
Conn            Connection among Nodes

Algorithm
Cooperative Communication (S, SNode, Dnode)
Begin
Initialize i, i ∈ S, where S = {s1,s2,s3,……,sn}
//  S is the list of nodes present in the network.
Set cur :=   Snode , where Snode ∈ S   // Set cur as
source node
while SNode<>Dnode  // Dnode is a destination node
for  i:= 1 to SLength Step 1 // Process all devices
if  Conn(Snode(i),cur)<>0    //  Check for neighbour Node
// Generate the list of neighbors based on distance adap-
tive analysis
neighborNode.Add(SNode(i))
```

```
End if
End for
For i:= 1 to S_Length
For j:= 1 to S_Length
Find  T_{a->b}(i, j) // Calculate link distance from a node to
its neighbour node.
Find P_{i,j}          // Calculate Packet loss of a node
End For
End For
For i:= 1 to S_Length
For j:= 1 to S_Length
// Find out a link with maximum load
Set Flag (i):=1
End For
End For
For i:= 1 to S_Length
For j:= 1 to S_Length
If P(i,j) > 20
Set Flag (i):=2
End if
End For
End For
For i := 1 to S_Length
if Flag(i)==1 && Flag (i)==2 //Conditions for checking
load  on a node
Node is congested.
Block the node
Else if  Flag(i)==1 && Flag (i)!=2
Distribute load of SNode(i) to SNode(j) // Load can be
distributed to neighbour nodes.
Else if  Flag(i)!=1 && Flag (i)==2
Block congested node
Else
Node can be used for normal operations.
End if

For i:= 1 to S_Length
For j:= 1 to S_Length
If T_{a->b}(i,j) > LOAD _LIMIT //  If load becomes greater than
link capacity
      CALL  ExponentialContentionControl(Wd)
End If
End For
End For
End while
End
```

Algorithm Description

Algorithm 1 defines network condition of a tree network. Various conditions have been taken for analyzing the node load factor. The communication optimization is achieved using contention window control mechanism applied using back-off algorithm. The contention window is responsible to decide the communication information size for the protocol stack for effective communication in the network. The contention window control mechanism analysis is done to get network failure estimation.

4.3 *Exponential Back-Off Algorithm*

In this method, the main requirement is to optimize the contention window processing using FDMA method. The content window control method is defined using back-off algorithm. The associated window-driven formula is defined to generate the slot specific value. The algorithmic formulation of this exponential algorithm is defined below. The node-level characterization and optimization are obtained using exponential algorithm. This algorithmic approach has optimized the window formation so that the effective communication will be drawn from the system.

Back-Off Algorithm

```
ExponentialContentionControl(Wd)
Begin
//  Wd is the standard contention window size calculated
based on load and time slot estimation
While Wd<> Null
For  i = 1 to TimeSlot(CommunicationData) step 1
if  IsIdle(Channel))
Set Wd := Wd-1
// Perform Communication
else
Set Wd := Exp(2,Wd)*SlotTime
End if
End for
End while
End
Result
```

For implementing HMCC, NS2 simulator tool is used. NS2 can be used for wired and wireless networks. This tool provides different traffic forms and connection control methods so that the communication in collaborative environment can be formed. Various parameters considered for evaluation are discussed in Table 1.

Network construction of switched wired network of 29 nodes is shown in Fig. 2.

Table 1 Network parameters

Parameters	Values
Number of nodes	29
Type of network	Switch hierarchal network
Simulation time	10 s
Packet size	512
Delay	0.01
Topology	Tree
Traffic	CBR

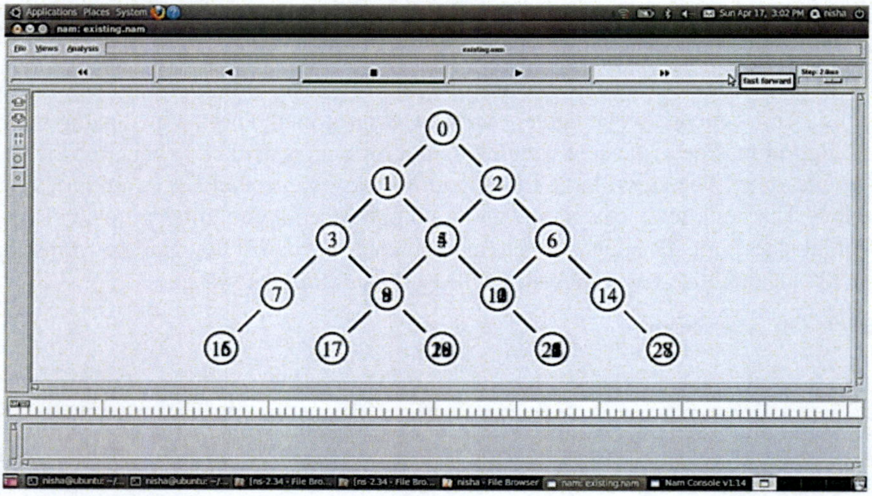

Fig. 2 Network scenario

The nodes are connected under tree topology. Here, 0th node is considered as the source node. All nodes are connected and transmission of packets from source node to all other nodes can take place in hierarchal order.

The result of the proposed approach is compared with existing (Gholami [10]) approach by constructing the same network scenario and results of both are compared on various defined parameters. Network communication from source node to all others is shown in Fig. 3. When load starts increasing on nodes, the bottleneck situation generated and it results in a communication loss.

Here, Fig. 4 shows communication flow in a switched network at time 1.5 s. It can be clearly seen that although the packets are already distributed over the network, but still due to heavy load lot of data will be lost.

Figure 5 shows packet communication in existing approach. Here, *x*-axis represents communication time and *y*-axis shows packet communication. As the network

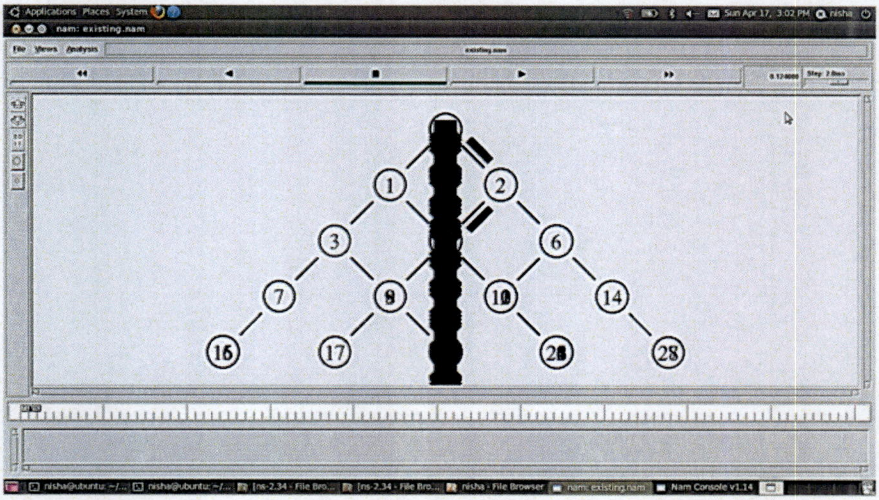

Fig. 3 Communication with network loss (existing approach)

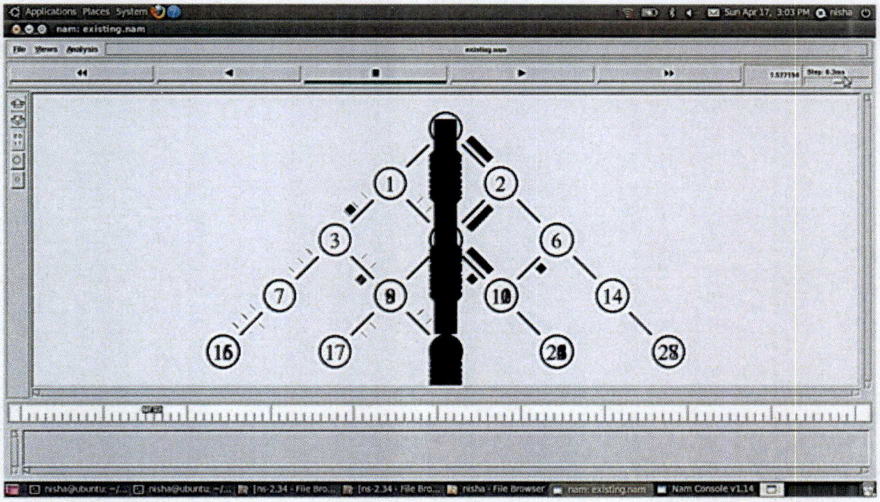

Fig. 4 Distributed communication with loss (existing approach)

is wired network, because of which once the flow weight is decided, the communication flows with constant rate. The figure is showing the communication flows at constant rate over the network.

Figure 6 shows the packet communication of HMCC. Here, x-axis represents communication time and y-axis is showing the packet communication. As the network is wired network, once a flow weight is decided, the communication flows with constant rate. The proposed protocol uses combined FDMA and TDMA-based

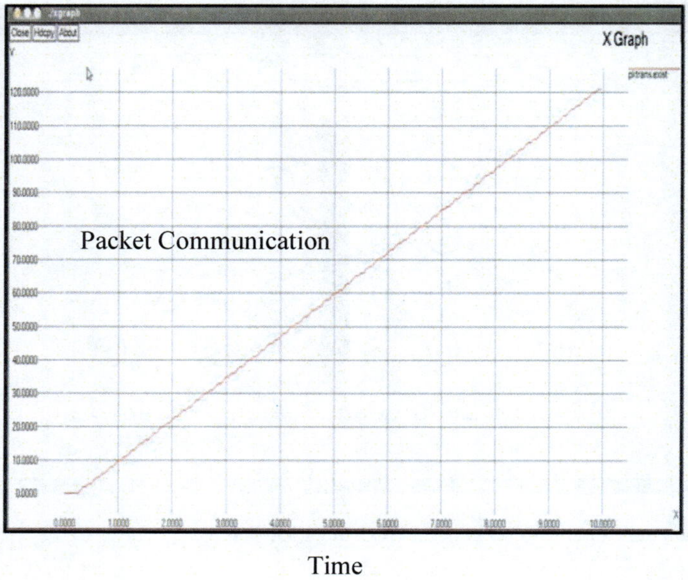

Fig. 5 Packet communication (Gholami [10] approach)

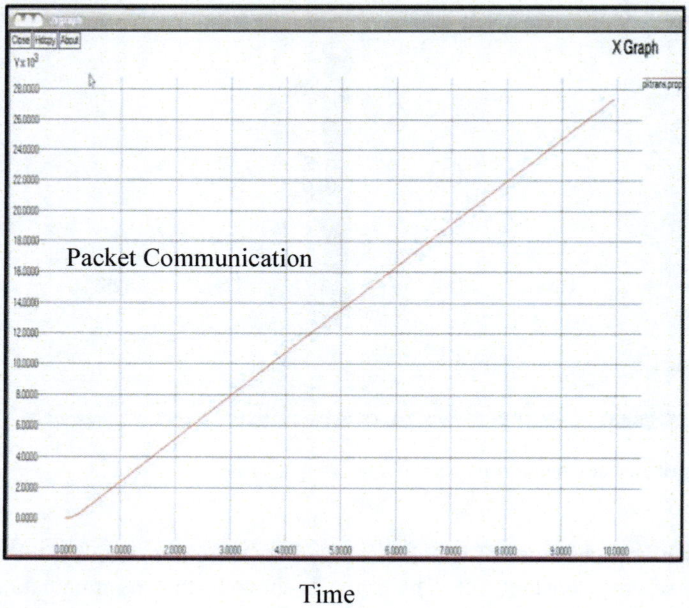

Fig. 6 Packet communication of HMCC

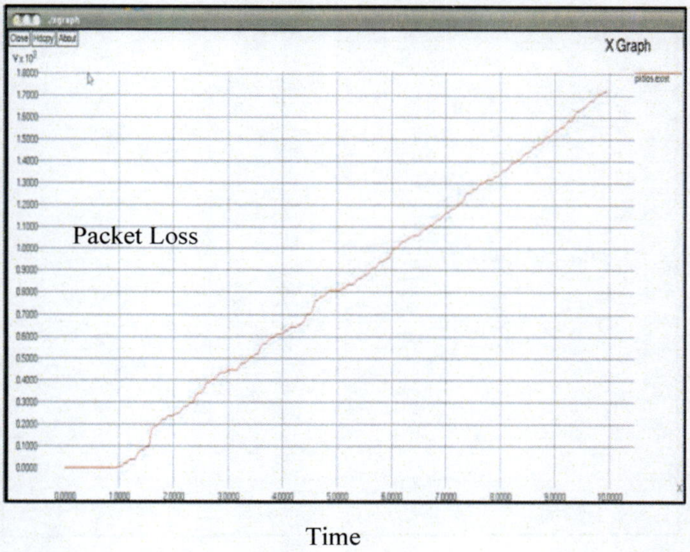

Fig. 7 Packet loss (Gholami [10] approach)

neighborhood analysis approach that results in improving communication up to an extent.

Figure 7 defines the packet loss in exiting protocol. It shows as simulation time increases, the loss of packets in the network also increases. This is mainly due to heavy load and bottleneck situation. As some nodes become more loaded and they start dropping packets.

Here, Fig. 8 shows communication loss occurred in HMCC. In this, TDMA and FDMA-based neighborhood analysis approach reduces the communication loss. As the communication begins, the loss occurs till the observation stage is not reached (as defined in Algorithm 1). Once reached, the load is distributed to neighboring nodes to minimize the traffic. Here, green line shows the proposed method scenario, and red line shows exiting network behavior.

Comparison between HMCC and Gholami [10] approach is shown in Fig. 9.

Here, Fig. 10 shows the packet delay analysis for Gholami [10] approach. Result shows that the communication delay is high when the network is established after that the communication delay is reduced.

Figure 11 shows the packet delay analysis for the proposed approach. The figure shows that the communication delay is high when the network is established after that the communication delay is reduced. The proposed FDMA and TDMA-based approach has improved the communication strength and reduced the communication delay when compared to exiting approach.

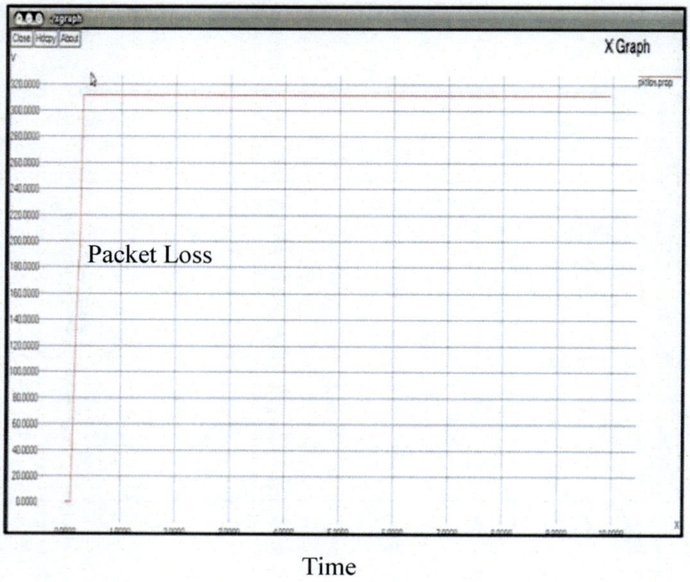

Fig. 8 Packet loss in HMCC

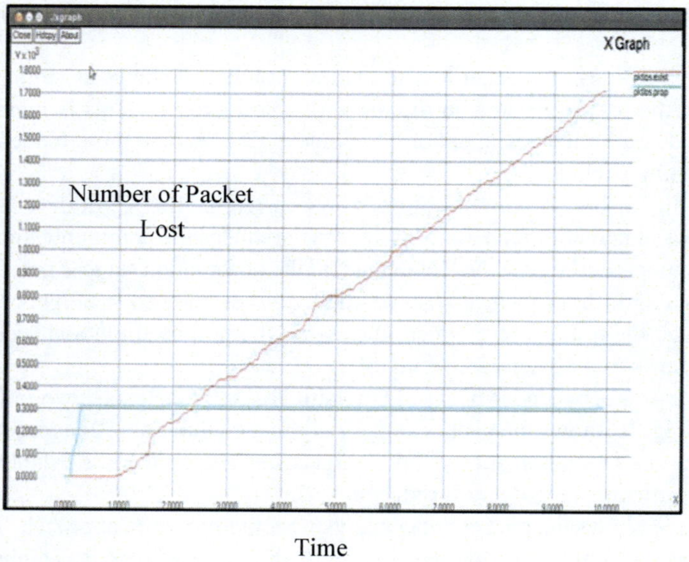

Fig. 9 Packet loss HMCC versus Gholami [10] approach

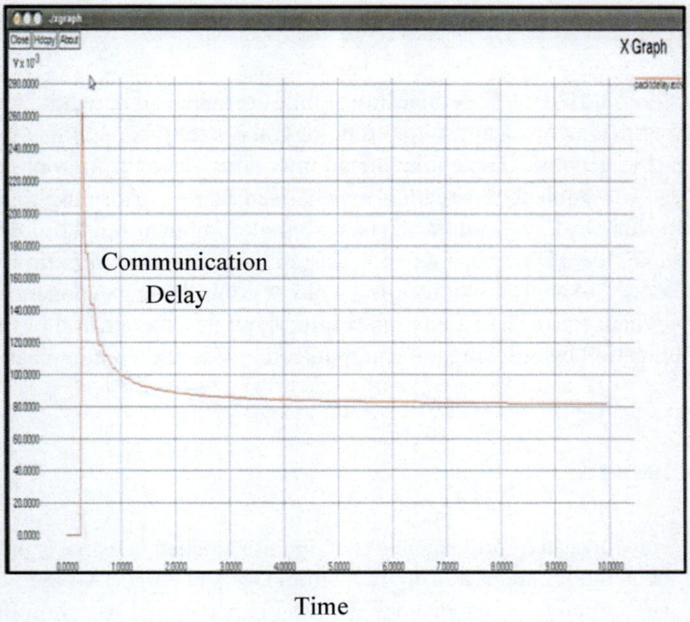

Fig. 10 Packet delay (Gholami [10] approach)

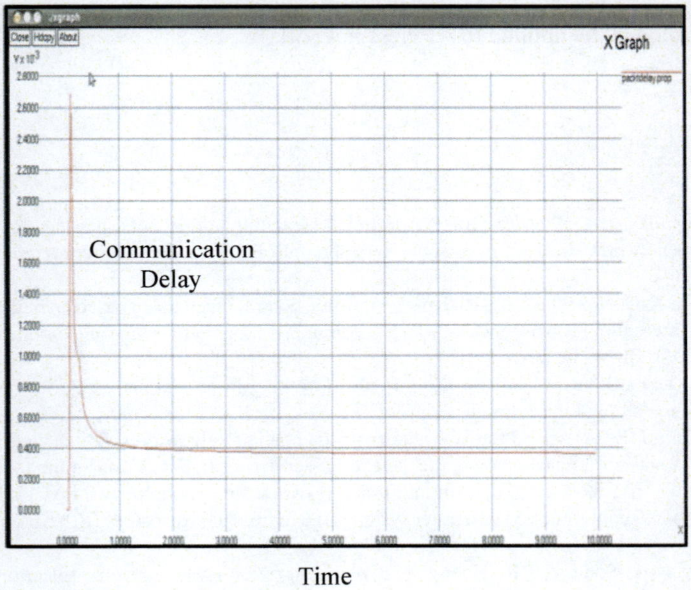

Fig. 11 Packet delay (HMCC)

5 Findings

The research is carried out for congestion control in switched network. A dedicated switch network is defined under tree topology and heavy communication flow is shown over the network. The communication is broadcasted from root node to all other nodes. The bottleneck situation is generated while performing this communication. In this work, a neighbor analysis-based combined model is defined. At the early stage, the TDMA is applied to assign the communication time slot. The FDMA is defined to apply the contention window control using exponential method during load conditions. The simulation results show that the method has improved the communication by reducing the communication loss and communication delay.

6 Conclusions

In this present work, an improvement to the circuit switched network is provided by observing the neighbor nodes and applying the TDMA and FDMA-based combined approach. The effective neighbor node analysis is performed based on delay, link observation and packet loss factors. The routing behavior of each node is defined to observe the previous load based on which the time slots are assigned. In the future, the work can be extended on other networks including star, mesh and hierarchical topology and can be applied to wireless network.

References

1. Mukhopadhyay A, Zhao ZJ (2009) Additional switching nodes: not a panacea for congested wireless networks. In: 18th Annual wireless and optical communications conference, Newark, NJ, pp 1–6
2. Liu Y, Li X, Chen S, Qin Z (2010) Link congestion control mechanism based on multi-topology. In: 6th International conference on wireless communications networking and mobile computing (WiCOM), Chengdu, pp 1–4
3. Khoo KL, Parthiban R (2010) Disjoint path re-routing in optical networks. In: International conference on photonics, Langkawi, pp 1–5
4. Li F, Chen J (2012) MPLS traffic engineering load balance algorithm using deviation path. In: International conference on computer science & service system (CSSS), Nanjing, pp 601–604
5. Nishijima H, Yakoh T (2012) Minimum delay switch for synchronous TDMA network. In: International conference on computer systems and industrial informatics (ICCSII), Sharjah, pp 1–6
6. Cheung MH, Southwell R, Huang J (2014) Congestion-aware network selection and data offloading. In: 48th Annual conference on information sciences and systems (CISS), Princeton, NJ, pp 1–6
7. Mora G, Garcia PJ, Flich J, Duato J (2007) RECN-IQ: A cost-effective input-queued switch architecture with congestion management. In: International conference on parallel processing (ICPP 2007), Xi'an, p 74

8. Girry KT, Ohzahata S, Wu C, Kato T (2014) A circuit switching method for improving congestion of Tor network. In: Ninth international conference on broadband and wireless computing, communication and applications (BWCCA), Guangdong, pp 416–421
9. Najah O, Seman K, Abdul Rahim K (2014) The performance of OCDM/WDM with buffering based on shared fiber delay line. In: International conference on information, communication technology and system (ICTS), Surabaya, pp 269–274
10. Gholami M, Akbari B (2015) Congestion control in software defined data center networks through flow rerouting. In: 23rd Iranian conference on electrical engineering, Tehran, pp 654–657
11. Wan Y et al (2007) A hybrid TDM-FDM MAC protocol for wireless sensor network using timestamp self-adjusting synchronization mechanism. In: International conference on wireless communications, networking and mobile computing, Shanghai, pp 2705–2709
12. Xie L, Zhang X (2007) TDMA and FDMA based resource allocations for quality of service provisioning over wireless relay networks. In: IEEE wireless communications and networking conference, Kowloon, pp 3153–3157
13. Tanoli UR, Abbasi R, Utmani QJ, Usman M, Khan I, Jan S (2012) Hybrid TDMA-FDMA based inter-relay communication in cooperative networks over Nakagami-m fading channel. In: International conference on emerging technologies (ICET), Islamabad, pp 1–5

Automated Object Detection System in Marine Environment

Vishal Gupta and Monish Gupta

Abstract Detection and classification of objects within images are an actively pursued field in computer vision. For more than a decade, a lot of work has been explored in the maritime surveillance domain but research in computer vision for maritime is still in the nascent support and several challenges remain open in this field. Although the tracking of ships widely explored, object detection of non-ship objects remains an unexplored part. The present work describes an automated novel method to detect different ship and non-ship objects. The proposed work is based on support vector machine (SVM), to train the algorithm with large data set. We have used the canny filter to detect the edges of an object and further extracted the features of images using bag of features. The algorithm is trained with the large set of extracted features of objects. Experimental results show the detected objects along with its classifications. The algorithm is implemented on MATLAB environment.

Keywords SVM · Computer vision · Surveillances · Tracking · Ships · Objects

1 Introduction

Maritime surveillance systems can be employed to increase the security of ports, airports, merchant and warships against pirates, terrorists or any hostile vessel attacks, to avoid collisions, to control maritime traffic at ports and channels and for coastal and oil platforms defence. Surveillance systems play an important role in management and monitoring of littoral and maritime areas by providing tools for situation awareness, threat assessment and decision-making. Video surveillance has been widely used for automatic monitoring in seaport security, maritime transportation and ocean engineering managements. Many applications have been developed for monitoring

V. Gupta (✉) · M. Gupta
University Institute of Engineering & Technology, Kurukshetra University, Kurukshetra, India
e-mail: v.vishu22@gmail.com

M. Gupta
e-mail: monish_gupta1976@kuk.ac.in

N. Marriwala et al. (eds.), *Mobile Radio Communications and 5G Networks*, Lecture Notes in Networks and Systems 140,
https://doi.org/10.1007/978-981-15-7130-5_17

ship waterways using stationary cameras. In order to control normal activities in these areas, ship detection and tracking play the important role in video surveillance systems [1–3]. Various sources of surveillance, monitoring, maritime safety information are available. These include the automatic identification system (AIS), automatic radar plotting aid (ARPA), vessel monitoring systems (VMS), air-and space-borne SAR systems, ship-and land-based radars, air-and space-borne optical sensors, harbour-based visual surveillance.

During the last decade, technology has been moving towards integration of several information sources, of which one essential component is vision. Fusion of vision with other information sources allows more accurate and descriptive monitoring of coastal areas, maritime borders and offshore assets. Examples of this trend include systems that integrate AIS/VMS with SAR imagery [4, 5], radar- and visual-based surveillance for ports [6, 7], land-radars with visual information from air-borne platforms, and ship-based systems that integrate visual and other sensor's information [8, 9]. A concept [10, 11] helps to integrate information from radars/AIS with vision sensors placed on autonomous buoys. Fefilatyev et al. [12] presented a novel algorithm for on-board processing of image-and video-data obtained by a buoy-based maritime surveillance system equipped with a low-sitting forward-looking camera.

There have been a number of attempts to address the problem of detecting and tracking objects at sea. In [13], several machine vision techniques that could help in easing the search tasks in maritime environment are investigated. In [14], temporal characteristics of sea clutter and that of a range of small boats are analysed using a comprehensive set of recorded data sets. Vicen-Bueno et al. [15] proposed neural networks-based signal processing techniques to reduce sea clutter. Zhang et al. [16] developed an algorithm for ship detection and tracking in image sequences, capable of handling in situ waterway surveillance. Frost and Tapamo [17] have investigated the use of prior knowledge of a shipshape to assist level set segmentation in video tracking for a maritime surveillance problem. Kocak et al. [18] proposed a stereo vision system to help navigation officers by detecting other ships and obstacles, measuring their locations and tracking them. Kaido [19] described a novel method to detect and tracking ships using a machine vision to reduce marine accidents. Szpak and Tapamo [20] presented an approach that attempts to track objects using a closed curve in the image (a method known as level set segmentation) after they had been detected using a motion-based detection system. Wang et al. [21] presented novel hierarchical saliency filtering (HSF) scheme, by taking both object- and contour-level cues into account, for efficient false alarm removal and precise ship target detection. Li et al. [22] proposed an integrated algorithm, based on multiscale saliency and surrounding contrast analysis, of ship detection and tracking. There were a number of prior works studying the problem of automatic vessel detection [23–25], as well as the related problem of vessel-type classification [26]. Buck et al. [23] developed a ship detection system that utilized high-resolution satellite imagery. Fefilatyev et al. [24] developed a vessel detection system using buoy-mounted cameras. Felzenszwalb et al. [25] developed a fairly general object detection approach that can be trained to detect different categories of objects including vessels.

The number of marine accidents with ships is decreasing because the quality of these electronic devices has been improved. However, the number of collision accident, especially with small ships, is not decreasing. The major reason is insufficient lookout around the ship by a navigation officer. Because small ships are not obligated to install the AIS and sometimes cannot be discriminated by radar, if a small ship enters a blind spot from a large ship while an officer does not watch it, collision of both ships will happen. In the same manner, the presence of other numerous objects in the sea may cause hindrance in the surveillance of coastal areas. These objects may be small boat, ferry, dynamic background, whale, water foams, human flesh, etc. Such objects are not easily detected by any electronics device, and using manual patrolling system, it is very difficult and a tedious job to detect these objects. There is a huge gap in the research work to detect the objects in coastal area, as the main focus of researchers has been laid down only on the detection and tracking of ships. In our proposed work, we have focussed on the detection of different objects which might be present in sea. Once the method of detection of objects successfully simulated, the tracking of ships can significantly be implemented.

2 Proposed Methodology

The final interest of the entire work being done on the marine areas is to make the coastline area safe and secure. Majorly only ships detection and tracking have been considered in all research work. There may be numerous other objects in the sea other than ships which may interfere in the detection of ships. A wide variety of objects may be encountered, ranging from small buoys and watercraft, to large commercial shipping tankers, so algorithms must be able to handle a broad range of object profiles. Therefore, the proposed work is focused on to detect the various objects/obstacles in the coastline area using training data sets. In our work, we have considered the 2D images along with the stereo vision data set a gap in [19].

(a) Binarization of Image

For the proper analysis of different objects, there is always a need to transfer an image into another represented form (binary representation), which can better be well represented by morphological operators because it provides mathematical as well as quantitative description of geometric structure. The morphological function includes erosion and dilation, opening and closing, labelling connected components, segmentation and finally the reconstruction of a binary image $B(x, y)$ to bounding the rectangle of the required region.

(b) Deep Learning with Training Data Set

The tracking of the ships can efficiently be done if the detection of different objects/obstructer in the sea finds out. For that reason, we have trained the algorithm using bag of features to extract the features of object. The feature samples are generated from the input image, and the bag of features returns a bag of features that use a custom feature extractor function to extract features.

The object boundaries of an image can be identified using edge detection algorithm.

1. **Edge Detection Algorithm**

The edge detection algorithms may be: Sobel, Prewitt, Roberts, Canny, Laplacian of Gaussian methods. The Laplacian method used in [19] may be replaced by Canny method as the later one can detect true weak edges without being fooled by "noise". The Canny filter works in the sequence of steps:

(a) First of all Gaussian filter is used to smooth the image in order to remove the noise.
(b) Calculating the intensity gradient of the image.
(c) Determining potential edges.
(d) Suppressing the weak edges which are not connected with the strong edges to finalize the detection of edges.

2. **Support Vector Machine (SVM)**

Support vector machine (SVM) is a machine learning algorithm which is used for classification and regression challenges. It is the data set of coordinates of individual observations. In the proposed work, support vector machine segregates the data set into two categories, "object or obstacle and sea or sky". Defining the data category helps in distinguishing the desired detection from the false detection. The set of frames are several times repeated to form the final candidate.

Algorithm Description:

(a) Defining RGB image $I(x, y)$ of an object
(b) Calculation of Gaussian filter coefficients
(c) Convolution of [image $I(x, y)$, Gaussian filter matrix]
(d) Applying edge detection.

In our present work, threshold values T_High = 0.50 and T_Low = 0.200 are taken into consideration and the Gaussian matrix obtained is:

$$\begin{bmatrix} 0.0126 & 0.0252 & 0.0314 & 0.0252 & 0.0126 \\ 0.0252 & 0.0566 & 0.0755 & 0.0566 & 0.0252 \\ 0.0314 & 0.0755 & 0.0943 & 0.0755 & 0.0314 \\ 0.0252 & 0.0566 & 0.0755 & 0.0566 & 0.0252 \\ 0.0126 & 0.0252 & 0.0314 & 0.0252 & 0.0126 \end{bmatrix}$$

(e) Features extraction of the object
(f) Training of the algorithm
(g) Matching with the preset database
(h) Repeat the steps (i–vii) if not matched

(i) Final output (Fig. 1).

The original images $I(x, y)$ of two different ships are shown in Figs. 2a and 3a. The images are then applied with the morphological operators to convert RGB image in binary form. The binary image $B(x, y)$ is then passed through Canny filter, the output

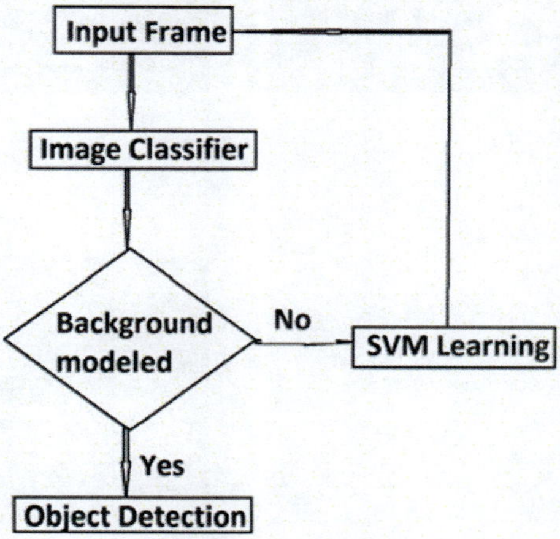

Fig. 1 Block diagram of object detection

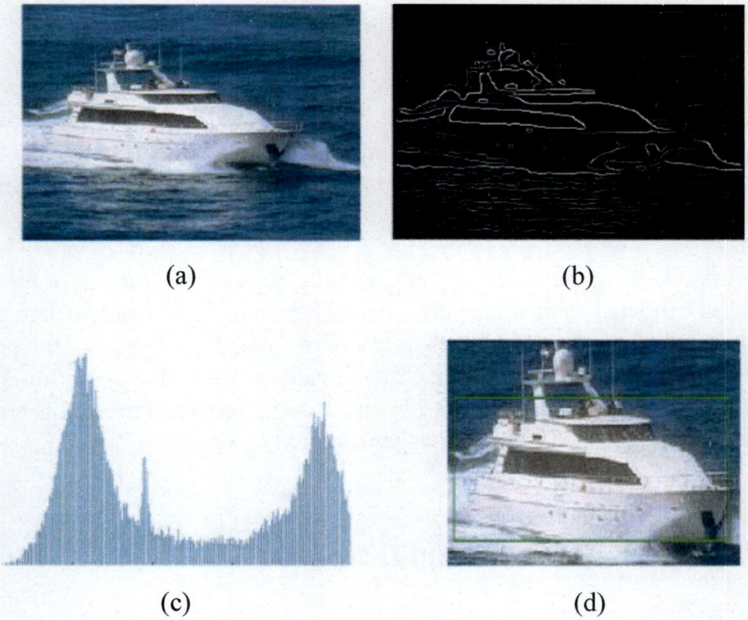

Fig. 2 **a** Original image, **b** Output of Canny filter, **c** Histogram of image, **d** Bounding rectangle of region

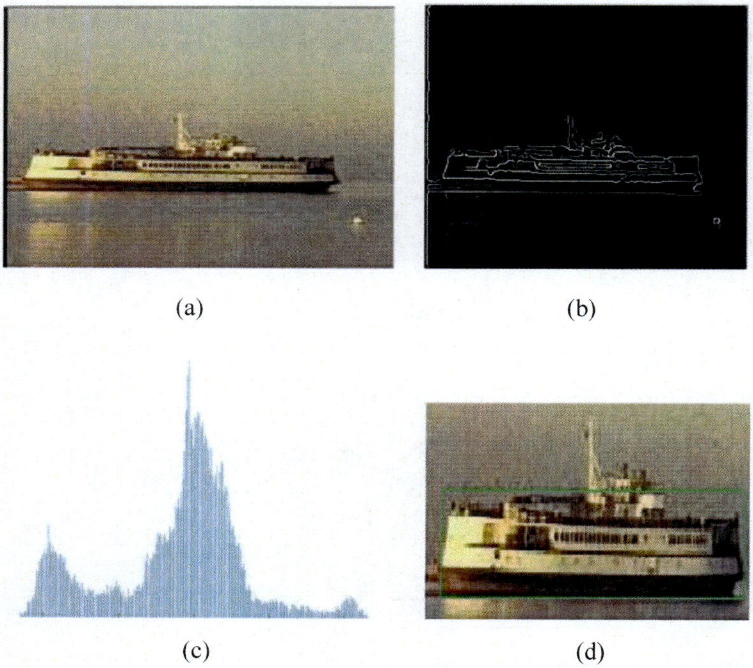

Fig. 3 **a** Original image, **b** Output of Canny filter, **c** Histogram of image, **d** Bounding rectangle of region

of which are shown in Figs. 2b and 3b. The Canny filter enhances the results by connecting the weak edges of the object. Figures 2c and 3c represent the histogram representation of images. The idea behind using the histogram is to prevent the over enhancement of the images which might have caused the overlapping of the unwanted region. Figures 2d and 3d represent the final output with the bounding region of the objects.

Before obtaining the bounding region of the objects, the features of the images have been extracted using image classifier as shown in Figs. 4 and 5. Here we have shown features of only two objects used in Figs. 2 and 3; otherwise, the proposed algorithm is supported with more than 2500 images data sets. The algorithm is trained with the extracted features of different objects. The proposed method is simulated on the data set collected from the images captured by on-board camera.

3 Experimental Results and Discussion

The data set for coastline areas are not easily available as most of the work is designed for the defence purpose only. For our work, we have gathered the data set from open-source readily available for the research purpose. The data set used as training

Fig. 4 Features extracted of image Fig. 2a

Fig. 5 Features extracted of image Fig. 3a

images includes images of various different categories. The images captured resized to 300 [pix] × 205 [pix]. Figure 6 shows images of different objects used for training purpose and the features of objects obtained after classifier. We have considered over 2500 images in the set of 16 different categories. But the algorithm is not limited to this number only; depending on the CPU utilization the number of images can be increased. The proposed work is simulated on the workstation with system configuration (Table 1).

Figure 7 shows the actual outcome of the method which clearly shows the objects and its classification with accuracy.

The accuracy is given by:

$$\textbf{Accuracy} = \textbf{mean}(\textbf{diag}(\textbf{confMatrix}))$$

where confMatrix is a confusion matrix which is a good initial indicator of how well the classifier is performing.

In our work, the accuracy comes out to be 0.6757 with an average computing time of 1.89 min. We compare the proposed algorithm with Koch et al. [27], Zhang et al. [28, 29] (Table 2).

Fig. 6 Data set of different Images for the training purpose

Table 1 Workstation configuration

Processor	i7
RAM	8 GB
Hard disk	1 TB
Software	MATLAB R2018a, Open CV 2.4.8
Tool box	Computer vision, image processing

4 Conclusion and Future Work

In this paper, we presented work on the detection of different objects in the coastline areas. The proposed algorithm is processed with above 2500 frames of images, and the computing confusion matrix gives enhanced results. The detection of objects plays a vital role in the coastline surveillance system. The proposed method is capable to handle large data set to detect and classify the objects. The futuristic work may be considered on the automation of tracking of objects and automation of analysing video frames.

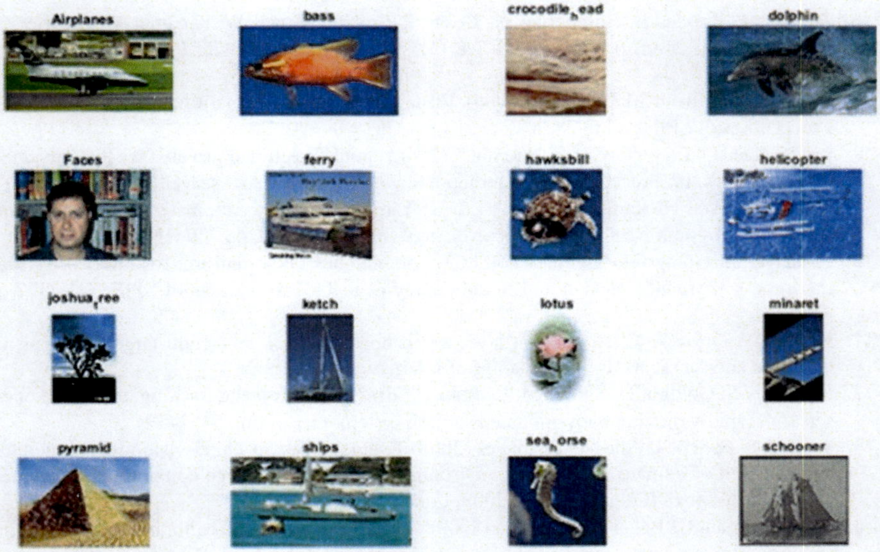

Fig. 7 Detection of objects with classification

Table 2 Processing parameters per set

Parameters	Koch's [27]	Zang [28]	Zhang [29]	Proposed
Time (min)	1356	58	39	1.89
Accuracy	0.2345	0.4569	0.5478	0.6757

References

1. Robert-Inacio F, Raybaud A, Clement E (2007) Multispectral target detection and tracking for seaport video surveillance. In: Proceedings of image and vision computing New Zealand, pp 169–174
2. Fefilatyev S, Goldgof D (2008) Detection and tracking of marine vehicles in video. In: International conference on pattern recognition (ICPR)
3. Rodriguez Sullivan MD, Shah M (2008) Visual surveillance in maritime port facilities. In: Proceedings of the SPIE 2008, vol 6978, pp 697811–697818
4. Lemoine G, Schwartz-Juste G, Kourti N, Shepherd I, Cesena C (2005) An open source framework for integration of vessel positions detected in space borne SAR imagery in operational fisheries monitoring and control. In: Proceedings of ENVISAT/ERS symposium, vol 572, pp 252–258
5. Saur G, Estable S, Zielinski K, Knabe S, Teutsch M, Gabel M (2011) Detection and classification of man-made off shore objects in terrasar-xand rapid eye imagery: selected results of the demarine-dekoproject. In: Proceedings of IEEE-Spain OCEANS, pp 1–10

6. Seibert M, Rhodes B, Bomberger N, Beane P, Sroka J, Kogel W, Kreamer W, Stauffer C, Kirschner L, Chalom E et al (2006) See Coast port surveillance. In: Proceedings of SPIE, 62040B-1
7. Rodriguez Sullivan M, Shah M (2008) Visual surveillance in maritime port facilities. In: Proceedings of SPIE, vol 6978, p 29
8. Liu H, Javed O, Taylor G, Cao X, Haering N (2008) Omni-directional surveillance for unmanned water vehicles. In: Proceedings of international workshop on visual surveillance
9. Wei H, Nguyen H, Ramu P, Raju C, Liu X, Yadegar J (2009) Automated intelligent video surveillance system for ships. In Proceedings of SPIE, vol 7306, pp 73061N
10. Fefilatyev S, Goldgof D, Lembke C (2009) Autonomous buoy platform for low- cost visual maritime surveillance: design and initial deployment. In: Proceedings of SPIE, vol 7317, p 73170A
11. Kruger W, Orlov Z (2010) Robust layer-based boat detection and multi-target- tracking in maritime environments. In: Proceedings of international waterside
12. Fefilatyev S, Goldgof D, Shreve M, Lembke C (2012) Detection and tracking of ships in open sea with rapidly moving buoy-mounted camera system. Ocean Eng 54, 1–12
13. Westall P, Ford J, O'Shea P, Hrabar S (2008) Evaluation of machine vision techniques for aerial search of humans in maritime environments. In: Digital image computing: techniques and applications (DICTA), 1–3 Dec 2008, Canberra, pp 176–183
14. PL Herselman, CJ Baker, de HJ Wind (2008) An analysis of X-band calibrated sea clutter and small boat reflectivity at medium-to-low grazing angles. Int J Navig Obs. https://doi.org/10.1155/2008/347518
15. Vicen-Bueno R, Carrasco-Alvarez R, Rosa-Zurera M, Nieto-Borge JC (2009) Sea clutter reduction and target enhancement by neural networks in a marine radar system. Sensors 2009(9):1913–1936
16. Zhang S, Qi Z, Zhang D (2009) Ship tracking using background subtraction and inter-frame correlation. IEEE, pp 1–4
17. Frost D, Tapamo J-R (2013) Detection and tracking of moving objects in a maritime environment using level set with shape priors. J Image Video Process 2013(42):1–16
18. Kocak G, Yamamoto S, Hashimoto T (2013) Detection and tracking of ships using a stereo vision system. Sci Res Essays 8(7):288–303
19. Kaido N, Yamamoto S, Hashimoto T (2016) Examination of automatic detection and tracking of ships on camera image in marine environment. IEEE, pp 58–63
20. Szapk ZL, Tapamo JR (2011) Maritime surveillance: tracking ships inside a dynamic background using a fast level-set. Expert Syst Appl 38(6):6668–6680
21. Wang S, Wang M, Yang S, Jiao L (2016) New hierarchical saliency filtering for fast ship detection in high-resolution SAR images. IEEE Trans Geosci Remote Sens, 1–12
22. Li H, Chen L, Li F, Huang M (2019) Ship detection and tracking method for satellite video based on multiscale saliency and surrounding contrast analysis. J Appl Remote Sens 13(2):026511
23. Buck H, Sharghi E, Guilas C, Stastny J, Morgart W, Schalcosky B, Pifko K (2008) Enhanced ship detection from overhead imagery. In Proceedings of SPIE, vol 6945, pp 69450U–69450U-12
24. Fefilatyev S, Goldof D, Lembke C (2010) Tracking ships from fast moving camera through image registration. In: Proceedings of international conference of pattern recognition, pp 3500–3503
25. Felzenszwalb P, Girshick R, McAllester D, Ramaman D (2010) Object detection with discriminatively trained part-based models. IEEE Trans Pattern Anal Mach Intell 32:1627–1645
26. Rainey K, Stastny J (2011) Object recognition in ocean imagery using feature selection and compressive sensing. IEEE applied imagery pattern recognition workshop, pp 1–6
27. Itti L, Koch C, Niebur E (1998) A model of saliency-based visual attention for rapid scene analysis. IEEE Trans Pattern Anal Mach Intell 20(11):1254–1259

28. Hou X, Zhang L (2007) Saliency detection: a spectral residual approach. In Proceedings of IEEE conference on computer vision and pattern recognition
29. Zhang Y, Zang F (2017) Ship detection for visual maritime surveillance from non-stationary platforms. J Elsevier Ocean Eng 141:53–63

Passive Authentication Image Forgery Detection Using Multilayer CNN

Sakshi Singhal and Virender Ranga

Abstract With the development of various image editing tools and techniques, the forgery has become a common aspect in the image domain, nowadays. We can now insert, delete, or transform a small part of an image very easily. We can also copy-paste an image into some other image or the same image frequently. The image forgery mainly focuses on passive-based approach, because it does not involve the addition of any information into the image unlike active-based approach. Our proposed method mainly focuses on copy-move forgery and splicing forgery detection which are types of passive authentication techniques. The proposed method uses four convolution and pooling layers in a succession with different parameters. A filter is applied on each convolution layer, and all the layers are initialized with some weights and a bias. Using a hidden layer after four convolutions and pooling layer and finally using output layer for decision making with a result of 86.4% accuracy, precision of 81%, recall of 79%, and F-measure of 84% as compared with state-of-the-art approaches.

Keywords Image forgery · Copy-move forgery · Splicing · Multilayer CNN · CASIA v1.0

1 Introduction

Due to the evolution of generation and technology, capturing moments through images and videos with the help of high-resolution cameras have become a crucial part of our life. Moreover, the images can be enhanced or manipulated using various softwares available today such as CorelDRAW [1], ACDSee Photo Editor [2], Adobe Photoshop [3] as required. The manipulation is done to gain political gain, to defame

S. Singhal · V. Ranga (✉)
Department of Computer Engineering, National Institute of Technology Kurukshetra, Kurukshetra, Haryana, India
e-mail: virender.ranga@nitkkr.ac.in

S. Singhal
e-mail: sakshisinghalmtech@gmail.com

N. Marriwala et al. (eds.), *Mobile Radio Communications and 5G Networks*, Lecture Notes in Networks and Systems 140,
https://doi.org/10.1007/978-981-15-7130-5_18

a person, to make false interpretation; in the medical healthcare, we can say that this can lead to the wrong diagnosis by enlarging the area of cancer [4]. The manipulation in the image can be referred to as integrity or authentication. Integrity refers to the change in the content of information [5]. Since one single image is a good carrier of information, tampering with such information can lead to various issues such as it can lead to misdiagnosis when the X-rays images are changed, can create violence when the social media images are tampered, can defame a person when the face of one person is changed with other [5, 6]. With the rise of such issues, there is a need to develop a highly efficient mechanism that can detect such tampering.

1.1 Types of Image Forgery

The integrity techniques are classified as active image integrity technique and passive image integrity technique [1–7]. The techniques are further divided based on the source image requirement as shown in Fig. 1. In active image integrity technique, source image is required to detect the forgery. In passive image integrity technique, requirement of the source image is not needed.

Active Approach. In active image integrity technique, some data is added to the image at the user's end which is extracted at the recipient's end to check the integrity of the data [4]. Active image integrity technique is further classified into digital watermarking [6–9] and digital signature [10–12].

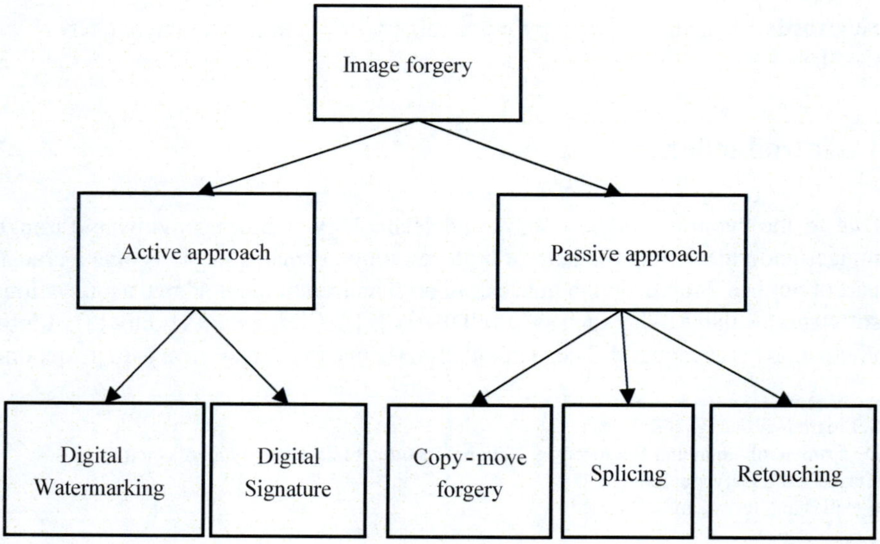

Fig. 1 Forgery techniques

Fig. 2 Copy-move image forgery example [7]. a The fake image, b The original image

Digital Watermarking. In digital watermarking, the watermark is added at the capturing end and this watermark will be used for forgery detection; later, the watermark is extracted from the source image at the receiver's end and if the watermark is found changed, then it can be detected that the forgery has taken place [13, 14].

Digital Signature. In case of digital signature, some algorithm is applied at the sender's end to extract the features from the image, and then the same algorithm is applied at the receiver' end, and extracted features are matched with that of sender's to check the integrity of the image [15].

Passive Approach. In passive image integrity technique, the copied region of the image is pasted into the same or some other image [16].

It is further classified as copy-move forgery, splicing, and retouching [17].

Copy-Move Forgery. As the name signifies in copy-move forgery, the copied part of the image is pasted into the same image [18]. The motive behind such a forgery hides some content behind an image or to highlight a portion of an image [19]. The components of the image being pasted remain the same such as texture, background color, noise.[20]. Hence, it becomes difficult to examine from the unaided eyes as shown in Fig. 2.

Splicing. In splicing, a part of the image is copied and pasted from one image to another image [21]. Due to which there is a change in the texture and background color composition. Hence, the detection of such forgery can be done by simply checking the components variation such as color palette, texture, and dynamic range [22]. The intuition behind such a forgery is to defame a person or to create a false impact. To make the forgery undetectable, the copied image undergoes geometric transformation such as scaling, rotation to fit the best into the image than to make the boundary of the image undetectable the post-processing operations are done [23].

Retouching. In case of retouching, the image is not copied and pasted rather the features such as brightness and color of the image are either improved or degraded to make it more pleasing as shown in Fig. 3. This is a popular technique used in films, newspapers. This type of passive-based image forgery does not have a major impact on the image. But it involves the manipulation with the image [24].

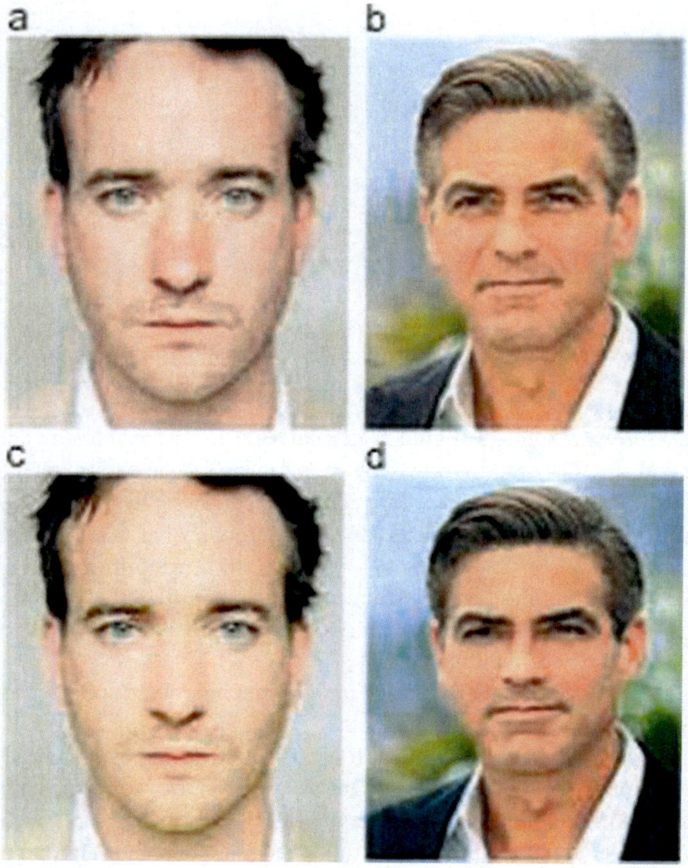

Fig. 3 Retouching examples [then (top) and after retouching (bottom)] [1]

1.2 Passive Image Forgery Attacks

Passive image forgery attacks are done either before or after the image which is copied and pasted to some other location [25]. It is classified into two categories: intermediate operations and post-processing operations.

Intermediate operations. Intermediate operations are done to transform the copied part before pasting it into the image. It is done so that the copied image should fit properly at some other place either in the same image or in the different image [26]. The various intermediate operations performed are:

- *Rotation*—changing the angle.
- *Scaling*—enhancing the image either by enlarging or reducing the size.
- *Reflection*—changing the image by replacing it with its reflection.
- *Chrominance Changing*—changing the blue or red component in the image.

- *Luminance Changing*—changing the contrast.

Post-processing operations. Post-processing operations are done after the copied image which is pasted at some other location [7]. This is done to hide any tampering clues. These operations can be:

- Compressing into JPEG
- Making blur
- Adding noise.

Further, paper is arranged as follows. Section 2 deals with the related work. Section 3 explains the proposed mechanism. Section 4 contains performance measures and results. Section 5 concludes the paper.

2 Related Work

There are various approaches implemented for detecting the forged images. Some of them are listed below.

In [27], the author used CASIA dataset (publically available dataset) to verify their approach. They first converted the image into YCrCb color component where Y stands for luminance channel and CrCb are the chrominance components where Cb and Cr are blue and red difference, respectively. The Cr and Cb components can detect the edge deformity in the image and shows good results in the literature to identify forged images that is why author has concentrated its work over the Cr and Cb components. Later Muhammad et al. [27], applied the SPT which is a multi-resolution technique consisting of various sub-bands on different scales. The zero scale contains high-pass and low-pass filters. In the first scale, the low-pass filter was again divided into four sub-bands and a low-pass sub-band has a frequency equal to half of the previous sub-band, and this is done until a required number of scales have been gain. Further, the local binary pattern (LBP) was applied for texture description in all the sub-bands, and the result from each LBP was stored as a feature vector, and then LLB and L_0-norm were applied for feature selection, and finally, the decision of the image to be forged or not is taken according to the support vector machine (SVM). The author does not use RGB color components because it does include relationship between the channels. The problem with this method is that the authors had used various sub-bands which are not feasible to implement in the real-time scenario.

In [28], the authors converted the image into the RGB components after that they applied Wiener filter for the noise removal and then they subtracted filtered image from the original image in order to get the noise image. Then the multi-resolution regression filter was applied over the estimated noise pattern to estimate the intensity of the central pixel. Then the normalization was done, and this input is feed into the SVM (support vector machine—which is use for binary classification) and ELM. ELM is a feedforward network with single hidden layer. The score of both the algorithm was sent as input to the BSR and the final decision of the image to be forged or not was measured using BSR. The proposed architecture was applied successfully over the smart healthcare system. The authors in this paper converted the image into RGB color components instead it could be converted into grayscale for lesser features or YCrCb component for edge deformity detection.

Wei et al. [29] devised a two-level CNN approach for detecting the deformity in the edges present in the forged image. To extract the difference between spliced regions and the image itself, they used two CNN layers: the refined convolutional neural network and the coarse convolutional neural network. They proposed method work on different scales by checking various intrinsic properties of the original image and the spliced image. Wei et al. [29] verified the effectiveness of their work on CASIA database which results in 0.70 weighted average of precision, 0.72 weighted average of recall, and 0.68 weighted average of F-measure.

3 Proposed Model

3.1 *Architecture of Model*

We used Keras library for our experiment which is powerful than TensorFlow. It runs over TensorFlow means it uses TensorFlow in the background. The basic flow of code in Keras is shown in Fig. 4.

3.2 *Preprocessing*

All the images present in CASIA v1.0 are of the same size but some are of 384 $\times$ 256 size while others are of 256 $\times$ 384 size so we reshape them into the same size before passing it through the convolution layer.

3.3 *Convolution Layer*

For our experiment, we used sequential model. Four layers of convolutional neural network were used in our approach each followed by one pooling layer. We used 64 units or neurons in the first convolutional layer with filter of depth 3 for RGB images and of size 3 $\times$ 3 with activation function as rectified linear unit (ReLU) which decides whether to fire a neuron or not. It fires a neuron if it crosses predefined threshold that is the function outputs the input if its value exceeds zero; otherwise, it outputs zero only. Let us consider i be the input and $f(i)$ be the function.

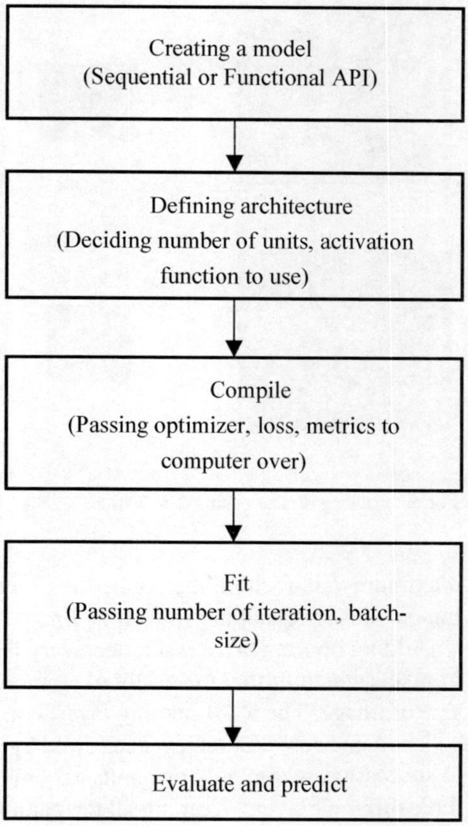

Fig. 4 General flow of Keras model

$$f(i) = \begin{Bmatrix} i, & i \geq 0 \\ 0, & i < 0 \end{Bmatrix} \tag{1}$$

The bias and the weights were chosen randomly at first afterward neural network will learn itself by applying multiple iterations doing forward propagation and backward propagation. Each unit in the convolution layer has unique weight and bias, and every unit performs weighted addition between pixels of an image and weights applied and adds bias to it.

The stride which signifies the number by which we want to shift our filter is set to 1 and padding this set to be the same.

3.4 Polling Layer

After convolution layer, we added pooling layer with pool size to be 2×2 and type to be max pooling. So the image will be traversed in 2×2 dimensional box and the

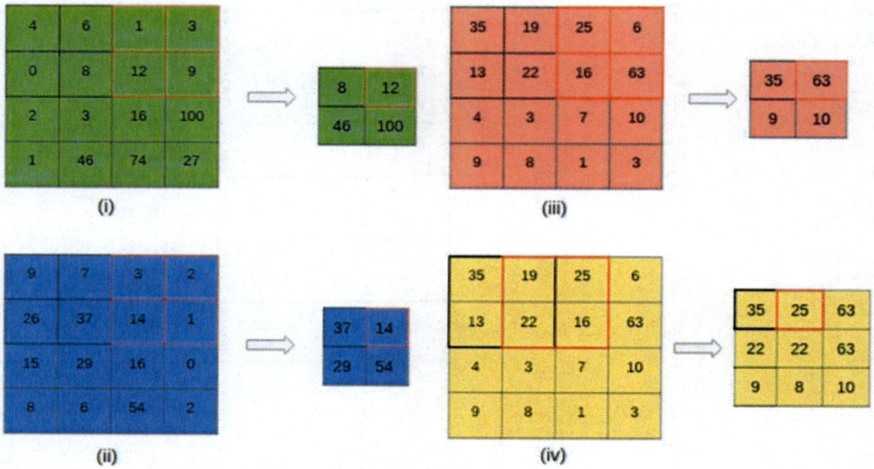

Fig. 5 Pooling process using 2 × 2 filter

maximum value within the applied region of an image will be stored in the output image, thus reducing the size of the image.

Adding pooling layer is not necessary but we had lot of images which means lot of pixels which further means lot of features. So we used pooling layer to reduce the size of image. The RGB pooling layer with pool size 2 × 2 is shown in Fig. 5.

The same convolution layer followed by pooling layer is applied three more times in succession with number of units at convolution layer to be 128, 256, and 512 for three different layers. Keeping all the parameters same for convolution layer as well as pooling layer.

3.5 Dense Layer

Before we pass our output from the above used four layers, we flattened the image then passing it through hidden layer consisting of 2048 units. Hidden layers are the intermediate layers, we add between input and output layer. We added single hidden layer for network with randomly chosen weights and bias. The output from hidden layer gets inputted into output layer. Since classifying whether the image is forged or not is a binary classification problem. So we used just a single unit at output layer with sigmoid as activation function. Sigmoid function return values between 0 and 1 and is s-shaped curved.

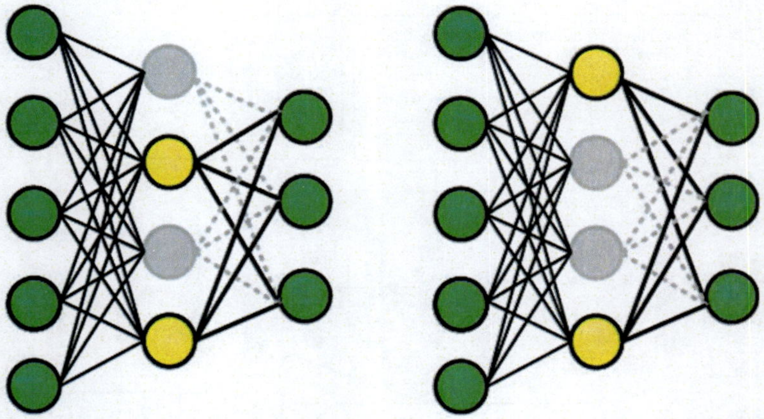

Fig. 6 Dropping random neurons output

3.6 Dropout Layer

To avoid overfitting, we added one dropout layer between hidden layer and output layer. Dropout layer will simply drop the output coming from some of the hidden layer neurons. This is done to avoid memorizing the complete training data and hence will efficiently handle overfitting (Fig. 6).

3.7 Completing the Network

While compiling our network, we passed optimizer as adam and loss to be the binary_crossentropy because our classification problem is binary in nature then used fit function to fit our model with the batch size of 32 which update the weights in the batch of 32 without waiting for complete input. Finally predicting the model over test set to verify the accuracy.

The proposed architecture is shown in Fig. 7.

3.8 Database Specification

In our experiment, we have used one publicly available database containing images out of which some are forged and some are authenticated. The forged images in the dataset are either copy-move forged or spliced images and have undergone geometric operations like scaling, rotation, chrominance change, reflection to make the edges of the image undetectable by naked eyes. All the images in the database are colored in nature. The database used is CASIA v1.0. CASIA v1.0 database contains more than

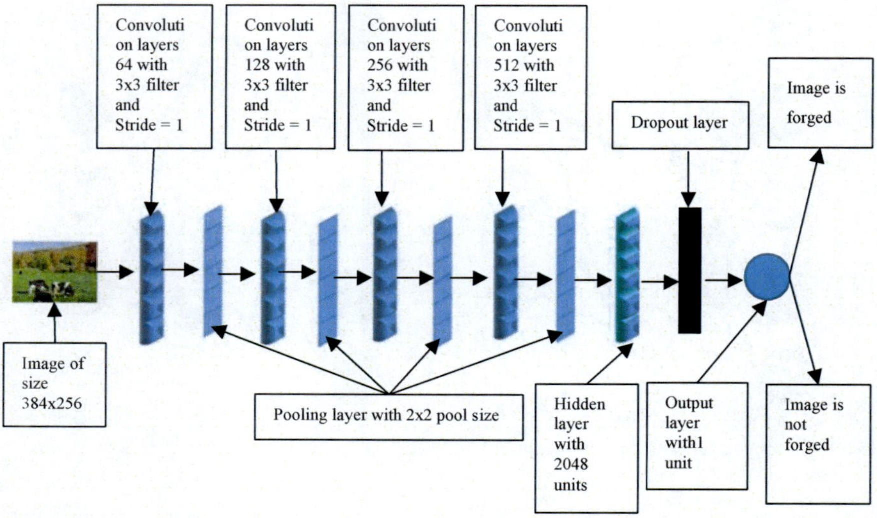

Fig. 7 Proposed architecture

900 forged color images and 800 authentic images out of which 459 are forged as copy-move, and the remaining are spliced. Most of the forged images have undergone various geometric operations to hide the tampered region. The images are in JPEG format and are of size either 384 × 256 pixels or 256 × 384.

4 Results

We verified the effectiveness of our model by comparing the performance of it with some other combinations of machine learning algorithms with or without filter. Filter is used to extracts out the noise from the image. The comparison is made between the classification reports of our proposed method with the existing techniques also and found that precision comes out 0.81, with recall value 0.79, and F-measure is shown 0.84 (Fig. 8). The values represent the weighted average of the precision, recall, and F-measure, respectively. We found that our proposed architecture performs better than other proposed schemes.

We calculate precision, recall, and F-measure by the following formulas:

$$\text{Precision}(P) = \frac{\text{TP}}{\text{TP} + \text{FP}} \tag{2}$$

$$\text{Recall}(R) = \frac{\text{TP}}{\text{TP} + \text{FN}} \tag{3}$$

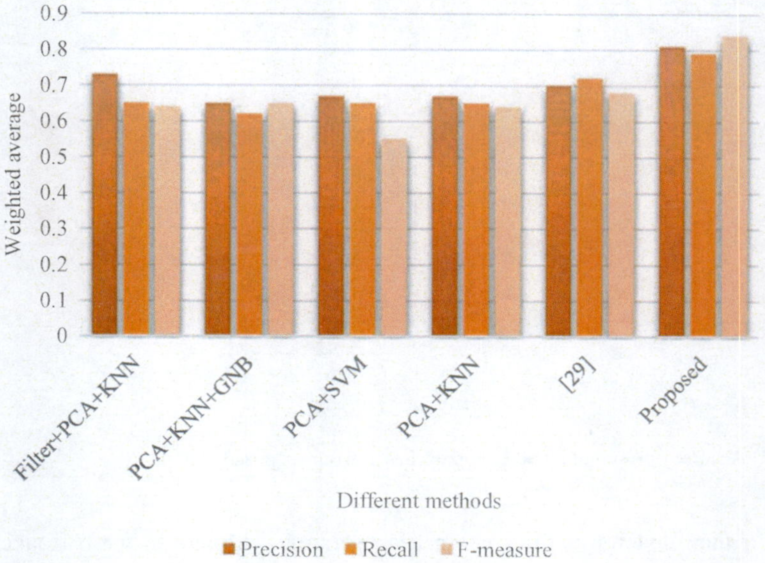

Fig. 8 Weighted average of the precision, recall, and F-measure

$$\text{F-measure} = \frac{2 * P * R}{P + R} \quad (4)$$

where TP is True Positive, FP is False Positive, and FN is False Negative.

F-measure is also known as F1-score which is the harmonic mean of precision and recall (Fig. 8).

We have also compared the accuracy of our model with proposed models, shown in [27, 28] and found that our model is producing 86.4% accuracy even when the image undergoes various geometric operations (Fig. 9).

The above-mentioned results are tested over CASIA v1.0 dataset which contains authentic as well-forged images. In both the figures, we can observe that our model is performing better.

5 Conclusion

This paper presents a convolutional neural network (CNN) approach to detect the forged images. The images can be either copy-move forged or spliced.

For the effectiveness of our method, we used publically available database CASIA v1.0. We compared our results with the previous proposed approaches used in [27, 28]. We have also tried and verified different combinations of machine learning algorithms and found that CNN approach works well since it is an image classifier. Our proposed method works well in detecting copy-move forged and spliced images

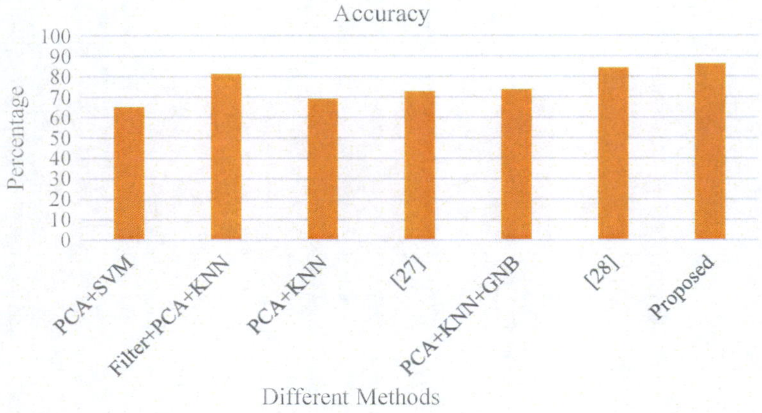

Fig. 9 Accuracy proposed by other approaches and our proposed method

even if they undergo various geometric operations. Moreover, we had taken care of overfitting by adding one dropout layer between hidden layer and output layer. Still much more work can be done in detecting the more efficient algorithm that can detect the forged images with higher accuracy or to develop a robust technique to detect forged images that will use for real-time applications, after detecting the forged images we the work can be done to localize the forged area.

References

1. Qureshi MA, Deriche M (2015) A bibliography of pixel-based blind image forgery detection techniques. Sig Process Image Commun 39:46–74
2. ACDSee. Accessed 2 Feb 2020. [Online]. Available https://www.acdsee.com/
3. Photoshop. Accessed 2 Feb 2 2020. [Online]. Available https://www.photoshop.com/
4. Lin X, Li JH, Wang SL, Cheng F, Huang XS (2018) Recent advances in passive digital image security forensics: a brief review. Engineering 4(1):29–39
5. Zheng L, Zhang Y, Thing VL (2019) A survey on image tampering and its detection in real-world photos. J Vis Commun Image Represent 58:380–399
6. Wang C, Zhang H, Zhou X (2018) A self-recovery fragile image watermarking with variable watermark capacity. Appl Sci 8(4):548
7. Al-Qershi OM, Khoo BE (2013) Passive detection of copy-move forgery in digital images: state-of-the-art. Forensic Sci Int 231(1–3):284–295
8. Zear A, Singh AK, Kumar P (2018) A proposed secure multiple watermarking technique based on DWT, DCT and SVD for application in medicine. Multimedia Tools Appl 77(4):4863–4882
9. Shehab A, Elhoseny M, Muhammad K, Sangaiah AK, Yang P, Huang H, Hou G (2018) Secure and robust fragile watermarking scheme for medical images. IEEE Access 6:10269–10278
10. Wang X, Xue J, Zheng Z, Liu Z, Li N (2012) Image forensic signature for content authenticity analysis. J Vis Commun Image Represent 23(5):782–797
11. Okawa M (2018a) Synergy of foreground–background images for feature extraction: offline signature verification using Fisher vector with fused KAZE features. Pattern Recogn 79:480–489

12. Okawa M (2018b) From BoVW to VLAD with KAZE features: offline signature verification considering cognitive processes of forensic experts. Pattern Recogn Lett 113:75–82
13. Kundur D, Hatzinakos D (1999) Digital watermarking for telltale tamper proofing and authentication. Proc IEEE 87(7):1167–1180
14. Singh N, Jain M, Sharma S (2013) A survey of digital watermarking techniques. Int J Mod Commun Technol Res 1(6):265852
15. Singh M, Kaur H, Kakkar A (2015) Digital signature verification scheme for image authentication. In: 2015 2nd international conference on recent advances in engineering & computational sciences (RAECS). IEEE, pp 1–5
16. Wang C, Zhang Z, Li Q, Zhou X (2019) An image copy-move forgery detection method based on SURF and PCET. IEEE Access 7:170032–170047
17. Mushtaq S, Mir AH (2014) Digital image forgeries and passive image authentication techniques: a survey. Int J Adv Sci Technol 73:15–32
18. Kuznetsov A, Myasnikov V (2017) A new copy-move forgery detection algorithm using image preprocessing procedure. Procedia Eng 201:436–444
19. Khudhur MH, Waleed J, Hatem H, Abduldaim AM, Abdullah DA (2018) An efficient and fast digital image copy-move forensic technique. In: 2018 2nd international conference for engineering, technology and sciences of Al-Kitab (ICETS). IEEE, pp 78–82
20. Shahroudnejad A, Rahmati M (2016) Copy-move forgery detection in digital images using affine-SIFT. In: 2016 2nd international conference of signal processing and intelligent systems (ICSPIS). IEEE, pp 1–5
21. Alahmadi AA, Hussain M, Aboalsamh H, Muhammad G, Bebis G (2013) Splicing image forgery detection based on DCT and local binary pattern. In: 2013 IEEE global conference on signal and information processing. IEEE, pp 253–256
22. Gupta CS (2016) A review on splicing image forgery detection techniques. IRACST-Int J Comput Sci Inf Technol Secur 6(2)
23. Patil B, Chapaneri S, Jayaswal D (2017) Improved image splicing forgery localization with first digits and Markov model features. In 2017 IEEE international conference on intelligent techniques in control, optimization and signal processing (INCOS). IEEE, pp 1–5
24. Shah H, Shinde P, Kukreja J (2013) Retouching detection and steganalysis. Int J Eng Innov Res 2(6):487
25. Ardizzone E, Bruno A, Mazzola G (2010) Copy-move forgery detection via texture description. In: Proceedings of the 2nd ACM workshop on multimedia in forensics, security and intelligence, pp 59–64
26. Ryu SJ, Lee MJ, Lee HK (2010) Detection of copy-rotate-move forgery using Zernike moments. In: International workshop on information hiding. Springer, Berlin, Heidelberg, pp 51–65
27. Muhammad G, Al-Hammadi MH, Hussain M, Bebis G (2014) Image forgery detection using steerable pyramid transform and local binary pattern. Mach Vis Appl 25(4):985–995
28. Ghoneim A, Muhammad G, Amin SU, Gupta B (2018) Medical image forgery detection for smart healthcare. IEEE Commun Mag 56(4):33–37
29. Wei Y, Bi X, Xiao B (2018) C2r net: the coarse to refined network for image forgery detection. In: 2018 17th IEEE international conference on trust, security and privacy in computing and communications/12th IEEE International conference on big data science and engineering (TrustCom/BigDataSE). IEEE, pp 1656–1659

Contiki Cooja Security Solution (CCSS) with IPv6 Routing Protocol for Low-Power and Lossy Networks (RPL) in Internet of Things Applications

Arun Kumar Rana and Sharad Sharma

Abstract Everyone is aware that existing approaches to IoT networks have nothing to do with certain security issues that may expose sensitive data to unauthorised user. Today, low-power and lossy networks (LLNs) speak as one of the most fascinating examination regions. They incorporate wireless personal area networks (WPANs), low-power line communication (PLC) systems and wireless sensor networks (WSNs). Such systems are frequently improved to spare vitality, bolster track designs not the same as the standard unicast correspondence, run steering conventions over connection layers with limited edge sizes and numerous others. The IoT is a quickly developing innovation. In IoT, the gadgets (device) are associated through the Internet and controlled from any remote territories. Before the approach of IoT, the association between the clients was distinctly through the Web. By 2020, there will be 75.4 billion gadgets interconnected through the Web. In IoT, we have routing protocol for low-power and lossy networks (RPL). RPL is a lightweight convention that has decent directing usefulness and mindful setting, and it underpins dynamic topology, having fundamental security usefulness. This paper explores the power-efficient and secure ipv6 routing protocol RPL and proposed the Caesar Cipher hash algorithm for the trust privacy and security of the information of IoT node with power full tool CONTIKI COOJA SIMULATOR security solution (CCSS).

Keywords Internet of things · Low-power and lossy network · RPL · IPv6 · Contiki Cooja

A. K. Rana (✉) · S. Sharma
Maharishi Markandeshwar (Deemed to be University), Mullana, India
e-mail: ranaarun1.ece@piet.co.in

S. Sharma
e-mail: sharadpr123@rediffmail.com

N. Marriwala et al. (eds.), *Mobile Radio Communications and 5G Networks*, Lecture Notes in Networks and Systems 140,
https://doi.org/10.1007/978-981-15-7130-5_19

1 Introduction

The Internet of Things (IoT) is anticipated hugely which affects the life of people. The report shows the enormous increment in the number of gadgets per individual, for example, in excess of eight gadgets for every individual by and large constantly 2020. This mostly gave to a value decrease of the gadgets which makes them progressively open for everybody. IoT is a multilayer design, partition into the four-layer as an observation layer, organizing layer, administration layer and application layer [1]. IoT is an overall heterogeneous system comprising of interconnected items and has a one of a kind location dependent on the standard correspondence conventions. In IoT, 'Internet' is an overall system of interconnected PC systems dependent on the (TCP/IP) correspondence conventions and 'Thing' is an item or any IoT gadget. IoT permits people to be associated whenever with the remote gadgets. A gadget can be associated with different gadgets utilizing any way/organization or by any assistance. Different sorts of correspondence can be used in IoT if the correspondence procedure transmits the data between the heterogeneous gadgets through heterogeneous systems. Distinctive directing conventions are required in IoT for gadget-to-gadget correspondence; however, we require versatile steering conventions in various situations to discover discretionary courses. RPL [1] is one of the institutionalized steering conventions in IoT systems. Right now, investigate the RPL convention by examining its security for various assaults in IoT (Fig. 1).

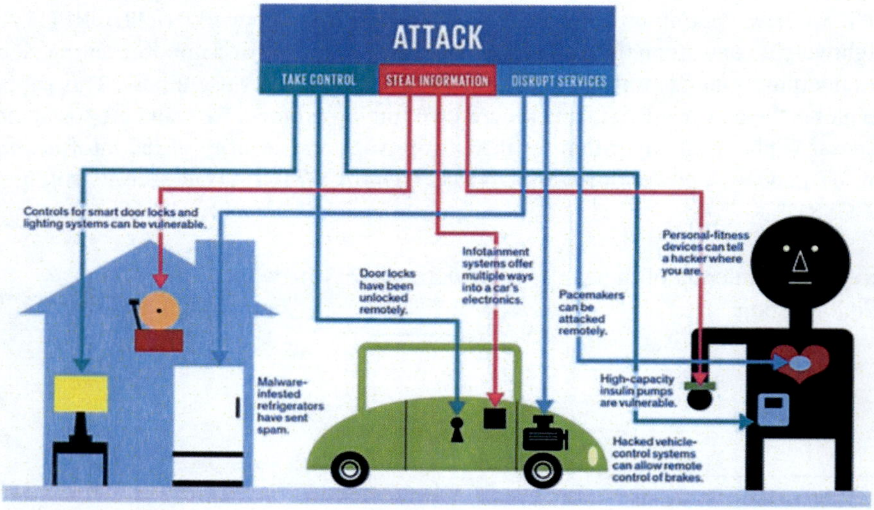

Fig. 1 Internet of Things security—unpredictable conduct

2 Literature Review

Lin et al. [1] have proposed the fog-based framework for IoT to empower processing administrations gadgets sent at the system edge, planning to improve the client's understanding and versatility of the administrations in the event of disappointments. With the benefit of circulated engineering and near end-clients, mist/edge figuring can give a quicker reaction and a more prominent nature of administration for IoT applications. The white paper by Atmel [2] has talked about the security issues for IoT. Right now, Web security advancements, for example, SSL and TLS were likewise talked about alongside basic issue happens in the system when the edge hub will be gotten to by the warning. Right now, benefits accessible to the hubs whose character has been checked are laid out. This paper has noticed the different kinds of pull in happen in the edge hub [3–5]. At last, they have given the keen answer for the encryption for the edge hub in this manner utilizing crypto quickening agent gadgets. Container Cheng et al. [6] have proposed a standard-based way to deal with the plan and actualize another mist figuring-based system, to be specific fog flow, for IoT shrewd city stages. They have likewise talked about how to progressively arrange and oversee information preparing undertakings over cloud and edges and how to advance errand allotment for insignificant dormancy and transmission capacity utilization. Travis Mick, Reza Tourani and Satyajayant Misra proposed lightweight validation and verified steering (LASeR) for NDN (Named Data Networking) IoT in Smart Cities [7], a protected onboarding and directing system for NDN-based IoT system. Right now, it is accomplished through a progressive system structure and almost no cryptographic or computational weight. Federico Montoriononi proposed an engineering [8] open to expansions in a few different ways and as yet permitting the concurrence of assorted information-gathering strategies. Joshua E. Siegel et al. propose an answer demonstrated [9] on human utilization of setting and discernment, utilizing cloud assets to encourage IoT on compelled gadgets. They present engineering application process information to give security through reflection and protection through remote information combination. The data proxy permits a framework's sensors to be examined to meet a predetermined 'Nature of Data' (QLD) focus with insignificant asset use. The productivity improvement of this design has appeared with a model vehicle following application. At long last, they think about future open doors for this design to diminish specialized, financial and conclusion obstructions to the selection of the IoT. Guan et al. [10] proposed a protected and proficient information obtaining plan for Cloud IoT in the shrewd network. In the proposed plan, the enormous information is divided into a few squares, and the squares are encoded/unscrambled and transmitted in a grouping. What's more, they receive the double-mystery sharing plan, which understands the protection safeguarding, the information honesty check and the traits check at the same time [11]. The examination shows that the proposed plan can meet the security necessities of information procurement in the brilliant framework, and it additionally diminishes reaction time overhead altogether contrasted with other mainstream plans. In this paper by Solapure et al. [12], different issues faced along with some solutions like power, storage and bandwidth are highlighted

The white paper by XILINX [13] inspects three key application zones that include the establishment of IoT network, cybersecurity and edge figure inside the setting of choosing an IoT edge stage that can adjust to the effect of market drifts after some time. In the paper, Sobral et al. [14] review the most recent approaches based on IoT with RPL presented. This paper will be based on the further enhancement in security scenarios using RPL routing protocol.

3 Methodology

IoT Contiki is an OS much like Microsoft Windows and Linux. Be that as it may, for a totally exact explanation and specifically, it concentrated on 'matters' inside the IoT. Different elements of an OS comprise programming/system control, valuable asset control, memory the executives and verbal trade the board. The objective of Contiki OS is to satisfy the necessities of the littlest devices together with shrewd soil. Things should have the option to talk a few bits of data to each other. Contiki OS is an IoT working framework intended to help arranged, asset-compelled gadgets. Actualized in C, Contiki organizes lightweight memory board and force proficiency, with a run of the mill set-ups being conveyed utilizing as small as 2 kB of RAM and 60 kB of ROM running at 1 MHz [10]. As indicated by the IoT engineer review of 2017 [5], Contiki is utilized in generally 13.4% of gadgets, and it is relied upon to develop consistently. Intended to associate little, battery-fuelled gadgets to the Web, Contiki gives lightweight usage to an assortment of mainstream correspondence principles, including IEEE 802.15.4, 6LoWPAN, CoAP, MQTT, TSCH and RPL. Also, Contiki highlights an equipment autonomous programming framework, with the moderate reflection being given by the centre framework. Given the undeniably application-driven nature of sensor gadgets, this encourages framework versatility, as extra stage backing can be actualized in libraries and administrations over Contiki's liquid design.

Function of each window in Cooja simulation in which I demonstrate the node distribution in form of name 'PIET' where presently working as assistant professor, shown in Fig. 2 (Fig. 3).

- **Network**—For the distribution of all nodes
- **Simulation Control**—This window is used for simulation control
- **Notes**—To write about simulation
- **Mote Output**—Status output of all nodes like power delay, etc.
- **Timeline**—Flow of message is noted by simulation timeline window.

Innovation has been improved so that the contraptions and gadgets can interface with one another without human collaboration. What's more, this kind of correspondence is known as the Web of things. IoT correspondence incorporates keen urban areas, business interchanges, barrier hardware, traffic frameworks, savvy workplaces, brilliant cost assortment, and satellite TV. These gadgets are for all intents and purposes associated and correspondence is performed naturally. IPv6 conventions

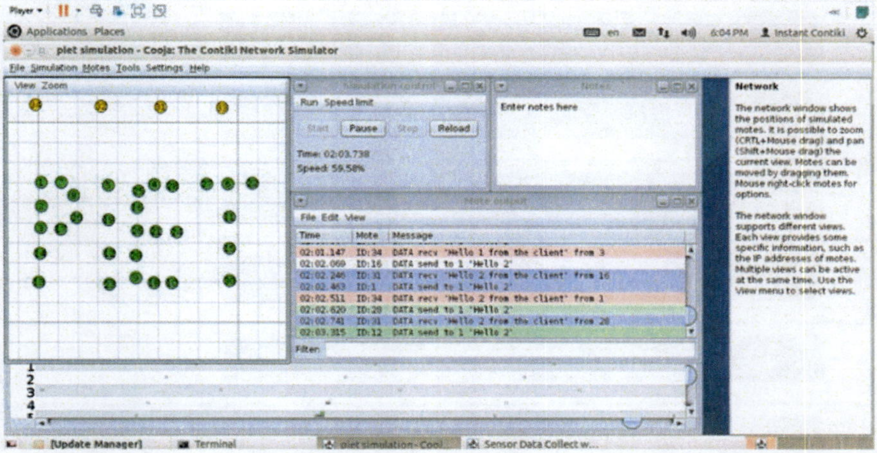

Fig. 2 Different window in Contiki Cooja simulator

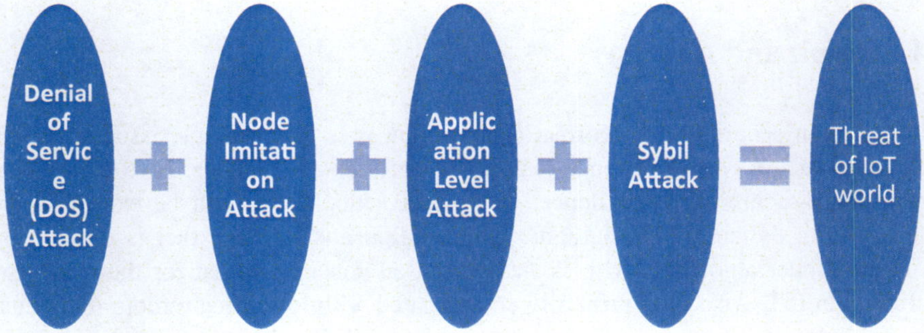

Fig. 3 Types of attack in IoT network

utilizing Contiki Cooja makes the system exceptionally secure. The DOS assault is finished by the noxious parcel or by the aggressor hub by obstructing the correspondence channel or data transmission of the gadgets. Hub imitation attack the assailant can make possess hub with a phony ID. Thus, the hub can get the messages that do not have a place with that hub. With the assistance of this assault, the aggressor can assault the entire situation of IoT. In application-level attack, the aggressor can assault the hub and alter the message and retransmit the message to the goal hub. By this assault, the aggressor can spread phony messages to the hubs. Sybil attack the assailant duplicates the source hub and controls the personality and powers different hubs to move away from the system (Fig. 4).

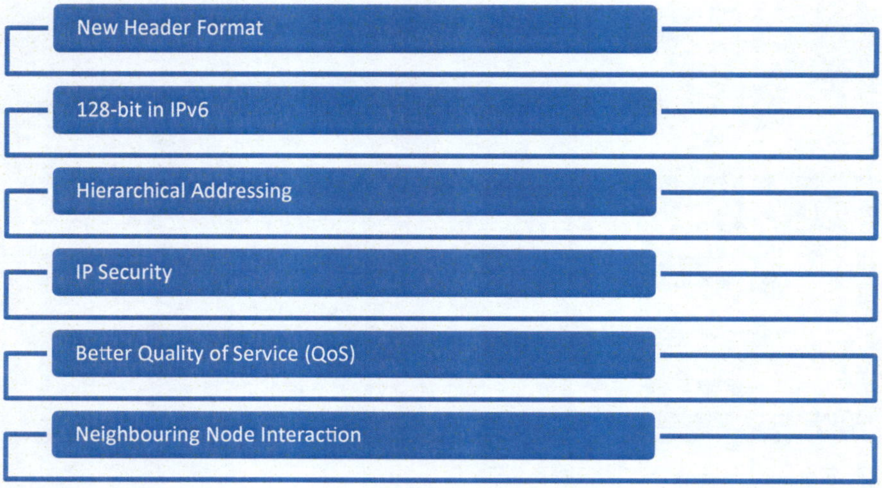

Fig. 4 Features introduced with the IPv6 protocol for RPL

4 Result and Analysis

Information security and information uprightness is the principle issue with IoT gadgets. Secure correspondence is required between the gadgets. IPv6 convention gives high-secure correspondence. IPv6-based calculations will be exceptionally secure with dynamic cryptographic confirmation and key age that is utilized by Caesar Cipher algorithm. RPL is an IPv6-based convention that for the most part utilized in IoT. Also, it is primarily co-ordinated with low-force remote individual region systems. Flow chart of security solution with IPv6 RPL routing is shown in Fig. 5 (Fig. 6).

Communication should be started between two nodes with authentication followed by message passes between them. The clients calculate the hash and send the hash to the server. Server calculates the hash from the information which it has saved earlier; if both have same, then I will send acknowledgement to the receiver. The receiver receiving the acknowledgement and then server will send co-ordinate of the key to the client these co-ordinate are chosen randomly of these values used for encrypting the message. Server match co-ordinates of key to save in his database when the key matches with the received co-ordinate then message decrypts successfully and if client have no message to send then server send a disconnection message. Encryption process and decryption process for the data security are shown in Figs. 7 and 8. Not at all like IPv4, is IPsec security ordered in the IPv6 convention detail, permitting IPv6 parcel validation or potentially payload encryption through the extension headers. In any case, IPsec is not naturally actualized; it must be arranged and utilized with a security key trade.

In our **mathematical model**, first translate all of our characters to numbers, '*a*' = 0, '*b*' = 1, '*c*' = 2 … '*z*' = 25 and demonstrate the Caesar Cipher encryption

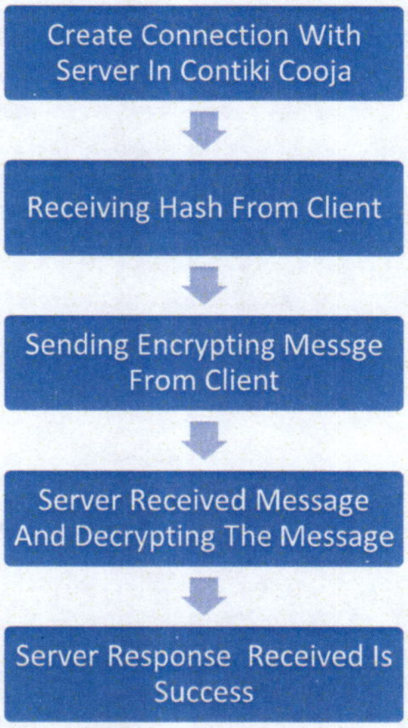

Fig. 5 Security solution with IPv6 RPL routing

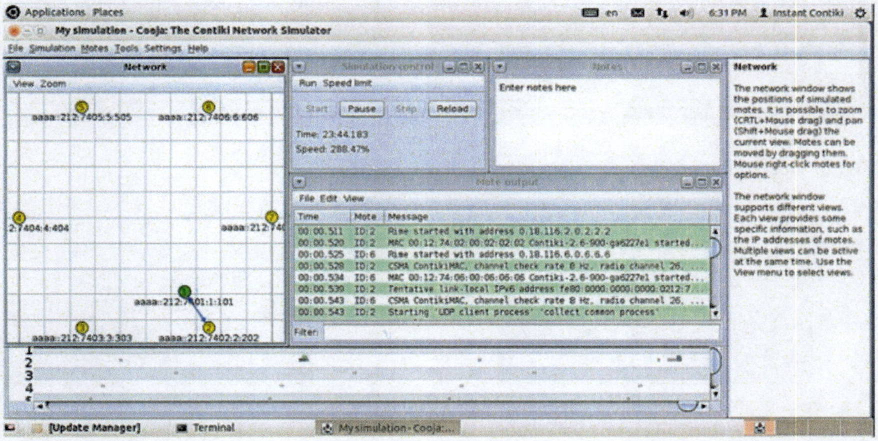

Fig. 6 Simulation scenarios with one sink (green colour) with six senders (yellow colour) with RPL routing

function, $e(x)$, where x is the character we are encrypting, as shown in Fig. 9:

$$e(x) = (x + k)(\bmod 26)$$

where k is the key (the shift) applied to each letter. The decryption function is:

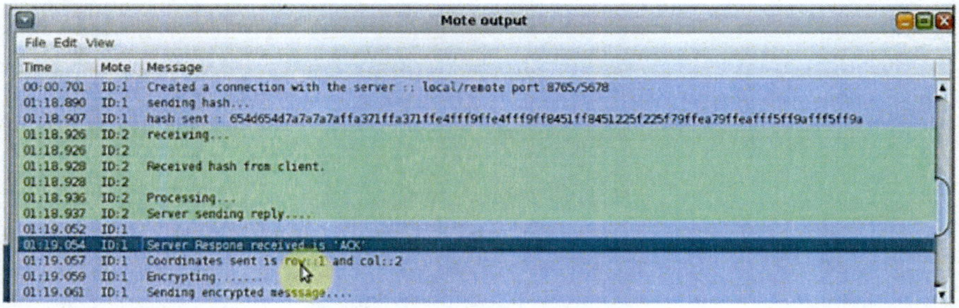

Fig. 7 Encryption process

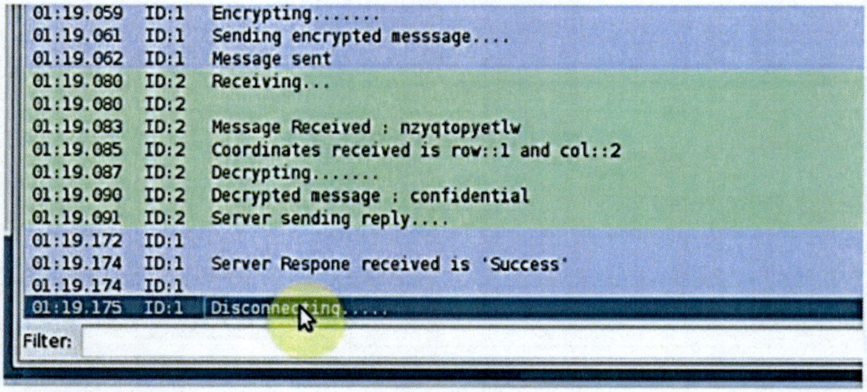

Fig. 8 Decryption process

Fig. 9 Caesar Cipher mathematical model

$$e(x) = (x - k)(\text{mod } 26)$$

5 Conclusions and Future Scope

IoT network is constrained network in terms of battery, processing power, memory, etc. Hence, it is very susceptible to the security attacks. RPL likewise permits the advancement of the system for various application situations and arrangements. For instance, it might consider the connection quality between hubs or their present measure of vitality which makes it a proficient answer for IoT organizations. This paper has talked about numerous assaults that influence the nature of correspondence in IoT and examined the alleviation procedures for different security assaults in IoT. With the help of Caesar Cipher mathematical model, trust, security and privacy are defined in the network. In future, RPL can be used for 5G technology so that 5G may offer a more reliable network that is extremely secure network for industrial IoT by integrating RPL security into the core network architecture.

References

1. Lin J, Yu W, Zhang N, Yang X, Zhang H (2015) A survey on internet of things: architecture enabling technologies security and privacy and applications. IEEE Internet Things J 4:2327–4662
2. Atmel-8994-Security-for-Intelligent-Connected-IoT-Edge-Nodes. Atmel Corporation./Rev.:Atmel-8994A-CryptoAuth-Security-for-Intelligent-Connected-IoT-Edge NodesWhitePaper_112015, 2015
3. Arun R, Salau A (2019) Recent trends in IoT, its requisition with IoT built engineering: a review. Springer, Singapore. ISBN 978-981-13-2553-3
4. Sharma S, Kumar A (2019) Enhanced energy-efficient heterogeneous routing protocols in WSNs for IoT application. IJEAT 9(1). ISSN: 2249-8958
5. Kumar K, Gupta S, Rana A (2018) Wireless sensor networks: a review on "challenges and opportunities for the future world-LTE". AJCS, 1(2). ISSN: 2456-6616
6. Cheng B., Solmaz G, Cirillo F, Kovacs E, Terasawa K, Kitazawa A (2017) FogFlow: easy programming of IoT services over cloud and edges for smart cities. IEEE Internet Things J 2:2327–4662
7. Mick T, Tourani R, Misra S (2017) LASeR: Lightweight authentication and secured routing for NDN IoT in smart cities. IEEE Internet Things J, 2327–4662
8. Montori F, Bedogni L, Bononi L (2017) A collaborative internet of things architecture for smart cities and environmental monitoring. IEEE Internet Things J 4:2327–4662
9. Siegel JE, Kumar S, Sarma SE (2017) The future internet of things: secure, efficient, and model-based. IEEE Internet Things J, 2327–4662
10. Guan Z, Li J, Wu L, Zhang Y, Wu J, Du X (2017) Achieving efficient and secure data acquisition for cloud-supported internet of things in smart grid. IEEE Internet Things J 4:2327–4662
11. Krishna R, Dhawan S, Sharma S, Kumar A (2019) Review on artificial intelligence with internet of things—problems, challenges and opportunities. In: 2nd International conference on power energy, environment and intelligent control (PEEIC). Greater Noida, India, pp 383–387
12. Solapure SS, Kenchannavar HH, Sarode KP (2020) Issues faced during RPL protocol analysis in Contiki-2.7. In: Tuba M, Akashe S, Joshi A (eds) ICT systems and sustainability. Advances in intelligent systems and computing, vol 1077. Springer, Singapore
13. ChetanKhona (2017) Key attributes of an intelligent IIoT edge platform. White Paper: All Programmable Devices XILINX. WP493 (v1.0) September 6
14. Sobral JVV (2019) Routing protocols for low power and lossy networks in internet of things applications sensors

Characterization of Thimbles Based upon Different Sensors

Harmeet Singh, Kamal Malik, and Anshul Kalia

Abstract Human can grasp and release any object based upon the tactile feedback. Without the feedback, the ability for fine control of a prosthesis is limited in the upper limb amputees. Based on the discrete event-driven sensory feedback control (DESC), there is a device that informs the users on the completion of the discrete events such as object contact and release in the form of vibrotactile feedback. The device (DESC-glove) comprises sensorized thimbles to be placed on the prosthesis digits to sense the contact events, a battery-powered electronic board, and vibrating units embedded in an arm-cuff. In this paper, we have presented the sensitivity in terms of the force applied on the thimble at different positions with different force rate. An experimental setup was designed in order to characterize the two thimbles with different sensors. The sensitivity was 0.82 ± 0.16 and 1.18 ± 0.05 N for two thimbles. The main motivation to present the sensitivity of the thimble in different position was to overcome the limitation of the previous design.

Keywords Thimble · DESC-glove · FSR sensor · Vibrator

1 Introduction

The loss of a limb is a dreadful event, and this is particularly apparent in case of the hand amputation. While a decent level of grasping function is often gained through active prostheses, the open challenge in this field is to restore the sensory function. Hand prosthetics can be classified into three main types. Firstly, conventional or body-powered prosthetic devices are those which work with the use of harness system. In this, user can control both their aperture (from the amount of the movement) and grip strength (through the reaction transmitted the by the control cable harness on their body). Second one is external power or myoelectric hand prosthetics. These devices are controlled by utilizing an electric motor that is powered by batteries instead of

H. Singh (✉) · K. Malik · A. Kalia
CT University, Ludhiana, India
e-mail: harmeet17333@ctuniversity.in

N. Marriwala et al. (eds.), *Mobile Radio Communications and 5G Networks*, Lecture Notes in Networks and Systems 140,
https://doi.org/10.1007/978-981-15-7130-5_20

the harness system. The electromyography (EMG) signals coming from the muscle are then transmitted to a processor to control the function of the motor. But user has to rely on the vision to regulate their actions because there is no feedback after the proper grasp [1–3]. Third is the cosmetic or passive functional hand prosthetics. This is just used to copy the appearance of the amputated hand and assist the user in simple carrying and balancing of the objects.

There is a significant progress that has been made in the fields of articulated prostheses, control interfaces, and control algorithms [4–7]. But the sensory feedback component and its practical effectiveness in daily activities are still an open issue. However, the sensory can be provided invasively or non-invasively. Invasive feedback can be provided using implanted electrodes targeting the nerves within the residual limb [8] and holds a potential of eliciting close to natural tactile sensations [9]. On the other hand, non-invasive simulation generally relies on the ability of the individual to learn to correctly interpret artificial sensory stimuli, although modality (e.g., pressure to pressure) and somatotopical matching (e.g., eliciting phantom sensations) can shorten the learning process [1]. Among the non-invasive, vibro, or electro-tactile feedback have been widely used, as they do not require any surgery and low power consumption and compatibly with EMG [10–12].

Discrete Event-Driven Sensory Feedback Control (DESC)-glove is a device that informs the users on the completion of the discrete events such as object contact and release as shown in Fig. 1. DESC-glove comprised of two parts. First part is the thimble, which is designed to fit digits of conventional myoelectric prostheses hand. Thimble is made from the silicon, so that it is easier to don and doff for the participants. Different sensors can be embedded into the thimbles, e.g., force-sensing resistor (FSR) sensor. The main function of the thimble is to sense the contact events. Second part of the DESC-glove is the arm-cuff attached to the user's arm containing a printed circuit board (PCB), a battery, and vibrator. The main functionality of the arm-cuff is to give the feedback to the user based on the signal sensed by the sensor in the thimble.

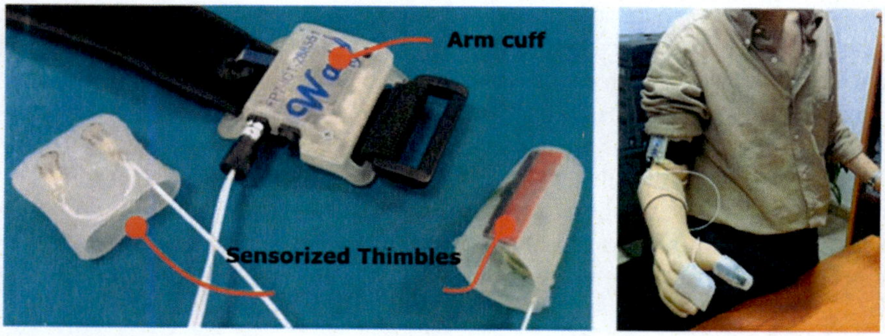

Fig. 1 DESC-glove: **a** The device is composed of two parts: sensorized digit thimbles that detect contact with objects and an arm-cuff used to stimulate the user via miniature vibrating motors, **b** the device fitted on the myoelectric prosthesis of an upper limb amputee (reprinted from Clemente et al. [13])

An experimental setup was developed in order to characterize the two thimbles with different sensors. As the working principle of both the thimbles are different, so.

Their data processing was also different. For the first thimble, the overall sensitivity was 0.82 ± 0.16 N, and for the second, it was 1.18 ± 0.05 N.

2 Architecture of the Thimbles

Two thimbles were designed with different sensors, and characterizations were done based upon the different forces applied with different force rates on different positions of the thimbles. The first thimble was made from the piezo film (thickness_28 μ) as shown in Fig. 2. First, the piezo film was cut according to the desired shape by the laser cutter. Then, to avoid the short circuit near the edges after the cut, edges were cleaned with help of acetone. With the help of conductive adhesive, the wires were connected on the different sides of the film. Then, the film was glued on the 1 mm inner layer of the thimble using the silicon glues (Sil-PoxyTM, Smooth-On Inc., Allentown, USA), and then, again the thimble was molded again with 1 mm of the outer layer of silicon (Smooth-SilTM 945, Smooth-On Inc., Allentown, USA).

The second thimble was made with the accelerometer (ADXL 335, Analog Devices Inc.) and the vibrator (vibrational frequency of 70 Hz) as shown in Fig. 3. I have placed the vibration on the center of the accelerometer as shown in Fig. 3.

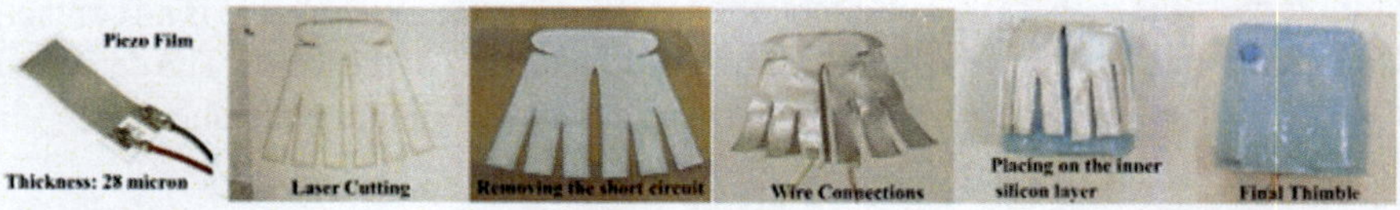

Fig. 2 Different stages in the implementation of piezo film thimble

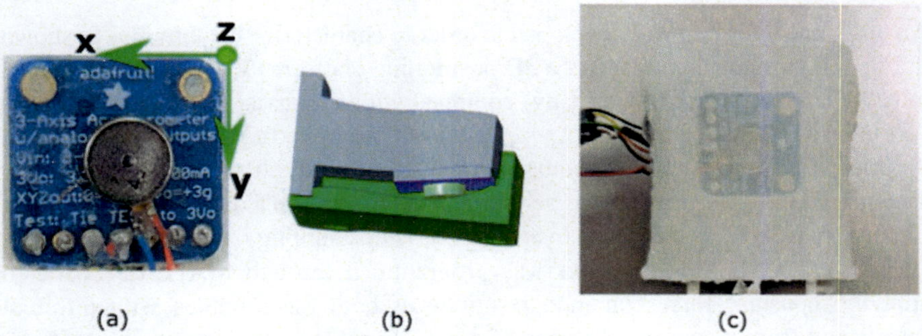

Fig. 3 Thimble made from accelerometer and the vibrator: **a** vibrator placed on the center, **b** mold used for the thimble, and **c** final thimble

Then, the thimble was made with the mold (Fig. 3), and then, the combination of accelerometer and the vibrator was glued with the silicon glue inside the thimble (Fig. 3).

3 Methods

Piezo film develops a voltage (potential difference) between its upper and the lower surfaces when the film is stressed. This voltage arises from the generation of the charge within the material. Piezo film is often used for dynamic tactile sensing [14]. The embedded strips are thin and flexible, having little effect on the elastomers in which they are embedded. The sensitivity of the film is largely due to the format of the piezo film material. The low thickness of the film makes, in turn, a very small cross-sectional area, and thus, relatively small longitudinal forces create very large stresses within the material. If the film is placed between two layers of complaint material, then any compressive forces are converted into much larger longitude extensive forces. In fact, this effect tends to predominate in most circumstances since most substances are compliant to some extent and the ratio of effective sensitivity in the 1 (length) versus 3 (thickness) directions is typically 1000:1. The signal was observed using 10 M input resistance and an amplifier (TL074IN, Texas Instruments).

On the other hand, the thimble with accelerometer and vibrator works on a concept of vibrating the thimble with forced vibration and measures the changes when the thimble touches some object [15]. For example, when there is contact between the thimble and object, there will be attenuation in the vibration. When there is no contact, that vibrational signal can be the baseline for the detection of the touch. The detection can be based upon the trained classifier or simply based upon the threshold.

4 Data Recording

An experimental setup was developed in order to characterize the thimbles as shown in Fig. 4. The setup comprised of a 3D positioning platform (VT-80 linear stages, PI miCos GmbH, Eschbach, Germany) equipped with a tri-axial load cell (NANO17, ATI Industrial Automation Inc., Apex, USA), which applied a known deformation on the thimbles' surface, while the embedded sensor and load cell outputs were recorded. The data from both thimbles was acquired by a PC through a data acquisition board (model NI-USB 6009, National Instruments), with sampling rate of 1 kHz.

The 3D positioning platform was programmed through MATLAB 2016a to applying constant forces on nine positions of both the thimbles with different speeds and from a constant distance of 4 mm (Fig. 5). At each position, one force was applied with five different speeds, and for each combination, there were ten

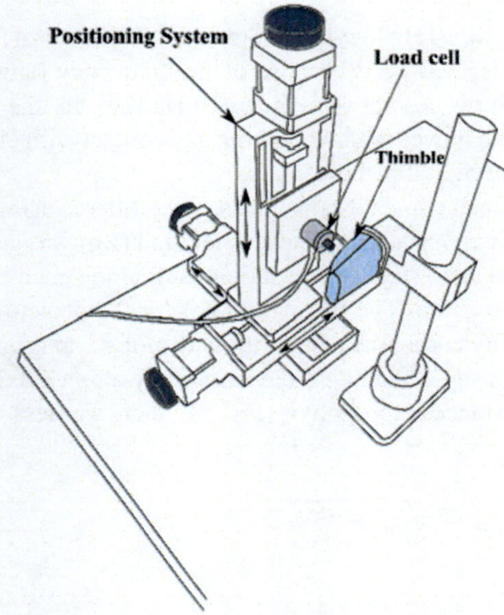

Fig. 4 Experimental setup used for the characterization of the thimbles

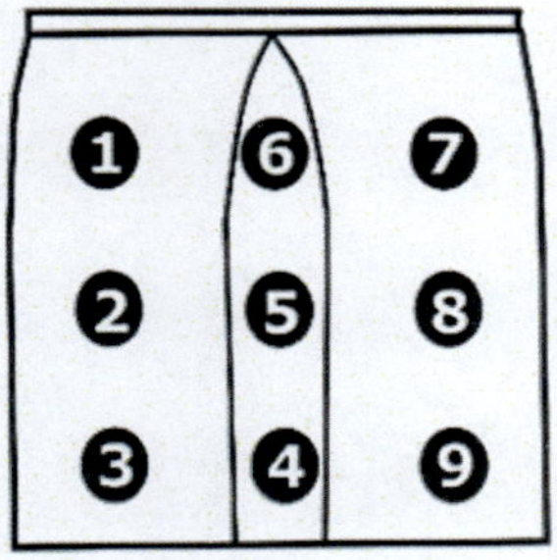

Force (N)	Speed (mm/s)
0.5±0.0018	0.5,1,2,5,10
0.1±0.0031	0.5,1,2,5,10
2.3±0.0039	0.5,1,2,5,10
3.2±0.031	0.5,1,2,5,10
4.5±0.017	0.5,1,2,5,10

Fig. 5 Different forces (mean ± SD) with different speeds applied on nine positions of thimble

trials. I have used five forces as shown in Fig. 5. So for one position, 250 trials (5 forces × 5 speeds × 10 trials) were performed, consisting of 2250 for all positions for one thimble.

5 Data Analysis

The data was first pre-processed using a Butterworth filter. Then, the relevant features were extracted for the further processing from both the fingers separately.

For the thimble with piezo film, 10 Hz low-pass third-order Butterworth filter was used to filter the data. This filter was used because of the frequency range of the human hand and fingers, which resides between 5 and 10 Hz [16]. In this way, the noise produced while the thimble moves without making any contact with object can be removed, and we will get the right thresholds.

For the thimble with accelerometer and vibrator, the data was filtered using a band-pass (5–150 Hz) third-order Butterworth filter. The acceleration (Rg) was calculated by combining all the axis (Eq. 1) and standard deviation (Std) along each axis with a window size 20 ms and 0% overlap. The threshold (Th) for the detection of the touch and release is calculated by taking the sum of the Std. along x-axis and y-axis (Eq. 2). The reason for considering the standard deviation only along x-axis and y-axis because as the vibrator was placed along z-axis (Fig. 3a), there was less variation along the z-axis.

$$\mathrm{Rg} = \sqrt{x^2 + y^2 + z^2} \tag{1}$$

$$\mathrm{Th} = stx + sty \tag{2}$$

where

x value of x-axis in g.
y value of y-axis in g.
z value of x-axis in g.
stx standard deviation along x-axis.
sty standard deviation along y-axis.

6 Results

The threshold for the thimbles is shown in Fig. 6. The upper and lower thresholds for piezo film thimble were set to 3.9 V and 3.1 V, respectively, for the detection of touch and release events (Fig. 6a). These thresholds were set by taking into account the noise threshold.

On the other hand, the upper and lower thresholds for the second thimble were set to 0.08 and 0.035, respectively (Fig. 6b). In addition to the noise threshold, the Rg was also calculated to avoid the wrong feedback when the thimble moved freely without any touch and release events. So Rg was set to be less than 1.2 g for the feedback, and if it was more than 1.2 g, the system would not give any wrong feedback to the users.

Figure 7 shows the response of the piezo thimble with respect to the thresholds. The response was in terms of force and force rate. The force rate was calculated according to the speed and the force applied. The positions 2, 4, 5, 7, and 8 are more sensitive as they reached the threshold with fore applied with less force rate as compared to the positions 1, 6, and 3. On the other hand, in case of the thimble with

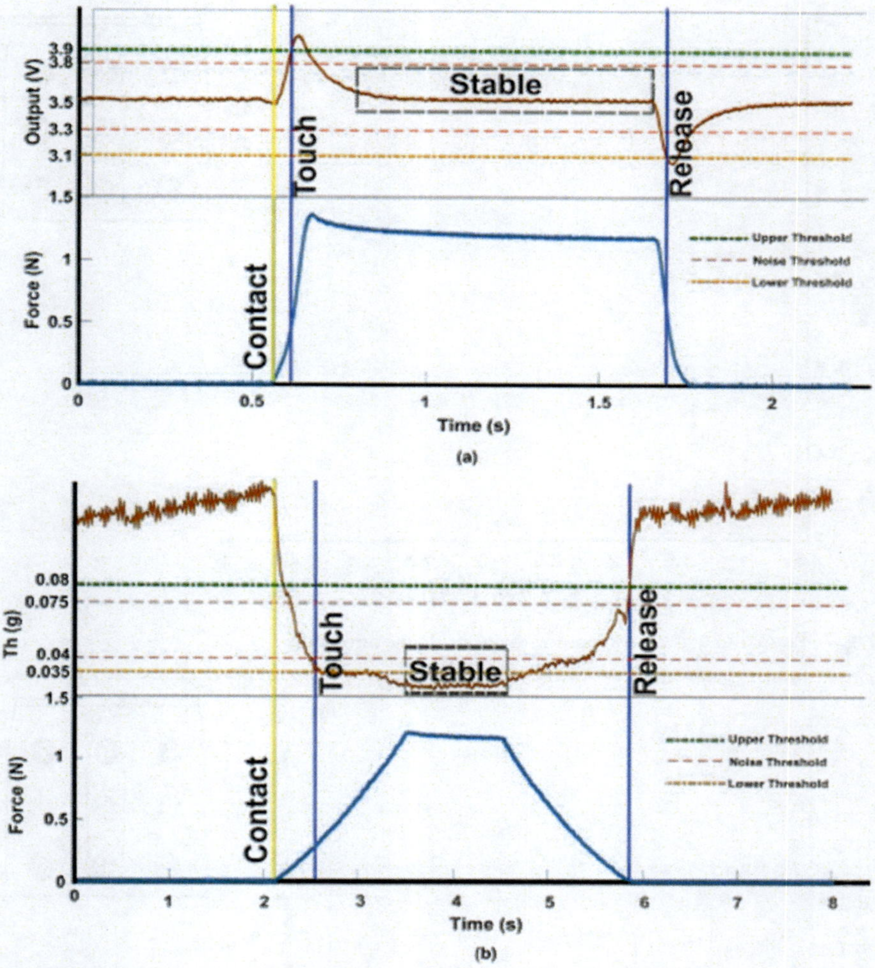

Fig. 6 Thresholds for: **a** piezo film thimble and **b** accelerometer and vibrator

the accelerometer and vibrator, all the positions are sensitive except the position 5 (Fig. 8).

7 Discussion

The two thimbles were designed with different sensors for the comparison. The working principle of both the thimbles is different. The piezo film develops a voltage between (potential differences) its upper and the lower when the film is stressed.

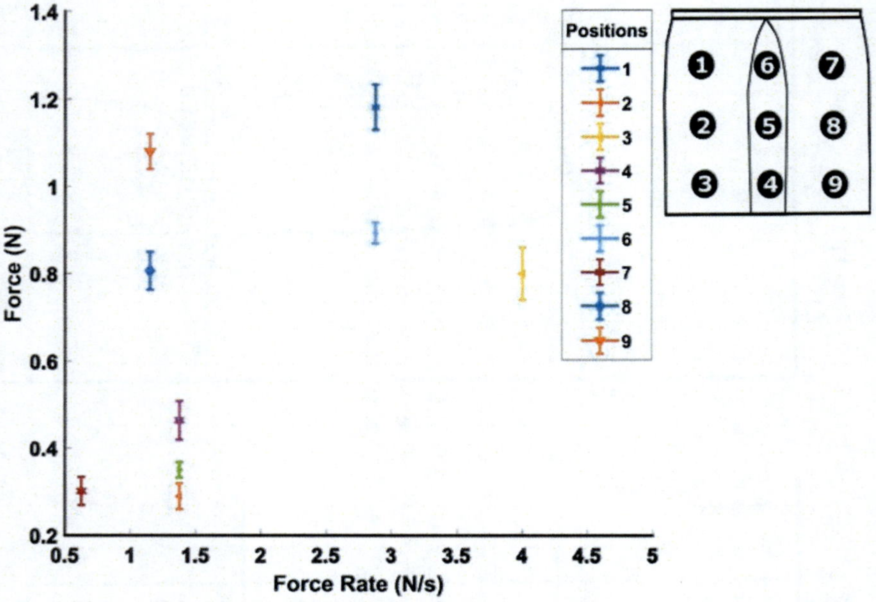

Fig. 7 Piezo film thimble response with respect to the thresholds

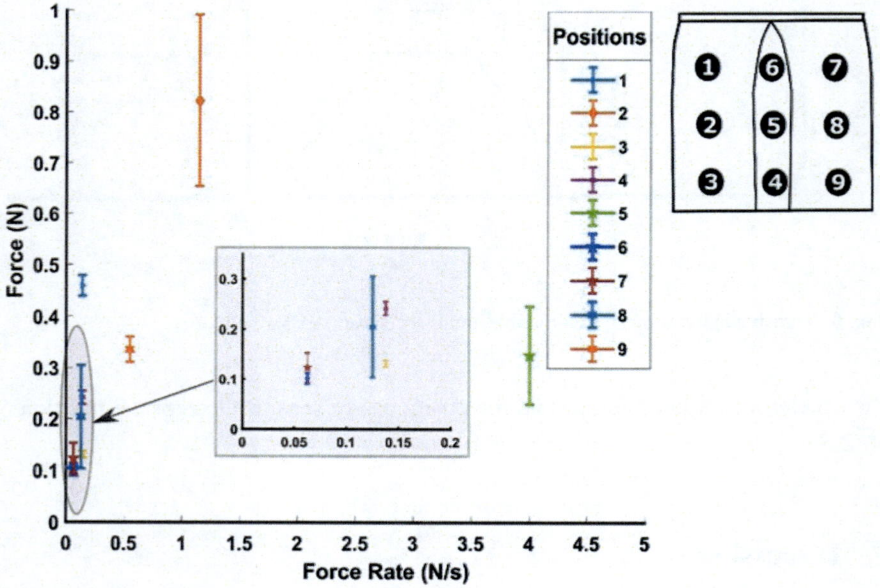

Fig. 8 Accelerometer and vibrator thimble response with respect to the thresholds

For the second thimble, the accelerometer recorded the vibration and measured the variation in the vibration when thimble touched or released the object.

As the working principles were different, their data processing was also different. For the piezo film, only one signal was used for the processing. But an amplifier was used to amplify the signal. Then, the filter was applied to remove the data with frequency more than 10 Hz. For the other thimble, the signals from three axis of accelerometer were collected and processed. Bandpass filter (5–150 Hz) was applied, and then Rg and Th were calculated. The delay of 10 ms (approx.) was observed in the piezo film thimble and 30 ms in the second thimble between the threshold reached and the feedback. The delay was more in the thimble with the accelerometer and vibrator because the window size of 20 ms was used to calculate the Std.

In terms of force rate, the thimble with accelerometer and vibrator seems to be more sensitive than the piezo film thimble. As you can see in Fig. 8, the force rate to reach the threshold for seven positions is less than 1 N/s. It implies that the thimble with accelerometer and vibrator was sensitive even if the speed of the myoelectric hand would be slow. On the other hand for the piezo film thimble, only one position is able to reach to threshold less than 1 N/s and four positions less than 1.5 N/s (Fig. 7).

In terms of force required to reach the threshold, the accelerometer thimble is more sensitive. In the thimble with accelerometer and the vibrator, eight positions required less than 0.5 N force to detect the events of touch and release, but in the piezo film only four positions.

For the overall sensitivity of the thimble, it is completely dependent on the contact between the thimble and the object, as some positions on the thimbles are more sensitive as compared to others. In our case, for the overall sensitivity of the thimble, I have selected the position with least sensitivity. So for the accelerometer and vibrator thimble, the overall sensitivity is 0.82 ± 0.16 N (second position), and for the piezo thimble, it is 1.18 ± 0.05 N (first position). So, in terms of the overall sensitivity, both thimbles are very similar.

In terms of the implementation, there are many factors that can affect the implementation of both the thimbles. The power consumption by the piezo film is very less as compared to the accelerometer and vibrator thimble, because the vibrator needs to be powered just before the detection of the events. Cutting the piezo film into the desired shape for the thimble is a challenging process because after cutting the film, if the edges are not cleaned properly, it would short circuit the film. There will be additional control required to active and deactivate the vibrator before and after the touch task. This can either be done by using the EMG control, i.e., vibrator could be activated with EMG signal or it could be activated if the thimble remains in certain position for fixed time and detect some changes in the signal with the help of accelerometer.

References

1. Antfolk C, D'Alonzo M, Rosén B, Lundborg G, Sebelius F, Cipriani C (2013) Sensory feedback in upper limb prosthetics. Expert Rev Med Devices 10(1):45–54
2. Childress DS (1980) Closed-loop control in prosthetic systems: historical perspective. Ann Biomed Eng 8(4–6):293–303
3. Biddiss EA, Chau TT (2007) Upper limb prosthesis use and abandonment: a survey of the last 25 years. Prosthet Orthot Int 31(3):236–257
4. Belter JT, Segil JL, Dollar AM, Weir RF (2013) Mechanical design and performance specifications of anthropomorphic prosthetic hands: a review. J Rehabil Res Dev 50(5):599–618
5. Ortiz-Catalan M, Hakansson B, Branemark R (2014) An osseointegrated human-machine gateway for long-term sensory feedback and motor control of artificial limbs. Sci Transl Med 6, 257:257re6–257re6
6. Pasquina PF et al (2015) First-in-man demonstration of a fully implanted myoelectric sensors system to control an advanced electromechanical prosthetic hand. J Neurosci Methods 244:85–93
7. Amsuess S, Goebel P, Graimann B, Farina D (2015) A multi-class proportional Myocontrol algorithm for upper limb prosthesis control: validation in real-life scenarios on amputees. IEEE Trans Neural Syst Rehabil Eng 23(5):827–836
8. Navarro X, Krueger TB, Lago N, Micera S, Stieglitz T, Dario P (2005) A critical review of interfaces with the peripheral nervous system for the control of neuroprostheses and hybrid bionic systems. J Peripher Nerv Syst 10(3):229–258
9. Riso RR (1999) Strategies for providing upper extremity amputees with tactile and hand position feedback–moving closer to the bionic arm. Technol Health Care 7(6):401–409
10. Dosen S, Schaeffer M-C, Farina D (2014) Time-division multiplexing for myoelectric closed-loop control using electrotactile feedback. J Neuroeng Rehabil 11(1):138
11. Kaczmarek KA, Webster JG, Bach-y-Rita P, Tompkins WJ (1991) Electrotactile and vibrotactile displays for sensory substitution systems. IEEE Trans Biomed Eng 38(1):1–16
12. Cipriani C, Zaccone F, Micera S, Carrozza MC (2008) On the shared control of an EMG-controlled prosthetic hand: analysis of user-prosthesis interaction. IEEE Trans Robot 24(1):170–184
13. Clemente F, D'Alonzo M, Controzzi M, Edin BB, Cipriani C (2016) Non-invasive, temporally discrete feedback of object contact and release improves grasp control of closed-loop myoelectric transradial prostheses. IEEE Trans Neural Syst Rehabil Eng 24(12):1314–1322
14. Balasubramanian R, Santos VJ (eds) Springer tracts in advanced robotics 95. The human hand as an inspiration for robot hand development
15. Singh H, Controzzi M, Cipriani C, Di Caterina G, Petropoulakis L, Soraghan J (2018) Online prediction of robot to human handover events using vibrations. In: 26th European signal processing conference
16. Brooks TL (1990) Telerobotic response requirements. In: 1990 IEEE international conference on systems, man, and cybernetics conference proceedings, pp 113–120

Application of the Optimization Technique in Analytical and Electrochemistry for the Anticorrosive and Complexing Activity of 2,4-Dihydroxyacetophenone Benzoylhydrazone

Navneet Kaur, Nivedita Agnihotri, Sonia Nain, Rajiv Kumar, and Rajesh Agnihotri

Abstract 2,4-Dihydroxyacetophenone benzoylhydrazone (DHABH) collectively has been used for the anticorrosive activity towards soft cast steel and as a complexing agent for formation of a coordination complex with tungsten (VI). The corrosion inhibition properties of the reagent are examined in acidic medium at variable concentrations and temperatures utilizing different techniques like gravimetric and electrochemical including polarization measurements and electrochemical impedance spectroscopy (EIS) and quantum chemical calculations. It has been found from the study that inhibition potency increased with increasing concentration of DHABH as indicated by weight loss measurement, polarization curves and EIS studies. However, the complexation conduct of the reagent with tungsten (VI) to form 1:3; W(VI): DHABH complex is studied spectrophotometrically. The produced method of determination has been observed to be exceedingly sensitive, selective, quick, reproducible and satisfactorily pertinent to a wide variety of technical and synthetic samples.

Keywords 2,4-dihydroxyacetophenone benzoylhydrazone · Corrosion · Spectrophotometry

---

N. Kaur · N. Agnihotri (✉)
Department of Chemistry, Maharishi Markandeshwar (Deemed to be University), Mullana, Ambala, Haryana 133207, India
e-mail: nivagni11@gmail.com

R. Agnihotri
Department of Applied Science, UIET, Kurukshetra University, Kurukshetra, Haryana 136119, India
e-mail: ragnihotri2015@kuk.ac.in

S. Nain · R. Kumar
Department of Chemistry, DCRUST, Murthal, Sonepat, Haryana, India

N. Marriwala et al. (eds.), *Mobile Radio Communications and 5G Networks*, Lecture Notes in Networks and Systems 140,
https://doi.org/10.1007/978-981-15-7130-5_21

1 Introduction

Corrosion is a naturally happening destructing phenomenon of a metal disintegration in view of metal ecological collaboration. A considerable lot of the ordinarily utilized metals particularly iron are helpless to consumption and subsequently lost. The process is undesirable, thus eradicating metallic properties and reducing their life [1]. Thus thinking about the wide applications of iron in its commercial steel alloy form, corrosion prevention and control are essentially required. Out of various techniques applied for the corrosion control process, a calculable decline in corrosion rate of iron is observed on supplementation of a chemical substance known as inhibitor. The inhibitors help in retarding the attacking tendency to metallic surface thereby to withstand environmental conditions. Corrosion control by application of inhibitors is observed to be the most efficacious, empirical and lucrative technique against erosion in acidic solution [2, 3].

Organic compounds generally act as effective corrosion inhibitors, and the most efficient are the compounds with π-bonds or heteroatoms like N, O and S in their structures [4]. Magnificient results are obtained in the past using hydrazones as the potential inhibitors [5, 6]. In continuation with the past studies [7–9], DHABH, another hydrazone, is used in the present work as a brilliant corrosion inhibitor. 2,4-Dihydroxyacetophenone benzoylhydrazone (DHABH) has been utilized for the anticorrosive action towards soft cast steel and as a complexing operator for formation of a coordination complex with W(VI). Alongside the utilization of DHABH as a powerful anticorrosive agent, it is found to have extensive use as a complexing agent in analytical chemistry for spectrophotometric determination of transition elements including tungsten (VI) [10–14] where DHABH goes about as a promising reagent for complexation of W(VI) in the present work.

2 Experimental

2.1 Preparation of 2,4-Dihydroxyacetophenone Benzoylhydrazone (DHABH)

Equimolar amounts (0.025 mol) of resacetophenone (3.8 g) and benzoyl hydrazine (3.4 g) are refluxed for two hours in ethanol. The product obtained, Fig. 1, is cooled, filtered and recrystallized with ethanol [15].

Molecular formula: $C_{15}H_{14}O_3N_2$; molecular mass: 270 g mol^{-1}; melting point: 197 °C.

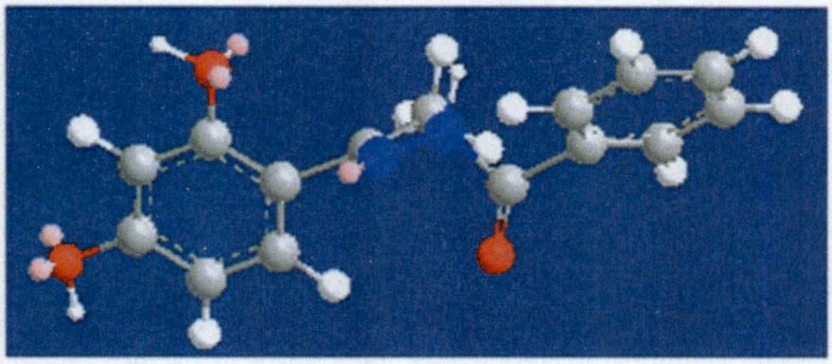

Fig. 1 Structure of 2,4-dihydroxyacetophenone benzoylhydrazone (DHABH)

2.2 Materials Required for Anti-corrosion Potential

Specimen Preparation. A ($3 \times 1.5 \times 0.028$ cm^3) rectangular soft cast steel specimen with the composition (wt%) C—0.14; Si—0.03; Mn—0.032; S—0.05; P—0.20; Ni—0.01; Cu—0.01; Cr—0.01 and the remaining iron was taken for all corrosion inhibition tests. All the samples and solutions required for gravimetric and electrochemical measurements were prepared as reported, and finally corrosion rate was computed in millimiles per year (mmpy) applying the formula [16]:

$$\text{Corrosion Rate} = (534 \times \Delta W)/\text{DAT} \quad (1)$$

where ΔW is loss in mass of soft cast steel in mg, D is the density of soft cast steel (7.86 g cm^{-3}), A is the total area in cm^2 and T is immersion time in hours of the soft cast steel. 'CR' values thus determined were applied to evaluate inhibition efficiency (η) using the equation:

$$\eta\% = \left[\left(\text{CR}^0 - \text{CR}\right)/\text{CR}^0\right] \times 100 \quad (2)$$

where CR^0 and CR are the corrosion rate values of soft cast steel in the absence and presence of the inhibitors, respectively.

Electrochemical Analysis. All types of electrochemical measurements including potentiodynamic polarization studies to record Tafel curves and electrochemical impedance spectroscopy (EIS) for Nyquist plots were performed using electrochemical work station, three-electrode cell assembly including soft cast steel sample as working electrode with 1 cm × 1 cm, exposed area, Ag/AgCl as reference electrode and graphite electrode as the counter electrode as reported [16].

The experimental measurements were carried out by immersing the steel sample in 1 M H_2SO_4 solution devoid of and in the presence of variable strengths of the inhibitor within the temperature range of 303–323 K.

All other testing conditions for electrochemical analysis were set as earlier similar reported experiments [16]. The polarization Tafel curves thus recorded in the absence and presence of inhibitor in the potential range ±250 mv with a scan rate of 0.001 Vs^{-1} were used to evaluate the percentage corrosion efficiency ($\eta\%$) as

$$\eta\%_{(\text{Tafel})} = \left[\left(I^0_{\text{corr}} - I_{\text{corr}}\right)/I^0_{\text{corr}}\right] \times 100 \quad (3)$$

$i^0{}_{\text{corr}}$ and i_{corr} denoted corrosion current densities of soft cast steel in 1 M H_2SO_4 in the absence and presence of DHABH, the inhibitor, respectively.

Similarly, based upon Nyquist curves obtained from EIS measurements and the EIS data including parameters like charge transfer resistance (R_{CT}), maximum frequency (F_{max}), double layer capacitance (C_{dl}) and % inhibition efficiency were determined applying the equation as:

$$C_{\text{dl}} = 1/2\pi F_{\text{max}} R_{\text{CT}} \quad (4)$$

$$\eta\% = \left[\left(R_{\text{CT}} - R^0_{\text{CT}}\right)/R^0_{\text{CT}}\right] \times 100 \quad (5)$$

R_{CT} and $R^0{}_{\text{CT}}$ indicated the values of charge transfer resistance in the presence and absence of inhibitor, respectively.

Quantum chemical parameters. The quantum chemical parameters were evaluated with the help of HyperChem Professional 8.0 packages (Hypercube, Inc., USA) [17].

2.3 *Instrumentation and Materials Required for Spectrophotometric Determination of Tungsten (VI)*

A UV–Vis spectrophotometer (2375; Electronics India) with 1 cm analogous quartz cuvettes was used for absorbance measurements and spectral studies.

A stock solution, 1 mg ml^{-1}, of tungsten (VI) was formulated by dissolving 0.179 g of sodium tungstate in distilled water and normalized by oxine method [18]. The employed solutions of lower concentration were prepared by suitable dilutions therefrom.

Solutions of other metal ions at mg ml^{-1} level were prepared by dissolving their commonly available sodium or potassium salts in deionized water or dilute acid which on appropriate dilution produced solutions of lower strength at μg level.

A 0.5% ethanolic solution of DHABH was prepared.

Distilled ethyl acetate throughout was used for extraction and determination.

2 M Hydrochloric acid (HCl) was prepared by appropriate dilution of concentrated HCl.

By homogenizing solutions of different metal ions in suitable proportions, the synthetic samples (some of them comparable to minargent, platinoid, tungsten alloy, heat-resistant steel and high-speed steel) were prepared.

The technical sample, reverberatory flue dust (0.1 g), was dried at 110–120 °C in a silica crucible after combining with 10 mg of tungsten solution and along with sodium peroxide (0.8 g) fused in a muffle furnace at 250–300 °C, dissolved in hot water, neutralized with concentrated hydrochloric acid and finally adjusted to 0.1 M acidity in 100 ml total volume. 0.5 and 0.4 ml fractions were applied to the determination of tungsten by the proposed method.

3 Result and Discussion

3.1 *Anticorrosive Behaviour*

Gravimetric Analysis/Weight Loss Measurements

Variation of corrosion rate with concentration of inhibitor. At divergent concentrations ranging between 200 and 1000 ppm of the inhibitor (without and with) in 1 M H_2SO_4, the weight loss measurements of rectangular steel sample were performed under stable aerated conditions and at variable temperature falling in the range of 303–323 K.

This is clearly evaluated from the weight loss studies as manifested from Table 1, and Figs. 2 and 3 that DHABH, the inhibitor, helps in increasing inhibition efficiency (η) with increase in concentration whereas the corrosion rate (CR) diminishes with increasing concentration of the studied inhibitor. The corrosion rate exhibited without DHABH was found to be 365.18 mmpy at 303 K which reduced to a much lower value of 30.74 mmpy on application of DHABH to the steel sample. Similarly, the inhibition efficiency increased significantly to a value of 91.58% at 303 K which is considerably a very high value of $\%\eta$ for corrosion inhibition of soft cast steel.

The results indicate effective adsorption of the hydrazone molecule with high concentration over the steel surface. The experiment of weight loss measurement was repeated at other temperatures of 313 and 323 K to observe the same trend in CR and $\%\eta$ with variable concentration as analysed at 303 K and shown in Table 1.

Table 1 Effect of concentration of DHABH and temperature on CR and $\%\eta$ from weight loss measurements

Inhibitor	Concentration (ppm)	CR at 303 K (mmpy)	$\eta\%$	CR at 313 K (mmpy)	$\eta\%$	CR at 313 K (mmpy)	$\eta\%$
Blank	–	365.18	–	598.44	–	1096.22	–
DHABH	200	88.70	75.71	315.13	47.34	677.68	38.18
	400	48.31	86.77	282.70	52.76	586.17	46.52
	600	35.27	90.34	257.32	57.00	548.00	50.01
	1000	30.74	91.58	215.55	63.98	475.21	56.65

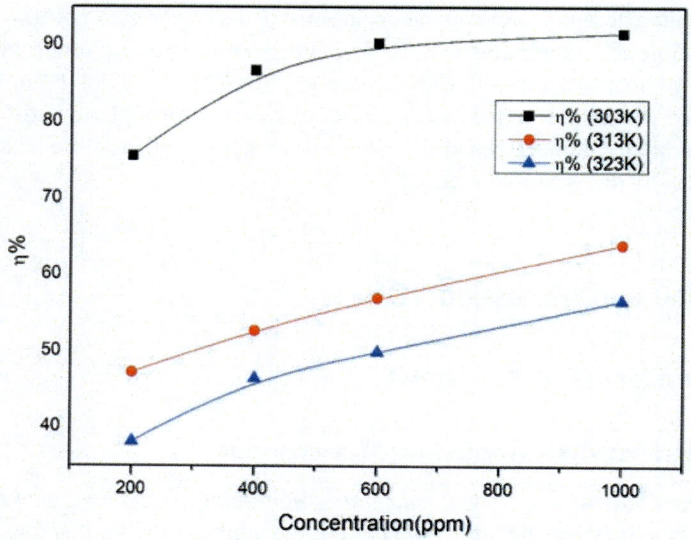

Fig. 2 Inhibition efficiency ($\eta\%$) with concentration of DHABH

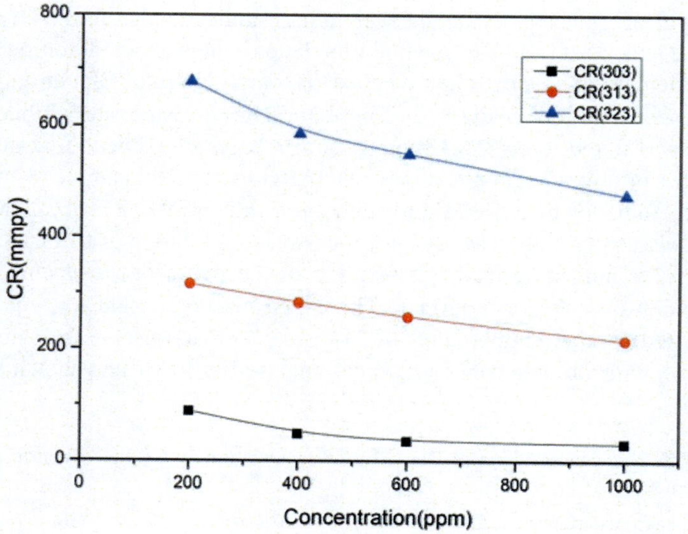

Fig. 3 CR with concentration of DHABH

Variation of corrosion rate with activation parameters. Table 1 from weight loss measurements indicates that rate of corrosion is highly affected with elevated temperature of the testing solution, thus influencing efficiency of the inhibitor. This is observed from the experiment that the corrosion rate enhances whereas $\%\eta$ diminishes with increased solution temperature showing maximum inhibition efficiency value of 91.58% at 1000 ppm, 303 K. The effect is ascribed to the enhanced corrosion process at elevated temperature, speeding up desorption of the absorbed DHABH (inhibitor) molecule from the steel surface.

The activation parameters showing dependency of corrosion rate on temperature can also be computed in terms of activation energy (E_a) employing Arrhenius equation as:

$$\mathrm{CR} = A\ \exp^{(-E_a/\mathrm{RT})}$$

$$\ln \mathrm{CR} = \ln A - \frac{E_a}{\mathrm{RT}}$$

$$\log \mathrm{CR} = \log A - \frac{E_a}{2.303\,\mathrm{RT}} \tag{6}$$

where CR stands for corrosion rate, A is pre-exponential factor, E_a is the apparent activation energy, R is gas constant (8.314 $\mathrm{JK^{-1}\ mol^{-1}}$) and T is the absolute temperature (K).

The value of activation parameter, i.e., E_a, is enumerated from the slope (E_a/2.303RT) of the plot drawn in Fig. 4 between log CR and 1/T at 303 K with 1000 ppm strength of the inhibitor in its presence and absence.

Analysis of the activation parameter data exhibited a much higher value of E_a of 111.57 kJ $\mathrm{mol^{-1}}$ for the solutions containing DHABH as inhibitor than to that of bare acid solution (E_a = 44.50 kJ $\mathrm{mol^{-1}}$). As is reported earlier that E_a with inhibitor is more than E_a without inhibitor [19] accordingly. The higher value of E_a energy shows increased energy barrier for the occurrence of corrosion slowing down the corrosion

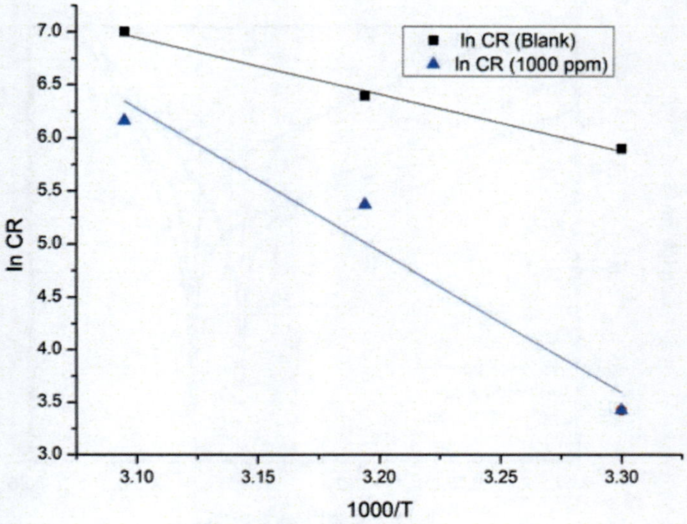

Fig. 4 Arrhenius plot for soft cast steel in 1 M H_2SO_4

process in 1 M H_2SO_4. The positive and increased value of E_a is also indicative of the process of physisorption between soft cast steel surface and DHABH molecule.

3.2 Electrochemical Measurements

Potentiodynamic polarization measurements. The polarization curves of the soft cast steel surface in 1 M H_2SO_4 solution without and in the presence of different concentrations of DHABH are represented in the form of Tafel polarization curves as shown in Fig. 5. From the potentiodynamic polarization curves thus obtained, the numerical values of all the electrochemical kinetic parameters like the anodic Tafel slope (β_a), the cathodic Tafel slope (β_c), the corrosion potential (E_{corr}) and corrosion current density (I_{corr}) at various concentrations of DHABH at 303 K were calculated and are given in Table 2. The corrosion inhibition efficiency ($\eta\%_{(Tafel)}$) was obtained from polarization measurements using Eq. 3. This is quite evident from Fig. 5 that DHABH influences both cathodic and anodic reactions changing shapes of the polarization curves to lower current densities indicating retarding nature of DHABH towards anodic dissolution as well as evolution of hydrogen takes place. The inhibition effect enhances with increasing concentration of the inhibitor indicating probably the respective enhancement of adsorption of DHABH over the surface. However, the change in E_{corr} value is less 85 mV, and it indicates the mixed-type nature of inhibitor.

Thus, the most probable mechanism expected for the process is adsorption of inhibitor through π-electrons of benzene ring and unshared pair of heteroatom

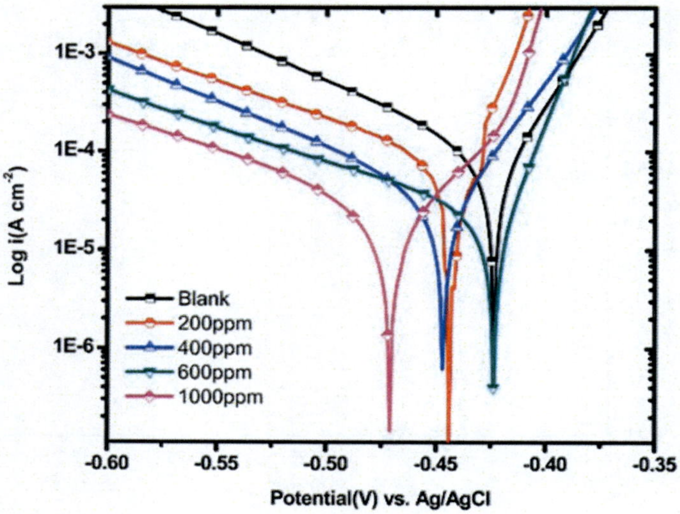

Fig. 5 Tafel polarization curves

Table 2 Polarization parameters of soft cast steel at different concentrations of DHABH

Inhibitor	C(ppm)	β_a (mV dec^{-1})	β_c (mV dec^{-1})	E_{corr} (mV vs. Ag/AgCl)	I_{corr} (μA cm^{-2})	η(%)	Θ
Blank	–	94.06	36.49	423	83.84	–	–
DHABH	200	40.56	19.89	444	28.08	66.50	0.66
	400	83.12	39.52	446	25.03	70.14	0.70
	600	100.72	24.59	425	18.28	78.19	0.78
	1000	86.70	56.35	471	15.90	81.03	0.81

(N and O) which blocks reactive sites of soft cast steel surface retarding the corrosion current density. Consequently with increased block fraction of steel sample by adsorption, inhibition efficiency increases. Further, the recorded values of corrosion potential (E_{corr}) do not show a significant change ($\Delta E_{corr} \leq 48$ mV) remaining less than 85 mV [20] confirming the nature of inhibitor under observation as a mixed type.

Electrochemical impedance spectroscopy (EIS). In 1 M H_2SO_4 solution at 303 K, after immersion of the steel sample for 60–65 min, the corrosion performance of soft cast steel was analysed by EIS studies (deprived of and with inhibitor). The investigational data thus procured is abstracted in Table 3, and the impedance spectra obtained for the sample without and in the presence of varying concentrations of DHABH are presented in the form of Nyquist plots in Fig. 6.

The total impedance remarkably rises up on applying inhibitor to the soft cast steel sample which is indicated in the results that an increase in concentration of DHABH affects positively on the impedance value.

The plots clearly reveal that each impedance spectrum contains one high frequency, an intermediate frequency and a low-frequency capacitive semicircle at different concentrations of the inhibitor. It can be interpreted on the basis of electrode behaviour and charge transfer process occurring on the interface of metallic surface. The variation in the semicircles shown in Nyquist plots may be ascribed to the surface heterogeneity that is its roughness and frequency variations at the steel

Table 3 Impedance parameters for soft cast steel at various DHABH concentrations

Inhibitor	C(ppm)	R_{CT} (Ω cm^2)	C_{dl} (μF cm-^{2})	Θ	η %
Blank	–	6.46	376.21	–	–
DHABH	200	71.56	33.9	0.791783	79.17831
	400	185.07	23.0	0.91949	91.94899
	600	215.62	19.8	0.930897	93.08969
	1000	342.54	16.5	0.956501	95.65014

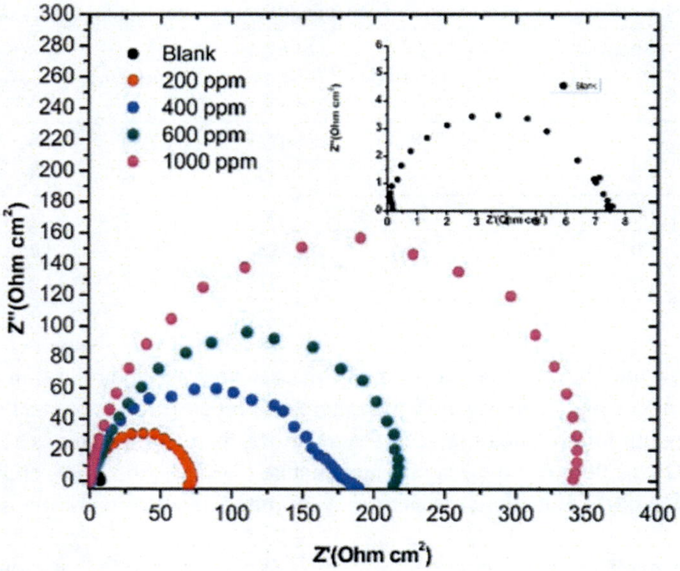

Fig. 6 Nyquist plots of DHABH

surface. The change observed in the impedance with the positively changed concentration of inhibitor may be due to enhanced adsorption of DHABH on metal surface and hereby obstructing its active sites.

R_{CT} (charge transfer resistance) was found to be maximum at the maximum concentration of inhibitor (1000 ppm) used for this study, while the corresponding C_{dl} value is found to be minimum as calculated by Eq. 4 indicating sleeving of the metal surface with protective film of the inhibitor. The same results were reflected in increased $\eta\%$ with rising concentration of DHABH as determined from Eq. 5.

3.3 Adsorption Isotherms

Corrosion inhibition, mainly due to adsorption, is generally studied through adsorption isotherm, i.e. relation between amounts of adsorbate adsorbed on the surface of adsorbent. Thus, it is significant to think about adsorption isotherm of soft cast steel (adsorbent) which can give essential information about the mechanism of corrosion hindrance for steel sample because of the examined compound. So as to portray the adsorption conduct of the inhibitor (DHABH), Langmuir adsorption isotherm [21] for soft cast steel was utilized which gave the best depiction of adsorption conduct of examined compound. The surface inclusion (θ) is characterized as given in Eq. 7

$$\theta = \frac{\eta}{100} \tag{7}$$

η denotes inhibition efficiency seen from weight loss estimations. The outcomes reveal that the plots of C/θ versus C provide a linear fit with the perfect linear correlation coefficient, $R^2 \geq 0.99$ as is given by Eq. 8, the general form of Langmuir equation

$$\frac{C}{\theta} = \frac{1}{K_{ads}} + C \tag{8}$$

where θ represents the extent of surface coverage/inclusion of soft cast steel the adsorbent, C is the concentration of DHABH present in bulk solution and K_{ads} is the equilibrium constant for adsorption–desorption process as determined from the reciprocal of intercept of the plot of C/θ versus C (Fig. 7).

K_{ads} is related to the free energy of adsorption ($\Delta G°_{ads}$) with Eq. 9

$$\Delta G = -\mathrm{RT}\ \ln\,(55.5\ K_{ads}) \tag{9}$$

where R stands for the ideal gas constant, T is the absolute temperature and the value 55.5 is concentration of water in bulk solution. The $\Delta G°_{ads}$ value as obtained from Eq. 15 at 303 K (30 °C) is -25.73 kJ mol^{-1} (Table 4). The value of $\Delta G°_{ads}$ signifies the nature of interaction between adsorbent and adsorbate that is physical adsorption ($\Delta G°_{ads}$ up to -20 kJ mol^{-1}) or chemical adsorption ($\Delta G°_{ads} \leq -40$ kJ mol^{-1})

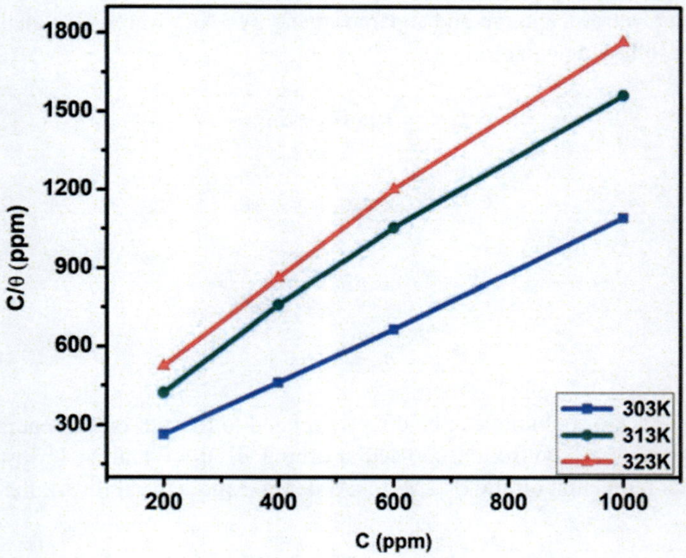

Fig. 7 Langmuir adsorption isotherm of DHABH for soft cast steel in 1 M H_2SO_4

Table 4 Thermodynamic parameters from Langmuir isotherm for DHABH in 1 M H_2SO_4

Temperature (K)	$1/K_{ads}$ (M)	K_{ads} (M^{-1})	R^2	$\Delta G°_{ads}$ (kJ mol^{-1})
303	0.002037551	490.7853	0.9995	−25.73
313	0.007026939	142.3095	0.9954	−23.36
323	0.009608571	104.0737	0.9958	−23.26

[22, 23]. The negative $\Delta G°_{ads}$ value is indicative of the spontaneity of the adsorption process [24] and stability of the adsorbed layer. Thus in the reported examination, the value slightly greater than −20 kJ mol^{-1} shows physicochemical nature of adsorption.

3.4 *Quantum Chemical Parameters*

Recently, quantum chemical calculation method has been utilized for assessing the corrosion inhibition mechanism with the aid of quantum chemical calculations [25]. The energy gap ΔE (HOMO–LUMO) is an important stability index and is applicable for developing the theoretical models which are able to explain conformational and structural barrier in various molecular systems [26]. Reactivity of the compound can be easily predicted on the basis of ΔE as represented in Table 5. Several other parameters like absolute hardness (h), global softness (σ) [26], absolute electronegativity (χ), electronic charge and separation energy (ΔE) were calculated with the help of the following equations:

$$\Delta E = E_{\mathrm{LUMO}} - E_{\mathrm{HOMO}} \tag{10}$$

$$\chi = -(E_{\mathrm{HOMO}} + E_{\mathrm{LUMO}})/2 \tag{11}$$

$$h = (E_{\mathrm{LUMO}} - E_{\mathrm{HOMO}})/2 \tag{12}$$

$$\sigma = 1/h \tag{13}$$

The tendency of bonding can be directly related to frontier orbital energy levels, that is, highest occupied frontier molecular orbital (HOMO) and the lowest unoccupied frontier molecular orbital (LUMO) in order to explore the ability of the inhibitor

Table 5 Computed quantum chemical parameters for DHABH

Quantum parameters	E_{HOMO}	E_{LUMO}	η	χ	ΔN	ΔE_T
DHABH	−8.545	−1.129	3.708	4.837	0.583	−0.927

[27]. HOMO represents the electron-donating ability, whereas LUMO is a representative of the electron acceptance ability of inhibitor molecule [28]. Greater the energy of HOMO, better is its ability to donate electrons to the *d* orbitals of metal while reverse is true for LUMO that is lower the energy of LUMO, excellently it can accept electrons from metal surfaces. Thus, energy gap between HOMO and LUMO is inversely varying with the inhibition efficiency. It could be observed from Fig. 8 that the electron density is mainly concentrated over heteroatoms, benzene ring pointing towards the electron-donating ability via lone pair and conjugated system to interact with the metal surface through donor–acceptor pathway.

The number of transferred electrons (ΔN) has been calculated by application of the Pearson method using Eq. 14 [29] as under

$$\Delta N = \frac{(\chi_{\mathrm{Fe}} - \chi_{\mathrm{inh}})}{2(h_{\mathrm{Fe}} - h_{\mathrm{inh}})} \tag{14}$$

where χ_{Fe} and χ_{inh} de note the absolute electronegativity of iron and inhibitor molecule. h_{Fe} and h_{inh} represent the absolute hardness of iron and the inhibitor molecule, respectively. However, the use of work function (ϕ) of the metal surface instead of electronegativity gives more appropriate results [30]. Thus, Eq. 15 is rewritten as follows:

$$\Delta N = \frac{(\phi_{\mathrm{Fe}} - \chi_{\mathrm{inh}})}{2(h_{\mathrm{Fe}} - h_{\mathrm{inh}})} \tag{15}$$

The derived quantum chemical value for Fe (1 1 0) surface is 4.26 eV [31]. ΔN is a measure of inhibitor's efficiency to transfer its electrons and hence its inhibition efficiency. In the presented work, ΔN value is positive demonstrating the potentiality of DHABH for adsorption to the soft cast steel surface. The presence of heteroatoms (N, O) and conjugated aromatic ring system in DHABH are the most probable modes of adsorption to the metal surface via electron transfer process. The experimental and theoretical calculations are in a good congruous with each other.

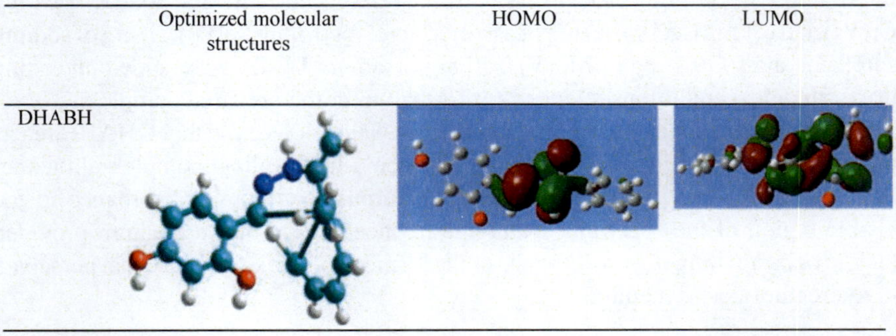

Fig. 8 Geometry optimized structure, HOMO and LUMO of DHABH

3.5 Spectrophotometric Determination of W(VI) Using DHABH as Complexing Agents

Another unconquerable significance of DHABH presented in the work is to go about as a complexing agent where it responds with tungsten (VI) in hydrochloric acid medium to create a 1:3 (M:L) yellow-coloured persistent complex quantitatively distinguishable by extraction into ethyl acetate and studied spectrophotometrically. The procedure for extraction and determination, experimental conditions being maintained while studying different parameters in a steady progression, the optical characteristics, stoichiometry of the formed complex and applications of the technique are referenced underneath.

3.6 Procedure Recommended for Extraction and Determination of Tungsten (VI)

1 ml of 1 M HCl and 1.5 ml of 0.5% (m/v) DHABH solution along with the sample solution containing $\leq$100 μg of W(VI) and appropriate amount of doubly deionized water were brought together into solution in a 125 ml separatory funnel to make up the volume of aqueous phase 10 ml. The contents were equilibrated once for 30 s by extracting out with 10 ml of ethyl acetate. After clear phase partition, the light yellow-coloured organic phase containing complex was passed through a Whatman filter paper (No. 41, 9-cm diameter, preprocessed with the solvent) to eradicate any traces of water left in the concentrate. At 400 nm, the analytical signals for absorbance of the light yellow complex were read against similarly treated reagent blank and from the standard curve thus obtained, concentration of the metal ion can be analysed under indistinguishable conditions.

However for the samples containing Cr(VI), Ce(IV), Se(IV) and Mo(VI), adjustment in the procedure is required to avoid co-extraction of these metals. In their respective specimens, masking or complexing agents were added to prevent the interference accordingly brought about by the referenced elements. For each of 1 mg Cr(VI) and 0.1 mg Ce(IV), 20 mg of ascorbic acid; for 0.1 mg Se(IV), 50 mg of sodium sulphate, and for 0.1 mg of Mo(VI), 50 mg disodium EDTA were added preceding use of the ideal conditions intended for formation of the prepared complex.

Effect of varying experimental conditions. This is observed that DHABH reacts with W(VI) under acidic conditions to produce a light yellow complex altogether extractable into ethyl acetate, indicating maximum and stable absorbance in the visible region of the spectrum. Nature and concentration of the medium provided (acidic or basic) influence absorbance of the formed complex with maximal perceived in hydrochloric acid medium.

The ideal, maximum and stable absorbance was accomplished for the specimens containing ≤100 μg of W(VI) in 10 ml aqueous conditions, 0.05–0.13 M HCl and 1.4–1.7 ml of 0.5% (m/v) DHABH solution in ethanol added progressively in a similar request. The resultant aqueous solution was then equilibrated once with an equivalent volume (10 ml) of ethyl acetate during 10–120 min as aggregated in Table 6. Various water incompatible solvents like chloroform, carbon tetrachloride, benzene, toluene, dichloromethane, 1,2-dichloroethane, isopentyl alcohol, isobutyl methyl ketone and cyclohexane could comfortably separate the investigated complex. However, ethyl acetate was noticed to be maximum acceptable imparting quick and clear phase partition.

To analyse selectivity of the method, tolerance limits of different foreign or diverse ions on complex formation were tested under ideal states of the proposed procedure. Various anions or complexing agents were examined by taking their generally accessible sodium or potassium salts and adding as their mg per 10 ml amounts along with 50 μg W(VI) per 10 ml aqueous phase. Nitrate, 100 mg; carbonate, 80 mg; disodium EDTA, thiocyanate, chloride, sulphate, ascorbic acid, acetate, hydrazine sulphate and iodide, 50 mg each; sulphite, bromide and dithionite, 40 mg each; citrate, 10 mg; tartrate and fluoride, 5 mg each; phosphate, 0.1 ml; H_2O_2, 30% (w/v); and glycerol, 0.5 ml each when added before reagent addition did not affect the determination. However, oxalate interferes seriously causing an error of more than 1%. Comparative investigation was performed of different cations as: Co(II), Hg(II), Ba(II), Mn(II), Mg(II), Pb(II) and OS(VIII), 1 mg each; Ru(III), 0.5 mg; Zr(IV) and V(V), 0.1 mg each; and Fe(III), Pd(II) and Th(IV), 0.08 mg each do not obstruct the absorbance of W(VI)–DHABH complex. Cr(VI), Ce(IV), Se(IV) and Mo(VI) do not influence the determination in the presence of respective masking or complexing agents as depicted in the procedure. The tolerance limit of a foreign ion is taken in the quantities as the maximum amounts causing an error in the absorbance value by ±1%.

Table 6 Effect of physical parameters on the absorbance of W(VI)–DHABH complex

HCl (M)[a]	0.01	0.03	0.04	0.05–0.13	0.14	0.15	0.20		
Absorbance	0.237	0.252	0.300	0.327	0.300	0.276	0.192		
DHABH (ml)[b]	0.2	0.4	0.6	0.8	1.0	1.2	1.4–1.8	1.9	2.0
Absorbance	0.325	0.388	0.453	0.497	0.514	0.540	0.610	0.560	0.542
Equilibration time (s)[c]	0	2	5	10–120					
Absorbance	0.420	0.550	0.590	0.610					

Conditions: [a]W(VI) = 50 μg; DHABH [0.5% (m/v) in ethanol] = 1.5 ml; aqueous volume = solvent volume = 10 ml; solvent = ethyl acetate; equilibration time = 30 s; λ_{max} = 400 nm
[b]HCl = 0.1 M; other conditions are the same as in (a) excepting variation in DHABH concentration; DHABH = 2,4-dihydroxyacetophenone benzoylhydrazone (DHABH)
[c]DHABH [0.5% (m/v) in ethanol] = 1.5 ml; other conditions are the same as in (b) excepting variation in equilibration time

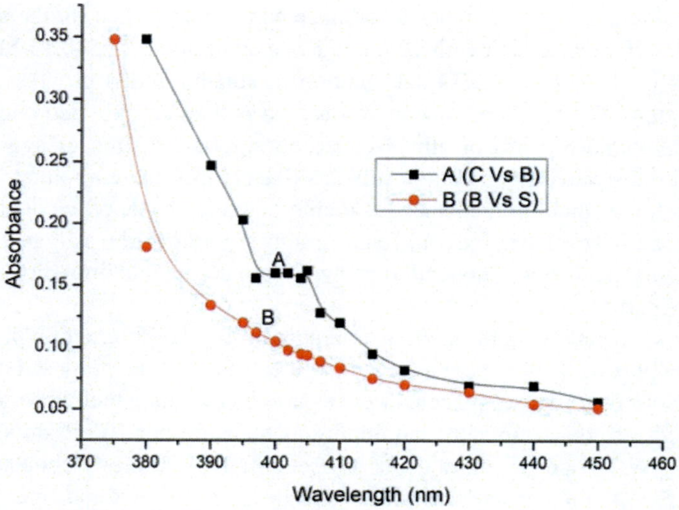

Fig. 9 Absorption spectrum

Optical characteristics and correlation data. At optimal conditions, the absorption spectrum of W(VI)–DHABH complex against reagent blank in ethyl acetate demonstrated the absorption maximum in the wavelength range 397–405 nm as is indicated in Fig. 9, Curve A. The spectrum of the reagent blank against pure ethyl acetate likewise demonstrated a little absorbance at this wavelength as is clearly assessed from Fig. 9, Curve B. The W(VI)–DHABH complex obeys linearity over the concentration range of 0.0–10.0 μg W(VI) ml^{-1}. However, the optimum range for precise determination of tungsten as analysed from Ringbom plot [32] is 1.17–6.98 μg ml^{-1}. Utilizing statistical techniques, various optical and statistical parameters were assessed and brought together in Table 7.

3.7 *Stoichiometry of the Complex*

The 1:3 (M:L) stoichiometry of the prepared complex was established by Job's continuous variations method [33] reformed by Vosburgh and Cooper [34] and certified by the mole ratio [35] and equilibrium shift [36] methods. Fixing two different concentrations (2.718×10^{-3} and 1.358×10^{-3} M) of the metal and ligand, solutions of equal molarity of W(VI) and DHABH were prepared and were employed to compute metal-to-ligand ratio. Quantifying absorbance value at 400 nm, the curves obtained absorbance against mole fraction of W(VI) with a maximum at a mole fraction of 0.25 of tungsten designating 1:3 (M:L) composition of the extracted complex. Validation of the 1:3 (M:L) composition was further made by mole

Table 7 Optical characteristics, precision and accuracy data

S. No.	Parameter	Value
1	λ_{max} (nm)	397–405
2	Beer's law limits (μg ml^{-1})	0–10
3	Molar absorptivity (l mol^{-1} cm^{-1})	2.244×10^4
4	Sandell's sensitivity (μg cm^{-2})	0.0082
5	Correlation coefficient (r)	0.992
6	Regression equation (*Y*)*	$Y = 0.620\,X + 0.015$
7	Slope (*b*)	0.620
8	Intercept (*a*)	0.015
9	Standard deviation	±0.0046
10	Relative standard deviation	0.756%
11	Limit of detection (μg ml^{-1})	1.233

*$Y = bX + a$, where Y is absorbance and X is concentration of W(VI) in μg ml^{-1}

ratio method utilizing two different concentrations of tungsten (2.718×10^{-3} and 1.358×10^{-3} M) kept constant per study and varying.

Concentration of DHABH. Measuring absorbance again at wavelength 400 nm of the variation between absorbance and mole ratio of the two components, a clear break at 1:3 metal to DHABH ratio and hence confirmed the composition. The stoichiometry of the complex was again verified by equilibrium shift method at 2.718×10^{-3} M tungsten concentration. The reagent concentration is varied from 5.436×10^{-4} to 6.795×10^{-4} M. The slope of the plot between log $A_I/(A_0 - A_I)$ and log C_L is found to be 3.2 supporting 1:3 (M:L) ratio as clearly shown in Fig. 10 (where A_0 denoted the maximum absorbance value attained at complete complex formation, A_I absorbance value at a reagent concentration C_L and C_L being the total molar concentration of the reagent added).

Thus, the feasible structure of the light yellow-coloured W(VI)–DHABH complex is proposed as shown in Fig. 11.

4 Computational Studies for Structural Elucidation

The energy optimized structure of W(VI)–DHABH complex (as shown in Fig. 12) was found to have octahedral geometry with an optimized energy −21.1525 kJ/mol. Three moieties of ligand are perpendicular to each other bound to the central metal through oxygen atom with a torsional strain of 1.4343. The equatorial M–O distance is 1.947 Å. The 1,4-VDW interaction is 14.3908 kJ/mol. The O–W–N and O–W–C bond angles are 84.472° and 78.296°, respectively.

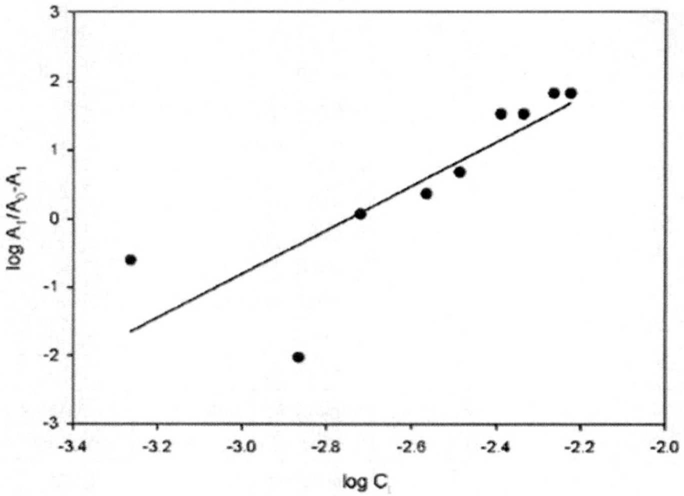

Fig. 10 Equilibrium shift method

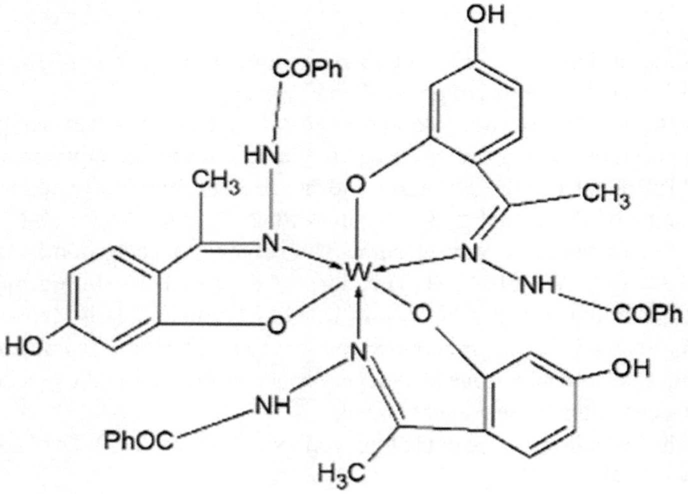

Fig. 11 W(VI)–DHABH complex

5 Conclusions

DHABH for the first time has been applied as anticorrosive as well as complexing agent. The anticorrosive studies of the reagent had been carried out with the help of gravimetric analysis, potentiodynamic polarization and EIS techniques. Theoretical methods were also taken into consideration to analyse corrosion inhibition efficiency. Theoretical and experimental results were found to be compatible with each other.

Fig. 12 Optimized structure of W(VI)–DHABH complex

This was clearly indicated from the obtained data that DHABH can act effectively as a mixed-type inhibitor against corrosion whose efficiency increases with increasing concentration of the reagent. Spectrophotometric studies offered a fruitful method for the micro-determination of W(VI) by the application of complexation behaviour of DHABH. The method was very simple, sensitive and fast (hardly taking 2–3 min for a single determination) and had much wider tolerance limits of foreign ions, thereby enhancing its scope of application. The validity of the method had further been tested by analysing several synthetic samples of varying compositions and the industrial reverberatory flue dust sample. The obtained results were in excellent agreement with the amount of metal ion initially added as is indicated in Table 8.

Table 8 Analysis of synthetic and technical samples by the proposed method

S. No.	Composition of sample*	*W* added (μg)	*W* found (μg)**
1	Cu(0.2), Ni(0.2), Pb(0.15)[a]	25	23.77
2	Cu(1.2), Ni(0.3), Zn(0.5)[a]	20	17.9
3	Cu(0.5), Zn(0.25), Al(0.4)[a]	50	46.5
4	Fe(0.08), Cu(0.02), Co(0.03), Mn(0.005), Ni(0.005)[a]	10	9.83
5	Fe(0.04), Cu(0.015), V(0.008)[a,b]	60	55.7
6	Ce(0.05), Ni(2), Mg(5)	50	48.6
7	Co(5), Hg(2), Ba(3)	50	46.9

(continued)

Table 8 (continued)

S. No.	Composition of sample*	*W* added (μg)	*W* found (μg)**
8	Co(5), Ba(3)	50	49.4
9	Ce(0.05), Ni(2), Mg(3)	25	24.9
10	V(0.008), Cu(0.5), Zn(0.5)[a]	50	46.9
11	Reverberatory flue dust	50 40	51.22 40.07

[a]1–5 correspond to minargent, platinoid, tungsten alloy, heat-resistant steel and high-speed steel, respectively; *mg amounts of metal ions in parentheses; ** average of triplicate analysis; [b]in the presence of 20 mg ascorbic acid

Acknowledgements Sincere thanks are due to the authorities, MMDU, Mullana; UIET, KU, Kurukshetra; and DCRUST, Murthal, for providing the required facilities.

References

1. Fontana MG (2005) Corrosion engineering, 3rd edn. Tata McGraw-Hill, New York
2. Saini N, Kumar R, Lgaz H, Salghi R, Chung Ill-M, Kumar S et al (2018) Minifieddose of urispas drug as better corrosion constraint for soft steel in sulphuric acid solution. J Mol Liq 269:371–380
3. Dahiya S, Lata S, Kumar R, Yadav O (2016) Comparative performance of uraniums for controlling corrosion of steel with methodical mechanism of inhibition in acidic medium: Part 1. J Mol Liq 221:124–132
4. Daoud D, Douadi T, Hamani H, Salah C, Al-Noaimi M (2015) Corrosion inhibition of mild steel by two new S-heterocyclic compounds in 1 M HCl: experimental and computational study. Corros Sci 94:21–37
5. Chaitra TK, Mohana KN, Tandon HC (2018) Evaluation of newly synthesized hydrazones as mild steel corrosion inhibitors by adsorption, electrochemical, quantum mechanical and morphological studies. Arab J Basic App Sci 25:45–55
6. Jandaly AL, Althagafi BA II, Abdallah M, Khairu KS, Ahmed SA (2016) Fluorenone hydrazone derivatives as efficient inhibitors of acidic and pitting corrosion of carbon steel. J Mater Environ Sci 7:1798–1809
7. Singh AK, Thakur S, Pani B, Ebenso EE, Quraishi MA, Pandey AK (2018) 2-Hydroxy-N'-((thiophene-2-yl)methylene) benzohydrazide: ultrasound assisted synthesis and corrosion inhibition study. ACS Omega 3:4695–4705
8. Mahgoub FM, Al-Rashdi SM (2016) Investigate the corrosion inhibition of mild steel in sulphuric acid solution by thiosemicarbazide. Open J Phys Chem 6:54–66
9. Bekheit MM, Allah MTG, Shobaky ARE (2012) Synthesis, antibacterial and corrosion activities of new thiophene-2-carboxaldehyde phenoxyacetylhydrazonecomplexes (HTCPA). ICAIJ 7:150–157
10. Agnihotri N, Dass R, Mehta JR (1998) Extractive spectrophotometric determination of vanadium(V) using 2′-hydroxyacetophenone benzoylhydrazone. J Ind Chem Soc 75:486–487
11. Agnihotri N, Dass R, Mehta JR (1999) 2′-hydroxyacetophenone benzoylhydrazone as an analytical reagent for the spectrophotometric determination of vanadium (III). J Ind Chem Soc 76:165–216

12. Agnihotri N, Dass R, Mehta JR (2003) Spectrophotometric determination of vanadium(III) as its ternary complex with 2′,4′-dihydroxyacetophenone benzoylhydrazone and pyridine. Chem Anal (Warsaw) 48:853–857
13. Jianning L, Zhang B, Wu B, Yongchun L, Xinqiao Y (2006) Spectrofluorimetric determination of trace amounts of cadmium with 2,4-dihydroxyacetophenone benzoylhydrazone. Rare Met 25:184–189
14. Dass R, Mehta JR (1994) A selective and rapid microdetermination of molybdenum(VI) using 2′,4′-dihydroxyacetophenone benzoylhydrazone as a new reagent. Indian J Chem 33A:438–439
15. Kotali A, Tsoungas PG (1987) Oxidation of N-aroylhydrazones of o-hydroxyarylketones with lead(IV) acetate: a facile route to aromatic o-diketones. Tetrahed Lett 28:4321–4322
16. Kumar R, Chahal S, Kumar S, Lata S, Lgaz H, Salghi R et al (2017) Corrosion inhibition performance of chromone-3-acrylic acid derivatives for low alloy steel with theoretical modeling and experimental aspects. J Mol Liq 243:439–450
17. Kumar R, Chahal S, Dahiya S, Dahiya N, Kumar S, Lata S (2017) Experimental and theoretical approach to exploit the corrosion inhibition activity of 3-formylchromone derivatives on mild steel in 1 M H_2SO_4. Corros Rev 35:95–110
18. Kodama K (1963) Methods of quantitative inorganic analysis. Interscience Publishers, New York
19. Verma C, Quraishi MA, Olasunkanmi LO, Ebenso EE (2015) L-proline promotedsynthesis of 2-amino-4-arylquinoline-3-carbonitriles as sustainable corrosion inhibitors for mild steel in 1 M HCl: experimental and computational studies. RSC Adv 5:85417
20. Zarrouk A, Hammouti B, Lakhlifi T (2015) New 1H-pyrrole-2,5-dione derivatives as efficient organic inhibitors of carbon steel corrosion in hydrochloric acid medium: electrochemical, XPS and DFT studies. Corros Sci 90:572–584
21. Zou CJ, Yan XL, Qin YB (2014) Inhibiting evaluation of β-Cyclodextrin-modified acrylamide polymer on alloy steel in sulfuric solution. Corros Sci 85:445–454
22. Verma C, Quraishi MA, Ebenso EE, Obot IB, El Assyry A (2016) 3-Amino alkylated indoles as corrosion inhibitors for mild steel in 1 M HCl: experimental and theoretical studies. J Mol Liq 219:647–660
23. Qu Q, Hao ZZ, Li L (2009) Synthesis and evaluation of Tris-hydroxymethyl-(2-hydroxybenzylidenamino)-methane as a corrosion inhibitor for cold rolled steel in hydrochloric acid. Corros Sci 51:569–574
24. Xu B, Ji Y, Zhang X, Jin X, Yang W, Chen Y (2015) Experimental and theoretical studies on the corrosion inhibition performance of 4-amino-N, N-di-(2-pyridylmethyl)-aniline on mild steel in hydrochloric acid. RSC Adv 5:56049–56059
25. Singh RN, Kumar A, Tiwari RK, Rawat P (2013) A combined experimental and theoretical (DFT and AIM) studies on synthesis, molecular structure, spectroscopic properties and multiple interactions analysis in a novel Ethyl-4-[2-(thiocarbamoyl)hydrazinylidene]-3,5-dimethyl-1H-pyrrole-2-carboxylate and its dimer. Spectrochim Acta A Mol Biomol Spectro 112:182–190
26. Yadav M, Sinha RR, Sarkar TK, Tiwari N (2015) Corrosion inhibition effect of pyrazole derivatives on Mild Steel in hydrochloric acid solution. J Adhes Sci Technol 29:1690–1713
27. Yüce AO, Telli E, Mert BD, Kardaş G, Yazıcı B (2016) Experimental and quantum chemical studies on corrosion inhibition effect of 5,5-diphenyl-2-thiohydantoin on Mild Steel in HCl solution. J Mol Liq 218:384–392
28. Lgaz H, Salghi R, Jodeh S, Hammouti B (2017) Effect of clozapine on inhibition of Mild Steel corrosion in 1.0 M HCl medium. J Mol Liq 225:271–280
29. Martinez S (2003) Inhibitory mechanism of mimosa tannin using molecular modeling and substitutional adsorption isotherm. Mater Chem Phys 77:97–102
30. Obot I, Macdonald D, Gasem Z (2015) Density functional theory (DFT) as a powerful tool for designing new organic corrosion inhibitors. Part 1: an overview. Corros Sci 99:1–30
31. Cao Z, Tang Y, Cang H, Xu J, Lu G, Jing W (2014) Novel benzimidazole derivatives as corrosion inhibitors of Mild Steel in the acidic media. Part II: Theoretical studies. Corros Sci 83:292–298
32. Ringbom A (1938) On the accuracy of Colorimeteric analytical methods I. Z Anal Chem 115:332–343

33. Job P (1928) Formation and stability of inorganic complexes in solution. Ann Chim 9:113–203
34. Vosburgh WC, Cooper GR (1941) The identification of complex ion in solution by spectrophotometric measurements. J Am Chem Soc 63:437–442
35. Yoe JH, Jones AL (1944) Colorimetric determination of iron with disodium-1,2-dihydroxybenzene-3,5-disulfonate. Ind Eng Chem (Anal Ed) 16:111–115
36. Tarasiewicz HP, Grudniewska A, Tarasiewicz M (1977) An examination of chlorpromazine hydrochloride as indicator and spectrophotometric reagent for the determination of molybdenum(V). Anal Chim Acta 94:435–442

A Comprehensive Overview of Sentiment Analysis and Fake Review Detection

Gurpreet Kaur and Kamal Malik

Abstract Sentiment analysis (SA) is based on natural language processing (NLP) techniques used to extract the user's feelings and opinions about any manufactured goods or services provided. Opinion mining is the other name for sentiment analysis. Sentiment analysis is very useful in the decision-making process. With greater Internet use, SA is a powerful tool for studying the opinions of customers about any product or services provided by any business organization or a company. Several approaches and techniques have came to existence in past years for sentiment analysis. Sentiment analysis is useful in decision making. In this paper, we offer an exhaustive description about techniques used for SA, approaches used for SA and applications of sentiment analysis.

Keywords SA · Opinion mining · Machine learning · Lexicon · NLP

1 Introduction to Sentiment Analysis

In this worldwide time, client reviews became relatively necessary to understand different aspects, especially the weaknesses of services and merchandise. The shortcomings or weaknesses can be overwhelmed by client criticism and expelling the bottlenecks. Sentiment analysis offers a tool to deal with client feedback expressed with the sort of text, which is formless data [1]. Opinion mining is the second name of SA [1, 2]. Sentiment analysis refers to processing of natural language and document analytics techniques used for the detection and retrieval of contextual knowledge in source material [3]. In general, the purpose of the sentiment analysis is to decide the speaker's, writer's and author's attitude toward any subject or the whole qualitative polarization of the manuscript. Opinion is either one's decision or the evaluation,

G. Kaur (✉) · K. Malik
CT University, Ludhiana, India
e-mail: gurpreetkaursandhu@gmail.com

K. Malik
e-mail: kamal.malik91@gmail.com

N. Marriwala et al. (eds.), *Mobile Radio Communications and 5G Networks*, Lecture Notes in Networks and Systems 140,
https://doi.org/10.1007/978-981-15-7130-5_22

emotional condition or emotional contact planned [3]. Sentiment analysis is widely used in many different areas like market prediction, recommender system education sector, box office detection, etc. Opinions are expressed with adjectives [4]. The number of adjectives used in conjunction with the product characteristic is called product rank [5]. An opinion could be of two types, one is direct and the other is relative (comparative) [6]. Direct sentiment signals about certain targets, entities like goods and services, events, subjects, people, for example, "Battery backup of Redmi pro 5 is amazingly good". Comparison opinion stated differences or similarities between the objects, e.g., "Bike X is cheaper than Bike Y" [6]. Review sites are increasingly met with either the increase in negativity, i.e., sentiment fraud aimed toward supporting or harming other potential clients through manipulating whether person who reads or computerized opinion mining and SA. Sentiment analysis approaches can be applied for spam review or fake reviews detection [7, 8].

1.1 Levels of Sentiment Analysis (SA)

The analysis of sentiments is conducted at three stages—one is document level, second is sentence level and third is at feature level. Sentence-level sentiment analysis is done via two jobs, subjectivity and objectivity [6] which are as follows:

> Objectivity defined for Dell laptop: I am using Dell laptop from last few years.
> Subjectivity for laptop: Dell laptop is very nice. Subjective sentences are classified into two types—one is positive and other is negative:
>
>> Positive sentence: Dell laptop is a very nice laptop.
>> Negative sentence: This laptop had a deprived battery backup.

In second-level mining, we get a review from the whole article or document and indicated that the whole document as positive sentiment value or negative sentiment value, [9] for example, "thumbs-up," "star ratings," etc. Opinions can also be of feature level, e.g., I bought a phone, it was such a nice phone, and the picture quality is too good. Each feature is classified, and overall sentiment is measured [6].

1.2 Architecture of Sentiment Analysis

In Fig. 1, the sentiment analysis method starts with data collection, which involves a collection of data from sources of opinion. The researcher takes only that data which is related to their research that was collected during the first step of the architecture. After that input document is analyzed. Identification of opinion is then performed to determine whether the text source includes opinions [10]. Yi et al. [11] defined a method that is used for opinion identification, opinion holders and subjects from sentences. They defined the process stepwise; in the first step, identification of

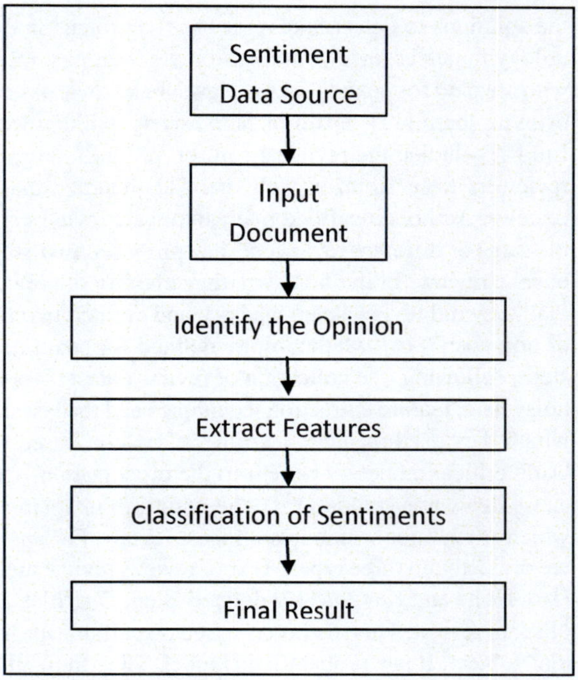

Fig. 1 Sentiment analysis architecture

opinion behavior words is (verbs and adjectives) categorized into positive, negative and neutral. In Fig. 1, after opinion identification, feature extraction is done using various methods or algorithms like lexicon approach and approaches to machine learning that we will discuss under approaches in the third section of this paper. Extracted features are fed to the classifier, and it is necessary to extract the exact features of sentiment. The main function of the classifier is to assign the sentiment polarity. Sentiment classifiers classify the sentiment either into neutral, positive or negative. And the final stage is just to summarize results [10]. The first step of the sentiment analysis architecture is that the reviews are taken and the data processing is done, and in the second step removing the unwanted data is done. In the third stage, review analysis is done with the help of POS (parts of speech tagging) and negation tagging. After that, learning models are applied for sentiment classification, and at the final stage, results are prepared [9].

2 Fake Review

Nowadays, online reviews of the product or service turn out to be the popular source of client reviews or comments. Those comments are utilized by people, companies or organizations to take decisions. Some of the companies or organizations, for profit, produce fake reviews, and customers are mislead by these reviews. But the categorization of the opinions or reviews as genuine or fake is the major problem. One of the solutions to detect fake reviews is sentiment analysis which is a powerful technology to detect reviews and customers' feelings. Opinion fraud includes reviewers who are paid for posting fake reviews. Fake reviews are of two types: defaming spam or hype spam [12]. Spam or fake review is like a web page fraud [13]. Jindal and Liu [13] studied the review spam for 5.8 million opinions and more than 2 million reviewers from Amazon web site. They studied that there exist a number of fake reviews. Authors identified many duplicate reviews and the same reviews posted by the same or different reviewer. They proposed two-step methodology to detect spam or fake review: In the first step, they classify the reviews into three types, and after that, they did review spam analysis and defined two classes for reviews either spam or non-spam [14]. Methodology defined for spam review detection consists of the steps, following the collection of review dataset—preprocess the dataset to remove noisy data, feature extraction techniques and the last step is to detect spam review for which they used machine learning or lexicon-based methods. Positive reviews can bring critical monetary benefit to the organization, and negative reviews frequently cause deals misfortune [15]. The main job in spam review detection is to identify which is spam or which is an honest review. For spam detection, data and features are divided into three types: Text of review, review meta data and review information [16]. Holla and Kavitha [17] defined latent Dirichlet allocation method for detection of spam review. This method is based on certain words or patterns related to a particular subject. It is a probabilistic model. They took 80% of training data and 20% of test data and chose evaluation metric for accuracy. Accuracy given by them is:

$$
\begin{aligned}
A = {} & \text{True Positive} + \text{True Negative}/\text{True Positive} \\
& + \text{True Negative} + \text{False Positive} + \text{False Negative}
\end{aligned}
$$

where true positive is the spam reviews, true negative is the truthful reviews, false positive and false negative reviews. Mukherjee et al. [18] in their paper defined method to detect group spammers in spam reviews. Group spamming is defined as number of consumers putting together spam or fake comments to support or downgrade aimed services or products. They proposed a relationship-based model called GSRank that can take into account the relationship between groups, users and services they reviewed to identify spammers.

Spam or fake review is a review that is not truthful review of the product. It is either posted by spammer or group of spammers to increase the profit of their organization, and customers are mislead by these reviews.

3 Literature Review

Pang et al. [19] defined machine learning methods to find the efficiency of documents classification by overall sentiment. Results showed that learning with machine gave improved results than human-created methods for opinion mining or SA on data of movie reviews. An experiment consists of more than 700 negative and positive reviews. The unigram and bigram features are used for classification. The research is carried out using support vector machine (SVM), Naïve Bayes and maximum entropy classification. The efficiency of machine learning methods was lower than topic-based methods.

Turney [20] used unsupervised learning algorithms to categorize the opinion as "suggested" and "not suggested". Thumbs-up is the example of suggested, and thumbs-down is the example of not suggested. They also used POS to categorize sentences which include adverbs and adjectives.

Yi and Niblack [21] defined a sentiment analyzer to take out a sentiments from given topic which was taken from online texts or documents. This analyzer used NLP techniques. This sentiment analyzer discovers every reference of the content or document and polarization of every reference. This paper gave good result for digital cameras and music.

Ghose [22] defined the product reviews focused on client-oriented and mechanism of producer ranking. The proposed method finds the most effective opinion or review. For aspect-based services or products, reviews containing validated details found in the product description were used, and the subjective point of view is useful for experienced products.

Hu et al. [23] defined removal and summarizing procedure to all the client opinions or reviews of a service. The following steps were used in the above said process:

1. Mined the product features which are written by client. Mining is done by NLP and techniques of data mining.
2. Opinions in the review are recognized and organized as positive and negative.
3. Finalize or summarize the results.

Several semi-supervised techniques have been proposed to provide the solution for the problem of lack of marked data for sentiment classification to solve the classification of opinions or feelings they used under-sampling methods [24].

Ravi et al. [1] explained sentiment analysis in the educational sector and analyzed text reviews on 270 training programs posted by 2688 participants in an organization. To find the efficiency of the proposed work, they found a correlation coefficient between sentiment score and numerical rating assigned by the participants. Pearson's coefficient calculated by them is 0.04, which indicates that sentiment score is nonlinearly related to program rating. Future scope of this paper was extended to find out sentiment expressed over the Internet connection, hospitality, etc.

Peng and Zhong [25] computed sentiment score by shallow parser dependency. It also defined the connection between score of sentiments and fake reviews. They developed a time sequence combined with some discriminative policy to find spam reviews efficiently. Classification accuracy of the different lexicons was given as SentiWordNet which has 56.4% and MPQA 58.5%.

This is the study of fake reviews in China [26]. Dataset was taken from the Chinese review hosting site Dianpoing.com. Firstly, they studied fake reviews of two classes, i.e., fake and unknown. But the unknown class has many fake reviews. To identify fake reviews form unknown class, they used positive unlabeled learning which identified a huge number of spam reviews contained in unmarked data.

The author studied all the features of fake hype reviews and discovered the text analysis was positive or negative [27]. They suggested a method to identify online fake hype reviews of a restaurant based on SA. In their paper, the reviews were of four types such as service, overall attitude, taste and environment, and the review is defined as hype, if the outcome of all types of reviews is constant. Their algorithm gave 74% accuracy.

Li et al. [7] defined machine learning technology to find spam or a fake review. They manually built a spam collection form their crawled reviews. Analyzed the impact of spam recognition features. They provide two methods to exploit unlabeled data. Methods used to exploit unlabeled data are semi-supervised methods and co-training. To identify review spam, they used spam corpus. For future scope, they said that the co-training model is also applied to find opinion mining in blogs or social networking sites.

Godsay [28] defined the process of SA. The first step in the process is to set the goal, and then do the text preprocessing and parsing the content based on their polarity. After that, do the text refinement by finding stop words, synonymous, etc. Last step of the process is to review the data and scoring. This process is applied for sentiment analysis in various areas.

Zvarevashe [29] proposed a sentiment analysis system for hotel customer reviews with an opinion mining. It is a sentiment polarity framework that automatically prepared dataset for training and testing. The analysis was done with the help of Naïve Bayes, a sequential minimal of optimization.

Krishna [30] analyzed customer complaints in Indian banks by using machine learning approaches on unprocessed textual data. For preprocessing raw data, author performed DTM driven by a term frequency—IDF and LIWC method. Raw data of grievances was marked either "Moderate" or "Extreme." They took their data from ICICI, AXIS, SBI, and HDFC banks. Results were good enough to use sentiment analysis in the future for complaints in the banking sector. The accuracy of their algorithm was 74%. This algorithm, in future, will use for various other services for review detection.

Wahyuni and Djunaidy [2] fake review detection using iterative computation framework. They proposed ICF++ for detecting spam reviews of any service that uses the text and ratings from a review. ICF calculates review's honesty value, reviewer's trustworthiness value and product's consistency value. To find review's honesty value, they used opinion mining, and to find fake review, they used algorithm for ratings. The ICF++ algorithm's accuracy value is 6 per cent higher than ICF.

4 Approaches Used for Sentiment Analysis

As discussed in fourth and fifth steps of architecture of sentiment analysis to extract features and classify sentiments, various classification techniques are used.

There are three approaches used for sentiment analysis [31]. For the classification of text, machine learning methods are used. It has two parts: One is training set, and the other is test set. For finding the different features of a document, we use training set, and to check the accuracy of the classifier, test set is used. These techniques are based on labeled data or unlabeled data that means supervised learning or unsupervised learning. Various approaches are used under this like SVM, maximum entropy, Naïve Bayes and neural networks. Naïve Bayes is a probabilistic classifier and employs for less size training data. The base of naïve Bayes algorithm is Bayes theorem given as under:

$$P(X/Y) = P(X)P(Y/X)/P(Y)$$

In this equation, probability of an event X arises given the proof Y. We use this equation to find a sentiment of any sentence as under [32–34]:

$$P(S/\mathrm{SN}) = P(S)P(S/\mathrm{SN})/P(S)$$

where S means sentiment and SN means sentence.

Support vector machine (SVM) is a binary classifier. SVM requires huge amount of training data. The objective of this technique is to locate a hyperplane that divides the tweets or reviews or opinions, and the high margin between the classes. It uses various univariate and multivariate techniques for characteristic selection by employing chi-square method to locate the appropriate text attributes. This method gives high accuracy of 85–88% in average for sentiment analysis [32, 34, 35]. Another probabilistic classifier based on exponential function is a maximum entropy classifier. It is used for NLP. It is based on the principle of maximum entropy which defines that the probability distribution of large entropy best represents the state of knowledge. The maximum entropy theory is that the scattering is supposed to be consistent when nothing is known that means it have maximum entropy [32, 34, 36, 37].

Second is the lexicon-based approach for sentiment analysis which is extensively used to classify text sentiments. Training dataset is not required for these techniques. Opinion words are the words which are commented by the client about any product, and opinion lexicons are the lexicons defined in those comments. This approach counts the number of words of positive and negative opinion in each review sentence near the characteristic of the product or service. To find sentiment score A_t on any day t, use the following equation of positive and negative words:

$$\begin{aligned} A_t &= \mathrm{count}\big(\text{positive words}^\wedge\text{topic}\big)/\mathrm{Count}\big(\text{Negative words}^\wedge\text{topic}\big) \\ &= p(\text{positive words/topic}, t)/p(\text{negative words/topic}, t) \end{aligned}$$

A major problem with the lexicon approach is that it does not operate on words depending on the meaning [32–34, 38, 39].

A hybrid approach is the mixture of both lexicon and machine learning approaches to enhance the performance of sentiment classification. Previous work shows that the hybrid approach increases the efficiency or accuracy over Naïve Bayes, SVM and lexicon-based methods [38, 40].

5 Real-Life Applications of Sentiment Analysis

The sentiment analysis (SA) is well known for multiple domains for studying the feelings and opinion of customers. In this part, we are discussing real-life applications of opinion mining or SA.

Sentiment analysis (SA) is commonly adopted technology to study the feelings and opinions of people about anything like product, service, etc. Therefore, sentiment analysis (SA) is used in many areas as under [41]: social media monitoring: Out of world's total population (i.e., 7.7 billion), 3.5 billion people have actively participated in social media, i.e., Facebook, Twitter and Instagram. Every day, people write lakhs of tweets on Twitter and comments on Facebook. The vast quality of data is a treasure for priceless information about the views, desires and feelings of people toward numerous things: From the food, they want to eat and to the candidate, they are selected for Prime Minister elections. Sentiment analysis (SA) analyzes this data and takes out the opinions and thoughts that underline social media discussion, to know about the thoughts of people about any service. Brand monitoring: In addition to social media, an online discussion will take place in blogs, Web site reviews, news Web sites and discussion forums. For example, for purchasers, product reviews are an important part in the decision process to buy something new. Before buying anything, they read at least 10–15 reviews. Sentiment analysis (SA) is a wonderful method for keeping an eye on the status of the brand, finding out what's correct or incorrect with one's business and understanding customers more. Customer support: The delivery of excellent customer service experiences should be the number one priority for a profitable firm. 96% of customers believe, after all, that excellent customer service is a key factor in choosing a brand and staying loyal. Sentiment analysis will help us make interactions with our customer support faster and more efficient. Customer feedback: Every businessman needs his/her customer feedback to evaluate the performance of their products or services, etc. For that they ask a customer to fill feedback form in which customer has to give a score to every service between 0 and 10. For processing these results, we use sentiment analysis and obtain sufficient results. Market research: If a company wants to introduce a new product and wishes to gather opinions of people's thoughts, perceptions and needs relevant to the product area of the business, we might use sentiment analysis to track various sources of online discussion regarding a particular subject. Analysis of emotions in sociology, psychology and political sentiment analysis: It helps individuals to get an opinion about things on a global scale such as film reviews, political reviews, global problem analysis, news inequality, etc., for example, people read from various blogs about which candidate is good for which post and they vote for that person accordingly [28]. Sentiments of social media images: [42] defined a framework for social media images known as unsupervised sentiment analysis (USEA). SEA provides them with a new perspective to better understand the semantic difference among graphical features and sentiment of picture. Movie review [43] for Turkish text documents they planned a structure for USEA. As part of their method, they modify the SentiStrength sentiment analysis library, by translating their lexicon into Turkish. The educational sector as already explained above in literature review, sentiment analysis is also used for the education and the banking sector [1, 30]. SA for stocks [44]: that paper explains the use of sentiment analysis in the finance and stock market to study the market sentiment, government policies and company announcements. Teraiya and Vohra [45] also defined usage of sentiment analysis (SA) to determine the strength and shortcomings of the government sector by observing group opinions.

6 Conclusion

We overall conclude that sentiment analysis (SA) which is a powerful technology used to evaluate customer's perceptions, feelings, emotions, opinions toward company's product or services provided by a company. It is a good technology for every business organization to know about its brand's strengths (positive opinion) and weaknesses (negative opinion). Sentiment analysis (SA) is broadly applied for different areas like brand monitoring, social media monitoring, customer feedback, finance, stock exchange, educational sector and banking sector. Sentiment analysis is also used to find out the spam reviews of various products or services using different techniques either lexicon or machine learning methods. Major challenge of the sentiment analysis is the fake review detection where the organization adds fake reviews and customers are mislead by those reviews. To overcome this, many techniques are used for detecting spam reviews. Machine learning methods are extensively employed for fake review detection, but some of the researchers used hybrid approach to detect fake reviews, but results are not accurate. We conclude that, in future, more work is to be done on hybrid approach to detect fake or spam reviews about the products or services provided by any organization.

References

1. Ravi K, Siddeshwar V, Ravi V, Mohan L (2016) Sentiment analysis applied to educational sector. In: 2015 IEEE International conference on computational intelligence and computing research ICCIC 2015, vol 500046. https://doi.org/10.1109/ICCIC.2015.7435667
2. Wahyuni ED, Djunaidy A (2016) Fake review detection from a product review using modified method of iterative computation framework. MATEC web of conferences, vol 58.https://doi.org/10.1051/matecconf/20165803003
3. Hovy EH (2015) What are sentiment, affect, and emotion? Applying the methodology of Michael Zock to sentiment analysis. Lang Prod Cogn Lex 48:13–24. https://doi.org/10.1007/978-3-319-08043-7

4. Baccianella S, Esuli A, Sebastiani F (2010) SENTIWORDNET 3.0: an enhanced lexical resource for sentiment analysis and opinion mining. In: Proceedings of 7th conference on language resources and evaluating LREC 2010, pp 2200–2204
5. Eirinaki M, Pisal S, Singh J (2012) Feature-based opinion mining and ranking. J Comput Syst Sci 78(4):1175–1184. https://doi.org/10.1016/j.jcss.2011.10.007
6. Jebaseeli P, Nisha A, Kirubakaran E (2012) A survey on sentiment analysis of (Product) reviews. Int J Comput Appl 47:0975–1888. https://doi.org/10.1007/s10462-017-9597-8
7. Li F, Huang M, Yang Y, Zhu X (2011) Learning to identify review spam. In: International joint conferences on artificial intelligence, pp 2488–2493. https://doi.org/10.5591/978-1-57735-516-8/IJCAI11-414
8. Fontanarava J, Pasi G, Viviani M (2017) Feature analysis for fake review detection through supervised classification. In: Proceedings—2017 international conference on data science and advanced analytics, DSAA 2017, vol 2018, Janua, pp 658–666. https://doi.org/10.1109/DSAA.2017.51
9. Behdenna S, Barigou F, Belalem G (2016) Sentiment analysis at document level. In: International conference on communications, computing and control, vol 628, CCIS, October, pp 159–168. https://doi.org/10.1007/978-981-10-3433-6_20
10. Tedmori S, Awajan A (2019) Sentiment analysis main tasks and applications: a survey. J Inf Process Syst 15(3):500–519. https://doi.org/10.3745/JIPS.04.0120
11. Yi J, Nasukawa T, Bunescu R, Niblack W (2003) Sentiment analyzer: extracting sentiments about a given topic using natural language processing techniques. In: Proceedings—IEEE international conference on data mining, ICDM, pp 427–434. https://doi.org/10.1109/icdm.2003.1250949
12. Akoglu L, Faloutsos C (2012) Opinion fraud detection in online reviews by network effects, pp 2–11
13. Jindal N, Liu B (2007) Analyzing and detecting review spam, pp 547–552. https://doi.org/10.1109/ICDM.2007.68
14. Hussain N, Mirza HT, Rasool G, Hussain I (2019) Spam review detection techniques: a systematic literature review, pp 1–26. https://doi.org/10.3390/app9050987
15. Ho-Dac NN, Carson SJ, Moore WL (2013) The effects of positive and negative online customer reviews: do brand strength and category maturity matter? J Mark 77(6):37–53. https://doi.org/10.1509/jm.11.0011
16. Heydari A, Tavakoli MA, Salim N, Heydari Z (2015) Detection of review spam: a survey. Expert Syst Appl 42(7):3634–3642. https://doi.org/10.1016/j.eswa.2014.12.029
17. Holla L, Kavitha KS (2019) Opinion spam detection and analysis by identifying domain features in product reviews. SSRN Electron J, pp 571–575. https://doi.org/10.2139/ssrn.3352429
18. Mukherjee A, Liu B, Glance N (2012) Spotting fake reviewer groups in consumer reviews. In: WWW'12—Proceedings of 21st annual conference on world wide web, pp 191–200. https://doi.org/10.1145/2187836.2187863
19. Pang B, Lee L, Rd H, Jose S (2002) Thumbs up ? Sentiment classification using machine learning techniques. July, pp 79–86
20. Turney PD (2002) Thumbs up or thumbs down? Semantic orientation applied to unsupervised classification of reviews, presented at the Association for Computational Linguistics 40th Anniversary Meeting, New Brunswick. In: ACL '02 Proceedings of 40th annual meeting on association for computational linguistics, pp 417–424, July, pp 417–424
21. Yi J, Niblack W (2003) Sentiment analyzer: extracting sentiments about a given topic using natural language processing techniques
22. Ghose A, Designing novel review ranking systems: predicting usefulness and impact of reviews categories and subject descriptors
23. Hu M, Liu B, Street SM (2004) Mining and summarizing customer reviews
24. Zhou G, Lee SYM, Li S, Wang Z (2012) Semi-supervised learning for imbalanced sentiment classification Shoushan. In: Proceeding of twenty-second international joint conferences on artificial intelligence semi-supervised, pp 13–16. https://doi.org/10.1109/IALP.2012.53

25. Peng Q, Zhong M (2014) Detecting spam review through sentiment analysis. J Softw 9(8). https://doi.org/10.4304/jsw.9.8.2065-2072
26. Liu B, Mukherjee A, Shao J (2014) Spotting fake reviews using positive-unlabeled learning, vol 18, no 3, pp 2–10
27. Deng X, Chen R (2014) Sentiment analysis based online restaurants fake reviews hype detection. Lecture notes in computer science (including its subseries lecture notes in artificial intelligence (LNAI) and lecture notes in bioinformatics, vol 8710 LNCS, pp 1–10. https://doi.org/10.1007/978-3-319-11119-3_1
28. Godsay M (2015) The process of sentiment analysis: a study. Int J Comput Appl 126(7):26–30. https://doi.org/10.5120/ijca2015906091
29. Zvarevashe K (2018) A framework for sentiment analysis with opinion mining of hotel reviews, pp 0–3
30. Krishna GJ (2019) Sentiment classification of Indian Banks' customer complaints. In: TENCON 2019—2019 IEEE Region 10 Conference, pp 427–432. https://doi.org/10.1109/TENCON.2019.8929703
31. Andrea AD, Ferri F, Grifoni P (2015) Approaches, tools and applications for sentiment analysis implementation, January 2016. https://doi.org/10.5120/ijca2015905866
32. Devika MD, Sunitha C, Ganesh A (2016) Sentiment analysis: a comparative study on different approaches. Procedia Comput Sci 87:44–49. https://doi.org/10.1016/j.procs.2016.05.124
33. Chaturvedi I, Cambria E, Welsch RE, Herrera F (2018) Distinguishing between facts and opinions for sentiment analysis: survey and challenges. Information Fusion, vol 44, June 2017, pp 65–77. https://doi.org/10.1016/j.inffus.2017.12.006
34. Bhuta S, Doshi U, Doshi A, Narvekar M (2014) A review of techniques for sentiment analysis of Twitter data. In: Proceedings of the 2014 international conference on issues and challenges in intelligent computing techniques, ICICT 2014, pp 583–591. https://doi.org/10.1109/ICICICT.2014.6781346
35. Haddi E, Liu X, Shi Y (2013) The role of text pre-processing in sentiment analysis. Procedia Comput Sci 17:26–32. https://doi.org/10.1016/j.procs.2013.05.005
36. Vryniotis V (2013) Machine learning tutorial: the max entropy text classifier. Datumbox [Online]. Available https://blog.datumbox.com/machine-learning-tutorial-the-max-entropy-text-classifier/
37. Nigam K, Lafferty J, Mccallum A, Using maximum entropy for text classification. Science (80)
38. Ding X, Liu B, Yu PS (2008) A holistic lexicon-based approach to opinion mining. In: WSDM'08—Proceedings 2008 international conference web search data mining, pp 231–239. https://doi.org/10.1145/1341531.1341561
39. O'Connor B, Balasubramanyan R, Routledge BR, Smith NA (2010) From Tweets to polls: linking text sentiment to public opinion time series Brendan. In: Fourth international AAAI conference on weblogs and social media
40. Khan FH, Bashir S, Qamar U (2014) TOM: Twitter opinion mining framework using hybrid classification scheme. Decis Support Syst 57(1):245–257. https://doi.org/10.1016/j.dss.2013.09.004
41. Sentiment Analysis Examples
42. Li B, Wang Y, Wang S, Tang J, Liu H (2015) Unsupervised sentiment analysis for social media images. In: 24th International joint conference on artificial intelligence, Buenos Aries, Argentina, 2015, pp 2378–2379. https://doi.org/10.1109/ICDMW.2015.142

43. Senkul P, Ozge Tokgoz Z, Gural Vural A, Barla Cambazoglu B (2012) A framework for sentiment analysis in Turkish: application to polarity detection of movie reviews in Turkish. In: Computer and information sciences III. https://doi.org/10.1007/978-1-4471-4594-3
44. Bapat P (2014) A comprehensive review of sentiment analysis of stocks. Int J Comput Appl 106(18):1–3. https://doi.org/10.5120/18702-9870
45. Teraiya J, Vohra S (2013) Applications and challenges for sentiment analysis: a survey. Int J Eng Res Technol 2(2)

Information Security in Software-Defined Network

Nidhi Kabta, Anjali Karhana, Neeraj Thakur, Soujanya, and Darpan Anand

Abstract For long, software-defined network (SDN) has been the new trend in the field of networking. Despite being way better than the traditional networking approach, security issues related to SDN are a major concern for technologists all over the globe. It has drawn great attention and is still a highly interesting field of study for network administrators. Among the various means suggested to overcome these security issues, working with securing Border Gateway Protocol (BGP) seems to be a feasible solution to securing SDN. BGP deals with transferring data between two autonomous systems by exchanging routing information between the systems. In this paper, we propose an efficient way to secure data transfer, through BGP, between two autonomous systems in SDN. This is achieved by encrypting data using Advanced Encryption Standard (AES) Encryption Algorithm. In order to prove its feasibility, AES data encryption algorithm is implemented on two hosts in SDN and data transfer is traced on Wireshark. Furthermore, we extend our study to conclude that SDN can be secured by encrypting data over BGP and then transferring it safely to the destination.

Keywords Software-defined network · SDN security · BGP security · AES · Data encryption

1 Introduction

Today, gathering information or transferring data to any nook and corner of the world has become a child's play. Everything can be done in a click, with minimum wastage of time. But earlier, things were different with the traditional networking approach.

In a traditional setup, one had to deal with a range of hardware devices; primarily routers, various types of switches and firewalls in order to control the movement of data. Figure 1 depicts the traditional networking architecture. Traditional network had

N. Kabta · A. Karhana · N. Thakur · Soujanya · D. Anand (✉)
Computer Science and Engineering, Chandigarh University, S.A.S. Nagar, Punjab, India
e-mail: darpan.e8545@cumail.in

N. Marriwala et al. (eds.), *Mobile Radio Communications and 5G Networks*, Lecture Notes in Networks and Systems 140,
https://doi.org/10.1007/978-981-15-7130-5_23

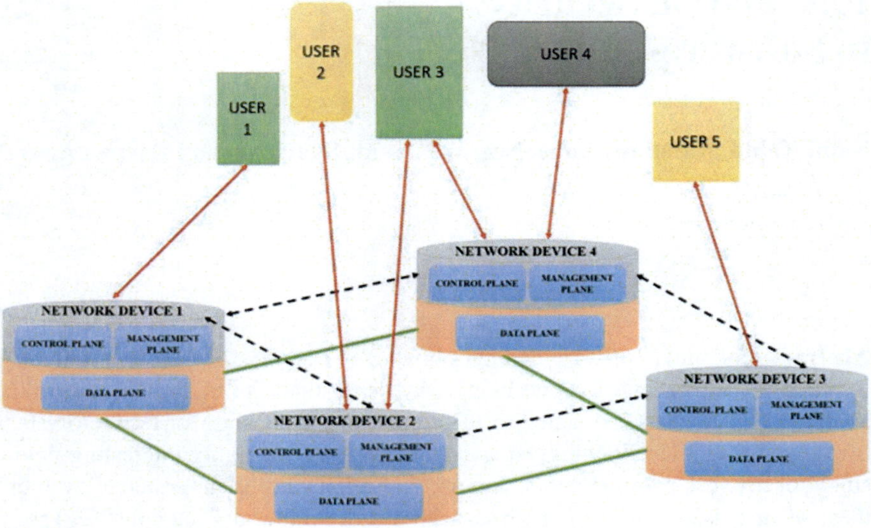

Fig. 1 Traditional networking architecture

control plane embedded with the data plane which provided decentralized control to the entire system. For many years this traditional networking approach was followed to meet the requirements of various organizations.

But as time elapsed and demands grew, the traditional networking approach failed to meet the needs. Many issues related to hardware, cost, security, programmability came up as obstacles to the growth of better networking conditions. A major reason behind the failure of the traditional network was that for a long time, networking was not uniform across the globe. Due to non-uniform networking paradigm, each vendor would create their own solution for a given request. This resulted in need for creating protocols that could operate across different vendors.

To overcome this, researchers thought of ways to minimize the complexity and cost of networking. The development of virtual machines by VMware, about a decade ago, brought about new transformations in the field of networking [1]. It gave rise to concepts like virtualized network functions (VNF) and network function virtualization (NFV).

Virtualized network functions dealt with virtualizing tasks that were carried out by dedicated hardware devices. VNF ensured that network functions were run as software applications, rather than hardware, on a device. VNFs not only increased network scalability and agility, but also ensured better use of network resources. VNF and NFV together not only optimized networking, but also resulted in promoting a better approach to networking through the software-defined network (SDN) [2].

Software-defined network is meant to address the dynamically changing, rapidly growing needs of computing and storage (of modern computing). Apart from being cost-effectiveness, it is flexible as it can be programmed (as per the need of the operator) and can also be implemented in an elastic computing environment in a

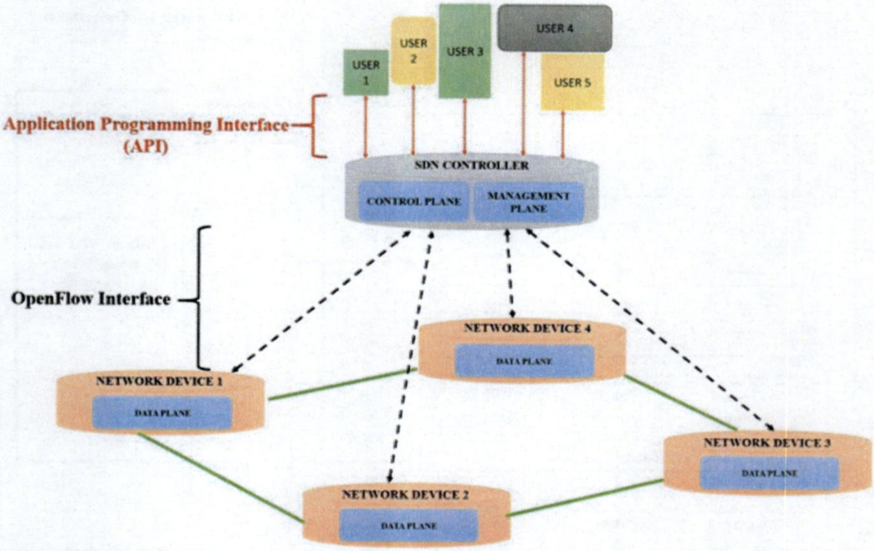

Fig. 2 Software-defined network (SDN) architecture

virtual machine. Software-defined network (SDN) architecture is shown in Fig. 2. SDN comes up with a centralized networking approach and dissociates the control plane from the data plane. The control plane along with the management plane forms the centralized controller of the SDN, which sends commands to data plane for data transfer. Controller maintains all networking paths and programs all networking devices (under it) with a set of commands which are described in OpenFlow protocol [3, 4].

Despite numerous benefits of SDN over the traditional networking approach, security of SDN has always been a matter of concern for the technologists. There are various network security challenges in SDN such as DoS attacks, trust between apps, controllers, and devices. These need to be overcome so as to obtain a robust, safe, and secure networking paradigm.

In this paper, we are to focus on securing Border Gateway Protocol (BGP) when data is transferred between controllers of SDN. BGP is an Exterior Gateway Protocol that deals with inter-domain transfer of data. BGP ensures transfer of data among various autonomous systems and also shares routing information among them. Our task is to secure this information so that it is not manipulated by unauthorized users. We will achieve it by using Advanced Encryption Standard (AES) Encryption Algorithm.

This paper is organized as follows: Section 2 gives a description and highlights the importance of VNF and SDN. Section 3 provides an overview about the related works. Section 4 explains in detail the work that we wish to propose and its implementation. Section 5 depicts the experimental results of the proposed work. Section 6 concludes the discussion. Section 7 highlights the future scopes of the work.

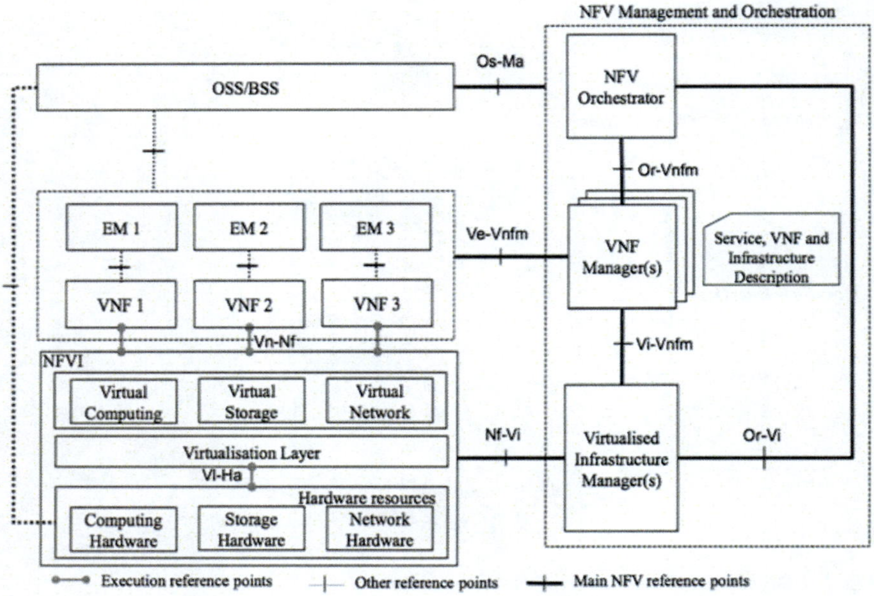

Fig. 3 Virtualized network functions (VNF) architecture

2 Virtualized Network Functions and Software-Defined Networks

Software-defined network (SDN) decouples control plane from data plane, which is programmable. Using SDN, you can monitor and/or configure traffic control and troubleshoot networking devices easily from controller, avoiding lots of manual computations. Therefore, it is both time saving and cost-effectiveness. SDN is agile and centrally managed and you can add new or remove element and devices. It allows network administrator to control traffic flow [5, 6] (Fig. 3).

Virtual network function (VNF) is a software consisting of one or more network functions that are used to provide security to networks created using virtual machines (VMs), without the use of hardware. Two or more VMs are connected to form a network. In VNF, multiple network functions work on single virtual machine using the boxes resources. Examples of VNF in security functions include firewalls, intrusion detection systems, virus scanners, and spam protection [7, 8] (Table 1).

3 Related Work

Although SDN is the future of networking architecture, but there are some limitations of SDN as well. The major limitation of SDN is its security issues such as controller hijacking, black-hole, and unauthorized data modification. Traditional system cannot fix these problems [9, 10].

Table 1 Difference between SDN and VNF

S. No.	SDN	VNF
1	SDN architecture decouples control plane and data plane	VNF architecture handles one or more network function that are built on one or more virtual machines
2	SDN is agile and centrally managed network traffic	VNF also has agile and centrally managed network functions
3	SDN has directly programmable network control	VNF has programmable network functions
4	SDN is a flexible and dynamic networking architecture	VNF runs on NFV infrastructure to provide programmable and dynamic security to NFV
5	Examples of SDN are load balancing using SDN, QoS support using SDN	Examples of VNF are firewalls, IPS, GGSN, SGSN, RNC, EPC, etc.

In SDN, the control of networks is centralized since it has separate data plane and control plane. SDN can be simulated by using different simulators such as NS3, Mininet, OpenDaylight, Floodlight, etc.

Casedo et al. [11] considered the idea of separating control plane and data plane, along with forwarding framework. His SANE architecture, comprising this idea, was proposed in 2006.

Later in 2008, SDN was made by using OpenFlow protocol. It is one of the key components of the SDN. It manages the direct traffic of packets among the network and the switches [12].

In 2015, Belema Agborubere and Erika Sanchez-Velazquez [13] implemented TLS protocol to protect confidential information communicated between the data plane and the control plane of SDN. For authentication, server had to confirm its identity to the client, and vice versa, using their private keys. MAC address was supplied by the sender, along with the messages sent, to the receiver in order to verify its identity at receiver side. TLS protocol protects the data from man-in-the-middle attack. But TLS protocol only ensures secure transmission of data, and it does not provide security to the data.

To provide security to the data, different encryption algorithms are used such as DES and AES algorithms. DES algorithm uses cryptographic key for any given block of code and converts the message into blocks of 64-bits and then encrypts these blocks with key. Decryption is done by implementing the encryption process in reverse order. On the other hand, AES algorithm uses same key to encrypt and decrypt messages [14, 15].

The threats in SDN are present in application layer, data plane, and the control plane. The threat in application layer is destruction of entire network by malicious applications. The control plane is a centralized unit and thus is an easy target for the attackers. These threats can be solved by using AES-256 algorithm [16]. This paper considers the security concerns related to SDN architecture simulated using Mininet. We then use AES algorithm to secure the network from the unauthorized users or attackers.

4 Proposed Work

4.1 Requirement Specification

Linux/Windows Operating System, 8 GB RAM, 15 GB ROM, Virtual box, Mininet, Wireshark.

4.2 Methodology

SDN involves various security challenges [17]. One of the challenges is that any third party can analyze the network traffic. This means they can analyze the source, destination, and type of protocols used to transmit the information between various hosts over SDN [18]. Also, disclosure of transmitted information to the third party can be possible. If less secure encryption technique is used for securing the data, then the possibility of decryption of encrypted data by the third party increases [19, 20]. The proposed scheme will prevent this insecurity involved in the process of transferring message from one host to another host in a software-defined network [21]. In this experiment, we have used Advanced Encryption Standard algorithm for securing the data over SDN [22]. AES: Advanced Encryption Standard algorithm provides high security with great hardware and software implementation speed. It can be implemented on various platforms, especially in small devices. AES uses 10 rounds for data block of 128-bit, 12 rounds for data block of 192-bit, and 14 rounds for data block of 256-bit [23]. In AES algorithm, the following steps are used to encrypt a 128-bit block of data:

1. A set of round keys is derived from the ciphertext.
2. The plain text is input into an array named as state array.
3. The initial round key is added to the starting state array.
4. Required number of rounds is performed on state array and data is manipulated.
5. The final state array obtained is our encrypted data or the ciphertext.

The series of steps involved in each round of encryption are as follows [24]:

- Substitute Bytes: Substitution operation is done to convert every byte into a different value.
- ShiftRows: Each and every row is rotated to the right by a certain number of bytes with respect to the row number.
- MixColumns: Each column of the state array is operated separately with the round key to produce a new column. This replaces the old columns in the state array.
- XorRoundKey: This XOR operation is done between the existing state array and the round key.

Decryption: In this, all the steps taken in encryption are reversed using inverse functions like InvSubBytes, InvShiftRows, InvMixColumns, and consequently receiver gets the actual data.

To perform this experiment, we need a simulation of SDN topology. Mininet allows you to simulate different SDN topologies. We can either use default topology by typing the command: sudo mn or we can create our own topology by using the following commands: sudo./miniedit.py opens up Miniedit interface where one can build different topologies by using OpenFlow switch, hosts, and reference controller. IPs should be assigned to each switch, host, and controller. Enable the flow of switches, save the topology and export the Mininet topology as Python file. To check whether end-to-end connection between hosts is established or not following commands can be used: pingall, h1 ping h2, h1 ping c4, etc. After establishing a successful connection between various hosts, we send information or messages from one host to another with encryption and without encryption by building a client–server socket. To analyze the difference between the two transmissions, we use a packet tracer, Wireshark. Advanced Encryption Standard (AES) algorithm is used to encrypt the data in the data plan. The main aim of the project is to ensure secure transmission of the data between the server and the client. It helps in combating unauthorized access of data by any third party.

Figure 4 represents the working of the project and can be explained as given below:

- Simulating software-defined network: For this, we need a simulator like Mininet. In Mininet, we simulate a network with number of hosts connected to each other using various switches. IPs are assigned to various hosts and switches. Path of data packets sent through these switches will be controlled by an OpenFlow reference controller [25].
- Enabling two-way communication between the two hosts: For this, we create a client–server socket program in Python. This helps in customized packet between the two hosts [26] as represented in Fig. 4.
- Tracing customized packets: For this, we require a packet tracing software like as Wireshark. Using Wireshark, we gather all information about the data packets, packet properties, source, and destination and we save that information for future reference [27] as shown in Fig. 4.
- Enabling secured two-way communication between the two hosts: For this, we create a client–server socket program, in which we encrypt data using AES algorithm. Later, an encrypted, customized data packet is exchanged between the two hosts as shown in Fig. 4.
- Tracing encrypted, customized packet: Using Wireshark, we gather all information about the data packet, other packet properties like source and destination and we save this information for further study and comparison.
- Analyzing the difference between the two captured files of Wireshark [28].

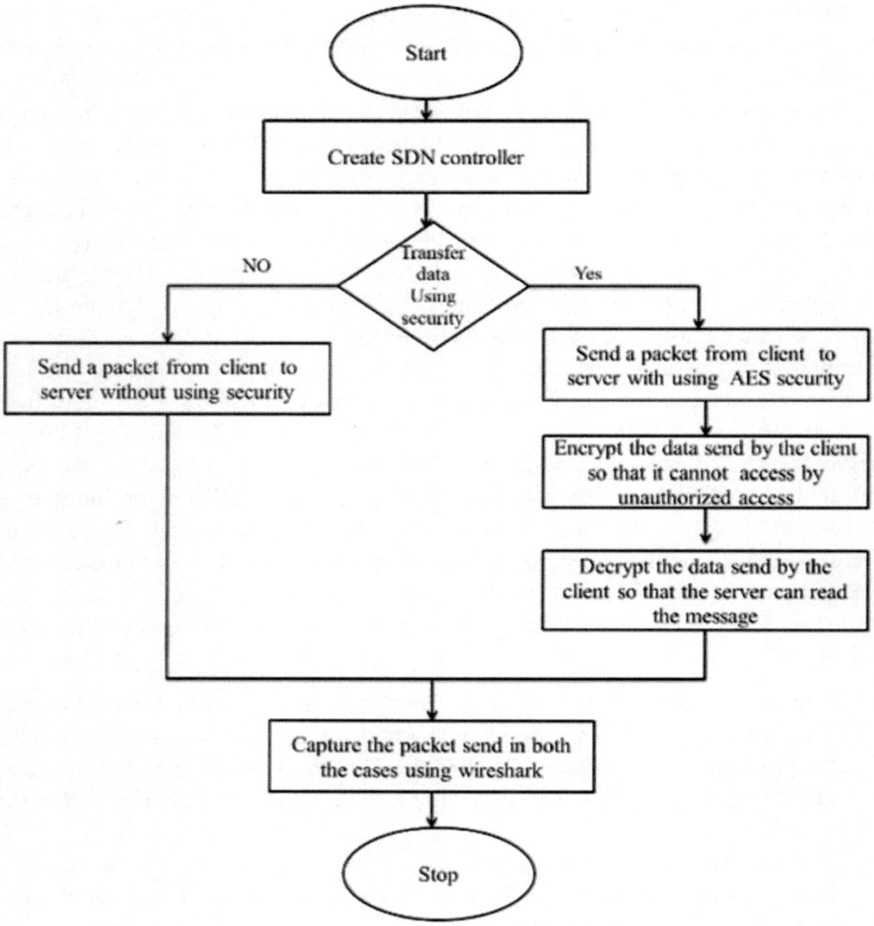

Fig. 4 Flowchart representing working of SDN with AES

5 Experimental Results

Software-defined network architecture is simulated with help of Mininet network emulator and data packets are then traced using Wireshark. Later, AES algorithm was implemented to encrypt data that was being transferred over the network. This helped in ensuring confidentiality of data being transferred over the network. As a result, our data was secure and free from attacking and hijacking.

5.1 Screenshots

Figure 5 shows the design of software-defined network topology created in Miniedit (Editor of Miniedit). It has one OpenFlow reference controller, Four Open vSwitches, and ten hosts connected to each other.

Figure 6 shows the data packets sent to all connected hosts. It checks if every host is reachable to every other host in the network. There is zero percent drop in data packets, indicating that all hosts are reachable.

Figure 7 shows the data packets from host h1 to host h10 are traced on Wireshark. Here, packet number 10 in the sequence is captured in Wireshark.

Figure 8 shows the exchange of customized data packets between two hosts in the network. For sending customized data packets, we create a client–server socket program. One host acts as a client, while the other becomes the server and the data is transferred between the two. Here, host 3 is the client and host 9 is the server.

Figure 9 shows captured data in Wireshark. It has information about customized data packets sent and received between the client and the server. In Wireshark, we note that from host 3 (with IP address 10.0.0.3) message is sent to host 9 (with IP

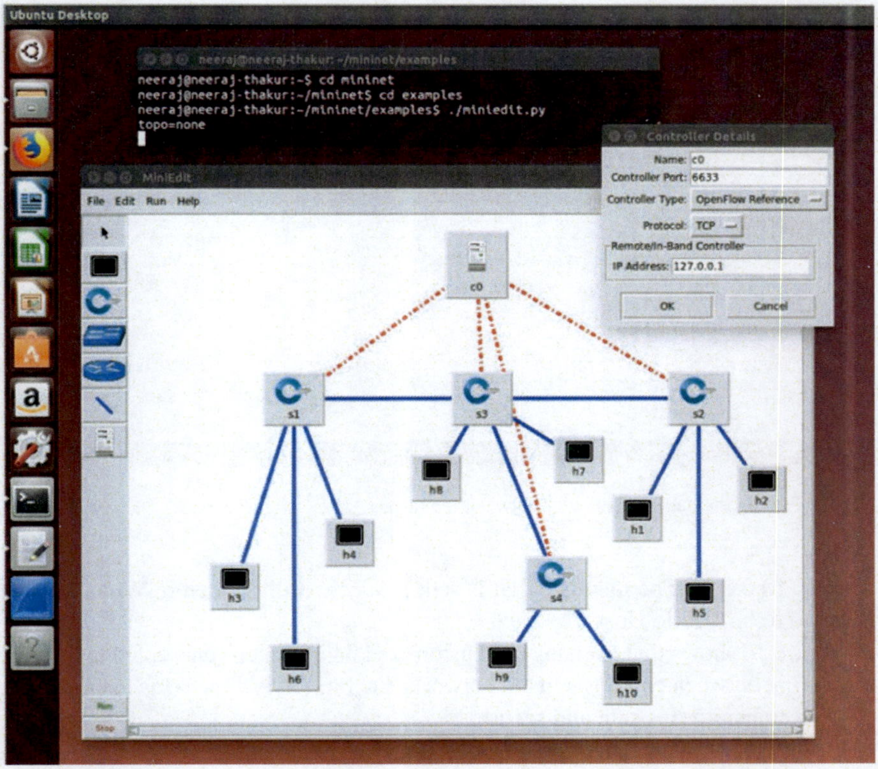

Fig. 5 SDN topology

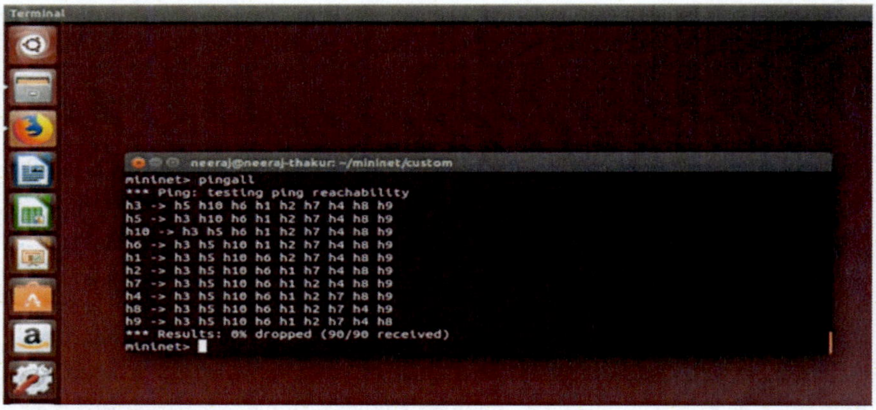

Fig. 6 Sending default data packets to all connected hosts

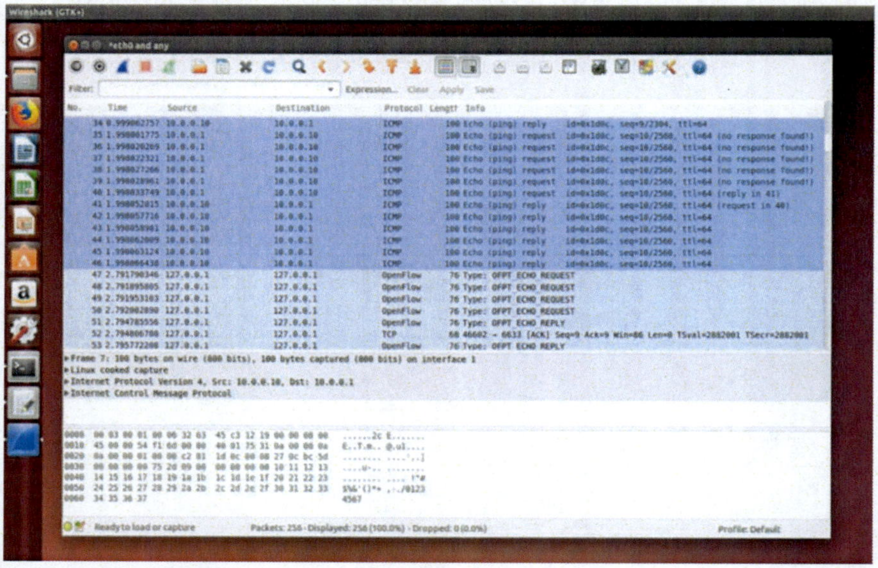

Fig. 7 Tracing the data packets sent between the hosts

address 10.0.0.9). The message that is sent is successfully traced in Wireshark and is found to be "Hello Host 9".

Figure 10 shows that exchange of customized data packets, encrypted using AES algorithm, between two hosts in the network. Encryption is done so as to ensure that the data transferred is safe and secure.

Figure 11 shows the encrypted data sent and received between the client and the server is captured in Wireshark. In Wireshark, we note that from host 3 (with IP address 10.0.0.3) message is sent to host 9 (with IP address 10.0.0.9).

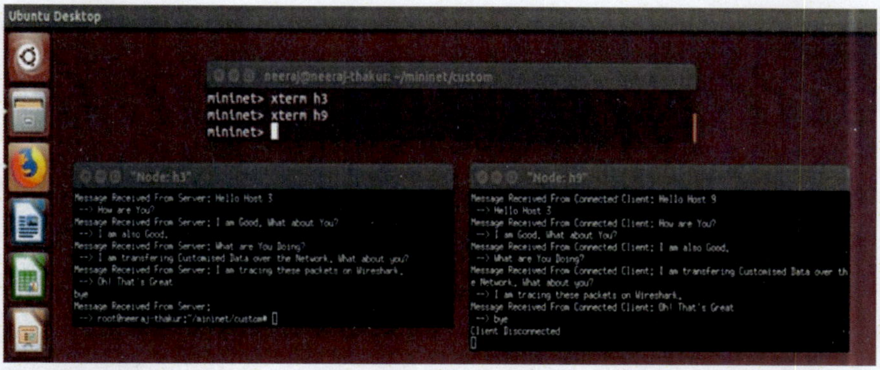

Fig. 8 Sending customized data packets between the hosts

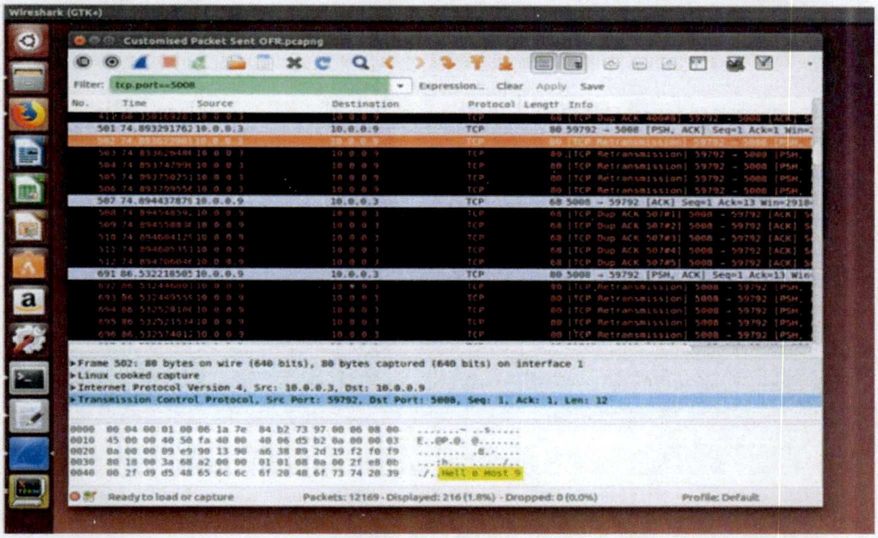

Fig. 9 Capturing customized packets sent between the hosts

The message that is sent is successfully traced (as highlighted in the screenshot above) in Wireshark and is found to be encrypted.

6 Conclusion

SDN is prone to various security risks due its network architecture design, which comprise of the control plane, application plane, and data plane layers.

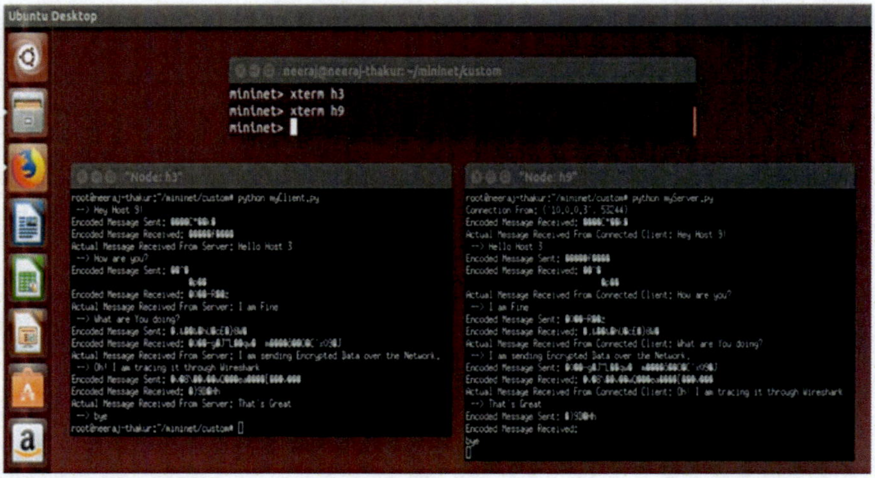

Fig. 10 Sending encrypted customized packet between server and client

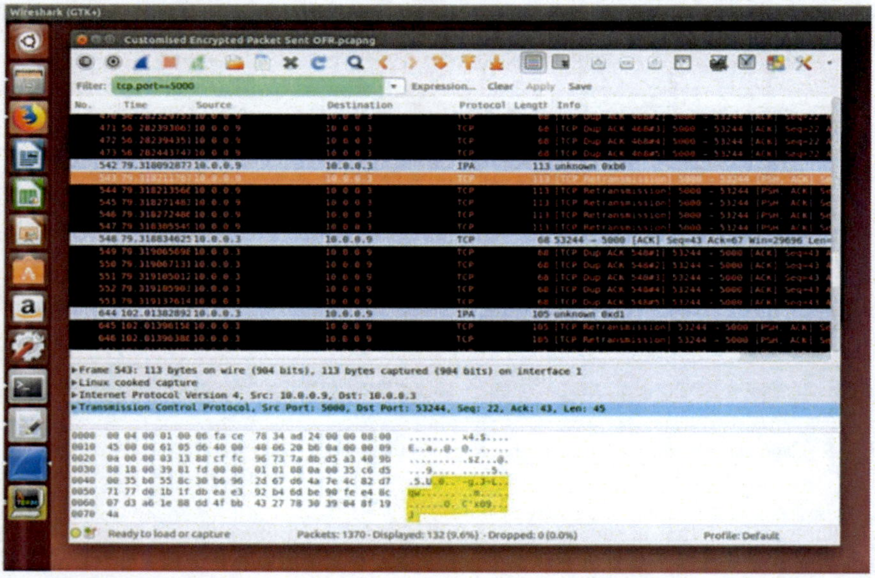

Fig. 11 Capturing encrypted customized data packets from server to client in Wireshark

Server virtualization mobility and cloud computing are becoming the new norms to meet changing business needs with SDN. With the rise of these technologies, the traditional network architecture is starting to fall short of meeting the significant network demands.

6.1 Proposed Work Achievement

SDN should able to meet the requirements of networking community and provide security to ensure integrity and confidentiality of data. Here, we have simulated a software-defined network on Mininet. We have used reference controller and Open-Daylight controller to handle control plane. For deploying security, we have tried to secure data plane as well as control plane by encrypting data, which is being communicated from one host to the other host in server and client environment, using AES algorithm. We have not introduced any new security feature but tried to secure controller by using existing data cryptography algorithm in software-defined network. As we can see that if data is not encrypted then attackers can easily steal information and our network is also prone to hijack attacks. On the other hand, if data is encrypted then attackers cannot have access to data. As a result, data cannot be manipulated as attackers cannot inject virus and worms in order to gain access to the controller and harm the network. Then, we have traced packets using Wireshark with security using ethn0 and any interface. To check the security issues, we have performed DoS attack on SDN and we found that network is secure enough. We worked on AES security to ensure secure transfer of data by each host connected to network and provide confidentiality to data being transferred on network. AES uses three main parameters—16-byte key which is to be shared with client for decryption, second one is AES vector, and third is AES MODE CBC. So, SDN can be secured in more efficient ways and has capacity to change the whole business networking scenario by making things easier and faster.

6.2 Findings

- Cryptography in software-defined network (SDN) using AES algorithm has enhanced security in network.
- While communicating within one switch, data was successfully shared from one host to other using private key.
- Encryption between one host to other across different controllers with minimal chance of data loss or modification is possible with our project model.
- We are able to maintain data integrity and data confidentiality successfully.

6.3 Advantages

- Does not require extra cost and effort.
- Effective and efficient with less time complexity.
- Include all AES advantage over other algorithms, for example, DES which is slow and can be decrypted by attackers.

- Any organization interested in having software-defined network as their network they can use AES in SDN to securely send data over network.

7 Future Scope

SDN can be integrated with various security features and protocols to make it secure. Certain security protocols of traditional networking can be incorporated to SDN with more research and efforts of networking scientists. They can be modified such that it can be applicable to SDN. SDN security already has certain issues which makes it prone to attackers, and can be modified to ensure validation of SDN communication link and content. Product evolution in both networking and virtual infrastructures by the incorporation of SDN as a network will make task easy as well as effective. The spread of micro-segmentation in pursuit of software-defined perimeters and zero-trust computing suggests that BGP can be incorporated in SDN. Various other encryption algorithms like DES can also be used to secure data. New security model can be designed to make controller more secure by authenticating hosts in LAN and WAN of SDN. In data centers as well as in clouds for supervising the whole network, which will decrease the cost and reduce time with strong security framework, SDN can bring wonders to the networking community.

References

1. Nadeau TD, Gray K (2013) SDN: Software defined network: an authoritative review of network programmability technologies, 1st edn. O'Reilly Media, Inc., Sebastopol, CA
2. Flauzac O, Gonzlez C., Hachani A, Nolot F (2015) SDN based architecture for IoT and improvement of the security. In: 2015 IEEE 29th International conference on advanced information networking and applications workshops. IEEE, pp 688–693
3. Cabaj K, Wytrebowicz J, Kuklinski S, Radziszewski P, Dinh KT (2014). SDN architecture impact on network security. In: FedCSIS position papers, pp 143–148
4. Yoon C, Park T, Lee S, Kang H, Shin S, Zhang Z (2015) Enabling security functions with SDN: a feasibility study. Comput Netw 85:19–35
5. https://www.oreilly.com/library/view/sdn-and-nfv/9780134307398/ch34.html. Last accessed 23 Sept 2019
6. Hu Z, Wang M, Yan X, Yin Y, Luo Z (2015) A comprehensive security architecture for SDN. In: 2015 18th International conference on intelligence in next generation networks. IEEE, pp 30–37
7. https://www.networkworld.com/article/3206709/what-s-the-difference-between-sdn-and-nfv.html. Last accessed 15 Oct 2019
8. https://www.cisco.com/c/en/us/solutions/software-de_ned-networking/sdn-vs-nfv.html. Last accessed 05 Nov 2019
9. Scott-Hayward S, O'Callaghan G, Sezer S (2013) SDN security: a survey. In: 2013 IEEE SDN for future networks and services (SDN4FNS). IEEE, pp 1–7
10. Banse C, Rangarajan S (2015) A secure northbound interface for SDN applications. In: 2015 IEEE Trustcom/BigDataSE/ISPA, vol 1. IEEE, pp 834–839

11. Casado M, Gar_nkel T, Akella A, Freedman MJ, Boneh D, McKeown N, Shenker S (2006) SANE: a protection architecture for enterprise networks. In: USENIX security symposium, vol 49, p 50
12. Kloeti R, OpenFlow: a security analysis, April 2013 [Online]. Available ftp://yosemite.ee.ethz.ch/pub/students/2012-HS/MA-2012-20signed.pdf
13. Agborubere B, Sanchez-Velazquez E (2017) OpenFlow communications and TLS security in software-defined networks. In: 2017 IEEE International conference on internet of things (iThings) and IEEE green computing and communications (GreenCom) and IEEE cyber, physical and social computing (CPSCom) and IEEE smart data (Smart-Data). Exeter, pp 560–566
14. Ertaul L, Venkatachalam K (2017) Security of software defined networks (SDN). In: Proceedings of the international conference on wireless networks (ICWN). The steering committee of the world congress in computer science, computer engineering and applied computing (WorldComp), pp 24–30
15. Nara R, Togawa N, Yanagisawa M, Ohtsuki T (2009) A scan-based attack based on discriminators for AES cryptosystems. IEICE Trans Fundam Electron Commun Comput Sci 92(12):3229–3237
16. Iqbal M, Iqbal F, Mohsin F, Rizwan M, Ahmad F, Security issues in software defined networking (SDN): risks, challenges and potential solutions
17. Li W, Meng W, Kwok LF (2016) A survey on OpenFlow-based software defined networks: security challenges and countermeasures. J Netw Comput Appl 68:126–139
18. Mishra S, AlShehri MAR (2017) Software defined networking: research issues, challenges and opportunities. Indian J Sci Technol 10(29):1–9
19. Sezer S, Scott-Hayward S, Chouhan PK, Fraser B, Lake D, Finnegan J, Vilijoen N, Miller M, Rao N (2013) Are we ready for SDN? Implementation challenges for software-defined networks. IEEE Commun Mag 51(7):36–43
20. Rana DS, Dhondiyal SA, Chamoli SK (2019) Software defined networking (SDN) challenges, issues and solution. Int J Comput Sci Eng 7:1–7
21. Padmavathi B, Kumari SR (2013) A survey on performance analysis of DES, AES and RSA algorithm along with LSB substitution. IJSR, India
22. Mahajan P, Sachdeva A (2013) A study of encryption algorithms AES, DES and RSA for security. Glob J Comput Sci Technol
23. Singh G (2013) A study of encryption algorithms (RSA, DES, 3DES and AES) for information security. Int J Comput Appl 67(19)
24. Abdullah AM (2017) Advanced encryption standard (AES) algorithm to encrypt and decrypt data. Crypt Netw Secur 16
25. De Oliveira RLS, Schweitzer CM, Shinoda AA, Prete LR (2014) Using Mininet for emulation and prototyping software-defined networks. In: 2014 IEEE Colombian conference on communications and computing (COLCOM). IEEE, pp 1–6
26. Elamaran V, Arunkumar N, Babu GV, Balaji VS, Gmez J, Figueroa C, Ramirez-Gonzlez G (2018) Exploring DNS, HTTP, and ICMP response time computations on brain signal/image databases using a packet sniffer tool. IEEE Access 6:59672–59678
27. Banerjee U, Vashishtha A, Saxena M (2010) Evaluation of the capabilities of WireShark as a tool for intrusion detection. Int J Comput Appl 6(7):1–5
28. https://hackertarget.com/wireshark-tutorial-and-cheat-sheet/. Last accessed 12 Nov 2019

Emotions Recognition Based on Wrist Pulse Analysis

Tanima, Akshay Kumar Dogra, Indu Saini, and B. S. Saini

Abstract In modern time, emotions affect people in many aspects of life. Long-term emotional problems lead to mental and physical problems such as depression. Wrist pulse contains the information regarding the physiological and pathological state of an individual. Overall mental and physical status of human can be checked with the wrist pulse through pulse examination or nadi parikshan. This paper discusses the parameters of wrist pulse in terms of Vata, Pitta, and Kapha energies for different emotions (anger, calm, fear). Manually, Vata, Pitta, and Kapha are sensed by the medical practitioners through three fingers. Here the pulse data is being acquired through the optical pulse sensor interfaced with the ATMEGA328 microcontroller. Signal processing techniques as approximate entropy, sample entropy, augmentation index, pulse rate are applied in MATLAB to extract the features of Vata, Pitta, and Kapha to recognize the emotions. It is found that Vata rate is high for fear emotion while Pitta rate is high for anger emotion and for calmness emotion, Kapha rate is high. Approximate entropy, sample entropy, augmentation index are more for anger emotion than other emotions.

Keywords Emotions · Wrist pulse · Pulse sensor · Microcontroller · Signal processing

Tanima (✉) · A. K. Dogra · I. Saini · B. S. Saini
Department of Electronics and Communication Engineering, Dr. B.R. Ambedkar National Institute of Technology, Jalandhar, India
e-mail: tanimaahuja@gmail.com

A. K. Dogra
e-mail: akshaykumarnit424@gmail.com

I. Saini
e-mail: indu.saini1@gmail.com

B. S. Saini
e-mail: sainibs@nitj.ac.in

N. Marriwala et al. (eds.), *Mobile Radio Communications and 5G Networks*, Lecture Notes in Networks and Systems 140,
https://doi.org/10.1007/978-981-15-7130-5_24

1 Introduction

The emotions are considered as a universal language as people interact socially with emotions. Facial expressions mainly express the emotions which are sometimes difficult to perceive. Basically, the mental state of the person is reflected through the emotions. This mental state determines the physical efforts and actions of the person. The automatic recognition of the emotions with the sensors is very useful in many areas as psychological studies, image processing, robotics, and virtual reality applications. Efforts in the area of emotion recognition are being made to gather the information from the ECG and EEG signals. Building an emotion recognition and face detection system is a great area of research these days [1]. Wrist pulse signal can also be used for emotion recognition [2]. Machine learning algorithms are being refined for emotion recognition in real time. Wrist pulse analysis is done through pulse examination which is known as nadi parikshan in Sanskrit language [3].

Arterial palpation is generated when the heartbeats. Blood flows from the left ventricles to aorta when the heart contracts. Pulse waveform is produced by the heart at that time. Due to the compliance of the arteries, the blood flows to other body parts of the body and the pulse pressure depends on it. The pulse wave comprised of forward wave which is formed during the systolic phase where the pulse wave travels away from the heart, and the reflected wave which is generated during diastolic phase. Figure 1 shows the standard pulse waveform.

Pulse is sensed through the nadi parikshan technique which is performed on the wrist with the help of index finger, middle finger, and ring finger. Normally, palm of the practitioners supports the wrist of the patient and pulse is examined using fingertips. This diagnosis requires long period of study and practice [5] without any physical recording and analysis. The radial artery gives the sensation of three bioenergies which are Vata, Pitta, and Kapha. The Vata is sensed by the index finger, Pitta by middle finger, and Kapha by ring finger. Figure 2 shows the pulse examination technique.

The characteristics of Vata, Pitta, and Kapha are different from each other, and dominance varies from person to person. Some people have one dominant energy while some have a combination of two or more energies. The imbalance in these energies is the root cause of many diseases. The characteristics of Vata, Pitta, and Kapha along with the physiological and pathological effects of these energies are listed in Table 1.

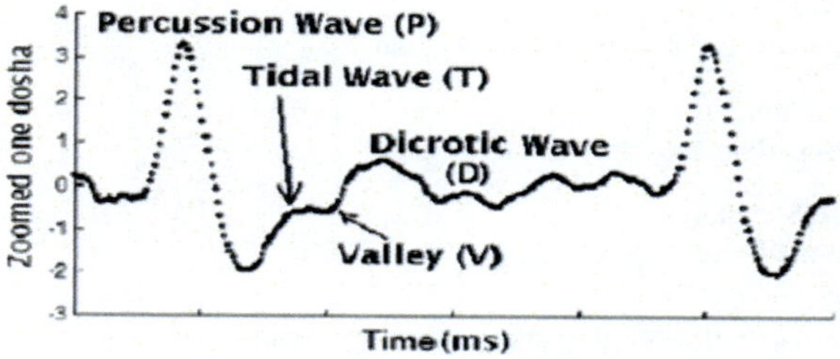

Fig. 1 Pulse waveform [4]

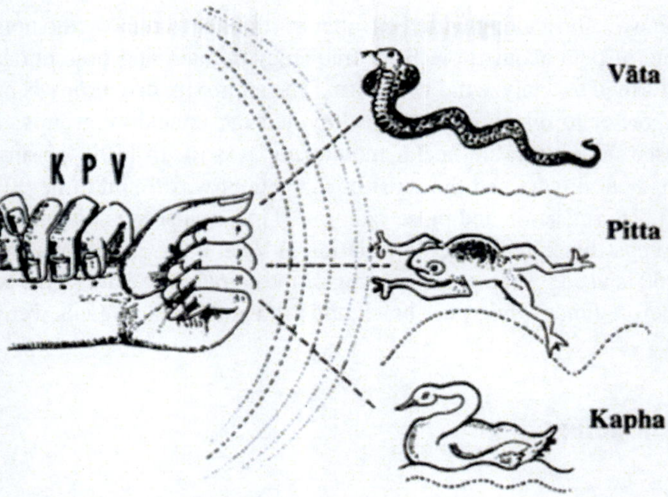

Fig. 2 Pulse examination [6]

Table 1 Characteristics, physiological, and pathological effect of Vata, Pitta, Kapha [6–8]

	Vata	Pitta	Kapha
Characteristics	Light, fast, thin, disappears on pressure	Forceful, strong as lifts up the palpating fingers	Regular, slow, deep
Sensation	Like a cobra	Like a frog	Like a swimming swan
Temperature	Cold	Hot	Cool or warm
Pulse Rate	80–95	70–80	50–60
Force	Less	High	Medium
Body Size	Slim	Medium	Large
Body Weight	Low	Medium	Overweight
Skin	Rough, thin, dry	Oily, smooth	Thick, oily
Chin	Thin	Triangular	Double chin
Eyes	Small, brown, black	Bright, sharp, gray	Big, beautiful
Nose	Uneven	Long	Short
Lips	Dry	Red	Oily
Digestion	Irregular, gas formation	Burning sensation	Mucous problem
Intellectually	Quick	Accuracy	Slow

In [9], the idea to diagnose the imbalance in Vata, Pitta, and Kapha is proposed with the help of machine learning algorithms. This is useful for the disease diagnoses. The system containing the transducer is used to record the pulse waveform for checking the imbalance in Vata, Pitta, and Kapha. In [3], the author presented the modern approach to design the Nadi diagnosis system. The photoelectric sensor was used to record the waveform along with the Butterworth filter to remove the noise produced at the time of data acquisition. Both frequency domain and time domain analyses were performed to analyze the waveform. The approximate entropy is proved as the main parameter to distinguish the healthy and the unhealthy groups. This system can be used as a wearable health monitoring system. In [10], the angle between the mechanical fingers and the wrist of the patient was found to be 90 degrees for accurate pulse sensation and pulse rate. In [11], the pulse was diagnosed before the lunch and post-lunch to check the variation in Vata, Pitta, and Kapha energies. Here Biopac 150 is used for the preprocessing of the signal. The Vata signal was found to be high before lunch while post-lunch, the Pitta and Kapha signals were high.

2 Methodology

For data acquisition, the optical pulse sensor is interfaced with the microcontroller to acquire the waveform of Vata, Pitta, and Kapha from the radial artery at the wrist. ATMEGA328 is used to convert the sensed signal of the pulse sensor from analog into digital form as the optical pulse sensor is of analog nature. The sensed signals are preprocessed with the Butterworth filter in the MATLAB. Necessary features are extracted from the Vata, Pitta, and Kapha waveform parameters. Methodology schematics is shown in Fig. 3.

2.1 Data Acquisition

For data acquisition, the emotions are simulated with the help of videos giving the sense of fear, anger, and calmness or peace. At that simulation, the band consisting of pulse sensor is placed on the wrist to acquire the required waveform. Figure 4 shows the wrist band used for data acquisition.

The optical pulse sensor works on the principle of photoplethysmography. This sensor has two components, i.e., LED and photodiode, embedded on it which are mainly responsible for pulse detection. LED emits the light on the radial artery. The blood flowing in the blood capillaries absorbs some amount of light, and rest is reflected back which is detected by the photodiode or photodetector. This reflection

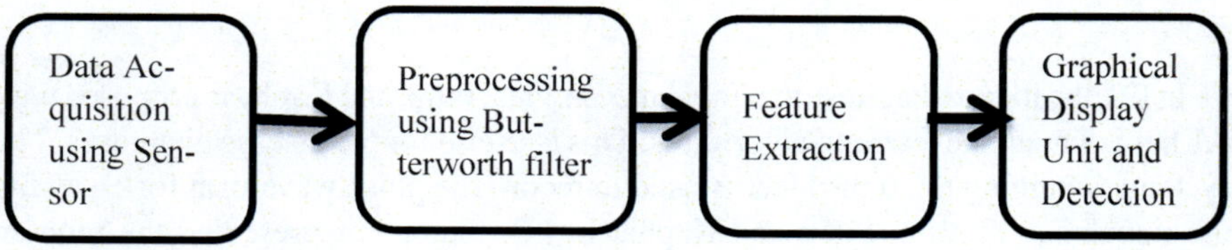

Fig. 3 Methodology

Fig. 4 Wrist band for data acquisition

is based on the volume of the blood inside the blood vessels. Figure 5 shows the setup for data acquisition.

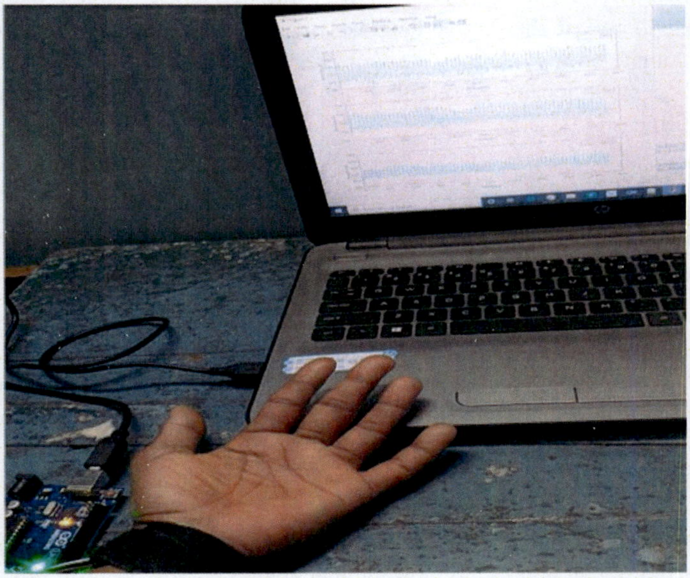

Fig. 5 Data acquisition setup

Pulse examination is done on the right hand for the male while on the left hand for female, as the nerve plexus is different for both male and female [12]. Proper position of the pulse sensor is necessary to acquire the precised data. To maintain the balanced pressure on the optical pulse sensor, the Velcro tape is being used.

2.2 Preprocessing

The pulse sensor has an inbuilt noise removal circuit below the LED. But due to interaction with the skin and the muscles, the signal is contaminated from noise. Here the low-pass Butterworth filter with cutoff frequency 200 Hz is designed in MATLAB to preprocess the acquired Vata, Pitta, and Kapha signals.

2.3 Feature Extraction

The preprocessed signals are obtained on MATLAB to extract the necessary features from different parameters. The Vata, Pitta, and Kapha rate, sample entropy, approximate entropy, augmentation index are mainly used to extract the features of Vata, Pitta, and Kapha signals. The approximate entropy quantifies the time domain signal based on the assumption that the signal which has more repetitive fluctuation patterns gives more predictable results than the signal with less repetitive patterns [13]. It has strong dependence on the sample length. Sample entropy being the modification of the approximate entropy defines the more regularity of the signal by less value of it [14]. It does not depend on sample length. Augmentation index is related to the arteries stiffness [15].

3 Result

With the data acquisition system, the data samples are being acquired. Videos were used to simulate the anger, fear, and calm emotions. Data is collected for 15 subjects where the health condition of person is also considered at the time of acquisition. Subjects are classified into three groups as fear emotion group, anger emotion group, calm emotion group. Table 2 lists the extracted features for these groups.

The approximate entropy, sample entropy, and augmentation index are found more for anger emotion group than other two groups which clearly indicates more repetitive pattern of fluctuation in pulse signal for anger emotion as approximate entropy is more than fear emotion group value which is followed by calm emotion group approximate entropy value. Similar trend is observed for sample entropy. As approximate entropy depends on the sample length, so total 5000 samples are considered for all the subjects. The augmentation index which measures arterial

Table 2 Extracted features for various emotions

Parameters	Fear emotion			Anger emotion			Calm emotion		
	Vata	Pitta	Kapha	Vata	Pitta	Kapha	Vata	Pitta	Kapha
Approximate entropy	0.192	0.202	0.156	0.200	0.220	0.211	0.150	0.182	0.136
Pulse rate	100.12	83.13	89.51	77.79	102.29	77.75	78.54	79.29	110.54
Sample entropy	0.154	0.165	0.129	0.170	0.178	0.183	0.109	0.144	0.100
Augmentation index	0.954	0.954	0.960	0.970	0.962	0.968	0.949	0.950	0.958

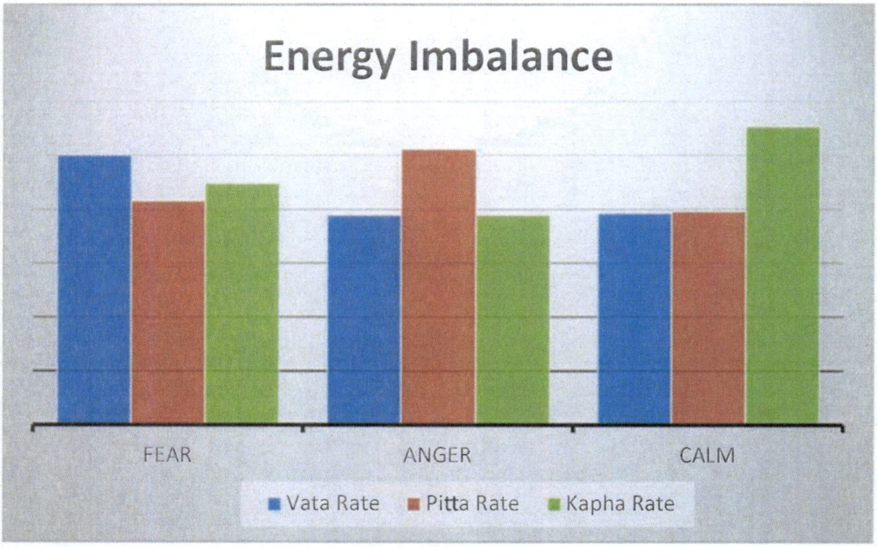

Fig. 6 Vata, Pitta, Kapha dominance for fear, anger, and calm emotion

stiffness is useful parameter to indicate the cardiovascular disease and is observed more in anger emotion. Fear emotion group has more augmentation index than the calm emotion group. For the anger emotion, the Pitta rate is high, and for fear emotion, the Vata rate is more while Kapha rate is high for calm emotion. The imbalance in Vata, Pitta, and Kapha energies are associated with fear, anger, and calm emotion, respectively. Figure 6 shows the Vata, Pitta, and Kapha imbalance for fear, anger, and calm emotion.

Figures 7, 8, and 9 show the Vata, Pitta, and Kapha waveform for fear, anger, and calm emotion group, respectively. More fluctuations in waveform are observed for fear and anger emotion group.

4 Conclusion

The study of the determination of emotion and imbalance in Vata, Pitta, and Kapha is carried out with the optical pulse sensor interfaced with the Arduino ATMEGA328. Low-pass Butterworth filter is used to remove the noise from the acquired signal. The features of Vata, Pitta, and Kapha are extracted from the required parameters. The subjects are classified into three groups based on simulated emotions, i.e., fear emotion group, anger emotion group, and calm emotion group. The approximate entropy and sample entropy are more for anger emotion group which indicates more similar pattern in the pulse waveform. The augmentation index is also more for anger emotion group which indicates more arterial stiffness and chances of cardiovascular

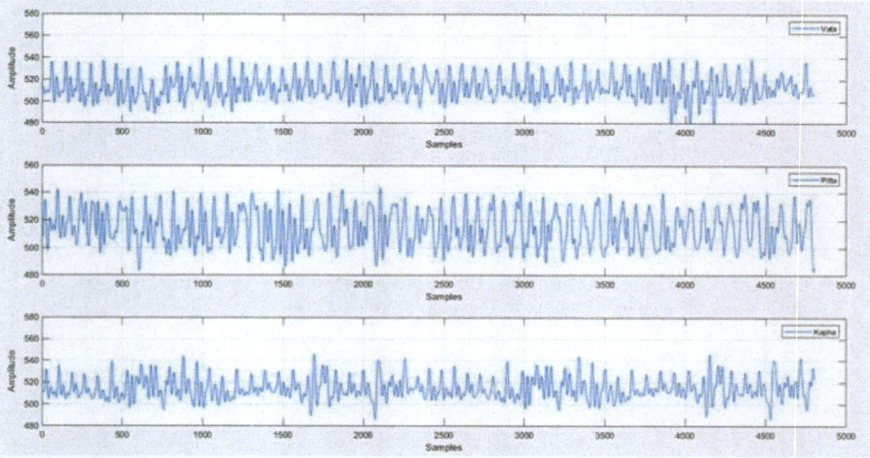

Fig. 7 Vata, Pitta, and Kapha waveform for fear emotion group

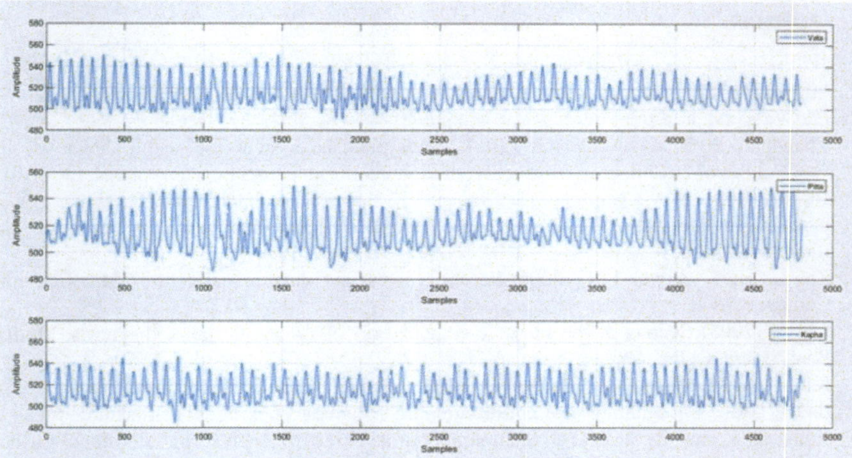

Fig. 8 Vata, Pitta, and Kapha waveform for anger emotion group

diseases. The dominance of imbalance in Vata is observed for fear emotion group while the imbalance of Pitta is associated with anger emotion group and Kapha imbalance for calm emotion group. The acquired waveform is plotted on MATLAB software package version R2018a. The imbalance in these energies is responsible for the change in the emotions of an individual.

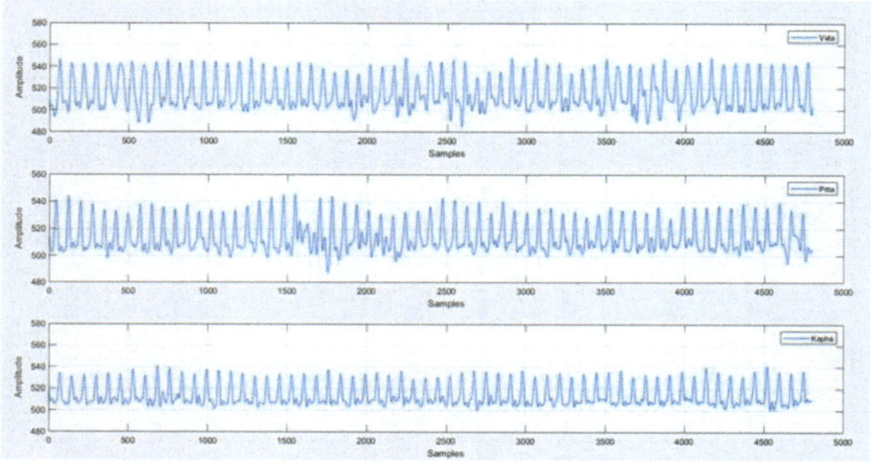

Fig. 9 Vata, Pitta, and Kapha waveform for calm emotion group

References

1. Singh D (2012) Human emotion recognition system. Int J Image Gr Sig Process 8:50–56
2. Lad V Ayurveda, a brief introduction and guide
3. Chaudhari S, Mudhalwadkar R (2017) Nadi pariksha system for health diagnosis. In: Proceedings of 2017 international conference on intelligent computing and control, I2C2 2017
4. Joshi A, Kulkarni A, Chandran S, Jayaraman VK, Kulkarni BD (2007) Nadi Tarangini: a pulse based diagnostic system. In: Proceedings of the 29th annual international conference of the IEEE EMBS, 2207–2210
5. Zhu L, Yan J, Tang Q, Li Q (2006) Recent progress in computerization of TCM. J Commun Comput 3:78–81
6. Lad V (2005) Secrets of the pulse: the ancient art of Ayurvedic pulse diagnosis. Motilal Banarsidass, Delhi
7. Kshirsagar M, Magno AC (2011) Ayurveda: a quick reference handbook. Lotus Press
8. Lad V, Lad U Determining your constitution. In: Ayurvedic cooking for self healing, p 1
9. Kallurkar P, Patil K, Sharma G, Sharma S, Sharma N (2015) Analysis of Tridosha in various physiological conditions. In: 2015 IEEE international conference on electronics, computing and communication technologies, CONECCT 2015
10. Chu YW, Chung YF, Chung CY, Chung S, Luo CH, Yeh CC, Si XC A new pulse pillow of traditional Chinese medicine—the wrist fixer system. In: ICOT 2013—1st international conference on orange technologies
11. Sareen M, Kumar M, Anand S, Salhan A, Santhosh J (2008) Nadi Yantra: a robust system design to capture the signals from the radial artery for non-invasive diagnosis. In: 2nd international conference on bioinformatics and biomedical engineering. IEEE, pp 1387–1390
12. Anand T, Deepali W, Kavita D, Mukund D (2018) Review of Ashtavidha Pariksha W.S.R. to Nadi Pariksha. Int J Ayurveda Pharma Res 6(1):40–43
13. Pincus SM, Gladstone IM, Ehrenkranz RA (1991) A regularity statistic for medical data analysis. J Clin Monitor Comput

14. Arunkumar N, Jayalalitha S, Dinesh S, Venugopal A, Sekar D (2012) Sample entropy based ayurvedic pulse diagnosis for diabetics. In: EEE-international conference on advances in engineering, science and management, ICAESM-2012
15. Wilkinson IB, MacCallum H, Flint L, Cockcroft JR, Newby DE, Webb DJ (2000) The influence of heart rate on augmentation index and central arterial pressure in humans. J Physiol 525(Pt 1):263–270

Trace Determination of Zirconium (IV) as its 3-Hydroxy-2-[2′-(5′-Methylthienyl)]-4*H*-Chromen-4-One Complex and Structural Elucidation by Quantization Technique

Chetna Dhonchak, Navneet Kaur, Rajesh Agnihotri, Urmila Berar, and Nivedita Agnihotri

Abstract Micro determination of Zirconium (IV) has been carried out using 3-Hydroxy-2-[2′-(5′-methylthienyl)]-4*H*-chromen-4-one (HMTC) as an analytical reagent. Zr (IV) forms a 1:4 (M:L) yellow coloured complex with HMTC extracted into dichloromethane from ammoniacal medium (pH 7.05–7.09). The complex system shows a maximum at 424–440 nm and follows Beer's law in the range 0.0–0.9 μg Zr (IV) ml^{-1} with an optimum range of determination as 0.27–0.79 μg Zr (IV) ml^{-1}as detected from Ringbom plot. Zr (IV)-HMTC complex has molar absorptivity of 8.22×10^4 $Lmol^{-1}$ cm^{-1}, specific absorptivity of 0.900 ml g^{-1} cm^{-1} and Sandell's sensitivity value 0.0011 μg Zr (IV)cm^{-2}; the linear regression equation being $Y = 0.981X - 0.036$ (Y = absorbance, X = μg Zr (IV) ml^{-1}) with the correlation coefficient 0.9987. Detection limit of the procedure is 0.0174 μg ml^{-1}. The repercussions obtained are highly consistent with the standard deviation of ± 0.0039 absorbance unit and has been confirmed by student's t-test with 0.5% limit. The proposed technique has been successfully applied in diverse synthetic and industrial samples.

Keywords Zirconium (IV) · 3-hydroxy-2-[2′-(5′-methylthienyl)]-4*H*-chromen-4-one · Spectrophotometric determination

C. Dhonchak · N. Kaur · N. Agnihotri (✉)
Department of Chemistry, Maharishi Markandeshwar (Deemed to be University), Mullana, Ambala, Haryana 133207, India
e-mail: nivagni11@gmail.com

R. Agnihotri · U. Berar
Department of Applied Science, UIET, Kurukshetra University, Kurukshetra, Haryana 136119, India

N. Marriwala et al. (eds.), *Mobile Radio Communications and 5G Networks*, Lecture Notes in Networks and Systems 140,
https://doi.org/10.1007/978-981-15-7130-5_25

1 Introduction

Zirconium (Zr) with atomic number 40 is a lustrous grey white transition element. It has no biotic role known so far. But depending on the nutritional custom human body contains 250 mg of Zr with a daily consumption of 4.1 mg, i.e., 3.5 mg from food and 0.65 mg from water, respectively. Zirconium is extensively scattered in nature and is found in all biological systems, for example: 2.86 $\mu g\ g^{-1}$ in whole wheat, 3.09 $\mu g\ g^{-1}$ in brown rice, 0.55 $\mu g\ g^{-1}$in spinach, 1.23 $\mu g\ g^{-1}$ in eggs, and 0.86 $\mu g\ g^{-1}$ in ground beef [1]. Additionally, zirconium has been frequently used in commercial products like deodorant sticks, aerosol antiperspirants such as the aluminium zirconium tetrachlorohydrex gly or AZG, is used as an antiperspirant in many deodorant products. It has the ability to obstruct pores in the skin and prevent sweat from leaving the body. Other use of zirconium is in water purification, for example, control of phosphorus pollution, bacteria and pyrogen-contaminated water [2]. Zirconium has not been assessed as health or environmental hazard and is not carcinogenic or genotoxic. However, at levels of 25 mg m^{-3}, zirconium has been detected to be immediately dangerous to life and health.

Zirconium bearing compounds have various biomedical applications, including dental implants and crowns, knee and hip replacements, middle-ear ossicular chain reconstruction and other restorative and prosthetic devices [2]. Sodium zirconium cyclosilicate has been used in the treatment of hyperkalemia [3].

Due to such vast applications of the element in commercial industry, it is needed that we develop a procedure which has good reliability, less time consuming and good sensitivity and selectivity for micro-determination of the same. Many approaches have been evaluated so far in this field where their pertinence has been diminished in terms of sensitivity, selectivity and rapidity using UV-VIS spectrophotometery [4–8]. The presented procedure manifests the complexation of new benzopyran named 3-hydroxy-2-[2′-(5′-methylthienyl)]-4*H*-chromen-4-one (HMTC) with Zr (IV) for its trace analysis, deriving preferable outcomes.

2 Experimental

2.1 *Equipments, Reagents and Solutions*

A UV-VIS spectrophotometer (2375; Electronics India) with 10 mm matched quartz cells was used for absorbance measurements and spectral analysis.

A stock solution of Zr (IV) containing 1 mg ml^{-1} of the metal ion was attained by solvation of promptly weighed amount of $ZrOCl_2.8H_2O$ (CDH® 'AR') in 2 M solution of HCl and making up the volume by same in 100 ml volumetric flask. Lower concentrations like that of $\mu g\ ml^{-1}$ level were obtained by appropriate dilutions therefrom.

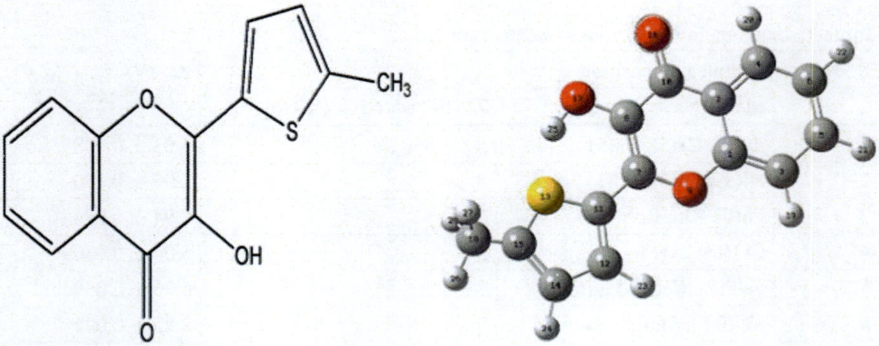

Fig. 1 Structure of HMTC

Similarly, stock solutions of other metal ions at the mg ml^{-1}level were prepared by dissolving their frequently available sodium or potassium salts ('chemically pure' grade) in de-ionized water or dilute acids. Dichloromethane (CDH® 'AR') and Ammonia (SDFCL 'AR') were used as such.

0.1% (m/v) solution of 3-hydroxy-2-[2′-(5′-methylthienyl)]-4*H*-chromen-4-one (HMTC) [9] [Molar Mass = 258 g mol^{-1}, Melting Point = 203 °C; Fig. 1) was prepared fresh in acetone.

2.2 *Synthetic and Technical Samples*

Synthetic samples were brought into solutions by mixing Zr (IV) and different metal ion solutions in desirable quantities with compositions as depicted in Table 1. However, technical samples like reverberatory flue dust and water (tap and well) were dissolved to a definite volume as per the earlier reported work [10]. Suitable aliquots were taken to determine zirconium as portrayed in the methodology.

2.3 *Procedure for Extraction and Determination*

To make final volume of aqueous phase as 10 ml, an aliquot containing ≤9 μg Zr (IV) solution and 0.5 ml of 0.1% HMTC solution were taken in a 125 ml separating funnel and enough amount of ammonia solution was added so as to adjust its pH 7.06. Ample volume of demineralized water was added to make the final volume of aqueous phase 10 ml. It was then equilibrated once with equal volume of dichloromethane for 30 s, releasing the pressure periodically through the stop cork. As the phase gets separated, the yellow coloured organic layer was filtered through Whatman filter paper no. 41 (pretreated with dichloromethane) in a 10 ml volumetric flask and was made up to mark with pure dichloromethane. Absorbance of the yellow coloured extract was

Table 1 Examination of various synthetic samples

S. No.	Sample composition		Zr (IV) found (μg 10 ml^{-1})**
	Matrix*	Zr (IV) added (μg 10 ml^{-1})	
1	Zn(5), Ca(5), Ba(5)	5	5.03 ± 0.008
2	Ni(3), Hg(3), Cd(1)	2	2.05 ± 0.100
3	Pb(5), Mn(3), Mg(2)	3	2.94 ± 0.084
4	Cr^{VI}(3), Ag(5), Co(2)	5	5.00 ± 0.006
5	Fe^{III}(2), Ce(2), Mg(5)[a]	7	6.96 ± 0.049
6	Mo(0.5), Sn(0.5), Ni(5)[b,c]	3	2.95 ± 0.022
7	Fe^{II}(2), Cu(0.5), Bi(0.5)[c]	5	5.02 ± 0.028
8	Pt(0.5), Nb(0.2), Au(0.5)[c]	7	6.98 ± 0.068
9	V(0.2), Co(2), Hg(5)[d]	3	2.96 ± 0.048
10	Ru(2), La(0.5), Cr^{III}(0.5)	5	5.16 ± 0.048
11	Ti(0.5), Pd(0.5), Se(2)[d]	2	2.04 ± 0.043
12	Ir(0.5), Al(2), Sr(0.5)	3	3.05 ± 0.021
13	As(0.5), W(2), Ba(5)[e]	7	7.00 ± 0.163
14	Water (i) Tap water (ii) Groundwater	5 5	5.11 ± 0.096 5.00 ± 0.100
15	Reverberatory flue dust[a]	5	4.95 ± 0.040

*Figure in bracket indicates the amount of metal ion in mg 10 ml^{-1}
**Average of triplicate analysis ± SD
[a]In presence of 10 mg fluoride
[b]In presence of 1 ml peroxide
[c]In presence of 30 mg oxalate
[d]In presence of 50 mg ascorbic acid
[e]In presence of 100 mg thiocyanate

measured at 430 nm against the reagent blank prepared in an equivalent manner. The amount of zirconium was determined from the calibration curve obtained by plotting a graph between variable amounts of Zr (IV) and the corresponding absorbance values obtained after applying optimum conditions of procedure.

3 Results and discussion

HMTC formed a very stable (stability >4 days) yellow coloured complex with Zr (IV) in an ammoniacal medium (pH 7.05–7.09). From the spectra taken for Zr (IV)—HMTC complex, the absorption maximum was observed in the range 424–440 nm in comparison with the reagent blank that also absorbed to a little extent in the reported range (Fig. 2). Due to this pretext entire measurements were taken at 430 nm against reagent blank. Though similar methodology as applied with ammonia was performed in Na_2CO_3, CH_3COOH, H_3PO_4, H_2SO_4, $NaHCO_3$ and HCl, the

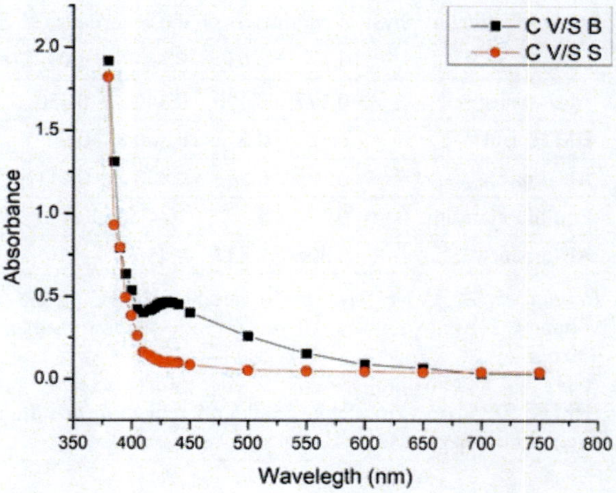

Fig. 2 Absorption spectra of Zr (IV)—HMTC complex

absorbance observed was comparatively less in the same order, and hence NH_3 medium was chosen over others where the complex not only showed high colour intensity but consistency over time.

Extraction conduct of Zr (IV)-HMTC complex was examined in various organic solvents and found to increases in the order: Ethyl acetate < isoamyl alcohol < methyl isobutyl ketone < toluene < benzene < carbon tetrachloride < chloroform < cyclohexane < iso amylacetate < 1,2dichloroethane < dichloromethane. In context to the mentioned order, it can be said that complex should be extracted in dichloromethane as it showed maximum absorbance followed by its stability for more than 4 days. Yellow coloured complex can be wholly uprooted into organic layer of dichloromethane in single step by quantitative (100%) extraction.

The influence of divergent physical variables like pH, HMTC concentration and equilibration time on the extraction of Zr (IV) was studied as summarized in Table 2. After examining Table 2, it can be said that to attain the optimum and constant intensity of the complex containing ≤ 9 μg Zr (IV) in 10 mL aqueous phase, 0.3–0.8 ml of 0.1% (w/v) HMTC solution in acetone was added maintaining pH 7.05–7.09 of the water solution by adding sufficient ammonia in the same order as per the applied procedure and equilibrating once with same amount (10 ml) of dichloromethane for 15–300 s, are appropriate for the Zr (IV)—HMTC complex formation.

3.1 *Effect of Anions/Complexing Agents and Cations*

To study the effect of diverse ions, the ions were added under ideal conditions of the suggested strategy to 5 μg concentration of Zr (IV) in a 10 ml aqueous volume. The impact of different anions, complexing agents or cations on the extraction and spectrophotometric determination of Zr (IV) was prospected to evaluate the selectivity

Table 2 Effect of physical parameters on the absorbance of Zr (IV)-HMTC complex

pH[a]	2.84	3.63	4.34	7.05–7.09	7.52	8.11	8.15	8.20
Absorbance	0.177	0.190	0.346	0.450	0.402	0.369	0.349	0.329
HMTC (ml)[b]	0.1	0.2	0.3–0.8	0.9	1.0	1.2	1.5	
Absorbance	0.133	0.233	0.450	0.411	0.391	0.361	0.301	
Equilibration time (s)[c]	0	5	10–300					
Absorbance	0.300	0.437	0.450					

Conditions [a]Zr (IV) = 5 μg; pH = variable; HMTC [0.1% (m/v) in acetone] = 0.5 ml; aqueous volume = solvent volume = 10 ml; solvent = dichloromethane; equilibration time = 30 s; λ_{max} = 430 nm

[b]NH_3 = 7.06; remaining parameters same as in (a) except for the variation in HMTC concentration
[c]HMTC [0.1% (m/v) in acetone] = 0.5 ml; remaining conditions same as in (b) excepting variation in equilibration time

and tolerance limit as shown in Tables 3 and 4. The tolerance limit was customized as the amount of diverse ion causing an error ≤1% in the extraction of Zr (IV). Some of the samples containing metal ions required addition of the appropriate masking/complexing agents before the addition of HMTC as depicted in Table 4. Out of 22 anions and 32 cations studied, none (alone or after masking) interfered with the determination of Zr (IV)-HMTC complex.

Table 3 Effect of anions or complexing agent on Zr (IV)-HMTC complex

S. No.	Anion or complexing agent	Zr (IV)-HMTC complex	
		Tolerance limit (mg 10 ml^{-1})	Absorbance
1	None	–	0.450
2	Chloride, Bromide, Iodide, Nitrate, Carbonate, Sulphite, Disodium EDTA, Thiocynate, Thiourea, Phosphate	100	0.450
3	Sulfosalicylic acid	80	0.450
4	Sulphate, Ascorbic acid, Dithionite, Tartarate, Sulphate	50	0.450
5	Acetate	40	0.450
6	Oxalate	30	0.450
7	Nitrite	20	0.450
8	Fluoride	10	0.450
9	Hydrogen peroxide (30%)*	0.5	0.450
10	Glycerol*	1	0.450

*Amount added in ml

Table 4 Effect of cations on Zr (IV)-HMTC complex

S. No.	Cation added	Zr (IV)-HMTC complex	
		Tolerance limit (mg 10 ml^{-1})	Absorbance
1	None	–	0.450
2	Ag(I), Zn(II), Pb(II), Hg(II), Ca(II), Ba(II), Mg(II), Mn(II), Ni(II), Cd(II)	10	0.450
3	Co(II), Al(III), Ru(III), Fe(III)[a], Se(IV), W(VI)[b]	5	0.450
4	Ce(IV)	4	0.450
5	Cr(VI)	2	0.450
6	Cu(II), Sn(II)[c] Sr(II), Pd(II)[d], Ir(III), As(III), Au(III), Pt(IV), La(III), Cr(III), Bi(III), Ti(IV), Mo(VI)[e]	1	0.450
7	V(V)[b], Nb(V)[c]	0.5	0.450
8	Fe(II)[c]	0.25	0.450

[a]In presence of 10 mg fluoride
[b]In presence of 50 mg ascorbic acid
[c]In presence of 30 mg oxalate
[d]In presence of 100 mg thiocyanate
[e]In presence of 1 ml hydrogen peroxide

3.2 *Spectral Characteristics*

The yellow complex of Zr (IV)-HMTC acquired under the optimum conditions of the mechanism showed linear response up to 0. 9 μg ml^{-1} of Zr (IV) with the optimum limit of determination as obtained from the Ringbom plot [11] falling in the range 0.27–0.70 ppm. Linearity of the calibration plot was confirmed by analysis of the correlation coefficient having the value 0.9987. Student's t-test was conducted and it was concluded that at 0.5% level, no biasness had occurred in the research process. Various optical and statistical characteristics as calculated by statistical methods are shown in Table 5.

3.3 *Stoichiometry of Zr (IV)-HMTC Complex*

The stoichiometric ratio of Zr (IV)-HMTC in the extracted species was established as 1:4 by the job's continuous variations method [12] as revised for a two-phase system by Vosburgh and Cooper [13] and portrayed in Fig. 3. The ratio 1:4 of the complex constituents was further confirmed by mole ratio method [14]. The stability constant as estimated by the mole ratio method was 3.76×10^{15}.

Thus, the probable structure of Zr (IV)-HMTC complex as determined by the two methods was suggested as shown in Fig. 4.

Table 5 Spectral characteristics, precision and accuracy data

S. No.	Parameters	Values
1	λ_{max} (nm)	424–440
2	Beer's law limits (μg mL^{-1})	0–0.9
3	Optimum range of determination (μg mL^{-1})	0.27–0.79
4	Molar absorptivity (L mol^{-1} cm^{-1})	8.212×10^4
5	Sandell's sensitivity (μg cm^{-2})	0.0011
6	Correlation coefficient (*r*)	0.9987
7	Regression equation (*Y*)*	$Y = 0.98X - 0.036$
8	Slope (b)	0.981
9	Intercept (a)	− 0.036
10	Standard deviation	±0.0039
11	Relative standard deviation (%)	0.861
12	Limit of detection (μg ml^{-1})	0.0174
13	Stability constant	3.76×10^{-15}

*where Y = absorbance and X = concentration of Zr (IV) in μg ml^{-1}

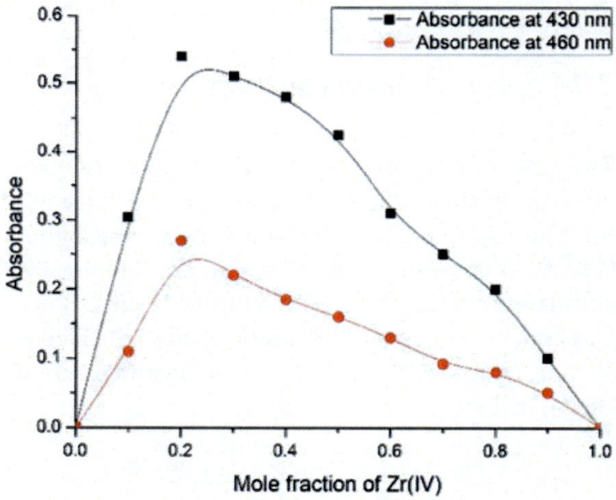

Fig. 3 Job's continuous variations method for Zr (IV)-HMTC complex (Total concentration: [Zr] + [HMTC] = 1.096×10^{-3} M; pH = 7.06)

3.4 Structural Elucidation by Optimization Technique

The metal complex was demonstrated with Avogadro 1.01 programme [15] and optimized using molecular mechanics. Various cycles of optimization were effectuated. Energy found for the complex was 4257.28 K J mol^{-1} before optimization and 2275.56 K J mol^{-1} after 500 cycles of optimization. The optimized structure has coordination number 8 and square antiprismatic geometry. Some optimized bond

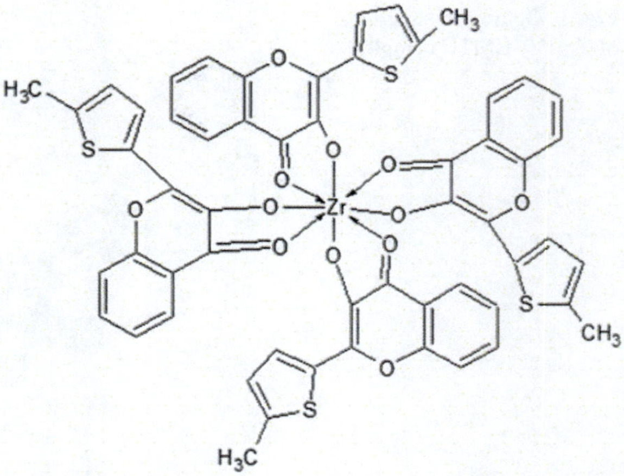

Fig. 4 Probable structure of Zr (IV)-HMTC complex

Table 6 Computational parameters of Zr (IV)-HMTC complex

S. No.	Bond length		Bond angle	
	Atom type	Bond length (A°)	Atom type	Bond angle (°)
1	Zr–O(1)	2.189	O(1)–Zr–O(8)	65.3
2	Zr–O(2)	1.843	O(2)–Zr–O(3)	62.4
3	Zr–O(3)	2.181	O(4)–Zr–O(5)	65.4
4	Zr–O(4)	2.177	O(6)–Zr–O(7)	62.3
5	Zr–O(5)	2.225		
6	Zr–O(6)	1.474		
7	Zr–O(7)	2.198		
8	Zr–O(8)	2.183		

lengths and bond angles are also calculated and are shown in Table 6. The optimized structure is shown in Fig. 5.

3.5 *Analytical Applications*

The proposed spectrophotometric method for the micro-determination of zirconium is rapid to operate (takes 2–3 min for a single determination), less equipment requiring, highly sensitive, reproducible and has a vast forbearance limit for the foreign ions including Fe, Cr, Mo, W, Nb, V, Sn and many other important elements including platinum group metals, thereby increasing the purview of application. The extensive usefulness of the technique is further tested by analysing a large variety of real samples and synthetic mixtures of varying composition as in Table 1. The obtained results are in great concurrence with the amount of metal ion initially added.

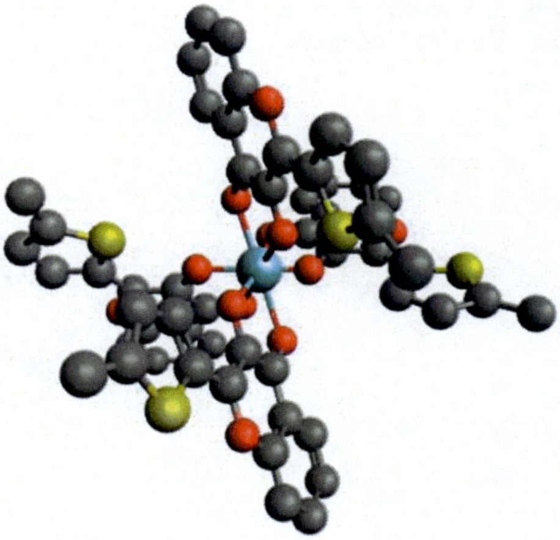

Fig. 5 Optimized structure of Zr (IV)-HMTC complex

The proposed method is reproducible and accurate with the relative standard deviation of 0.861% for 10 replicates containing 0.5 μg Zr (IV) ml^{-1} every time. The procedure preferred has been compared with existing methods in respect of rapidity, selectivity and sensitivity as summarized in Table 7.

4 Conclusion

3-Hydroxy-2-[2′-(5′-methylthienyl)]-4*H*-chromen-4-one has been used as an analytical reagent for the spectrophotometric determination of zirconium. Zirconium (IV) in presence of several cations and anions/complexing agents forms a yellow 1:4 (M:L) complex with HMTC which is extractable into dichloromethane in ammoniacal medium of pH 7.06 and 0.3–0.8 ml of 0.1% HMTC solution in acetone and is stable for more than 4 days. The complex shows an absorption maximum at 424–440 nm with a molar absorptivity of 8.212×10^4 l mol^{-1} cm^{-1} and Sandell's sensitivity equal to 0.0011 μg Zr cm^{-2}. The linear regression equation was $\boldsymbol{Y = 0.981X - 0.036}$ and the correlation coefficient, $r = 0.9987$; detection limit of the method being 0.0174 μg ml^{-1}. The optimum range of determination was 0.27–0.79 μg Zr (IV) ml^{-1}. The method has been adopted in account for its simplicity, selectivity, rapidity and precision and has been applied for the determination of zirconium in synthetic and technical samples.

Table 7 Comparison of the proposed method with reported methods

S. No.	Aqueous conditions	λ_{max} (nm), Solvent	Molar absorptivity (l mol^{-1} cm^{-1})	Interference	References
1	Zr (IV), 6-Chloro-3-hydroxy-2-(2′-hydroxyphenyl)-4-oxo-4*H*-1-benzopyran	400, Propane-1-ol-H_2O	1.78×10^4	EDTA 'disodium salt', tartrate, oxalate, phosphate, Nb (V) and Pb (II)	[5]
2	Zr (IV), 6-Chloro-3-hydroxy-2-phenyl-4*H*-chromen-4-one, In HCl medium	415, Dichloromethane	5.930×10^5	Oxalate, phosphate, EDTA 'disodium salt'	[6]
3	Zr (IV), Pyrazolo (1, 5-a) quinazolin-6-one, pH 3.2–4.7 in aqueous medium, 15 min standing time	368, –	1.68×10^5	–	[7]
4	Zr (IV), 2-Hydroxynaphthaldehyde-p-hydroxybenzoic hydrozone, pH 1.0, acidic medium, heating for 15 min at 60° C	415, –	9.86×10^3	–	[8]
5	Zr (IV), 5,7-Dibromo-8-hydroxyquinoline in the presence of thiocyanate	416, Chloroform	1.05×10^4	Oxalate, EDTA disodium salt, fluoride	[9]
6	Zr (IV), 3-Hydroxy-2-[2′-(5′-methylthienyl)]-4*H*-chromen-4-one, pH 7.06	430, Dichloromethane	8.212×10^4	None (22 anions and 32 cations studied)	Proposed method

References

1. Schroeder AH, Balassa JJ (1966) Abnormal trace metals in man: zirconium. J Chronic Diseases 19(5):573–586
2. Lee DBN, Roberts M, Bluchel CG, Odell RA (2010) Zirconium: biomedical and nephrological. Appl ASAIO J 56(6):550–556
3. Ingelfinger JR (2016) A new era for the treatment of hyperkalemia. J Med 372(3):275–277
4. Chawaria M, Sharma HK (2018) A non-extractive spectrophotometric determination of Zirconium with6-Chloro-3-hydroxy-2-(2'-hydroxyphenyl)-4-oxo-4 H-1-benzopyran using propan-1-ol-H2O mixture as solvent. Internat J Green Herbal Chem 7(2):230–236
5. Chawaria M, Kumar A (2017) Extractive spectrophotometric determination of zirconium with 6-chloro-3-hydroxy-2-phenyl-4H-chromen-4-one as an analytical reagent. J Chem Bio Phy Sci 7(3):600–604
6. Lasheen TA, Hussein GM, Khawassek YM, Cheira MF (2013) Spectrophotometric determination of zirconium (IV) and hafnium (IV) with pyrazolo (1, 5-a) quinazolin-6-onederivative reagent. Anal Chem Indian J 12(10):368–376
7. Sumathi G, Sreenivasulu Reddy RT Direct and derivative spectrophotometric determination of zirconium(IV) with 2-hydroxynaphthaldehyde-phydroxybenzoichydrazone. J Appl Chem 2(1):81–85
8. Jain A, Prakash O, Kakkar LR (2010) Spectrophotometric determination of zirconium with 5,7-dibromo-8-hydroxyquinoline in presence of thiocyanate. J Anal Chem 65(8):820–824
9. Algar J, Flynn JP (1934) A new method for the synthesis of flavones. Proc Royal Irish Acad 42B:1–8; Oyamada T(1934) A new general method for the synthesis of flavonol derivatives. J Chem Soc Jpn 55:1256–1261
10. Agnihotri N, Mehta JR (2006) A highly sensitive extractive spectrophotometric determination of zirconium(IV) using 2-(2'-furyl)-3-hydroxy-4-oxo-4H-1-benzopyran. J Indian Chem Soc 83(8):846–848
11. Ringbom A (1938) On the accuracy of colorimetric analytical methods I. Z Anal Chem 115:332–343
12. Job P (1928) Formation and stability of inorganic compexes in solution. Ann Chim 9:113
13. Vosburgh WC, Cooper GC (1941) The identification of complex ion in solution by spectrophotometric measurements. J Am Chem Soc 63:437
14. Yoe JH, Jones AL (1944) Colorimetric determination of iron with disodium-1,2-dihydroxybenzene-3,5-disulfonate.Ind. Eng Chem (Anal.Ed.) 16:111
15. Hanwell MD, Curtis DE, lonie DC, Vandermeersch T, Zurek E, Hutchison GR (2012) Avogadro: an advanced semantic chemical editor, visualization and analysis platform. J Cheminform 4(17)

Barriers Which Impede the Implementation of an Effective Deep Web Data Extraction in VBPS

Meena Chaudhary and Jyoti Pruthi

Abstract Semantic block detection is an approach to deal with mining of data from Web pages and Web applications. Conventional strategies cannot perform better, as the new Web site configuration develops with new technologies. Extraction of Web information from the full Web page will be the intensive assignment to recover the substantial info because they are Web site programming language subordinate. A "layer tree" is developed to replace various levels of irregularities between the DOM tree portrayal and the visual format of the Web page. There are various limitations and barriers arise for the different scenarios and to find out and resolve the issues accordingly with the degree of complexity of the problem. As the flow of information has been upgraded and updated on the Web pages exponentially on the daily bases and to find out the correct information in this bulk of data is the real task. VBPS helps in finding the deep data among the huge Web pages but as the technology has been updated dynamically this approach has faced with some limitations. In this paper, we are discussing about these limitations and its comparison with the other techniques in this field.

Keywords DOM · VBPS · Semantic nodes

M. Chaudhary (✉) · J. Pruthi
Manav Rachna University, Faridabad, Haryana, India
e-mail: meena@mru.edu.in

J. Pruthi
e-mail: jyoti@mru.edu.in

N. Marriwala et al. (eds.), *Mobile Radio Communications and 5G Networks*, Lecture Notes in Networks and Systems 140,
https://doi.org/10.1007/978-981-15-7130-5_26

1 Introduction

1.1 The Recent Trends

In recent information retrieval from Web, Web crawler restores similar outcomes for various third party clients. This technique is awkward and improper in light of the fact that clients' interests are different from each other, while query output on the Web crawler is not separate among users. As a rule, this technique for data recovery may not be material since in our quick changing Web and pattern of customized data prerequisite, data recovery should supply more insightful, more precise what is more, client aim data benefit. To take care of these issues, Web indexes must be ready to express and manage the interpreted data in the HTML pages and return clients a clever and describes query items. VIPS is a procedure got from individuals' recognition on site. A site HTML page is not just a single dissent; instead, the impression of different things consisting of linguistic starts. It is in a perfect world that particular limit program will be located at specific territories. Exactly when a site page displayed, customers would isolate into a couple of linguistic normally. It is possible to interface a site page with using location and vision data. VIPS has extensively been used as a piece of Web mining. Then, once more, it is all around exhibited that similar semantic limit arena reflecting similar HTML marks.

1.2 The Working Architecture

The issue of mining can be settled for the most part by separating the Web organization and significant names. In this, we use VIPS figuring to extract the specific content from Web lists. The issue of Web information extraction has accumulated a considerable measure of consideration as of late, and the vast majority of the proposed solutions depend on breaking down the HTML DOM program or the DOM trees of the pages. A DOM arrangement takes the accompanying principle impediments: First, the programming dialects which are HTML site pages' subordinate. As most Web site pages are composed in HTML, it is not amazing that every single past arrangement relies on investigating the HTML source code of pages. Be that as it may, HTML itself is as yet developing and when upcoming forms or different labels are presented, the past activity should be changed more than once to adjust to new forms or new labels, assist more, HTML is not any more the selective site page programming dialect, and different dialects have been presented, for example, XHTML and XML, the past arrangements now confront the accompanying problem, would it be advisable for them to be essentially modified or even surrendered? Or, then again should different methodologies be suggested to suit the new dialects? Second, they are unequipped for taking care of the regularly expanding many-sided quality of HTML DOM code. Past works have not connected the contents, like.js and.css, in the HTML content. Nowadays, Web site page architects are utilizing extremely complex JavaScript, jquery or css, in light of our examination and got to numerous sites, and we came into conclusion that present structure of pages is more muddled then before rendition of site page outline this make out more troublesome for existing answers for induce the flow of the structure for the pages the Web sites contains.

With the help of this, we can separate the blocks of information from the Web pages according to the layout of the page and locations, but the revolution of updating of techniques in this decades has put up the limitations on these techniques as the development of the Web pages has been updated from <HTML> to <HTML.5.> and now <HTML> with JavaScript and <XML>. To find out the barrier in searching, the information in this update has put the real effort and this is what we have tried to find out.

2 Objectives

(i) To review and identify the list of challenging factors (barriers) for impede, the effective Deep Web Data Extraction in VBPS.
(ii) To analyze and identify the inter-relationship between barriers for the implementation of an effective Deep Web Data Extraction in VBPS.
(iii) To identify the influential barrier with the help of DEMATEL approach which is suitable to find the cluster of the barriers in finding out the knowledge.

3 Research Background

Authors: Liu and Meng [1]: Profound Web substance is gotten to by questions presented in Web site database and revert information archives enclosed in progressively created Web site pages. Extracting organized information from profound Web site sheets and testing issue due to the basic complicated organizations of such sheets. As not long ago a substantial number of methods been suggested to statement this issue, yet every one of them has inborn impediments in light of the fact that they are pages are constantly shown consistently for client to peruse.

Authors: Cai and Yu [2]: Another Web site content structure examination in view to visual portrayal is suggested. Several Web site applications, for example, data recovery, data extraction and programmed page adjustment can profit by this organization. This paper offers a programmed top-down label hierarchy autonomous way to deal with distinguish Web site content organization. It exaggerates a client comprehends Web site design organization in light of his visual discernment. Contrasting with other existing techniques, for example, DOM tree.

Authors: Silambarasan and Mohan [3]: Many Web applications, for example, data recovery, data extraction and programmed page adjustment can profit by semantic content structure. This paper exhibits a programmed top-down, tag-tree free way to deal with recognize Web content structure. It reenacts how a client comprehends Web format structure in view of his/her visual recognition. This paper diagnostically talks about VIPS calculation to remove the semantic organization of a Web site sheets.

Authors: Akpınar et al. [4]: Site pages are regularly intended for visual cooperation. They incorporate numerous visual components, for example, header, footer, menu and so forth that guide the per user. One can undoubtedly take a gander at the visual rendering and can separate the portion which regularly varies in foundation shading, textual style styles, fringes or edges around the portions. Then again, the basic source code regularly does not give such sort of clear division or, on the other hand design. In this manner, when Web site pages are naturally prepared by assistive innovations or, then again adjusted for cell phones this sort of data is not accessible.

Authors: Singhal [5]: Site page portrayal is a subject of worry for little screen gadgets, similar to, versatile, palm and so on. In a Web site page, greater part of insignificant information including ads and other boisterous data's make get to bother. Web site page division is a method which settles this issue by legitimately separating a site sheet into fragments. Sheet portions could be made by utilizing DOM and Visual Page Segmentation methods. This research includes, a half breed technique for Web site page division has been composed utilizing mix of DOM strategy and VIPS calculation for creating portions from a Web site sheet. Here, both the basic and graphic parts of a Web site sheet to make a section been measured. A portion is such a fundamental part of Web site sheet that could not be additionally partitioned. This is finished by preparing a sheet over a BCA which would further discuss.

Authors: Doan et al. [6]: This paper examines the issue of removing information records on the reaction pages came back from Web databases or Web indexes. Existing answers for this issue are constructing fundamentally with respect to breaking down the HTML DOM trees and labels of the reaction pages. While these arrangements can accomplish great outcomes, they are too intensely reliant on the specifics of HTML.

4 Vision-Based Page Segmentation (VBPS)

To encourage perusing and draw in consideration, pages more often than not enclose much graphic data in the markers in HTML. Run of the mill visual indications incorporate lines, clear regions, hues, pictures, textual styles and so on. Visual prompts are exceptionally useful to identify the semantic locales in Web site sheets. Document object model all in all gives a valuable arrangement to a page. Be that as it may, labels, for example, HTMLTABLE and HTMLP are utilized for content association, as well as for format introduction. Much of the time, DOM has a tendency to uncover introduction organization other than content organization and is frequently not sufficiently exact to segregate distinctive semantic squares in a site page. Many times, subjects or goals can be recognized with visual intimations. For example, position, anchor, textual style, images and so on. With the improvement of Hyper

Text Markup Language Calibration, as Extensible Hyper Text Markup Language, the pattern Hyper Text Markup Language tag must steady with the semantic squares.

VIPS is the sort of calculation which extricates the semantic arrangement of a site page in view of its visual introduction. The semantic structure of VIPS is a tree structure. In view of the VIPS calculation, it is helpful to remove the fundamental or expectation data from the Web page. Clear from the tree structure of the site page, commotion data is expelled naturally. The belief system of VIPS calculation is the direction of Web mining. We utilize the strategy to creep the normal pursuit motor outcomes. Its points of interest are self-evident. Right off the bat, VIPS is to discover the DOM structure and from the site page. A level of coherence (DOC) is characterized for each square. The permitted degree of coherence (PDOC) can be pre-characterized to accomplish diverse granularities for the content structure. Also, it is a helpful strategy for Web content mining and particularly helpful to separate the Web indexes' query items. These particular data is gathered to do next seeking step, such as report division, tf-idf count and expectation words weight estimation, bunching and positioning calculations.

4.1 Deep Web Data Extraction in VBPS

The browser can provide the visual data for any Web page with the support of the logics and interfaces provided on it. In this paper, we apply the VIPS concept to change the visual prospects of the Web page in to the knowledge information in the form of a visual structure tree. A visual structured block tree is transformation of the segmented content of a Web site. The node represents the root block as a complete page, and each block in the tree sync as a region of the Web page. The least differentiated nodes are represented as leaf blocks, such as the paragraph of texts or images. Figure 1a shows a Web page presented on a Web site which considered as main source data and Fig. 1b gives its correlated visual block tree. In reality, the Visual structured block the structured tree of a deep information Web page contains numbers of blocks that may be in huge numbers. Visual structured block tree has consist of many nodes, the highest node is considered as the root node with the highest degree, and at the level one, the leaf nodes appear with the visual structured blocks, the more we go down the levels the more depth of Web page will arise with correlated blocks. There are three segments of representation in this tree, the root node is considered as the main node and the leaf nodes are successor of this ancestor node. The sub-level nodes will be further divided as leaf node can be further categorized as root, left child and right child. In reference to Fig. 1a, which shows the real-time example of a Web site, and it consists of multiple blocks of pages here the top block representing the advertisement image which can be converted as node. The second and third blocks are representing the leaf nodes of left and right child data. Each internal blocks can be represented by = (Root, Left, Right) and continuously with the different levels. Each leach node can be further divided into the same way

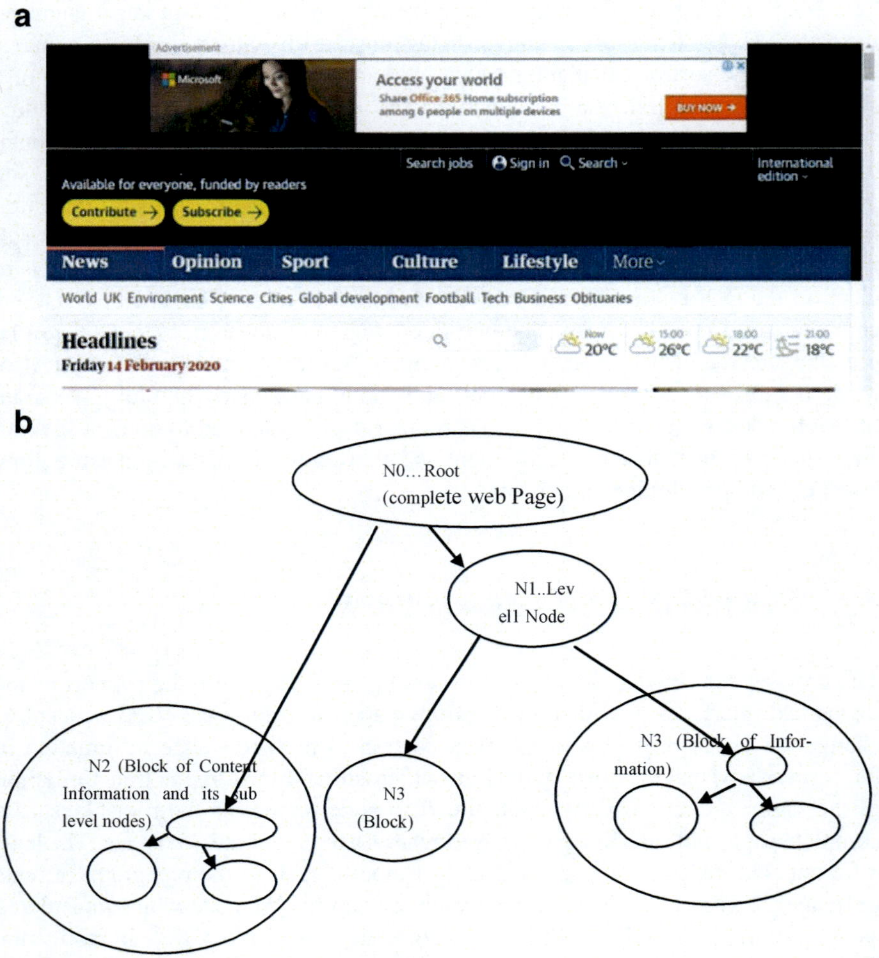

Fig. 1 **a** Page layout. **b** DOM tree of the page

as Root, Left, Right child nodes. The same set of blocks is arranged on every level. On every level, we categorized the fonts, images and contents of the blocks.

Figure 1b is representing the DOM structure of the tree which consists of root as whole Web page with node level N0 at the highest node and further sub-level N1, N2, N3 are sub-levels of these nodes. Each node is representing its corresponding blocks of node in the structured tree. These blocks consist of further images, different fonts of different text used (Fig. 2).

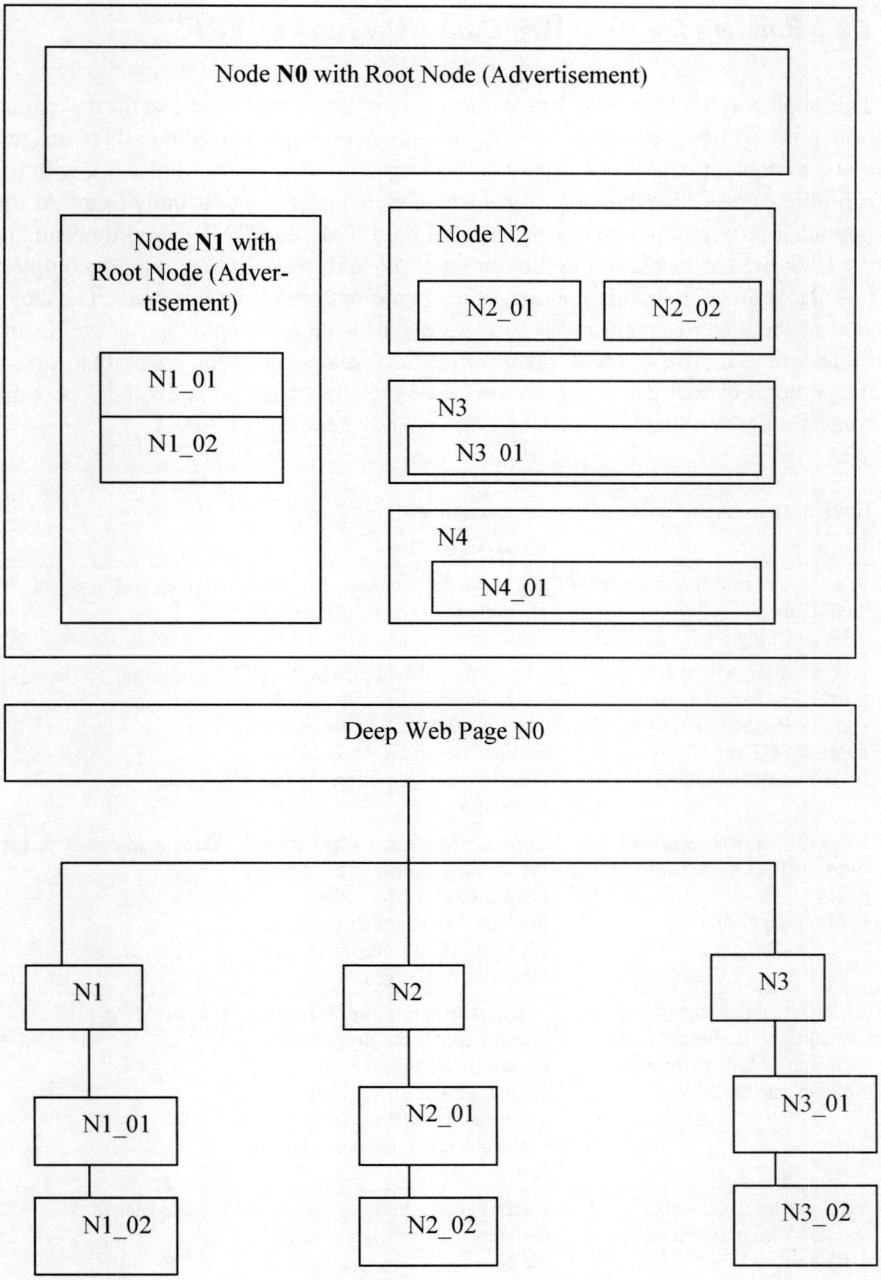

Fig. 2 **a** Block Web page structure and **b** its visual structured block tree

4.2 Barriers for Deep Web Data Extraction in VBPS

This approach for deep Web data extraction in VBPS is very effective for the extraction of the Web pages as it works with the content of pages which consists of images, text and content. But it consists of some barriers in it as it does not work with the run-time pages or the dynamic pages which were updated on the daily basis. As the upgraded programming of the pages from the HTML to HTML 5 and inclusion of the CSS and the JavaScript in the design of the Web pages makes it more complex, it is also tedious in finding the rank of the pages with run-time threshold. The inclusion of the advertisement in these pages makes it more complex as we are unable to detect the motive of these advertisements. The algorithm can check that it is an image but it cannot detect for what purpose the advertisement is floated. There are some ambiguous situations faced by these parameters also (Table 1).

Table 1 Barriers of deep web data extraction in VBPS

Barriers	Short description	References
Rule set and thresholds are not well defined • Web Mining	The set of rules defined for visual structured are not clearly defined	Akpınar and Yesilada [7]
Programming wise it is insufficient to handle the complexity generated by the upgraded HTML 5 • Web Content Mining	The barrier of the upgradation of the technology from the initial Tags of <HTML > to the <HTML5> makes it more complex and hinders the algorithm	Akpınar and Yesilada [7]
VIPS cannot deal work with the run-time dynamism of the Web pages • Web Usage Mining	The real-time Web pages contain the dynamic portions in it which is updated and creates the hinderers' in applying the VIPS. This algorithm is mostly covers the static Web pages	Akpınar and Yesilada [7]
VIPS is unable to handle the cascading style sheet and the inclusion of JavaScript in it • Web Structure Mining	The appearance of the Web pages means a lot for the users which creates the application of JavaScript and CSS in this, hence, the complexity of the pages increased and it creates the barrier	Liu and Meng [1]
VIPS cannot work with the updated Web pages in and on daily basis • Web Server Logs • App Server Logs • App Level Logs	As the exponentially growth of the Web sites for every ideality the generation of pages as been increased and the updated technology has put it in more complex while applying for the search on the deep Web pages	Liu and Meng [1]

5 Conclusions and Recommendations

The booming growth of the deep Web gives the scope to get benefit from abundant information lying beneath it. Generally, the knowledge or the useful information is just waiting to get extracted from the queries generated by the end users. So, it is very important to fetch the desired information from the huge stack and pile of information. That is why this paper is useful in defining the relevant requirement and problem arises on the way of the destined knowledge. As every Web page has been structured in a particular way to visualize the relevant information in it, we categorize this in a structural form.

References

1. Liu W, Meng X (2010) A Vision-based approach for deep web data extraction. IEEE Trans Knowl Data Eng 22(3)
2. Cai D, Yu S (2012) A vision based page segmentation algorithm. Microsoft Research Lab, Asia
3. Silambarasan G, Mohan K (2016) Perceiving syntactic content structure of web page using VIPS. Int J Eng Sci Comput
4. Akpınar ME, Yesilada Y, Kadam KS, Majgave AB, Kamble YM (2016) Extracting content structure from web pages by applying vision based approach. Int J Comput Sci Inf Technol 7(4):2011–2016.
5. Singhal S (2013) Hybrid web-page segmentation and block extraction for small screen terminals. In: International journal of computer applications (0975–8887) 4th international IT summit confluence 2013
6. Kadam KS, Majgave AB, Kamble YM (2016) Extracting content structure from web pages by applying vision based approach. Int J Comput Sci Inf Technol 7 (4):2011–2016
7. Akpınar ME, Yesilada Y (2013) Vision based page segmentation algorithm: Extended and perceived success. In: International conference on web engineering. Springer, Cham
8. Zhao H, Meng W, Yu CT (2006) Automatic extraction of dynamic record sections from search engine result pages. In: Proceedings of the international conference on very large data bases (VLDB), pp 989–1000
9. Zeleny J, Burget R, Zendulka J (2017) Box clustering segmentation: a new method for vision-based web page preprocessing. Inf Process Manag (Science Direct)
10. Liu W, Meng X (2010) A holistic solution for duplicate entity identification in deep web data integration. In: 2010 sixth international conference on semantics knowledge and grid (SKG). IEEE, pp 267–274
11. Sun F, Song D, Liao L (2011) DOM based content extraction via text density. In: Proceedings of the 34th international ACM SIGIR conference on research and development in information retrieval
12. Kadam KS, Majgave AB, Kamble YM (2016) Extracting content structure from web pages by applying vision based approach. Int J Comput Sci Inf Technol 7(4):2011–2016
13. Ma W, Chen X, Shang W (2012) Advanced deep web crawler based on Dom. In: Fifth international joint conference on computational sciences and optimization (CSO). IEEE, pp 605–609
14. M. Sahami, S. Dumais, D. Heckerman, and E. Horvitz, "A Bayesian approach to filtering junk e-mail," in Learning for Text Categorization: Papers from the 1998 workshop, vol. 62, pp. 98–105, 1998.

15. Chang C-H, Kayed M, Girgis MR, Shaalan K (2006) A survey of web information extraction systems. IEEE Trans Knowl Data Eng 18(10):1411–1428
16. Arasu A, Garcia-Molina H (2003) Extracting structured data from web pages. In: Presented at the ACM SIGMOD conference. San Diego, CA
17. Dean T, Ruzon M, Segal M, Shlens J, Vijayanarasimhan S, Yagnik J (2013) Fast, accurate detection of 100,000 object classes on a single machine. In: IEEE conference on computer vision and pattern recognition (CVPR)
18. Doan A, Ramakrishnan R, Halevy AY (2011) Crowdsourcing systems on the world-wide web. Commun ACM 54(4):86–96
19. Singh J, Nene MJ (2013) A survey on machine learning techniques for intrusion detection systems. Int J Adv Res Comput Commun Eng 2(11):4349–4355
20. Blanzieri E, Bryl A (2008) A survey of learning-based techniques of email spam filtering. Artif Intell Rev 29(1):63–92

The Prediction of Stock Market Trends Using the Hybrid Model SVM-ICA-GA

Kamal Malik and Manisha Malik

Abstract In this paper, the trends and the timings of stock market of Japanese Candlestick are predicted and analyzed empirically by developing the hybrid model that uses the three prominent techniques of artificial intelligence, i.e., SVM, ICA and genetic algorithms. In order to conduct the effective technical analysis of stock market—support vector machines (SVM) are used with genetic algorithms and imperialist competition algorithms (ICA). ICA is used to indicate the stock market timing and to optimize the SVM parameters, whereas GA is used to select the best features in addition to SVM parameters optimization. The input data of a model is generated using the two very important approaches—raw-based and signal-based. The results of the paper indicate that the performance of SVM-IC-GA is far better than the existing feed forward static neural network techniques of the existing literature.

Keywords SVM · ICA · GA · ANN

1 Introduction

To predict, the timing and the behavior of the stock market prices have become the biggest challenge. Nowadays, for all the financial scientists and investors, the right information about the forecasting trends and stock changes is the outmost requirement for investing in the stock market. So, there must be a powerful mechanism that deals with this forecasting. To predict the stock prices, the fluctuations in stock market work in a nonlinear dynamic system, and hence their prediction becomes a quite complicated tasks in all aspects and respects. In order to handle nonlinear complexities and the uncertainities of these dynamic systems, the modern approach of artificial intelligent systems has been discovered that raises the accuracy in predicting

K. Malik (✉) · M. Malik
CT University, Ludhiana, India
e-mail: kamal.malik91@gmail.com

M. Malik
e-mail: Malikmanisha2306@gmail.com

N. Marriwala et al. (eds.), *Mobile Radio Communications and 5G Networks*, Lecture Notes in Networks and Systems 140,
https://doi.org/10.1007/978-981-15-7130-5_27

this stock market effectively. The new methods of AI are quite reliable and up to mark as compared to the traditional methods and these methods have their own pros and cons. For instance, while taking into consideration, the technical analysis of Japanese Candlestick as the investment technique applied for the stock market timing, a supervised feed forward neural network is used. One of the most prominent and supervised learning methods is support vector machine, i.e., SVM that identifies the fluctuations occurred in the rules of the time series and predicts the different outcomings. The accuracy of the back-up vectors depends upon the setting up of the parameters of SVM. In this paper, along with SVM, the two metaheuristic approaches are also used to predict the changes in stock prices. These two metaheuristics algorithms are genetic algorithms (GA) and imperialism competition algorithm (ICA) that helps us to modify and optimize the various features of the models. The aim of the study is to evaluate SVM over GA and ICA to solve the various problems regarding forecasting the different predictions of stock market funds.

The structure of paper is as follows: Section II contains the review of the literature. In section III, the overview of SVM, GA and ICA is discussed which is further used to develop the proposed model. Section 4 presents the proposed methodology of the model indicating the step by step procedure. Section 5 deals with the results obtained for the real data on different aspects. Section 6 presents the final discussions of the entire study.

2 Literature Review

Stock market researchers and anyone who wants to sell or buy the stocks is keenly interested in knowing its right time at predicting the behavior of stock market but due to the nonlinearity and noisy nature of stock process, the accurate prediction is quite challenging and cumbersome. There is significant impact of some factors like general economic conditions, rate of interest, political events, psychological factors, etc., on the stock market and the different methods are used by the traders for the decision making in the stock market. Basically, the analysis of stock market can be further categorized into two types.

i. Fundamental analysis
ii. Technical analysis.

Fundamental analysis involves the economy and the industrial conditions, financial conditions, various quantitative and qualitative factors of secure investigations, whereas the technical analysis includes the previous prices to predict the future prices of the stock. In this research, the analysis of Japanese Charts and their patterns are used as the technical information. Due to the nonlinearity of nature, soft computing methods are widely used so as to deal with the stock market issues. ANN, i.e., artificial neural network and SVM have been used to resolve the problems of forecasting and financial time series predictions of stock market trends. Although there are many evolutionary mechanisms of classification techniques in neural networks, yet they lack in the learning process that completely depends on the limited reproducibility of the process [1]. Due to which new approaches and modern statistical techniques like SVM are used. Software vector machine is one of the advanced applications of regression and classification methods. There are several studies [2] that use the SVM to predict the time series and SVM was initially developed and introduced by Vapnik [2]. SVM is a machine learning technique that is widely used in pattern recognition and time series forecasting due to its excellent features and high performance [3]. Apart from this [4] uses SVM to determine the air quality and finally concluded that the two kernel parameters C and Ψ have the significant impact on the accuracy and precision of the SVM [5]. Pai and hang [5] used genetic algorithms and gradual annealing algorithm. Sequential parameter optimization (SPO) was given by Bartz [6]. Reference [7] Hong Dong Chen and Wei have used the continuous ant colony algorithm and genetic algorithm to achieve SVR parameters. For numerical analysis and to set the other parameters, some other important methods of genetic algorithms are also available like PSO, i.e., particle swarm optimization methods, etc.

3 Overview of SVM

Support vector machines are one of the prominent techniques of supervised learning models that help us to analyze the data for classification and regression analysis. Support vectors are shown in Fig. 1, SVM is a representation of instances as points in spaces so that the instances can be separately categorized with a clear gap [7]. Further, some new instances can be added to it and are again categorized and the process goes on for the no. of iterations required as per the threshold. Mathematically, SVM is a binary classifier that separates the two classes using a linear boundary.

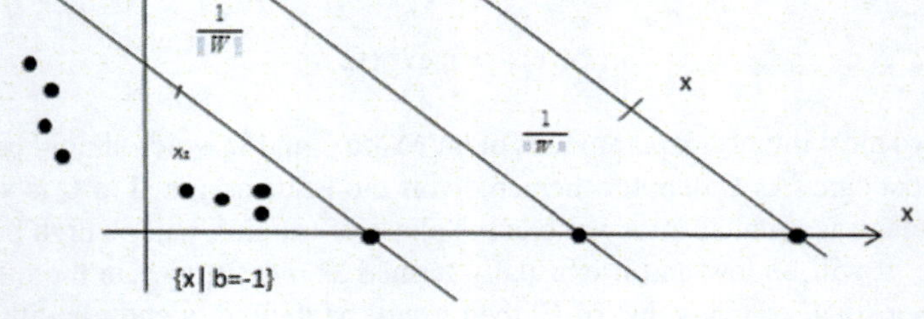

Fig. 1 SVM model

An optimization algorithm is used to create the boundary classes. These are called support vectors. Figure 1 indicates that the input feature space includes two classes and the classes holds xi educational points where $i = 1, 2, 3, 4 \ldots N$. These two classes are tagged in $yi = m$ 1 to -1. The method of optimal margins given by is used to calculate the decision boundary of the two absolutely separate classes and it is boundary line can be given by

$$wx + b = 0 \tag{1}$$

where x represents the support vector on decision boundary and w is an n-dimensional vector which is perpendicular to this decision boundary. $\frac{b}{w}$ denotes the distances between the origin and decision boundary and $w.x$ is the inner product of the two vectors.

Consider the mapping $\mu{:}R^n \rightarrow R^m$, i.e., it is started using the initial data from R^n dimensions to R^m and in these dimensions, the classes usually have less interference.

Mathematically, its optimal decision boundary is as follows:

$$\text{Max:}\, \alpha_1, \alpha_2, \alpha_3, \alpha_4 \ldots \alpha_N$$

$$\frac{1}{2}\sum_{i=1}^{N} \alpha_i \sum_{j=1}^{N} \alpha_j y_i y_j (\mu(x_i)\mu(x_j)) + \sum_{i}^{N} \alpha_i$$

where $0 \leq \alpha_i \leq c_i$, where $i = 1, 2 \ldots N$

$$\sum \alpha_i y_i = 0 \tag{2}$$

Here, α_i and α are Lagrange's multiplier and C is a constant. Now, in Eq. 2, if we apply the core function

$$K(x_i, x_j) = \mu(x_i)\mu(x_j) \tag{3}$$

After calculating the exact value of $K(x_i, x_j)$, we can replace the cost function $\mu(x_i, x_j)$ by the calculated value. One of the sigmoidal kernel equations formed after that will be

$$k(x_i, x_j) = n\,\exp(\Omega, c) \tag{4}$$

The two most important parameters of SVM are C and Ω which should be selected with utmost care. As C denotes the penalty, if the value assigned to C is very high, then the classification, accuracy rate is the phase of training will be high but in case of testing, it will be low and it is usually termed as overfitting. On the other hand, if there is a small or low value of C, then it will be definitely end up with the poor accuracy of classification that will give the wrong results as shown Fig. 1. Similar scenario applies for the parameter Ω while it has more adverse effects that C in the results as it directly affects the feature space.

4 SVM with ICA and GA

Segregation of SVM with ICA and GA

i. In this section, we aim to develop a model for determining the timing of stock market by using SVM with imperialist competitive algorithm (ICA) and genetic algorithms (GA) on the business strategy and technical analysis with the help of Japanese Candlesticks. ICA is applied to optimize the SVM parameters. The segregation of ICA with multiclass SVM is used to determine the amount of SVM parameters helps in the accurate timing of the stock market.
ii. Secondary, GA is used for the feature selection and to choose the optimal parameters from SVM. The purpose of these selected features is to provide us with the useful and unique features by selecting the most essential attributes having the less no. of features will decrease the overall computational cost, SVM features are also the kernel function parameters. Genetic algorithms are basically used to optimize the feature selection done by SVM.

SVM with ICA

An optimization of kernel parameters is using imperialist competitive algorithm. First of all, to select the appropriate centroid or kernel for the SVM is one of the most important tasks as an appropriate kernel determines the data effectively from the initial state to the state at which we want to map our data in new space. For accomplishing this, the linear classifier has to be chosen effectively and we usually use the polynomial kernels. There are two ways to select the appropriate kernels, one is to use the spatial data, another one is to select a different kernel and then classify the data and on the basis of the tested data, the final appropriate kernel is decided, and the Gaussian and sigmoidal kernels are most popularly used kernels and along with the kernels it is also important to have the different kernel parameters.

Polynomial kernel with parameters p

$$P(a, b) = (a.b + 1)^p$$

Polynomial kernel with parameter Ω

$$K(a, b) = 2\exp\left(\frac{-(a - b)}{2\Omega}\right)$$

Polynomial kernel with parameter μ and k is

$$(a, b) = \tan\, h(ka.b - \mu)$$

RBF kernel is basically a benchmark for the optimization of parameters. ICA which is used for optimization is a social human phenomenon but not a natural phenomenon.

SVM and GA

Genetic algorithms are actually based on Darwin's principle "*Struggle for Survival and the Survival of the fittest*" and can be obtained after the series of iterations. In genetic algorithms, different solutions are provided to successive populations until the acceptable results are achieved. In evaluation step, the quality fitness function is calculated and the operations that randomly affects its fitness values as shown in Fig. 2. In every iteration, more suitable chromosomes have been chosen for reproducing the better results. Crossover is an important operator that actually takes place by one-point, two-point or multipoint crossovers and this evolutionary process continuously repeats till the stop condition is satisfied.

Fig. 2 Data flow in ICA

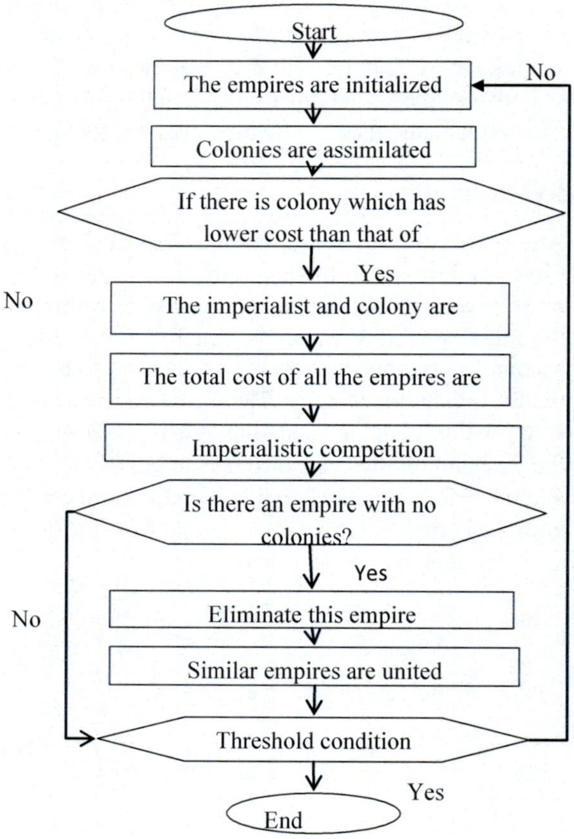

5 Proposed Methodology

The main aim of this study is to create an appropriate model for predicting stock market trading signals with very high precision. In this paper, the model is developed by segregating ICA and GA with SVM, and then appropriate results have been obtained. The following figure shows the flow of the data in it.

In Table 1, the input dataset that we have used is actually based on two approaches []. The stock market's price is categorized into low, high, open and closed prices and turn into 15 and 24. Indicators Oi, Mi, Li and Ci, respectively, denote open, high, low and closed prices on the ith day. 48 datasets are described on the daily basis and are classified into two groups—training and testing. Each dataset includes the daily stock prices, for example, data of the year 2002 is used for testing, whereas the dataset of 2000 is used for training. In one dataset, the data of the year 2000 and 2001

Table 1 Input data set

No	Training period	Test period	No	Training period	Test period
1	2000	2001	25	2000–2003	2007
2	2000	2002	26	2000–2003	2008
3	2000	2003	27	2001	2002
4	2000	2004	28	2001	2003
5	2000	2005	29	2001	2004
6	2000	2006	30	2001	2005
7	2000	2007	31	2001	2006
8	2000	2008	32	2001	2007
9	2000–2001	2002	33	2001	2008
10	2000–2001	2003	34	2001–2002	2003
11	2000–2001	2004	35	2001–2002	2004
12	2000–2001	2005	36	2001–2002	2005
13	2000–2001	2006	37	2001–2000	2006
14	2000–2001	2007	38	2001–2002	2007
15	2000–2001	2008	39	2001–2002	2008
16	2000–2002	2003	40	2001–2003	2004
17	2000–2001	2004	41	2001–2002	2005
18	2000–2001	2005	42	2001–2002	2006
19	2000–2001	2006	43	2001–2002	2007
20	2000–2001	2007	44	2001–2002	2008
21	2000–2001	2008	45	2001–2004	2005
22	2000–2003	2004	46	2001–2004	2006
23	2000–2003	2005	47	2001–2004	2007
24	2000–2003	2006	48	2001–2004	2008

are combined together as a single training data as shown in Fig. 3. The algorithm SVM-ICA-GA is as follows.

SVM-ICA-GA Model

A hybrid approach is followed where a support vector machine is used with ICA[] and genetic algorithms(GA). The following notations can be used in the paper:

N_{pop}: It indicates no. of primary population.

N_{Imp}: No. of imperialist.

THETA: It indicates the "cost on Empires". As it represents the impact on coefficient of colonies.

Imperial cost: Imperialist cost.

Imperial fitness: Imperialistic fitness.

Fig. 3 Model of SVM-ICA-GA

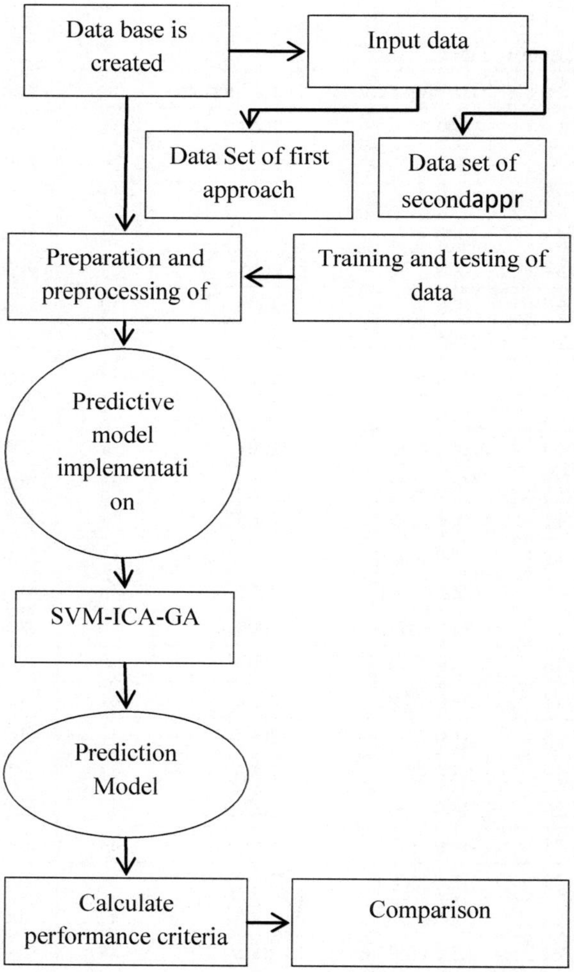

Pre-evolution: The probability of revolution.

$Heus_{Ratio}$: The impact co-efficient, the difference between parent and child.

Imp colonies cost: It is the cost of imperialist's colonies.

Max decades: It indicates the maximum no. of decades as a factor for stopping in ICA.

In this study, RBF as kernel function has been used for the following reasons:

i. The kernel functions can scale the boundaries of the linear space to another very higher dimensional space.
ii. By considering RBF kernel, the linear space with one parameter (C) has entirely the same performance for the two parameters (C,Ω) for higher dimensional space.

Algorithm

Step 1: Firstly, the input dataset of Japanese Candlestick has been taken into consideration. After that an optimal hyperplane is chosen that clearly categorizes the data points. Our objective is to find out the plane that has the maximum margin for the different classes. Maximizing the margin distances provides some reinforcement so that the future data points can be classified with confidence. Using support vector machines, the data is categorized into three classes $H_{\rm i}$, $M_{\rm i}$ and $L_{\rm i}$ denoting high, medium and low chances of the particular data points using two hyperplanes.

Step 2: $H_{\rm i}$, $M_{\rm i}$ and $L_{\rm i}$, the different data points are achieved and obtained in support to these three classes and with the help of SVM, we are having the data points with high prediction ($H_{\rm i}$), low prediction ($L_{\rm i}$) and medium prediction ($M_{\rm i}$) classes.

Mathematically, for data points, $x1,\ x2\ldots xi$ belongs to $H_{\rm i}$, $L_{\rm i}$, $M_{\rm i}$ where $x = 1-48$ in the dataset of Japanese Candlestick and if the soft margin is taken, we wish to minimize the following:

$$\frac{1}{n}\sum_{i=1}^{n}(\max(0, 1 - y_i(w.x_i - b))) + \mu||w||$$

Here, μ is the trade-off between increasing the margin size and ensures that x_i lies on the correct side of the margin.

Step 3: The ICA, i.e., imperialist competitive algorithm is used in this step.

i. The primary population of the colonies is created randomly. Then, the cost function is evaluated for each colony. In this paper, the data points which are missed and are used as a cost function.

$$\text{Miss\_points} = 100 * \left(\sum\nolimits_{(Y=1)}^{N} X_{\text{predicted}} - X\right)$$

Here, N is the no. of outputs. $X_{\text{predicted}}$ is the predicted output by SVM and x is the real output that is given as an input to SVM.

ii. From the above cost evaluation function, the strongest colonies are selected as an imperialist and the colonies move toward the different directions. This step of ICA is usually known as assimilation. Then, the residual colonies are allocated to the imperialists based on the strength of the empires.
iii. Revolution in some colonies takes place, i.e., the location of one colony and imperialist changes if the empire's cost is less.
iv. The objective functions of all the empires are compared and the weakest colony is assigned to the best empire.
v. Weak empires are eliminated and the algorithm is stopped when the stopping condition is reached which accomplishes the specified grounds, otherwise, step ii is repeated.

In this way, the best SVMs parameters are selected using ICA so, in each iteration, it will have the co-ordinates for Miss_points.

Step 4: On the selected parameters of SVM, genetic algorithm is applied. The basic notion of using genetic algorithms is to apply the reinforcement learning. In the selected SVM parameters, GA is used in the three steps:

i. Firstly, the initial population is generated by the system and it is used to find out the optimal parameters and features. Each chromosome contains all the necessary information that is required to select the optimal features and important parameters.
ii. Training
After the initial population is created, system uses support vectors with the help of process parameter values in the chromosomes and the performance of each chromosome is calculated. In this paper, the optimal or near-optimal parameters are calculated and predicted in order to make the predictions more accurate and precise. Then, the fitness function is applied for testing the accuracy of prediction.
iii. Genetic Operations.

By applying the genetic operations, such as selection, crossover and mutation, a generation of new population is created. The best chromosomes are selected depending upon the fitness value of each chromosome and then the crossover as the mutation operator is applied on it with a very small mutation rate. With the new generation again the training phase will be executed to calculate the fitness values. This process is repeated until the stopping condition is achieved, i.e., the no. of predetermined generations. The chromosomes that indicates the best performance in the last selected population are considered as the final result.

In this way, support vectors with very high predictions and medium predictions are achieved in an optimal way.

Table 2 For raw approach with complete details of SVM-ICA-GA

	Total	Raw approach	Signal approach	1-day raw approach	1-day signal approach
Feed Forward Neural Network	74.2	74.8	73.6	45	43.4
SVM-ICA	76	70	79	19	27
SVM-GA	60	59	61	33	33
SVM-ICA-GA	78	62	80	42	33

6 Results and Discussions

The parameters taken for ICA are as follows:

MaxDecades = 100;

$N_{\text{Pop}} = 30$;

$N_{\text{Imp}} = 5$;

$N\text{col} = n_{\text{Pop}} - n\text{Imp}$;

Prevolution = 0.7;

Zeta = 0.1;

Heuristic ratio = 0.5;

With the implementation of SVM-ICA-GA model for the two approaches—raw and signal, the results of 48 datasets have been shown in Table 1. The mean of the predictions with the optimal parameters (C) with raw and signal approaches is 63. 84% and 64.75%, respectively. As the prediction accuracy of both the approaches are quite close to each other and the no. of the features do not make a major change in the forecast accuracy. However, the second approach seems to be better (Tables 2 and 3).

In Table 4, the comparative analysis of the two approaches are discussed. 9% and 61%, respectively while 1-day hit rate is almost same.

7 Conclusion

This study has proposed SVM-ICA-GA model for determining the stock market trends in the way that SVM acts as a classifier, ICA will figure out the strongest imperialist colonies and then genetic algorithms are applied to select the optimum features for the exact and better predictions. In previous studies, SVM-ICA and SVM-GA have already been done but both of these algorithms overall enhances the space and time complexities and their prediction trend was less accurate, but in this paper, three of the approaches are merged and SVM-ICA-GA got the fantastic hit rates of 80% which is quite approachable.

It is to be noted that 1-day hit rate of this base study is quite dominant on others under all the circumstances and provide a good reason for the deeper studies.

Table 3 For signal approach with complete details of SVM-ICA-GA

No.	1	2	3	4	5	6	7	8
1	33	16	6	2	2	0	62	102
2	21	8	6	3	3	1	42	113
3	41	15	6	2	3	2	69	72
4	53	10	3	3	2	1	72	81
5	52	11	3	2	1	1	81	80
6	41	12	6	3	1	2	80	73
…								
…								
45	18	11	8	5	4	3	72	73
46	25	12	6	2	1	0	112	112
47	38	13	5	2	1	1	72	72
48	9	6	4	4	3	2	43	43

Table 4 Comparative analysis of the two approaches

No	1	2	3	4	5	6	7	8
1	52	16	6	2	2	0	62	102
2	40	8	6	3	3	1	42	113
3	62	15	6	2	3	2	69	72
4	83	10	3	3	2	1	72	81
5	51	11	3	2	1	1	81	80
6	40	12	6	3	1	2	80	73
…								
…								
…								
…								
45	33	17	8	5	4	3	72	73
46	28	12	6	2	1	0	112	112
47	33	13	5	2	1	1	72	72
48	24	6	4	4	3	2	43	43

References

1. Ahmadi E, Abooie MH, Jasemi M, ZareMehrjardi Y (2016) A nonlinear au-toregressive model with exogenous variables neural network for stock market timing: the candlestick technical analysis. Int J Eng 29(12):1717–1725
2. Xie H, Zhao X (2012) A comprehensive look at the predictive infor-mation in Japanese candlestick. Proc Comput Sci 9:1219–1227

3. Tay FE, Cao L (2001) Application of support vector machines in financial timeseries forecasting. Omega 29(4):309–317
4. Barak S, Heidary J, Dahooie TT (2015) Wrapper ANFIS-ICA method to do stock market timing and feature selection on the basis of Japanese Candlestick. Expert Syst Appl 42(23):9221–9235
5. Boutte D, Santhanam B (2009) A hybrid ICA-SVM approach to continuous phasemodulation recognition. IEEE Signal Process Lett 16(5):402–405
6. Kuo SC, Lin CJ, Liao JR (2011) 3D reconstruction and face recognition using Kernel-based ICA and neural networks. Expert Syst Appl 38(5):5406–5415
7. Cristianini N, Shawe-Taylor (2000) An introduction to support vector machines. Cambridge University Press

Bone Fractured Detection Using Machine Learning and Digital Geometry

Ashish Sharma, Abhishek Mishra, Aashi Bansal, and Achint Bansal

Abstract The use of technology in the field of medical sciences has increased a lot since the last decade. Nowadays, you can see many technologies like computers and cameras used in medical sciences. They not only help in detecting the disease or the cause but also help in the curing process by maintaining records. The use of computers for image processing has provided an upper hand to the physicians. By using this technique, anybody can tell if there is a fracture in the bone or not. Fractures are common these days, so fracture detection is a crucial part of orthopedic X-ray image analysis. The automatic fracture detection technique helps the doctor to start medical care immediately. We propose a new technique using machine learning and digital geometry. The method can detect bone fractures by bone contours by removing discontinuity processed by segmentation. It overcomes the shortcoming of the previous method which only works on texture analysis. In this, several digital X-ray images are taken as input, and the machine will give the output whether there is a fracture or not based on particular algorithms.

Keywords Bone fracture detection · Medical imaging · Digital image processing · Classification

A. Sharma (✉) · A. Mishra · A. Bansal · A. Bansal
Department of Computer Engineering & Applications, GLA University, NH#2, Delhi Mathura Highway, Post Ajhai, Mathura, UP, India
e-mail: ashishs.sharma@gla.ac.in

A. Mishra
e-mail: abhishek.mishra_cs16@gla.ac.in

A. Bansal
e-mail: aashi.bansal_cs16@gla.ac.in

A. Bansal
e-mail: achint.bansal_cs16@gla.ac.in

N. Marriwala et al. (eds.), *Mobile Radio Communications and 5G Networks*, Lecture Notes in Networks and Systems 140,
https://doi.org/10.1007/978-981-15-7130-5_28

1 Introduction

This research work proposed a new technique of detecting fractures from X-ray images which are normally examined manually. It takes a lot of time to do it manually and is prone to errors. The paper is based on many preprocessing techniques to remove noise from X-rays because these images are more prone to noise (dark spots and patches). The system can detect fractures more efficiently. First, noise is removed from the image, and then the clearer image is used for further process. All unwanted and small objects are removed by the system. Connected components are used by the system to detect the fracture. Orthopedic surgeons must provide ray images as input. Then the system will do preprocessing to remove noise from the scanned image. It will also eliminate all the undesirable objects from the image. Then the machine will give output whether the given input image has a fracture or not. It will do this on the basis of cracks detected in the image. This system saves time and can help a doctor to treat his patients faster.

Here, the proposed model performs considerably well with the digital geometry technique. The introduction is done in Sect. 1. In Sect. 2, literature review is presented; in Sect. 3, the proposed model is presented; Sect. 4 covers implementation details. Section 5 covers results and discussions. Finally, Sect. 6 presents the conclusion and future work.

2 Literature Review

Bone fracture detection is implemented by various researchers, and still, the detection is in progress to identify the bone fracture, its type and severity. The literature available provides a real path to identify the work done in this area as well as gives a systematic view to provide a solution.

In this paper [1], the author has developed a classification of detection of long bone fracture. The system works on two stages, the first stage is basically for preprocessing, where the shape and size of bone fracture is identified and in the second stage the classification is done using back propagation technique.

In another paper [2], the authors use basic image processing techniques to detect multiple fractures in an X-ray image. First, it takes an X-ray as an input, then it preprocesses it to remove noise, then edge detection is done to extract the edges of the bone, then segmentation and feature extraction techniques are executed after which the resultant image is classified. The authors of the paper [3] are using MATLAB as a basic tool for image loading and preprocessing. Also, the gray level co-occurrence matrix (GLCM) is used for feature detection. Also, decision trees and artificial neural networks and meta-classifiers are used for classification.

Another article presented by [4] the author has elaborated the Gaussian filter to remove noise from an image. Also, it uses Canny edge detection and Sobel edge

Table 1 Summary of the accuracy for bone fracture

Algorithm used	Accuracy (%)	Result
Canny edge detection, support vector machine (SVM) [4]	84.7	Morphological operators are used along with Canny edge detection method. After that the classification is done on the basis of SVM
Simple image processing with two performance assessment methods [5]	80	This uses image processing as a basic technique, and then one performance method is used in evaluation of bones and other is used to reduce error
Back propagation, SIFT [1]	100	This uses image processing and neural network technique and SIFT to classify the images
Combination of decision tree and neural network which is named as meta-classifier [3]	85	In this, Sobel operator is used, after that area of fracture is calculated followed by GLCM. It classifies the image as fractured or unfractured

detector to detect edges in an image. It uses support vector machine (SVM) to distinguish the fractured bones from the unfractured bone.

In this paper [5, 6], the author has proposed the idea of classification of bone fractures into fractured and un-fractured ones. In this, the output of the system is measured by two performance assessment methods. The first one is the performance evaluation of bones and the second is for accurate analysis of fracture type within error conditions.

This investigation develops a programmed model [7, 8] for distinguishing and characterizing blacktop asphalt split. Various situations of highlight choice have been endeavored to make informational indexes from computerized pictures. These informational collections are then utilized to prepare and confirm the presentation of AI calculations including the support vector machine. The list of capabilities that comprises the properties got from the projective indispensable and the properties of split articles can convey the most attractive result.

The following table contains the summary of the research work done for bone fracture (Table 1).

3 Methodology

First of all, the types of bones are identified, so that the proper classification can be done, based on the features of different types of the bone fracture, one can easily identify the fracture in the bone as well as its type for the better prediction (Table 2).

Bones are the most sophisticated part of the body. This paper has approached to detect deformities in bones through algorithms using digital image processing. The proposed approach mainly focuses on detecting minor or hairline fractures through

Table 2 Types of the fractures

Simple fracture	This fracture occurs when the broken bone does not penetrate into the skin. It is also called as closed fracture
Open fracture	This type of fracture occurs when the bone is fragmented into small parts which then penetrates through the skin. Another name is compound fracture
Hairline fracture	In this type of fracture, large amount of force stress is exerted on one leg or on one part of the body. It generally occurs when a person does rigorous jogging and running exercise
Greenstick fracture	In this, the bone bends and cracks. In this, the bone is not broken it just simply bends and cracks from its original position
Complicated fracture	In this fracture, the bones adjacent to the fractured bone are also damaged. It may result in severe injuries like in the arteries
Avulsion fracture	In this fracture, the connecting tissues between bone and muscles called tendons are affected. This type of fracture generally occurs in knees and shoulders
Communicated fracture	This type of fracture will take more time to heal. In this, the bone is broken into several pieces which penetrate through the skin
Compression fracture	This type of fracture happens when two bones are pressed against each other. Aged people are more prone to this type of fracture because of osteoporosis

the above techniques. First, the input image is preprocessed so that we can extract the required objects from the image. This is done by blurring the background details and processing the highlighted foreground details. Mathematical morphological techniques like opening and edge detection are used to do these types of operations. The foreground objects are highlighted by edge detection and smoothening technique. At last, the image is classified into two categories whether it is a fractured bone or non-fractured bone. The techniques used in these operations are image preprocessing, noise removal, segmentation, feature extraction and then the classification of the image whether it is fractured or not. The steps involved in these operations are as follows.

Image Preprocessing: In this stage, the features of the image are enhanced. This stage consists of particular set of procedure that helps to highlight the foreground features and smoothening background to increase the performance of subsequent stages of the system. The unwanted features like, noise are removed from the image and contrast is improved for feature extraction. A median filter and average filters are used to remove noise. Median filters are more effective than any other filter because it does not decrease the image sharpness. Also for image enhancement, logarithmic operators are used. The feature extraction is performed by using erosion and dilation techniques and the Sobel edge detection technique is also used.

Noise Removal: Noise is the unwanted pixels that are present in an image. It decreases the quality of image. Noise can be represented as.

$h(a, b) = s(a, b) + i(a, b)$, where $h(a, b)$ is noisy image, $s(a, b)$ is original image, and $i(a, b)$ is noise present in an image.

There are two types of noise present in an image-Gaussian noise and Salt and Pepper noise. Salt and Pepper is more common in X-ray images. It is caused during transmission and appears as light and black dots in the X-rays. Salt and Pepper noise can be removed by mathematical transformation because this method helps us to preserve the edges. To remove it generally median filters are used.

Edge Detection: It determines the boundary of the objects present in the image thus saving its structure. It is basically a technique to find a period in an image on which the intensity of the image shifts suddenly or abruptly. The points on which the illumination of the image changes are then organized into a group of curved portion known as edges. Gradient and Laplacian are the two approaches which are used in edge detection. In the gradient method, the first derivative of the image is used to detect the edges, whereas in Laplacian second derivative of the image is used.

Segmentation: Image segmentation is one of the most important steps because in this step we basically extract data from an image. In this, the regions of the image are homogenously divided with respect to some parameters like color, texture, brightness, etc. It is a necessary step as it locates objects and their boundaries. Through this, an image can be partitioned into a collection of pixels that are connected. It helps in the annotation of the object scene. Many approaches can be used like region approach, boundary approach, edge approach and many more.

Feature Extraction: It is considered the next step in many image processing applications. It is a type of dimension reduction. Parts of an image are represented as compact. This is useful when images are of large size, and feature extraction technique is used to complete tasks such as retrieval very quickly. It is often used with feature detection so that it can work efficiently. Many techniques like scale-invariant feature transform (SIFT) technique are used to detect features from the digital image.

Classification: In this, the data are categorized into a number of categories. It is basically a study of data and then categorizes it accordingly. Each category has its own properties, and the data which belong to that category have some common properties. In this, the classifiers divide the input image into two categories—the fractured and the non-fractured one.

4 The Model

The approach in the current paper is basically an intelligent system that consists of two phases: processing and classification, i.e., done by digital geometry and machine learning. In the first stage, we extract the content of the image by using the Haar wavelet transform and SIFT. SIFT method is used to detect those points which are of interest from the grayscale image. Image intensities which are present in the image structure are used in this process. In this system, we used the SVM because it is very efficient in identifying different types of bone fractures by extracting the depth of the fracture. These features are used to increase the character of the image and also to excerpt useful parts from the bone (fractured bone). After that, the processing phase of the system image is set to the next phase for classification, which is done by a machine learning classification system, and in the current approach SVM is used for the classification.

Haar wavelet transform: As stated in 2013, one of the best ways to do image preprocessing is image transforms because they are extensively used in filtering, image processing, along with denoising. Haar wavelet can be used for several distinct image transform and preprocessing methods. The compression method is basically used for the investigation of medical images. In studies, the Haar wavelet transform technique is used to strengthen the features of the image. Firstly, input image is transformed into grayscale so to decrease the processing time, then image filtering is done to make image balance. Smoothing which is also called blurring is a method that means to reduce noises in an image for further processing. Wavelet transform has an advantage that this is not Fourier-based, therefore better results can be obtained by handling the discontinuities of the image.

Scale-Invariant Feature Transform (SIFT): SIFT is a method used for feature extraction. In these design stats of gradient directions of image, depths are used. These depths are compiled in the image structure of each interest point to match each relevant interest point descriptor is used. This algorithm is used to extract features and to identify interest points, and it uses different kinds of images. A method is performed which uses stats of gradient direction of image depths.

This approach is used to extract the feature of the image for processing of image so that the system can classify the image as fractured and non-fractured. For extracting features from the image, an image is converted into a "large collection of local feature vectors" by the algorithm. Scaling invariance, rotation and translation can define properties of these local feature vectors. The system is trained by the back propagation network through the features of images that are extracted from images by SIFT. After training, images are input into the neural network to check if it is working properly and can it pass the feature extraction phase (Fig. 1).

5 Result

Data Set: The data set is taken from the TCIA [7] and USI [10]. MATLAB 16.0 and Python are used for carrying out this system for real-time use and for detecting fractured bones.

The Table 3, illustrate that the bone fracture detection is performed by various approaches like Canny Edge Detection, SVM (Support Vector Machine), Simple Image Processing with two performance assessment methods, Back Propagation, SIFT, Combination of Decision Tree and Neural Network, which is named as Meta classifier. Finally, the proposed approach i.e. Machine Learning and Digital Geometry based technique provides better accuracy level than the state of art approaches. The results are based on overall accuracy of bone fractured as calculated so far.

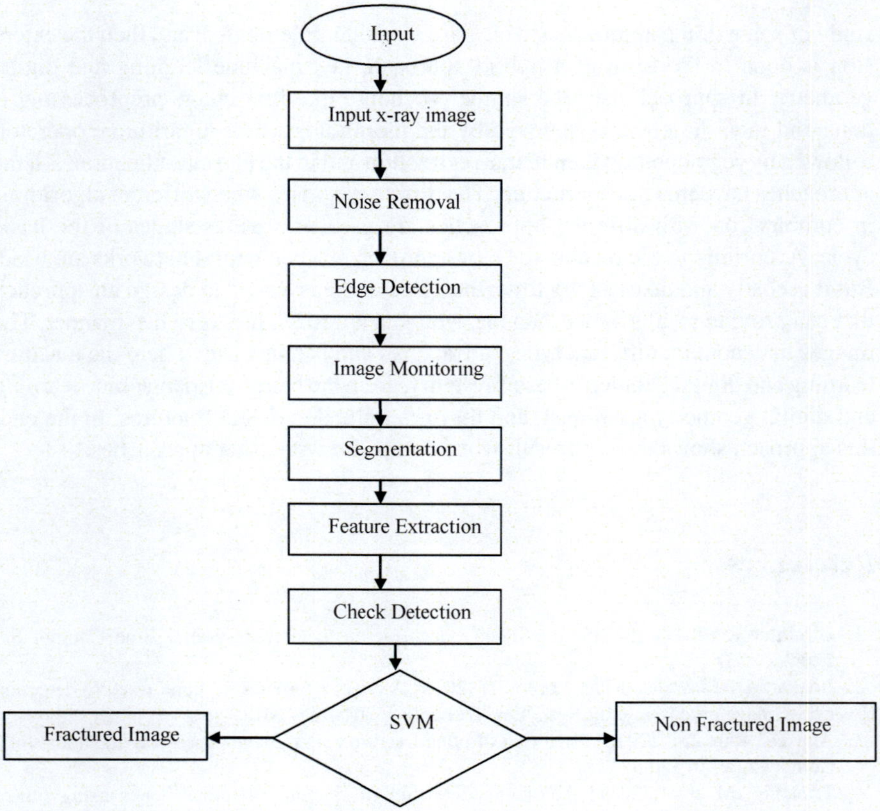

Fig. 1 Flowchart of the proposed approach

Table 3 Results of the accuracy for bone fracture

Algorithms	Accuracy (%)
Canny edge detection, support vector machine (SVM) [4]	84.7
Simple image processing with two performance assessment methods [5]	80
Back propagation, SIFT [1]	90
Combination of decision tree and neural network, which is named as meta-classifier [3]	85
Machine learning and digital geometry	**92**

6 Conclusion

The approach is based on two things, i.e., machine learning and digital geometry; therefore, it uses the advantages of these approaches. The image processing techniques are very useful in the medical field. The work is based on the comparative study of some of the approaches which are already implemented, and then the extension is done in the form of a hybrid approach, i.e., machine learning and digital geometry. In approach uses the simple procedure like first image preprocessing is done and then the noise is removed by the morphological or logarithmic operator, followed by segmentation then feature extraction and in the end classification. All the approaches have this basic structure. The difference arises when different algorithms in combination with different approaches are used on various stages of the basic cycle. According to the results, the best approach is when neural networks are used. But it is costly and takes a long time; therefore, there is a need to design an approach that can provide results better than the available approach in a very fast manner. The images may contain different types of fractures and patches which help the machine learning and digital geometry-based system to train the better. Also, machine learning and digital geometry are robust, and they are trained to detect fractures. In the end, this approach shows the best result when compared with other approaches.

References

1. Dimililer K (2017) IBFDS: intelligent bone fracture detection system. Proc Comput Sci 120:260–267
2. Bhakare DB, Jawalekar PA, Korde SD (2018) A novel approach for bone fracture detection using image processing. Int Res J Eng Technology (IRJET) (2018)
3. Anu TC, Raman R (2015) Detection of bone fracture using image processing methods. Int J Comput Appl 975:8887
4. Tripathi AM et al (2017) Automatic detection of fracture in femur bones using image processing. In: 2017 international conference on innovations in information, embedded and communication systems (ICIIECS), IEEE
5. Bandyopadhyay O, Biswas A, Bhattacharya BB (2016) Long-bone fracture detection in digital X-ray images based on digital-geometric techniques. Comput Methods Programs Biomed 123:2–14
6. Bandyopadhyay O, Biswas A, Bhattacharya BB (2014) Long-bone fracture detection in digital X-ray images based on concavity index. International workshop on combinatorial image analysis. Springer, Cham
7. Hoang N-D, Nguyen Q-L (2019) A novel method for asphalt pavement crack classification based on image processing and machine learning. Eng Comput 35(2):487–498
8. Gopalakrishnan K et al (2017) Deep convolutional neural networks with transfer learning for computer vision-based data-driven pavement distress detection. Construc Build Mater 157(2017):322–330
9. https://www.cancer.net/cancer-types/bone-cancer/statistics. Last accessed on 26 Jan 2019
10. https://archive.ics.uci.edu/ml/machine-learning-databases/. Last accessed on 26 Jan 2019

Assessment of Microgrid Communication Network Performance for Medium-Scale IEEE Bus Systems Using Multi-Agent System

Niharika Singh, Irraivan Elamvazuthi, Perumal Nallagownden, Gobbi Ramasamy, and Ajay Jangra

Abstract Microgrids help to achieve power balance and energy allocation optimality for the defined load networks. In this paper, the focus is to enhance the intelligence of the microgrid network using a multi-agent system and validation is using network performance metrics such as delay, throughput and jitter. Network performance is analyzed for the medium-scale microgrid using two IEEE test systems, i.e., IEEE 34 and IEEE 39. In this paper, Bellman–Ford algorithm is incorporated to calculate the shortest path to a given destination. The algorithm is defined for the distributed nature of the microgrid. From this model, researchers have achieved up to 30% improvement in the network performance of a microgrid.

Keywords Distributed energy resources (DERs) · Microgrid · Multi-agent system (MAS) · Network performance · Renewable energy sources (RES) · Smart grid

N. Singh (✉) · G. Ramasamy
Smart Assistive and Rehabilitative Technology (SMART) Research Group, Seri Iskandar, Malaysia
e-mail: niharika.academics@gmail.com

I. Elamvazuthi
Department of Electrical and Electronics Engineering, Universiti Teknologi PETRONAS, 32610 Perak, Malaysia
e-mail: irraivan_elamvazuthi@utp.edu.my

P. Nallagownden
Faculty of Engineering, Multimedia University, Persiaran Multimedia, 63100 Cyberjaya, Selangor, Malaysia

A. Jangra
University Institute of Engineering and Technology, Kurukshetra University, Kurukshetra, India

N. Marriwala et al. (eds.), *Mobile Radio Communications and 5G Networks*, Lecture Notes in Networks and Systems 140,
https://doi.org/10.1007/978-981-15-7130-5_29

1 Introduction

Microgrids are small electric power systems that consist of generation, transmission, distribution and load. These are the single, independent and controllable power systems incorporating various distributed generators, energy storage devices, sensing and controlling devices, etc., connected to the users [1, 2]. Microgrids help to achieve power balance and energy allocation optimality for the defined load networks. Microgrids are the distribution networks helpful to handle grid variability and energy harvesting using high penetrations of bidirectional power flow. Microgrids have two working modes, i.e., grid-connected mode and islanded mode [3]. In the grid-connected mode, microgrid is connected to the national grid for transmitting/receiving power and information through their respective networks, but in islanded mode, microgrid is in the self-sufficient mode that generates/distributes energy and information as per their defined area. In islanded mode, power sharing is local though the information/data sharing is not only limited to local but to global network as well [4, 5]. This global connection to the Internet builds a strong communication between microgrid and national grid in terms of economical, commercial, technical and social decision making [6]. To share energy optimally and efficiently, energy information is exchanged within installed energy routers in the data network [7]. The microgrid data network is distributed either using wired or wireless connections. This distribution varies based on the size of microgrids' geographical network area (such as small, medium and large). As per the network size, the routers may use different routing protocols. These protocols are applied to maintain optimization, misrouting, of packets, re-routing of data packets, packet dropping, fabrication, security and modification [8, 9].

Traditional microgrids are the electrical systems that have a major role of energy routing than the information/data routing [10, 11]. To make the microgrids smarter, various techniques are required that can equip traditional microgrids with intelligence and vision. To aim a certain level of intelligence in the modern microgrids, it should have a robust communication network. Modern microgrids use communication technologies to collect and respond to the information, as represented in Fig. 1.

Figure 1 describes the environment between inter- and intra-microgrid communication. In the scenario, collected information considers the behavior of the consumer and supplier, production and distribution stability, transmission and monitoring operations, etc. The requirements for power routing are quite different from the data packet routing (such as best effort delivery is suitable for data routing but not for energy routing since lost power signals are difficult to be resent) [1, 4]. Today, the most challenging task for the microgrid is decision making. Hence, it seeks upgradation for the microgrid control and monitoring systems. Several methods are proposed based on the physical model of the microgrid considering the physical layer, but nowadays communication and networking demand more focus. The futuristic approaches can be listed as Internet of Things (IoT), artificial intelligence, multi-agent systems, game theory, etc. [2, 6]. Multi-agent and game theories widely have been used for the energy

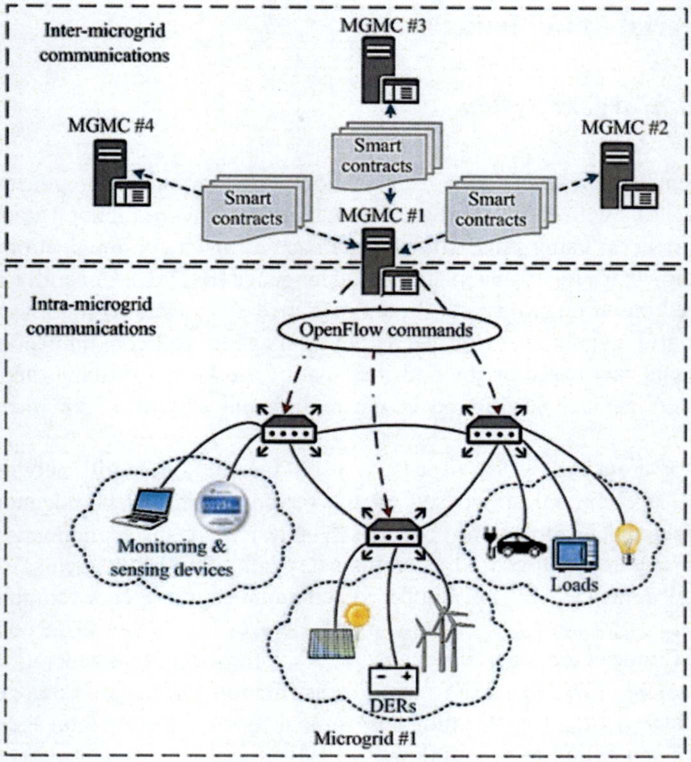

Fig. 1 Pictorial representation of inter- and intra-microgrid communications [1]

management and decision making of the AC and DC microgrids. Several intelligent approaches are explored for changing electricity consumption behavior and energy demand response. Previously, researchers only focused on formulating optimization problems for achieving economic efficiency with the centralized approach. Regardless in this paper, researchers are focusing on robust communication designing for the smart microgrid in order to make the microgrid wiser than other proposed approaches to get a better network performance based on the decentralized distribution network [2, 12].

The uniqueness of the paper is the network performance analysis for medium-scale microgrids. The paper is organized in a manner that explains outline through introduction followed by material and methods that describes basic building blocks of the work. Further, the simulation environment is explained followed by the results and discussion section.

2 Materials and Methods

2.1 Multi-Agent System

Multi-agent system refers to a team of homogeneous and heterogeneous agents communicating with each other based upon their intelligence (defined algorithm and past experiences) using standard language such as agent communication language (ACL), knowledge query and manipulation language (KQML). Generally, agents are categorized based on two types: i) administrative agents and ii) runtime agents. In the microgrid, agents are classified as physical agents and communication agents. These agents vary based on their attribute nature, i.e., static attributes and dynamic attributes. A general description of the multi-agent system is explained through Table 1.

Table 1 describes a list of agents used for the microgrid network model designing. It communicates based on the heterogenous and homogenous nature of the agents. The framework considers N physical agents (combination of Σn_j generator agents deployed to communicate with Σn_j load agents) communicating and controlled by M number of communication agents (combination of $m_{master}, m_{gen\_host}, m_{dis\_tran\_host}, m_{load\_host}$). Here, m_{master} is the master communication node for the communication agents m_{gen_host} (host agent for generation controlling agents), $m_{dis\_tran\_host}$ (host agent for distribution and transmission controlling agents) and m_{load_host} (representing host for load agents). Hence, total agents for the framework can be followed as follows:

$$\text{Number of agents } A = [N + M] \tag{1}$$

$$A = \left[\left(\sum n_i + \sum n_j\right) + \left(m_{master} + m_{gen_{host}} + m_{dis_{tran_{host}}} + m_{load_{host}}\right)\right] \tag{2}$$

Based on this equation, researchers have analyzed microgrid performance for IEEE 34 and IEEE 39.

Table 1 Agent response table for the multi-agent system

Agent type	Responsibility
User agent	Represents a user by expressing its information request
Information agent	Represents an information source and answers users' requests
Query agent	Follows up a request from the user
Support agent	Set of agents providing internal service and coordination among multiple agents
Communication agent	Transmits messages to various agents
Communication area	Provides mailbox to each agent's disposal
Mailboxes	Accumulates message addressed to one agent

2.2 IEEE Bus Systems for Microgrid

Microgrids are considered as the whole grid with a significantly smaller geographical extent. Several testing models are provided by the IEEE such as IEEE 4, IEEE 9, IEEE 14, IEEE 33, IEEE 34, IEEE 39, IEEE 72, IEEE 118 and IEEE 123. These systems are proposed for the electricity grid testing but have been used for microgrid testing as well [13]. To test for the microgrids, electrical lengths are considered smaller in comparison with national grid testing. Based on the power production, demands and line length microgrid testing systems are classified into three categories, i.e., small, medium and large. The major differences of using these testing systems are seen for the monitoring, response times, system handling, voltage fluctuations, less inertia but not on the topology. Hence, the scale of microgrids can be categorized as small microgrids as IEEE 4, 5, 9, 14, etc., medium microgrids as IEEE 23, 33, 34, 39, etc., and large microgrids as IEEE 72, 118, 123, etc.

2.3 Network Performance

In a microgrid, communication and network performance play an important role. The network communication in a microgrid is supported by 100 Mbps Ethernet, DNP3 over TCP/IP that may lead to end-to-end delay. Network communication in a microgrid environment can be demonstrated as shown in Fig. 2 [14].

Figure 2 encapsulates the power and communication network of the microgrid. Here, it is not limited to the local microgrid network but also shows how microgrids interact with the outside environment. It has network nodes, actuator nodes and sensor nodes making coordination among distributor, operator and user. Environment perception and local control system of microgrids are integrated into the LAN network. It is responsible to handle distributed power and load based on the energy storage and distributed power acquisition system [3, 4].

Network performance can be evaluated based on various parameters, such as packet loss, throughput, bandwidth, delay or latency. These measures vary under different stressed conditions. In this paper, the focus is to evaluate microgrid networks based on the end-to-end delay in the network. It is analyzed for the medium-scale microgrid using two IEEE test systems, i.e., IEEE 34 and IEEE 39. The protocol and algorithm used for the demonstration are mentioned in the below section.

2.4 Protocol and Algorithm Description

In a microgrid network, routing plays an important role as the network layer protocol to guide the packets from the communication source to their designated destinations. These packets contain information related to energy flowing in the circuit [6, 15].

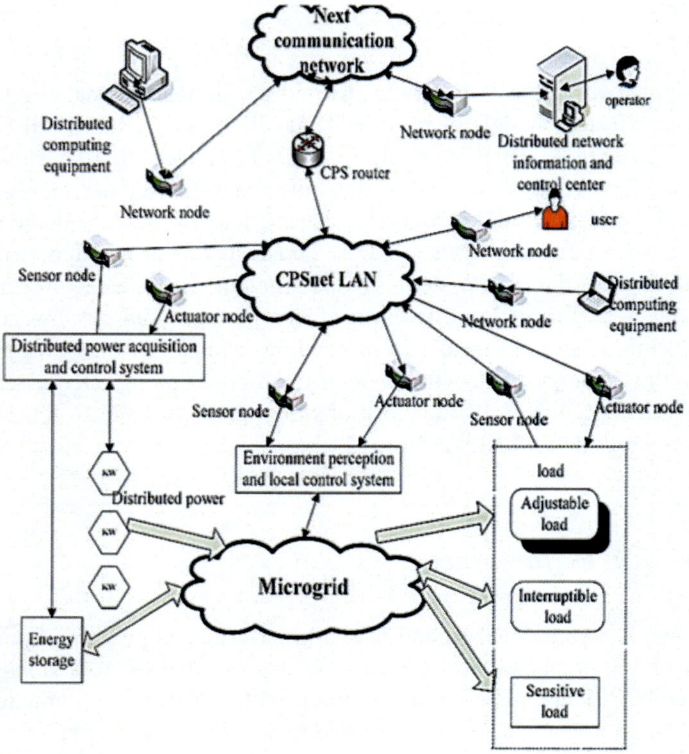

Fig. 2 Interactive environment of microgrid communication network [14]

Routing in the microgrid involves complex calculation of algorithms supported by each other proposing service or information exchange. The complexity of algorithms relates to three reasons. First, it requires coordination among all microgrid network nodes based on the module as well as their subnet. Second, microgrid routing needs to cope with node failures and links in order to redirect packets and update the databases maintained and third, for achieving high performance over congested nodes in the microgrid network. In microgrid, mainly two performance measures are affected by the routing algorithm, i.e., average packet delay and throughput [16]. Average packet delay refers to the quality of service, and throughput refers to the quantity of service. A good routing is responsible for increasing throughput for the same value of average delay per packet during high load demand and a decrement of average delay per packet for the low or moderate load conditions. In this paper, Bellman–Ford algorithm is incorporated to calculate the shortest path to a given destination. The algorithm is defined for the distributed nature of the microgrid. Bellman–Ford algorithm is as follows:

$$D_i = \min_j \left[d_{ij} + D_j\right]$$

Here, D_i represents estimation of the shortest distance of the node i to a definite destination. Length between the link (i, j) is designated to d_{ij}. Node i executes the iteration periodically by considering minimum from neighbor $j.d_{ij} + D_j$ is the estimated shortest distance from node i to the destination passing through j and $\min_j[d_{ij} + D_j]$ is the shortest distance estimated through the best neighbor [6].

Several algorithms are proposed for optimal routing computation of the smart grid but very few for the microgrid data network. In this paper, authors have successfully analyzed data networks for medium-scale microgrids based on the described material and methods.

3 Performance Metrics

To assess the performance of microgrid networks, several baselines are used. The default baseline is the multi-agent system. In addition, the Bellman–Ford algorithm is used as a second baseline which provides routing optimality to the microgrid network. For the simulation network, data/packet streaming uses unicast and broadcast communication channels. The network uses Ethernet with 1 Gbps network bandwidth (channel capacity). For network performance evaluation, following metrics are used:

(a) *Throughput*: This is the average amount of packets received by the node per second.
(b) *Delay*: This is the function value of travel and processing time of the signal/data packet traversing between sender and receiver.
(c) *Jitter*: This is the delay inconsistency between each packet. Jitter occurs due to inconsistent delay pacing during packet transmission.

The message transmission in the network uses various protocols. Snapshots from two different timescales have been recorded and shown in Figs. 3 and 4.

Figures 3 and 4 describe behavior of various protocols used during the communication process. The snapshots are taken for two different scenarios where in Fig. 3 one can observe the involvement of TCP protocols along with ARP and ICMP, whereas through Fig. 4 one can notice the communication using UDP protocol. The communication starts with a broadcast packet and receives information through ARP protocol. For TCP protocol, it uses frame information including port address, status of SYN or ACK, window size, packet length, etc. For UDP protocol, it uses information about acknowledgement, flag status, length of the packet, etc.

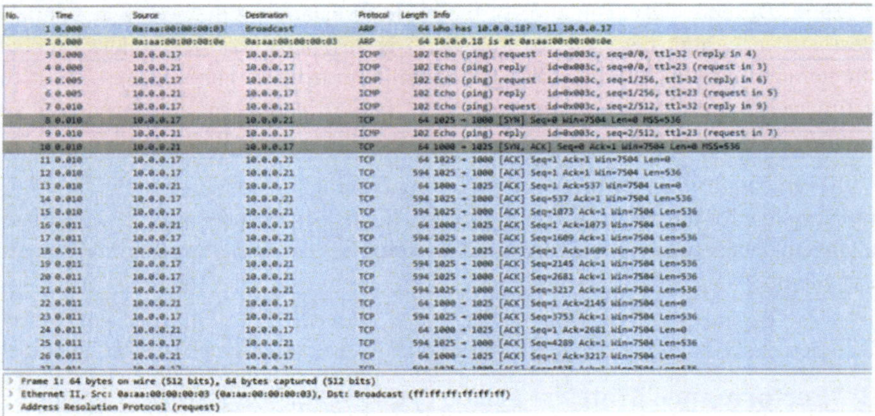

Fig. 3 Packet flow for TCP protocol in the microgrid network

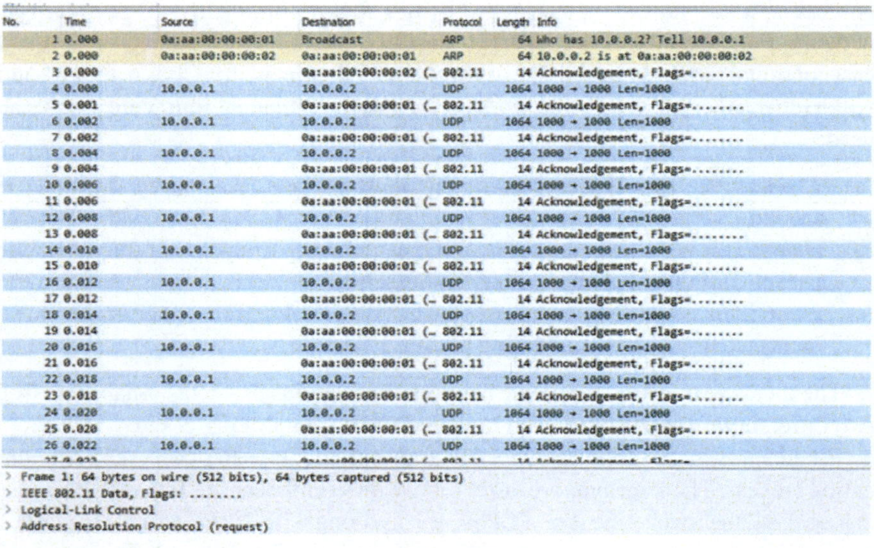

Fig. 4 Packet flow for UDP protocol in the microgrid network

4 Results and Discussion

Microgrids are small electric power systems that consist of generation, transmission, distribution and load. These are the single, independent and controllable power systems incorporating various distributed generators, energy storage devices, sensing and controlling devices, etc.

The research is run for two medium-scale microgrid testing models, i.e., IEEE 34 and IEEE 39. In order to test the communication performance of the microgrid,

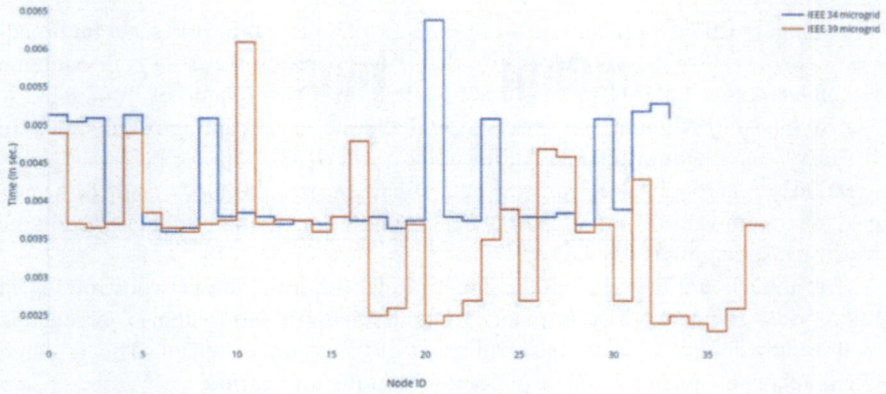

Fig. 5 Average delay of IEEE 34 and IEEE 39 microgrid test model

communication delay is considered as the first performance metric. The average delay of IEEE 34 and IEEE 39 microgrid test models is shown in Fig. 5.

In Fig. 5, there is an overlap for the delay sequence. It is observed that delay varies from one node to another node. Here, the total no. of nodes are 34 and 39 for IEEE 34 and IEEE 39 microgrid, respectively. The highest average delay for IEEE 34 is 0.0064 s, and for IEEE 39 it is 0.0061 s. The average delay for the microgrid should be less than 0.01 s (10 ms) for the high priority information. This resulting value for each node mentioned in Fig. 5 has experienced very less delay. After the average delay, communication performance is checked using throughput and jitter. These performance results are analyzed and compared in Table 2.

For network simulation, microgrid is using two ways for data streaming, i.e., unicast and multicast along receiving traffic from CBR mode. Hence, in Table 2, three comparisons are made based on the unicast, CBR and broadcast packets. For unicast, no jitter was observed, but the case is different for broadcast data

Table 2 Comparison of microgrid-based network performance metrics

Microgrid bus system (100 Mbps bandwidth)	UDP: Unicast		CBR server		UDP: Broadcast		
	Throughput (bits/s)	Delay (in s)	Unicast end to end throughput (bits/s)	Unicast end to delay (bits/s)	Throughput (bits/s)	Delay (in s)	Jitter (in s)
IEEE bus 34	3450	0.0145	4300	0.014 s	11,600	0.0064	0.00065
IEEE bus 39	3290	0.0132	4200	0.0132 s	11,200	0.0061	0.00051

streaming. In IEEE 34, maximum average delay for unicast is 0.0145 s, but for broadcast delay is 0.0064 s. Another metric that is compared in the table is throughput. For unicast, it is 3450 bits/s, and for broadcast maximum value of throughput is 11,600 bits/s. Maximum jitter observed in the broadcast streaming is 0.00065 s. In IEEE 39, maximum average delay for unicast is 0.0132 s, but for broadcast delay is 0.0061 s. Throughput for unicast data streaming is 3290 bits/s, and for broadcast maximum value of throughput is 11,200 bits/s. Maximum jitter observed in the broadcast streaming is 0.00051 s.

The models are using the Bellman–Ford algorithm for the network routing in microgrids. The network performance for both microgrid bus systems is acceptable as it is showing better performance in terms of the reference results. This research opens scope to enhance network performance of the microgrid.

References

1. Li Z, Shahidehpour M, Liu X (2018) Cyber-secure decentralized energy management for IoT-enabled active distribution networks. J. Mod. Power Syst, Clean Energy
2. Coelho VN, Weiss Cohen M, Coelho IM, Liu N, Guimarães FG (2017) Multi-agent systems applied for energy systems integration: State-of-the-art applications and trends in microgrids. Appl Energy
3. Bartolucci L, Cordiner S, Mulone V, Rocco V, Rossi JL (2017) Renewable sources integration through the optimization of the load for residential applications. Energy Proc 142:2208–2213
4. Giaouris D et al (2013) Performance investigation of a hybrid renewable power generation and storage system using systemic power management models. Energy 61:621–635
5. Yoldaş Y, Önen A, Muyeen SM, Vasilakos AV, Alan İ (2017) Enhancing smart grid with microgrids: challenges and opportunities. Renew Sustain Energy Rev
6. Anvari-Moghaddam A, Rahimi-Kian A, Mirian MS, Guerrero JM (2017) A multi-agent based energy management solution for integrated buildings and microgrid system. Appl Energy
7. Nguyen TL et al (2018) Multi-agent system with plug and play feature for distributed secondary control in microgrid—controller and power hardware-in-the-loop implementation. Energies 11(12):1–21
8. Kamal Z, Mohammed A, Sayed E, Ahmed A (2017) Internet of things applications, challenges and related future technologies. World Sci, News
9. Kuzlu M, Pipattanasomporn M, Rahman S (2014) Communication network requirements for major smart grid applications in HAN, NAN and WAN. Comput Netw
10. Supriya S, Magheshwari M, Sree Udhyalakshmi S, Subhashini R, Musthafa (2015) Smart grid technologies: communication technologies and standards. Int J Appl Eng Res
11. Galli S, Scaglione A, Wang Z (2011) For the grid and through the grid: the role of power line communications in the smart grid. Proc IEEE
12. Anvari-Moghaddam A, Monsef H, Rahimi-Kian A, Guerrero JM, Vasquez JC (2015) Optimized energy management of a single-house residential micro-grid with automated demand response. In: 2015 IEEE Eindhoven PowerTech, PowerTech 2015
13. Hirsch YPA, Guerrero J (2018) Microgrids: a review of technologies, key drivers, and outstanding issues. Renew Sustain Energy Rev. 90(April):402–411
14. Hirsch A, Parag Y, Guerrero J (2018) Microgrids: a review of technologies, key drivers, and outstanding issues. Renew Sustain Energy Rev

15. Simmhan Y et al (2013) Cloud-based software platform for big data analytics in smart grids. Comput Sci Eng
16. Jaradat M, Jarrah M, Bousselham A, Jararweh Y, Al-Ayyoub (2015) The internet of energy: smart sensor networks and big data management for smart grid. Proc Comput Sci

Optimum Digital Filter Design for Removal of Different Noises from Biomedical Signals

Shailu Srivastava and Shruti Jain

Abstract The denoised signal with enhanced quality is utilized for correct analysis of biomedical signal. In real-time applications, these signals are mostly affected by distinct types of artefacts. The objective of this research paper is to design the optimum digital filter for the elimination of noises lies in high- and low-frequency bands. The proposed methodology is implemented using approximations and windowing techniques. The results of frequency responses of both the designed filters are evaluated and compared on the basis of pole–zero plot and gain. It is interpreted from the simulated results that the design technique with the help of Kaiser window results in −61.75 and −56.89 dB for high-frequency and low-frequency signals, respectively, while assuming shape parameter as 0.5. The proposed design method can be used in different digital signal processing applications.

Keywords Digital filters · Low-frequency noise · High-frequency noise · Window technique · Butterworth filter

1 Introduction

Signal processing is performed in most of the systems for biosignal interpretation and analysis [1]. Nowadays, biomedical signal processing has been close towards objective or the quantitative analysis of physiological systems and phenomena through signal evaluation [2]. These signals contain valuable clinical information in real time which is hampered by different noises such as electrode contact noise, instrumental noise, baseline noise, motion artefacts, power line interference and electrosurgical noise [3]. Different types of digital filters are used for the removal of artefacts from signal components. It is very tedious to reduce random noises having fixed coefficients.

S. Srivastava · S. Jain (✉)
JUIT, Solan, Himachal Pradesh, India
e-mail: jain.shruti15@gmail.com

N. Marriwala et al. (eds.), *Mobile Radio Communications and 5G Networks*, Lecture Notes in Networks and Systems 140,
https://doi.org/10.1007/978-981-15-7130-5_30

The situations which include motion of the patient (running or walking), breathing, interplay among the electrodes pores and skin are the reasons of occurrence of low-frequency noises/baseline wandering [4]. The signal gets drifted with a high degree from baseline and is known as baseline wander. The presence of muscle noise is due to high-frequency noise/electromyography (EMG) noise that represents a fast rate trouble in most of the applications, particularly in recordings acquired during exercise, for the reason that low amplitude waveforms may turn out to be absolutely obscured. EMG noise occurs due to the shrinkage of muscles other than the cardiac muscular tissues [3, 4]. The degree of crosstalk is always in proportion to the amount of muscular contraction that takes place either due to the probe quality or movement. The frequency of the EMG noise lies in between 100 and 500 Hz or further taken to the higher frequencies. Muscle noise, baseline wander and 50/60 Hz noise are not always removed by using narrowband filtering. This affords a much extra filtering trouble because of the spectral content of muscle activity.

To remove the artefacts from the biomedical signals, different techniques were proposed in the last twenty years. To get the important data which is present in a specific frequency range, digital filters may be used [5]. In general, these techniques can be categorized into adaptive and non-adaptive filtering [6]. The non-adaptive filtering consists of infinite impulse response (IIR) filter, finite impulse response (FIR) filter and the notch filter. The FIR filters are considered for removing the noise and spectral effectively from the biomedical signal. An FIR filter includes multipliers, delays and adders to get the output [7–9]. As the order of the filter increases, there is an increase in the execution time and complexity. Theoretical and experimental results using the windowing technique are almost similar due to the availability of a well-structured equation [10].

In this paper, authors have studied distinct research papers on specific filtering strategies [11]. Alarcon et al. implemented algorithms for high-pass and low-pass three-pole recursive Butterworth filters at particular cut-off frequency [12]. Gaikwad and Chavan suggested that digital IIR filtering approach is best suited for the removal of high-frequency noise from ECG Signal [13]. Rahman et al. [14] remove the baseline wandering by using the wandering path finding algorithm. Authors in [15] worked on FPGA-based FIR low-pass filter for the removal of EMG noise from electrocardiogram. Authors in [16] discussed the various noises present in biomedical signals and different techniques to remove these noises.

The main utility of a filter is to discard unnecessary signal components [17, 18], like baseline wander noise or EMG noise. The main aim of this research work is to design an optimum digital filter for removal of noises lies in high- and low-frequency bands. For the smoother transitions, better gain and ability to control the ripple content, authors have implemented different digital filters using LabVIEW. The main requirement of this method is to maintain physiological characteristics after denoising. Gain (in dB) is calculated for different digital filters. The pole–zero plot and frequency responses of both the filters were also analysed.

The rest of the paper is structured as follows: Sect. 2 explains the methodology segment, Sect. 3 explains the results and discussion which are followed by the conclusion and future scope in Sect. 4.

2 Methodology

Various types of filters are categorized primarily based on signal processing, elements type, construction filters, impulse response and frequency range. The transfer function for a time-invariant, linear digital filter known as the recursive filter can be expressed in z- domain and is represented by Eq. (1).

$$H(z) = \frac{B(z)}{A(z)} = \frac{b_0 + b_1 z^{-1} + b_2 z^{-2} + \ldots b_N z^{-N}}{1 + a_1 z^{-1} + a_2 z^{-2} + \ldots a_M z^{-N}} \quad (1)$$

where $a_0, a_1, \ldots a_n$ and $b_0, b_1, \ldots b_n$ are filter coefficients. If the denominator in Eq. (1) is made equal to unity (no feedback), then the filter will act as an FIR filter [6]. The impulse response is a measurement of how a filter will respond to the delta function and is expressed by Eq. (2).

$$\delta_{ij} = \begin{cases} 1 & i = j \\ 0 & i \neq j \end{cases} \quad (2)$$

Window functions are a category of time-domain functions. Gibb's oscillations are decreased by using a suitable window function [1, 2]. Window functions $[w(nT_s)]$ are used to restrict the impulse response $[h(nT_s)]$ with certain value. $H(z)$ is the z transform of $[h(nT_s)]$ expressed by Eq. (3), $W(z)$ is the z transform of $w(nT_s)$ expressed by Eq. (4), and $H_w(z)$ is the z transform of $[w(nT_s)h(nT_s)]$ expressed by Eq. (5).

$$H(z) = \sum_{n=-\infty}^{\infty} h(n\,T_s)\,z^{-n} \quad (3)$$

$$W(z) = \sum_{n=-\infty}^{\infty} w(n\,T_s)\,z^{-n} \quad (4)$$

$$H_w(z) = \sum_{n=-\infty}^{\infty} [w(n\,T_s).\,h(n\,T_s)]\,z^{-n} \quad (5)$$

Different filtering approaches are implemented for denoising of the different noises from biomedical signals. Initial step is the identification of noises present in the signal followed by the selection of the filter. Selection of optimal filter is a challenging task as in literature there are two types of filter IIR and FIR, and both have its benefits and downsides. In this paper, authors have stressed on FIR filter using windowing technique and IIR using approximate filter. When a digital FIR filter is designed using window techniques, it is essential to specify window function and the order of the filter as shown in Fig. 1. Gain (in dB) is calculated for different

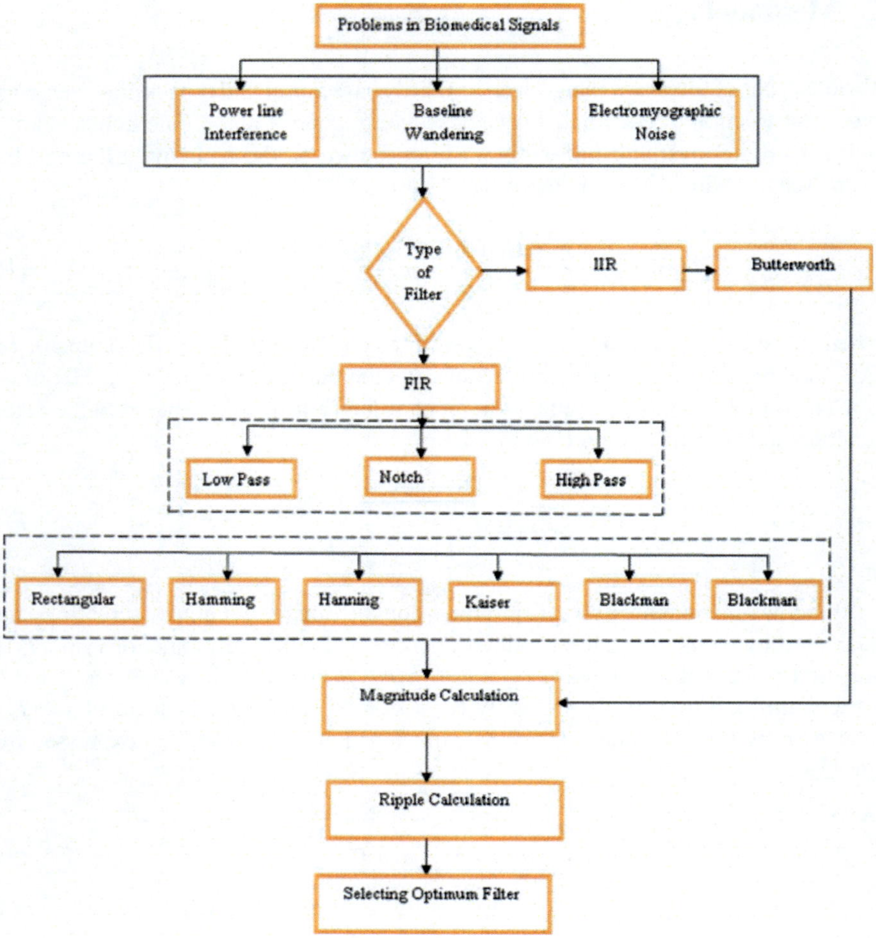

Fig. 1 Filter design steps

digital filters. The pole–zero plot and frequency responses of both filters were also analysed.

The fallacious electrodes placing, patient's motion and respiration (breathing) are the foremost reason of baseline wandering. The baseline signal is a low-frequency signal, with a cut-off frequency of 0.5 Hz FIR high-pass zero phase filters is used for the removal of the low-frequency noise from the biomedical signal [4–6]. Likewise for the removal of the high-frequency signal, low-pass filter with cut-off frequency of 100 Hz is designed.

3 Results and Discussion

There are some random noises such as high-frequency and low-frequency noises present in the biomedical signal. To remove these noises, optimum digital filters were designed using FIR and IIR filtering techniques. FIR and IIR filters are designed using different windowing techniques and approximate techniques, respectively. All the simulation was carried out in Intel 2.4 GHz, 64-bit operating system using LabVIEW and MATLAB software.

Removal of low-frequency noises: The high-pass FIR and IIR filter with cut-off frequency 0.5 Hz are designed (as shown in Fig. 2) to eliminate the low-frequency noises; this helps in removing the signal component of frequency beyond 0.5 Hz.

Table 1 tabulates the effect of order on gain using Butterworth approximate filtering (for IIR).

The gain of −36.53 dB is evaluated for the designed filter of order 2. It is observed that there is a sharp transition from stopband to passband. The pole–zero plot of the designed filter is also studied which is shown in Fig. 3.

The pole and zero both lie on the unit circle (as shown in Fig. 3) that signifies the filter is stable but the impulse response is infinite (as shown in Fig. 4) which is difficult to handle.

To overcome the problem of infinite response of IIR filter, FIR filters using different windows, namely Kaiser, Hamming, Hanning, Blackman, Rectangular and Gaussian are considered [6, 18]. Among various filters, Kaiser window shows the remarkable results. Kaiser window function is expressed by Eq. (6).

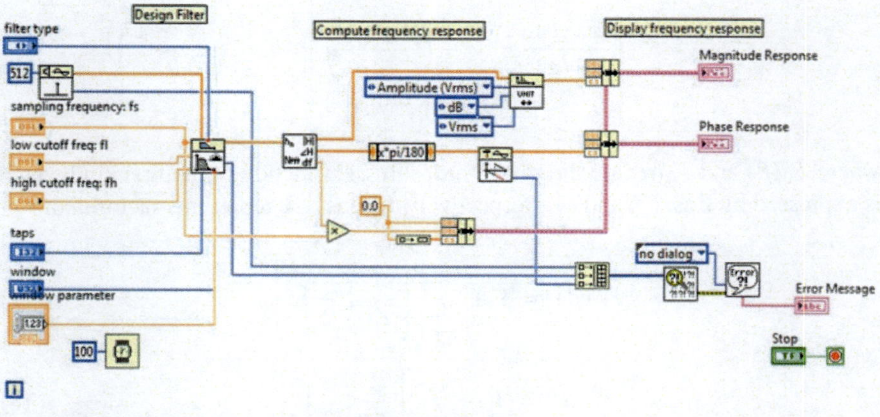

Fig. 2 Implementation of digital filter using LabVIEW

Table 1 Effect of order on gain for Butterworth HPF

Order	Gain (dB)
1	−18.33
2	−36.53

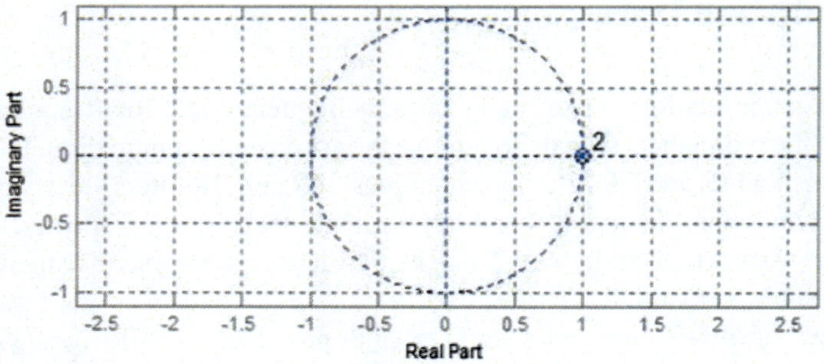

Fig. 3 Pole–zero plot of Butterworth high-pass filter of order 2

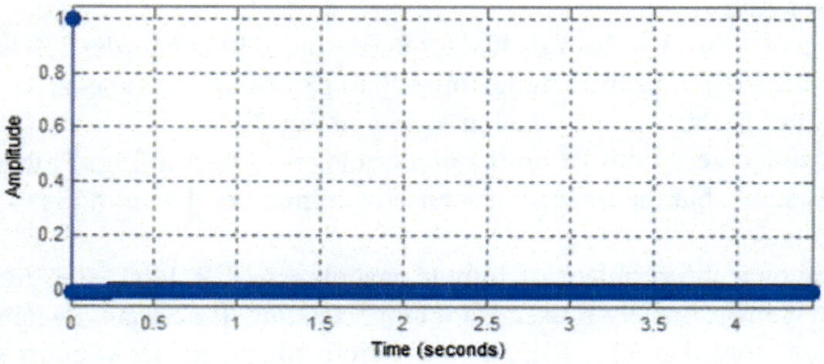

Fig. 4 Impulse response of Butterworth high-pass filter

$$W_k\,(n\,T_s) = \begin{cases} \dfrac{F_o\,(\beta)}{F_o\,(\alpha)} & -\left(\dfrac{N-1}{2}\right) \le n \le \left(\dfrac{N-1}{2}\right) \\ 0 & \text{otherwise} \end{cases} \tag{6}$$

where $F_0(\beta)$ and $F_0(\alpha)$ are the zeroth-order Bessel function of the first kind which is expressed by Eqs. (7) and (8), respectively. In Eqs, $\angle k$ represents factorial of k.

$$F_o\,(\beta) = 1 + \sum_{k=1}^{\infty}\left[\frac{1}{\angle k}\left(\frac{\beta}{2}\right)^k\right]^2 \tag{7}$$

$$F_o\,(\alpha) = 1 + \sum_{k=1}^{\infty}\left[\frac{1}{\angle k}\left(\frac{\alpha}{2}\right)^k\right]^2 \tag{8}$$

α is an independent parameter, and β is a dependent parameter which depends upon α and is expressed by Eq. (9).

$$\beta = \alpha\sqrt{1 - \left(\frac{2n}{N-1}\right)} \tag{9}$$

The ripples found in the window can be removed by varying the shape parameter (β). The effect of β on Kaiser window is tabulated in Table 2.

The effect of odd and even order of filters using windowing technique (for FIR) on gain has been studied, and the results are tabulated in Tables 3 and 4, respectively. If the number of taps is odd, then the delay of the filter is an integer number of samples, which is desirable for some applications while if the number is even, then it leads to half-sample delay.

From Table 4, it is interpreted that for odd number order, the gain value decreases, but the transition is sharper for a higher-order filter which is desirable for some applications. From the results, it is interpreted that for linear phase FIR filter, the order of the filter should be odd. Figures 5 and 6 represent the impulse response and pole–zero plot of FIR high-pass filter, respectively, of order 11. It is interpreted that if the length of the impulse response is even, anti-symmetric and has zero at $z = 1$, it represents that the filter is of Type 4 and is best suited for the high-pass filtering.

From all the simulated results, it is analysed that Kaiser window (shown in Fig. 7) yields better results in comparison with other methods because of its sharp transition and better gain value.

The main reason for analysing any filter using windowing technique is its simplicity of design, and the design method is very well understood with basic DSP knowledge.

Removal of EMG/high-frequency noises: The low-pass IIR and FIR filter with cut-off frequency 100 Hz are designed and simulated. For the Butterworth IIR filter of

Table 2 Effect of β on Kaiser window HPF

β	Gain (dB)
0.5	−2.98
1	−2.79
2	−2.33
3	−1.97
4	−1.72
5	−1.539
6	−1.4

Table 3 Effect of even order on gain for HPF Kaiser window

Order	Gain (dB)
50	−0.45
100	−0.894
150	−1.39
200	−1.88
250	−2.409
300	−2.98

Table 4 Effect of odd order on gain for HPF Kaiser window

Order	Gain (dB)
11	−56.89
51	−44.13
101	−38.31
201	−32.48
251	−30.64
301	−29.16

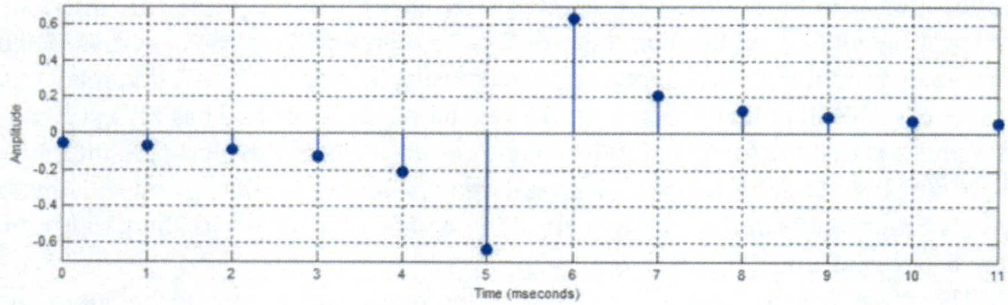

Fig. 5 Impulse response of FIR HPF of order 11

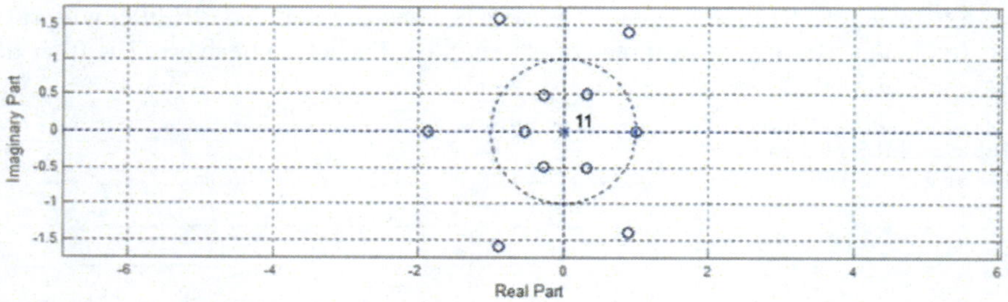

Fig. 6 Pole–zero plot of FIR HPF of order 11

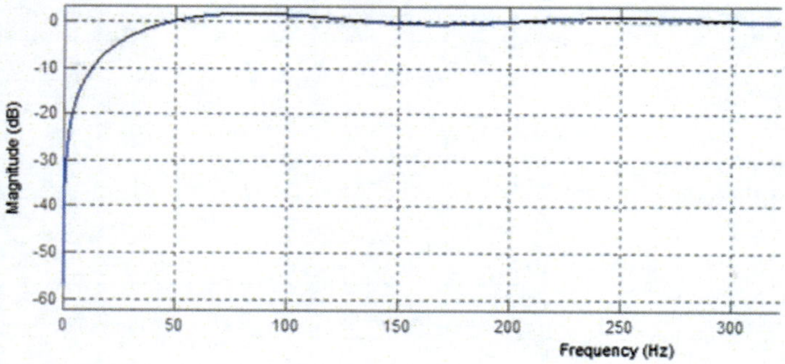

Fig. 7 FIR Filter design at 0.5 Hz using Kaiser window

order 2, it is interpreted that the frequency response is not good because the transition is very poor as shown in Fig. 8.

The pole–zero plot of the filter is also studied as shown in Fig. 9 which illustrates that the pole and zero both lie inside the unit circle that signifies the filter is stable but the impulse response is infinite which is difficult to handle. To overcome this problem, FIR filters using different window techniques are realized, and the results are tabulated in Table 5.

The choice of the window depends on the requirement like characteristics of noise, region of transition and the number of coefficients required. For the different windows, the Blackman windowing technique delivers significant results but have some limitations like large passband width (frequency of 135 Hz), while for the

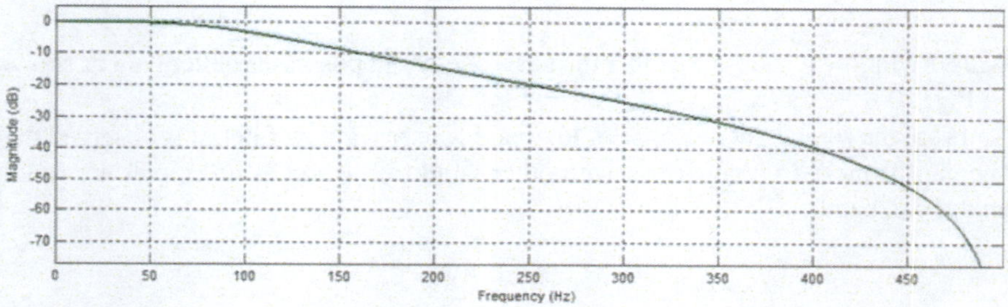

Fig. 8 Frequency response of Butterworth low-pass filter

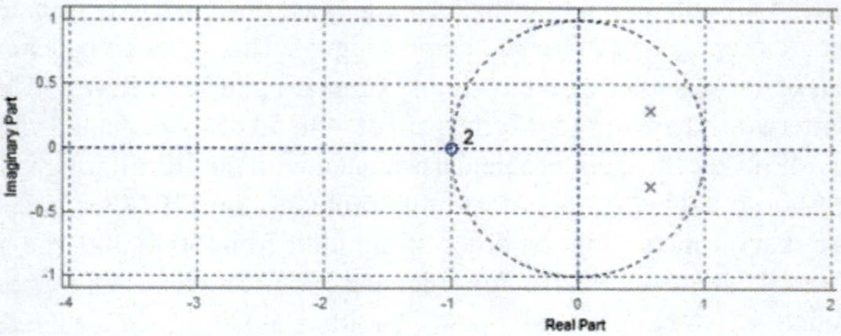

Fig. 9 Pole–zero plot of Butterworth low-pass filter of order 2

Table 5 Effect of gain on different window using LPF

Window	Gain (dB)
Hamming	−80
Hanning	−79.59
Kaiser	−61.75
Blackman	−108.21
Rectangular	−60.28

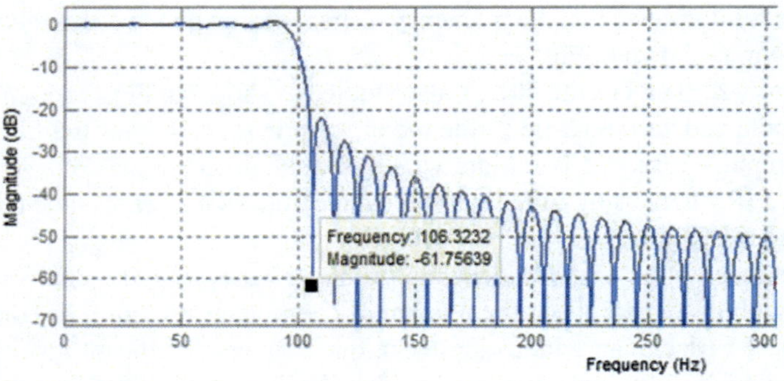

Fig. 10 Frequency response of low-pass FIR filter using Kaiser window

Kaiser window it is observed that the passband width possesses better gain as shown in Fig. 10.

From the frequency response of low-pass Kaiser window filter, it is observed that the gain is more in comparison with other window method as no ripples are found in the passband.

4 Conclusion

In this paper, optimum digital filtering techniques are designed to eliminate different noises that are propagating in the biomedical signal. This research paper carried out the analysis of various filtering methods and simulating in LabVIEW and MATLAB. For IIR Butterworth HPF of order 2, the gain of −36.53 dB is evaluated which shows a sharp transition but the main problem associated with the IIR filtering is its infinite impulse response nature. To overcome this problem, the FIR filter is designed. It is observed that on increasing the order of the filter from 50 to 300, the gain value changes from −0.45 to −2.98 dB. For order eleven, FIR high-pass filter, the value of gain is around −56.89 dB. The low-pass FIR filters are designed for the removal of high-frequency noises that results in −61.75 dB of gain. It is analysed that the FIR filters are best suited for the removal of noises from any biomedical signals because of its finite impulse response and no recursion which is mainly found in the IIR filtering. The Kaiser window is best suited with the appropriate value of β for the removal of different noises because of its better magnitude value and sharp transition from passband to stopband. In future, authors will try to filter the biomedical signals with the help of field-programmable gate array (FPGA).

References

1. Manolakis D, Proakis GJ. Digital signal processing principles, algorithms, and applications, 4th edn. Northeastern University Boston College
2. Jackson LB (2013) Digital filters and signal processing with MATLAB exercises, 3rd edn. Springer Science & Business Media
3. Prashar N, Dogra J, Sood M, Jain S (2018) Removal of electromyography noise from ECG for high performance biomedical systems. Netw Biol 8(1):12–24
4. Prashar N, Jain S, Sood M, Dogra J (2017) Review of biomedical system for high performance applications. In: 4th IEEE International conference on signal processing and control (ISPCC 2017), Jaypee University of Information technology, Waknaghat, Solan, HP, India, pp 300–304, 21–23 Sept 2017
5. Kirti, Sohal H, Jain S (2019) FPGA implementation of power-efficient ECG pre-processing block. Int J Recent Technol Eng 8(1):2899–2904
6. Prashar N, Sood M, Jain S (2019) Design and performance analysis of cascade digital filter for ECG signal processing. Int J Innov Technol Exploring Eng 8(8):2659–2665
7. Pun CKS, Chan SC, Yeung KS, Ho KL (2002) On the design and implementation of FIR and IIR digital filters with variable frequency characteristics. IEEE Trans Circ Syst—II: Analog Digital Signal Proc 49(11):689–703
8. Barnela M, Kumar S, Kaushik A, Satvika (2014) Implementation and Performance Estimation of FIR Digital Filters using MATLAB Simulink Int J Eng Adv Technol 3(5):62–65
9. Wang X, Meng X, He Y (2006) A novel neural networks-based approach for designing FIR filters. In: Proceedings of the 6th world congress on intelligent control and automation, 21–23 June 2006, Dalian, China, pp 4029–4032
10. Schaumann R, Xiao H, Mac VV (2009) Design of analog filters, 2nd edn. The Oxford Series in Electrical and Computer Engineering
11. Rahul (2019) Signal processing techniques for removing noise from ECG signals. J Biomed Eng Res 1–9
12. Alarcon G, Guy CN, Binnie CD (2000) A simple algorithm for a digital three pole Butterworth filter of arbitrary cut-off frequency: application to digital electroencephalography. J Neurosci Methods 104(1):35–44
13. Gaikward KM, Chavan MS (2014) Removal of high frequency noise from ECG signal using digital IIR butterworth filter. IEEE Glob Conf Wireless Comput Netw (GCWCN) 2014 121–124
14. Rahman MA, Milu1 MMH, Anjum A, Khanam F, Ahmad M (2017) Baseline wandering removal from ECG signal by wandering path finding algorithm. In: 3rd International conference on electrical information and communication technology (EICT), 7–9 Dec 2017
15. Bhaskara PC, Uplaneb MD (2016) High frequency electromyogram noise removal from electrocardiogram using FIR low pass filter based on FPGA. Global colloquium in recent advancement and effectual researches in engineering, science and technology (RAEREST 2016) pp 497–504
16. Aswathy V, Soniy P. Noise analysis and different denoising techniques of ECG signal—a survey. IOSR J Electron Commun Eng (IOSR-JECE) pp 40–44
17. Bhogeshwar SS, Soni MK, Bansal D (2014) Design of simulink model to denoise ECG signal using various IIR and FIR filters. In: 2014 international conference on reliability, optimization and information technology—ICROIT 2014. India. 6–8 Feb 2014, pp 477–483
18. Subhadeep C (2013) Advantages of Blackman window over hamming window method for designing FIR filter. Int J Comput Sci Eng Technol 4(8):1181–1189

5G Inset Feed Antenna Array for 28 GHz Wireless Communication

Rohit Yadav, Leeladhar Malviya, and Dhiraj Nitnaware

Abstract Millimeter wave technology is the solution of the current generation of the mobile users to provide uninterrupted signaling and high data rate. In this paper, a compact 1 × 6 antenna array is designed at 28 GHz millimeter wave frequency application with 2:1 VSWR. For the impedance matching, inset feed is utilized with the circular patch. The proposed antenna occupies 37.60 × 8.45 mm^2 space on the Rogers RT/duroid 5880 dielectric substrate. The designed array antenna achieves 27.654–28.291 GHz bandwidth, 12 dBi gain, 88.04% radiation efficiency, at the resonant frequency.

Keywords 5G · Antenna array · 28 GHz · Millimeter wave · Microstrip patch antenna

1 Introduction

Present technology uses shared resources dynamically to support multiple users based on IP packet switching network [1]. Many researchers and scientists are trying their best to increase data rate and to reduce delays in worst conditions and to provide uninterrupted connectivity due to the increasing demand of wireless devices. The 4G has many of the limitations. Therefore, 5G has come in urgent need [2]. Also the bandwidth in 4G is limited and has only 20 MHz scalable bandwidth. This is also one of the reasons to use 5G millimeter wave (mmw) communication technology, which

R. Yadav (✉)
Swami Vivekanad College of Engineering, Indore, M.P, India
e-mail: yadu86rohit@gmail.com

L. Malviya
Shri G. S. Institute of Technology and Science, Indore, M.P 452003, India
e-mail: ldmalviya@gmail.com

D. Nitnaware
Institute of Engineering & Technology, Devi Ahilya University, Indore, M.P, India
e-mail: dnitnawwre@ietdavv.edu.in

N. Marriwala et al. (eds.), *Mobile Radio Communications and 5G Networks*, Lecture Notes in Networks and Systems 140,
https://doi.org/10.1007/978-981-15-7130-5_31

attracted researchers and industry to work over these bands for high data range, low latency, etc., for wireless communications systems [3].

The main goals of the 5G technology are to provide the solutions of the enhanced wireless services, coverage at low cost, high data rate, reuse/utilization of allotted frequency band, low power consumption, etc. Frequency bands of 28 and 43 GHz are being deployed in Korea, Japan and China, and 64 and 71 GHz are also proposed by FCC for 5G application [4]. For modern communication, low-cost microstrip antennas are the better choice. Taconic TLY-5, Rogers RT/duroid, FR-4 and polyimide, etc., are used to develop the 5G antennas [5].

Microstrip antenna is also known as the printed antenna because it is fabricated on printed circuit boards (PCBs). Bidirectional PCBs are used for the creation of ground and patch, where the substrate is sandwiched between these two layers. Varieties of shapes are available to create the ground and patches nowadays [6]. In the rectangular patch, the length is responsible for resonant frequency, and width is responsible for impedance matching at resonant frequency [7].

Various 5G antennas have been designed to achieve high data rate, high gain and high directivity, with different polarizations. A dual band antenna operated on 28 and 38 GHz was designed with such concept [8]. Similarly, a 15×15 mm^2 antenna was designed for 28 GHz using planar-inverted frequency antenna (PIFA) approach [9]. To cover the wide range from 26.6–40 GHz, a T-shaped antenna was designed [10]. In another design, an antenna was operated on 23.92–43.8 GHz frequency and occupied $10 \times 12 \times 1.48$ mm^3 size on AgHT-8 substrate [2]. To cover X band of 5G technology, an antenna was designed on polyethylene terephtalete substrate, which occupied 60.0×75.0 mm^2, and had average gain of 5.0 dBi [11]. An antenna covered frequency range from 27.5–28.25 GH and occupied 5 mm $\times$ 5 mm size on Rogers RT/duroid 6002 substrate [12]. A 28 GHz antenna was investigated with different orientation and excitation phases for the beamforming and beam steering operations [13]. An array antenna was designed with meta-material unit cell approach and occupied 11.3×31 mm^2 on Rogers RT/duroid 5880 dielectric substrate. The design had 10 dBi gain, better than 21 dB isolation [14]. Three antenna designs were investigated on the size of $55 \times 110 \times 0.508$ mm^3 and operated at 28/38 GHz. The maximum gain of 9.49 dBi was achieved with the designs [15].

2 Antenna Design

The proposed array antenna of 1×6 is designed on the Rogers RT/duroid 5880 dielectric substrate and occupied 37.60×8.45 mm^2 space. The Rogers RT/duroid has dielectric constant of 2.2, 0.0009 tangent loss and thickness 0.79 mm. An inset feed with circular patch is selected for design to proper impedance matching. The proposed array antenna resonates at 27.966 GHz frequency in 27.654–28.291 GHz frequency band. The circular patch has the freedom that only radius is required to control the frequency of operation [16]. The dimension of the circular patch is given by Eq. (1) [17].

$$R = \frac{F}{\sqrt{1 + \frac{2h}{\pi \varepsilon_r F\left[\ln\left(\frac{F\pi}{2h}\right) + 1.7726\right]}}} \tag{1}$$

where

$$F = \frac{8.791 \times 10^9}{f\sqrt{\varepsilon_r}}$$

R is radius of circular patch, *h* is substrate height, ε_r is effective permittivity of substrate, and *f* is resonant frequency.

The design consists of two steps. In the first step, a circular patch with microstrip feed is designed, which resonates at 28.09 GHz frequency. In step two, an inset feed is selected with the circular patch. Each patch is connected to the independent feed with and feed lengths for the desired results. The schematic view of the proposed array antenna is shown in Fig. 1. All the optimized design parameters are given in Table 1.

3 Result and Discussion

3.1 Resonant Frequency

All the design steps are compared in Fig. 2. In design step 1 with 1 × 6 array, antenna resonates at 28.09 GHz and covers 27.73–28.46 GHz 10 dB return loss bandwidth. In design step 2, 1 × 6 array antenna with inset feed array antenna resonates at 27.966 GHz and occupies 27.654–28.291 GHz bandwidth in −10 dB return loss band. The inset feed with circular patch provides better impedance matching in comparison with the microstrip feed-based 1 × 6 array antenna.

3.2 Gain

The gain and efficiency of the proposed antenna array are shown in Fig. 3, and the 3D gain pattern is shown in Fig. 4. The proposed array antenna has 11.57–12.24 dBi gain in whole band. Also, the radiation efficiency in the band varies from 89.47–88.32%. The total efficiency in the band varies from 77.66–78.77%. At resonant, the gain is 12 dBi, and the radiation efficiency is 88.02%.

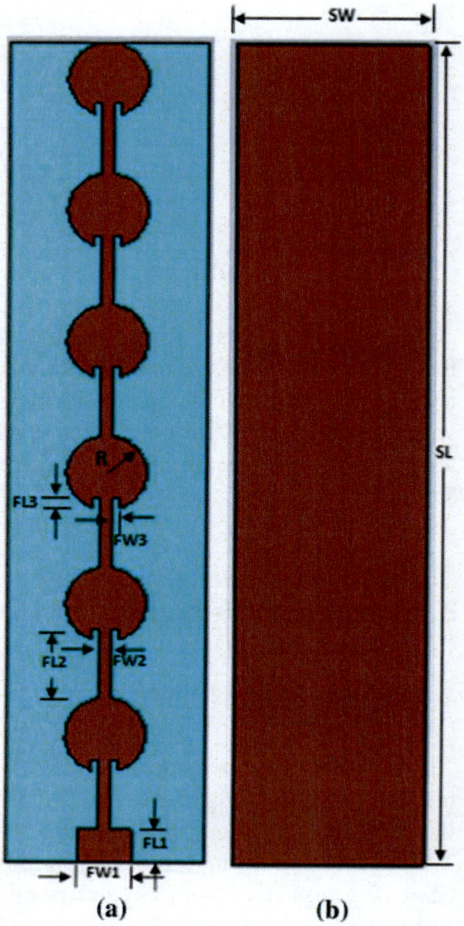

Fig. 1 Schematic views of the proposed design. **a** Front view. **b** Back view

Table 1 Dimension of proposed circular patch antenna

Parameter	Value (mm)
SL	37.60
FL1	1.59
FL2	3.17
FL3	0.50
SW	8.45
FW1	2.28
FW2	0.52
FW3	0.33
R	1.72

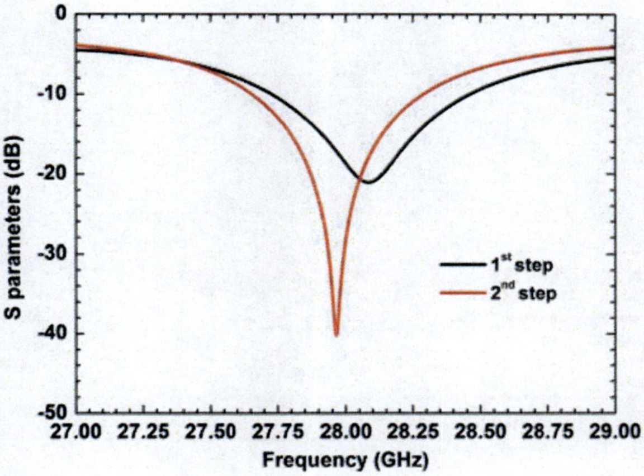

Fig. 2 S_{11} parameter

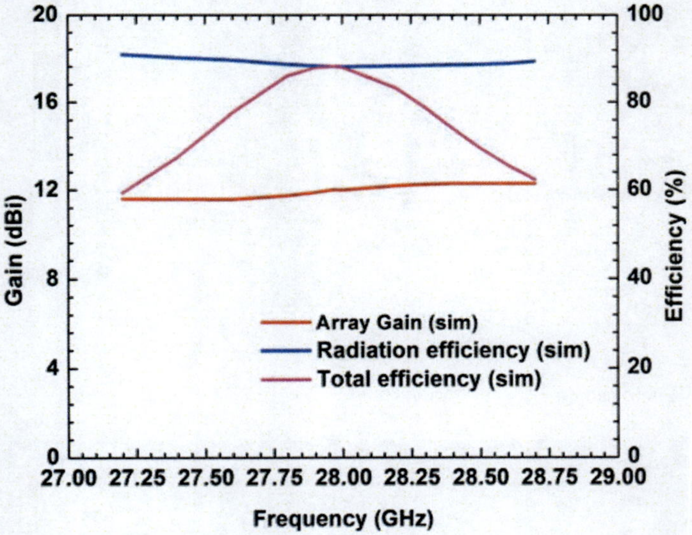

Fig. 3 Gain and efficiency

3.3 Radiation Pattern

The three-dimensional E-field and H-field radiation patterns are shown in Figs. 5 and 6. Similarly, the normalized E-field and H-field patterns are shown in Figs. 7 and 8. The proposed antenna array has 26.7 dBV/m value of E-field at resonant frequency and −24.8 dBA/m value of H-field. The antenna array has 62.8° beamwidth, and the main lobe directions of E-field is 0° and 340° for H-field.

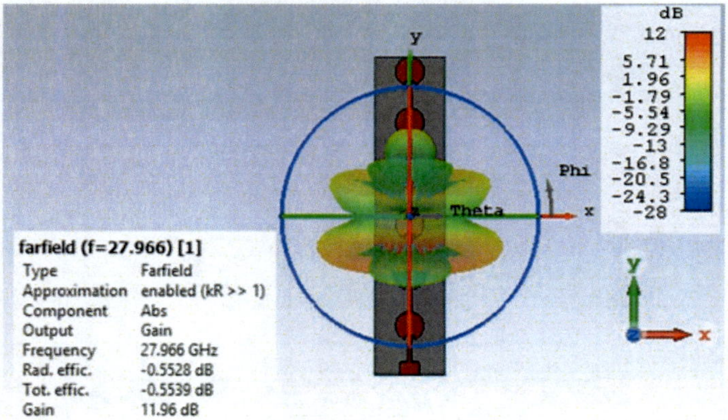

Fig. 4 3D gain pattern

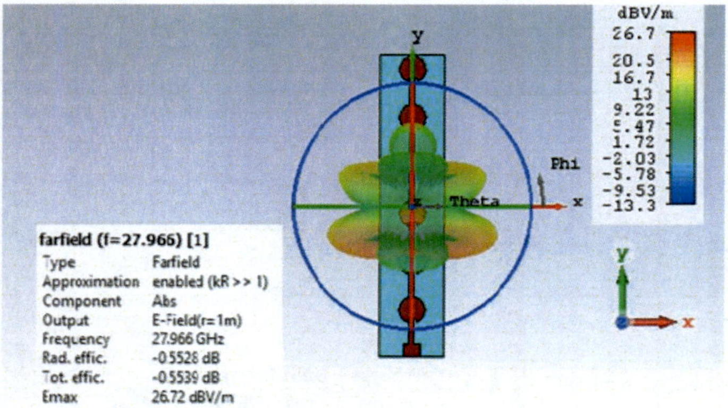

Fig. 5 3D E-field pattern

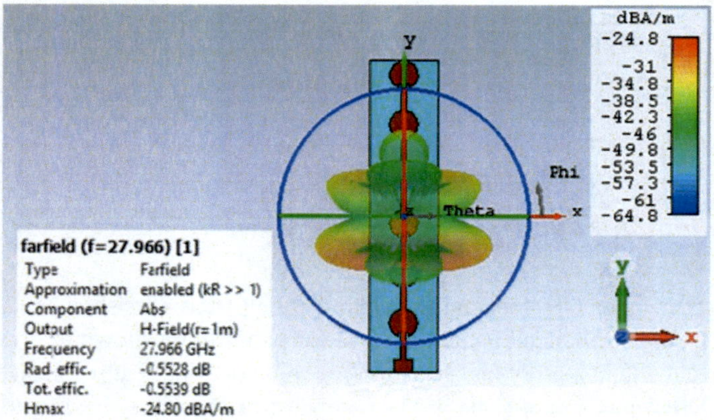

Fig. 6 3D H-field pattern

Fig. 7 Normalized E-field pattern

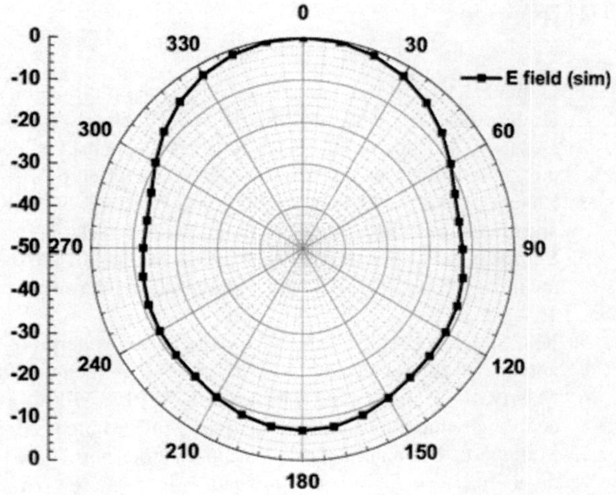

Fig. 8 Normalize H-field pattern

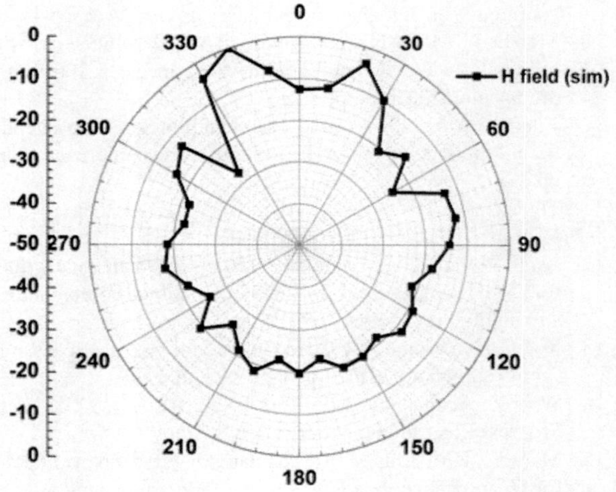

4 Conclusion

A 1 × 6 array antenna has been presented with inset feed and circular patch to resonate at 28 GHz millimeter wave wireless application. The design was fabricated on low loss tangent Rogers RT/duroid substrate. The design occupied the 37.60 × 8.45 mm^2 size on dielectric substrate. The bandwidth in 2:1 VSWR band is 27.654–28.291 GHz. The minimum gain in design is 11.57 dBi, and the radiation efficiency is 88.01%. The proposed array antenna is suitable for high data rate 5G technology.

References

1. Gupta P, Malviya L, Charhate SV (2019) 5G multi-element/port antenna design for wireless applications: a review. Int J Microw Wirel Technol 11:918–938
2. Desai A, Upadhyaya T, Patel R (2019) Compact wideband transparent antenna for 5G communication systems. Microw Opt Technol Lett 61:781–786
3. Zhang J, Ge X, Li Q, Guizani M, Zhang Y (2017) 5G millimeter-wave antenna array: design and challenges. IEEE Wirel Commun 24(2):106–112
4. Shorbagy ME, Shubair RM, AlHajri MI, Mallat NK (2016) On the design of millimetre-wave antennas for 5G. In: 2016 16th mediterranean microwave symposium (MMS), Abu Dhabi, pp 1–4
5. Kim S, Tentzeris MM (2018) Parylene coated waterproof washable inkjet-printed dual-band antenna on paper substrate. Int J Microw Wirel Technol 10(7):814–818
6. Malviya L, Panigrahi R, Kartikeyan M (2017) MIMO antennas with diversity and mutual coupling reduction techniques: a review. Int J Microw Wirel Technol 9(8):1763–1780
7. Malviya L, Chouhan S (2019) Multi-cut four-port shared radiator with stepped ground and diversity effects for WLAN application. Int J Microw Wirel Technol 11(10):1044–1053
8. Aliakbari H, Abdipour A, Mirzavand R, Costanzo A, Mousavi P (2016) A single feed dual-band circularly polarized millimeter-wave antenna for 5G communication. In: 2016 10th European conference on antennas and propagation (EuCAP), pp 1–5
9. Morshed KM, Esselle KP, Heimlich M (2016) Dielectric loaded planar inverted-f antenna for millimeter-wave 5G hand held devices. In: 2016 10th European conference on antennas and propagation(EuCAP), pp 1–3
10. Jilani SF, Alomainy A (2016) Planar millimeter-wave antenna on low cost flexible PET substrate for 5G applications. In: 2016 10th European conference on antennas and propagation (EuCAP), pp 1–3
11. Tighezza M, Rahim SKA, Islam MT (2017) Flexible wideband antenna for 5G applications. Microw Opt Technol Lett 60:38–44
12. Awan WA, Zaidi A, Baghdad A (2019) Patch antenna with improved performance using DGS for 28 GHz applications. In: 2019 international conference on wireless technologies, embedded and intelligent systems (WITS), pp 1–4
13. Yu LC, Kamarudin MR (2016) Investigation of patch phase array antenna orientation at 28 GHz for 5G applications. Proc Comput Sci 86:47–50
14. Wani Z, Abegaonkar MP, Koul SK (2018) A 28-GHz antenna for 5G MIMO applications. Progress Electromagnetic Res Lett 78:73–79
15. Marzouk HM, Ahmed MI, Shaalan AA (2019) Novel dual-band 28/38 GHz MIMO antennas for 5G mobile applications. Progress Electromagnetic Res C 93:103–117
16. Balanis CA (1997) Antenna theory: analysis and design, 2nd ed., Wiley
17. Singh M, Basu A, Koul SK (2006) Circular patch antenna with quarter wave transformer feed for wireless communication. In: 2006 annual IEEE India conference, pp 1–5

Comparison of Various Attacks on WSN Layers—Based on the Site of the Attacker and Access Level

Rekha Rani and Narinder Singh

Abstract In the earlier few years, more concern has been focused on wireless sensor networks with its large range of applications in different areas even in critical situations like commercial applications, battlegrounds, pollution sensing, habitat observing of animals, smart buildings and homes, secure homeland, military monitoring, hospitals, and other various locations; thus, there are many chances of attacks in WSNs. Security is major concern to control of these attacks for a secure data as well as secure transmission, in wireless sensor network. Our main focus in this work is to provide a detailed comparison of various attackers site-based attack and access level attacks on various network layers of WSNs.

Keywords WSN · Security · Integrity · Privacy

1 Introduction

A WSN is a self-configuring network of little nodes exchange sense data between them with radio signals, and install in capacity to sense, monitor and recognize the real world. WSN gives a connection among the real world and virtual worlds [1].

Sensor units of homogeneous or heterogeneous types can be deployed at random or at preset positions with a deterministic method. Sensor deployed nodes are mostly static, while mobile nodes can be deployed according to application requirements. One or several static or mobile base stations (BSs) are deployed together with the network. Sensor units continue to monitor the network region after being installed. Later than an action of interest happens, one of the surrounding sensor nodes can sense it, make a report and send this report to a base station with multi-hop wireless communications. Association can be accepted out in case multiple nearest nodes sense the similar event. In case, each sensor node generates final report after communicating with the other nearest nodes. The base station can develop the report and

R. Rani (✉) · N. Singh
Department of Computer Science, Guru Nanak College, Budhlada, Punjab, India
e-mail: rekha_nskalra@yahoo.co.in

N. Marriwala et al. (eds.), *Mobile Radio Communications and 5G Networks*, Lecture Notes in Networks and Systems 140,
https://doi.org/10.1007/978-981-15-7130-5_32

then transforms it through either high-quality links; it may be wired or wireless to the external world for more dispensation. The sensor network authority can transmit queries to a base station, which spreads those commands within the sensor network. Thus, a base station works as a gateway among the wireless sensor network as well as the external world.

2 WSN Architecture

It consists of microcontroller unit antenna, receiver and transmitter sensor, and control units are powered by small batteries and voltage sensor sense the input to the analog-to-digital converter to convert the sense data into digital form and send to the microcontroller [2] (Fig. 1).

2.1 Organization of Wireless Sensor Network

Any wireless sensor network may be constructed as a five-layered structure as described following with their working:

- **Physical layer**. Responsibility of physical layer is to modulate and encrypt the sensed data.
- **Data link layer**. It works for multiplexing of data streams, detection of data frame, error control and medium access control.
- **Network layer** [3]. This layer is used for data transmission through transport layer by using multi-hop routing algorithms among sensor units and base stations.
- **Transport layer** [4]. It controls the flow of sensed data.
- **Application layer**. Constructs both software and hardware of last layers visible to user.

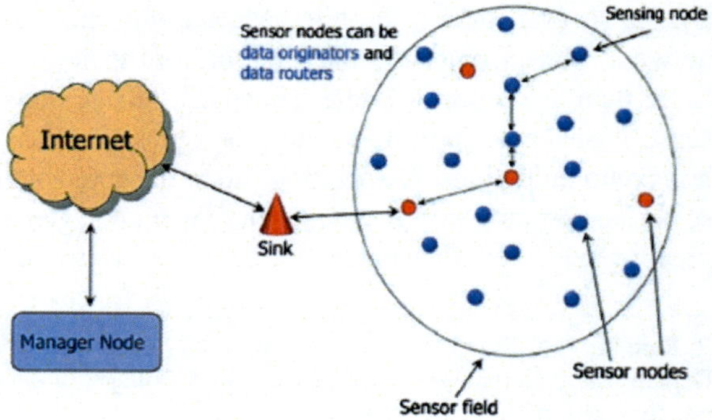

Fig. 1 A block diagram of a wireless sensor network

3 Limitations of Wireless Sensor Networks

- It has little storage capacity due to very small storage unit
- It has modest processing control
- It utilizes more memory because it has little communication range and needs number of sensor nodes
- It has limited of fixed lifespan batteries.

4 Security in WSN

Sensor networks affectation limited challenges, thus traditional security schemes which are used for traditional networks are not be appropriate for today wireless sensor network. The sensor units are insufficient in their power, processing, and communication abilities. While WSNs are settled in unfriendly surroundings, security is very important, as they have various kinds of attacks. For example, an attacker can effortlessly eavesdrop to traffic, copy the data of nodes, or purposely supply deceptive data to other sensor nodes. Wireless sensor network together closely with their physical surroundings, affectation new security problems. As outcomes, sensor nodes are installed in an unsecure open and unattended environment, so:

- Mechanisms are insufficient, and novel ideas are needed. WSN involves a large number of nodes in the network implementing security in overall.
- An enemy can simply infuse cruel node into sensor network.
- Wireless sensor network utilizes wireless communication among nodes, so eavesdrop on it is predominantly easy.
- Security is essential for such network nodes are source limitation in phrase of storage, power, sending the data packets.
- The levels are essential and too difficult.
- Symmetric cryptography technique is utilized as substitute.

Wireless sensor networks common security objectives are efficiency, integrity, confidentiality, availability, authentication, survivability, scalability and freshness as shown in Table 1. Wireless sensor network is vulnerable to a lot of attacks as of its unattended deployment in unrestrained surroundings. To make sure security facilities.

In WSN, several cryptographics like symmetric and asymmetric techniques are proposed. To accomplish safety in WSNs, it is essential to be capable to authenticate as well as encrypt data transform among sensor nodes.

Table 1 Security services of WSN

Access control	Restrict to resources accessing
Revocation	Repudiation of approval
Survivability	The lifespan of sensor unit has to be extensive and even node is falsified
Non-repudiation	Secure or prevent denial of an earlier assurance
Availability	High accessibility networks with plan to stay accessible at every time avoiding service interruption owing to energy wastages, hardware collapse and organization upgrades. Ensuring accessibility also occupy prevent DoS attacks
Data freshness	Data freshness ensures about the messages freshness, stands for that are in correct order as well as haven't been use again
Confidentiality	Having sensor node information secret from other nodes however only authorized users can see it
Device authentication	Validation of individuality of the sensor node
Integrity	Achievable for the destination sensor node of message to verify the originality of data means; it has not been altered during transmission
Validation	To give accuracy of authorization to utilize or control resources
Message authentication	Validation source of information

5 Classification of Attacks in WSN

A range of attacks is potential in wireless sensor network. As per to different criteria, for example, methods utilized in attacks and domain of enemies, security attacks can be classified in different forms. In WSN, security attacks are classified in the following categories: [5].

5.1 *Based on the Site of the Attacker*

According to [6], attackers can be inside (internal attack) or outside (external attack). In these types of attack, attackers classify the location in the WSNs.

External attacker (outsider). These attacks are real one, where the person does not have any facts regarding the structure and protection measures of sensor network. It utilizes the cooperated node to assault sensor network that is able to obliterate or interrupt the sensor network effortlessly.

Some of the most general characteristics of such types of attack are:

- Committed by illegal person, for financial or personal harms.
- Attacks without being valid authentication (Fig. 2).
- Utilization of resources in sensor networks.
- Jamming the whole communication in networks.
- DOS attacks.

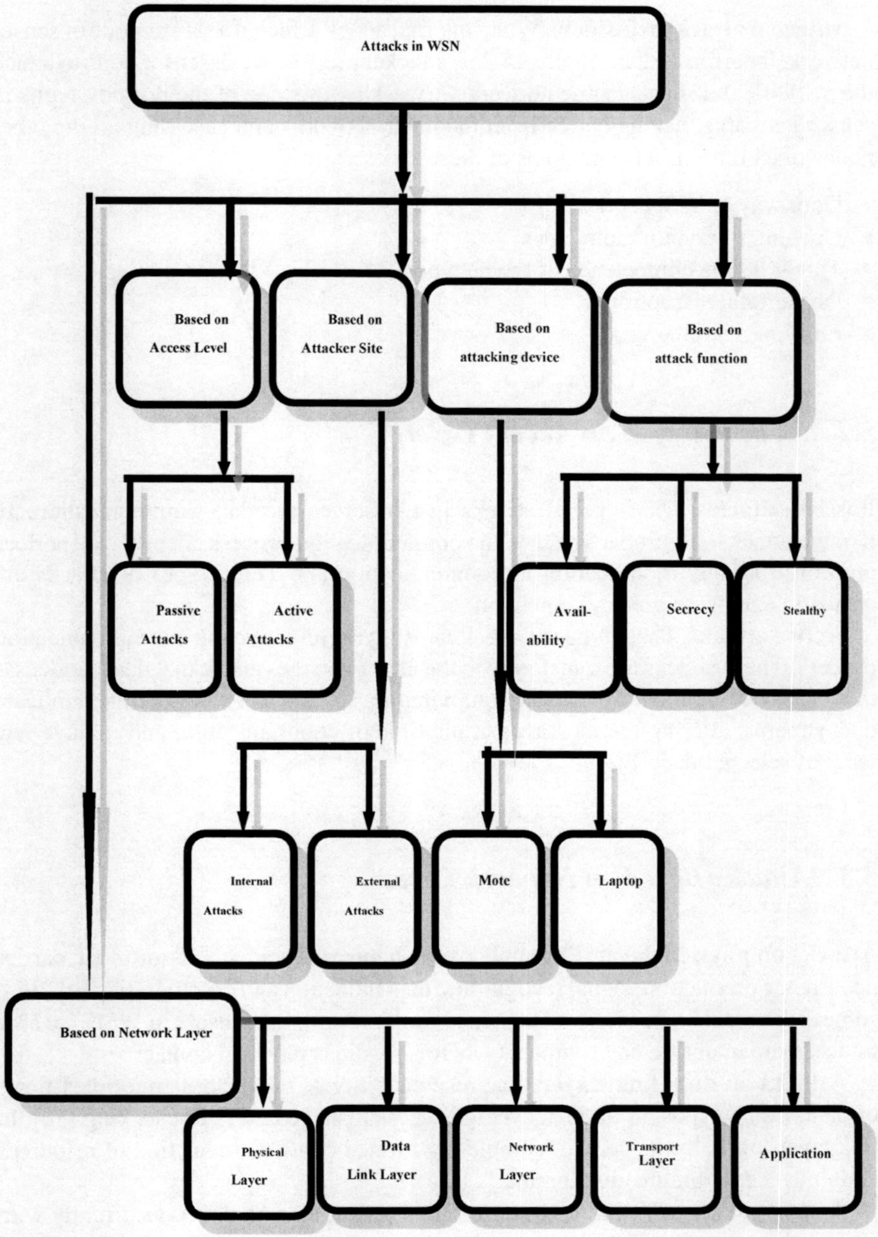

Fig. 2 Classification of attacks in WSN

Internal attacker (insider). When internal attack affected a sensor node of sensor network, it performs abnormally. In this, attacker uses the weak sensor node to attack the network and suspends the network very easily. It is one of the primary faults in sensor networks, having entrée from the inner networks and accessing all the other nodes in its limit. Important goals of these types are:

- Door way to WSN codes
- Opening to cryptography keys
- Threat: to the competence of the network
- Partial/total disruption
- Edifying secret keys.

5.2 *Attacks Based on Access Level*

Passive attacks. This type of attacks just observes the data communication. In passive attack, an intruder watches the communication process silently except does not construct any modification in communication [7]. These types of attacks are basically against to privacy of network.

Active attacks. These types of attacks are playing an active role in communication process. These attacks are transferred to the alter messages and actual data condensation or producing the false facts in communiqué. An attacker can generate duplicate data streams, altering the messages at the time of communication and remove few parts of selected data [8].

5.3 *Attacks Based on Network Layer*

Attacks on physical layer. Physical layer is liable for incidence assortment, carrier occurrence production, signal recognition, data inflection and cryptography [9]. With some radio media, chance of interference, furthermore, the sensors in WSN are able to distribute in unsafe environments where a rival has physical contact.

Attacks on data link Layer. The responsibility of this layer is multiplex flows of data, frame exposure of data, average access manage and fault checking [10]. In link layer level attacks contain completely formed conflicts, depletion of resources and injustice within the distribution.

A conflict happens while two sensor nodes effort to send data concurrently with similar frequency [11]. After that, packets are remaining and must be resend. An enemy can tactically reason collisions in particular packages, for example ACK manages communication. A potential outcome of this conflict is expensive exponential setback. The enemy can basically break the working algorithm and always sends messages in an effort to create conflict. Repetitive conflicts are able to utilize by an enemy to motivate the collapse of resources [11].

Table 2 Types of attack

Layers of network	Types of attack	Site of attack	Access level
Physical layer	Jamming	External	Active
Data link layer	Exhaustion, Unfairness	External External	Active Active
Network layer	Sinkhole Selective forwarding Spoofed Sybil, Wormhole Hello flood	Internal Internal Internal Both Both External	Active Active Active Active Active Active
Transport layer	Flooding De-synchronization	External Both	Active Active
Application layer	Elective message forwarding Eavesdropping	Internal Both	Active Passive

Such as an implementation of the naive link level can constantly try to resend infected packets. If the resending process is sensed in advance, the power levels of sensors will speedily run out. The erroneousness is fragile variety of denial-of-service attack. An enemy must origin evil from time to time using previous attacks on the link level. In this situation, the enemy causes the degradation of simultaneous appliance operating on another sensor by irregularly interrupting their edge communications.

Attacks on network link layer. This layer of WSNs is susceptible for various kinds of assaults like choosy packets forward, Sybil and imitation recognition which are defined as follows (Table 2).

Spoofed routing information: The purpose of this attack in routing algorithm is to execute steering information from sensor network. An enemy can read, modify or reproduce the path [12].

Acknowledgements spoofing: Several algorithms of WSN routing have need to communication of recognition packet. An attacker sensor node can detect data communication from neighbouring nodes along with falsify send recognition, and thus, giving such false information to sensor nodes [12].

Attacks at the transport layer level. Several attacks like de-synchronization and flooding are usually on effect on this layer attacks; through this manner, the rival can broadcast incorrect data regarding position of sensors.

Flooding. Routing protocol collapses through flooding when it tries to maintain status at either ending of a link [13]. Rival deliberately sends incessantly request to establish connection with nodes to exhaust the network energy. In either case, legal request will be suffered.

De-synchronization. An enemy sends repetitively spoof message to the end-user forcing host node appeal resending of miss borders. To prevent the end host to exchange the real data, and unnecessary indulged in attempt to recover the errors which really never exists and led them to waste their energy [14].

Attacks at the application layer. The data are composed and managed in this layer. It is essential to make sure the consistency of sensed data than to send it to lower levels. The main problem is that it can occur a security attack on the de-synchronization of transfer data.

6 Conclusion

The use of wireless sensor network involves almost all phases' in our routine life. Wireless sensor network has achieved implicit significance in last few years. Increasing the exploit of wireless sensor network also presents the way to enlarged security threats and attacks to steal the sensed data by the sensor network before it reaches to destination for further process. This paper summarizes the attacks based on attacker location and various network layers of wireless sensor networks. This paper will hopefully inspire upcoming researchers to come up with smarter and further robust security schemes to make a secure network.

References

1. Zhang J, Stojmenovic I (2005) Cellular networks, pp 654–662
2. Apoorva SK, Sreerangaraju MN (2017) Energy efficient routing protocol for wireless sensor networks. Int Res J Eng Technol (IRJET) 4(6):2505–2509
3. Mauri K, Hännikäinen M, Hämäläinen T (2005) A survey of application distribution in wireless sensor networks. EURASIP J Wirel Commun Network 774–788
4. Pereira P, Grielo A, Rocha F, Nunes M, Casaca C, Almsrtrom P, Johansson M (2007) End-To-end reliability in wireless sensor networks: survey and research challenges. In: EuroFGI workshop on IP QoS and traffic control, Lisbon
5. Hunt R (2004) Network security: the principles of threats, attacks and intrusions, part1 and part 2, APRICOT
6. Mohammadi S, Jadidoleslamy H (2011) A comparison of link layer attacks on wireless sensors network. Int J Appl Graph Theory Wirel Ad hoc Netw Sensor Netw 3(1):35–56
7. Singh SK, Singh MP, Singh DK (2011) A Survey on network security and attack defense mechanism for wireless sensor networks. Int J Comput Trends Technol 1–10
8. Lupu TG (2009) Main types of attacks in wireless sensor networks. In: International conference in recent advances in signals and systems, pp 114–118
9. Sharma R, Diwakar C (2012) Security analysis of wireless sensor networks. IJREAS 2(2):774–786
10. Akyildiz IF, Su W, Sankarasubramaniam Y, Cayirci E (2002a) Wireless sensor networks: a survey. Comput Netw Int J Comput Telecommun Network 38(4):393–422
11. Akyildiz IF, Su W, Sankarasubramaniam Y, Cayirci E (2002b) A survey on sensor networks. IEEE Commun Mag 40(8):102–114
12. Wood A, Stankovic J (2002) Denial of service in sensor networks. Computer 35(1):54–62
13. Karlof C, Wagner D (2003) Secure routing in wirelesss sensor networks: Attacks and countermeasures. In: Proceedings of the 1st ieee international workshop on sensor network protocols and applications, pp 113–127

14. Carman DW, Krus PS, Matt BJ (2000) Constraints and approaches for distributed sensor network security, Technical report, NAI Labs, Network Associates, Inc., Glenwood, MD, pp 00–010

Non-conventional Energy Source-Based Home Automation System

Rashmi Vashisth, Rahul Verma, Lakshay Gupta, Harsh Bansal, and Vishal Kharbanda

Abstract This paper aims to design and execute the advanced development in home automation system based on non-conventional energy source for conserving energy. Nowadays, people are too engaged in their busy schedule and are unable to switch off the lights when not in use. The present system is like: the lights will get turn ON/OFF only when person entering/exiting the room presses the switch manually. Also, the inverter battery gets charged from the main AC supply, which further leads to more energy consumption. This paper gives the best solution to reduce high energy consumption by appliances. Also, the manual handling of the lighting system and charging of inverter battery through the main AC supply are completely eliminated. The main purpose of this paper is to provide continuous power supply to an appliance, by selecting the supply from any of the source, namely DC battery charged through solar panel and main AC supply automatically, in case if one of the sources is absent, and this is done with the help of microcontroller. Finally, the objectives of the prototype, which are to be implemented, have successfully achieved.

Keywords Solar panel · Inverter · Switching circuit · Solar inverter · IR sensor · Microcontroller · LCD display · Main AC supply · DC battery · Relay · Inverter transformer · Step-down transformer · Energy conservation · Circuit design

1 Introduction

The idea of implementing this project or prototype comes as by seeing the great amount of energy to be used by the people and that energy is mostly referred as conventional energy which is used by the people in their homes for running the home appliances. Thought behind this idea also leads to the easy lifestyle and convenience

R. Vashisth (✉) · R. Verma · L. Gupta · H. Bansal · V. Kharbanda
Department of Electronics and Communication Engineering, Amity School of Engineering and Technology (ASET), Guru Gobind Singh Indraprastha University (GGSIPU), New Delhi, India
e-mail: rashmiapj@gmail.com; er_jangra@yahoo.co.in

N. Marriwala et al. (eds.), *Mobile Radio Communications and 5G Networks*, Lecture Notes in Networks and Systems 140,
https://doi.org/10.1007/978-981-15-7130-5_33

for the people. As there are two types of energy, i.e., conventional energy and non-conventional energy.

Conventional energy refers to the energy source which is obtained from the fixed reserves present in nature like coal, gas, and oil. But non-conventional energy refers to the energy which is generated by using the wind, tides, solar geothermal heat, etc. [1]. These sources are renewable and inexhaustible and even does not cause any harm or pollute the environment. There are various types of non-conventional sources—geothermal energy, wind energy, tidal energy, solar energy. The non-conventional energy used in this project is solar energy and this energy used to charge the battery of inverter, and rest of the details is described in below sections.

2 System Model

In this section, the system model designed will be discussed in detail. Each component has its own crucial role to play in. Each component goes in systematic order and generates signal which acts as either an input or an output to other component. Figure 1 shown below is a block diagram of prototype.

Main AC supply of 220 V is used to provide input to a step-down transformer of 220–12 V, which step downs the AC supply from 220 to 12 V, and then this 12 V AC is supplied to rectifier circuit [2]. A rectifier is a device or circuit which handles the conversion of alternating current (AC) into direct current (DC). This is then supplied to the coil of Relay 1 as an input. The coil of Relay 1 is getting supply from the rectifier circuit, i.e., the supply which is transformed from AC to DC. The normally closed has positive of solar battery and normally open has positive of rectifier circuit, and the common is supplied as input to Relay 2 at normally open terminal. Then, the common terminal of this Relay 2 is supplied as input to inverter circuit. This inverter circuit converts the DC into AC over which the appliances are operable. Due to this, when the main AC supply is provided, then appliance will run over that

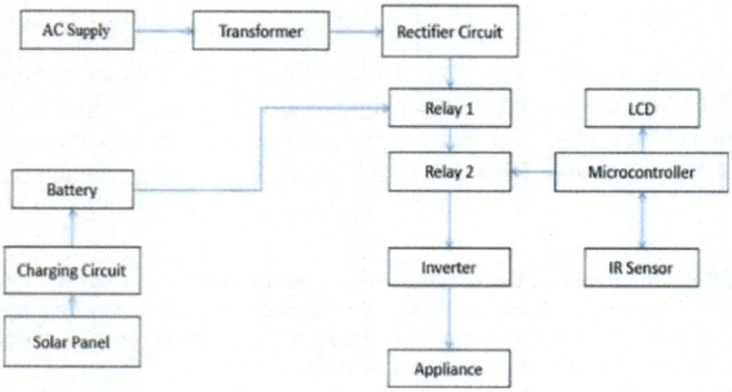

Fig. 1 Block diagram

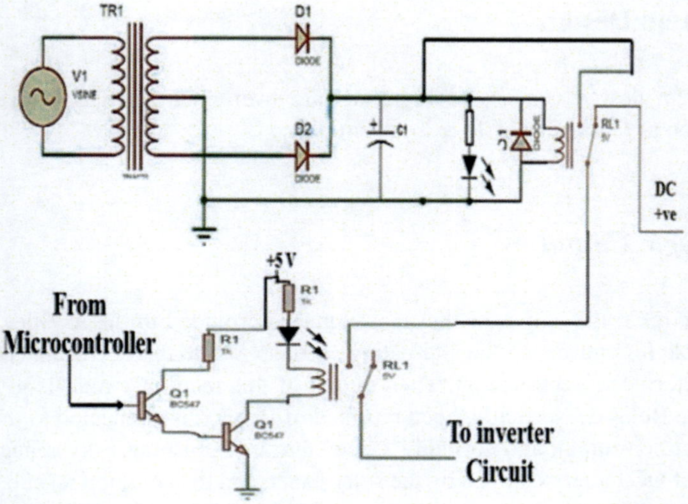

Fig. 2 Logic circuit diagram

supply according to the signal given by the microcontroller but whenever, the main AC supply cuts off, the appliances will work on that supply which is provided by the DC battery charged by solar energy and at that time also, appliance will run only when signal is provided by the microcontroller. Microcontroller receives the input from the IR sensor, and it counts the number of person entering and exiting the room and calculates accordingly, and accordingly, it generates signal for the Relay 2 to operate the appliance [3]. It also displays the number of people inside the room on the LCD screen connected to its pin. When there is no person inside the room, the IR sensor generates no signal and LCD screen shows number of persons to be zero. But when even a single person entered a room, the LCD shows the total number of persons inside the room, and the microcontroller generates a signal for the Relay 2 to provide further signal; the inverter circuit is used to convert the DC input coming from the Relay 2 into AC output, and this AC output of inverter circuit is provided to the home appliances. In inverter, the inductor is used to block the AC Repulse so that, pure DC is converted as AC output. Solar battery charging refers to charging a battery using solar energy but directly connecting the solar panel to battery may increase the chances of reverse current, i.e., from battery to solar panel, so to avoid this reverse current, the charging circuit is used which is in between solar panel and battery which charges the inverter battery from solar energy [4]. The solar panel converts the solar energy into the electrical energy which is given to the charging circuit. This circuit makes use of a diode which prevents the backward flow of electrical energy so the chances of reverse current reduces and battery gets charged, and it also reduces the chances of overcharging of battery. This battery is connected to the Relay 1 but only positive terminal, whereas negative terminal of battery is common with the center tapping of transformer and given to the inverter circuit [5].

3 System Design

The system design consists of logic circuit, inverter circuit and microcontroller connection to IR sensor, LCD & logic circuit.

3.1 *Logic Circuit*

In logic circuit, the output of the transformer is provided to the rectifier, which is responsible for converting the alternating current (AC) to direct current (DC) which is also referred as rectification. Then output of this rectifier circuit is given to the coil of the Relay 1. Normally, open terminal of Relay 1 is connected to the positive of the rectifier output, and normally, closed terminal of Relay 1 is connected to the positive of DC battery charged by the solar panel, and the common terminal of relay is connected to the normally open terminal of Relay 2. In Fig. 2, the circuit diagram is shown. Microcontroller provides signal to the coil of Relay 2 but it is not enough to drive a relay so Darlington pair is used to amplify the current and then provide it to the coil of relay. A Darlington pair is referred to two transistors which act as a single transistor but produces a much higher current gain and due to the high current gain a tiny amount of current from a sensor, microcontroller or similar can be used to drive a larger load [6]. The Darlington pair is used in order to have high current gain because the current gain provided by the Darlington pair is much higher as compared to a single transistor, and due to this high current gain, we are able to drive various electrical devices [7]. So microcontroller provides signal to the Darlington pair, and then output of it is provided to the coil of Relay 2. At last, the common terminal of Relay 2 is provided as input to the inverter circuit. In Fig. 3, implemented logic circuit is shown. In this, PCB board is used, on which the components are soldered using soldering iron and soldering wire, and further connection is made between the components.

3.2 *Inverter Circuit*

As we know that inverter refers to the transformation of DC into AC, as most of the appliances do not work on DC supply. So, to drive them AC supply is required. But in the case when the main AC supply gets OFF, inverter circuit plays a crucial role by generating AC from DC input. In inverter circuit of this project, the output of logic circuit or common of Relay 2 is used as an input to the inverter along with a signal which is common for both, negative of battery charged by solar panel and center tapping of transformer to which main AC supply is provided [8].

The circuit shown in Fig. 4 consists of two switching transistor (CTC1061) and a push–pull transformer [9]. The base feedback of two switching transistor is collected

Fig. 3 Implemented logic circuit

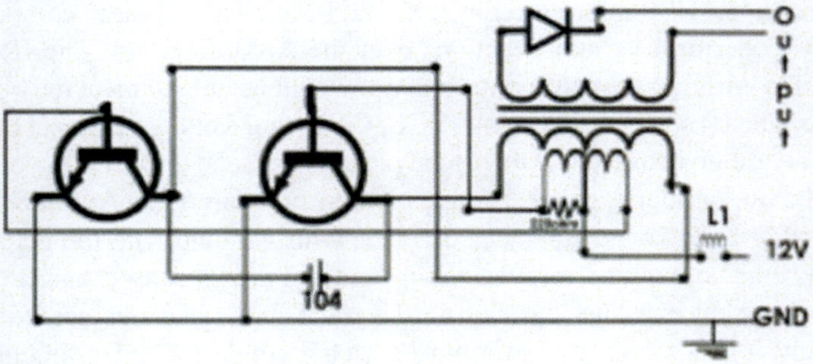

Fig. 4 Inverter circuit diagram

from the transformer itself. This one has output power of up to 18Watt stated at 12 V; though by seeing Fig. 4, it can be observed that the output of the circuit is coupled, and this is coupled by a series capacitor/diode here. The transformer has 5 lines at input section. Central connector goes to positive supply, through a L1 inductor coil. It absorbs the spikes due to the switching of transformer [6]. There are two power stage pins, and two feedbacks go to each of transistor's collector, base. The initial base feeding is done by a 220 Ω resistor to the base of any of the transistor, having a capacitor of 104 J (0.1uF) to the other transistor [7]. In this, the inductor is used to block the AC Repulse so that pure DC is converted as AC output as shown in Fig. 5. The key circuit or the most basic circuit which is to be needed to construct the prototype can be referred as inverter circuit.

Fig. 5 Implemented inverter circuit

3.3 Microcontroller Connections

This project consists of Arduino UNO board which is based on microcontroller ATmega328P. The main components of the Arduino UNO board are—USB connector, power port, microcontroller, analog input pins, digital pins, reset switch, crystal oscillator, USB interface chip, TX, and RX LEDs. All these components are necessary to perform various functions over the Arduino board. The IR sensor is connected to Arduino in such a way so that it would be able to meet the objective of prototype. The IR sensor has 3 pins—VCC, GND, and Vout. So connect the VCC to 5 V pin of Arduino board, GND to ground pin of board, and the last pin, Vout to A0 of Arduino board. Similarly, connect another IR sensor to pin A5 on Arduino. By seeing Figs. 6 and 7, LCD connections can be made with Arduino with the potentiometer of 10 K [10]. The Arduino receives the input from the IR sensor, and it counts the number of person entering and exiting the room and calculates accordingly, and accordingly, it generates signal over pin 2 which is connected to Darlington pair and then that Darlington pair connects to Relay 2 to operate the appliance. It also displays the number of people inside the room on the LCD screen connected to its pin. When there is no person inside the room, the IR sensor generates no signal and LCD screen shows number of persons to be zero. But when even a single person entered a room, the LCD shows the total number of persons inside the room and the microcontroller generates a signal for the Relay 2 to provide further signal.

4 Results

The research aims were to make a prototype of non-conventional energy source-based home automation system. In this, solar energy is used as non-conventional energy resource which in turn used to charge a DC battery connected to inverter via a logic circuit. The system itself decide that whether the appliance would work on

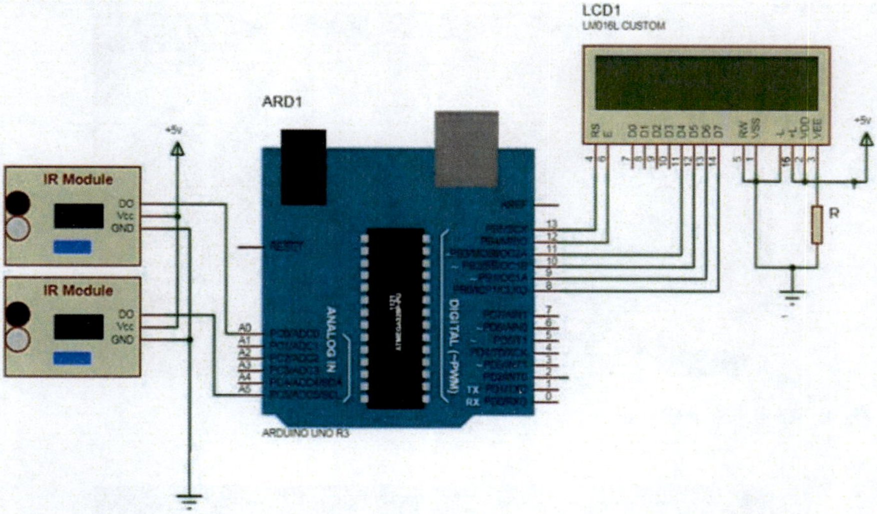

Fig. 6 Arduino board connections

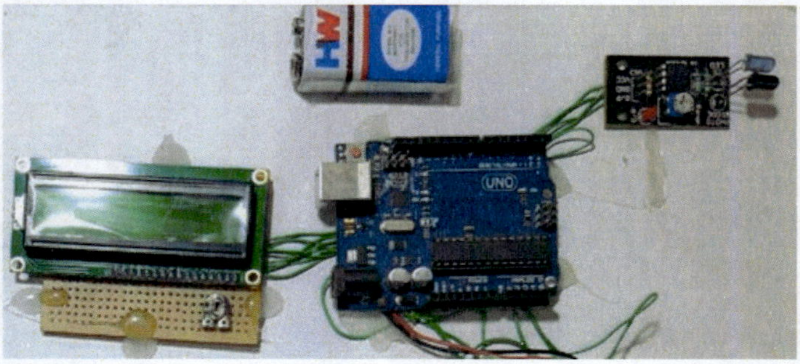

Fig. 7 Implemented connections

the main AC supply or the supply provided by the DC battery charged by using solar panels and also making the appliance (bulb) work only if the person is present at that place.

In Fig. 8, the bulb is in off state because no person entered in the room or we can say that nothing signal detected by the first IR sensor or that IR sensor which is present at the entry gate of room, due to which the IR sensor sends no signal to the microcontroller and the LCD display shows that there is no person in the room or zero on the display and microcontroller does not send any signal to Relay 2. And if Relay 2 does not get the signal, then circuit gets uncompleted and bulb remains OFF.

In Fig. 9, the bulb is in ON state because one person entered in the room and IR sensor which is present at the entry gate of room, sends signal to the microcontroller, and the LCD display shows there is one person in the room on the display and

Fig. 8 Appliance in OFF state

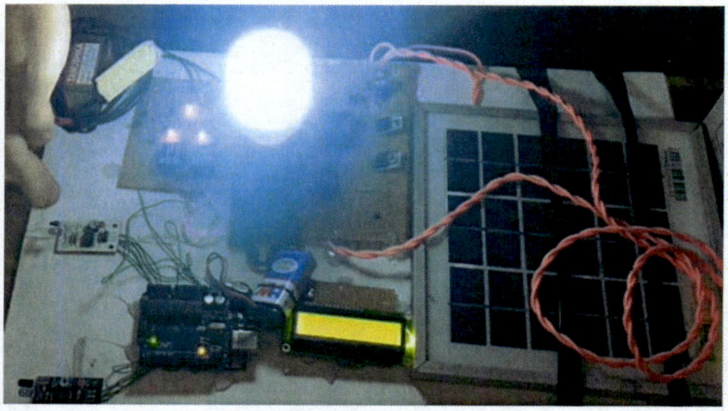

Fig. 9 Appliance in ON state

microcontroller sends a signal to Relay 2 passing through Darlington pair. And when Relay 2 gets the signal, then circuit gets completed, and the input is provided to inverter circuit and then output of inverter circuit is given to the appliance or bulb which makes the bulb ON. When the IR sensor 2 detects any motion, it sends signal to the microcontroller, and the LCD display shows there is one person left and accordingly decrement the value shown in the display, and if the display shows zero, then microcontroller stop sending signal to Relay 2 passing through Darlington pair. And then again bulb comes to the OFF state. As stated above, the switching of logic with help of relay and microcontroller shows the execution of system which is able to provide signal as an input to the load, and in this way, execution of the prototype is done successfully.

5 Conclusion

This paper elaborates the design and construction of non-conventional energy source-based home automation system. The objective is to make a system which is able to conserve energy as well as to provide such type of convenience which reduces human efforts by making the home appliance to work automatically. Finally, this system can be used in homes, industries, and for various commercial purposes.

Circuit works properly to turn ON/OFF the appliance (LED bulb) on the main AC supply according to the motion detected by IR sensor or else the person should be present in a room. If the main AC supply cuts off, then the appliance will be provided with a supply of DC battery charged by solar energy, which is converted into AC output with the help of inverter circuit, after designing the logic circuit and inverter circuit which controls the shift between the supplies and converting the DC input into AC output, respectively, as illustrated in the previous sections. Two IR sensors are the main conditions behind the working of the circuit. If the conditions have been satisfied, the circuit will do the desired work according to specific program. One IR sensor for entering and other for leaving a room controls the turning ON\OFF of appliance. This prototype of home automation system based on non-conventional energy source is successfully controlled by microcontroller. Due to the commands given by the controller, the light will get ON as person move inside the room and controller gives command to get light off when there is no person in the room.

References

1. Khan BH Non-conventional sources of energy, 5/e, Mc Graw Hill Education (India)
2. Reinisch C (2007) Wireless communication in home and building automation. Master thesis, Viennia University of Technology
3. Gauger M, Minder D, Wacker A, Lachenmann A (2008) Prototyping sensor—actuator networks for home automation. In: REALWSN'08, Glasgow
4. Chandane A, Narkhade D, Nageshingale A, Yadde S (2018) Renewable energy based home automation using IoT. Int J Innov Eng Res Technol [IJIERT], ISSN: 2394–3696
5. Sriskanthan N, Tan F, Karande A (2002) Bluetooth based home automation system. Microprocess Microsyst J 26:281–289. Elsevier Science B.V
6. Gupta JB (2009) Electronics devices & circuit, 3/e, S.K. Kataria & Sons
7. Hussain A Electrical machines, 2nd edn, Dhanpat Rao Publications
8. Yang X, Zhai Y, Ai X (2011) Design of inverter power supply for household solar power generation in pastoral area. In: 2011 IEEE power engineering and automation conference, Wuhan, pp 279–282
9. Milliman J, Halkias C Integrated electronics, 2/e, Mc Graw Hill Education (India)
10. Yu H, Chen Z (2002) Charge pump for LCD driver used in cell phone. Tsinghua Sci Technol 7(5):517–520

Resource Scheduling on Basis of Cost-Effectiveness in Cloud Computing Environment

Rupali and Neeraj Mangla

Abstract Cloud computing is a computing software that enables a user to manipulate, configure, and provide access to the applications over the network. It is a model that allows the user to take advantage of computing facilities over the web as per user demand and to share resources such as network, data storage, servers, and applications without the need of actually installing it on their device. Thus, it can be said that it is a computing model that helps manage cost and saves time. Cloud computing has its applications over many fields like health, education or banking due to its special features. Cloud computing is an Internet-based software, and thus, it is the responsibility of cloud service providers (CSPs) to maintain data stored by users at data centers. Scheduling in cloud plays a very important role to achieve maximum utilization, and user satisfaction resources need to be allocated effectively. In this paper, we have discussed some optimum scheduling technique to enhance the performance of cloud. We consider cost as one of the major attributes that result in enhancing the performance (Cloud computing tutorial tutorials point), (Armbrust et al. in Commun ACM 53:50–58, 2010).

Keywords Cloud computing · Resource scheduling · Cost-based scheduling · DAG

1 Introduction

The word cloud computing comes from two terms cloud and computing where the word cloud refers to storage at remote location and computing refers to calculation or processed data; thus, we can say that cloud computing is a software application that

Rupali (✉) · N. Mangla
M. M. Engineering College, Maharishi Markandeshwar (Deemed to be University) (MMDU), Mullana, Ambala, India
e-mail: rupali.sharma514@gmail.com; er_jangra@yahoo.co.in

N. Mangla
e-mail: erneerajynr@gmail.com

N. Marriwala et al. (eds.), *Mobile Radio Communications and 5G Networks*, Lecture Notes in Networks and Systems 140,
https://doi.org/10.1007/978-981-15-7130-5_34

allows to store, access, or process data from a remote location. Cloud offers an online space that allows a user to store data, infrastructure, and their application. Cloud computing is defined as a computer service which allows changing, configuring, and using the application over network. Cloud computing is an advancement of all forms of computing such as parallel computing, distributed computing, and grid computing [1, 2]. Cloud computing has now become an emerging trend through which technical services, and information is provided online. It is one of the fastest new trends that deliver services on demand over the network and is purely Internet-based software. It is an enhanced form of utility computing that has gained prominence in no time. The cost of using the cloud services is less than setting up a datacenter. CSPs such as Google and Amazon provide users with services on pay-per-use basis without charging a high premium value. Cloud services can be accessed on both private and public networks. These service providers have their own schemes to provide computing facilities with varying prices. For instance, Amazon EC2 provides micro, small, medium, large and extra-large packages, whereas Google computed charges on monthly basis [2]. The management of user's data is also the responsibility of these service providers. But to make an effective use of its capabilities, better scheduling algorithms are required. These are practiced by cloud resource manager to optimally communicate tasks to cloud resources. In cloud computing environment, it becomes mandatory to follow such a technique to schedule resources so that cloud users can efficiently use resources. To decrease the execution time and accessing cost, various scheduling algorithms are implemented in cloud computing environment to enhance the usage or resources by scheduling them. There exists a large variety of scheduling algorithms that are capable of minimizing the total completion time of tasks. This minimization process is carried out by matching the task with the most suitable resources. Scheduling in cloud computing is attaining attention day by day. Generally, it can be said that scheduling is a process of matching or mapping tasks to existing resources based on features and requirements of a task. It is one of the important fragments that help in effective working of cloud computing as many task parameters require to be considered for proper scheduling. The issue in scheduling is that the logical sequence of task must be preserved. The scheduler must be capable of upholding total execution time as well as the monetary cost, which can only be attained by using a better scheduling algorithm. These applications are commonly represented using a directed acyclic graph (DAG) [3]. DAGs have the capability to represent real-life situations well. Heuristic methods and approximation algorithms might be well-suited to solve problems like NP-complete problem to get optimal values. Thus, heuristic-based scheduling algorithms can be recommended to achieve optimal solutions for DAG structure.

2 Related Work

In order to select the related work to this survey, various papers have been chosen from databases (IEEE, ACM, Elsevier, Springer, and Google Scholar) based on some below-listed keywords:

I. Virtual machine allocation
II. Monetary cost
III. Makespan
IV. Trade-off cost
V. Task completion cost
VI. Total cost
VII. Computation cost
VIII. Datacenter cost
IX. Virtualization
X. SLA.

The author in [4] has deliberated a heterogeneous earliest finish time (HEFT) algorithm for mapping jobs to resources using a Directed_Acyclic_Graph (DAG), with less number of virtual machines. Various methods to calculate weight of nodes and edges of a graph have been implemented to achieve less cost and makespan.

The author in [5] has offered a job scheduling method in cloud and proposes an improved cost-based scheduling algorithm in which a feasible method has been chosen to allocate resources to the processes. In this work, the algorithm is capable to calculate both cost and performance, and also, it improves the calculation ratios by mapping user jobs to their respective resources.

The author in [6] has offered hybrid cloud optimized cost (HCOC) scheduling algorithm which helps in taking in consideration that which resources can be borrowed from public cloud to get enough handling related to the execution of a workflow in an agreed period of time. This algorithm also helps in lessening the cost while completing the job at mentioned execution time.

The author in [7] has presented a cost-based resource allocation theory in cloud environment where according to the market trend, the resources are mapped to the jobs as per the user demand. The resources with the lower prices are mapped to jobs according to availability of resources with suppliers and minimum associated cost.

The author in [8] deliberates deadline as one of the important restriction and offers a rank-based deadline controlled workflow scheduling algorithm. Here they have used the idea of score which signifies the abilities of the hardware assets. This worth of rank is used when assigning assets to jobs of workflow application which in result reduces the rate of failure and also have less execution time. Thus, it can be concluded that algorithm executes workflow in controllable cost while meeting resource requirements.

The author in [9] has proposed an algorithm based on deadline in a hybrid cloud scenario. In this algorithm, cost has been optimized by taking resources on contract from open cloud to cover the workflow with deadline. This level centered algorithm implements jobs level-wise and also uses the principles of sub-deadline that helps in searching the finest assets on free cloud that acts as an aid in saving cost and complete the task on time.

The author in [10] offered a customer-facilitated cost-based scheduling (CFCSC) algorithm to help the cloud users to enjoy less cost. It compares its implemented algorithm to HEFT algorithm on the basis of balancing the load, minimizing the cost and choosing a less complex cost function. In this work, it was also found that both HEFT and CFCSC have similar makespan.

The author in [11] has discussed scheduling in cloud computing environment and has recorded a survey on various scheduling algorithms implemented so far. The author discussed four types of scheduling strategies based on different parameters, i.e., based on virtualization, energy conservation, SLA, and cost-effectiveness. The author compares various algorithms on the basis of above-mentioned criteria.

The author in [12] has aimed upon task scheduling in cloud computing environment on basis of cost. The proposed algorithm calculates the monetary cost and checks the chances of completing the task within the referred time and cost. When related with deterministic scheduling algorithm, this algorithm enhances performance by executing task on time.

The author in [13] has proposed a hybrid type of particle swarm optimization (PSO) as nearest neighbor cost-aware PSO to evaluate various parameters like makespan, energy consumption, utilization, and cost efficiency.

The author in [14] has discussed various issues related to scheduling methods and their limitations. The author also has discussed various scheduling methods based on different parameters to cover their characteristics. The survey further is based on three different strategies methods, applications, and measurements on the basis of some parameters.

The author in [15] offers a scheduling algorithm which calculates the jobs in line and executes them in order. On the basis of a fixed goal, jobs are evaluated, i.e., minimizing the cost of job completion or minimizing the time of job completion. This algorithm follows a greedy approach to choose a suitable resource and map various tasks to resources. This also improves the utilization of resources and also improves efficiency.

The author in [16] has presented a hybrid algorithm known as CR-AC which combines both chemical reaction and ant colony optimization (ACO) algorithms to solve this problem. The chosen algorithm has achieved better results as compared to PSO and CCGA algorithms in terms of total cost, time complexity, and schedule length.

The author in [17] has offered a SLA-RALBA algorithm that is being compared to different schedulers, i.e., MCT, Profit-MCT, SLA-MCT, execution min-min, profit min-min, and SLA min-min in relation to average source consumption, finishing time, and price of cloud facilities, the proposed algorithm provides even balance among the execution time and cost of services.

The author in [18] has discussed the task scheduling for iteration process-based situation that works on less cost and less time. The author has used two load balancing techniques which are compared to Round-robin scheduling technique which comes to be successful in achieving less cost and time period when equally related to Round-robin scheduling algorithm.

3 Resource Scheduling

Job scheduling is a technique used to assign resources to the process as per demand of their usage such as processing, network usage, storage in such a way that the resources can be utilized at maximum. In cloud computing environment, it becomes mandatory to follow such a technique to schedule resources so that cloud users can efficiently use resources. To decrease the execution time and accessing cost, various scheduling algorithms are implemented in cloud computing environment to enhance the usage or resources by scheduling them. Scheduling in cloud computing remains a problem for fair allocation of resources to the cloudlets. Most scheduling algorithms are based upon parameters like load balancing, execution time, fault tolerance, cost-effectiveness, makespan, and migration of resources.

Scheduling in cloud computing is attaining attention day by day. Generally, it can be said that scheduling is a process of matching or mapping tasks to existing resources based on features and requirements of a task. It is one of the important fragments that help in effective working of cloud computing as many task parameters require to be considered for proper scheduling. Thus, it focuses on the fact that available resources must be properly used without affecting the service quality of cloud. Scheduling process in cloud computing environment is divided into three phases that include discovering resource, filtering, selecting resources, and allocating tasks to resources.

Workflows have been used to solve a range of engineering applications that involve high processing and storage capabilities. Thus to please these applications, cloud computing can be well-thought-out as a new computing model. These workflow models can be symbolized using a DAG model. Research work conducted in line of workflow scheduling mostly concentrates on minimizing the scheduling length, whereas the existing research in scheduling algorithms focuses on economic cost factors. Scheduling in cloud computing can be either cost-based or credit-based scheduling [3].

Resource scheduling in cloud computing can be classified into various categories. Some of the commonly used strategies are Fig. 1:

I. Based on energy conservation.
II. Based on virtualization.
III. Based on service-level agreement (SLA).
IV. Based on cost-effectiveness.

Fig. 1 Resource scheduling in cloud computing environment

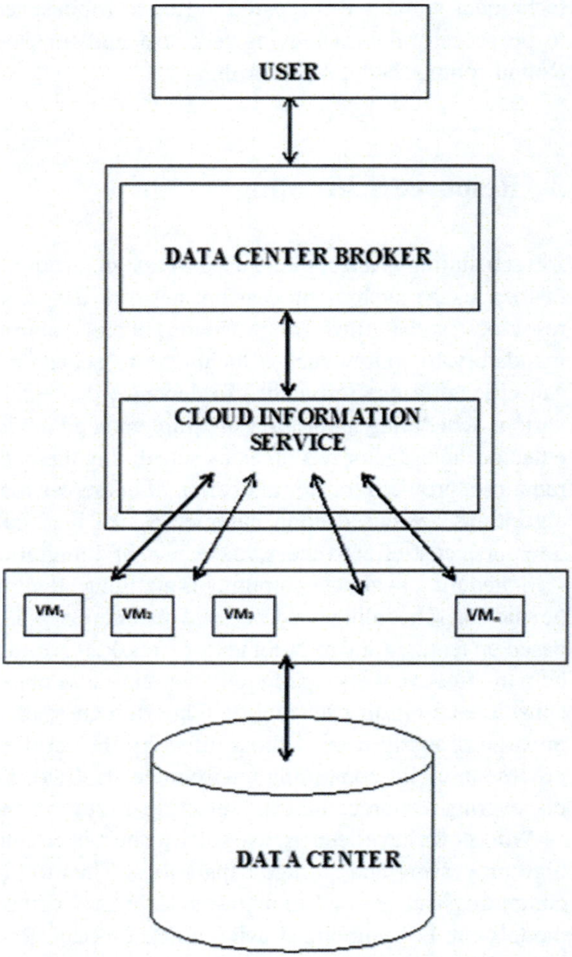

3.1 Resource Scheduling Algorithms Based on Energy Conservation in Cloud Computing

A large amount of energy is required to perform high-level tasks in a huge cloud computing environment. The consumption of energy can be lessened by executing applications or processes on less number of servers and putting the rest on sleep or power-off mode. Since the workload in cloud computing scenario is different at different time, therefore real-time virtual machine scheduling can control energy utilization for calculating while servers get decreased, and more virtual machines should be allocated when load is high [19, 11].

3.2 Resource Scheduling Algorithms Based on Virtualization in Cloud Computing

Virtualization can be defined as a way through which cloud users share same data present at cloud-like network, application software. It actually maintains a virtual space in cloud environment which can be anything like a software, hardware, or anything. Virtualization allows a task to migrate from a physical machine (PM) to virtual machine (VM) so that resources can be utilized efficiently and energy consumed and cost related to a cloud service provider could be reduced. But as per the work recorded till yet allocation of tasks over virtual machine from physical machine; i.e., migration is not easy. Various algorithms have been proposed and implemented so far to increase efficiency and reduce the capital cost [11].

3.3 Resource Scheduling Algorithms Based on Service-Level Agreement (SLA) in Cloud Computing

As the name suggests SLA is a major parameter of cloud computing which means to assure the customer to deliver services as per requirement. SLA is an agreement between the service providers and customers; therefore, SLA is based on some properties. Properties on basis of customers SLA are based on properties such as request-type, product-type, account-type, contract-length, number of accounts, number of records, response time, and on the properties based on service provider are VM types, service initiation time, VM price, data transfer, data transfer speed. The main issue that arises is violation of these properties of SLA. To overcome this issue, various algorithms have been proposed and implemented [20].

3.4 Resource Scheduling Algorithms Based on Cost in Cloud Computing

Two major QoS checks of scheduling for workflows cloud computing are processing time and execution time. Normally, one would like the task to be completed within required time at the lowest cost. In this paper, cost-based workflow scheduling method is represented that allows the workflow management system to minimize the execution cost while delivering results within a desired period of time [11]. Some important terms required to be known are discussed below:

(a) **Makespan**: Makespan can be defined as the total span of schedule or the time when all tasks have finished processing. Practically, this problem can be termed as an online problem or dynamic scheduling that is the decision of scheduling a job can only be made online.

(b) **Monetary cost**: Monetary cost can be well-defined as the cost necessary to purchase the facility; it may also include the time and expenses involved in shopping and the risk taken in spending money to obtain the estimated benefit from the good's features [10].

In Table 1, we have discussed various cost-based scheduling algorithms proposed and implemented so far with an aim to reduce the cost and utilize the cloud resources more efficiently.

4 Directed Acyclic Graph

Directed acyclic graph (DAG) can be defined as a directed graph with no cycles in it. It is the most widely type of graph used in the field of mathematics and computer science. It can be said that it is a directed graph that consists of finite number of edges and nodes, where each edge is directed from one node to another node. There are no cycles formed in this graph which means one cannot loop back to the starting node when following a steady directed path. A DAG is a directed graph that is topologically sorted; it consists of a series of nodes in such a way that each edge has a direction from previous node to next node. In engineering and scientific fields, many problems can be made easy using DAGs, such as game evaluation, expression tree evaluation, and path analysis [3].

In computer science, DAG is a data structure that may be used to solve a large number of problems. A DAG comprises of following main components:

(a) *Nodes*: in DAG, each node represents some object or piece of data or a task.
(b) *Directed edges*: the edges in a DAG are directed and have a direction marked with arrow from one node to another node. The edge represents the relationship between the nodes or their dependencies. Generally in a DAG, the arrows are pointed from child to parent that is from new version to the version from which it was inherited.
(c) *Parent node/Root node*: in each graph, there exists a node that may have no parents, and this node is called the root node or parent node.
(d) *Child node/Leaf nodes*: like the root node, there exist some nodes in the graph that may not have any children. These are called leaf nodes or leaves or child node.

5 Conclusion

Cloud computing has enlarged the attention of a great number of users through its several services. The pay-per-use basis improved the cloud customer population terrifically in limited number of years. Researches indicate that there is a giant market awaiting due to the significance of a business model that bids high performance with

Table 1 Comparison of various cost-based scheduling strategies used in cloud computing

Technique proposed	Performance matrices	Platform used	Results	Related work
HEFT: heterogeneous earliest finish time algorithm	Computation cost	Java-based simulator	In the results, it was found that there is a huge difference between the performance of heterogeneous earliest finish time(HEFT) based upon DAG for calculating weights	[4]
Improved cost-based algorithm for task scheduling in cloud computing	Task computation time, cost	Cloudsim 1.0b	The results indicate that the proposed algorithm is more efficient than activity based on cost-effective algorithm	[5]
HCOC: a cost optimization algorithm for workflow scheduling in hybrid clouds	Makespan, cost	Cloudsim and workflows montage cost and montage makespan	From the results of implementation, it is concluded that the proposed algorithm, HCOC speeds up the execution of workflow within desired time and also decreases cost	[6]
A cost-based resource scheduling cost paradigm in cloud computing	Cost	Java Cloudware, Pure Java-based platform in cloud	This algorithm evaluates cost and is first such algorithm used in cloud environment	[7]
Deadline and budget distribution-based cost–time optimization workflow scheduling algorithm for cloud	Execution time, cost	Java and randomly generated workflows	The proposed algorithm reduces the cost and execution time and manages deadline and overall budget simultaneously	[21]
Deadline and cost-based workflow scheduling in hybrid cloud	Makespan, cost	Cloudsim	The proposed algorithm results to the conclusion that this level-based hybrid scheduling algorithm is more efficient than existing workflows based upon level-free scheduling and min–min algorithm	[9]

(continued)

Table 1 (continued)

Technique proposed	Performance matrices	Platform used	Results	Related work
Customer-facilitated cost-based scheduling (CFCSC) in cloud	Makespan, monetary cost	Cloudsim	From the results, it is recorded that the proposed algorithm is helpful in load balancing and minimizing the total monetary cost when compared with HEFT	[10]
Resource scheduling base on fuzzy clustering in cloud computing	Response time, waiting time, running cost	Java-based simulator	The proposed algorithm is cost-based clustering (CBFCMP) is a fuzzy algorithm and is more efficient and gives more accurate results when run in different iterations	[22]
Cost-effective scheduling precedence constrained tasks in cloud computing	Makespan, monetary cost	MATLAB	The method adopted in this work has chosen multi-population genetic algorithm to obtain the optimal scheduling strategy so that slot time can be utilized efficiently. An improved cost-effective task copying is considered as result of which the required schedule length and minimum cost can be acquired. The results based upon DAG have much better performance based upon monetary cost. The algorithm is capable of reducing the monetary cost along with schedule length	[12]
Cost-effective task scheduling using hybrid approach in cloud	Makespan, energy consumption, cost	MATLAB 7.1	This paper proposed a novel heuristics called nearest neighbor (NN) and a variant of PSO called NNCA_PSO. NN heuristic achieved aligned allocation of tasks to VMs by reducing the difference between execution time requirements of tasks and execution time characteristics of VMs. This minimization is validated through reduction in variance. NN proved to be an effective scheduling algorithm for set of independent tasks over other traditional heuristics. The proposed NNCA_PSO also provides an edge over other cost-aware PSO variants. Multiple issues of cloud related to makespan, energy usage, economic effectiveness, and resource utilization have been addressed through this research work	[13]

(continued)

Table 1 (continued)

Technique proposed	Performance matrices	Platform used	Results	Related work
Optimization of task scheduling and cloudlets cost scheduling algorithms on cloud using cloud simulator	Execution cost, task completion	Cloudsim2.1.2	Both the cost and time of completion are lessened by the proposed algorithm, thus proving advantageous over the sequential logic. Minimization of turnaround time and cost of each job on an individual basis in turn minimizes the average turnaround time and cost of execution. As the number of tasks executed based on these criteria increase, the efficiency gained over the sequential approach also increases	[15]
Minimizing cost and makespan for workflow scheduling in cloud using fuzzy dominance Sort-based HEFT	Makespan, cost-trades off	Real-world and synthetic workflows Iaas model, Cloudsim	In this work, FDHEFT scheduling algorithm has been proposed that merges together the fuzzy sorting method with heuristic HEFT method. The experiments were performed on two simulators: one is real-world workflow-based and another is a synthetic application so that performance of FDHEFT can be validated. The experiments performed are based on parameters like cost and elements of Amazon EC2. Thus, it is concluded that this algorithm is capable to get better makespan, cost tradeoffs with less CPU run time	[23]
An efficient cost-based algorithm for scheduling workflow tasks in cloud computing systems	Schedule length, monetary cost	Cloudsim	the paper analyze the improving this work ICTS, improved cost task scheduling algorithm is analyzed for mapping tasks to resources in cloud computing environment. This algorithm is capable of minimizing both time and monetary cost when using cloud resources. The algorithm is a three phase algorithms—sorting the levels, allocating priority to task, and selection of virtual machine. From the results, it was found that the proposed algorithm is superior than hybrid algorithm. The percentage of improvement in cost and time is different for different DAGs	[24]

(continued)

Table 1 (continued)

Technique proposed	Performance matrices	Platform used	Results	Related work
Cost-effective algorithm for workflow scheduling in cloud computing Under deadline constraint	Makespan, cost in montage and inspiral workflows	Cloudsim toolkit and Java-based discrete event simulator	This paper presented a two-phase hybrid CR-AC algorithm to solve scheduling problem under deadline constraint. The algorithm is based on two famous algorithms, CRO and ACO. In the first phase, a modified CRO algorithm is applied to find a near-optimal solution in low-time complexity under deadline constraint. While, in the second phase, the ACO algorithm is applied to minimize the total cost and improve solution quality. The proposed CR-AC approach can schedule large number of tasks Into the available VMs considering several resource constraints. It enhances the overall system performance at low-time complexity. By comparing the new CR-AC approach with the CRO, the ACO, modified PSO and CEGA algorithms, the new approach is more efficient than them. It can achieve a high-quality near-optimal solution with low-cost under the deadline constraint	[16]
SLA-RALBA: cost-efficient and resource-aware load balancing algorithm for cloud computing	Makespan, gain_cost	Cloudsim 3.0.2	A comprehensive investigation of the state-of-the-art SLA-based heuristics fallouts in reduced resource utilization and increased execution cost due to the improper allocation of jobs to VM and straggling jobs in scheduling. The proposed novel technique SLA-RALBA ensures enhanced resource utilization with reduced execution time and cost. The performance of SLA-RALBA has been compared using benchmark HCSP and GoCJ datasets with the scheduling algorithms in terms of execution time, cost, and resource utilization. The exhaustive examination of pragmatic results has shown that SLA-RALBA has presented an improved balance among the resource utilization, execution time, and cost than existing state-of-the-art algorithms (i.e., Execution MCT, Execution Min-Min, Profit-MCT, Profit Min-Min, SLA-MCT, and SLA Min-Min)	[17]
Distributed cost-based scheduling in cloud computing environment	Monetary cost, makespan	Cloudsim 3.0.3	The result is that the throttled method of scheduling, when utilized under reconfigure dynamically service is better than Round-robin and active VM load balancing strategy	[18]

low cost. In this paper, we have taken an overview of what is scheduling and types of scheduling in cloud computing environment. The major focus of the paper is toward cost-based scheduling techniques. In a tabular format, we have differentiated various cost-effective scheduling strategies implemented and proposed so far. We also focused on the data structure required to represent workflows in cloud computing in the form of the DAG.

References

1. Cloud computing tutorial tutorials point. https://www.tutorialspoint.com/cloud_computing/
2. Armbrust M, Fox A, Griffith R, Joseph AD, Katz R, Konwinski A, Lee G, Patterson D, Rabkin A, Stoica I, Zaharia M (2010) A view of cloud computing. Commun ACM 53(4):50–58
3. Directed Acyclic Graph: https://ericsink.com/vcbe/html/directed_acyclic_graphs.html
4. Zhao H, Sakellariou R (2003) An experimental investigation into the rank function of the heterogeneous earliest finish time scheduling algorithm. In European conference on parallel processing. Springer, Berlin, pp 189–194
5. Selvarani1 S, Sudha Sadhasivam G (2010) Improved cost-based algorithm for task scheduling in cloud computing
6. Bittencourt LF, Madeira ER (2011) HCOC: a cost optimization algorithm for workflow scheduling in hybrid Clouds. J Internet Serv Appl 2(3):207–227
7. Yang Z, Yin C, Liu Y (2011) A cost-based resource scheduling paradigm in cloud computing. In: 12th international conference on parallel and distributed computing, applications and technologies
8. Singh R, Singh S (2013) Score based deadline constrained workflow scheduling algorithm for Cloud systems. Int J Cloud Comput Serv Architec (IJCCSA) 3(6)
9. Chopra N, Singh S (2013) Deadline cost based workflow scheduling in hybrid cloud. In: International conference on advances in computing, communications and informatics (ICACCI)
10. Amalarethinam DIG, Beena LA (2015) Customer facilitated cost based scheduling (CFCSC) in cloud. In: international conference on information and communication technologies. Elsevier, p 46
11. Mangla N, Singh M, Rana SK (2016) Resource scheduling in cloud environment: a survey. Adv Sci Technol Res J 10(30):38–50
12. Wang B, Li J, Wang C (2017) Cost effective scheduling precedence constrained tasks in cloud computing. In: 2017 the 2nd IEEE international conference on cloud computing and big data analysis. IEEE
13. Thaman J, Singh M (2017) Cost effective task scheduling using hybrid approach in cloud. Int J Grid Utility Comput 8(3)
14. Arunarani AR, Manjula D, Sugumaran V (2018) Task scheduling techniques in cloud computing: a literature survey. Future Gener Comput Syst 0167–7390
15. Kumar D, Jiaswal N, Gill SK, Saharma P, Singh VP (2018) Optimization of task scheduling algorithms on cloud simulator. Int J Adv Stud Sci Res (IJASSR). ISSN 2460-4010
16. Nasr AA, El-Bahnsawy NA, Attaya G, EL-Sayed (2018) Cost effective algorithm for workflow scheduling in cloud computing under deadline constraint. Arabian J Sci Eng
17. Hussain A, Iqbal MA, Islam MA (2019) SLA-RALBA: cost efficient and resource—aware load balancing algorithm for cloud computing. J Super Comput
18. Rupali, Jaiswal AK (2019) Distributed cost-based scheduling in cloud computing environment. World Acad Sci Eng Technol Int J Comput Syst Eng 13(1)
19. Sarkar AMJ, Adhikary T, Das AK, Razzaque MA (2013) Energy—efficient scheduling algorithms for data center resources in cloud computing. In: 2013 IEEE international conference

on high performance computing and communications and 2013 IEEE international conference on embedded and ubiquitous computing. IEEE
20. Wu L, Garg SK, Buyya R (2011) SLA-based resource allocation for software as a service provider (SaaS) in cloud computing environment. In: 11th IEEE/ACM international symposium on cluster, cloud and grid computing. IEEE
21. Verma A, Kaushal S (2012) Deadline and budget distribution based cost-time optimization workflow scheduling algorithm- international conference on recent advances and future trends in information technology. Int J Comput Appl
22. Xiaojun W, Yun W, Zhe H, Juan D (2015) The research on resource scheduling based on fuzzy clustering in cloud computing. In: 8th international conference on intelligent computation technology and automation Jiangsu Open University, Nanjing, Jiangsu, 210036, China
23. Zhou X, Zhang G, Sun J, Zhou J, Wei T, Hu S (2018) Minimizing cost and makespan for workflow scheduling in cloud using fuzzy dominance sort based HEFT. Future Gener Comput Syst Future 4551
24. Amaan M, El-Bahnasawy N, Elkazaz M (2018) An efficient cost- based algorithm for scheduling workflows tasks in cloud computing systems. J Neural Comput Appl

Optimized Multi-level Data Aggregation Scheme (OMDA) for Wireless Sensor Networks

Shilpy Ghai, Vijay Kumar, Rajneesh Kumar, and Rohit Vaid

Abstract Data aggregation plays an important role over WSN as aggregated data is utilized for decision making/analysis purpose; but due to complex aggregation computations, sensors may consume excessive energy and thus may reduce the network lifespan. So there is requirement to optimize the aggregation process. In this paper, an optimized data aggregation scheme, called optimized multi-level data aggregation scheme (OMDA), is introduced using LEACH protocol. Its performance is analyzed using different performance parameters (throughput/end-to-end delay/energy consumption/network lifespan) under the constraints of sensor node density that varies from 50 to 200.

Keywords WSN · Data aggregation · Data redundancy · LEACH

1 Introduction

The process of summarization of collected data from the various sources called data aggregation that is quite complex and inefficient aggregation may degrade the network performance. It is used to filter out the redundant data as well as number of transmission cycle can be reduced. Following are the data aggregation types:

S. Ghai (✉) · V. Kumar · R. Kumar · R. Vaid
Department of Computer Science and Engineering, MMEC, Maharishi Markandeshwar Deemed to be University, Mullana, Ambala, India
e-mail: ghai.shilpy2010@gmail.com

V. Kumar
e-mail: katiyarvk@mmumullana.org

R. Kumar
e-mail: drrajneeshgujral@mmumullana.org

R. Vaid
e-mail: rohitvaid@mmumullana.org

N. Marriwala et al. (eds.), *Mobile Radio Communications and 5G Networks*, Lecture Notes in Networks and Systems 140,
https://doi.org/10.1007/978-981-15-7130-5_35

- **Lossless and Lossy data aggregation**: In case of lossless approach, large packet size is used for data storage and processing, whereas lossy data aggregation reduces the packet size.
- **Structured and non-structured data aggregation**: Structured-based aggregation is suitable for the applications that use a fixed data pattern, whereas non-structured-based aggregation can handle the dynamic traffic patterns.

1.1 Requirements of Data Aggregation Over WSN

- **Minimization of redundancy**: Sensors collect and forward the data that may contain duplicate values and transmission of redundant data consumes excessive resources. So there must be a provision to filter out this type of data.
- **Optimization of energy consumption**: Energy can be preserved by regulating the transmission interval and thus may reduce the cycle of data aggregation process as well as network life span can be extended.
- **Efficient utilization of shared channel resources**: Sensors utilize the shared channel. Accuracy of data aggregation can be achieved using optimized slot allocation.

1.2 Constraints for Data Aggregation Over WSN

- Dynamic sampling rate may cause the congestion, buffer overflow and delay and thus may reduce the efficiency/accuracy of aggregation process.
- Optimal size of the cluster may enhance the overall output of the aggregation function but it is complex to define an idle size for the cluster.
- Packet collision may degrade the accuracy of the aggregation process so there is a need to regulate the schedule of channel access.
- Cluster head acts as aggregator and excessive computations over large-scale data may cause energy depletion.
- Secure aggregation is another major concern for end users [1–5].

2 Existing Solutions for Data Aggregation Over WSN

Zhang et al. [6] introduced an aggregation method using the combination of tree/entropy and gradient deployment, etc. Analysis identifies few factors that those can affect the network performance as well as its lifespan, i.e., deployment method, sink position and the sensor density, etc.

Wen et al. [7] identified the relationship between compression ratio and its effect over data developed a scheme to optimize the data aggregation process through different phases, i.e., optimal selection of cluster head, compression, and mutual reformation. Cluster head filters the common data for current cluster, and after that residual data is further process by intermediate member nodes. Data reformation is done through a matching pursuit method. Simulation results show its outcome in terms of minimum delay, low redundancy, and less resource consumption as compared to the traditional compression method. It can be further utilized over real-time WSN.

Sarangi et al. [8] investigated various data aggregation schemes and found different methods that can enhance the WSN performance, i.e., random data collection points, cluster selection through neural networks, and bio-inspired target locking, etc. This study can be implemented to conserve the resources of WSN.

Tolani et al. [9] developed an aggregation scheme that uses multiple layers for data processing. First of all, aggregation is performed at intermediate member, and later on *cluster head* performs the aggregation. Simulation results show that this method extends the network lifespan by reducing the overall resource consumption during transmission as compared to existing schemes (spatial–temporal correlation/aggregation window function).

Tamiji et al. [10] developed used automata for data aggregation over WSN. Sensors adjust their sensing rate to forward the data to current cluster head that is responsible to redirect the data to base station. Simulation results show that dynamic sensing rate improves the node's lifespan by reducing the energy consumption.

Akila et al. [11] introduced a secure data aggregation method for WSN. Session keys are used to initiate the communication, and aggregation is performed over encrypted data, and finally, data is collected by multiple sinks. Simulation results show that it outperforms in terms of less overhead/resource consumption and also ensures the security goals.

Egidius et al. [12] introduced an aggregation scheme for software-defined WSN. It performs aggregation at packet level using flow admissions. Simulation results show that it can reduce the overall resource consumption by aggregating the incoming/outgoing traffic streams.

Goyal et al. [13] investigated the influence of data aggregation/non-data aggregation over the performance and resource consumption of WSN. Simulation-based analysis indicates that cluster-based aggregation methods are more efficient as compared to non-cluster based methods. Outcomes show that various performance factors (i.e., packet drop/packet delivery ratio/energy consumption/collision level, etc.) are affected by variable packet size.

Hadi [14] explored the data aggregation issues related to geographical routing over WSN and introduced a scheme that can filter out the redundant data at node level, and it does not depend over cluster head and thus results in the reduction of control overhead. Simulation results show that it outperforms in terms of optimal resource consumption and higher accuracy of data aggregation as compared to non-aggregated data processing.

Galkin et al. [15] introduced an ant-based data aggregation method that collects the data using the shortest paths under the constraints of semaphore, and finally, it is aggregated at node level. Simulation results show its performance in terms of enhanced network lifetime/optimal resource consumption, etc. Its capabilities can be further extended using blockchain theology.

Nguyen et al. [16] investigated that collision and delay both act as a fence for data aggregation and developed a distributed scheduling method to avoid collisions. It transmits the data through time slots using multiple channels and thus reduces the probability of collisions. Simulation results display that aggregation accuracy can be achieved by minimizing the collision level as well as delay/resource consumption can be reduced.

Mortada et al. [17] developed a resource-efficient method to minimize the redundant data at intermediate node level, and after that processed data is forwarded to cluster head (CH) at the end of periodic interval. Finally, CH aggregate the received data using Euclidean distance algorithm. Simulation results show its performance in terms of efficient resource consumption. It can be further extended to adapt variable sampling rate over different periodic intermissions.

Merzoug et al. [18] developed a localized data aggregation method that builds the one hop routes for aggregation to avoid the link failure/topology dynamics. Simulation results indicate that it offers scalable/collision free transmission and uses optimal resources as compared to traditional aggregation approaches (depth-first search/peeling algorithm/greedy–boundary traversal). It can be further optimized by reducing extra control overhead.

Chen et al. [19] merged the tree-based approach with scheduling method to achieve higher aggregation rate using optimal resources. It uses the intermediate nodes on the basis of their residual energy and delay factor to build a schedule for aggregation. Simulation results show that it outperforms in terms of extended battery life/minimum delay as compared to existing scheme (energy-collision aware data aggregation scheduling).

Mosavvar et al. [20] introduced a scheduling-based aggregation technique for WSN. It organizes the sensors in cluster groups, and in each set, sensors are marked active/inactive on the basis of residual energy, area and distance, etc. Aggregation is scheduled only for activated sensors thus reduce the overall energy consumption. Simulation results show its performance in terms of optimal resource consumption as compared to existing schemes (shuffled frog/hierarchical clustering). It can be further enhanced using machine learning algorithms.

Yuvaraj et al. [21] introduced a location-based scheduling method for data aggregation over WSN. Reliable routes are selected on the basis of different factors, i.e., residual energy/hop, etc. Simulation results show its outcomes in terms of extended lifespan/energy efficiency/optimal delay/minimal packet drop as compared to multiple sink positioning and relocation scheme. It can be further extended by embedding a security provision.

Sudha et al. [22] developed a load balancing scheme for efficient data aggregation by selecting intermediate nodes randomly on the basis of their residual energy over the current interval. It also performs inter-cluster switching for load balancing. Simulation results indicate its performance in terms of optimal delay/energy consumption. It can be further implemented for hierarchical cluster networks.

Sarode et al. [23] introduced a query-based efficient data aggregation scheme for WSN. It executes an optimized group search over collected data for aggregation and thus reduces the overall processing time and resources. Simulation results show that its performance in terms of the highest throughput/minimal delay as compared to traditional schemes (swarm optimization and genetic algorithm).

Padmaja et al. [24] developed a scheme for secure data aggregation over WSN. First of all, trusted aggregation is initiated at node level, and data is forwarded to base station through cluster head. Base station verifies the authenticity of the cluster head along with the trust value of the source. Simulation results show its performance in terms of the extended lifespan of network, efficient aggregation with optimal resource consumption/delay as compared to existing scheme (secure data aggregation).

Lee et al. [25] presented an energy-efficient MAC-based aggregation scheme that can estimate the wakeup interval of the sensors for aggregation purposes. Simulation results show its outcomes in terms of less control overhead/delay/higher throughput and optimal resource consumption.

Le et al. [26] proposed an aggregation method that uses interlinked set to form a tree. As per the transmission interval, schedule is created for aggregation. Study displays that it can employ the transmission interval efficiently, and it is capable to reduce aggregation delay by avoiding collision/extra control overhead and thus improves the overall network performance. It can be further extended by using subtree and interference model.

Jothiprakasam et al. [27] introduced a MAC-based aggregation scheme for multi-hop WSN. It estimates the number of required slots to forward the data to sink and then aggregation is executed. Analysis shows that sensors can efficiently utilize the allocated slots and thus results in extended network life span of network as compared to cluster/chain-based schemes. It can be further employed for distributed WSN also.

John et al. [28] investigated various data aggregation schemes, i.e., structureless, cluster, and tree-based, etc., for WSN. Study found the merits and demerits of all these schemes. In case of structureless aggregation, maintenance cost is minimal but it does not support efficient aggregation/routing, in case of cluster-based scheme; it reduces the cost of resource consumption but it is challenging to overcome from the node failure, and in case of tree-based scheme, it consumes optimal resources for transmission and topology maintenance is required to cope with node failure. Study claims that tree-based aggregation is more efficient as compared to others.

Idrees et al. [29] developed a distributed aggregation approach using K-means algorithm for WSN. Each sensor collects the data periodically, and finally, after eliminating the redundancy using K-means, only the selected cluster forwards the data to sink. Simulation results show its performance in terms of higher accuracy of data aggregation, improved network lifespan and optimal resource consumption. It can be further optimized by correlating the sampling and aggregation together.

Jhuang et al. [30] proposed a collision-free efficient aggregation scheme using overlapping algorithm. It reduces the overall cycle required for transmission, and later on these cycles are overlapped for aggregation purpose. Simulation results show its performance in terms of minimal usage of time slots and higher accuracy of aggregation.

Harb et al. [31] introduced a frequency-based filtering approach that defines prefix/suffix for a given set to minimize the redundancy over aggregation. Simulation results show that it can optimize the delay factor for aggregation as well energy consumption as compared to the existing scheme (prefix frequency filtering). Its efficiency can be further extended by embedding the dynamic sampling rate.

Fissaoui et al. [32] presented a mobile agent-based data aggregation method that defines the minimum spanning tree for data collection and only the cluster head(s) over that tree is selected by agents for aggregation purpose. Simulation results show that it outperforms in terms of resource consumption and delay as compared to traditional method (global closest first method).

3 Optimized Multi-level Data Aggregation Scheme (OMDA) for WSN

Wireless Sensor network WSNi
Coverage Area Ca
Sensor Node Sn
Cluster Head CH
Last Sensed LSnd
Last SentdataLStd
Current Data Cd
Data Type Dt
Source Sr
Destination Dt
Link Distance Ld
Duplicate Data Threshold dTH
//assumption
ifCHi has m member & each sends n bytes then total data size == m*n for CHi

Phase-I: Redundancy Check at Sensornode

At sensor level, it is enforced that sensor should not be able to forward the duplicate data, so last sent data is matched with the current data (to be sent) as given below:
if (Sr->LStd == Cd) //Last forwardeddata is similar to the current data to be sent.
Discard(Cd) //ignore duplicate data at initial stage
Else if (Sr->LStd!=Cd)
addtolist(Lst, Cd) //otherwise add to list
end if
If data is already in the list but it is not forwarded to CH till time, then check it for redundancy in the given list. Data redundancy is verified using multiple parameters, i.e., source of data, its type/size, and destination, etc. If it is found in the current list, then it is discarded otherwise it is added to the list to be forwarded to CH.
Procaddtolist(Lst, Cd)

If (Find ((Sr.Cd->Type, Sr.Cd->data_size, Sr->id, Dt->id, Sr->dst), Lst))==FALSE)
Update Msg_list (Lst, Cd)
Else
Discard(Cd) //discard duplicates
end if
send (sr->Msg, CH)
Update (Sr->LStd, Cd) // update last sent data

Phase-II: Redundancy Check at ClusterHead

As CH may receive same data from multiple sources so need to prepare a master message list to be forwarded to base station.
For each Sri->val in Lsti
If (Find_Duplicates (Sri->val, CH->Lst)==TRUE)
Discard(Sr->Val)
Sri->dTH++
Else
Update (Mmsgl)
End for
Forward (Mmsgl, Bs, True)
Set sleep mode forcefully, as sensor node is producing redundant data, in order to reduce resource consumption.
If (Sri->dTH>1, Sri)
set sleep (Sri, interval)
End if

Phase-III: Redundancy Check at Base Station

Base station may collect duplicate data values from multiple CHs, so there is a need to filter final data for analysis purpose as given below:
For each CHi->Mmsgl in Lsti
If (Find_Duplicates (CHi->Mmsgl, Bs->Lst)==TRUE)
Discard(CHi->Mmsgl)
Else
Accept (Mmsgl)
End for.

4 Simulation Scenario

As per Table 1, NS-2.34 was used for analysis purpose with different parameters. Routing protocol is LAECH, and its performance was analyzed under various simulation scenarios, i.e., Terrain size is 1000 × 1000, MAC Protocol is Mac/Sensor, Node Density is 50/100/ 200, Propagation Model is TwoRay Ground, Data Type is CBR, Sampling Interval is 1.0 ms, Simulation Time is 600 s, Initial Energy 10.0j, rxPower/txPower is 1, IFQ 100, Antenna Type is Omni.

Table 1 Simulation configuration

Simulation parameters	Parameter values
Routing protocol	LEACH
Terrain	1000 × 1000
MAC protocol	Mac/Sensor
Node density	50/100/200
Propagation model	TwoRay ground
Data type	CBR
Sampling interval	1.0 ms
Simulation time	600 s
Network simulator	NS-2.34
Initial energy	10.0j
rxPower	1
txPower	1
IFQ	100
Antenna type	Omni
Simulation scenario (s)	a. Traditional data aggregation scheme (TD-DA) b. Optimized multi-level data aggregation scheme (OMDA)

5 Simulation Results and Performance Analysis

Performance of the OMDA is analyzed under the constraints of different parameters (Throughput/Delay/Energy consumption/Alive sensors and their lifespan over the simulation interval) using LEACH protocol with the sensor density that varies from 50–200 sensors.

5.1 Throughput

Figure 1 shows the throughput of LEACH protocol using different data aggregation schemes under the constraints of sensor node density that varies from 50 to 200.

In case of traditional data aggregation scheme (TD-DA), it is 869.37 Kbps using 50 sensors and it is slightly degraded (865.49 Kbps) with 100 sensors, and finally, it reaches to its lowest level (798.43) with 200 sensors.

In case of optimized multi-level data aggregation scheme (OMDA), it is 946.8 Kbps using 50 sensors and it is slightly degraded (913.29 Kbps) with 100 sensors, and finally, it reaches to its lowest level (805.5 Kbps) with 200 sensors.

It can be analyzed that sensor node density affects the throughput of LEACH using both schemes. However, OMDA delivers the highest throughput as compared to TD-DA under the constraints of sensor node density variations.

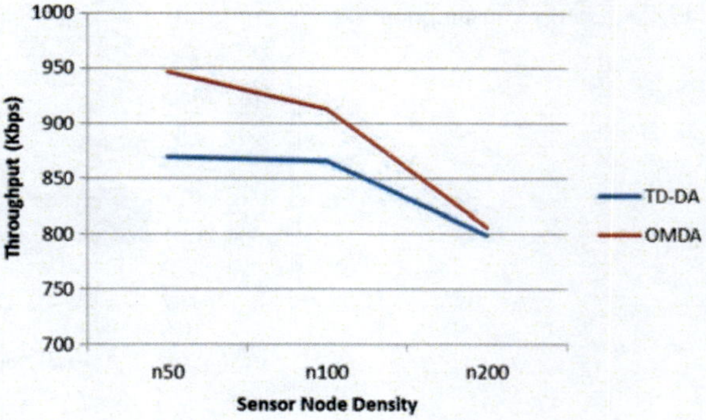

Fig. 1 Throughput

5.2 *End-To-End Delay*

Figure 2 shows the end-to-end delay of LEACH protocol using different data aggregation schemes under the constraints of sensor node density that varies from 50 to 200.

In case of traditional data aggregation scheme (TD-DA), it is 823.018 ms using 50 sensors and it is slightly increased (1695.49 ms) with 100 sensors, and finally, it reaches to its peak level (3245.92 ms) with 200 sensors.

In case of optimized multi-level data aggregation scheme (OMDA), it is 819.943 ms using 50 sensors and it is slightly increased (1617.23 ms) with 100 sensors, and finally, it reaches to its peak level (3034.58 ms) with 200 sensors.

It can be analyzed that sensor node density affects the end-to-end delay of LEACH using both schemes. However, OMDA offers less delay as compared to TD-DA under the constraints of sensor node density variations.

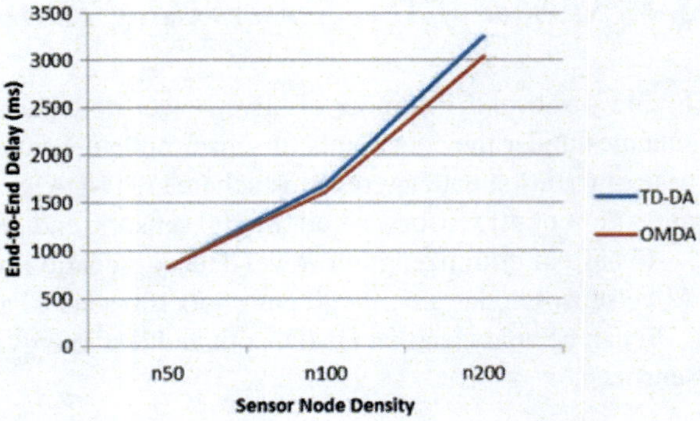

Fig. 2 End-to-end delay

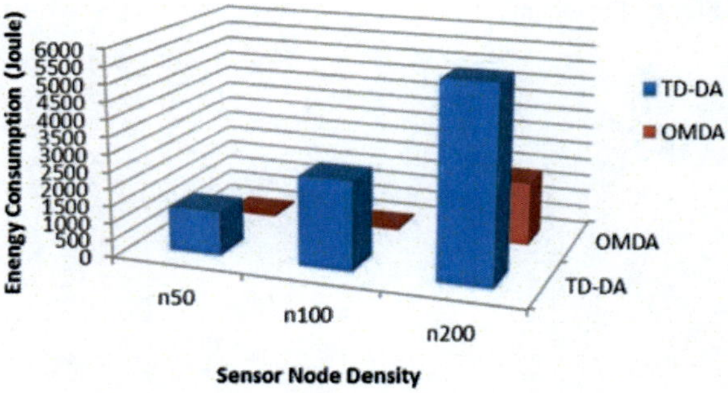

Fig. 3 Energy consumption

5.3 *Energy Consumption*

Figure 3 shows the energy consumption of LEACH protocol using different data aggregation schemes under the constraints of sensor node density that varies from 50 to 200.

In case of traditional data aggregation scheme (TD-DA), it is 1278.851j using 50 sensors and it is slightly increased (2594.561j) with 100 sensors, and finally, it reaches to its peak level (5615.457j) with 200 sensors.

In case of optimized multi-level data aggregation scheme (OMDA), it is 65.40449j using 50 sensors and it is slightly decreased (61.06243j) with 100 sensors, and finally, it reaches to its peak level (1868.837j) with 200 sensors.

It can be analyzed that sensor node density affects the energy consumption of LEACH using both schemes. However, OMDA consumes less energy as compared to TD-DA under the constraints of sensor node density variations.

5.4 *Number of Alive Sensors Using TD-DA and OMDA*

Figure 4 shows the number of alive sensor nodes using different data aggregation schemes under the constraints of sensor node density that varies from 50 to 200. In case of tradition data aggregation scheme (TD-DA), a number of alive sensor nodes are 34 out of 50 sensors, 80 out of 100 sensors, and 181 out of 200 sensors.

In case of optimized multi-level data aggregation scheme (OMDA), a number of Alive sensor nodes are 50/100 out of 50/100 sensors and 188 out of 200 sensors.

It can be analyzed that OMDA utilized less energy thus enhanced the lifespan of sensors.

5.4.1 Lifespan of Sensors Over a Time Interval with Sensor Density 50

Figure 5 shows the lifespan of sensor nodes over time interval with sensor node density 50 using different data aggregation schemes. In case of TD-DA, lifespan

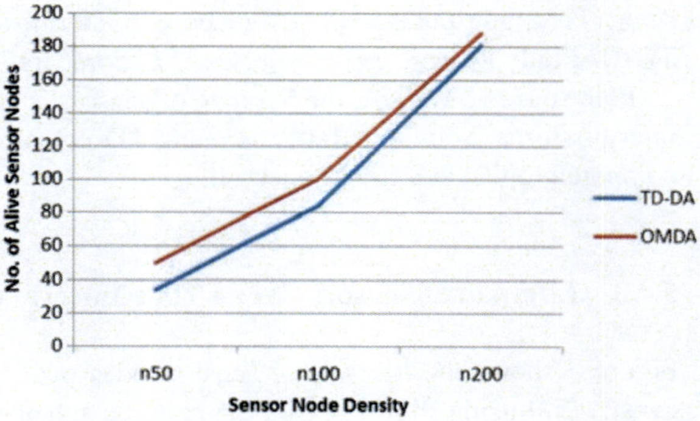

Fig. 4 Number of alive sensor nodes

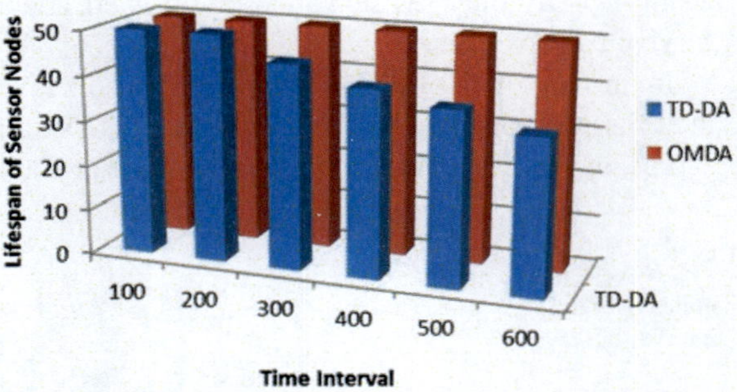

Fig. 5 Lifespan of sensor nodes over time interval-n50

Table 2 Lifespan of sensor nodes over time interval-n50

Time interval (sec.)	Data aggregation schemes	
	TD-DA	OMDA
100	50	50
200	50	50
300	45	50
400	41	50
500	38	50
600	34	50

of sensors is exhausted gradually over a given interval as shown in Table 2. Using TD-DA, only 34 sensors were survived as compared to OMDA at 600 interval.

In the case of TD-DA, the lifespan of sensors is exhausted gradually over a given interval as shown in Table 2.

5.4.2 Lifespan of Sensors Over a Time Interval with Sensor Density 100

Figure 6 shows the lifespan of sensor nodes over time interval with sensor node density 100 using different data aggregation schemes. In case of TD-DA, lifespan of sensors is exhausted steadily over a certain intermission as shown in Table 3. Using TD-DA, only 84 sensors were survived as compared to OMDA at 600 interval.

In the case of TD-DA, the lifespan of sensors is exhausted steadily over a certain intermission as shown in Table 3. Using TD-DA, only 84 sensors were survived as compared to OMDA at 600 interval.

5.4.3 Lifespan of Sensors Over a Time Interval with Sensor Density 200

Figure 7 shows the lifespan of sensor nodes over time interval with sensor node density 200 using different data aggregation schemes. In case of TD-DA, lifespan of sensors is exhausted steadily over a certain intermission as shown in Table 4. Using TD-DA, only 181 sensors were survived, and using OMDA, 188 sensors were survived at 600 interval.

In the case of TD-DA, the lifespan of sensors is exhausted steadily over a certain intermission as shown in Table 4. Using TD-DA, only 181 sensors were survived and using OMDA 188 sensors were survived at 600 interval.

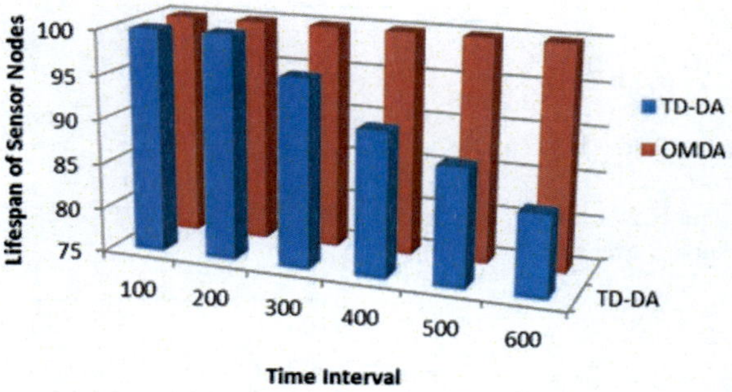

Fig. 6 Lifespan of sensor nodes over time interval-n100

Table 3 Lifespan of sensor nodes over time interval-n100

Time interval	Data aggregation schemes	
	TD-DA	OMDA
100	100	100
200	100	100
300	96	100
400	91	100
500	88	100
600	84	100

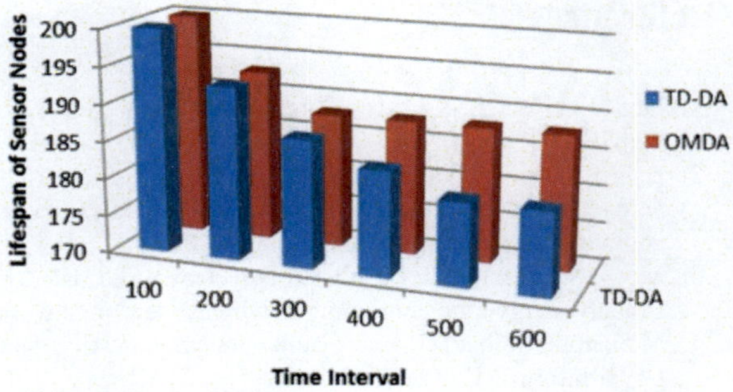

Fig. 7 Lifespan of sensor nodes over time interval-n200

Table 4 Lifespan of sensor nodes over time interval-n200

Time interval	Data aggregation schemes	
	TD-DA	OMDA
100	200	200
200	193	193
300	187	188
400	184	188
500	181	188
600	181	188

6 Conclusion

In this paper, issues and solutions related data aggregation over WSN were discussed and an optimized data aggregation scheme was introduced to enhance the overall performance and lifespan of the network using LEACH protocol. Simulation analysis was performed using different parameters, i.e., throughput/delay/energy consumption/number of alive sensors/sensor's lifespan over interval, etc., under the constraints of sensor node density that varies from 50 to 200.

As per the simulation outcomes, it can be observed that TD-DA could not perform well as compared to OMDA.

TD-DA has the lowest throughput with the highest energy consumption/end-to-end delay, and it has the minimum alive nodes with respect to sensor node density.

OMDA offers the highest throughput with optimal energy consumption and thus extends the overall lifespan of sensors. However, there is need to reduce the end-to-end Delay that increases with respect to sensor node density.

Finally, it can be concluded that the optimization of data aggregation can enhance the performance of routing protocol. Currently, it is implemented only for LEACH protocol, and it can be further extended for other routing protocols.

References

1. Dhand G, Tyagi SS (2016) Data aggregation techniques in WSN: survey. Proc Comput Sci 92:378–384
2. Bala Krishna M, Vashishta N (2013) Energy efficient data aggregation techniques in wireless sensor networks. In: 5th international conference and computational intelligence and communication networks. IEEE, pp 160–165
3. Movva P, TrinathaRao P (2019) Novel two-fold data aggregation and MAC scheduling to support energy efficient routing in wireless sensor network. IEEE Access 7 (In Press)
4. Mohanty P, Kabat MR Energy efficient structure-free data aggregation and delivery in WSN. Egypt Inform J 17(3), 273–284
5. Sarode P, Nandhini R (2018) Intelligent query-based data aggregation model and optimized query ordering for efficient wireless sensor network. Wirel Press 1405–1425
6. Zhang J, Lin Z, Tsai P, Xu L (2020) Entropy-driven data aggregation method for energy-efficient wireless sensor networks. Inform Fusion 56:103–113 (In Press)
7. Wen XD, Liu CW (2019) Decentralized distributed compressed sensing algorithm for wireless sensor networks. Proc Comput Sci 154:406–415
8. Sarangi K, Bhattacharya I (2019) A study on data aggregation techniques in wireless sensor networkin static and dynamic scenarios. Innov Syst Softw Eng 15:3–16
9. Tolani M, Singh RK (2019) Lifetime improvement of wireless sensor network by information sensitive aggregation method for railway condition monitoring. Ad Hoc Netw 87:128–145
10. Tamiji M, Nasri S (2019) Data aggregation for increase performance of wireless sensor networks using learning automata approach. Wirel Personal Commun 108(1), 187–201
11. Akila V, Sheela T (2019) Secure data aggregation to preserve data and key privacy in wireless sensor networks with multiple sinks. In: 3rd international conference on computing and communications technologies (ICCCT). IEEE, pp 86–93
12. Egidius PM, Abu-Mahfouz AM, Hancke GP (2019) A comparison of data aggregation techniques in software-defined wireless sensor network. In: 28th international symposium on industrial electronics (ISIE). IEEE, pp 1551–1555
13. Goyal N, Dave M, Verma AK (2019) Data aggregation in underwater wireless sensor network: Recent approaches and issues. J King Saud Univ Comput Inf Sci 31(3):275–286
14. Hadi K (2019) Analysis of exploiting geographic routing for data aggregation in wireless sensor networks. Proc Comput Sci 151:439–446
15. Galkin P, Klyuchnyk I (2019) An optimal concepts for aggregation of data in wireless sensor network. In: 2nd Ukraine conference on electrical and computer engineering (UKRCON). IEEE, pp 1162–1166
16. Nguyen NT, Liu BH, Chu SI, Weng HZ (2019) Challenges, designs, and performances of a distributed algorithm for minimum-latency of data-aggregation in multi-channel WSNs. IEEE Trans Netw Service Manag 16(1):1–14
17. Mortada M, Makhoul A, Jaoude CA, Harb H, Laiymani D (2019) A distributed processing technique for sensor data applied to underwater sensor networks. In: 15th international wireless communications & mobile computing conference (IWCMC). IEEE, pp 979–984
18. Merzoug MA, Boukerche A, Mostefaoui A, Chouali S (2019) Spreading aggregation: a distributed collision-free approach for data aggregation in large-scale wireless sensor networks. J Parallel Distrib Comput 125:121–134
19. Chen K, Gao H, Cai Z, Chen Q, Li J (2019) distributed energy-adaptive aggregation scheduling with coverage guarantee for battery-free wireless sensor networks. In: INFOCOM-IEEE conference on computer communications. IEEE, pp 1018–1026
20. Mosavvar I, Ghaffari A (2019) Data aggregation in wireless sensor networks using firefly algorithm. Wirel Personal Commun 104(1):307–324
21. Yuvaraj R, Chandrasekar A, Jothi S (2018) Time orient LEAD based polling point selection algorithm for efficient data aggregation in wireless sensor networks. Cluster Comput 22:3339–3346

22. Sudha L, Thangaraj P (2018) Improving energy utilization using multi hop data aggregation with node switching in wireless sensor network. Cluster Comput 22:12749–12757
23. Sarode P, Nandhini R (2018) Intelligent query-based data aggregation model and optimized query ordering for efficient wireless sensor network. Wirel Personal Commun 100:1405–1425
24. Padmaja P, Marutheswar GV (2018) Energy efficient data aggregation in wireless sensor networks. Mater Today Proc 5(1):388–396
25. Lee H, Park I (2018) Performance of a receiver-initiated MAC protocol with aggregation for event-driven wireless sensor networks. In: International conference on information networking (ICOIN). IEEE, pp 853–856
26. Le DT, Lee T, Choo H (2018) Delay-aware tree construction and scheduling for data aggregation in duty-cycled wireless sensor networks. EURASIP J Wirel Commun Network 1–15
27. Jothiprakasam S, Muthial C (2018) A method to enhance lifetime in data aggregation for multi-hop wireless sensor networks. AEU-Int J Electron Commun 85:183–191
28. John N, Jyotsna A (2018) A survey on energy efficient tree based data aggregation techniques in wireless sensor networks. In: International conference on inventive research in computing applications. IEEE, pp 461–465
29. Idrees AK, Yaseen WLA, Taam MA, Zahwe O (2018) Distributed data aggregation based modified K-means technique for energy conservation in periodic wireless sensor networks. In: Middle East and North Africa communications conference (MENACOMM). IEEE, pp 1–6
30. Jhuang YY, Lin T, Nguyen TD, Chu S, Liu BH, Pham VT (2017) A new overlap circle technique for reducing data aggregation time in wireless sensor networks. In: international conference on system science and engineering (ICSSE). IEEE, pp 546–549
31. Harb H, Makhoul A, Tawbi S, Zahwe O (2017) Energy efficient filtering techniques for data aggregation in sensor networks. In: 13th international wireless communications and mobile computing conference (IWCMC). IEEE, pp 693–698
32. Fissaoui ME, Hssane AB, Saadi M (2017) Mobile agent protocol based energy aware data aggregation for wireless sensor networks. Proc Comput Sci 113:25–32

Potentiality of Nanotechnology in Development of Biosensors

Deepika Jain, Bikram Pal Kaur, and Ruchi Pasricha

Abstract Biosensors have been the most alluring research area since long times back due to their highly correlation with human beings, health issues and environment. Constraints in the way of fabrication and design of biosensors for commercial applications require constant attention. Certain materials can exhibit different properties based on its shape and size and have been realized with the advent of an interdisciplinary and integrated present-day science profoundly called as "Nanotechnology." Intervention of nanotechnology in this biosensor field provides some exceptional electronic, optical and biological properties of these nanomaterials that find their use in variety of applications like glucose monitoring, estimation of harmful diseases, hazardous chemicals detection and drug discovery. Nanomaterials have empowered lower working potentials and facilitate electron transfer and lower detection limits, thereby improving the sensing capacity of the device. This paper mainly highlights the concept of biological sensors, their applications and different ranges of nanomaterials used in conjunction with the biosensor field to get a faster and reliable detection of electrical signals. Reducing dimensions in nano range opens up various opportunities in the biosensor field for significant improvement in their characteristic parameters.

Keywords Analyte · Biosensor · Nanobiosensor · Nanotechnology · Nanomaterials

D. Jain (✉)
ECE, IKGPTU Jalandhar, Punjab, India
e-mail: jain.deepika2010@gmail.com

B. P. Kaur
IT Department, Chandigarh Engineering College, Mohali, Punjab, India

R. Pasricha
Chandigarh Group of Colleges, Mohali, Punjab, India

N. Marriwala et al. (eds.), *Mobile Radio Communications and 5G Networks*, Lecture Notes in Networks and Systems 140,
https://doi.org/10.1007/978-981-15-7130-5_36

1 Introduction

Every human being has inbuilt sensors that assist in recognizing and perceive the environment. These sensors work by responding to a noticeable activity, trailed by response to that activity by developing a signal that can be estimated. Being highly selective and sensitive toward numerous analytes, biosensors are among the intriguing areas of research in real-time applications when compared with conventional standard chromatographic techniques to determine the concentration of different pollutants, hazardous chemicals and various other biological parameters. Interlacing of nanotechnology in the field of biosensors made the recognition of target analytes easier, proficient and cost effective, thereby stimulating the development of more sophisticated sensors. Diverse range of nano-based biosensors has already been developed, and some are under development [1]. Biosensors have immense number of applications in different areas such as agriculture, defense, food additives, industrial, environmental and pharmaceutical because of phenomenal properties of nanomaterials utilized in this field. Nanotechnology plays an imperative role in growth of biosensors since they have massive capacities to provide high mechanical strength, stability, reduction in interferences, lower detection potential and increase surface area during interaction with bioreceptors on the transducer surfaces. Biosensors utilizing nanomaterials can provide detection limits up to femto level with reduced response time [2].

Biosensor research plays a vital role in the growth of modern electronics. Biosensor, a device for detecting target analytes, comprises bioreceptor element that perceives the analyte to produce a signal that can be measured like antibody/antigen interactions, nucleic acid and protein–protein interactions, etc., while transducer converts a physicochemical change into electrical signal based on analyte–bioreceptor interactions. These devices are associated with the user-friendly electronic circuits like signal processors for the display of the results [3]. Immobilization of biological elements to the surface of the sensor is an inherent characteristic of biosensor that decides the different parameters of the biosensor [4]. High selectivity toward the target analyte is a key necessity of the bioreceptor. Biosensors can detect variety of analytes based on choice of recognition elements at the biological input, immobilization method and transduction working principle as shown in Fig. 1.

Electrochemical biosensors emerge as the most commonly used biosensors due to ease of construction and inexpensive. They can be potentiometric, amperometric or

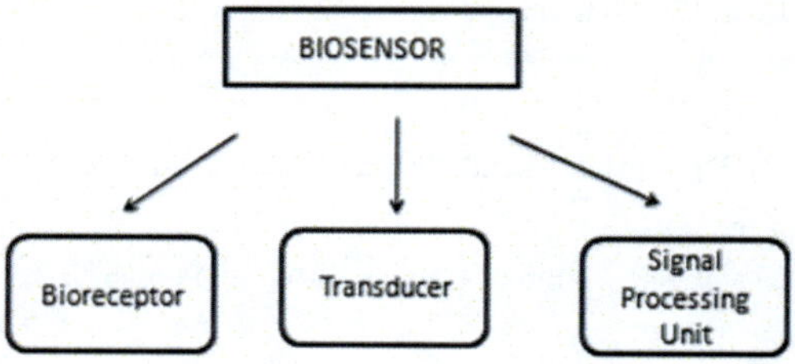

Fig. 1 Components of biosensor

conductometric based on measured parameter such as voltage, current and conductance at the working electrodes. Fabrication and design of electrodes are the ultimate steps in the growth of electrochemical sensors [5]. The fundamental prerequisites in biosensor development are the bioreceptors availability related to specific target analytes, proper immobilization of bioreceptors on transducer surfaces followed by disposable and portable detection devices when compared with traditional laboratory-based chromatographic techniques. Most common applications of biosensors involve human health care, e.g., glucose monitoring in diabetes patients, biomarker detection for easy diagnosis and disease prevention, environmental applications like heavy metal contamination detection, pesticide recognition, drug discovery, crime, defense, etc. [6, 7].

2 Intertwining Nanotechnology in Biosensors

Nanotechnology rose as an attractive field for detection of pollutants due to their unprecedented electronic, chemical and physical properties. Working potential required for generation of current in biosensors can be reduced by utilization of numerous redox mediators like Prussian blue, cophthalocyanin, ferricyanide, etc. Other than the diminishing of the working potential, utilization also improves reproducibility and sensitivity of thiocholine detection. Potential can be lowered by an alternative way through utilization of nanomaterials that enhance the surface region, enhancing the rate of movement of electrons. Use of nanomaterials improves sensitivity and enhances conductivity, repeatability and stability.

Recently, there is a developing enthusiasm for nanomaterial-based biosensors. Nano-biosensors and its applications play a vital role in roads of biosensor advancement, which has been conceivable due to the marvels of nanotechnology as shown in Fig. 2. Integration of nanomaterials with various electrical systems gives rise to nanoelectromechanical systems, with highly specific transduction mechanisms and biological signaling [8].

Nanotechnology and nanomaterials are interlaced in the growth of most of the biosensor devices. Nanomaterials involving carbon nanotubes, metal nanoparticles

Fig. 2 Applications of nano-biosensors

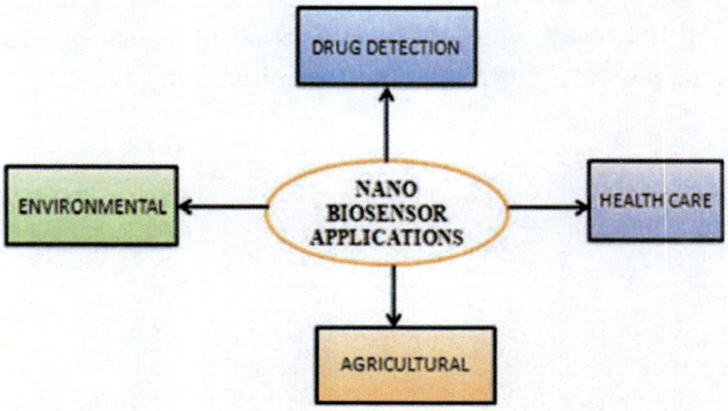

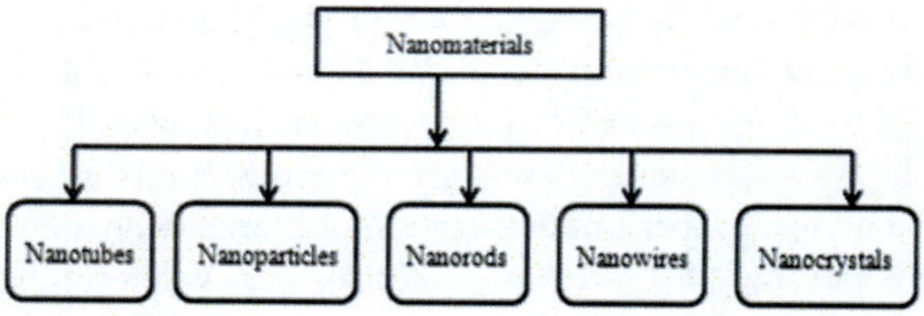

Fig. 3 Different types of nanomaterials

(MNPs), CdTe quantum dots (QDs), SiO_2 nanosheets, etc., have been widely used in variety of applications [9]. The utilization of nanocomposites in the biosensor field allows for target analyte detection at extremely lower detection limits but at the expense of overpotential. In contrast, nanotubes allow catalysis at lower working potentials but offer poor enzyme loading. In spite of advancements in biosensor technology, selectivity still appears as the challenge for these biosensors. Different varieties of nanomaterials are shown in Fig. 3.

3 Different Components of Nanotechnology

3.1 Nanotubes

Carbon nanotubes (CNTs) are tubes made of carbon with diameters generally measured in range of nanometers. Their properties are useful in different technological areas such as electronics, optical, mechanical and material sciences. Carbon nanotubes might be single-walled (SWCNTs) or multi-walled (MWNTs) [10]. Characteristic properties of carbon nanotubes are smaller diameters, sharpness, unique composition, geometry, better electrical conductivity even at moderate voltages, exceptional tensile strength, good thermal conductivity, adequate electron emission, lower enzyme loading and higher aspect ratios, etc.

Sun et al. [11] developed a biosensor for detection of dichlorvos organophosphate compound with aniline and multi-wall carbon nanotubes (MWNTs)-modified electrode surfaces. The lower working potential in this electrochemical biosensor can be achieved with the assistance of carbon nanotubes. Cyclic voltammetry was used to study the electrochemical behavior of carbon nanotube-modified electrode surface, and the results verified that the modified surfaces decrease the working potential, thus increasing the sensitivity of biosensor [11].

3.2 Nanoparticles

These particles are having size running somewhere in the range between 1 and 100 nm (nm) encompassed by interfacial layer that contain ions, organic and inorganic molecules. Most of the properties of these particles are contributed by the presence of this layer. As indicated by diameter size and morphology, these particles can be metal-based, ceramic, carbon-based, polymeric and semiconductor ones [12]. Examples of semiconductor nanoparticles involve Si, Ge, **GaP, InP, ZnO, CdS, CdTe**, etc. These particles due to their biocompatibility and outstanding electronic, optical properties enable fast catalysis of electrochemical reactions, lower working potential and increase surface area by facilitating electron transfer on transducer surfaces. Nanoparticle-based biosensors can provide detection limits up to 10^{-7}–10^{-13}. These nanoparticles due to their simplicity, shape variability and large surface area, when attached to the QCM surface, can be used to estimate the malignancy of cancer cells. According to research studies, it is found that exactly 58 nm-sized PHEMA nanoparticles are required on QCM surface for estimating in vitro breast cancer.

Upadhyay et al. [13] developed a pesticide recognition bienzymatic biosensor having platinum- and gold-modified electrode surfaces. Conventionally, platinum (Pt) electrodes have been used for the electrochemical oxidation of H_2O_2. Nowadays, platinum nanoparticles are used in conjunction with gold- or tin-based nanoparticles to resolve issues associated with high working potential. The combined capacity of gold and platinum nanoparticles facilitated the recognition of hydrogen peroxide at 0.4 V, thus improving electroactive surface area, potential sensitivity and dynamic range [13]. Yang et al. [14] presented another acetylcholinesterase (AChE) biosensor utilizing nickel oxide nanoparticles combined with carboxlic graphene and nafion for detection of carbofuran, methyl parathion and chlorpyrifos. These nanocomposites require lower potential for enzyme catalysis, thereby improving the sensitivity, stability and reproducibility of biosensor [14].

3.3 Nanocrystals

These are the tiny crystalline particles that exhibit size-dependent optical and electronic properties. These can be made by top–bottom or bottom-up technology. Exceptional fluorescence, unique electronic and optical properties and quantum confinement of charge carriers are the inherent features of these particles. They can emit light of different colors when ultraviolet rays strike their surfaces.

Sahub et al. [15] developed a biosensor that combines graphene quantum dots and enzyme for biomonitoring of organophosphate exposure in environment, water and food commodities. The working principle of the proposed biosensor involves the catalytic hydrolysis of acetylcholine by ACHE followed by oxidation of choline to produce hydrogen peroxide.H_2O_2 generated reacts with graphene quantum dots to produce photoluminescence changes that correspond to the amount of pesticide present in the sample. Oxoform of orangophosphates exhibits much stronger inhibitory effect than thioform of organophosphates. To validate the biosensor, results obtained were compared with traditional analytical method UHPLC-MS. The proposed biosensor then used for estimating interference due to mercury ions and found that their presence profoundly influences the fluorescence quenching of graphene quantum dots [15]. Wang et al.[16] presented another novel immunoassay using amorphous magnetic particles for the recognition of phosphorylated ACHEadducts. A pair of antibodies, anti-phosphoserinepolyclonal antibodies (Ab1) and anti-human AChEmonoclonal antibodies (Ab2) coupled with quantum dots, was used for capturing of adducts from biological samples followed by recognition. This highly sensitive novel immunoassay was then tested for detection of OP-ACHE adducts in plasma samples spiked with paraxon and could yield a linear response of 0.3–300 ng/mL [16].

3.4 Nanowires

A nanowire is a nanostructure having diameter as small as 10^{-9} m. Nanowires may exist as superconducting, metallic semiconducting, insulating, etc. These nanowires are exceptionally versatile, better charge conduction capacity and good electrical communication capabilities.

Song et al. [17] proposed an ultrasensitive biosensor utilizing palladium–copper nanowires coupled with chitosan (CS) and enzyme acetylcholinesterase for organophosphate recognition in variety of fruits and vegetables. The use of Pd-Cu NWs in the proposed biosensor exhibits unprecedented performance such as large active surface area for enzyme loading, excellent reproducibility, electrocatalytic activity, electron transport capability and anti-interference ability. The proposed biosensor showed remarkable performance for detection of malathion with lower limits of detection be 4.5 pM [17].

3.5 Nanorods

Nanorods are produced by direct chemical synthesis. Nanorods assimilate in near IR, and produce heat when energized with IR light. This natural property has made their usage in malignant growth therapy. When a person is infected with cancer-like disease, generally infrared exposure is given; because in IR light, nanorods acquire energy, thereby devastating the malignant cells, while healthy cells remain flawless. Their inherent properties make it suitable for various emerging applications like miniaturization of sensors, biomedicine, detectors, fabrication of solar cells, etc (Table 1).

Table 1 Summary of nanomaterial-based biosensors for different applications

Analyte	Bioreceptor	Principle	Nanomaterial	Parameter LOD	References
Carbofuran	ACHE Enzyme	Amperometry	Gold nanoparticles (AuNPS)	Nm	[19]
Triazophos	ACHE enzyme	Amperometry	Carbon nanotubes (CNTs)	0.01 μM	[20]
Blood glucose	Glucose oxidase enzyme	Amperometry	Multi-walled carbon nanotubes (MWNTs)	High	[21]
Antibiotic penicillin G	Penicillinase enzyme	Amperometry	Gold nanoparticles (AuNPS)	4.5 nM	[22]
Carcino-embryonic antigen (CEA)	Adamantine-modified antibody	Antigen antibody	Cu @Ag NPs	20 fg/ml	[23]

Lang et al. [18] proposed highly sensitive ACHE biosensor utilizing gold nanorods for recognition of pesticide organophosphate paraxon and dimethoate. The use of gold nanorods-modified electrodes was found to be an excellent platform for achieving a lower oxidation potential of thiocholine at an anodic peak at 0.55 and generate larger amperometric current due to their excellent electrocatalytic ability when compared with heterogeneous nanorods. The proposed biosensor exhibits good operational stability and reproducibility at gold nanorods-modified electrodes [18].

4 Conclusion

In the present revolutionary era, nanotechnology made a significant contribution in the biosensor field to achieve the targets of excellent specificity, stability and selectivity. The transduction mechanisms to convert a physicochemical change into electrical signal have been greatly improved with the utilization of various nanomaterials. Their addition to the biosensor field makes the overall system cheap, user friendly, smarter, proficient and faster for the recognition of variety of target analytes. Day-by-day advancements in the miniaturization of electronic circuits and features of nanomaterials have increased several gateways toward the commercialization of biosensors on large scale. Despite of exceptional improvement of biosensors with the addition of nanotechnology, yet there are some technical complexities that are still a pelat in the road of growth.

References

1. Zhang A, Lieber CM (2016) Nano-bioelectronics. Chem Rev 116(1):215–257
2. Lahir YK, Samant M, Dongre PM (2016) Role of Nanomaterials in the development of biosensors. Global J Biosci Biotechnol 5(2):146–163
3. Murugaiyan SB, Ramasamy R, Gopal N, Kuzhandaivelu V (2014) Biosensors in clinical chemistry: an overview. Adv Biomed Res 3(67):1–9
4. Sassolas A, Blum LJ, Leca-Bouveir BD (2012) Immobilization strategies to develop enzymatic biosensors. Biotechnol Adv 3(3):489–511
5. Yang H (2012) Enzyme-based ultrasensitive electrochemical biosensors. Curr Opin Chem Biol 16(3–4):422–428
6. Gruhl FJ, Rapp B, Lange K (2013) Biosensors for diagnostic applications. Adv Biochem Eng/Biotechnol 2013:115–148
7. Bahadir EB, Sezginturk MK (2015) Applications of commercial biosensors in clinical, food, environmental, and biothreat/biowarfare analyses. Anal Biochem 478:107–120
8. Jianrong C, Yuqing M, Nongyue H, Xiaohua W, Sijiao L (2004) Nanotechnology and biosensors. Biotechnol Adv 22(7):505–518
9. Kumar S, Ahlawat W, Kumar R, Dilbaghi N (2015) Graphene, carbon nanotubes, zinc oxide and gold as elite nanomaterials for fabrication of biosensors for healthcare. Biosens Bioelectron 70:488–503
10. Hirlekar R, Yamagar M, Garse H, Vij M, Kadam V (2009) Carbon nanotubes and its applications: a review. Asian J Pharm Clin Res 2(4):17–27
11. Sun X, Li Q, Wang Q (2011) AChE biosensor based on aniline-MWNTs modified electrode for the detection of pesticides. In: Proceedings of international conference on CME. IEEE, pp 441–444
12. Heera P, Shanmugam S (2015) Nanoparticle characterization and application: an overview. Int J Curr Biol Appl Sci 4(8):379–386
13. Upadhyay S, Rao GR, Sharma MK, Bhattacharya BK, Rao VK, Vijayaraghavan R (2009) Immobilization of acetylcholineesterase–choline oxidase on a gold–platinum bimetallic nanoparticles modified glassy carbon electrode for the sensitive detection of organophosphate pesticides, carbamates and nerve agents. Biosens Bioelectron 25(4):832–838
14. Yang L, Wang G, Liu Y, Wang M (2013) Development of a Biosensor based on immobilization of NiO nanoparticles-carboxylic graphene-nafion modified electrode for detection of pesticides. Talanta 113:135–144
15. Sahub C, Tuntulani T, Nhujak T, Tomapatanaget B (2018) Effective biosensor based on grapheme quantum dots via enzymatic reaction for directly photoluminescence detection of organophosphate pesticide. Sens Actuators B Chem 258:88–97
16. Wang H, Wang J, Timchalk C, Lin Y (2008) Magnetic electrochemical immunoassays with quantum dot labels for detection of phosphorylated acetylcholinesterase in plasma. Anal Chem 80(22):8477–8484
17. Song D, Li Y, Lu X, Sun M, Liu H, Yu G, Gao F (2017) Palladium-copper nano-wires based biosensor for the ultrasensitive detection of organophosphate pesticides. Anal Chimica Acta 982:168–175
18. Lang Q, Han L, Hou C, Wang F, Liu A (2016) A sensitive acetylcholinesterase biosensor based on gold nanorods modified electrode for detection of organophosphate pesticide. Talanta 156–157:34–41
19. Shulga O, Kirchoff JR (2007) An Acetylcholinesterase enzyme electrode stabilized by an electrodeposited gold nanoparticle layer. Electrochem Commun 9(5):935–940
20. Du D, Huang X, Cai J, Zhang J (2007) Amperometric detection of triazophos pesticide using acetylcholinesterase biosensor based on multiwall carbon nanotube-chitosan matrix. Sens Actuators B Chem 127:531–535
21. Wang SG, Zhang Q, Wang R, Yoona SF (2003) A novel multi-walled carbon nanotube-based biosensor for glucose detection. Biochem Biophys Res Commun 311:572–576

22. Moreira Goncalves L., Callera WFA, Sotomayor MDPT, Bueno PR (2014) Penicillinase based amperometric biosensor for penicillin G. Electrochem Commu 38:131–133
23. Gao J, Gao Z, Su F, Gao L, Pang X, Cao W, Du B, Wei Q (2015) Ultrasensitive electrochemical immunoassay for CEA through host–guest interaction of β-cyclodextrin functionalized graphene and Cu@Ag core–shell nanoparticles with adamantine-modified antibody. Biosens Bioelectron 63:465–471

Microstrip Patch Antenna for Future 5G Applications

Nikhil Kalwit, Piyush Pawar, and Piyush Moghe

Abstract In this paper, a fifth-generation microstrip patch antenna has proposed. The proposed antenna design is working on 10.04 GHz with a return loss of −56.65 dB. The patch antenna has a compact structure of 18 mm × 18 mm with a FR4 glass epoxy substrate of 1.6 mm thickness. The results are simulated using Computer Simulation Technology Microwave Studio.

Keywords 5G · Patch antenna · Low return loss · High gain

1 Introduction

The speedy decrease in the dimensions of the mobile phone has led to the evolution of compact antenna structures. The conventional antennas are replaced by different antenna structure used in mobile communication [1–3]. The microstrip patch antenna has various advantages such as low cost, lightweight and easy to manufacture; despite various advantages, a major drawback that microstrip patch antenna is its narrow bandwidth.

The communication system moves on the next generation. In fifth-generation communication system, it has improved data rates and speed as compared to 4G. Various different fields have already adopted the 5G technology such as Internet of Things (IOT) and advance MIMI structure [4–6].

It has been discovered that millions of devices can be connected and operated using 5G technologies. Some 5G future systems are smart grids, smart cities, smart transportation, telemedicine and smart communication [7].

N. Kalwit (✉) · P. Pawar · P. Moghe
Swami Vivekanand College of Engineering, Indore, MP, India
e-mail: ec16.nikhilkalwit@svceindore.ac.in

P. Moghe
e-mail: piyushmoghe@gmail.com

N. Marriwala et al. (eds.), *Mobile Radio Communications and 5G Networks*, Lecture Notes in Networks and Systems 140,
https://doi.org/10.1007/978-981-15-7130-5_37

Based on the requirements for 5G, antennas with lightweight, low profile, low-cost mass production, ease of installation, conformal to planar surface and also non-planar surface, mechanically robust when mounted on a rigid surface and compatible with monolithic integrated circuits are quite important [8, 9]. Despite its narrow bandwidth, microstrip patch antenna can be perfect coordinate to meet all the above requirements.

A microstrip patch antenna is proposed for 5G communication. The proposed antenna is designed to resonate at 10.04 GHZ and has a low profile structure with the dimensions of 18 mm × 18 mm × 1.6 mm.

2 Antenna Structure and Dimensions

The proposed small patch antenna using a microstrip line for feeding is given in Fig. 1. The patch antenna has rectangular patch of 10 mm × 6 mm with a rectangular slot on it. Various parameters, such as dielectric constant ($\varepsilon_r = 4.4$), resonant frequency ($f_r = 10.04$ GHz) and thickness of substrate ($h = 1.6$ mm), are considered while designing the proposed antenna. We have used FR4 substrate material for the design of the proposed antenna. In this structure, a waveguide port is used to excite the antenna. The precise dimensions of the proposed patch antenna are summarized in Table 1. The proposed structure works on one of the proposed frequency bands, i.e., 9.7–10.2 GHz for future 5G communication (Fig. 2).

Designing and simulation of the proposed patch antenna are performed using a commercially available simulation tool called Computer Simulation Technology (CST) Microwave Studio. The antenna is particularly designed for one of the frequency bands which may get considered for future 5G wireless communication.

Fig. 1 3D view of the proposed antenna in CST

Table 1 Dimensions of the proposed antenna

Parameter	Value (in mm)
Ground plane length	18
Ground plane width	18
Patch length	10
Patch width	6
Feed line length	6
Feed line width	1.40
Substrate thickness	1.6
Patch slot length	3
Patch slot width	1

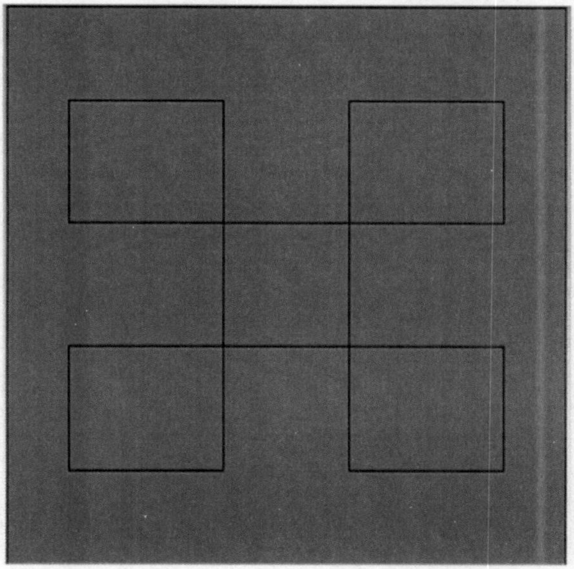

Fig. 2 Back view of the proposed antenna in CST

3 Simulation Results

3.1 Plot of Return Loss

Using wave port configuration, S11 parameters are obtained as antenna return loss. A value of −10 dB is taken as the base value which is considered fairly good in case of mobile communication. The proposed antenna works at the proposed band for 5G wireless standard. This antenna resonates at 10.044 GHz with a return loss of −56.65122 dB, covering a band from 9 to 12 GHz. Figure 3 represents return loss or S11 plot of the antenna.

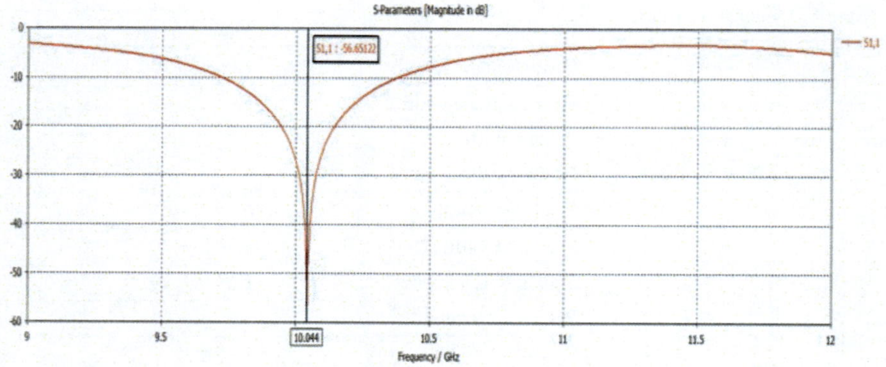

Fig. 3 Plot of return loss (in dB) of the proposed patch antenna

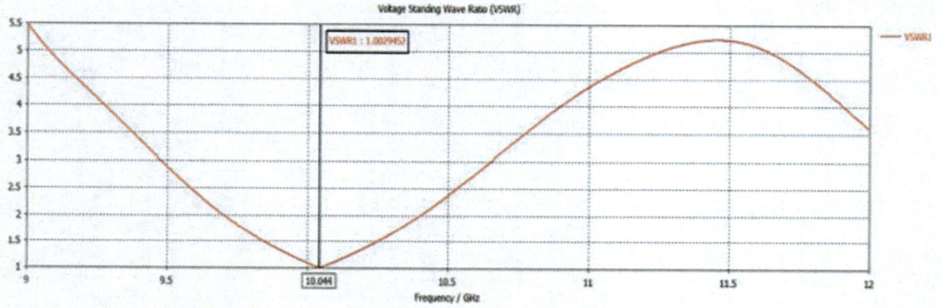

Fig. 4 Simulated VSWR plot of the proposed antenna

3.2 *Plot of VSWR*

The voltage standing wave ratio (VSWR) plot of the antenna is presented in Fig. 3. The consent level of VSWR for most of the wireless applications should not be more than 2.5 and it should be 1 ideally. As seen in Fig. 4, the VSWR value achieved at resonant frequency of 10.04 GHz is 1.0029 which is acceptable for its use in wireless applications.

3.3 *Gain Plot*

The 3D gain plot determines the antenna efficiency. The proposed patch antenna achieved moderate gain of 5.039 dB which is considered fairly well in terms of a compact antenna design. Figure 5 presents the 3D gain plot for the proposed antenna.

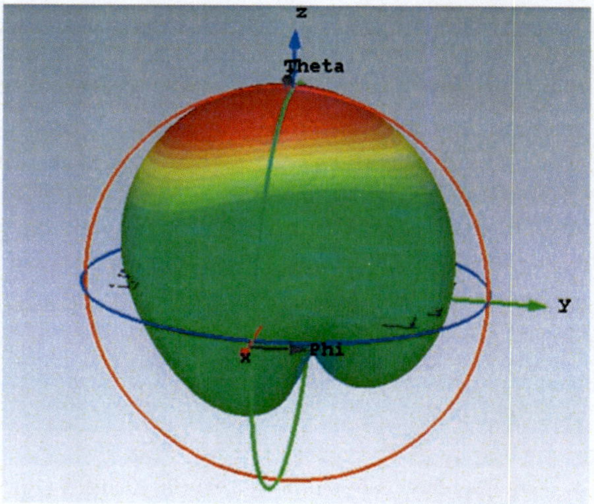

Fig. 5 3D gain plots

3.4 Radiation Pattern

Two-dimensional radiation pattern of the patch antenna is presented in Fig. 6. An omnidirectional pattern has been shown by the proposed antenna which is desirable for mobile communication.

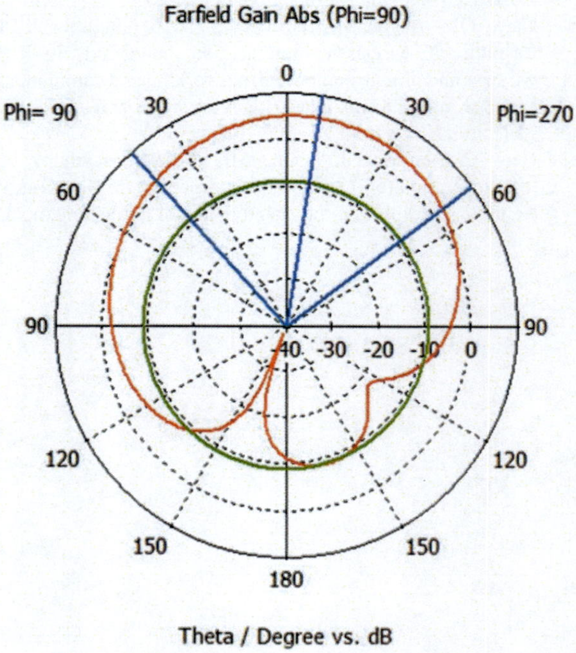

Fig. 6 Radiation pattern

4 Conclusion

A small and compact microstrip patch antenna has been proposed for 5G wireless standard. The stupendous increase in mobile speed and technologies is approaching from fourth generation (4G) to fifth generation (5G). The antenna resonates at 10.04 GHz with a return loss of −56.65 dB and can be used in future 5G wireless devices. The proposed patch antenna shows the gain of 5.039 dB. The structure of the antenna is small and can be easily integrated in devices where space is a major issue.

References

1. Verma S, Mahajan L, Kumar R, Saini HS, Kumar N (2016) A small microstrip patch antenna for 5G applications. In: 5th international conference on reliability, infocom technologies and optimization (trends and future directions) (ICRITO)
2. Chen S, Zhao J (2014) The Requirements, challenges, and technologies for 5G of terrestrial mobile telecommunications. IEEE Commun Mag 52(5):36–43
3. Wong K-L (2003) planar antennas for wireless communication. In: Chapter 2. Wiley, pp 26–65
4. Wong H, Luk K-M, Chan CH, Xue Q, So KK, Lai HW (2012) Small antennas in wireless communications. Proc IEEE J 100(7):2019–2021
5. Mak KM, Lai HW, Luk KM, Chan CH (2014) Circularly polarized patch antenna for future 5G mobile phones. IEEE Access 2:1521–1529
6. Lai HW, Wong H (2015) Substrate integrated magneto-electric dipole antenna for 5G Wi-Fi. IEEE Trans Antennas Propog 63(2):870–874
7. Haraz OM, Ashraf M, Alshebeili S (2015) Single-band PIFA MIMO antenna system design for future 5G wireless communication applications. In: IEEE 11th international conference on wireless and mobile computing, networking and communications (WiMob), pp 608–612, 19–21
8. Puri S, Kaur K, Kumar N (2014) A review of antennas for wireless communication devices. Int J Electron Electr Eng 2(3):199–201
9. Neto ASS, de Macedo Dantas ML, dos Santos Silva J, Fernandes HCC (2015) Antenna for fifth generation (5G) using a EBG structure. In: New contributions in information systems and technologies, vol 2. Springer International Publishing, pp 33–38

Multilayer Perceptron and Genetic Algorithm-Based Intrusion Detection Framework for Cloud Environment

Parul Singh and Virender Ranga

Abstract The attractive characteristics of the cloud computing environment encourage its growth and penetration in various sections of society like government, education, entertainment, etc. The large-scale adoption of cloud computing not only provides services to users but also presents a wide attack landscape to the attackers and intruders in order to perform sophisticated attacks. Widespread implementation of cloud computing and its distributed and decentralized existence makes this computing paradigm prone to intrusion and attacks. Thus, the creation of network intrusion detection framework using anomaly detection method for cloud computing network with a better assault identification level and less false positives is essential. This paper discusses an effective network-based intrusion detection model utilizing artificial neural network strategies such as multilayer perceptron programmed with a genetic algorithm, as well as compares it using other machine learning techniques. The genetic algorithm was incorporated in multiple layer perceptron to predict the connection weights. Standard IDS dataset, namely CICIDS 2017, was used for simulation and testing of the suggested model. The results of implementation demonstrate the ability of the proposed model in the identification of intrusions in the cloud environment with a higher rate of detection and generation of minimal false alarm warnings, which suggests its dominance relative to state-of-the-art approaches. The implementation results show an accuracy of 90%.

Keywords Cloud computing · Intrusion detection system · CICIDS 2017 · Multiple perceptron · Genetic algorithm

---

P. Singh · V. Ranga (✉)
Department of Computer Engineering, National Institute of Technology Kurukshetra, Kurukshetra, Haryana, India
e-mail: virender.ranga@nitkkr.ac.in

P. Singh
e-mail: singhparul291@gmail.com

N. Marriwala et al. (eds.), *Mobile Radio Communications and 5G Networks*, Lecture Notes in Networks and Systems 140,
https://doi.org/10.1007/978-981-15-7130-5_38

1 Introduction

1.1 Cloud Computing Environment

The emergence of the cloud computing world has changed the information technology sector entirely. Cloud computing is a paradigm that provides users with tools, applications, and services on a pay-as-go basis according to their need [1]. Characteristics of the cloud are as follows on-demand self-service, access to a broad network, multi-tenancy and pooling of resources, faster scalability, and measured service. These attractive features of cloud computing are motivating its adoption in different sectors such as government, education, and entertainment. There are three ways in which applications can be deployed are public cloud for multiple users, for individual organizations or users private cloud and hybrid cloud for both at the same time. Cloud services can be accessed through the provision of infrastructure, services, and execution platforms. It has emerged as a computing paradigm which has proved to be a driving force for small-scale IT companies. Cloud computing eliminates the need for maintaining data centers, continuous up-gradation of applications, etc. Cloud computing led to the separation of operating systems and hardware [2].

Intrusion detection framework classification depending on their placement in Cloud. Classification of the cloud computing intrusion detection framework by location is hypervisor-based cloud IDS, network-based cloud IDS, host-based cloud IDS, and distributed cloud IDS [3]. Network-based cloud IDS tracks incoming and outgoing traffic in network, network contracts, and software operations to detect suspicious behaviors and attacks such as a bot, heartbeat, and denial of service or even tries to unlock and intrude into the computers [4]. An intelligent agent is installed on the hypervisors installed on the servers maintained in the data centers of cloud infrastructure in hypervisors-based cloud IDS. Smart software is installed over the monitored host in the host-based cloud IDS. Distributed IDS is deployed in the network as well as on the host.

Motivation. The virtualized machines running on hypervisors installed on servers in the cloud network can be compromised by zombie hosts or hackers bypassing the security inspection conducted by firewall safeguarding the network at the entry point in order to get unauthorized access to data centers that threaten the security of sensitive information uploaded on it. Attackers have developed advanced techniques and are exploiting new vulnerabilities to bring down cloud services [5]. 2011, witnessed a data breach in which an intruder used services of Amazons Elastic Cloud service to perform an attack on online entertainment systems of Sony entertainment by registering and opening an anonymous Amazon Elastic Computer Cloud service account. This attack was the largest data breach in which more than a hundred million account details of customers belonging to Sony Company were stolen [6]. In the year 2013 in the USA, Stratford announced a security breach in which credit card numbers of seventy-five thousand users were leaked [6].

The above-mentioned incidents show that security is one of the major challenges in cloud infrastructure and the importance that intrusion detection systems hold in safeguarding user's information stored in the cloud and in maintaining the trust of consumers. Although designing a network-based cloud intrusion detection framework has become sophisticated, it is understood that the network-based cloud intrusion detection framework should perform analysis on a large amount of information and should be able to determine attack patterns with the highest possible accuracy and minimal false positive. Conventional intrusion detection techniques pose overhead on cloud infrastructure in terms of processing resources and cost involved. Therefore new techniques, such as machine learning, need to be implemented [7]. The paper comprises five sections. The first provides a brief overview of the cloud computing paradigm and the importance of IDS. The second section describes the already existing solutions. The third section outlines the techniques used in the proposed framework. The fourth section gives a detailed description of the suggested IDS. The fifth section describes the implementation details and results. The fifth section provides a conclusion and future scope.

2 Literature Review

In [8], a new CS-PSO-based feature selection system was proposed and implemented to develop an effective IDS that reliably and rapidly classify attacks in the cloud infrastructure. The two components of the suggested IDS framework are a CS-PSO algorithm and a logistic regression classifier. The CS-PSO algorithm integrates CS and PSO methods to select the best features. The second component is a logistic regression classifier that was used to categorize instances selected from the NSL-KDD dataset as a normal or an assault pattern. PSO was incorporated to find the local best in search space, and CS was used to determine the global best value. The combination of these two evolutionary models inspired by nature produces greater accuracy compared to the other methods.

Besharati et al. proposed a cloud intrusion detection system for protecting virtual machines and need to be installed on a particular machine [9]. Logistic regression algorithm is used in significant feature extraction of each class, and the values obtained are enhanced further which are using regularization techniques. The classification of cyber-attacks was done by using three different classification algorithms of machine learning: linear discriminate analysis, decision tree, and artificial neural network with ensemble machine learning technique: bagging algorithm. The proposed solution was demonstrated using a benchmark intrusion detection system dataset NSL-KDD. The results of the implementation of the solution illustrate its ability in the determination of intrusion within cloud computing environment with high accuracy. The simulation results show its superiority on other methods for detecting an attack.

In [10], the authors suggested an intrusion detection framework using an anomaly-based detection technique for the environment of the cloud computing paradigm via the integration of different machine learning techniques. The authors have applied the proposed system to the cloud computing environment's hypervisor layer. The combination of artificial neural network and fuzzy c-means machine learning algorithm was utilized for better categorization. Simulation of the proposed algorithm was done using benchmark intrusion detection dataset DARPA's KDD1999 cup dataset. The proposed model achieved high accuracy and minimal false positive alarms. The algorithm outperformed the other classification approach adopted in machine learning-based intrusion detection framework using algorithms such as classic neural networks and the Naive Bayes classifier.

In [11], the authors proposed a dimensionality reduction method in order to eliminate redundant and irrelevant features. Efficient correlation-based feature selection (ECOFS) technique was used to handle both linear and nonlinear dependent features. The effectiveness of the proposal is evaluated by employing it in Libsvm-IDS. The proposed solution is demonstrated using two benchmark intrusion detection system dataset, namely DARPA's KDD1999 cup and NSL-KDD. The implementation results indicate that the developed technique picks the smallest number of features after the deletion of obsolete features. When the algorithm was applied Libsvm-IDS, it produced accurate results in less computational time and cost. The accuracy obtained through this technique proved to better than the other two existing techniques.

Mehibs et al. suggested a network-based intrusion detection framework to protect cloud services and networks from various assaults using a fuzzy c-means clustering algorithm [12]. Fuzzy c-means technique was used to split the dataset into two clusters, one for anomalous patterns and the other for regular patterns. The proposed framework comprises two steps; the first step is training in which best cluster centers are identified. The cluster centers obtained from the training phase were used to assess the cluster of fresh instances that are not seen. Training and testing of the proposal were done using the KDD cup 99 datasets. The authors' findings indicate that the proposed system has a good detection rate with a small false positive warning.

In [13], the authors have suggested a distributed intrusion detection system for the environment of cloud computing paradigm based on machine learning techniques. The system proposed is intended to be implemented side by side in the cloud with the edge network components of the service provider. The suggested IDS comprises five modules: cloud network data storage module, a preprocessing module for network traffic, module for detection of an anomaly, module for the synchronization of network traffic, and attack reporting module. Ensemble learning-based random forest classifier was used for the classification of instances. The new IDS was reviewed using the benchmarked dataset Coburg Intrusion Detection Dataset-001 and built on the Google Cloud Platform. The results obtained by applying the regular random forest on Coburg Intrusion Detection Dataset-001 are outperformed by the result obtained on the proposed system in terms of precision and runtime. The system obtained an overall precision of ninety percent.

3 Methodologies Used

3.1 Multilayer Perceptron

Multilayer perceptron (MLP) is one of the neural network architectures consisting of more than one perceptron in the deep artificial neural network. They consist of an input layer to obtain input signals, an output layer to make a decision or decision about the input data obtained and an unspecified number of hidden layers to do all the computation. Multilayer perceptron is applied to problems dealing with supervised learning. They are trained on a series of pairs of input–output and practice modeling the association between those inputs and outputs. Training involves modifying the features of the model, including weights and bias in order to eliminate the misclassification. The error concerned may be calculated using such as several methods root mean square error (RMSE). Adjustment in connection weights and bias relative to error is made by applying a back-propagation neural network (bpnn). The multilayer perceptron works in two stages, namely forward pass and backward pass [14]. The input signal travels via the neurons of hidden layers from the neurons of the input layer to the neurons of the output layer, and the output layer decision is evaluated in the forward pass. In the backward pass, the partial derivative of the error function is propagated back over a multilayer perceptron using a back-propagation neural network (bpnn) with respect to different biases and weights. Parameters for multilayer perceptron neural networks may be modified to bring it one step nearer to the lowest error value that could be done using and gradient-based optimization technique. The stage at which the decrement of error stops is known as the convergence stage.

Genetic algorithm. In the 1970s, John Holland developed an algorithm based on the fundamentals of natural selection, biological evolution, and recombination of genes which were named genetic algorithms (GAs). GAs are artificial intelligence techniques that are often used for the optimization of problems. All the solutions to a particular are encoded in a chromosome. The features contained in a chromosome are called a gene. The group of chromosomes generated by genes is known as the population. The fitness function is used to determine the quality of each chromosome according to the solution required. Genetic algorithm involves a series of steps which are as follows: An initial population is created using a random selection of solutions. Each solution is assigned a fitness value, depending on its proximity to the problem solution. Chromosomes with good fitness value are retained while chromosomes with bad fitness value are discarded. Higher fitness value chromosomes have a higher probability to produce new offspring. If the newly generated chromosomes contain a solution near enough to the optimal solution, then the target has been achieved. The new generation will undergo the same process as their parents if the target is not achieved (Fig. 1).

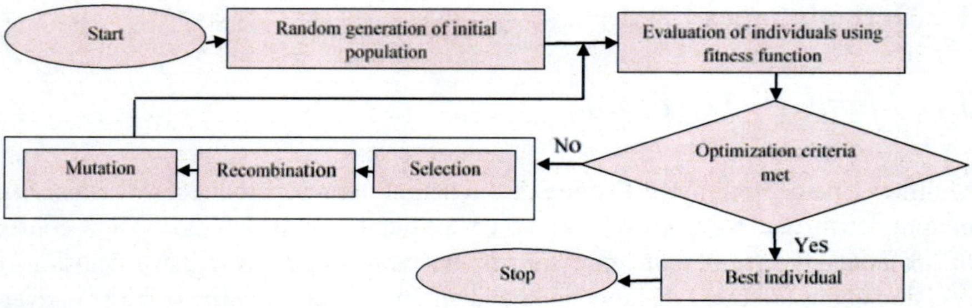

Fig. 1 Genetic algorithm workflow

4 The Proposed System

This section provides a detailed overview of the proposed IDS. In this paper, we have used an artificial neural network technique, namely multilayer perceptron because of the flexibility it provides in the detection of intrusion, the ability to process nonlinear data, high speed of processing, excellent generalization capacity, and exceptional classification performance [13].

4.1 Workflow of the Proposed System

A hybrid framework is employed in proposed IDS that is obtained by the integration of two artificial intelligence techniques, namely multilayer perceptron and genetic algorithm. The CICIDS 2017 dataset was preprocessed using the following operation categorical encoding, handling NAN values and missing values, and normalization. The preprocessed dataset is then split into a training dataset and a test dataset. The IDS was trained using a multilayer perceptron, and a genetic algorithm was used for weights and bias adjustment of the model. Our artificial neural network is a multilayer perceptron, with one layer of input neurons, two layers of hidden layer neurons, and one layer of output neurons. The number of attributes extracted from the dataset and presented to the input layer specifies the number of neurons in the input layer. The output layer is comprised of one neuron gives value 1 in case of an instance of dataset classified as abnormal pattern and value 0 in case of classification of an instance as a normal pattern. Genetic algorithm (GA) is utilized to find optimal values for connection weights and bias which reduces convergence time, the execution time, and processing power of the system. The trained model is tested with test dataset instances. The suggested model predicts the class to which test input instance belongs that is to an abnormal class or to a normal class. Figure 2 shows workflow of the proposed IDS.

Framework of IDS based on multilayer perceptron and genetic algorithm. Our proposed framework comprises three modules, data preprocessing module,

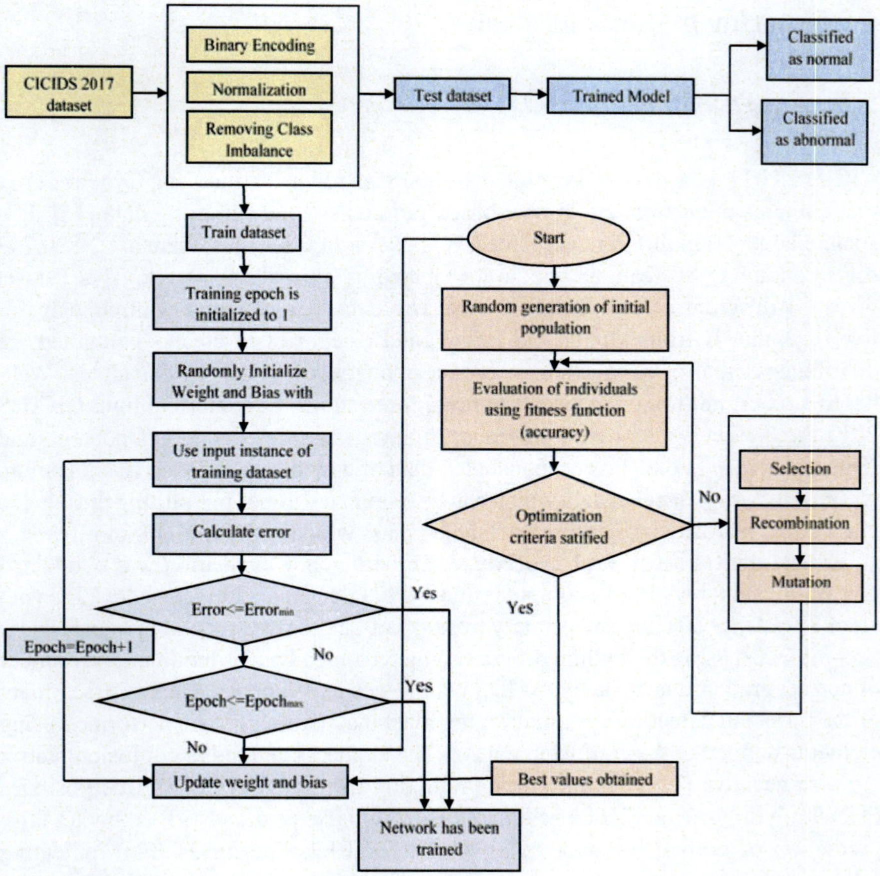

Fig. 2 Functioning of the proposed IDS

training classifier, and attack detection module. Module for preprocessing data includes two categorical encoding operations which process assigning numerical values to string values in a given dataset and normalization which refers to value scaling of dataset instance between 0 and 1. Categorical encoding was performed using the binary encoding method, and for normalization, min–max normalization was used. In the training classifier module, the IDS was trained using a multilayer perceptron and genetic algorithm, and in the attack detection module, the trained model is used to test the proposed IDS using test input instance to classify them as benign or attack instance.

5 Experimentation and Results

5.1 Experimental Dataset

CICIDS 2017 is a dataset generated by the Canadian Institute for Cybersecurity that contains the actions of 25 user-based protocols while capturing data [15]. It is spanned over eight different CSV files. In those eight CSV files, there are 2,830,743 rows containing 80 features which are labeled as normal and attack. This dataset gives 14 different categories of attacks. The data was captured continuously for five days that is from Monday to Friday, and categories of attacks contained are distributed denial-of-service attack, port scan attack, denial-of-service attack, Web-based attacks, infiltration attacks, and brute force attack. Few shortcomings CICIDS 2017 the dataset are scattered presence, irrelevant features, large data content, and large imbalance in the class of the labeled data of which classification result is more inclined to benign labeled data which can be eliminated either by splitting the labeled class in the majority or merging the labeled class which is present in minority.

Experimental setup and procedure. The efficiency evaluation was conducted on a 64-bit Windows 10 Pro computer fitted with Intel ® i7-7700 four-core CPU with clock speed of 3.60 GHz and primary memory of 12 GB. Preprocessing of datasets is accomplished using the Python programming language Pandas module. The number of correct predictions made by the IDS determines its efficiency. Quality assessment of the intrusion detection system utilizing machine learning can be performed using elements defined in the confusion matrix. The elements defined in confusion matrix are true negative (TN) meaning true prediction of normal behavior, true positives (TP) implying true prediction of attack behavior, false positives (FP) showing false prediction of normal behavior as an assault, and false negatives (FN) indicating false prediction of attack as normal. The performance metrics generated using the confusion matrix which will be used for the evaluation of the proposed IDS are accuracy rate, F-score, and detection rate.

$$\text{Accuracy} = \frac{\text{True Positive} + \text{True Negative}}{\text{True Positive} + \text{True Negative} + \text{False Positive} + \text{False Negative}}$$

$$\text{Detection Rate} = \frac{\text{True Positive}}{\text{True Positive} + \text{False Negative}}$$

$$f - \text{score} = \frac{2 * \text{Detection Rate} * \text{Precision}}{\text{Detection Rate} + \text{Precision}}$$

Results and Analysis. To test the proposed approach, certain assessment requirements are specified. The performance metrics consisting of accuracy, the detection rate, and F-score are used to evaluate the system as well as compare it with other IDS. Figure 3 illustrates comparative results of the proposed IDS with respect to other popular IDS such as network-based intrusion detection system using gradient boosting and clustering [7] and an intrusion detection framework using CS-PSO for

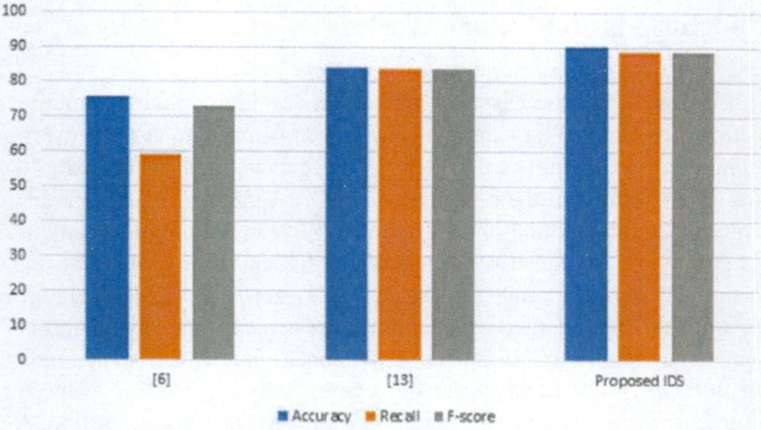

Fig. 3 Comparison of the proposed IDS and other popular IDS

the cloud computing environment [8]. The suggested approach outperformed other IDSs with regard to the evaluation parameters. The proposed IDS shows a significant improvement over the IDSs used for comparison in terms of accuracy, rate of detection or recall, and f-score. Figure 4 displays a comparison of hybrid machine learning used in the proposed and other machine learning techniques which shows that the hybrid algorithm used in the implementation of the proposed IDS outperformed other single and hybrid algorithms (Figs. 3 and 4).

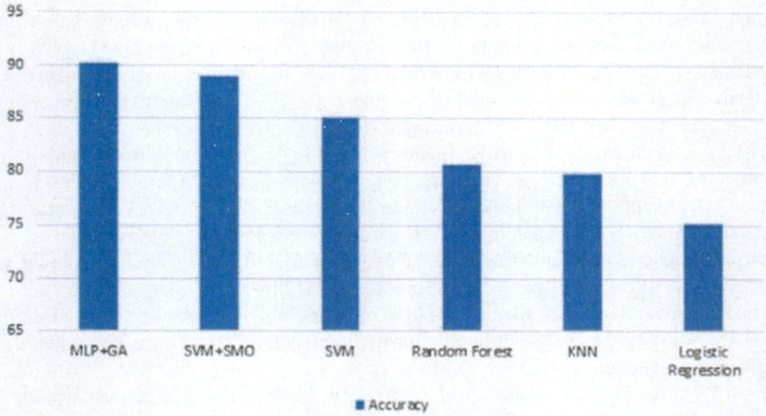

Fig. 4 Comparison of the proposed system utilizing different algorithms of machine learning

6 Conclusions and Future Work

Network intrusion detection systems are very important for cloud computing since it is a platform on which a large amounts of user sensitive data uploads every second. Soft computing are being used widely in the development of intrusion detection systems (IDS) because of their ability to learn different patterns existing in the data, the evolution of the learning model, and flexibility in learning the patterns. In this paper, a machine learning network-based cloud IDS has been proposed using an artificial neural network technique (multilayer perceptron) and a genetic algorithm. The genetic algorithm was merged with a multi-layer perceptron to obtain optimal values for connection weights and bias. The implementation results obtained indicate that several recent experiments have been outperformed by our IDS. In fact, performance enhancement approaches also reduced convergence and execution time. Although hybrid machine learning algorithm requires more computational time, they provide more accurate results. In the future, we plan to use a combination of other neural network models with met heuristic techniques such as whale optimization, particle swarm optimization, and crow search algorithm in order to obtain minimal false positive and higher precision.

References

1. Deshpande P, Sharma SC, Peddoju SK, Junaid S (2018) HIDS: a host based intrusion detection system for cloud computing environment. Int J Syst Assur Eng Manag 9(3):567–576
2. Singh DAAG, Priyadharshini R, Leavline EJ (2018) Cuckoo optimisation based intrusion detection system for cloud computing. Int J Comput Network Inf Secu 10(11):42
3. Ghanshala KK, Mishra P, Joshi RC, Sharma S (2018) BNID: a behavior-based network intrusion detection at network-layer in cloud environment. In: 2018 first international conference on secure cyber computing and communication (ICSCCC). IEEE, pp 100–105
4. Patel A, Taghavi M, Bakhtiyari K, JúNior JC (2013) An intrusion detection and prevention system in cloud computing: a systematic review. J Netw Comput Appl 36(1):25–41
5. Hatef MA, Shaker V, Jabbarpour MR, Jung J, Zarrabi H (2018) HIDCC: a hybrid intrusion detection approach in cloud computing. Concurr Comput Pract Exp 30(3):e4171
6. Mozumder DP, Mahi JN, Whaiduzzaman M, Mahi MJN (2017) Cloud computing security breaches and threats analysis. Int J Sci Eng Res 8(1):1287–1297
7. Verma P, Anwar S, Khan S, Mane SB (2018) Network intrusion detection using clustering and gradient boosting. In: 2018 9th international conference on computing, communication and networking technologies (ICCCNT). IEEE, pp 1–7
8. Ghosh P, Karmakar A, Sharma J, Phadikar S (2019) CS-PSO based intrusion detection system in the cloud environment. In: Emerging technologies in data mining and information security. Springer, Singapore, pp 261–269
9. Besharati E, Naderan M, Namjoo E (2019) LR-HIDS: logistic regression host-based intrusion detection system for cloud environments. J Ambient Intell Humanized Comput 10(9):3669–3692
10. Pandeeswari N, Kumar G (2016) An anomaly detection system in cloud environment using fuzzy clustering-based ANN. Mob Networks Appl 21(3):494–505
11. Wang W, Du X, Wang N (2018) Building a cloud IDS using an efficient feature selection method and SVM. IEEE Access 7:1345–1354

12. Mehibs SM, Hashim SH (2018) Proposed network intrusion detection system based on fuzzy c mean algorithm in cloud computing environment. J Univ Babylon Pure Appl Sci 26(2):27–35
13. Idhammad M, Afdel K, Belouch M (2018) Distributed intrusion detection system for cloud environments based on data mining techniques. Procedia Comput Sci 127:35–41
14. Mehibs SM, Hashim SH (2018b) Proposed network intrusion detection system in cloud environment based on back propagation neural network. J Univ Babylon Pure Appl Sci 26(1):29–40
15. Wu SX, Banzhaf W (2010) The use of computational intelligence in intrusion detection systems: a review. Appl Soft comput 10(1):1–35
16. Manickam M, Rajagopalan SP (2019) A hybrid multi-layer intrusion detection system in cloud. Cluster Comput 22(2):3961–3969

Position Falsification Misbehavior Detection in VANETs

Ankita Khot and Mayank Dave

Abstract VANETs stands for vehicular ad hoc networks. In VANETs, alerts like post-crash notification (PCN), beacon messages, etc., (with sender id, position, speed and timestamp) are exchanged between vehicles in order to improve road safety so that the driver is previously alerted of the hazard or crash that she/he could face ahead. This technology has a great potential to reduce the number of accidents that are happening every year. If the driver is alerted few seconds before the accident about the hazard, then the accident could be prevented from happening. But, in VANETs, there is a possibility that due to selfish or malicious reasons, some attacker might send false alerts and falsified information in beacon leading to change in driver's behavior and entire network. This could result in accidents in the network or long-distance travel of driver. Hence, it is very much necessary to detect the false messages that are communicated in vehicular network.

Keywords Vehicular ad hoc networks · Misbehavior detection · Vehicular security · Position verification and machine learning

1 Introduction

VANETs stands for vehicular ad hoc networks. It is a variant of mobile ad hoc networks (MANETs). Mobile ad hoc networks are self-configuring, infrastructure-less network of mobile devices that are connected wireless. In MANETs, each device is free to move independently in any direction, and hence, its wireless link with other devices keeps changing frequently. Each device forwards or relays the traffic to other devices, hence acting as a router.

A. Khot (✉) · M. Dave
Department of Computer Engineering, National Institute of Technology, Kurukshetra, India
e-mail: ankitakhot001@gmail.com

M. Dave
e-mail: mdave@nitkkr.ac.in

N. Marriwala et al. (eds.), *Mobile Radio Communications and 5G Networks*, Lecture Notes in Networks and Systems 140,
https://doi.org/10.1007/978-981-15-7130-5_39

Vehicular ad hoc networks are used for communication between vehicles (V2V) for relying alert messages and between vehicles and infrastructure (V2I) for reporting some event to RSU/CA or road conditions and speed limit alerts from RSU or for communication between vehicle to pedestrian (V2P) and vehicle to network (V2N). Together, everything forms vehicle to everything (V2X). Vehicles are equipped with radio communication. The communication between vehicles is arranged in ad hoc manner with an IEEE 802.11 g-based WiFi communication system, supported with wireless roadside base stations. Each vehicle is equipped with short- and medium-range wireless communication (DSRC—Dedicated Short-Range Communication System).

Vehicles are referred to as nodes in the network. Vehicles are embedded with OBUs. OBUs follow ad hoc communication and can be used to get vehicle position, predict driving behavior and detect traffic violations. Additional functionalities can be added to OBUs, for instance, traffic, road conditions ahead or alerting emergency services in case of an accident. RSUs are stationary units located on the roadside that provide connectivity support to passing vehicles. RSUs can follow both ad hoc and WLAN communications. Certifying authorities are authorities that issue certificates, sign the message digitally and provide the private and public keys. The CAs are government agencies that maintain record of vehicles and their owners and issue unique identities as license plate and secret credentials like pseudonyms, public/private keys and certificates. MA is headed by government of state/province. Each province is divided into several smaller regions each having a local authority CAs (Fig. 1).

Applications [1] of VANETs are classified into three major groups like commercial-oriented applications which include advertisement, internet access, passenger infotainment, etc.; convenience-oriented applications like route diversion, electronic toll collection, parking discovery, etc.; and safety-oriented applications like accident notification, road hazard notification, forward collision warning, etc (Fig. 2).

VANETs are infrastructure-less. They do not have fixed infrastructure. They are ephemeral; network lasts for a very short time. They have dynamic topology; topology changes very frequently. Density also changes frequently. They are self-organizing, distributed networks where mobility of vehicles is very high.

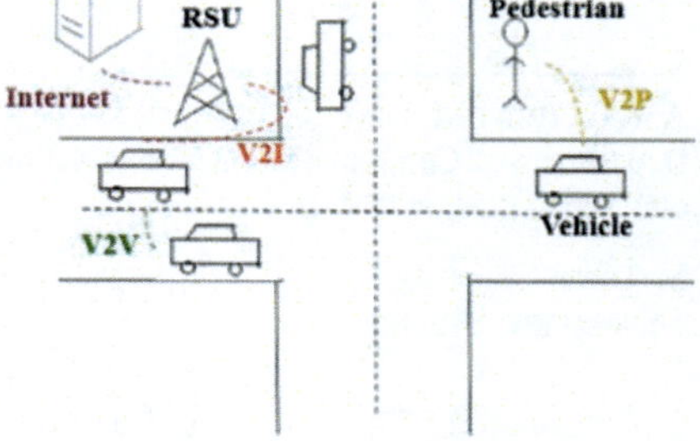

Fig. 1 VANETs communications model

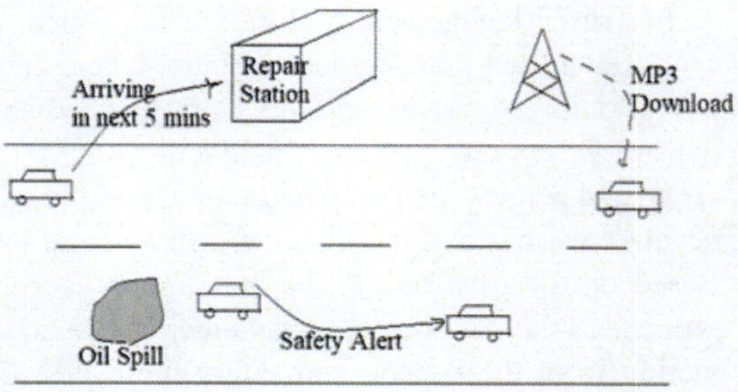

Fig. 2 Applications of VANETs

Different research areas of VANETs are broadly classified into routing, broadcasting and security [2–4]. Different research areas within 'security' domain of VANETs are broadly classified into security attacks, security challenges and security requirements. Different security attacks possible on VANETs are illusion attack, bogus attack, DoS attack, Sybil attack, false alert generation attack, insider attack, eavesdropping, jamming, replay attack, etc [5]. Different security requirements include integrity, confidentiality, privacy, non-repudiation, authentication and availability. And security challenges include certificate management, misbehavior detection and data trust [6].

Misbehavior in VANETs is sending/transmission of false information [7]. Vehicles can send false alerts either due to internal failure (faulty nodes) or due to selfish reasons (malicious nodes). False alerts could disturb normal functioning of network and could also change drivers' behavior and create disastrous situations in network. In order to prevent this, misbehavior detection is essential and various misbehavior detection schemes are proposed by various researchers. Misbehavior detection schemes (MDS) are broadly classified into two types. The first is data-centric MDS, where false data is identified. And the second is entity-Centric MDS where the misbehaving node is identified. The messages exchanged in vehicular communication network are of two types. The first is the beacon message that specifies the location and speed information of the sender vehicle. And the second is alert message to ensure safety of vehicles on the road. This alert message specifies information like collision warning, violation warning, road hazards, post-crash notification, emergency vehicle approaching, etc.

The rest of paper is organized as follows. In Sect. 2, we present the related works. In Sect. 3, the methodology used is explained in brief followed by Sect. 4 where the implementation results are evaluated and Sect. 5, where conclusions are drawn.

2 Related Work

There is a lot of active research being carried out in order to detect different misbehaviors in vehicular network.

In Trust Management and Fake Data Detection Scheme [8], author et.al. have proposed a trust management scheme, where each neighboring vehicle is assigned a trust rating by calculating the average speed and average density from speed and density values received from neighboring nodes' beacon. These values of average speed and density are compared with threshold values. If the difference is high, then negative trust is assigned, and if difference is low, then positive trust is assigned. Based on the trust rating, the message is accepted or rejected. Author et.al. also proposed fake data detection scheme, where he used the position and speed information from the beacon and safety messages. Using this information, the author calculated acceleration and hence the distance by using motion equations. Later by using the position data of beacon and safety messages, the distance is calculated. If there is a huge difference between both the distance values, then the message is considered fake and is rejected.

The proposed 'fake data detection scheme' filters out only the false messages that are transferred between nodes and not between roadside units (RSUs) and node. It might be possible that the RSU is compromised by the attackers. The proposed technique has to calculate average speed and average density every second due to dynamic characteristics of VANETs which result in lot of computations and power usage.

In detection of malicious node algorithm [9], author et.al. have proposed a detection of malicious node (DMN) algorithm, where there is one verifier node elected in the network that can verify the activities of neighboring nodes. If neighboring node takes time which is greater than a certain predetermined value, then verifier node detects it to be dropping the packets. Every node is associated with a 'distrust value.' Now, the verifier node increases the distrust value by 1 and broadcasts this value to other nodes in the network. If distrust value increases the threshold, then the verifier node sends the ID of that particular node to Certifying Authority (CA). The CA then broadcasts the ID of that particular node in the network of being a malicious node and blacklists it and isolates from the network. The author does not talk about how to handle the scenario, wherein the cluster head in itself is compromised. The cluster head is also a vehicle; hence, the verifier would be monitoring its behavior but how will it identify misbehavior of cluster head (like whether if it does not take any action of misbehaving nodes) and whom will the verifier report in this case?

On data-centric misbehavior detection in VANETs [10], author et.al. have proposed a scheme that can detect whether the alert message received by a node in vehicular network is valid or not. First, they check whether the message is having valid signature using Efficient Certificate Management for Vehicular network (ECMV) scheme. Then, they check the freshness of messages by using threshold value. Later checks if the event location, sender node and receiver nodes location is in particular order or not (event—ns—nr or nr—ns—event). If not, then message is rejected. If correct, then the position of the sender node and the location of event are compared. If found contradicting, then reported to the CA. CA maps the entity and imposed negative points and fines the entity. If large numbers of such reports are detected against a particular node, then its certificate is revoked and is isolated from the network. The paper does not discuss how RSU will check whether ni gives wrong report about nj to RSU. It is possible that ni falsely accuses nj.

In verification-based authentication scheme for bogus attacks in VANETs [11], author et al. have proposed a verification scheme in order to detect bogus attacks on vehicular networks. Here, they used three types of sensors: (1) ART sensor acceptance range threshold sensor for observation, (2) proactive sensor for proactive exchange of neighboring table and (3) reactive position request sensor. Now, if information from two nodes is different, then it requests information from other neighbors, and based on majority, the decision is taken, and thus, an attacker is detected. The information about this attacker is distributed over the network by the RSU. Thus, the other nodes in the network drop the other packets from the attacker node.

All types of bogus attacks (false speed, density, position, etc.) are not detected by the proposed scheme, and only, the falsified position information is detected and requires deployment of many different types of sensors at each node, thereby increasing cost.

In [12], author et.al. have proposed a machine learning approach to detect false alerts and position falsification attacks. Here, the author have used various information from the alert messages, beacons, neighboring table and historic data to extract features and applied them on various machine learning algorithms and have analyzed the results. In false alert verification scheme, attacker has use speed and density information and calculated flow value. Difference in flow value received and calculated is derived as a feature. In position falsification verification, the author has used information from the beacon and compared it with the previous beacon and local information of nodes sensor. Likewise, features are extracted. This information along with basic information is fed to its local MDS. MDS which is trained based on extensive simulations can verify the attack.

In integrated plausibility check and machine learning for misbehavior detection in VANETs [13], author et.al. have proposed a framework where they perform location plausibility check and movement plausibly check to address the issue of location spoofing in VANETs. In order to enhance their results, they have further used SVM and KNN algorithms.

In VANET alert endorsement using multi-source filters [14], author et.al. have proposed a framework to endorse alert messages by applying multi-source filter to check whether the message is authentic or malicious. Their filtering model for misbehavior detection is based on threshold curve and certainty of event (CoE) curve. Threshold curve gives the importance of event according to drivers' relative position to the driver, and certainty of event curve represents the probability of correctness of alert message. This misbehavior detection model prevents unnecessary warnings and alerts from disturbing the driver. The author has used six complimentary sources of information, and the results from these sources are aggregated, and only, when the aggregated result indicates that the message is valid, the driver is alerted.

In position verification for vehicular networks via analyzing two-hop neighbors information [15], author et.al. have used information from its direct neighbors and neighbors at one hop position in order to make judgment of position announcement made by sender vehicle. This scheme defines the plausible area where the vehicle should exist by setting the lower bound and higher bound by taking the last observation of sender from its direct neighbors. If sender is outside the boundaries of this area, then it is suspected as malicious. In case of sparse traffic vehicle analysis, the consistence in receives signal strength and said location.

In predicting vehicle's position using roadside units, a machine learning approach [16], author et.al. have proposed a mechanism to locate position of vehicle using the reception power of packets that are send by the vehicles to RSU. The author has used the relation between reception power of packets received at RSU and distance between vehicle and RSU. Then, they train a model using machine learning algorithms like KNN, SVM and random forest. Training is done where the vehicle sends its location in packet and later testing is done, wherein the vehicle asks for its current location.

3 Methodology

3.1 Statistical Approach

Beacon message received by a receiver vehicle from sender vehicle in VANETs contains sender id, position (Px, Py, Pz), speed (Vx, Vy, Vz) and time stamp. Assuming beacon is received at regular interval of 300 ms. Figure 4 shows the position validation scheme (Fig. 3).

Let vehicle A receives beacon (Pbt1, Vbt1, t1) at time t from Vehicle B.

where
Pbt1: position of Vehicle B at time t1
Vbt1: speed of Vehicle B at time t1
t1: send time of beacon
t: receive time of beacon
c: speed of light
Pat: position of Vehicle A at time t.

$$\text{Distance} = \text{Speed} * \text{time}$$
$$|\text{dist(veh A, Veh B)}| = \text{Speed of Beacon} * (t - t1)$$
$$\text{Pbt1} - \text{Pat} = c * (t - t1)$$
$$\text{Pbt1} = c * (t - t1) + \text{Pat}$$

Now, receiver can calculate what the location of sender would be after 300 ms using.

Distance = Speed * time.

Let sender position at time t1 + 300 be P′. This new position could be obtained as follows.

$$300\text{ ms} = (P' - P\text{ bt1}) / V\text{ bt1}.$$

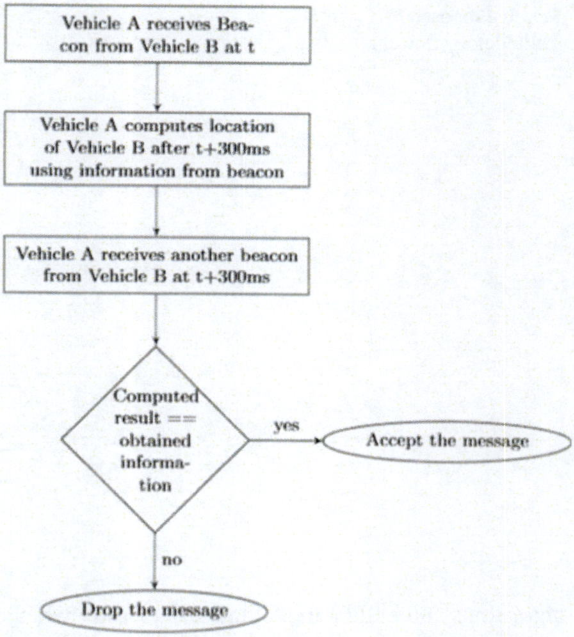

Fig. 3 Proposed validation flowchart

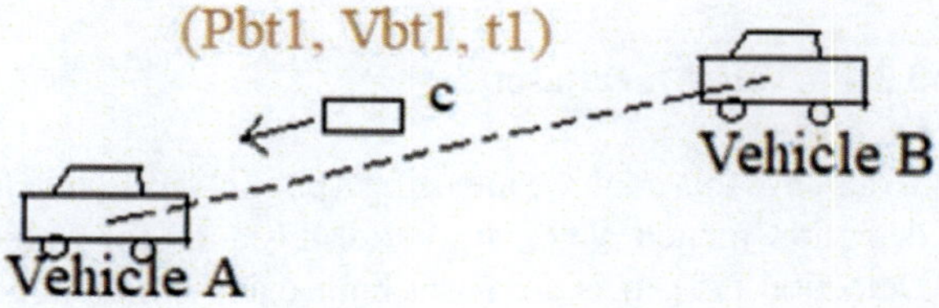

Fig. 4 VANETs communication scenario

i.e., $P' = 300$ ms $*V$ bt1 $+ P$ bt1
where P' is the calculated position of sender after 300 ms.

If the next received beacon from sender does not approximately have the same location as predicted, the message is rejected, else accepted.

3.2 *Machine Learning Approach*

Position falsification attacks are those wherein false position information is passed in the beacon message. In order to detect these position falsification attacks, machine learning-based solution is proposed. The proposed methodology flowchart is given in Fig. 1. In the proposed scheme, we use data within the beacon message from neighboring vehicles in order to generate various features like distance between sender and receiver, etc. Then, the features are used to train different machine learning

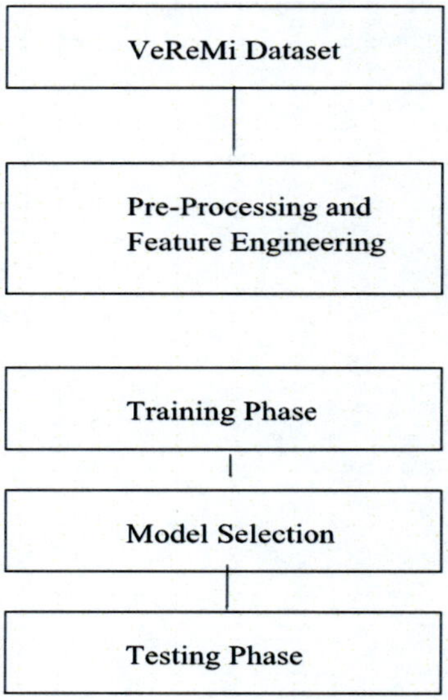

Fig. 5 Proposed methodology flowchart

algorithms and build a model that can classify whether the message is falsified with respect to its position data or not (Fig. 5).

3.2.1 VeReMi Dataset

To achieve this goal, we are using a publicly available labeled dataset for misbehavior detection named VeReMi. VeReMi stands for Vehicular Reference Misbehavior Detection [17]. It is a dataset built using simulation tools like SUMO, OMNET++ and Veins. Using these tools, the attack is simulated and the message logs that are received bj Onboard Units (OBUs) of vehicles are used to build the dataset. This dataset is built using five different vehicular densities, five different attacks and five different attacker densities. The five different attacks in dataset are (1) constant position attack (Type = 1), where the attacker transmits a fixed location. (2) Constant offset attack (Type = 2), where the attacker transmits fixed offset added to the real position. (3) Random attack (Type = 4), attacker sends random position within the simulated area. (4) Random offset attack (Type = 8), where attacker sends random position within the simulation area. (5) Eventual stop (Type = 16), where attacker behaves normally for some time and then attacks by transmitting the same position repeatedly.

VeReMi dataset is a log of around 2257 files. Each file has approximately 300 data points. The message logs include two types of data points. Type 2 indicates that the information is obtained from Global Positioning System (GPS), and Type 3 indicates the information of beacon obtained from neighboring vehicles. Refer Fig. 4 for distribution of data points according to the type of message. Other information in log is receive signal strength information (RSSI), sender ID, claimed sender position, message ID, claimed sender speed, noise, position noise, speed noise, GPS position and speed of receiver.

The message logs downloaded are in the form of JSON files with file names in the format of 'JSONlog-0-7-A0.json,' where '0′ stands for 0th vehicle, '7′ stands for 7th OMNET++ module and 'A0′ states that vehicle is not an attacker. The dataset zip files are downloaded, extracted, merged and uploaded into Google Cloud Platform (GCP). Using Python, the data points from all files are read into pandas data frame for further operations.

3.2.2 Preprocessing and Feature Engineering

Feature analysis involves analyzing the importance of a feature in the dataset. Some of the features are important, but some are not. Excluding the least important features will help in better classification rate. The results of the dataset are highly dependent on the type and combination of features used for classification. There are total 3,412,471 data points and six class labels. Figure 3 describes the distribution of class labels (Fig. 6).

The data needs to be preprocessed and cleaned. The features with all zero values like noise, position noise and speed noise are dropped. The VeReMi dataset was analyzed, and various features like receiver position and receiver speed at the time of reception of beacon, distance between sender and receiver, etc., are extracted.

The plausible location feature is analytically extracted from data points available. The beacon message consists of data like sender's position Ps, sender's speed Ss, time stamp or send time ts. And through GPS, the receiver can locate its position that is receiver's position Pr. Let tr be the time at which beacon message was received. Beacon message will travel with speed of light (c).

Since distance = speed * time.

$$(Ps' - Pr) = c * (ts - tr)$$

$$Ps' = c * (ts - tr) + Pr$$

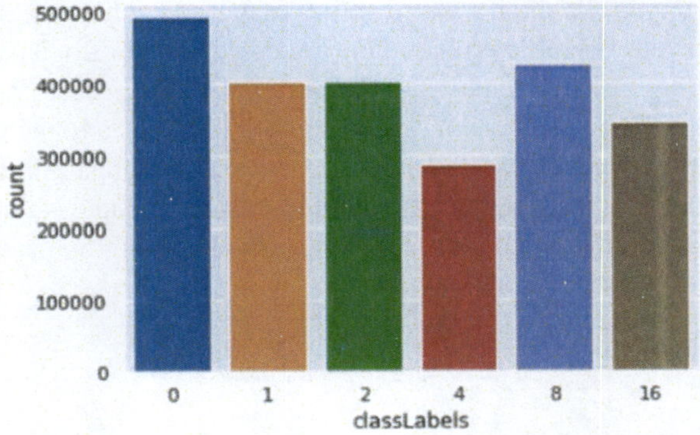

Fig. 6 Distribution of class labels

where Ps′ is the calculated sender vehicles' position according to time received in beacon message. If Ps and Ps′ are equal, then it is an authentic message with correct position information; else, its position information is falsified.

This data along with basic features is together used to train models using different machine learning algorithms.

Feature scaling, hyperparameter tuning and cross-validation techniques are performed. Feature scaling is the method to reduce down the value scale of features between zero and one. This is done to bring the entire data in a fixed range when there are highly varying magnitudes or units. If feature scaling is not done, then a machine learning algorithm can weigh higher magnitudes to be higher and lower magnitudes to be lower irrespective of their units. For example, 5000 m will be considered higher than 8 km and will predict wrong results.

Hyperparameter tuning parameter is the technique where the parameters of algorithm are chosen in such a way that it gives optimal solution. Parameters which define such a model are called hyperparameters, and the process of searching for this ideal model is called hyperparameter tuning. For example, deciding the maximum height of leaf in decision tree, number of layers in neural networks, number of neurons in neural network, etc., that gives optimized model.

Cross-validation is the method where the dataset is divided into different sets, and one of the sets is for testing, whereas the remaining others are for training. Once done with this, then one of the other sets is used for testing and remaining others are used for training and likewise repeated. So, here, all the data is being used for training and all the data is being used for testing. It is primarily used to estimate the skill of machine learning model on unseen data.

3.2.3 Training, Model Selection and Testing Phase

In this study, the data is split into training and test data using 80:20 split. For position falsification misbehavior detection scheme, different machine learning algorithms were implemented like Gaussian Naive Bayes classifier, logistic regression, decision tree, random forest, SVM, KNN, KNN + Random forest + Naive. Gaussian Naive Bayes classifier is based on Bayes theorem where the probability of occurrence of an event is predicted based on prior knowledge. Logistic regression is a classification algorithm which uses statistical approach for analyzing a dataset and determines the outcome. A decision tree is a graph that uses a branching method to illustrate every possible outcome of a decision. Random forest builds multiple decision trees and merges them together to get a more accurate and stable prediction. KNN, SVM and random forest are used together for pipelining by means of stacking classifiers.

4 Results

The metrics used for evaluating model performance are precision, recall, f1 score and accuracy. Precision defines the percentage of samples with a certain predicted class label actually belonging to that class label. Recall defines the percentage of samples of a certain class which were correctly predicted as belonging to that class. The f1 score is defined as the harmonic mean of precision and recall. Table 1 shows accuracy metric obtained for different machine learning algorithms.

We see random forest gives the highest of accuracy; hence, we use it for comparison with results obtained in [13]. Table 2 shows the precision, recall and f1 score metric comparison between the proposed works with [13].

The bias and the weights were chosen randomly at first; afterward, neural network will learn itself by applying multiple iterations doing forward propagation and backward propagation. Each unit in the convolution layer has unique weight and bias, and every unit performs weighted addition between pixels of an image and weights applied and add bias to it (Figs. 7 and 8).

5 Conclusion

The falsified position information in beacon can change drivers' behavior, can lead to accidents and could disrupt the network. Overall, it has a potential to cause hazardous situation. In proposed work, we have analyzed the beacon message information and proposed solutions using statistical approach and machine learning approach. Beacon information is obtained through VeReMi dataset, performed different computations

Table 1 Results: falsified position detection scheme

Model	Accuracy
Naive Bayes	0.23
Logistic regression	0.24
KNN	0.55
Random forest	0.94

Table 2 Results: comparison of the proposed scheme with SVM used in [13]

Attack	Proposed		[13] SVM	
	Precision	Recall	Precision	Recall
Type1	0.94	0.95	1	0.82
Type2	0.93	0.92	0.57	0.17
Type4	0.91	0.90	0.81	0.88
Type8	0.95	0.95	0.80	0.87
Type16	0.95	0.95	0.81	0.46

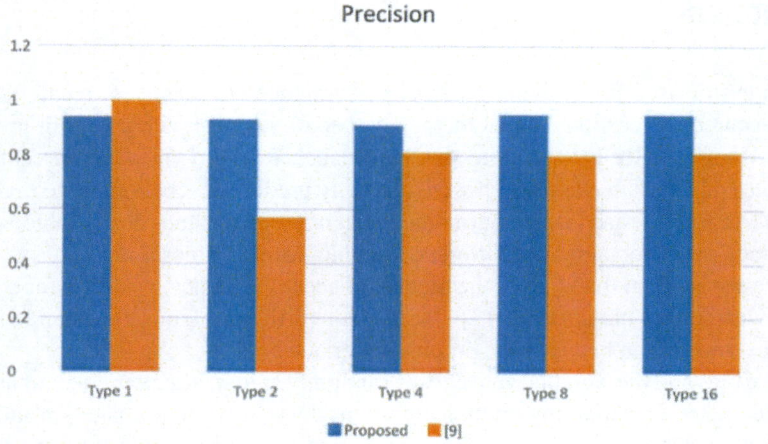

Fig. 7 Comparison of precision

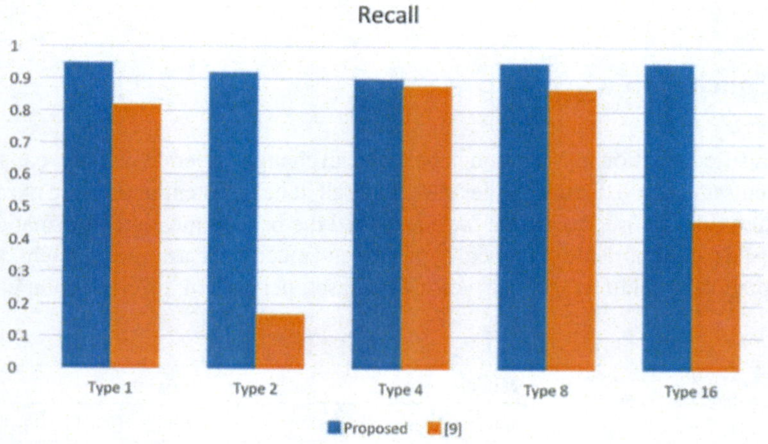

Fig. 8 Comparison of recall

to get additional features and then applied different machine learning algorithms for classifying the packet to be normal or malicious. We see that random forest performs best among the models used in the study. In comparison with [13], we see that this model with extracted features outperforms. It is powerful model which can be applied for position falsified misbehavior detection in VANETs.

As future work, we will use the same machine learning approach for detection of Sybil attack in VANETs. Sybil attacks can be simulated in VANET environment using simulation tools like SUMO, OMNET++ and Veins. From which, we can obtain a labeled dataset on which we can apply the above detection models for classification.

References

1. Zhang J (2011) A survey on trust management for vanets. In: 2011 IEEE international conference on advanced information networking and applications. IEEE, pp 105–112
2. Jakubiak J, Koucheryavy Y (2008) State of the art and research challenges for Vanets. In: 2008 5th IEEE consumer communications and networking conference. IEEE, pp 912–916
3. Qu F, Wu Z, Wang FY, Cho W (2015) A security and privacy review of VANETs. IEEE Trans Intell Transp Syst 16(6):2985–2996
4. Rawat A, Sharma S, Sushil R (2012) VANET: security attacks and its possible solutions. J Inf Oper Manag 3(1):301
5. Luckshetty A, Dontal S, Tangade S, Manvi SS (2016) A survey: comparative studyof applications, attacks, security and privacy in vanets. In: 2016 international conference on communication and signal processing (ICCSP). IEEE, pp 1594–1598
6. Wex P, Breuer J, Held A, Leinmuller T, Delgrossi L (2008) Trust issues for vehicular ad hoc networks. In: VTC Spring 2008-IEEE vehicular technology conference. IEEE, pp 2800–2804
7. Le A, Maple C (2019) Shadows don't lie: n-sequence trajectory inspection for misbehaviour detection and classification in VANETs. In: 2019 IEEE 90th vehicular technology conference (VTC2019-Fall). IEEE, pp 1–6
8. Arshad M, Ullah Z, Khalid M, Ahmad N, Khalid W, Shahwar D, Cao Y (2018) Beacon trust management system and fake data detection in vehicular ad-hocnetworks. IET Intel Transport Syst 13(5):780–788
9. Khan U, Agrawal S, Silakari S (2015) Detection of malicious nodes (dmn) in vehicularad-hoc networks. Proc Comput Sci 46(9):965–972
10. Ruj S, Cavenaghi MA, Huang Z, Nayak A, Stojmenovic I (2011) On data-centric misbehavior detection in VANETs. In: 2011 IEEE vehicular technology conference (VTC Fall). IEEE, pp 1–5
11. Celes AA, Elizabeth NE (2018) Verification based authentication scheme for bogusattacks in vanets for secure communication. In: 2018 international conference on communication and signal processing (ICCSP). IEEE, pp 0388–0392
12. Gyawali S, Qian Y (2019) Misbehavior detection using machine learning in vehicularcommunication networks. In: ICC 2019–2019 IEEE international conference on communications (ICC). IEEE, pp 1–6
13. So S, Sharma P, Petit J (2018) Integrating plausibility checks and machine learning for misbehavior detection in VANET. In: 2018 17th IEEE international conference on machine learning and applications (ICMLA). IEEE, pp 564–571
14. Kim THJ, Studer A, Dubey R, Zhang X, Perrig A, Bai F, Bellur B, Iyer A (2010) VANET alert endorsement using multi-source filters. In: Proceedings of the seventh ACM international workshop on vehicular internetworking, pp 51–60
15. Abu-Elkheir M, Hamid SA, Hassanein HS, Elhenawy IBM, Elmougy S (2011) Position verification for vehicular networks via analyzing two-hop neighbors in-formation. In: 2011 IEEE 36th conference on local computer networks. IEEE, pp 805–812
16. Sangare M, Banerjee S, Muhlethaler P, Bouzefrane S (2018) Predicting vehicles' positions using roadside units: A machine-learning approach. In: 2018 IEEE conference on standards for communications and networking (CSCN). IEEE, pp 1–6
17. van der Heijden RW, Lukaseder T, Kargl F (2018) Veremi: a dataset for comparableevaluation of misbehavior detection in vanets. In: International conference on security and privacy in communication systems. Springer, pp 318–337

Artificial Intelligence Fuelling the Health Care

Sahil Jindal, Archit Sharma, Akanksha Joshi, and Muskan Gupta

Abstract Artificial intelligence (AI) is defined as the power of intellectual human mind that designs an intelligent working system which proves in terms of computational power designed by the human intelligence. AI is designed in such a way that it commences to simulate the thought of brain, their thinking pattern, analysing approach and way of computing the problem. AI is one of the most notable fields in the current scenario of Fourth Industrial Revolution. AI is not a concept which finds its application in single field, but it can be used in many domains such as health care, medicines, evolutionary computation, security purposes, diagnosis and evaluation, image classification, accounting databases, transportation and smart cities. AI explores new routes of computation and follows heuristic approaches in order to solve biological problems. The present paper defines applications and current role of AI in health care and how it relates in studies including diagnosis process, image classification for diagnostic sciences, measuring the tendency of congestive heart failure, genetic analysis, drug discovery and much more. New techniques and methods are always a great tool to learn more, analyse more, as well as extract some useful information. Finally, a broad perception of this emerging topic is mentioned here to prove the positive role of AI with biotechnology and health care.

Keywords Artificial intelligence · Biotechnology · Diagnosis and health care

---

S. Jindal (✉) · A. Sharma
University Institute of Engineering and Technology, Kurukshetra University, Kurukshetra, India
e-mail: sahiljindal8991@gmail.com

A. Sharma
e-mail: architulet18@gmail.com

A. Joshi
Department of Applied Sciences, Limerick Institute of Technology, Limerick, Ireland

M. Gupta
MSM Institute of Ayurveda, BPSMV, Sonipat, India

N. Marriwala et al. (eds.), *Mobile Radio Communications and 5G Networks*, Lecture Notes in Networks and Systems 140,
https://doi.org/10.1007/978-981-15-7130-5_40

1 Introduction and Impact of AI

AI is a track in the field of computer technology that stresses on the curation of intelligent machines that perform tasks and think relatively just like humans as per the report present in brookings.edu. Innovative word like AI was first coined by John McCarthy in 1956 during his first academic conference. AI is not just a word, but a tool that forces the people to integrate the information that comes in their mind or enables them to rethink for enhancement purposes of previous given information [1]. AI improves the concept of decision-making by analysing the data and exerts the resulting insights in terms of transforming every walk of life [2]. AI is the analytic approach in the development of computer systems that matches the perception level of the human intelligence.

There are lots of working model to prove that level such as speech recognition by intelligent machine, visual perception in behalf of natural intelligence, decision-making skill and some more [3]. Human brain is considered as the supreme source of intelligence, and this is absolutely true. For the sake of simplicity, it is assumed naturally that AI can be accomplished by constructing a brain-like structure, matching with the concept of neuron like processing units working in collaborative manner. Intelligence is a term that is present in both human beings and the word AI that specifies the cognitive ability of human beings in contrary to the computational ability of the intelligent system.

AI is a technology and as you know human study comes under biology, so it is really a good approach when combine both terms and study the role of AI in biotechnology [4]. In simple words, AI has taken root in different areas, increases the efficiency of computation, provides efficient result and cures a problem in most statistical way. AI also increases the employability status that reduces depression level of unemployability in the candidates. Some of the common form of AI is machine learning, neural networks, deep learning and natural language processing.

There are lots of divisions where AI is used such as AI techniques smartly working in the power system stabilizer (PSS) design. It is used to add damping to electromechanical oscillations [5]. It is also used for the security purposes for protecting computer and communication networks from intruders. Also, there are lots of factors where AI shows their potential in almost every field of medical area. From cure to treatment, it shows miracles.

AI is not stopped there, and it follows broad prospect and shows their capability and effectiveness in terms of computational power. From accounting databases to large volumes of data, an accurate approach with or without direct participation of the decision maker to be possible with the investigation role of intelligent system such as AI. It changes the theme and visualizing pattern of gaming world. AI adds the features to the games such as path finding for the non-playing character, decision-making and real-time learning and some more. AI does not give a command to see a future, and it is a realistic term which is here today and established into a variety of domains. These include fields of finances, securities, health care, transportation, smart cities, etc. AI makes an impact on the world by following the natural intelligence and reinforcing human potentials in extreme ways [6].

2 Scenario of AI with Health Care

Historically, biology field is used in correlation with the deep learning which is one of the parts where AI shows their existence and to be more meticulous where the whole scenario is revolving around statistics and genetics. It is noted that Gregor Johann Mendel's experiment of pea breeding provided decisive data for the expansion of the statistical theories given by Pearson and Fischer. Heredity laws uncovered by Mendel provided crucial data which gives the pea plant genetic factors that are further moved to the offspring as stated by a certain direction to determine the traits of offspring [7]. Relationship between genetic factor and the observed traits is described on the collaborative studies on statistical models defined by Mendel, Pearson and Fischer.

As compared with the image or text data sets, biological data is much more networked and multidimensional. Indeed of this, you can say that a simple statistical model for biological data sets is more appropriately represented using AI model. You can consider to uncover the genetic factors as the key objective of biological research. Based on the automated robots, results of high-speed DNA synthesis technology created by the J. Craig Venter Institute [8] are calculated on the basis of first artificial micro-organism and the artificial yeast synthesis project [9].

In health care, AI plays differently such as AI tools helps designers to improve computational sophistication. Let us take an example of Merantix, a German company [1] that applies deep learning concepts to originating medical issues. This includes lymph node detection using computer tomography (CT) images in the human body [10]. This one is not a such type of computation that humans cannot perform but the problem is that a radiologist may be able to carefully read only four images an hour and charge $50 per hour as per their need. Let us consider 50,000 images, the cost of this process would be $625,000, which is extremely high, and the casualty offer in terms of increase in cost and decrease in computation power. Now the question is that how AI solves this situation; basically, it trains intelligent machine on given data sets which is created on the basis of previous computation that learns to differentiate between the lymph node regularities. So, here AI plays their role and increases the computation power and correctness of the labelling. Further, radiological with the help of this intelligent system apply imaging constraints to patient and calculated the thoroughness of lymph nodes either is at risk of cancer or not.

In AI, fuzzy logic is a concept which is used for data handling methodology but sometimes it permits ambiguity and hence this type of concept plays their role in the medical applications. AI role is there in a way that it uses the conviction of fuzziness in a computationally effective manner. Second role is the evolutionary computation, in which natural selection method is taken into consideration for the best fit in unfold real-world problems. Genetic algorithms are the most useful example for medical applications used as an evolutionary computation. In the medical area, AI is used for the enhancement purposes and improves the steps of decision-making. And the example of such area where AI is used such as Clinical decision support system, i.e. one of the first successful application of AI. As well as analysis of model-based intelligent system and tools for decision-making are important in medical imaging for computer-assisted diagnosis and evaluation. Here, AI techniques are used in the biomedical image classification for diagnostic sciences. Some common examples are artificial neural networks focusing on diagnostic science, endoscopic images and MRI brain tumour analysis.

One of the big roles of AI has been applied to measure the tendency of congestive heart failure, an ailment that is mainly seen in the senior citizens. In order to cure a disease, a prevention on time is necessary, so AI tools are benevolent here because they estimate the potential challenges in advance and dispense the proper resources to patient education, sensing and dynamic interventions that keep patients out of the hospital [11]. Currently, genetic analysis is one of the most obvious implementations of ML techniques for diagnostics named as Sophia Genetics [12], a Swiss-based start-up that exemplifies the state of the art. They take a blood sample from the patient and, with the help of their powerful analytical AI algorithms, process it and then analyse the data. Nowadays, gene editing or data analysis done in the laboratory is handed to AI programs as a form of secretarial work [13].

Using CRISPR, desktop genetics that works through AI designed constructs for gene editing. In health care, a booming business is created by AI with the development of drugs around 21% of global demand for pharmaceuticals [14]. In the last decade, the emergence of biotechnologies seen in many areas such as technology is used to convert any blood type into universal donor type O, human organs printed by SD printers, monitoring fetal genetic abnormalities using non-invasive techniques & now infamous CSISPR/cas9 gene-editing technique have inspired a serious ingress of interested investors into biotechnologies. In medical field, case sheets of each and every patient are maintained. These case sheets carry all information related to the patient's disease. Some of the disease like STDs has a marked social stigma attached with them. Because of such reason, patients do not want to disclose them. Nowadays, AI offers 100% privacy to data. Fusion of AI with biotechnology I shown in Fig. 1.

3 AI is the Future of Health Care

Whether it was Google, Microsoft and any other big MNCS, everyone was trying to explore the new era of artificial intelligence in each and every field. No matter how big the deal is, but each and everyone was trying to grab the new in AI. But above all this, use of AI in biotechnology proved to be a bloom for mankind. In the discovery and development of drugs, AI plays a crucial role. Some sort of new chemical entities can be recognized that based on the discovery of drug metabolites. The conventional method used in drug discovery was very obsolete as it is very much time consuming and fleeting as it takes several years and a lot of investment averaging over $2.5B or more [15].

Apart from all these, major drawbacks are they are labour- and resource-intensive. Some of the drugs failed due to economic and technical reasons and the wrong combination of compounds. AI resolves all these problems. Its effectiveness was amazing,

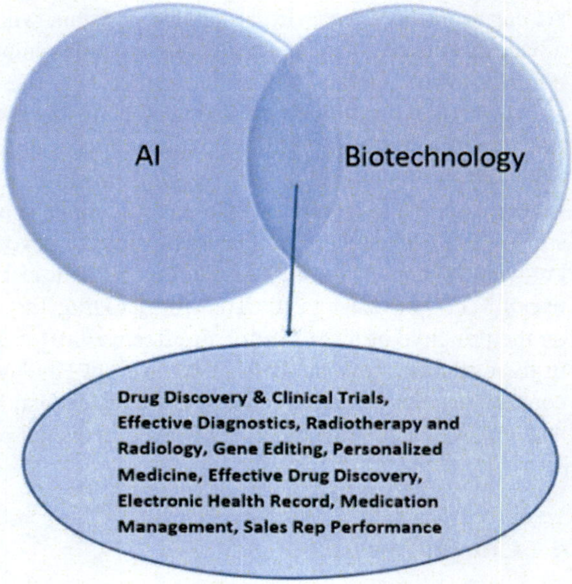

Fig. 1 Fusion of AI with biotechnology. From the discovery to development when AI works with the biotechnology, the fusion or the role of both increases the experiment role of many applications in the area of healthcare

for example, deep learning algorithm, in which model is defined that calculates 70% accurate reaction types. Result demonstrates AI has certain predictive ability and application value. Another exclusive feature of AI is its assistance to people with information gathering, data crunching, routine customer service and physical labour, thereby freeing them for higher-level tasks that require leadership, creative thinking, judgement and other human skills. All of the above-mentioned facts prove that AI had an extreme impact on humanity at large.

Humans have a tendency to make things smarter and smarter. Another example is where humans enabled a computer to ingest data, process it and provide an outcome. In layman terms, this is cognition and technologies that enable cognition are AI in short. Surely, this will impact human race at a large scale. Even today, we have AI complimenting humans and not replacing them; for example, AI assesses an X-ray and helps a doctor to arrive at better diagnosis. This fascination with biotechnological advancement makes life comfortable for all. Humans, their expectations and field associated with them are vast, and they continuously tried their best to chase them to make life more comfortable and easy. Amongst all these fields, health care is one of the most highlighted and star marked. AI plays a vital role in promoting and improving health care. In many fields like oncology, neurology and immunology, works of AI were really fruitful [15]. From hours to minutes, then confinement to just mini seconds was really amazing. Primary aim of health-related AI application is to analyse the input and patient outcome.

Another most important aspect of AI in medical field is its nursing assistance. Hospitals cannot be run alone by single member, and all the working members have to work as a single unit. A single mistake can deteriorate the condition of patient.

AI can work here as a virtual nursing assistant which can give accurate medical advice and observe patient's health based on the symptoms that a person can initially describe to an AI model (i.e. virtual nurse).

AI also plays an important role in inventing a new drug, which was not an easy task to perform. It takes millions of expenditure, and along with these it was a time-consuming process. But in the emerging situations like pandemics or epidemics, it is very essential to launch a new drug in a limited span of time. Currently, a disease spreads like a fire such as coronavirus. Scientists have no idea related to its treatment, but with the help of AI they can do this difficult task in a short time span with the use of MTDTI (i.e. Molecule Transformer Drug Target Interaction) [16]. It is based on the principle of identification of effective drug based on the compounds present in them. Finally with the help of AI, we got Favilavir, first effective drug against coronavirus. So, we can say that AI can execute healthcare tasks better than humans and also AI can shorten the time involved in overall process.

4 Conclusion

Artificial intelligence is not just two or more words or phrases that can be concluded. It is a perspective that can be explored on a large scale. In every field, it works like multi-specialist. Its accuracy, precision in work and time-saving qualities make it different from others. This is the main reason for the flourishment of AI in all sectors specially biotech of biomedical. Right from the diagnosis to treatment, it fulfils all the tasks with such an accuracy that was impossible for human being to achieve it alone. Right from the emergence of existence of human being on the earth, evolution was taken place. It is an ongoing process. From human beings to animals, a lot of major and minor changes were going to happen for the sake of betterment, to adjust themselves according to changing external factors like temperature, weather, etc. They want to live their life in a standardized manner. There is no chance of mistake. And with the advent of AI in every field specifically in biotechnology, our dream of being perfect looks some mile away from us. AI has taken root in biotechnology and increase the efficiency of computation provides efficient result. AI agents, in turn, can assist people with information gathering, data crunching and physical labour so that AI will flourish well. Early enthusiasm for the application of AI must be needed. So, it can be concluded that AI can suppress the large-scale automation of healthcare professional jobs for a considerable period. Hence, we cannot deny its existence in our life. It is the need of the hour that we all work in collaboration with the flourishment of AI to make our present secure and future brighter.

References

1. West DM, Allen JR (2018) How artificial intelligence is transforming the world. ABA J 202.238.3507
2. Irizarry-Nones A, Palepu A, Wallace M (2020) Artificial Intelligence (AI). https://cisse.info/pdf/2019/rr_01_artificial_intelligence.pdf. Last accessed 24 Feb 2020
3. Wang P (2008) What do you mean by AI. Front Artifi Intell Appl J 171(1):09226389
4. Pharma B (2018) The robots are coming: is AI the future of biotech? https://www.labiotech.eu/features/ai-machine-learning-biotech/. Last accessed 24 Feb 2020
5. Pannu A (2015) artificial intelligence and its application in different areas. Certified Int J Eng Innov Technol (IJEIT) 4(10):2277–3754. M. Tech Student, Department of Computer Science & Engineering, DAV Institute of Engineering and Technology, Jalandhar, India
6. West DM (2018) The future of work: robots, ai and automation. Brookings Institution Press, pp 1–205
7. Kim H (2019) AI, big data, and robots for the evolution of biotechnology. Genomics Inform J 17(4):0742–2234
8. Gibson DG, Glass JI, Lartigue C, Noskov VN, Chuang RY, Algire MA et al (2010) Creation of a bacterial cell controlled by a chemically synthesized genome. Science 329:52–56
9. Richardson SM, Mitchell LA, Stracquadanio G, Yang K, Dymond JS, DiCarlo JE et al (2017) Design of a synthetic yeast genome. Science 355:1040–1044
10. Rothe R (2017) applying deep learning to real-world problems. Medium J 1–17
11. Horvitz E (2016) Reflections on the status and future of artificial intelligence. In: Testimony before the U.S. senate subcommittee on space, science, and competitiveness, p 5
12. Crunch E (2020) Swiss data analytics company sophia genetics could be Switzerland's next unicorn. https://techcrunch.com/2017/01/02/swiss-data-analytics-company-sophia-genetics-could-be-switzerlands-next-unicorn/. Last accessed Feb 24 2020
13. San Francisco AS (2016) The software secretaries. Econ J 2–5
14 Rue N (2019) Trending AI articles: 1. Making a simple neural network 2. From perceptron to deep neural nets 3. Neural networks for solving differential equations 4. Turn your Raspberry Pi into homemade Google home biotechnology is big business. J
15. Hernandez D (2017) How AI is transforming drug creation. Wall Street J 1–5
16. Beck BR, Shin B, Choi Y, Park S, Kang K (2020) Predicting commercially available antiviral drugs that may act on the novel coronavirus (2019-nCoV), Wuhan, China through a drug-target interaction deep learning model. BioRxiv J. 2020.01.31.929547

A Systematic Review of Risk Factors and Risk Assessment Models for Breast Cancer

Deepti Sharma, Rajneesh Kumar, and Anurag Jain

Abstract Breast cancer is the utmost frequently occurring as well as the most common reason for cancer-related deaths among women community worldwide. In Indian females, breast cancer ranks with the highest rate as 25.8 out of 100,000 with the mortality rate of 12.7 per 100,000 women. Early detection and accurate diagnose will facilitate the clinicians to fight against this deadly disease worldwide. To differentiate between the patients at higher risk and lower risk of breast cancer, various risk factors and risk analysis models have been developed. Machine learning-based models help in the categorization of high-risk and low-risk patients. Once categorized properly, high-risk patients require more surveillance, prophylactic count, and other preventive measures like chemoprevention or surgery. Patients with low risk should also be kept under surveillance to minimize the probability to turn in high-risk patients. In this paper, the authors have identified the key risk factors for breast cancer. The authors have done a systematic review of different risk assessment models for breast cancer.

Keywords Breast cancer · Breast cancer gene (BRCA) · Risk assessment · Breast cancer gene carrier probability (BRCAPRO) · Breast and ovarian analysis of disease incidence and carrier estimation algorithm (BOADICEA)

1 Introduction

In contemporary health challenges in women all over the world, breast cancer is responsible for significant morbidity and mortality. Breast cancer constitutes about

D. Sharma (✉) · R. Kumar
Department of Computer Science and Engineering, Maharishi Markandeshwar (Deemed to be University), Mullana, Ambala, India
e-mail: deeptisharma85@gmail.com

A. Jain
Department of Virtualization, School of Computer Science, University of Petroleum and Energy Studies, Dehradun, India

N. Marriwala et al. (eds.), *Mobile Radio Communications and 5G Networks*, Lecture Notes in Networks and Systems 140,
https://doi.org/10.1007/978-981-15-7130-5_41

27% of cancers affecting humans and about 25% of cancers affecting females in developed countries. Although it can affect both males and females, it is about a hundred times more commonly found in females compared to males [1].

Among the women in the age bracket of twenty to thirty-nine, breast cancer is the subsequent reason for cancer-associated deaths worldwide. In breast cancer history, both clinical and pathological factors are involved. Usually, breast cancer starts with the normal proliferation of the outer skin of the breast to carcinoma, carcinoma to carcinoma in situ, and ultimately carcinoma in situ to metastatic diseases. Breast lesions can be of two types, malignant and non-malignant [2].

Increasing age cases of breast cancer is an important issue. Timely detection of breast cancer improves the survival rate of patients, and overall, the key factor in improving the mortality rate in cancer is its early detection and accurate diagnosis. However, the increase in patient numbers and experienced clinicians or pathologists is a big challenge for inaccurate diagnosis. The imbalance of medical resources in allocation also increases the chance of the wrong diagnosis. In this scenario of increase incidence of cases and challenges for providing a base for early diagnosis and treatment, a machine learning-based reliable program helps the doctor in breast cancer diagnosis in a fast and reliable manner [3, 4].

1.1 Overview of Breast Cancer Risk

Breast cancer is a collection of various breast diseases in which uncontrolled divisions and changes in the tissues of breast cells occur. This division and change of tissues result in the formation of mass and lumps. The main origination of cancer in the breast is the lobules (milk glands) or in the ducts of the cells. Breast cancer can be diagnosed by Clinical Breast Examination (CBE) or by screening through mammograms or MRIs. Breast cancers are categorized into two types: In-Stu (localized) and invasive or infiltrating, means the walls or glands of the abnormal cells have broken up and spread to the neighboring cells. Timely detection of cancer helps in the increase of the possibility of survival and cure of patients. Proper risk assessment and the study of factors associated with these risks are significant tools for this [5].

1.2 Significance of Risk Assessment

Risk assessment is a term used to understand the comprehensive methods to calculate the risks that may be the potential to cause some dangerous results. Breast cancer is a crucial challenge for healthcare professionals on both diagnosis and treatment part. The prevalence of breast cancer is increasing worldwide, which ultimately results in the mortality of the patients. Breast cancer diagnose at late stages led to an increase in death rates. Early detection of cancer helps in reducing the risks of diagnosing and improves the survival and cure rate of patients. Timely detection of cancer is important in both cases, early diagnosis in the significant women and screening in the insignificant women [6, 7].

1.3 Risk Assessment Modeling

Different statistical models represent the breast cancer risk assessment that has been designed and validated for this purpose. The main constraint that is related to these models is sensitivity, specificity, and accuracy. Variable results are received when we independently validate all the existing models. Some models predict risks in mutation carriage in breast cancer, but the accuracy is the main discriminatory factor. Receiver operating characteristic curves best represent a balance between specificity and sensitivity. An area under the curve (AUC) determines a model's accuracy. AUC is also known as C-statistics. A perfect, accurate model has AUC of 1.0, while 0.8 or 0.7 is as of good accuracy, and, at last, AUC of 0.5 is a matter of chance for a model.

This paper has been structurally organized as follows: Section 2 lays down the various risk factors involved with breast cancer, Sect. 3 discusses the various existing models involved with risk assessment in breast cancer, Sect. 4 is about the comparison of different models according to the risk factors involved in them, Sect. 5 describes the significance of machine learning in breast cancer prediction, and Sect. 6 presents the paper conclusion and future possibilities.

2 Risk Factors for Breast Cancer

Risk factors involved in a disease are something that inclined the feasibility of developing a disease in an individual. Broadly, the breast cancer risk factors can be of two types: environmental and host factors. Environmental factor in the presence of previous incidence of breast cancer in family members is strongly related to new incidences. Apart from it, host factors associated with risk are endocrine imbalance, excess body weight, benign breast proliferation, poor nutrition, and sedentary lifestyle with habits like smoking and alcohol. These together or in isolation damage DNA, leading to the mutation in the BRCA family of the cancer suppressor gene. This genomic cause, although harder to diagnose and manage, is the future of breast cancer management [5].

To estimate the risks of breast cancer efficiently, we need to identify and quantify the risk factors involved in breast cancer. There are various aspects involved in the increased incidences of breast cancer. Many of them are summed up here.

2.1 Risk Factors

Family background related to the breast cancer: Despite fewer incidences of mortality from breast cancer every year, breast cancer is still a deadly disease and burden for the doctors [8]. Many personal and family constraints involve in this like:

(i) Age/gender: Growing age is a crucial risk aspect for breast cancer. Two-third cases of breast cancer rise and appear after the age of fifty-five. Also, breast cancer frequently appears in women as compared to men [9].
(ii) Unilateral or bilateral diseases.
(iii) Degree of relationship: Younger age diagnosis of breast cancer in the family may increase the breast cancer risk thrice in their\first-degree relatives (mother, sister, and daughter).
(iv) History of previous family cases of breast cancer.
(v) Related early tumors diagnosed.
(vi) Personal history of malignancy.
(vii) Hormonal and reproductive factors of breast cancer risk: Various aspects like early onset of menarche (less than twelve years) or prolonged age of menopause (more than 55 years) are concerned with the rise in the risks of breast cancer. Every year, 4% of cases of total breast cancer cases are by cause of decrease age at menarche and 3% cases with the association of late menopause [10].
(viii) Late age at first pregnancy (age greater than 35) also inclined the concern for breast cancer. In females having first pregnancy at age of 30 years is more vulnerable for the risk for breast cancer than the women giving birth around age of 20. Breastfeeding is linked with reduced breast cancer risks; therefore, an increased no. of years of breastfeeding were necessary for the impact on risks [11].
(ix) Mammography density: Important risk factor mammography density is defined as the breast tissue proportion on a mammogram, which is denser than the surrounding tissues. The mammogram can rapidly and reliably measure breast density, and such mammogram data could be pre-owned accurately in risk estimation along with breast cancer after-treatment prognosis [12].
(x) The proliferation of breast diseases: Many types of breast proliferations are linked with breast cancer risk. Diseases like LCIS and DCIS or benign condition of the breast are ten times more prone to breast cancer. Ductal hyperplasia or lobular hyperplasia has immense possibility of breast cancer [13].
(xi) Oral contraceptive pills: Continuous intake of birth control pills (BCP) is also related to the rise in breast cancer incidences. The risk seemed to disappear after stop taking the pills [8, 14].
(xii) Ionizing radiation exposure: Atomic bomb survivors in Japan are seen with increased breast cancer's risk. Patients suffering from Hodgkin's disease below 20 years age also have chance of developing breast cancer. Therapeutic radiation in breast zone in the first thirty years of the lifetime also inclined the risk of breast cancer [15].

2.2 *Lifestyle-Related Risk Factors*

Many epidemiological studies conclude some lifestyle factors are also responsible for breast cancer. Sedentary lifestyle and inactivity in daily life also led to increasing the risk.

(i) Diet/nutrition: Excess intake of dietary fat is directly proportional to increased incidence of breast cancer.
(ii) Excess body weight and weight gain: Postmenopausal weight gain is strongly linked with the risks of incidence of breast cancer.
(iii) Smoking and alcohol consumption: Smoking and alcohol intake are strongly linked to increased risks of breast cancer.
(iv) Physical inactivity: Regular physical activity, especially postmenopausal, has effective protection against breast cancer. Brisk walking of minimum 10 h in a week by women benefits them against fighting cancer, according to the WHI study.

2.3 *Other Risk Factors*

According to recent studies, the role of vitamin D and calcium is associated with the risk of breast cancer. If during premenopausal period females have taken low dairy fat products, vitamin D, and calcium, it is associated with increased breast cancer risk. Various other factors like genetic predisposition, height, breast density, menstruation cycles, breastfeeding, environmental pollution, etc., are some other factors involved in risk [16] (Table 1).

3 Risk Factor Incorporation in Models for Risk Estimation

Models for cancer risk estimation are the combination of risk aspects involved in breast cancer and their prediction as an output of the model over a specific time or lifetime of the patient. The frequent risk aspects responsible are pedigree and its advanced stage. The lifetime risk (LTR) is defined by many risk assessment models [17]. The risk assessment models are classified into two classes:

(i) Empirical models.
(ii) Genetic risk prediction models or statistical models.

These models are used by the doctors to decide the patient is surveillance or not. Some models are implanted on the pedigree of patients, while some include other related factors. The main task of the models is basically to find the BRCA1/2 gene mutation of the patients to ultimately find the aspects of breast cancer.

For healthy women, those who want to find about their breast cancer risk in the future, the risk estimation is calculated by examining their pedigree. The most difficult task for clinicians is to decide the correct model for an individual from a large diversity of models.

To find organized analysis of all the available risk assessment models, different models have a variable set of risk association, pedigree, and methods to calculate the risk in an individual. There is a variety of scoring systems, and models have derived the possibility of BRCA1 or BRCA2 mutation in an individual [18].

(i) *Empirical Models*: These models detect the BRCA1 and BRCA2 mutation possibility on genetic testing. The empirical models consist of some early models like the Shattuck-Eidens model and Couch model. The Couch model further improvised as the Penn II model. The tabular scoring system is also a part of empirical models. To simplify the time-consuming scoring system, two similar screening models developed. These are family history assessment tool (FHAT) and the Manchester model. Another group of empirical models, such as the Australian LAMBDA model and National Cancer Institute model, both are used for the calculation of risk in a specified population [19].

(ii) *Genetic Risk Prediction Model*: The models that used for emphatic conclusion about the genes convoluted and frequency of alleles carried out by an individual and the aspects of breast cancer carried out by these alleles are under this category of model. These illustrations use parentage background, personal history, and the pedigree of the individuals. Mainly, they calculate the cancer risk. The models are Gail model, BRCAPRO, Breast and Ovarian Analysis of Disease Incidence and Carrier Estimation Algorithm (BOADICEA), Yale University model, International Breast Cancer Intervention (IBIS) model) also called as Tyrer–Cuzick model, Claus model. These models are summarized in more detail below [20].

(iii) *Gail Model*: The utmost common model is Gail model, and model is known for the risk assessment in breast cancer. It was initially designed in 1989 and then further modified in the year 1999 and called as NCI Gail model. The Gail model discussed six main risk factors. These were the age of the patient, hormonal and pattern of reproduction (age at menarche, age at menopause, age at first live pregnancy), history of breast disease, and family history (first-degree relatives in the family) [21]. Gail model has been approved in large no. of community and provides accurate results. The Gail model provides low results when tested individualized risk assessment in the high-risk populations.

(iv) *Claus Model*: The Claus model contained data from cancer patients and steroids. Breast cancer sufferer enrolled in Surveillance, Epidemiology, and End Results Region (SEER), and the authentic model includes data from ovarian cancer. Claus model uses family background to predict breast cancer risk. It also includes the first-and second-degree relatives who have breast cancer and the age of those relatives at the time of cancer diagnosed. The Claus model also included paternal lineage with breast cancer.

(v) Major drawbacks of Clause model are: This model did not include hormonal and reproductive patterns. The calculated values and the computerized results of this model have shown the difference in their results. The calculated values give a higher figure of results as compared to the computerized results. This discrepancy has resulted because the calculated values do not make adjustments with the unaffected relatives [21].

(vi) *Breast Cancer Risk Assessment Tool (BCRAT)*: BCRAT is also called as modified Gail model (Gail 2). It is adjusted particularly for the population of African–Americans in 2007, which is further modified again for the Asian populations. It is validated using WHI data [22].

(vii) *BRCAPRO Model*: BRCAPRO is used to calculate mutation of BRCA1, BRCA2, or both based upon suffering of ovarian cancer in the first- or second-degree relatives or the person's personal history. Bayes' rule of determination of a mutation calculated overall risks for family history. The main advantage of this model is that it is used for affected or unaffected patients both. It helps the clinicians in the decision to refer the patients further for genetic testing. The drawback of this model is that none of the non-transmissible risk aspect have been used yet for calculation; therefore, it underestimates the risk in women having nonhereditary risk factors [23].

(viii) *Jonker Model*: Jonker model published a genetic model for breast cancer prediction using family history and ovarian cancers. Jonker's model is an expansion of Claus model in the combination of the BRCAPRO model. BRCA1, BRCA2, and a hypothetical gene BRCAu explained the familial breast and ovarian cancer. Hypothetical gene showed all other familial clustering not included in the BRCA1 and BRCA2. Jonker's model does not involve the personal risk factor data; it is also called a Claus extended model. Limitations of the Jonker model are that it is unable to estimate the risks in women with complicated parentage [24].

(ix) *IBIS (Tyrer–Cuzick) Model*: International Breast Intervention Study integrates family history, measures of endogenous estrogen, or benign cancer diseases. Unlike the Jonker model, these models are used for the calculation of BRCA1 or BRCA2 mutations in an individual for ten-year period or lifetime estimation of breast cancer risk. This tool helps in the decision about the genetic counseling and testing of a patient. If the prediction for the mutation is 10% or greater, patient is advised for the genetic counseling [18].

(x) *BOADICEA Model*: Breast and Ovarian Analysis of Disease Incidence and Carrier Estimation Algorithm, this algorithm predicts the probability of mutations and the association of cancer in an individual depending upon the family history. Familial transmission of the disease is calculated and not accounted for BRCA1 or BRCA2 mutations [22].

Fig. 1 Prediction system in health care

4 Comparative Analysis of Model Accuracy

Analysis of prediction models depends upon the accuracy of predicting the breast cancer risk in patients in for 5 years, 10 years, or lifetime. Few models perform with good results than other in certain prospects and availability of data. Specificity and sensitivity are also the main concerns in providing the ROC for a model [18] (Table 2).

There are some practical limitations of the models, that are computer-based programs needs time to input all the relevant data, adoption of any particular model is also a concern, small family size, insufficiency of data about the parentage background also decreases the efficiency of any model up to the same extent as a case of missing value [18].

5 Machine Learning in Breast Cancer Prediction

Diagnosis of breast cancer and the classification of patients into malignant or non-malignant are the main concerns nowadays. Artificial intelligence and machine learning play an important role in critical applications, such as image recognition, natural language processing, time series forecasting, regression, and prediction. The use of an accurate machine learning algorithm for early detection could save precious lives. Feature selection and identification from obscure breast cancer datasets are important aspects. Machine learning (ML) is a preference in breast cancer pattern classification and forecast modeling. Machine learning or deep learning models have provided significant results from the previous studies for the classification of breast cancer.

Prediction system in health care helps in the automation of the processes and in the prediction of the disease. Machine learning-based prediction models predict the disease in an early stage of disease to provide more accurate treatment [25]. Figure 1 explains the process of healthcare prediction system using machine learning in detail.

6 Conclusion

Many different types of risk estimation models are validated to find BRCA1 and BRCA2 mutations in the patients to decide the associated risk of breast cancer in any individual. Several studies have independently validated or compared and resulted that no one type of model is good. Empirical and genetic models can differentiate based on mutation carriers and non-carriers. Accuracy, sensitivity, and specificity of risk models may vary, based upon the test population. An accurate, individualized risk prediction model is necessary so that the appropriate patients may be selected for further treatments. Daily new researches for the improvements in the models are carrying out, and we need to revalidate those changes.

Table 1 Risk factors involved in the breast cancer

Category	Risk factors
History of cancer involved in family	One or more first-degree relatives suffering from cancer BRCA1 and BRCA2 mutations
Hormonal and reproductive factors	Early menarche (onset age less or equal to 12 years) Late menopause (onset age greater or equal to 55 years) Delayed childbearing or first pregnancy (age after 35) Hormone replacement therapy
Mammographic density	Mammographic screening of breast
The proliferation of breast disease	A proliferative lesion showing atypical features or could be without it
Oral contraceptive pill	Birth control pills
Ionizing radiation exposure	Exposure to radiation in the first three decades of life in the breast tissue region
Personal history of malignancy	Earlier cases of breast cancer in a family
Factors related to lifestyle	Obesity Smoking and alcohol Physical inactivity Fat intake

Table 2 Assessment of models with risk factors

Personal/hormonal factors	Breast disease	Hereditary
Factors: • Menarche age • Menopausal age • Hormone therapy • Obesity • Alcohol and smoking • Exercise	Factors • LCIS • DCIS • Breast density • Breast biopsies • Radiation exposure	Factors: • Family history breast cancer • Ovarian cancer • Mutation carrier
Models: • Gail model • BCRAT		Models • Claus model • BRCAPRO model • BOADICEA
Models • IBIS (Tyrer—Cuzick) model		

References

1. Global Health Observatory (2018). Geneva: sWorld Health Organization; https://www.who.int/gho/database/en
2. Siegel RL, Miller KD, Fedewa SA, Ahnen DJ, Reinier GS Meester, Afsaneh Barzi, Ahmedin Jemal (2017) Colorectal cancer statistics. CA: A Cancer J Clin 67(3):177–193
3. Burstein HJ, Polyak K, Wong JS, Lester SC, Kaelin CM (2004) Ductal carcinoma in situ of the breast. N Eng J Med 350(14):1430–1441
4. Easton DF, Pooley KA, Dunning AM, Pharoah PD, Thompson D, Ballinger DG, Struewing JP (2007) Genome-wide association study identifies novel breast cancer susceptibility loci. Nature 447(7148):1087–1093
5. Breast Cancer Facts and Figures (2019–2020), https://www.cancer.org/content/dam/cancer-org/research/cancer-facts-and-statistics/breast-cancer-facts-and-figures/breast-cancer-facts-and-figures-2019–2020.pdf
6. Tryggvadottir L, Sigvaldason H, Olafsdottir GH, Jonasson JG, Jonsson T, Tulinius H, Eyfjörd JE (2006) Population-based study of changing breast cancer risk in Icelandic BRCA2 mutation carriers 1920–2000. J Natl Cancer Inst 98(2):116–122
7. Evans DG, Shenton A, Woodward E, Lalloo F, Howell A, Maher ER (2008) Penetrance estimates for BRCA1 and BRCA2 based on genetic testing in a clinical cancer genetics service setting: risks of breast/ovarian cancer quoted should reflect the cancer burden in the family. BMC Cancer 8(1):1–9
8. Calle EE, Heath CW, Miracle-McMahill HL, Coates RJ, Liff JM, Franceschi S, Talamini R (1996) Breast cancer and hormonal contraceptives: further results: Collaborative group on hormonal factors in breast cancer. Contraception 54(3):1–106
9. Jemal A, Siegel R, Ward E, Hao Y, Xu J, Murray T, Thun MJ (2008) Cancer statistics. CA A Cancer J Clin 58(2):71–96
10. Carey LA, Perou CM, Livasy CA, Dressler LG, Cowan D, Conway K, Karaca G (2006) Race, breast cancer subtypes, and survival in the carolina breast cancer study. JAMA 295(21):2492–2502
11. Jones LA, Chilton JA (2002) Impact of breast cancer on African American women: priority areas for research in the next decade. Am J Public Health 92(4):539–542
12. Boyd NF, Guo H, Martin LJ, Sun L, Stone J, Fishell E, Jong RA (2007) Mammographic density and the risk and detection of breast cancer. N Engl J Med 356(3):227–236
13. Dupont WD, Parl FF, Hartmann WH, Brinton LA, Winfield AC, Worrell JA, Schuyler PA, Plummer WD (1993) Breast cancer risk associated with proliferative breast disease and atypical hyperplasia. Cancer 71(4):1258–1265
14. Marchbanks PA, McDonald JA, Wilson HG, Folger SG, Mandel MG, Daling JR, Bernstein L (2002) Oral contraceptives and the risk of breast cancer. N Engl J Med 346(26):2025–2032
15. Kelsey JL, Gammon MD, John EM (1993) Reproductive factors and breast cancer. Epidemiol Rev 15(1):36–47
16. Thiebaut AC (2007) Victor kipnis: dietary fat underreporting and risk estimation. Public health nutrition 10(2):212–213
17. Dhahri H, Al Maghayreh E, Mahmood A, Elkilani W, Faisal Nagi M (2019) Automated breast cancer diagnosis based on machine learning algorithms. J Healthc Eng
18. Amir E, Freedman OC, Seruga B, Evans DG (2010) Assessing women at high risk of breast cancer: a review of risk assessment models. JNCI J Natl Cancer Inst 102(10): s680–691
19. Alghunaim S, Al-Baity HH (2019) On the scalability of machine-learning algorithms for breast cancer prediction in big data context. IEEE Access 7:91535–s91546
20. Cintolo-Gonzalez JA, Braun D, Blackford AL, Mazzola E, Acar A, Plichta JK, Griffin M, Hughes KSHughes (2017) Breast cancer risk models: a comprehensive overview of existing models, validation, and clinical applications. Breast Cancer Res Treat 164(2):263–284
21. Tirona MT, Sehgal R, Ballester O (2010) Prevention of breast cancer (part I): epidemiology, risk factors, and risk assessment tools. Cancer Invest 28(7):743–750

22. Ming C, Viassolo V, Probst-Hensch N, Chappuis PO, Dinov ID, Katapodi MC (2019) Machine learning techniques for personalized breast cancer risk prediction: comparison with the BCRAT and BOADICEA models. Breast Cancer Res 21(1):75
23. Berry DA, Iversen ES Jr, Gudbjartsson DF, Hiller EH, Garber JE, Peshkin BN, Lerman C (2002) BRCAPRO validation, the sensitivity of genetic testing of BRCA1/BRCA2, and prevalence of other breast cancer susceptibility genes. J Clin Oncol 20(11):2701–2712
24. Jonker JW, Smit JW, Brinkhuis RF, Maliepaard M, Beijnen JH, Schellens JH, Schinkel AH (2000) Role of breast cancer resistance protein in the bioavailability and fetal penetration of topotecan. J Natl Cancer Inst 92(20):1651–1656
25. Agarap AFM (2018) On breast cancer detection: an application of machine learning algorithms on the wisconsin diagnostic dataset. In: Proceedings of the 2nd international conference on machine learning and soft computing pp 5–9

Reflection of Plane Harmonic Wave in Transversely Isotropic Magneto-thermoelastic with Two Temperature, Rotation and Multi-dual-Phase Lag Heat Transfer

Parveen Lata, Iqbal Kaur, and Kulvinder Singh

Abstract The aim of the present investigation is to study the propagation of plane harmonic waves in transversely isotropic homogeneous magneto-thermoelastic rotating medium with multi-dual-phase lag heat transfer. In the current research, it is observed that there are three types of coupled longitudinal waves (quasi-longitudinal, quasi-transverse and quasi-thermal) for the 2D assumed model. Different thermoelasticity theories (coupled theory (CTE), Lord–Şhulman (L–S) theory, Green–Naghdi (G-N) theory and multi-dual-phase lag (MDPL) theory) are used to study the propagation of plane harmonic waves. The characteristics of various reflected waves such as amplitude ratios, energy ratios, phase velocities, specific loss, penetration depth and attenuation coefficients are computed and depicted graphically. The conservation of energy at the free surface is verified. The characters of wave with different theories of thermoelasticity are represented graphically.

Keywords Thermoelastic · Transversely isotropic · Rotating medium · Multi-dual-phase lag heat transfer · Plane harmonic wave propagation

1 Introduction

The study of the plane wave propagation in a thermoelastic solid gained considerable attention of different researcher from various fields interested in its different behaviors because of its applications in the areas of geophysics, nuclear fields and related topics. Multi-dual-phase lag heat transfer has been incorporated by a

P. Lata · I. Kaur (✉)
Department of Basic and Applied Sciences, Punjabi University, Patiala, Punjab, India
e-mail: bawahanda@gmail.com

P. Lata
e-mail: parveenlata@pbi.ac.in

K. Singh
UIET, Kurukshetra University, Kurukshetra, Haryana, India
e-mail: kshanda@rediffmail.com

N. Marriwala et al. (eds.), *Mobile Radio Communications and 5G Networks*, Lecture Notes in Networks and Systems 140,
https://doi.org/10.1007/978-981-15-7130-5_42

number of researchers in generalized theory for thermoelastic materials. It may also be mentioned that modern laminated media which are being used more and more in engineering and other applications behave anisotropically locally (thermally and elastically). Thus, there is an imperative need to consider the anisotropic media particularly transversely isotropic. Ting [29] discovered a surface wave propagation in an anisotropic rotating medium. Othman and Song [24] presented different hypotheses about magneto-thermoelastic waves in a homogeneous and isotropic medium. Kumar and Chawla [9] discussed the plane wane propagation in the anisotropic three-phase lag model and two-phase lag model. Kumar and Gupta [13, 14] studied the effects of reflection and refraction at the boundary of elastic and a thermoelastic diffusion media for plane waves by expanding the Fick law with dual-phase lag diffusion model with delay times of both mass flow and the potential gradient. Besides, Kumar et al. [15] had depicted the effect of time and thermal and diffusion phase lags for axisymmetric heat supply in a ring for the dual-phase lag model.

Lata [16, 18] studied the effect of energy dissipation on plane waves in sandwiched layered thermoelastic medium with two temperature, rotation and Hall current in the context of GN type-II and type-III theory of thermoelasticity. Alesemi [2] validated the efficiency of the thermal relaxation time depending upon LS theory, Coriolis and centrifugal forces on the reflection coefficients of plane waves in an anisotropic magneto-thermoelastic rotating medium with stable angular velocity medium. Zenkour [30] discussed the refined microtemperature multi-phase lag theory for plane wave propagation in thermoelastic medium. Lata and Kaur [8] studied the effect of Hall current on propagation of plane wave in transversely isotropic thermoelastic medium with two temperature and fractional-order heat transfer. Lata and Kaur [10] investigated Rayleigh wave propagation in transversely isotropic magneto-thermoelastic medium with three-phase lag heat transfer and diffusion. Despite this, several researchers worked on different theories of thermoelasticity such as Mahmoud [22], Kumar and Chawla [11], Kumar and Devi [12], Bijarnia and Singh [4], Othman et al. [26], Ezzat and El-Barrry [6], Allam et al. [3], Zenkour [31], Abo-Dahab et al. [1], Othman and Marin [27], and Lata and Kaur [18–21].

In spite of these, not much work has been carried out in harmonic plane wave propagation with multi-dual-phase lag heat transfer in transversely isotropic thermoelastic rotating medium with two temperature. Keeping these considerations in mind, plane harmonic wave propagation problem in transversely isotropic magneto-thermoelastic rotating medium with two temperature is studied under different theories of thermoelasticity by using reflection techniques.

2 Basic Equations

The simplified Maxwell's linear equation of electrodynamics for a slowly moving and perfectly conducting elastic solid is

$$\operatorname{curl}\vec{h} = \vec{j} + \varepsilon_0 \frac{\partial \vec{E}}{\partial t}, \tag{1}$$

$$\operatorname{curl}\vec{E} = -\mu_0 \frac{\partial \vec{h}}{\partial t}, \tag{2}$$

$$\vec{E} = -\mu_0 \left(\frac{\partial \vec{u}}{\partial t} + \vec{H}_0 \right), \tag{3}$$

$$\operatorname{div}\vec{h} = 0. \tag{4}$$

The constitutive relations for a transversely isotropic thermoelastic medium are given by

$$t_{ij} = C_{ijkl}e_{kl} - \beta_{ij}T. \tag{5}$$

Equation of motion was described by Schoenberg and Censor [25] for a transversely isotropic thermoelastic medium rotating uniformly with an angular velocity $\mathbf{\Omega} = \Omega \boldsymbol{n}$, where $\mathbf{n}$ is a unit vector representing the direction of the axis of rotation and taking into account Lorentz force

$$t_{ij,j} + F_i = \rho\{\ddot{u}_i + (\Omega \times (\Omega \times \mathrm{u})_i + (2\Omega \times \dot{u})_i\}, \tag{6}$$

where $F_i = \mu_0 \left(\vec{j} \times \vec{H}_0 \right)$ are the components of Lorentz force, $\vec{H}_0$ is the external applied magnetic field intensity vector, $\vec{j}$ is the current density vector, $\vec{u}$ is the displacement vector, μ_0 and ε_0 are the magnetic and electric permeabilities, respectively, and t_{ij} is the component of Maxwell stress tensor. The terms $\Omega \times (\Omega \times \mathrm{u})$ and $2\Omega \times \dot{u}$ are the additional centripetal acceleration due to the time-varying motion and Coriolis acceleration, respectively.

and $\beta_{ij} = C_{ijkl}\alpha_{ij}$,

$$\begin{gathered} e_{ij} = \tfrac{1}{2}\left(u_{i,j} + u_{j,i}\right), \quad i, j = 1, 2, 3. \\ T = \varphi - a_{ij}\varphi_{,ij}, \end{gathered} \tag{7}$$

$\alpha_{ij} = \alpha_i \delta_{ij}, \beta_{ij} = \beta_i \delta_{ij}, K_{ij} = K_i \delta_{ij}$, i is not summed.

Here, $C_{ijkl}\left(C_{ijkl} = C_{klij} = C_{jikl} = C_{ijlk}\right)$ are elastic parameters, β_{ij} is the thermal elastic coupling tensor, e_{ij} are the components of strain tensor, u_i are the displacement components, ρ is the density, C_E is the specific heat, K_{ij} is the thermal conductivity, α_{ij} is the coefficient of linear thermal expansion, δ_{ij} is the Kronecker delta, Ω is the angular velocity of the solid.

Following Zenkour (2018), heat conduction equation with multi-phase lag heat transfer is:

$$K_{ij}\mathcal{L}_\theta \nabla^2 \varphi_{,ij} = \mathcal{L}_q \frac{\partial}{\partial t}\left(\beta_{ij} T_0 u_{i,j} + \rho C_E T\right), \tag{8}$$

where $\mathcal{L}_\theta = 1 + \sum_{r=1}^{R_1} \frac{\tau_\theta^r \partial^r}{r! \partial t^r}$, and $\mathcal{L}_q = \varrho + \tau_0 \frac{\partial}{\partial t} + \sum_{r=2}^{R_2} \frac{\tau_q^r \partial^r}{r! \partial t^r}$.

The thermal relaxation time parameters τ_q, τ_θ and τ_0 are the thermal memories in which τ_q is the phase lag (PL) of the heat flux ($0 \leq \tau_\theta < \tau_q$), while τ_θ is the PL of the temperature gradient. For example, L–S theory will be appearing when $\tau_\theta = \tau_q = 0$ and $= 1$. T is the absolute temperature, T_0 is the reference temperature, and φ is the conductive temperature. Generally, the value of $R_1 = R_2 = R$ may reach five or more according to refined multi-dual-phase lag theory required while is a non-dimension parameter ($= 0$ or 1 according to the thermoelasticity theory).

3 Formulation and Solution of the Problem

We consider a homogeneous transversely isotropic magneto-thermoelastic rotating medium with angular velocity $\boldsymbol{\Omega}$ initially at a uniform temperature T_0, permeated by an initial magnetic field $\vec{H}_0 = (0, H_0, 0)$ acting along the y-axis. The rectangular Cartesian coordinate system (x, y, z) having origin on the surface ($z = 0$) with z-axis pointing vertically into the medium is introduced. In addition, we consider that

$$\boldsymbol{\Omega} = (0, \Omega, 0).$$

From the generalized Ohm's law,

$$E = -\mu_0 H_0(-\ddot{w}, 0, \ddot{u}), \tag{9}$$

$$j = \left(-\frac{\partial Hy}{\partial z} - \varepsilon_0 \dot{E}_x, 0, -\frac{\partial Hy}{\partial x} - \varepsilon_0 \dot{E}_z\right), \tag{10}$$

$$\vec{F} = \left(\mu_0 H_0^2\left(\frac{\partial e}{\partial x} - \varepsilon_0 \mu_0 \ddot{u}\right), 0, \mu_0 H_0^2\left(\frac{\partial e}{\partial z} - \varepsilon_0 \mu_0 \ddot{w}\right)\right). \tag{11}$$

In addition, the equations of displacement vector $(\vec{u}, \vec{v}, \vec{w})$ and conductive temperature φ for transversely isotropic thermoelastic solid are

$$\vec{u} = u(x, z, t), \vec{v} = 0, \vec{w} = w(x, z, t) \text{ and } \varphi = \varphi(x, z, t). \tag{12}$$

Now using the transformation on Eqs. (1)–(3) following Slaughter [28] Eqs. (7)–(8) with the aid of Eqs. (10)–(12), yield

$$C_{11}\frac{\partial^2 u}{\partial x^2} + C_{13}\frac{\partial^2 w}{\partial x \partial z} + C_{44}\left(\frac{\partial^2 u}{\partial z^2} + \frac{\partial^2 w}{\partial x \partial z}\right) - \beta_1 \frac{\partial}{\partial x}\left\{\varphi - \left(a_1 \frac{\partial^2 \varphi}{\partial x^2} + a_3 \frac{\partial^2 \varphi}{\partial z^2}\right)\right\}$$

$$+\mu_0 H_0^2\left(\frac{\partial e}{\partial x}-\varepsilon_0\mu_0\ddot{u}\right)=\rho\left(\frac{\partial^2 u}{\partial t^2}-\Omega^2 u+2\Omega\frac{\partial w}{\partial t}\right), \tag{13}$$

$$(C_{13}+C_{44})\frac{\partial^2 u}{\partial x\partial z}+C_{44}\frac{\partial^2 w}{\partial x^2}+C_{33}\frac{\partial^2 w}{\partial z^2}-\beta_3\frac{\partial}{\partial z}\left\{\varphi-\left(a_1\frac{\partial^2\varphi}{\partial x^2}+a_3\frac{\partial^2\varphi}{\partial z^2}\right)\right\}$$
$$+\mu_0 H_0^2\left(\frac{\partial e}{\partial z}-\varepsilon_0\mu_0\ddot{w}\right)=\rho\left(\frac{\partial^2 w}{\partial t^2}-\Omega^2 w-2\Omega\frac{\partial u}{\partial t}\right), \tag{14}$$

$$K_1\left(1+\sum_{r=1}^{R_1}\frac{\tau_\theta^r\partial^r}{r!\partial t^r}\right)\frac{\partial^2\varphi}{\partial x^2}+K_3\left(1+\sum_{r=1}^{R_1}\frac{\tau_\theta^r\partial^r}{r!\partial t^r}\right)\frac{\partial^2\varphi}{\partial z^2}$$
$$=\left(\varrho+\tau_0\frac{\partial}{\partial t}+\sum_{r=2}^{R_2}\frac{\tau_q^r\partial^r}{r!\partial t^r}\right)\left[T_0\left(\beta_1\frac{\partial\dot{u}}{\partial x}+\beta_3\frac{\partial\dot{w}}{\partial z}\right)\right.$$
$$\left.+\rho C_E\left\{\dot{\varphi}-a_1\frac{\partial^2\dot{\varphi}}{\partial x^2}-a_3\frac{\partial^2\dot{\varphi}}{\partial z^2}\right\}\right] \tag{15}$$

and

$$t_{11}=C_{11}e_{11}+C_{13}e_{13}-\beta_1 T, \tag{16}$$

$$t_{33}=C_{13}e_{11}+C_{33}e_{33}-\beta_3 T, \tag{17}$$

$$t_{13}=2C_{44}e_{13}, \tag{18}$$

where

$$\beta_1=(C_{11}+C_{12})\alpha_1+C_{13}\alpha_3,$$

$$\beta_3=2C_{13}\alpha_1+C_{33}\alpha_3,$$

$$e=\frac{\partial u}{\partial x}+\frac{\partial w}{\partial z}$$

To simplify the solution, the following dimensionless quantities are used:

$$(x',z')=\frac{1}{L}(x,z),\ (u',w')=\frac{\rho c_1^2}{L\beta_1 T_0}(u,w),$$
$$\Omega'=\frac{\mathrm{L}}{c_1}\Omega,\ a_1'=\frac{a_1}{L^2},\ a_3'=\frac{a_3}{L^2},\ \rho c_1^2=c_{11}$$

$$\varphi' = \frac{\varphi}{T_0}, (t'_{xx}, t'_{xz}, t'_{zz}) = \frac{1}{\beta_1 T_0}(t_{xx}, t_{xz}, t_{zz}), (\tau'_0, \tau'_\theta, \tau'_q, t') = \frac{c_1}{L}(\tau_0, \tau_\theta, \tau_q, t). \tag{19}$$

Making use of Eq. (19) in Eqs. (13)–(15), after suppressing the primes, yields

$$\begin{aligned}&(1+\delta_5)\frac{\partial^2 u}{\partial x^2} + (\delta_4+\delta_5)\frac{\partial^2 w}{\partial x \partial z} + \delta_2\left(\frac{\partial^2 u}{\partial z^2} + \frac{\partial^2 w}{\partial x \partial z}\right)\\ &- \frac{\partial}{\partial x}\left\{\varphi - \left(a_1\frac{\partial^2\varphi}{\partial x^2} + a_3\frac{\partial^2\varphi}{\partial z^2}\right)\right\}\\ &= \left(\frac{\varepsilon_0\mu_0^2 H_0^2}{\rho} + 1\right)\frac{\partial^2 u}{\partial t^2} - \Omega^2 u + 2\Omega\frac{\partial w}{\partial t},\end{aligned} \tag{20}$$

$$\begin{aligned}&(\delta_1+\delta_5)\frac{\partial^2 u}{\partial x \partial z} + \delta_2\frac{\partial^2 w}{\partial x^2} + (\delta_3+\delta_5)\frac{\partial^2 w}{\partial z^2}\\ &- \frac{\beta_3}{\beta_1}\frac{\partial}{\partial z}\left\{\varphi - \left(a_1\frac{\partial^2\varphi}{\partial x^2} + a_3\frac{\partial^2\varphi}{\partial z^2}\right)\right\}\\ &= \left(\frac{\varepsilon_0\mu_0^2 H_0^2}{\rho} + 1\right)\frac{\partial^2 w}{\partial t^2} - \Omega^2 w - 2\Omega\frac{\partial u}{\partial t},\end{aligned} \tag{21}$$

$$\begin{aligned}&K_1\left(1 + \sum_{r=1}^{R_1}\frac{\tau_\theta^r \partial^r}{r!\partial t^r}\right)\frac{\partial^2\varphi}{\partial x^2} + K_3\left(1 + \sum_{r=1}^{R_1}\frac{\tau_\theta^r \partial^r}{r!\partial t^r}\right)\frac{\partial^2\varphi}{\partial z^2}\\ &= \left(\varrho + \tau_0\frac{\partial}{\partial t} + \sum_{r=2}^{R_2}\frac{\tau_q^r \partial^r}{r!\partial t^r}\right)\left[\frac{\partial}{\partial t}\left(\delta_6\frac{\partial u}{\partial x} + \delta_8\frac{\partial w}{\partial z}\right)\right.\\ &\left. + \delta_7\frac{\partial}{\partial t}\left\{\varphi - a_1\frac{\partial^2\varphi}{\partial x^2} - a_3\frac{\partial^2\varphi}{\partial z^2}\right\}\right],\end{aligned} \tag{22}$$

where

$$\delta_1 = \frac{c_{13}+c_{44}}{c_{11}}, \delta_2 = \frac{c_{44}}{c_{11}}, \delta_3 = \frac{c_{33}}{c_{11}}, \delta_4 = \frac{c_{13}}{c_{11}}, \delta_5 = \frac{\beta_1 T_0 \mu_0 H_0^2}{L\rho^2 C_1^4},$$
$$\delta_6 = \frac{L\beta_1^2 T_0}{\rho c_1}, \quad \delta_8 = \frac{L\beta_1\beta_3 T_0}{\rho c_1}, \delta_7 = \rho C_E c_1 L.$$

The stress–strain relations after applying dimensionless quantities defined by and suppressing primes give

$$t_{11}(x, z, t) = \frac{\partial u}{\partial x} + \delta_4\frac{\partial w}{\partial z} - \left\{\varphi - \left(a_1\frac{\partial^2\varphi}{\partial x^2} + a_3\frac{\partial^2\varphi}{\partial z^2}\right)\right\} \tag{23}$$

$$t_{33}(x, z, t) = \delta_4\frac{\partial u}{\partial x} + \delta_3\frac{\partial w}{\partial z} - \frac{\beta_3}{\beta_1}\left\{\varphi - \left(a_1\frac{\partial^2\varphi}{\partial x^2} + a_3\frac{\partial^2\varphi}{\partial z^2}\right)\right\}, \tag{24}$$

$$t_{13}(\mathrm{x,z,t}) = \delta_2\left(\frac{\partial u}{\partial z} + \frac{\partial w}{\partial x}\right), \quad (25)$$

4 Plane Wave Propagation

We pursue plane wave to be a time-harmonic solution of the form

$$\begin{pmatrix} u \\ w \\ \varphi \end{pmatrix} = \begin{pmatrix} U \\ W \\ \varphi^* \end{pmatrix} e^{i(\xi(xn_1+zn_3)-\omega t)} \quad (26)$$

where n_1, n_3 denotes the projection of wave normal on to the x-z plane, and ξ and ω are the wave numbers and angular frequency of plane waves propagating in x-z plane, respectively.

Upon using Eq. (26) in Eqs. (20)–(22), we get

$$U\left[\zeta_1\xi^2 + \zeta_2\right] + W\left[\zeta_3\xi^2 + \zeta_4\right] + \varphi^*\left(\zeta_5\xi + \zeta_6\xi^3\right) = 0,$$

$$U\left[\zeta_7\xi^2 - \zeta_4\right] + W\left[\zeta_8\xi^2 + \zeta_2\right] + \varphi^*\left(\zeta_9\xi + \zeta_{10}\xi^3\right) = 0,$$

$$\zeta_{11}U\xi + \zeta_{12}W\xi + \varphi^*\left[\zeta_{13}\xi^2 + \zeta_{14}\right] = 0.$$

and then eliminating U, W and φ^* from the resulting equations yields the following characteristic equation

$$\left(A\xi^6 + B\xi^4 + C\xi^2 + D\right) = 0, \quad (27)$$

where

$$A = \zeta_1\zeta_{13}\zeta_8 - \zeta_1\zeta_{12}\zeta_{10} - \zeta_3\zeta_{13}\zeta_7 + \zeta_3\zeta_{11}\zeta_{10} + \zeta_6\zeta_{12}\zeta_7 - \zeta_6\zeta_{11}\zeta_8,$$

$$\begin{aligned} B = {} & \zeta_2\zeta_8\zeta_{13} + \zeta_1\zeta_2\zeta_{13} + \zeta_8\zeta_1\zeta_{14} - \zeta_{12}\zeta_1\zeta_9 - \zeta_{12}\zeta_{10}\zeta_2 - \zeta_4\zeta_7\zeta_{13} \\ & + \zeta_3\zeta_4\zeta_{13} - \zeta_3\zeta_7\zeta_{14} + \zeta_3\zeta_{11}\zeta_9 + \zeta_4\zeta_{11}\zeta_{10} + \zeta_5\zeta_7\zeta_{12} \\ & - \zeta_6\zeta_2\zeta_{11} - \zeta_6\zeta_4\zeta_{12} - \zeta_{11}\zeta_8\zeta_5, \end{aligned}$$

$$\begin{aligned} C = {} & \zeta_2\zeta_8\zeta_{14} + \zeta_2\zeta_2\zeta_{13} + \zeta_1\zeta_{14}\zeta_2 - \zeta_2\zeta_{12}\zeta_{19} + \zeta_4\zeta_4\zeta_{13} \\ & - \zeta_7\zeta_{14}\zeta_4 + \zeta_3\zeta_4\zeta_{14} + \zeta_4\zeta_9\zeta_{11} - \zeta_4\zeta_{12}\zeta_5 - \zeta_5\zeta_2\zeta_{11}, \end{aligned}$$

$$D = \zeta_2\zeta_2\zeta_{14} + \zeta_4\zeta_4\zeta_{14},$$

$$\zeta_1 = -(1+\delta_5)n_1^2 - \delta_2 n_3^2,$$

$$\zeta_2 = \left(\frac{\varepsilon_0\mu_0^2 H_0^2}{\rho} + 1\right)\omega^2 + \Omega^2,$$

$$\zeta_3 = -(\delta_4 + \delta_5 + \delta_2)n_1 n_3,$$

$$\zeta_4 = -2\omega\Omega i,$$

$$\zeta_5 = -in_1,$$

$$\zeta_6 = -n_1 i\left(a_1 n_1^2 + a_3 n_3^2\right)$$

$$\zeta_7 = -(\delta_1 + \delta_5)n_1 n_3,$$

$$\zeta_8 = -\delta_2 n_1^2 - (\delta_3 + \delta_5)n_3^2,$$

$$\zeta_9 = -i\frac{\beta_3}{\beta_1}n_3,$$

$$\zeta_{10} = -\frac{\beta_3}{\beta_1}n_3 i\left(a_1 n_1^2 + a_3 n_3^2\right)$$

$$\zeta_{11} = \left[\varrho - i\omega\tau_0 + \sum_{r=2}^{R_2}\frac{\tau_q^r}{r!}(-i\omega)^r\right]\delta_6\omega n_1,$$

$$\zeta_{12} = \left[\varrho - i\omega\tau_0 + \sum_{r=2}^{R_2}\frac{\tau_q^r}{r!}(-i\omega)^r\right]\delta_8\omega n_3,$$

$$\zeta_{13} = \left[1 + \sum_{r=1}^{R_1}\frac{\tau_\theta^r}{r!}(-i\omega)^r\right]\left[-K_1 n_1^2 - K_3 n_3^2\right] - \delta_7 i\omega\left[\varrho - i\omega\tau_0 + \sum_{r=2}^{R_2}\frac{\tau_q^r}{r!}(-i\omega)^r\right]\left(a_1 n_1^2 + a_3 n_3^2\right),$$

$$\zeta_{14} = -\left[\varrho - i\omega\tau_0 + \sum_{r=2}^{R_2}\frac{\tau_q^r}{r!}(-i\omega)^r\right]\delta_7 i\omega,$$

The six nonzero roots of Eq. (27) give six roots of ξ, that is, $\pm\xi_1$, $\pm\xi_2$ and $\pm\xi_3$, in which we are interested in those roots whose imaginary parts are positive. Corresponding to these roots, there exist three waves corresponding to descending order of their velocities, namely a quasi-longitudinal (QL), quasi-transverse (QTS) and quasi-thermal (QT) waves. The phase velocities, attenuation coefficients, specific loss and penetration depth of these waves are obtained by the following expressions.

4.1 Phase Velocity

The phase velocities are given by

$$V_i = \frac{\omega}{\mathrm{Re}(\xi_i)}, i = 1, 2, 3$$

where V_1, V_2, V_3 are the velocities of QL, QTS and QT waves, respectively.

4.2 Attenuation Coefficient

The attenuation coefficient is defined as

$$Q_i = \mathrm{Img}(\xi_i), \mathrm{i} = 1, 2, 3.$$

where Q_1, Q_2, Q_3 are the attenuation coefficients of QL, QTS and QT waves, respectively.

4.3 Specific Loss

The specific loss is the ratio of energy (ΔS) dissipated in taking a specimen through the cycle, to elastic energy (S) stored in a specimen when the strain is maximum. The specific loss is the most direct method of defining internal friction for a material. For a sinusoidal plane wave of small amplitude, it was shown by Kolsky [7] that specific loss $\frac{\Delta S}{S}$ equals 4π times the absolute value of the imaginary part of ξ to the real part of ξ, i.e.

$$S_i = \left(\frac{\Delta S}{S}\right)_i = 4\pi\left|\frac{\mathrm{Img}(\xi_i)}{\mathrm{Re}(\xi_i)}\right|, \mathrm{i} = 1, 2, 3.$$

where S_1, S_2, S_3 are specific losses of QL, QTS and QT waves, respectively.

4.4 Penetration Depth

The penetration depth is defined by

$$D_i = \frac{1}{\mathrm{Img}(\xi_i)}, \mathrm{i} = 1, 2, 3.$$

where D_1, D_2, D_3 are penetration depths of QL, QTS and QT waves, respectively (Fig. 1).

5 Reflection at the Boundary Surfaces

We consider a homogeneous transversely isotropic magneto-thermoelastic half-space occupying the region $z \geq 0$. Incident quasi-longitudinal or quasi-transverse or quasi-thermal waves at the stress-free, thermally insulated surface ($z = 0$) will generate reflected QL, reflected QTS and reflected QT waves in the half-space $z > 0$. The total displacements and conductive temperature are given by

$$u = \sum_{j=1}^{6} A_j e^{-iM_j},$$

$$w = \sum_{j=1}^{6} d_j A_j e^{-iM_j},$$

$$\varphi = \sum_{j=1}^{6} l_j A_j e^{-iM_j}, \; j = 1, 2, 3, \ldots, 6 \tag{28}$$

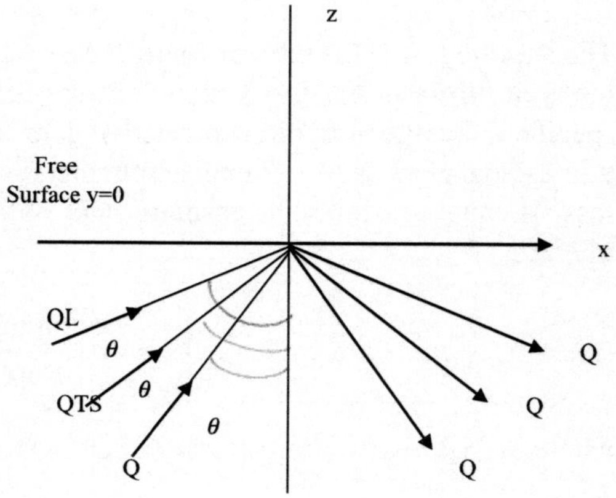

Fig. 1 Geometry of the problem

where

$$M_j = \omega t - \xi_j\big(xn_{1j} - zn_{3j}\big),\ j = 1, 2, 3,$$

$$M_j = \omega t - \xi_j\big(xn_{1j} + zn_{3j}\big),\ j = 4, 5, 6.$$

Here, subscripts $j = 1, 2, 3$, respectively, denote the quantities corresponding to incident QL, QTS and QT modes, whereas the subscripts $j = 4, 5, 6$ denote the corresponding reflected waves, ξ_j are the roots obtained from Eq. (24), and $n_{1j} = \sin\theta_j; n_{3j} = \cos\theta_j$.

$$d_j = \frac{(\zeta_2\zeta_{14}) + (\zeta_2\zeta_{13j} + \zeta_{1j}\zeta_{14} - \zeta_{11j}\zeta_{5j})\xi_j^2 + \zeta_{1j}\zeta_{13j}\xi_j^4}{\zeta_{8j}\zeta_{13j}\xi_j^4 + (\zeta_2\zeta_{13j} + \zeta_{8j}\zeta_{14j} - \zeta_{12j}\zeta_{9j})\xi_j^2 + (\zeta_2\zeta_{14} + \zeta_{12j}\zeta_{10j})},$$
$$j = 1, 2, 3.$$

$$l_j = \frac{(\zeta_2^2 + \zeta_4^2) + (\zeta_2\zeta_{1j} + \zeta_2\zeta_{8j} - \zeta_4\zeta_{3j} + \zeta_4\zeta_{7j})\xi_j^2 + (\zeta_{1j}\zeta_{8j} + \zeta_{7j}\zeta_4)\xi_j^4}{\zeta_{8j}\zeta_{13j}\xi_j^4 + (\zeta_2\zeta_{13j} + \zeta_{8j}\zeta_{14} - \zeta_{12j}\zeta_{9j})\xi_j^2 + (\zeta_2\zeta_{14} + \zeta_{12j}\zeta_{10j})},$$
$$j = 1, 2, 3.$$

$$d_j = \frac{(\zeta_2\zeta_{14}) + (\zeta_2\zeta_{13j} + \zeta_{1j}\zeta_{14} - \zeta_{11j}\zeta_{5j})\xi_j^2 + \zeta_{1j}\zeta_{13j}\xi_j^4}{\zeta_{8j}\zeta_{13j}\xi_j^4 + (\zeta_{2j}\zeta_{13j} + \zeta_{8j}\zeta_{14} - \zeta_{12j}\zeta_{9j})\xi_j^2 + (\zeta_2\zeta_{14} - \zeta_{12j}\zeta_{10j})},$$
$$j = 4, 5, 6.$$

$$l_j = \frac{(\zeta_2^2 + \zeta_4^2) + (\zeta_2\zeta_{1j} + \zeta_2\zeta_{8j} + \zeta_4\zeta_{3j} - \zeta_4\zeta_{7j})\xi_j^2 + (\zeta_{1j}\zeta_{8j} - \zeta_7 j\zeta_4)\xi_j^4}{\zeta_{8j}\zeta_{13j}\xi_j^4 + (\zeta_2\zeta_{13j} + \zeta_{8j}\zeta_{14} - \zeta_{12j}\zeta_{9j})\xi_j^2 + (\zeta_2\zeta_{14} - \zeta_{12j}\zeta_{10j})},$$
$$j = 4, 5, 6.$$

6 Boundary Conditions

The dimensionless boundary conditions at the free surface $z = 0$ are given by

$$t_{33} = 0, \tag{29}$$

$$t_{31} = 0, \tag{30}$$

$$\frac{\partial\varphi}{\partial z} = 0. \tag{31}$$

Making use of Eq. (28) into the boundary condition Eqs. (29)–(31), we obtain

$$\sum_{j=1}^{3} A_j e^{-i(\omega t-\xi_j(x\sin\theta_j))}\Big[\delta_4 i\xi_j \sin\theta_j - \delta_3 i d_j \xi_j \cos - \frac{\beta_3}{\beta_1} l_j\big(1 + a_1\xi_j^2 \sin\theta_j^2 + a_3\xi_j^2\cos\theta_j^2\big)\Big] + \sum_{j=4}^{6} A_j e^{-i(\omega t-\xi_j(x\sin\theta_j))}\Big[\delta_4 i\xi_j \sin\theta_j + \delta_3 i d_j \xi_j \cos\theta_j - \frac{\beta_3}{\beta_1} l_j\big(1 + a_1\xi_j^2 \sin\theta_j^2 + a_3\xi_j^2\cos\theta_j^2\big)\Big] = 0, \tag{32}$$

$$\sum_{j=1}^{3} A_j e^{-i(\omega t-\xi_j(x\sin\theta_j))}\big[\xi_j\cos\theta_j - d_j\xi_j\sin\theta_j\big] - \sum_{j=4}^{6} A_j e^{-i(\omega t-\xi_j(x\sin\theta_j))}\big[\xi_j\cos\theta_j + d_j\xi_j\sin\theta_j\big] = 0, \tag{33}$$

$$\sum_{j=1}^{3} A_j e^{-i(\omega t-\xi_j(x\sin\theta_j))}\big[i l_j \xi_j \cos\theta_j\big] - \sum_{j=4}^{6} A_j e^{-i(\omega t-\xi_j(x\sin\theta_j))}\big[i l_j \xi_j \cos\theta_j\big] = 0, \tag{34}$$

Equations (32)–(34) are satisfied for all values of x; therefore, we have

$$M_1(x,0) = M_2(x,0) = M_3(x,0) = M_4(x,0) = M_5(x,0) = M_6(x,0) \tag{35}$$

From Eqs. (28) and (35), we obtain

$$\xi_1\sin\theta_1 = \xi_2\sin\theta_2 = \xi_3\sin\theta_3 = \xi_4\sin\theta_4 = \xi_5\sin\theta_5 = \xi_6\sin\theta_6 \tag{36}$$

which is the form of Snell's law for the stress-free, thermally insulated surface of transversely isotropic magneto-thermoelastic medium with rotation. Equations (32)–(34) and (36) yield

$$\sum_{j=1}^{3} X_{ij}A_j + \sum_{j=4}^{6} X_{ij}A_j = 0, \quad (i = 1, 2, 3) \tag{37}$$

where for $j = 1, 2, 3$ we have

$$X_{1j} = \Big[\delta_4 i\xi_j\sin\theta_j - \delta_3 i d_j\xi_j\cos\theta_j - \frac{\beta_3}{\beta_1} l_j\big(1 + a_1\xi_j^2\sin\theta_j^2 + a_3\xi_j^2\cos\theta_j^2\big)\Big]$$
$$X_{2j} = \xi_j\cos\theta_j - d_j\xi_j\sin\theta_j,$$
$$X_{3j} = i l_j\xi_j\cos\theta_j$$

And for $j = 4, 5, 6$, we have

$$X_{1j} = \left[\delta_4 i\xi_j \sin\theta_j + \delta_3 i d_j \xi_j \cos\theta_j - \frac{\beta_3}{\beta_1} l_j \left(1 + a_1 \xi_j^2 \sin\theta_j^2 + a_3 \xi_j^2 \cos\theta_j^2\right)\right],$$
$$X_{2j} = -\xi_j \cos\theta_j - d_j \xi_j \sin\theta_j,$$
$$X_{3j} = -i l_j \xi_j \cos\theta_j$$

Incident QL Wave

In the case of a quasi-longitudinal wave, the subscript p takes only one value, that is $p = 1$, which means $A_2 = A_3 = 0$. Dividing the set of Eq. (34) throughout by A_1, we obtain a system of three homogeneous equations in three unknowns which can be solved by Cramer's rule, and we have

$$A_{1i} = \frac{A_{i+3}}{A_1} = \frac{\Delta_i^1}{\Delta}, \tag{38}$$

Incident QTS Wave

In the case of a quasi-transverse wave, the subscript q takes only one value, that is $q = 2$, which means. Dividing the set of Eq. (34) throughout by, we obtain a system of three homogeneous equations in three unknowns which can be solved by Cramer's rule, and we have

$$A_{2i} = \frac{A_{i+3}}{A_2} = \frac{\Delta_i^2}{\Delta}, \tag{39}$$

Incident QT Wave

In the case of a quasi-thermal wave, the subscript q takes only one value, that is $q = 3$, which means. Dividing the set of Eq. (34) throughout, we obtain a system of three homogeneous equations in three unknowns which can be solved by Cramer's rule, and we have

$$A_{3i} = \frac{A_{i+3}}{A_3} = \frac{\Delta_i^3}{\Delta}, \tag{40}$$

where A_{ji} ($i,j = 1, 2, 3$) is the amplitude ratios of the reflected QL, reflected QTS, reflected QT waves to that of the incident QL (QTS or QT) waves, respectively.

Here,

$$\Delta = \left|A_{i(i+3)}\right|_{3X3},$$

$$\Delta_i^p, \ (i = 1, 2, 3).$$

can be obtained by replacing, respectively, the 1st, 2nd and 3rd columns of Δ by $\left[-X_{1p}, -X_{2p}, -X_{3p}\right]'$.

Following Achenbach[32] , the energy flux across the surface element, which is the rate at which the energy is communicated per unit area of the surface, is represented as

$$P^* = t_{lm} n_m \dot{u}_l, \tag{41}$$

where t_{lm} is the stress tensor, n_m are the direction cosines of the unit normal and $\dot{u}_l$ are the components of the particle velocity.

Plane waves in an anisotropic thermoelastic, where is the stress tensor, are the direction cosines of the unit normal and are the components of the particle velocity.

The time average of P^* over a period, denoted by $\langle P^* \rangle$, represents the average energy transmission per unit surface area per unit time and is given at the interface $z = 0$ as

$$\langle P^* \rangle = \langle Re(t_{13}).Re(\dot{u}_1) + Re(t_{33}).Re(\dot{u}_3) \rangle, \tag{42}$$

Following Achenbach (1973), for any two complex functions f and g, we have

$$\langle Re(f) \rangle \langle Re(g) \rangle = \frac{1}{2} Re(f\dot{g}), \tag{43}$$

The expressions for energy ratios $E_i (i = 1, 2, 3)$ for reflected QL, QTS, QT waves are given as

(i) In case of incident QL wave

$$E_{1i} = \frac{\langle P^*_{i+3} \rangle}{\langle P^*_1 \rangle}, i = 1, 2, 3. \tag{44}$$

(ii) In case of incident QTS wave

$$E_{2i} = \frac{\langle P^*_{i+3} \rangle}{\langle P^*_2 \rangle}, i = 1, 2, 3. \tag{45}$$

(iii) In case of incident QT wave

$$E_{3i} = \frac{\langle P^*_{i+3} \rangle}{\langle P^*_3 \rangle}, i = 1, 2, 3. \tag{46}$$

where $\langle P^*_i \rangle, i = 1, 2, 3$ are the average energy transmissions per unit surface area per unit time corresponding to incident QL, QTS, QT waves, respectively, and $\langle P^*_{i+3} \rangle, i = 1, 2, 3$ are the average energy transmissions per unit surface area per unit time corresponding to reflected QL, QTS, QT waves, respectively.

7 Numerical Results and Discussion

To demonstrate the theoretical results and effect of multi-phase lag, the physical data for cobalt material, which is transversely isotropic, is taken from Dhaliwal and Singh [5] given as

$$
\begin{aligned}
&c_{11} = 3.07 \times 10^{11}\,\mathrm{Nm}^{-2}, c_{33} = 3.581 \times 10^{11}\,\mathrm{Nm}^{-2}, c_{13} = 1.027 \times 10^{10}\,\mathrm{Nm}^{-2},\\
&c_{44} = 1.510 \times 10^{11}\,\mathrm{Nm}^{-2}, \beta_1 = 7.04 \times 10^{6}\,\mathrm{Nm}^{-2}\mathrm{deg}^{-1},\\
&\beta_3 = 6.90 \times 10^{6}\,\mathrm{Nm}^{-2}\mathrm{deg}^{-1}, \rho = 8.836 \times 10^{3}\,\mathrm{Kgm}^{-3},\\
&C_E = 4.27 \times 10^{2}\,\mathrm{j\,Kg}^{-1}\mathrm{deg}^{-1}, K_1 = 0.690 \times 10^{2}\,\mathrm{Wm}^{-1}\,\mathrm{K\,deg}^{-1},\\
&K_3 = 0.690 \times 10^{2}\,\mathrm{Wm}^{-1}\,\mathrm{K}^{-1}, K_1^* = 1.313 \times 10^{2}\,\mathrm{W\ sec},\\
&\mathrm{K}_3^* = 1.54 \times 10^{2}\,\mathrm{W\ sec}, T_0 = 298\,\mathrm{K},\\
&H_0 = 1\,\mathrm{Jm}^{-1}\,\mathrm{nb}^{-1}, \varepsilon_0 = 8.838 \times 10^{-12}\,\mathrm{Fm}^{-1}, L = 1.
\end{aligned}
$$

The values of frequency, rotation Ω magnetic effect H_0 are taken as 0.03, 0.5 and 10, respectively. The software MATLAB 8.0.4 has been used to determine the amplitude ratios of reflected QL, QTS and QT waves with respect to incident QL, QTS and QT waves, respectively. The variation of the magnitude of amplitude ratios has been plotted in Figs. 2, 3, 4, 5, 6, 7, 8, 9 and 10 with respect to the angle of incidence. A comparison has been made to show the effect of multi-phase lag on the various quantities.

1. The black line with square symbol represents $a_1 = 0.0, a_3 = 0.0, \alpha = 0.5$
2. The red line with circle symbol represents $a_1 = 0.02, a_3 = 0.04, \alpha = 0.5$
3. The green line with triangle symbol represents $a_1 = 0.0, a_3 = 0.0, \alpha = 1.5$
4. The blue line with diamond symbol represents $a_1 = 0.02, a_3 = 0.04, \alpha = 1.5$

Fig. 2 Variations of amplitude ratio A_{11} with angle of incidence θ

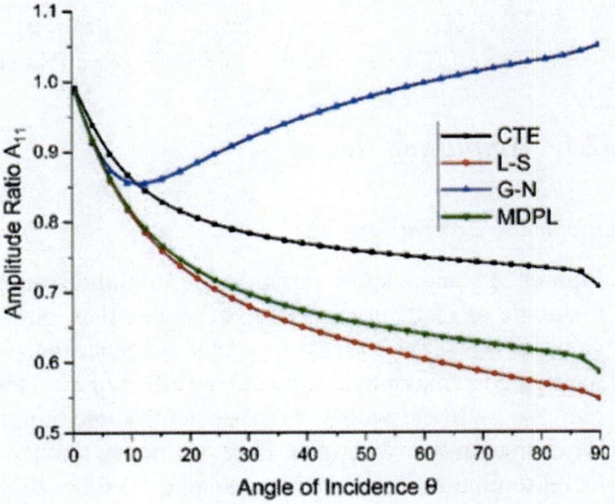

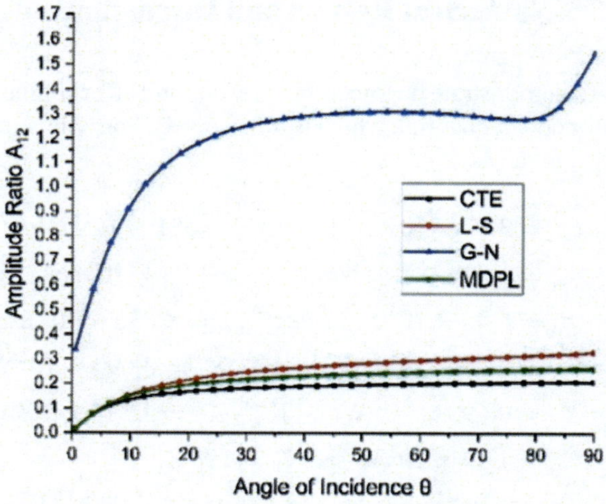

Fig. 3 Variations of amplitude ratio A_{12} with angle of incidence θ

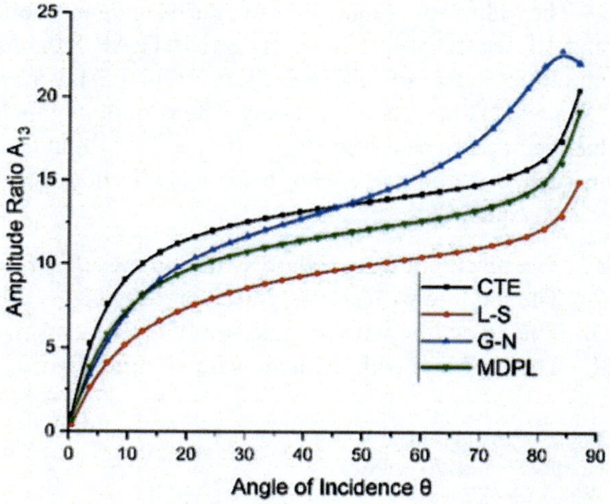

Fig. 4 Variations of amplitude ratio A_{13} with angle of incidence θ

7.1 Amplitude Ratios

Incident QL Wave

Figures 2, 3 and 4 show variations of amplitude ratios A_{11}, A_{12}, A_{13} with respect to the angle of incidence θ. Here, we notice that, initially, there is a decrease in the values of A_{11} for all the cases angle of incidence θ 40 and then increases sharply with the change in magnitude showing the effect of α nd two temperature. The amplitude ratio A_{12} increases with an increase in the angle of incidence θ and shows variations for change in the value of α. Here, we notice that the values of amplitude ratio A_{13} increase monotonically for the range $0 < \theta < 40$ and after achieving maximum

Fig. 5 Variations of amplitude ratio A_{21} with angle of incidence θ

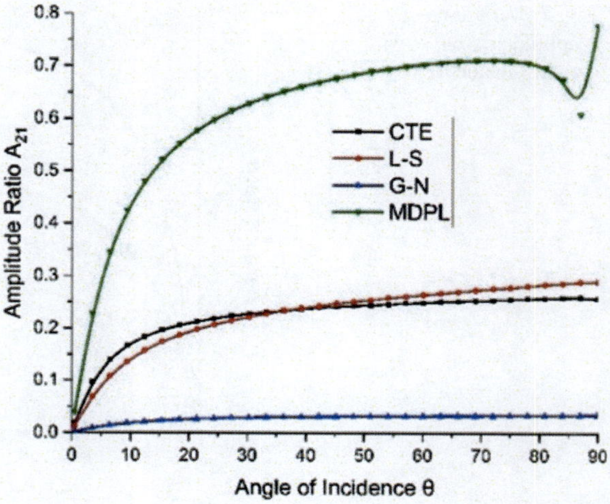

Fig. 6 Variations of amplitude ratio A_{22} with angle of incidence θ

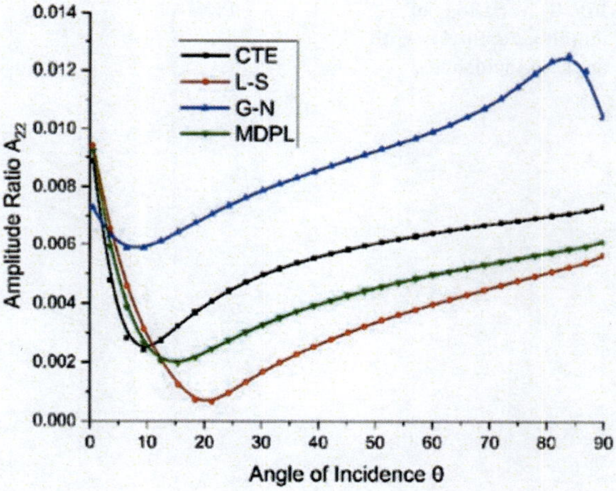

value at 40, the values start decreasing with the difference in magnitude for all the cases with a small change in amplitude.

Incident QTS Wave

Figures 5, 6 and 7 depict the variations of amplitude ratios A_{21}, A_{22}, A_{23} with respect to the angle of incidence θ. Here, we notice a sharp increase in the values of amplitude ratios A_{21} for without two temperature while it shows a small increase for the initial angle of incidence and again decreases for rest of the range for with temperature case. While A_{22} shows opposite behavior than A_{21}, A_{23} shows oscillatory behavior for all the cases of α with a change in the magnitude of amplitude ratio with the change in the value of fractional-order parameter α and two temperature. These variations

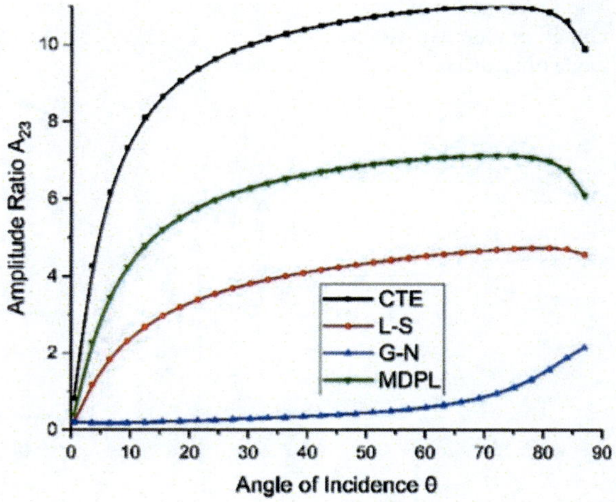

Fig. 7 Variations of amplitude ratio A_{23} with angle of incidence θ

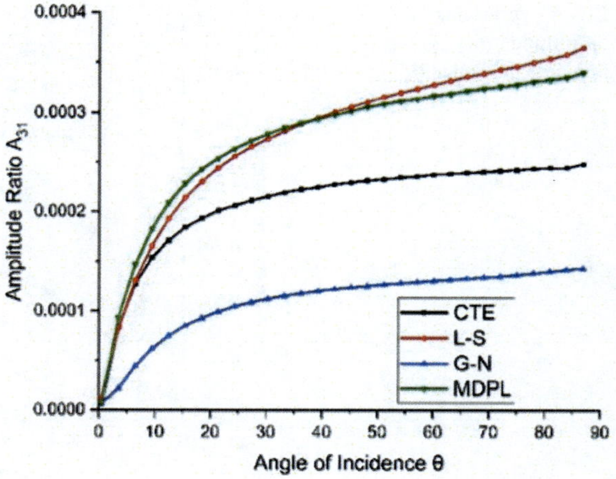

Fig. 8 Variations of amplitude ratio A_{31} with angle of incidence θ

show the effect of fractional-order parameter α and two temperature on the amplitude ratios A_{21}, A_{22}, A_{23}.

Incident QT Wave

Figures 8, 9 and 10 show the variations of amplitude ratios A_{31}, A_{32}, A_{33} with respect to the angle of incidence θ For all the cases, they show a sharp increase and then after attaining maximum value decrease with a small amplitude difference showing the effect of two temperature and fractional-order parameter α on the amplitude ratios A_{31}, A_{32}, A_{33}.

Fig. 9 Variations of amplitude ratio A_{32} with angle of incidence θ

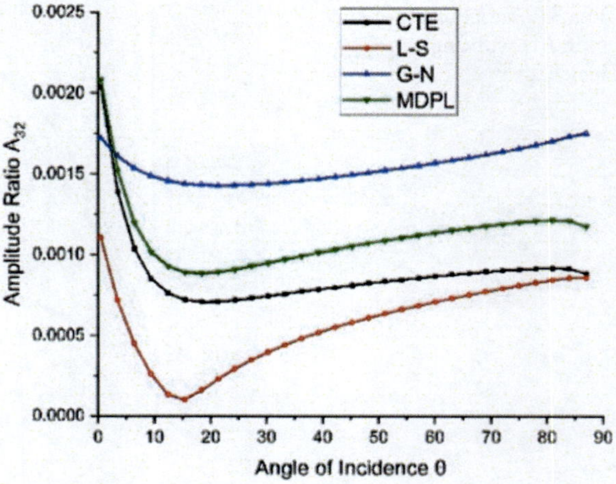

Fig. 10 Variations of amplitude ratio A_{33} with angle of incidence θ

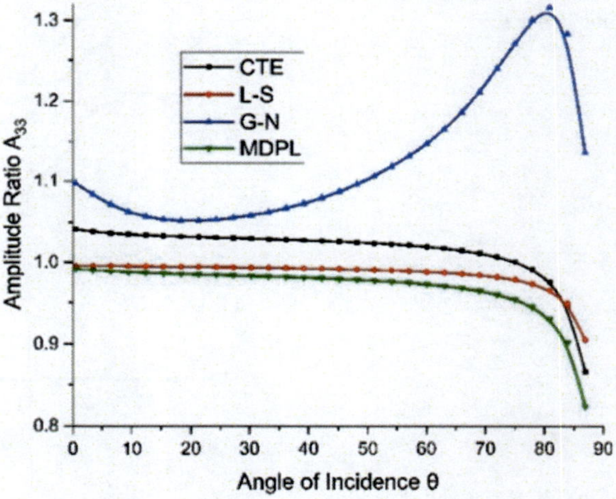

7.2 Energy Ratios

Incident QL Wave

Figures 11, 12 and 13 show variations of energy ratios E_{11}, E_{12}, E_{13} with respect to the angle of incidence θ. Here, we notice that, initially, there is a decrease in the values of E_{11} for all the cases angle of incidence $\theta <= 40$ and then remains same with a difference in magnitude showing the effect of α and two temperature. The energy ratio E_{12} decreases with increase in the angle of incidence θ for without two temperature case, whereas it remains almost same for with two temperature case and shows little variations in magnitude for change in the value of α. Here, we notice that the values of energy ratio E_{13} decrease monotonically for the range $0 < \theta < 10$

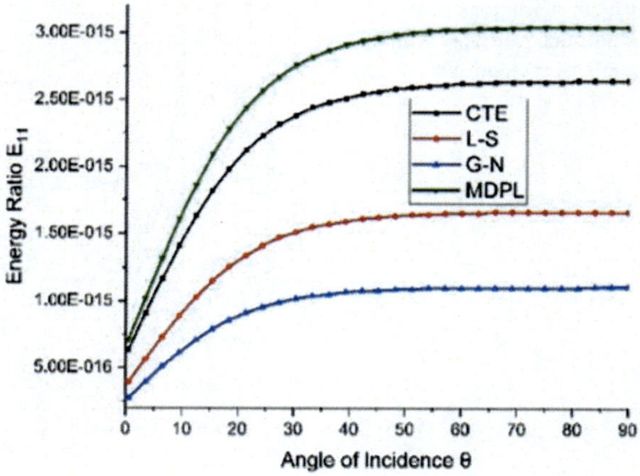

Fig. 11 Variations of energy ratio E_{11} with angle of incidence θ

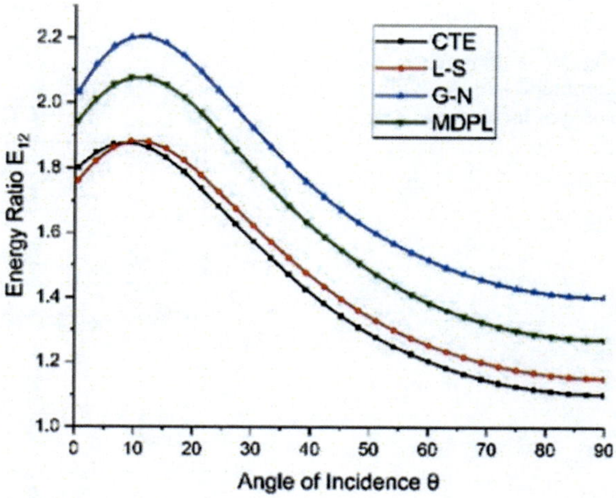

Fig. 12 Variations of energy ratio E_{12} with angle of incidence θ

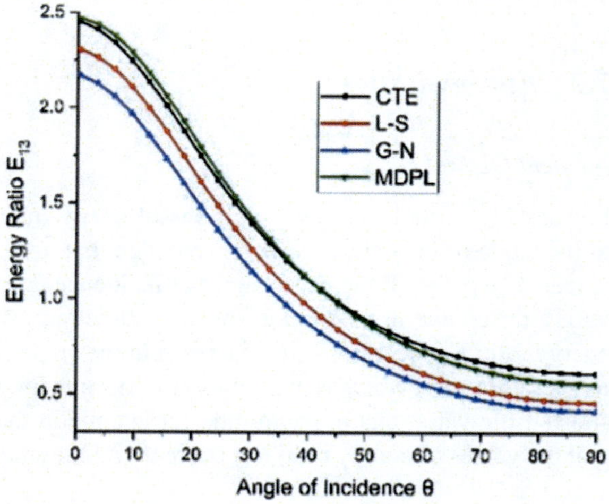

Fig. 13 Variations of energy ratio E_{13} with angle of incidence θ

and the values start increasing with the difference in magnitude for all the cases with a small change in amplitude.

Incident QTS Wave

Figures 14, 15 and 16 depict the variations of energy ratios E_{21}, E_{22}, E_{23} with respect to the angle of incidence θ. Here, we notice a decrease in the values of energy ratio E_{21} for all the case except for with two temperature and $\alpha = 1.5$ where it shows little variation in magnitude. E_{22} increases for two temperature case, decreases for without two temperature case and shows variation in magnitude for different values of α, while E_{23} increases with the difference in magnitude for all the four cases. These variations show the effect of fractional-order parameter α and two temperature on the energy ratios E_{21}, E_{22}, E_{23}.

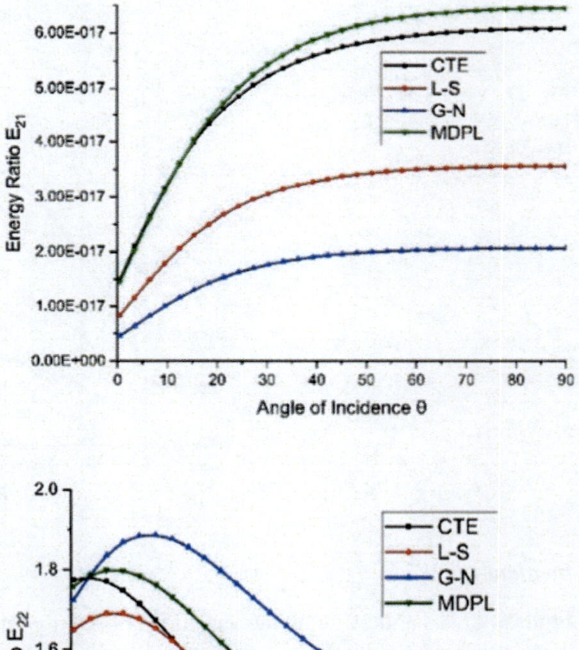

Fig. 14 Variations of energy ratio E_{21} with angle of incidence θ

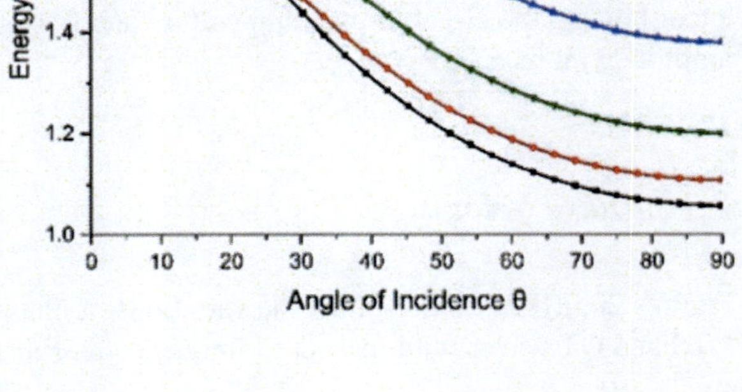

Fig. 15 Variations of energy ratio E_{22} with angle of incidence θ

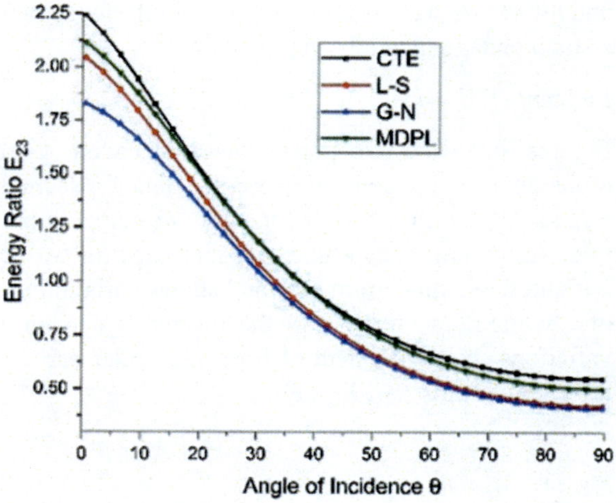

Fig. 16 Variations of energy ratio E_{23} with angle of incidence θ

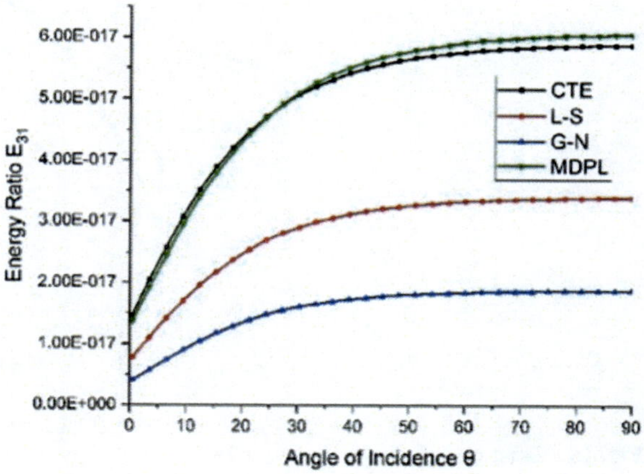

Fig. 17 Variations of energy ratio E_{31} with angle of incidence θ

Incident QT Wave

Figures 17, 18 and 19 show the variations of energy ratios E_{31}, E_{32}, E_{33} with respect to the angle of incidence θ respectively. For all the cases, they show a sharp increase and then after attaining maximum value decrease with a small amplitude difference showing the effect of two temperature and fractional order parameter α on the amplitude ratios E_{31}, E_{32}, E_{33}.

7.3 Phase Velocity

Figures 20, 21, 22 and 23 show the variations of phase velocities V_1, V_2, V_3 of QL, QTS and QT waves with respect to frequency ω. From the graphs, we observe that

Fig. 18 Variations of energy ratio E_{32} with angle of incidence θ

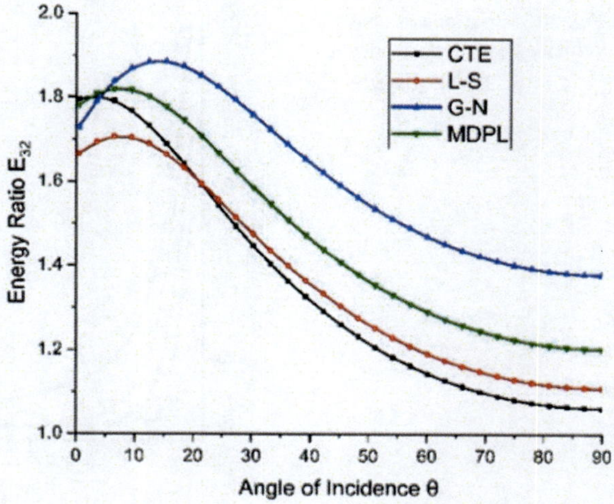

Fig. 19 Variations of energy ratio E_{33} with angle of incidence θ

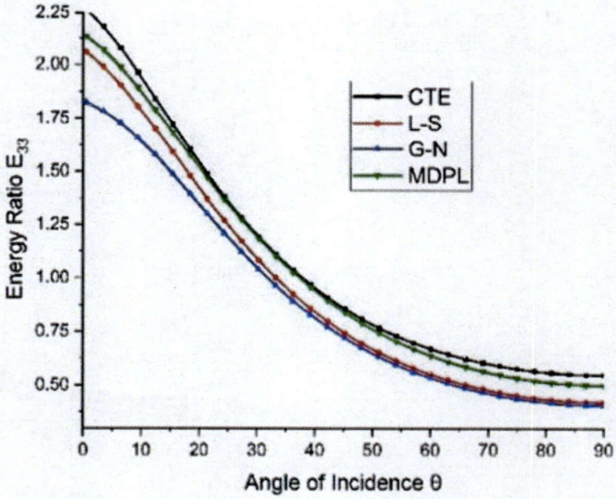

the phase velocities V_1, V_2, V_3 of QL, QTS and QT waves decrease with increase in the value of frequency in all the four cases with a little difference in magnitude.

7.4 Attenuation Coefficients

Figures 24, 25 and 26 show the variations of attenuation coefficients Q_1, Q_2, Q_3 of QL, QTS and QT waves with respect to frequency ω. From the graphs, we observe that the attenuation coefficients Q_1, Q_2, Q_3 of QL, QTS and QT waves increase sharply with the increase in frequency ω with variations in magnitude for different values of α.

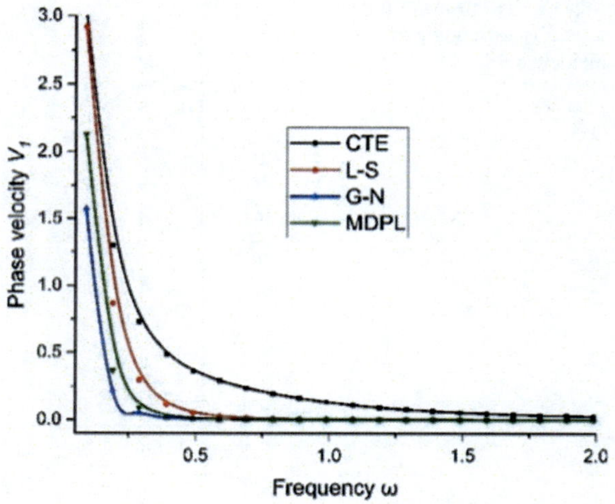

Fig. 20 Variations of phase velocity V_1 with frequency ω

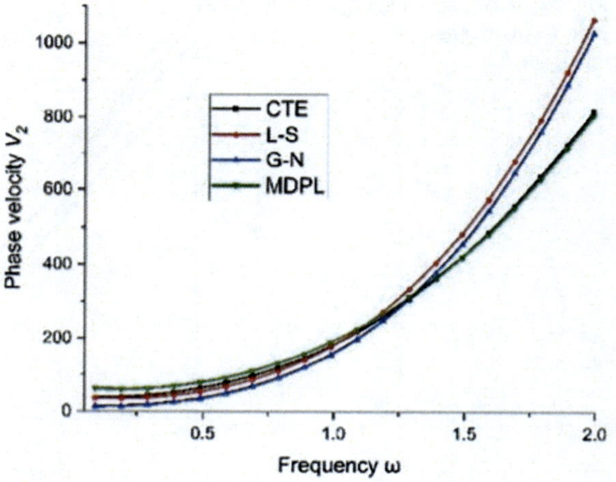

Fig. 21 Variations of phase velocity V_2 with frequency ω

7.5 Specific Loss

Figures 27, 28 and 29 show the variations of specific losses W_1, W_2, W_3 of QL, QTS and QT waves with respect to frequency ω. From the graphs, we observe that specific losses W_1, W_2, W_3 of QL, QTS and QT waves decrease with increase in frequency ω with variations in magnitude for different values of fractional-order parameter α.

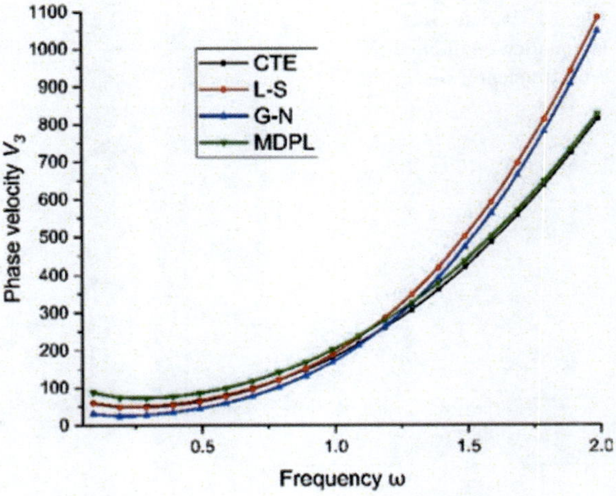

Fig. 22 Variations of phase velocity v_3 with frequency ω

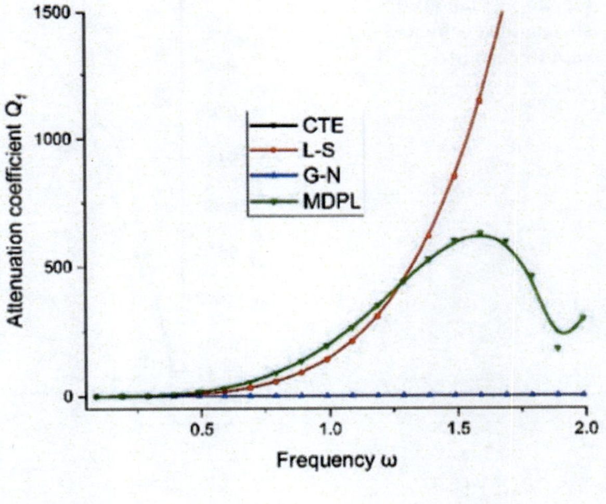

Fig. 23 Variations of attenuation coefficient Q_1 with frequency ω

7.6 Penetration Depth

Figures 30, 31 show the variations of penetration depths S_1, S_2, S_3 of QL, QTS and QT waves with respect to frequency ω. From the graphs, we observe that the penetration depths S_1, S_2, S_3 of QL, QTS and QT waves show an increase in values with a small difference in magnitude for the four different cases showing the effect of two temperature and fractional-order parameter α.

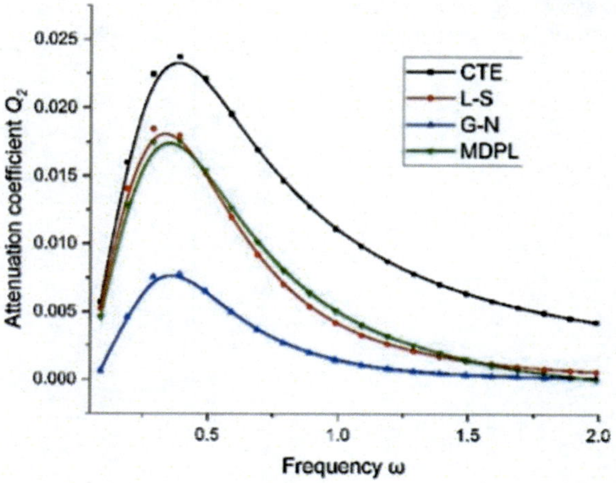

Fig. 24 Variations of attenuation coefficient Q_2 with frequency ω

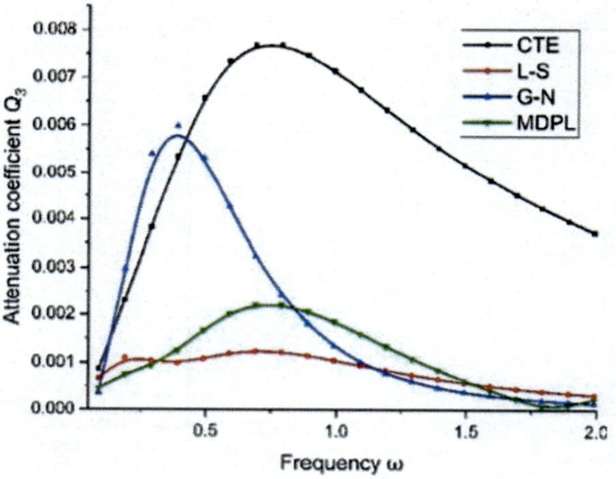

Fig. 25 Variations of attenuation coefficient Q_3 with frequency ω

8 Conclusions

The propagation of plane harmonic wave's inhomogeneous transversely isotropic magneto-thermoelastic rotating medium with multi-dual-phase lag heat transfer, rotation and two temperature has been studied. From the graphs, we observe the following concluding remarks:

(i) Different theories show the significant difference in the amplitude ratios, energy ratios, phase velocities, specific loss, penetration depth, attenuation coefficients of the plane harmonic wave.
(ii) The amplitude ratios, energy ratios, phase velocities, specific loss, penetration depth and attenuation coefficients also show maximum variations with GN and MDPL theories in the plane harmonic wave.

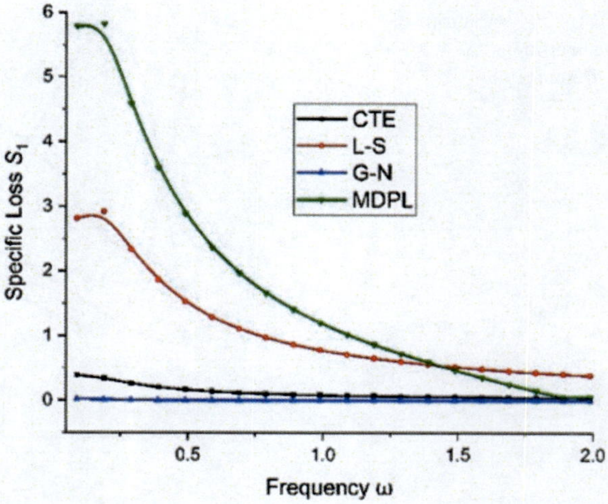

Fig. 26 Variations of specific loss S_1 with frequency ω

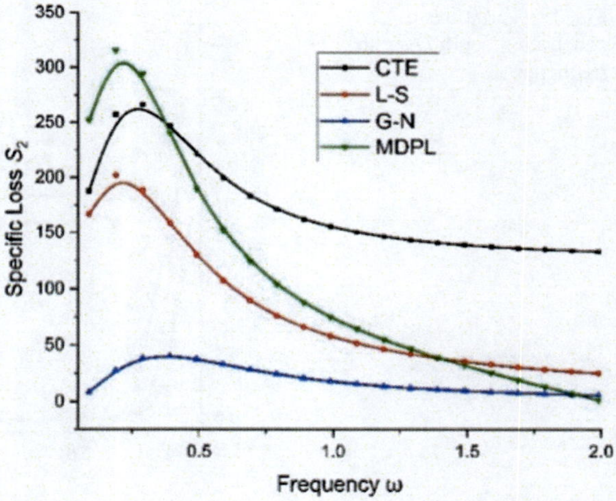

Fig. 27 Variations of specific loss S_2 with frequency ω

(iii) Study of these waves is not only helpful in providing information about the internal structures of the earth but also helpful in geophysics, for understanding the effects of the earth's magnetic field on seismic waves, damping of acoustic waves in a magnetic field, emission of electromagnetic radiations from nuclear devices, etc.

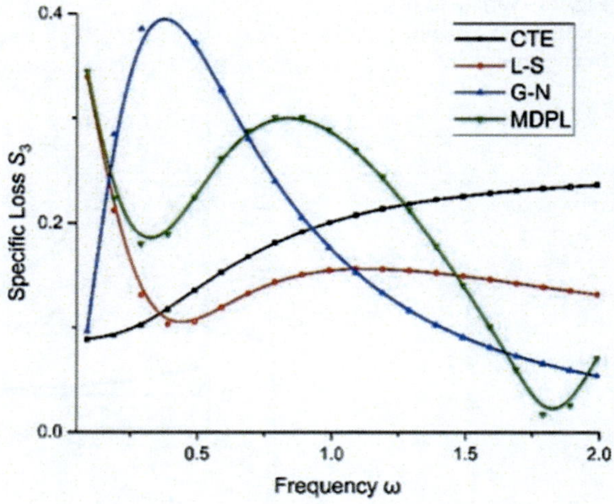

Fig. 28 Variations of specific loss S_3 with frequency ω

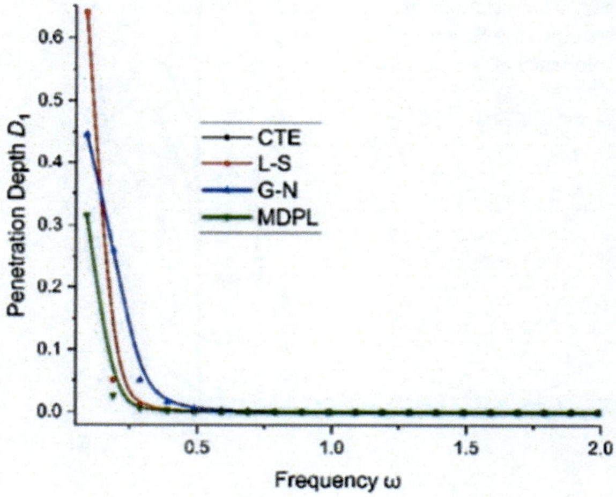

Fig. 29 Variations of penetration depth D_1 with frequency ω

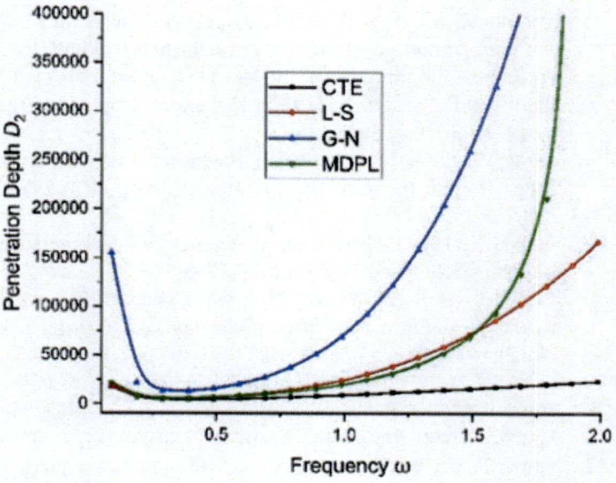

Fig. 30 Variations of penetration depth D_2 with frequency ω

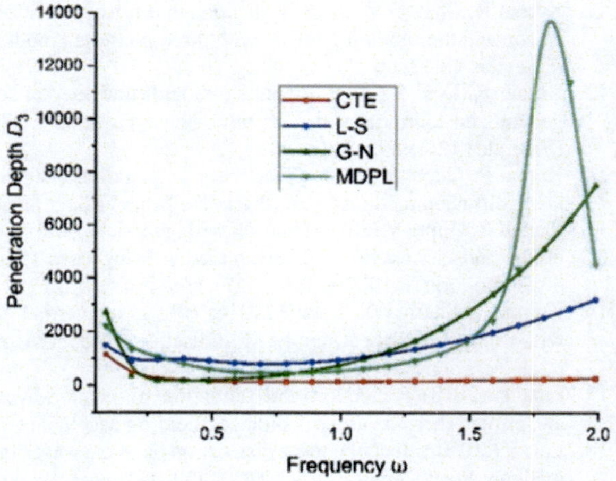

Fig. 31 Variations of penetration depth D_3 with frequency ω

References

1. Abo-Dahab SM, Jahangir A, Abd-alla A-e-nN (2018) Reflection of plane waves in thermoelastic microstructured materials under the influence of gravitation. Continuum Mech Thermodyn 1-13. https://doi.org/10.1007/s00161-018-0739-2
2. Achenbach J (1973) Wave propagation in elastic solids, North-Holland, Amsterdam, Elsevier, The Netherlands
3. Alesemi M (2018) Plane waves in magneto-thermoelastic anisotropic medium based on (L-S) theory under the effect of Coriolis and centrifugal forces. International Conference on Materials Engineering and Applications 348:01–12. IOP Conf. Series: Materials Science and Engineering. https://doi.org/10.1088/1757-899x/348/1/012018
4. Allam M, Tantawy R, Zenkour A (2018) Magneto-thermo-elastic response of exponentially graded piezoelectric hollow spheres. Adv Comput Des 3(3):303–318. https://doi.org/10.12989/acd.2018.3.3.303

5. Bijarnia R, Singh B (2016) Propagation of plane waves in a rotating transversely isotropic two temperature generalized thermoelastic solid half-space with voids. Int J Appl Mech Eng 21(1):285–301. https://doi.org/10.1515/ijame-2016-0018
6. Dhaliwal R, Singh A (1980) Dynamic coupled thermoelasticity. Hindustan Publication Corporation, New Delhi, India
7. Ezzat M, El-Barrry aA (2017) Fractional magneto-thermoelastic materials with phase-lag Green-Naghdi theories. Steel Compos Struct 24(3):297–307. http://dx.doi.org/10.12989/scs.2017.24.3.297
8. Kaliski S (1963) Absorption of magnetoviscoelastic surface waves in a real conductor in a magnetic field. Proc Vibr Prob 4:319–329
9. Kaur I, Lata P (2019) Rayleigh wave propagation in transversely Isotropic magneto thermoelastic medium with three phase lag heat transfer and diffusion. Int J Mech Mater Eng 14(12):1–11. https://doi.org/10.1186/s40712-019-0108-3
10. Kaur I, Lata P (2019) Effect of hall current on propagation of plane wave in transversely isotropic thermoelastic medium with two temperature and fractional order heat transfer. SN Appl Sci 1:900. https://doi.org/10.1007/s42452-019-0942-1
11. Kumar R, Chawla V (2011) A study of plane wave propagation in anisotropic three phase-lag model and two-phae-lag model. Int Commun Heat Mass Transfer 38(9):1262–1268. https://doi.org/10.1016/j.icheatmasstransfer.2011.07.005
12. Kumar R, Chawla V (2013) Reflection and refraction of plane wave at the interface between elastic and thermoelastic media with three-phase-lag model. Int Commun Heat Mass Transfer (Elsevier) 48:53–60
13. Kumar R, Devi S (2016) Plane waves and fundamental solution in a modified couple stress generalized thermoelastic with three-phase-lag model. Multidiscipline Model Mater Struct (Emerald) 12(4):693–711
14. Kumar R, Gupta V (2014) Plane wave propagation in an anisotropic dual-phase-lag thermoelastic diffusion medium. Multidiscipline Model Mater Struct (Emerald) 10(4):562–592
15. Kumar R, Gupta V (2015) Dual-phase-lag model of wave propagation at the interface between elastic and thermoelastic diffusion media. J Eng Phys Thermophys, Springer 88(1):252–265. https://doi.org/10.1007/s10891-015-1188-4
16. Kumar R, Sharma N, Lata P (2016) Effects of thermal and diffusion phase-lags in a plate with axisymmetric heat supply. Multidiscipline Model Mater Struct (Emerald) 12(2):275–290. https://doi.org/10.1108/MMMS-08-2015-0042
17. Lata P (2018) Reflection and refraction of plane waves in layered nonlocal elastic and anisotropic thermoelastic medium. Struct Eng Mech 66(1):113–124
18. Lata P (2018) Effect of energy dissipation on plane waves in sandwiched layered thermoelastic medium. Steel Compos Struct 27(4), 439–451. http://dx.doi.org/10.12989/scs.2018.27.4.439
19. Lata P, Kaur I (2019) Plane wave propagation in transversely isotropic magnetothermoelastic rotating medium with fractional order generalized heat transfer. Struct Monit Maintenance 6(3):191–218. https://doi.org/10.12989/smm.2019.6.3.191
20. Lata P, Kaur I (2019) Axisymmetric thermomechanical analysis of transversely isotropic magneto thermoelastic solid due to time-harmonic sources. Coupled Syst Mech 8(5):415–437. https://doi.org/10.12989/csm.2019.8.5.415
21. Lata P, Kaur I (2019) Effect of rotation and inclined load on transversely isotropic magneto thermoelastic solid. Struct Eng Mech 70(2):245–255. http://dx.doi.org/10.12989/sem.2019.70.2.245
22. Lata P, Kaur I (2019) Thermomechanical interactions in transversely isotropic magneto thermoelastic solid with two temperatures and without energy dissipation. Steel Compos Struct 32(6):779–793. https://doi.org/10.12989/scs.2019.32.6.779
23. Mahmoud S (2012) Influence of rotation and generalized magneto-thermoelastic on Rayleigh waves in a granular medium under effect of initial stress and gravity field. Meccanica, Springer 47:1561–1579. https://doi.org/10.1007/s11012-011-9535-9
24. Mahmoud SR, Marin M, Al-Basyouni KS (2015) Effect of the initial stress and rotation on free vibrations in transversely isotropic human long dry bone. Versita 171–184. https://doi.org/10.1515/auom-2015-0011

25. Othman M, Marin M (2017) Effect of thermal loading due to laser pulse on thermoelastic porous medium under G-N theory. Results Phys 7:3863–3872
26. Othman MI, Song YQ (2008) Reflection of magneto-thermoelastic waves from a rotating elastic half-space. Int J Eng Sci 46:459–474. https://doi.org/10.1016/j.ijengsci.2007.12.004
27. Othman MI, Abo-Dahab SM, S Alsebaey, ON (2017) Reflection of Plane waves from a rotating magneto-thermoelastic medium with two-temperature and initial srtress under three theories. Mech Mech Eng 21(2):217–232
28. Schoenberg M, Censor D (1973) elastic waves in rotating media. Q Appl Math 31:115–125
29. Slaughter W (2002) The linearised theory of elasticity. Birkhausar
30. Ting TC (2004) Surface waves in a rotating anisotropic elastic half-space. Wave Motion 40:329–346. https://doi.org/10.1016/j.wavemoti.2003.10.005
31. Zenkour AM (2018) Refined microtemperatures multi-phase-lags theory for plane wave propagation in thermoelastic medium. Results Phys 11:929–937. https://doi.org/10.1016/j.rinp.2018.10.030
32. Zenkour AM (2018) Refined two-temperature multi-phase-lags theory for thermomechanical response of microbeams using the modified couple stress analysis. Acta Mechanica 229(9):3671–3692. https://doi.org/10.1007/s00707-018-2172-9

Deformation in Generalized Transversely Isotropic Magneto-Thermoelastic Rotating Solid Due to Inclined Load and Thermal Laser Pulse

Parveen Lata, Iqbal Kaur, and Kulvinder Singh

Abstract The present research deals with the study of deformation in generalized transversely isotropic magneto-thermoelastic solid with two temperature (2T), rotation, due to inclined load and laser pulse. Generalized thermoelasticity theory has been considered for this mathematical model. The entire thermoelastic medium is rotating with uniform angular velocity and subjected to thermally insulated and isothermal boundaries. The inclined load is supposed to be a linear combination of a normal load and a tangential load. The Fourier and Laplace transform techniques have been used to find the solution to the problem. The displacement components, conductive temperature distribution and stress components with the horizontal distance are computed in the transformed domain and further calculated in the physical domain using numerical inversion techniques. The effect of laser pulse in different theories of thermoelasticity is depicted graphically on the resulting quantities.

Keywords Thermal laser pulse · Transversely isotropic generalized thermoelastic · Rotation · Inclined load · Magneto-thermoelastic solid

1 Introduction

The laser technology has dynamic applications in testing and analysis of materials. When a solid is exposed with a laser pulse, it absorbs some energy which results in thermal deformation and generates ultrasonic waves in the material. The change at some point of the medium is beneficial to detect the deformed field near mining

P. Lata · I. Kaur
Department of Basic and Applied Sciences, Punjabi University, Patiala, Punjab, India
e-mail: parveenlata@pbi.ac.in

I. Kaur
e-mail: bawahanda@gmail.com

K. Singh (✉)
UIET, Kurukshetra University, Kurukshetra, Haryana, India
e-mail: ksingh2015@kuk.ac.in

N. Marriwala et al. (eds.), *Mobile Radio Communications and 5G Networks*, Lecture Notes in Networks and Systems 140,
https://doi.org/10.1007/978-981-15-7130-5_43

shocks, seismic and volcanic sources, thermal power plants, high-energy particle accelerators and many emerging technologies. The study of a time-harmonic source is one of the broad and dynamic areas of continuum dynamics.

Youssef and Al-Felali [1] described the generalized thermoelasticity problem of material with thermal loading due to laser pulse. Every et al. [2] calculated the laser thermoelastic generation in metals above the melt threshold. Youssef and El-Bary [3] discussed the thermoelastic material response due to laser pulse heating with four different theorems of thermoelasticity. Sharma et al. [4] investigated the 2D deception in a transversely isotropic homogeneous thermoelastic solids in the presence of two temperatures in GN-II theory with an inclined load (linear combination of normal load and tangential load). Kumar et al. [5] discussed the thermomechanical interactions due to laser pulse in microstretch thermoelastic medium. Kumar [6] studied the effect of laser pulse in micropolar thermoelastic medium with three-phase lag model. Kumar and Kumar [7] illustrated the elastodynamic response of thermal laser pulse in micropolar thermoelastic mass diffusion medium. Kumar et al. [8] depicted the elastodynamic interactions of laser pulse in microstretch thermoelastic mass diffusion medium with dual-phase lag. Ailawalia et al. [9] discussed the laser pulse heating in thermo-microstretch elastic layer overlying thermoelastic half-space. Othman and Marin [10] described the effect of heat laser pulse on generalized thermoelasticity for micropolar medium. Othman et al. [11] showed the effect of thermal loading due to laser pulse on 3D problem of micropolar thermoelastic solid with energy dissipation. Abbas and Marin [12] provided the analytical solutions of a 2D diffusion problem for generalized thermoelastic due to laser pulse. Despite this, several researchers worked on different theories of thermoelasticity such as Allam et al. [13], Lal et al. [14], Kumar et al. [15], Hassan et al. [16], Lata and Kaur [17, 18], Lata and Kaur [19, 20] and Kaur and Lata [21, 22].

Irrespective of these, not much work has been carried out in magneto-thermoelastic transversely isotropic solid with rotation, for inclined load with two temperature in generalized thermoelasticity. In this paper, we have attempted to study the deformation in transversely isotropic magneto-thermoelastic solid with the combined effects of rotation for inclined load with two temperature by considering the laser pulse effect. The expressions of displacement components, conductive temperature and stress components are calculated in the transformed domain by using the Laplace and Fourier transformation. Numerical inversion technique is used to find the resulting quantities in the physical domain, and different thermoelastic theories have been represented graphically.

2 Basic Equations

Following Kumar and Kumar [7] for a generalized anisotropic thermoelastic medium, the constitutive equation is given by

$$t_{ij} = C_{ijkl}e_{kl} - \beta_{ij}\left(1 + \tau_1 \frac{\partial}{\partial t}\right)T. \tag{1}$$

and equation of motion as described by Schoenberg and Censor [23] for a uniformly rotating medium with angular velocity and Lorentz force which governs the dynamic displacement u is

$$t_{ij,j} + F_i = \rho\{\ddot{u}_i + (\boldsymbol{\Omega} \times (\boldsymbol{\Omega} \times \mathbf{u})_i + (2\boldsymbol{\Omega} \times \dot{\boldsymbol{u}})_i\}, \quad (2)$$

where $\boldsymbol{\Omega} = \Omega\,\hat{\boldsymbol{n}}$, $\boldsymbol{n}$ is a unit vector representing the direction of the axis of rotation, the term $\boldsymbol{\Omega} \times (\boldsymbol{\Omega} \times \mathbf{u})$ is the additional centripetal acceleration due to the time-varying motion only and the term $2\boldsymbol{\Omega} \times \dot{\boldsymbol{u}}$ is the Coriolis acceleration.

$$F_i = \mu_0(\mathbf{J} \times \mathbf{H_0}).$$

Following Kumar and Kumar [7], the generalized heat conduction equation is

$$K_{ij}\varphi_{,ij} + \rho\big(Q + \tau_0\dot{Q}\big) = \beta_{ij}T_0\big(\dot{e}_{ij} + \varepsilon\tau_0\ddot{e}_{ij}\big) + \rho C_E\big(\dot{T} + \tau_0\ddot{T}\big), \quad (3)$$

where

$$\beta_{ij} = C_{ijkl}\alpha_{ij}, \quad (4)$$

$$\begin{gathered} e_{ij} = \tfrac{1}{2}\big(u_{i,j} + u_{j,i}\big), \quad i, j = 1, 2, 3. \\ T = \varphi - a_{ij}\varphi_{,ij} \end{gathered} \quad (5)$$

$\beta_{ij} = \beta_i\delta_{ij}$, $\quad K_{ij} = K_i\delta_{ij}$, i is not summed.

3 Formulation and Solution of the Problem

We consider a homogeneous transversely isotropic magneto-thermoelastic medium, permeated by an initial magnetic field $\mathbf{H_0} = (0, H_0, 0)$ acting along y-axis. The rectangular Cartesian coordinate system (x, y, z) having origin on the surface $(z = 0)$ with z-axis pointing vertically into the medium is introduced. The surface of the half-space is subjected to an inclined load acting at $z = 0$.

In addition, we consider that

$$\boldsymbol{\Omega} = (0, \Omega, 0).$$

From the generalized Ohm's law,

$$J_2 = 0.$$

The density components J_1 and J_3 are given as

$$J_1 = -\varepsilon_0 \mu_0 H_0 \frac{\partial^2 w}{\partial t^2}, \tag{6}$$

$$J_3 = \varepsilon_0 \mu_0 H_0 \frac{\partial^2 u}{\partial t^2}. \tag{7}$$

In addition, the equations of displacement vector (u, v, w) and conductive temperature φ for transversely isotropic thermoelastic solid in the presence of two temperature are

$$u \equiv u(x, z, t), v = 0, w \equiv w(x, z, t) \text{ and } \varphi \equiv \varphi(x, z, t). \tag{8}$$

Following Slaughter [24], using the proper transformation on Eqs. (1)–(3) with the aid of (8), yields:

$$C_{11}\frac{\partial^2 u}{\partial x^2} + C_{13}\frac{\partial^2 w}{\partial x \partial z} + C_{44}\left(\frac{\partial^2 u}{\partial z^2} + \frac{\partial^2 w}{\partial x \partial z}\right) - \beta_1\left(1 + \tau_1\frac{\partial}{\partial t}\right)\frac{\partial}{\partial x}\left\{\varphi - \left(a_1\frac{\partial^2 \varphi}{\partial x^2} + a_3\frac{\partial^2 \varphi}{\partial z^2}\right)\right\} - \mu_0 J_3 H_0 = \rho\left(\frac{\partial^2 u}{\partial t^2} - \Omega^2 u + 2\Omega\frac{\partial w}{\partial t}\right), \tag{9}$$

$$(C_{13} + C_{44})\frac{\partial^2 u}{\partial x \partial z} + C_{44}\frac{\partial^2 w}{\partial x^2} + C_{33}\frac{\partial^2 w}{\partial z^2} - \beta_3\left(1 + \tau_1\frac{\partial}{\partial t}\right)\frac{\partial}{\partial z}\left\{\varphi - \left(a_1\frac{\partial^2 \varphi}{\partial x^2} + a_3\frac{\partial^2 \varphi}{\partial z^2}\right)\right\} - \mu_0 J_1 H_0 = \rho\left(\frac{\partial^2 w}{\partial t^2} - \Omega^2 w - 2\Omega\frac{\partial u}{\partial t}\right), \tag{10}$$

$$K_1\frac{\partial^2 \varphi}{\partial x^2} + K_3\frac{\partial^2 \varphi}{\partial z^2} + \rho(Q + \varepsilon\tau_0\dot{Q}) = \rho C_E(\dot{T} + \tau_0\ddot{T}) + T_0\frac{\partial}{\partial t}\left\{\beta_1\left(1 + \varepsilon\tau_0\frac{\partial}{\partial t}\right)\frac{\partial u}{\partial x} + \beta_3\left(1 + \varepsilon\tau_0\frac{\partial}{\partial t}\right)\frac{\partial w}{\partial z}\right\}, \tag{11}$$

and

$$t_{xx} = C_{11}e_{xx} + C_{13}e_{xa} - \beta_1\left(1 + \tau_1\frac{\partial}{\partial t}\right)T, \tag{12}$$

$$t_{zz} = C_{13}e_{11} + C_{33}e_{zz} - \beta_3\left(1 + \tau_1\frac{\partial}{\partial t}\right)T, \tag{13}$$

$$t_{xz} = 2C_{44}e_{xz}, \tag{14}$$

where

$$T = \varphi - \left(a_1 \frac{\partial^2 \varphi}{\partial x^2} + a_3 \frac{\partial^2 \varphi}{\partial z^2} \right),$$

$$\beta_1 = (C_{11} + C_{12})\alpha_1 + C_{13}\alpha_{3,}$$

$$\beta_3 = 2C_{13}\alpha_1 + C_{33}\alpha_{3.}$$

where τ_0 and τ_1 are thermal relaxation times with $\tau_0 \geq \tau_1 \geq 0$.

According to Marin et al. [10], the surface of transversely isotropic thermoelastic solid is illuminated by laser pulse given by the heat input:

$$Q = I_0 f(t) g(x) h(z).$$

where I_0 is the energy absorbed, i.e. the laser intensity which is defined as the total energy carried by a laser pulse per unit area of the laser beam, and $f(t)$ is a temporal profile given as

$$f(t) = \frac{t}{t_0^2} e^{-(t/t_0)}.$$

here, $t_0 = 2ps$ is the pulse rise time. The pulse is also assumed to have a Gaussian spatial profile in x

$$g(x) = \frac{1}{2\pi r^2} e^{-(x^2/r^2)}.$$

where r is the beam radius and as a function of the depth, z, the heat deposition due to the laser pulse is assumed to decay exponentially within the solid:

$$h(z) = \gamma e^{-\gamma z}.$$

Therefore,

$$Q = \frac{\gamma I_0 t}{2\pi r^2 t_0^2} e^{-\left(\frac{x^2}{r^2} + \frac{t}{t_0} + \gamma z\right)}.$$

We consider that the medium is initially at rest. Therefore, the preliminary and symmetry conditions are given by

$$u(x, z, 0) = 0 = \dot{u}(x, z, 0),$$

$$w(x, z, 0) = 0 = \dot{w}(x, z, 0),$$

$$\varphi(x,z,0)=0=\dot{\varphi}(x,z,0)\ \text{for}\ z\geq 0,\ -\infty<x<\infty,$$

$$u(x,z,t)=w(x,z,t)=\varphi(x,z,t)=0\ \text{for}\ t>0\ \text{when}\ z\to\infty.$$

To facilitate the solution, the following dimensionless quantities are introduced:

$$\begin{aligned}
&x'=\frac{x}{L},\ z'=\frac{z}{L},\ t'=\frac{c_1}{L}t,\ u'=\frac{\rho c_1^2}{L\beta_1 T_0}u,\ w'=\frac{\rho c_1^2}{L\beta_1 T_0}w,\\
&T'=\frac{T}{T_0},\ t'_{11}=\frac{t_{11}}{\beta_1 T_0},\ t'_{33}=\frac{t_{33}}{\beta_1 T_0},\ t'_{31}=\frac{t_{31}}{\beta_1 T_0},\ \varphi'=\frac{\varphi}{T_0},\\
&a'_1=\frac{a_1}{L^2},\ a'_3=\frac{a_3}{L^2},\ h'=\frac{h}{H_0},\ \Omega'=\frac{\mathrm{L}}{C_1}\Omega,\ Q'=\frac{1}{\varpi T_0 C_E}Q,\\
&\varpi=\frac{\rho C_E C_1^2}{K_1},\ \tau'_0=\frac{c_1}{L}\tau_0,\ \tau'_1=\frac{c_1}{L}\tau_1,\ \rho C_1^2=C_{11}
\end{aligned}\tag{15}$$

Making use of (15) in Eqs. (9)–(11), and the stress–strain relations (12)–(14) after suppressing the primes, yields

$$\begin{aligned}
&\frac{\partial^2 u}{\partial x^2}+\delta_4\frac{\partial^2 w}{\partial x\partial z}+\delta_2\left(\frac{\partial^2 u}{\partial z^2}+\frac{\partial^2 w}{\partial x\partial z}\right)-\left(1+\tau_1\frac{\partial}{\partial t}\right)\frac{\partial}{\partial x}\left\{\varphi-\left(a_1\frac{\partial^2\varphi}{\partial x^2}+a_3\frac{\partial^2\varphi}{\partial z^2}\right)\right\}\\
&=\left(\frac{\varepsilon_0\mu_0^2H_0^2}{\rho}+1\right)\left(\frac{\partial^2 u}{\partial t^2}\right)-\Omega^2 u+2\Omega\frac{\partial w}{\partial t},
\end{aligned}\tag{16}$$

$$\begin{aligned}
&\delta_1\frac{\partial^2 u}{\partial x\partial z}+\delta_2\frac{\partial^2 w}{\partial x^2}+\delta_3\frac{\partial^2 w}{\partial z^2}-\frac{\beta_3}{\beta_1}\left(1+\tau_1\frac{\partial}{\partial t}\right)\frac{\partial}{\partial z}\left\{\varphi-\left(a_1\frac{\partial^2\varphi}{\partial x^2}+a_3\frac{\partial^2\varphi}{\partial z^2}\right)\right\}\\
&=\left(\frac{\varepsilon_0\mu_0^2H_0^2}{\rho}+1\right)\left(\frac{\partial^2 w}{\partial t^2}\right)-\Omega^2 w+2\Omega\frac{\partial u}{\partial t},
\end{aligned}\tag{17}$$

$$\begin{aligned}
&\frac{\partial^2\varphi}{\partial x^2}+\frac{K_3}{K_1}\frac{\partial^2\varphi}{\partial z^2}+Q_0 f(x,t)e^{-\gamma z}=\delta_5\frac{\partial}{\partial t}\left(1+\tau_0\frac{\partial}{\partial t}\right)\left[\varphi-a_1\frac{\partial^2\varphi}{\partial x^2}-a_3\frac{\partial^2\varphi}{\partial z^2}\right]\\
&+\delta_6\frac{\partial}{\partial t}\left(1+\varepsilon\tau_0\frac{\partial}{\partial t}\right)\left[\beta_1\frac{\partial u}{\partial x}+\beta_3\frac{\partial w}{\partial z}\right],
\end{aligned}\tag{18}$$

$$t_{xx}(x,z,t)=\frac{\partial u}{\partial x}+\delta_4\frac{\partial w}{\partial z}-\left(1+\tau_1\frac{\partial}{\partial t}\right)\left\{\varphi-\left(a_1\frac{\partial^2\varphi}{\partial x^2}+a_3\frac{\partial^2\varphi}{\partial z^2}\right)\right\},\tag{19}$$

$$t_{zz}(x,z,t)=\delta_4\frac{\partial u}{\partial x}+\delta_3\frac{\partial w}{\partial z}-\frac{\beta_3}{\beta_1}\left(1+\tau_1\frac{\partial}{\partial t}\right)\left\{\varphi-\left(a_1\frac{\partial^2\varphi}{\partial x^2}+a_3\frac{\partial^2\varphi}{\partial z^2}\right)\right\},\tag{20}$$

$$t_{xz}(x,z,t)=\delta_2\left(\frac{\partial u}{\partial z}+\frac{\partial w}{\partial x}\right),\tag{21}$$

$$t_{xz}(x, z, t) = \delta_2\left(\frac{\partial u}{\partial z} + \frac{\partial w}{\partial x}\right), \tag{21}$$

where

$$\delta_1 = \frac{c_{13} + c_{44}}{c_{11}}, \delta_2 = \frac{c_{44}}{c_{11}}, \delta_3 = \frac{c_{33}}{c_{11}}, \delta_4 = \frac{c_{13}}{c_{11}}, \delta_5 = \frac{\rho C_E C_1 L}{K_1}, \delta_6 = -\frac{T_0 \beta_1 L}{\rho C_1 K_1},$$

$$Q_0 = \frac{L^2 \rho \varpi C_E}{K_1} \frac{\gamma I_0}{2\pi r^2 t_0^2},$$

$$f(x, t) = \frac{L}{c_1} t\left[1 + \varepsilon \tau_0 \frac{\partial}{\partial t}\right] e^{-\left(\frac{x^2}{r^2} + \frac{L}{c_1}\frac{t}{t_0}\right)}.$$

Laplace transforms with respect to variable t with s as a Laplace transform variable are defined by

$$\mathcal{L}[f(x, z, t)] = \int_0^\infty e^{-st} f(x, z, t)\mathrm{d}t = \tilde{f}(x, z, s). \tag{22}$$

Fourier transforms with respect to variable x with ξ as a Fourier transform variable are defined by

$$\hat{f}(\xi, z, s) = \int_{-\infty}^{\infty} f(x, z, s)e^{i\xi x}\mathrm{d}x \tag{23}$$

Using Laplace and Fourier transforms defined by Eqs. (22)–(23) on Eqs. (16)–(18), we obtain a system of equations

$$\begin{aligned} &[-\xi^2 + \delta_2 D^2 - \delta_7 s^2 + \Omega^2]\hat{u}(\xi, z, s) + [\delta_4 Di\xi + \delta_2 Di\xi - 2\Omega s]\hat{w}(\xi, z, s) \\ &+ (-i\xi)(1 + \tau_1 s)\left[1 + a_1\xi^2 - a_3 D^2\right]\hat{\varphi}(\xi, z, s) = 0, \end{aligned} \tag{24}$$

$$\begin{aligned} &[\delta_1 Di\xi - 2\Omega s]\hat{u}(\xi, z, s) + \left[-\delta_2\xi^2 + \delta_3 D^2 - \delta_7 s^2 + \Omega^2\right]\hat{w}(\xi, z, s) \\ &- \frac{\beta_3}{\beta_1}(1 + \tau_1 s)D\left[1 + a_1\xi^2 - a_3 D^2\right]\hat{\varphi}(\xi, z, s) = 0, \end{aligned} \tag{25}$$

$$\begin{aligned} &[\delta_6 s\delta_8\beta_1 i\xi]\hat{u}(\xi, z, s) + [\delta_6 s\delta_8\beta_3 D]\hat{w}(\xi, z, s) \\ &+ \left[\xi^2 - \frac{K_3}{K_1}D^2 + \delta_5\delta_8 s\left(1 + a_1\xi^2 - a_3 D^2\right)\right]\hat{\varphi}(\xi, z, s) \\ &= Q_0 f(\xi, s)e^{-\gamma z}, \end{aligned} \tag{26}$$

where

$$\delta_7 = \frac{\varepsilon_0 \mu_0^2 H_0^2}{\rho} + 1, \delta_8 = (1 + \varepsilon \tau_0 s),$$

$$f(\xi, s) = \left[1 - \varepsilon \tau_0 \frac{L}{C_1 t_0} \left(\frac{C_1 t_0}{C_1 t_0 s + L} \right)^2 \right] r \sqrt{\pi} e^{-\left(\frac{r^2 \xi^2}{4} + \frac{\frac{L}{C_1} t}{t_0} \right)}.$$

By taking $\hat{Q}(\xi, z, s) = 0$, i.e. no external heat is supplied, the non-trivial solution of (24)–(26) yields

$$(AD^6 + BD^4 + CD^2 + E)(\hat{u}) = f_1(x, \gamma, t)e^{-\gamma z} \tag{27}$$

$$(AD^6 + BD^4 + CD^2 + E)(\hat{w}) = f_2(x, \gamma, t)e^{-\gamma z} \tag{28}$$

$$(AD^6 + BD^4 + CD^2 + E)(\hat{\varphi}) = f_3(x, \gamma, t)e^{-\gamma z} \tag{29}$$

where

$$A = \delta_2 \delta_3 \zeta_7 - \zeta_5 \delta_2 \frac{\beta_3}{\beta_1} a_3,$$

$$\begin{aligned} B = {} & \delta_3 \zeta_1 \zeta_7 - a_3 (1 + \tau_1 s) \zeta_1 \zeta_5 \frac{\beta_3}{\beta_1} + \delta_2 \delta_3 \zeta_6 + \delta_2 \zeta_7 \zeta_3 - \zeta_5 \zeta_9 \delta_2 \\ & - \zeta_8 \delta_1 i \xi \zeta_7 + \zeta_8 \zeta_4 \frac{\beta_3}{\beta_1} (1 + \tau_1 s) a_3 \\ & - a_3 (1 + \tau_1 s) \xi^2 \zeta_5 \delta_1 - a_3 (1 + \tau_1 s) \delta_3 \zeta_4 i \xi, \end{aligned}$$

$$\begin{aligned} C = {} & \delta_3 \zeta_1 \zeta_6 + \zeta_1 \zeta_3 \zeta_7 - \zeta_1 \zeta_5 \zeta_9 + \delta_2 \zeta_6 \zeta_3 + \zeta_4 \zeta_8 \zeta_9 \\ & - \zeta_8 \delta_1 i \xi \zeta_6 + 4 \Omega^2 s^2 \zeta_7 \\ & + \zeta_2 \delta_1 i \xi \zeta_5 - \zeta_2 \zeta_4 \delta_3 - a_3 (1 + \tau_1 s) \zeta_4 i \xi \zeta_3, \end{aligned}$$

$$E = \zeta_3 \zeta_1 \zeta_6 + 4 \Omega^2 s^2 \zeta_6 - \zeta_2 \zeta_4 \zeta_3,$$

$$\zeta_1 = -\xi^2 - \delta_7 s^2 + \Omega^2,$$

$$\zeta_2 = -i\xi (1 + \tau_1 s) \left(1 + a_1 \xi^2\right),$$

$$\zeta_3 = -\delta_2 \xi^2 - \delta_7 s^2 + \Omega^2,$$

$$\zeta_4 = \delta_6 \delta_8 s \beta_1 i \xi,$$

$$\zeta_5 = \delta_6 \delta_8 s \beta_3,$$

$$\zeta_6 = \xi^2 + \delta_5\delta_8 s\left(1 + a_1\xi^2\right),$$

$$\zeta_7 = -\frac{K_3}{K_1} - a_3\delta_5\delta_8 s,$$

$$\zeta_8 = \delta_1 i\xi,$$

$$\zeta_9 = -\left(1 + a_1\xi^2\right)(1 + \tau_1 s)\tfrac{\beta_3}{\beta_1},$$
$$f_1(\xi,\gamma,t) = Q_0 f(\xi,t)\Big[(\gamma\zeta_8 + 2\Omega s)\Big(\zeta_9\gamma + \tfrac{\beta_3}{\beta_1}a_3\gamma^3\Big) - \left(\zeta_2 + a_3 i\xi\gamma^2\right)\left(\zeta_3 + \delta_3\gamma^2\right)\Big],$$
$$f_2(\xi,\gamma,t) = Q_0 f(\xi,t)\Big[\left(\zeta_1 + \delta_2\gamma^2\right)\Big(\zeta_9\gamma + \tfrac{\beta_3}{\beta_1}a_3\gamma^3\Big) + (-\gamma\delta_1 i\xi + 2\Omega s)\left(\zeta_2 + a_3 i\xi\gamma^2\right)\Big],$$
$$f_3(\xi,\gamma,t) = Q_0 f(\xi,t)\left[\left(\zeta_1 + \delta_2\gamma^2\right)\left(\zeta_3 + \delta_3\gamma^2\right) + (-\gamma\delta_1 i\xi + 2\Omega s)(\gamma\zeta_8 + 2\Omega s)\right],$$
$$f_4(\xi,\gamma,t) = A\gamma^6 + B\gamma^4 + C\gamma^2 + E.$$

The roots of the characteristic equation of Eq. (27) are $\pm\,\lambda j$, (j = 1, 2, 3), and the solution of Eqs. (27)–(29) is calculated by using the radiation condition of $\left(\hat{u}, \hat{w}, \hat{\varphi}\right) \to 0$ as $z \to \infty$ and can be written as

$$\hat{u}(\xi, z, s) = \sum_{j=1}^{3} A_j e^{-\lambda_j z} + \frac{f_1}{f_4}e^{-\gamma z}, \tag{30}$$

$$\hat{w}(\xi, z, s) = \sum_{j=1}^{3} d_j A_j e^{-\lambda_j z} + \frac{f_2}{f_4}e^{-\gamma z}, \tag{31}$$

$$\hat{\varphi}(\xi, z, s) = \sum_{j=1}^{3} l_j A_j e^{-\lambda_j z} + \frac{f_3}{f_4}e^{-\gamma z}, \tag{32}$$

where $A_j(\xi, s)$, $j = 1, 2, 3$ being undetermined constants and d_j and l_j are given by

$$d_j = \frac{\delta_2\zeta_7\lambda_j^4 + (\zeta_7\zeta_1 - a_3\zeta_4 i\xi + \delta_2\zeta_6)\lambda_j^2 + \zeta_1\zeta_6 - \zeta_4\zeta_2}{\left(\delta_3\zeta_7 - \frac{\beta_3}{\beta_1}a_3\zeta_5\right)\lambda_j^4 + (\delta_3\zeta_6 + \zeta_3\zeta_7 - \zeta_5\zeta_9)\lambda_j^2 + \zeta_3\zeta_6},$$

$$l_j = \frac{\delta_2\delta_3\lambda_j^4 + (\delta_2\zeta_3 + \zeta_1\delta_3 - \delta_1\zeta_8 i\xi)\lambda_j^2 + 4\Omega^2 s^2 + \zeta_3\zeta_1}{\left(\delta_3\zeta_7 - \frac{\beta_3}{\beta_1}a_3\zeta_5\right)\lambda_j^4 + (\delta_3\zeta_6 + \zeta_3\zeta_7 - \zeta_5\zeta_9)\lambda_j^2 + \zeta_3\zeta_6}.$$

And the stress relations (19)–(21) after using Laplace and Fourier transforms defined by Eqs. (22)–(23) and then putting values of $\hat{u}, \hat{w}, \hat{\varphi}$ from Eqs. (30)–(32) become

$$\hat{t}_{xx}(\xi, z, s) = i\xi\left[\sum_{j=1}^{3} A_j e^{-\lambda_j z} + \frac{f_1}{f_4}e^{-\gamma z}\right] + \delta_4 D\left[\sum_{j=1}^{3} d_j A_j e^{-\lambda_j z} + \frac{f_2}{f_4}e^{-\gamma z}\right]$$
$$-(1+\tau_1 s)\left\{1-\left(a_1\xi^2 + a_3 D^2\right)\right\}\left[\sum_{j=1}^{3} l_j A_j e^{-\lambda_j z} + \frac{f_3}{f_4}e^{-\gamma z}\right], \quad (33)$$

$$\hat{t}_{zz}(\xi, z, s) = \delta_4 i\xi\left[\sum_{j=1}^{3} A_j e^{-\lambda_j z} + \frac{f_1}{f_4}e^{-\gamma z}\right] + \delta_3 D\left[\sum_{j=1}^{3} d_j A_j e^{-\lambda_j z} + \frac{f_2}{f_4}e^{-\gamma z}\right]$$
$$-\frac{\beta_3}{\beta_1}\left(1+\tau_1\frac{\partial}{\partial t}\right)\left\{1-\left(a_1\xi^2 + a_3 D^2\right)\right\}\left[\sum_{j=1}^{3} l_j A_j e^{-\lambda_j z} + \frac{f_3}{f_4}e^{-\gamma z}\right], \quad (34)$$

$$\hat{t}_{xz}(\xi, z, s) = \delta_2\left(D\left[\sum_{j=1}^{3} A_j e^{-\lambda_j z} + \frac{f_1}{f_4}e^{-\gamma z}\right] + i\xi\left[\sum_{j=1}^{3} d_j A_j e^{-\lambda_j z} + \frac{f_2}{f_4}e^{-\gamma z}\right]\right). \quad (35)$$

4 Boundary Conditions

We consider a normal line load F_1 per unit length acting in the positive z-axis on the plane boundary $z = 0$ along the y-axis and a tangential load F_2 per unit length, acting at the origin in the positive x-axis. The appropriate boundary conditions are

$$t_{33}(x, z, t) = -F_1\psi_1(x)H(t), \quad (36)$$

$$t_{31}(x, z, t) = -F_2\psi_2(x)H(t) \quad (37)$$

$$h_1\frac{\partial\varphi}{\partial z}(x, z, t) + h_2\varphi(x, z, t) = 0, \quad (38)$$

where $h_2 \to 0$ corresponds to insulated boundaries and $h_1 \to 0$ corresponds to isothermal boundaries. F_1 and F_2 are the magnitude of the forces applied, $\psi_1(x)$ and $\psi_2(x)$ specify the vertical and horizontal load distribution functions along the x-axis, and $H()$ is the Heaviside function and is given by

$$H(t) = \begin{cases} 1, & t \geq 0 \\ 0, & t < 0 \end{cases}. \tag{39}$$

$$H(s) = L(H(t)) = \frac{1}{s}. \tag{40}$$

Applying dimensionless conditions defined by Eq. (15) and then suppressing primes and then taking Laplace and Fourier transforms defined by Eqs. (22) and (23) on the boundary conditions (36)–(38), and solving the resultant equations for $A_j,\ j = 1, 2, 3$, by Cramer's rule and then using $A_j,\ j = 1, 2, 3$ in Eqs. (30)–(32) and (33)–(35), we obtain the components of displacement, normal stress, tangential stress and conductive temperature as

$$\hat{u} = \frac{F_1\hat{\psi}_1(\xi)}{\Lambda}\left[\sum_{j=1}^{3}\Gamma_{1j}e^{-\lambda_j z}\right] + \frac{F_2\hat{\psi}_2(\xi)}{\Gamma}\left[\sum_{j=1}^{3}\Gamma_{2j}e^{-\lambda_j z}\right] + \frac{f_1}{f_4}e^{-\gamma z}, \tag{39}$$

$$\hat{w} = \frac{F_1\hat{\psi}_1(\xi)}{\Gamma}\left[\sum_{j=1}^{3}d_j\Gamma_{1j}e^{-\lambda_j z}\right] + \frac{F_2\hat{\psi}_2(\xi)}{\Gamma}\left[\sum_{j=1}^{3}d_j\Gamma_{2j}e^{-\lambda_j z}\right] + \frac{f_2}{f_4}e^{-\gamma z}, \tag{40}$$

$$\hat{\varphi} = \frac{F_1\hat{\psi}_1(\xi)}{\Gamma}\left[\sum_{j=1}^{3}l_j\Gamma_{1j}e^{-\lambda_j z}\right] + \frac{F_2\hat{\psi}_2(\xi)}{\Gamma}\left[\sum_{j=1}^{3}l_j\Gamma_{2j}e^{-\lambda_j z}\right] + \frac{f_3}{f_4}e^{-\gamma z}, \tag{41}$$

$$\widehat{t_{xx}} = \frac{F_1\hat{\psi}_1(\xi)}{\Gamma}\left[\sum_{j=1}^{3}S_j\Gamma_{1j}e^{-\lambda_j z}\right] + \frac{F_2\hat{\psi}_2(\xi)}{\Gamma}\left[\sum_{j=1}^{3}S_j\Gamma_{2j}e^{-\lambda_j z}\right] + g_1e^{-\gamma z}, \tag{42}$$

$$\widehat{t_{xz}} = \frac{F_1\hat{\psi}_1(\xi)}{\Gamma}\left[\sum_{j=1}^{3}N_j\Gamma_{1j}e^{-\lambda_j z}\right] + \frac{F_2\hat{\psi}_2(\xi)}{\Gamma}\left[\sum_{j=1}^{3}N_j\Gamma_{2j}e^{-\lambda_j z}\right] + g_2e^{-\gamma z}, \tag{43}$$

$$\widehat{t_{zz}} = \frac{F_1\hat{\psi}_1(\xi)}{\Gamma}\left[\sum_{j=1}^{3}M_j\Gamma_{1j}e^{-\lambda_j z}\right] + \frac{F_2\hat{\psi}_2(\xi)}{\Gamma}\left[\sum_{j=1}^{3}M_j\Gamma_{2j}e^{-\lambda_j z}\right] + g_3e^{-\gamma z}, \tag{44}$$

where

$$\Gamma_{11} = -N_2R_3 + R_2N_3,$$

$$\Gamma_{12} = N_1R_3 - R_1N_3,$$

$$\Gamma_{13} = -N_1R_2 + R_1N_2,$$

$$\Gamma_{21} = M_2R_3 - R_2M_3,$$

$$\Gamma_{22} = -M_1R_3 + R_1M_3,$$

$$\Gamma_{23} = M_1R_2 - R_1M_2,$$

$$\Gamma = -M_1\Gamma_{11} - M_2\Gamma_{12} - M_3\Gamma_{13},$$

$$N_j = -\delta_2\lambda_j + i\xi d_j,$$

$$M_j = i\xi - \delta_3 d_j\lambda_j - \frac{\beta_3}{\beta_1}(1+\tau_1 s)l_j\big[\big(1+a_1\xi^2\big) - a_3\lambda_j^2\big],$$

$$R_j = -h_1\lambda_j l_j + h_2 l_j,$$

$$S_j = -i\xi - \delta_4 d_j\lambda_j - l_j(1+\tau_1 s)\big[\big(1+a_1\xi^2\big) - a_3\lambda_j^2\big],$$

$$g_1 = i\xi\frac{f_1}{f_4} - \delta_4\gamma\frac{f_2}{f_4} - \big(1+a_1\xi^2 - a_3\gamma^2\big)\frac{f_3}{f_4},$$

$$g_2 = \delta_2\left(\gamma\frac{f_1}{f_4} + i\xi\frac{f_2}{f_4}\right),$$

$$g_3 = i\xi\frac{f_1}{f_4} - \delta_3\gamma\frac{f_2}{f_4} - \frac{\beta_3}{\beta_1}\big(1+a_1\xi^2 - a_3\gamma^2\big)\frac{f_3}{f_4}.$$

5 Applications

Suppose an inclined load, F_0 per unit length, is acting on the *y*-axis and its inclination with *z*-axis is θ, we have (see Fig. 1),

$$F_1 = F_0\cos\theta \text{ and } F_2 = F_0\sin\theta. \tag{45}$$

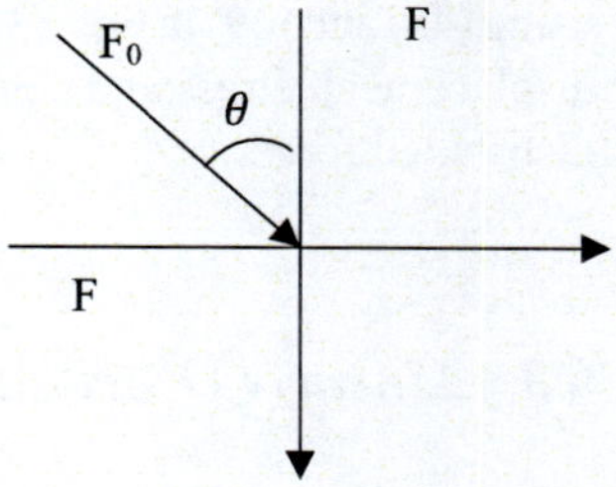

Fig. 1 Inclined load over a transversely isotropic magneto-thermoelastic solid

6 Special Cases

6.1 Concentrated Force

The solution due to the concentrated normal force on the half-space is obtained by setting

$$\psi_1(x) = \delta(x),\ \psi_2(x) = \delta(x), \tag{46}$$

where $\delta(x)$ is Dirac delta function.

Applying Fourier transform defined by (23) on (46), we obtain

$$\hat{\psi}_1(\xi) = 1,\ \hat{\psi}_2(\xi) = 1. \tag{47}$$

Using Eqs. (45) and (47) in Eqs. (39)–(44), the components of displacement, stress and conductive temperature for concentrated force and inclined load are obtained.

6.2 Uniformly Distributed Force

The solution due to uniformly distributed force applied on the half-space is obtained by setting

$$\psi_1(x),\ \psi_2(x) = \begin{cases} 1\ \text{if}|x| \le \text{m}, \\ 0\ \text{if}|x| > \text{m}. \end{cases} \tag{48}$$

The Fourier transforms of $\psi_1(x)$ and $\psi_2(x)$ with respect to the pair (x, ξ) for the case of a uniform strip load of non-dimensional width 2 m applied at origin of co-ordinate system $x = z = 0$ in the dimensionless form after suppressing the primes become

$$\hat{\psi}_1(\xi) = \hat{\psi}_2(\xi) = \left\{\frac{2\sin(\xi m)}{\xi}\right\}, \xi \ne 0. \tag{49}$$

using (45) and (49) in Eqs. (39)–(44), we obtain the expressions for the components of displacement, stress and conductive temperature for an inclined load, for uniformly distributed force.

6.3 Linearly Distributed Force

The solution due to linearly distributed force applied on the half-space is obtained by setting

$$\{\psi_1(x), \psi_2(x)\} = \begin{cases} 1 - \frac{|x|}{m} & if|x| \leq m, \\ 0 & if|x| > m. \end{cases} \tag{50}$$

Here, 2*m* is the width of the strip load, and applying the transform defined by (23) on (50), we get

$$\hat{\psi}_1(\xi) = \hat{\psi}_2(\xi) = \left\{\frac{2\{1 - \cos(\xi m))}{\xi^2 m}\right\}, \xi \neq 0. \tag{51}$$

Using Eqs. (45) and (51) in Eqs. (39)–(44), the components of displacement, stress and conductive temperature for an inclined load for linearly distributed force are obtained.

7 Inversion of the Transformation

For obtaining the result in the physical domain, we need to invert the Fourier transforms in Eqs. (39)–(44), using

$$\tilde{f}(x, z, s) = \frac{1}{2\pi} \int_{-\infty}^{\infty} e^{-i\xi x} \hat{f}(\xi, z, s) d\xi = \frac{1}{2\pi} \int_{-\infty}^{\infty} |\cos(\xi x) f_e - i \sin(\xi x) f_o| d\xi, \tag{52}$$

where f_o is odd and f_e is the even parts of $\hat{f}(\xi, z, s)$, respectively [25]. Following Honig and Hirdes [26], the Laplace transform function $\widetilde{f}(x, z, s)$ can be inverted to $f(x, z, t)$ for problem I by

$$f(x, z, t) = \frac{1}{2\pi i} \int_{e-i\infty}^{e+i\infty} \widetilde{f}(x, z, s) e^{-st} ds. \tag{53}$$

The last step is to calculate the integral in Eq. (53). The method for evaluating this integral is described in Press et al. [27]. It involves the use of Romberg's integration with adaptive step size. This also uses the results from successive refinements of the extended trapezoidal rule followed by extrapolation of the results to the limit when the step size tends to zero.

8 Particular Cases

- If $Q_0 = 0$ we obtain relations for displacement temperature distribution and stress for without laser pulse from Eqs. (39)–(44), in transversely isotropic magneto-thermoelastic solid with two temperature (2T) and rotation due to inclined load.
- If $\varepsilon = 0$ we obtain relations for displacement temperature distribution and stress for two relaxation times (for Green–Lindsay (G-L) model) from Eqs. (39)–(44), in transversely isotropic magneto-thermoelastic solid with two temperature (2T), rotation due to inclined load and laser pulse.
- If $\tau_1 = 0$ and $\varepsilon = 1$ we obtain relations for displacement temperature distribution and stress for one relaxation time (for Lord–Shulman (L-S) model) from Eqs. (39)–(44), in transversely isotropic magneto-thermoelastic solid with two temperature (2T), rotation due to inclined load and laser pulse.
- If $\tau_1 = \tau_0 = 0$ we obtain relations for displacement temperature distribution and stress for coupled thermoelastic (CT) half-space from Eqs. (39)–(44), in transversely isotropic magneto-thermoelastic solid with two temperature (2T), rotation due to inclined load and laser pulse.
- If $\theta = \frac{\pi}{2}$ we obtain relations for displacement temperature distribution and stress for tangential load thermoelastic half-space from Eqs. (39)–(44), in transversely isotropic magneto-thermoelastic solid with two temperature (2T), rotation due to inclined load and laser pulse.
- If $\Omega = 0$ we obtain relations for displacement temperature distribution and stress from Eqs. (39)–(44), in transversely isotropic magneto-thermoelastic solid with two temperature (2T), without rotation due to inclined load and laser pulse.
- If $a_1 = a_3 = 0$ we obtain relations for displacement temperature distribution and stress from Eqs. (39)–(44), in transversely isotropic magneto-thermoelastic solid without two temperature (2T), with rotation due to inclined load and laser pulse.

9 Numerical Results and Discussion

In order to illustrate our theoretical results in the proceeding section and to show the effect of inclined load and laser pulse, we now present some numerical results. Following Dhaliwal and Singh [28], cobalt material has been taken for thermoelastic material as

$$c_{11} = 3.07 \times 10^{11}\,\mathrm{Nm}^{-2},\ c_{33} = 3.581 \times 10^{11}\,\mathrm{Nm}^{-2},\ c_{13} = 1.027 \times 10^{10}\,\mathrm{Nm}^{-2},$$
$$c_{44} = 1.510 \times 10^{11}\,\mathrm{Nm}^{-2},\ \beta_1 = 7.04 \times 10^{6}\,\mathrm{Nm}^{-2}\,\mathrm{deg}^{-1},\ \beta_3 = 6.90 \times 10^{6}\,\mathrm{Nm}^{-2}\,\mathrm{deg}^{-1},$$
$$\rho = 8.836 \times 10^{3}\,\mathrm{K\,gm}^{-3},\ C_E = 4.27 \times 10^{2}\,\mathrm{j\,Kg}^{-1}\,\mathrm{deg}^{-1},$$
$$K_1 = 0.690 \times 10^{2}\,\mathrm{Wm}^{-1}\,\mathrm{K\,deg}^{-1},$$
$$K_3 = 0.690 \times 10^{2}\,\mathrm{Wm}^{-1}\,\mathrm{K}^{-1},\ T_0 = 298\,\mathrm{K},\ H_0 = 1\,\mathrm{Jm}^{-1}\,\mathrm{nb}^{-1},$$
$$\varepsilon_0 = 8.838 \times 10^{-12}\,\mathrm{Fm}^{-1},\ L = 1,\ \gamma = 0.01,\ \tau_0 = 0.02,$$
$$\tau_1 = 0.01,\ r = 0.01,\ I_0 = 1 \times 10^{11}\,\mathrm{Jm}^{-2},\ a_1 = 0.02 \text{ and } a_3 = 0.04$$

Using the above values, the graphical representations of displacement component u, normal displacement w, conductive temperature φ, stress components t_{xx}, t_{xz} and t_{zz} for transversely isotropic thermoelastic medium have been investigated and the effect of inclination with two temperature has been depicted.

9.1 *Concentrated Force Due to Inclined Load and with Laser Pulse*

Figures 1, 2, 3, 4, 5 and 6 show the variations of the displacement components (u and w), conductive temperature φ and stress components (t_{xx}, t_{xz} and t_{zz}) for a transversely

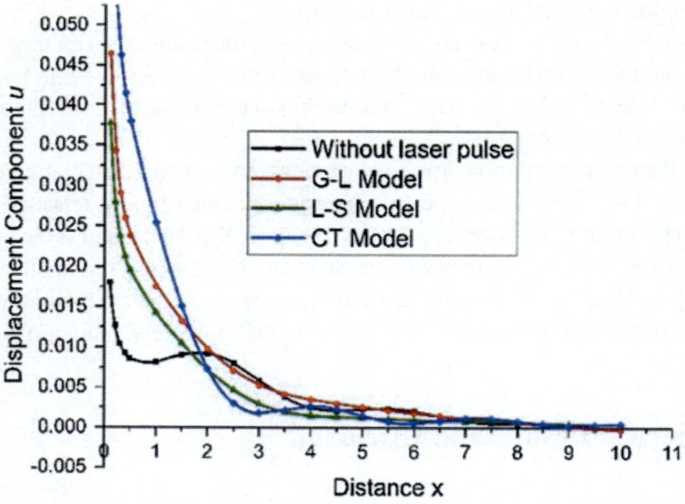

Fig. 2 Variations of displacement component u with distance x

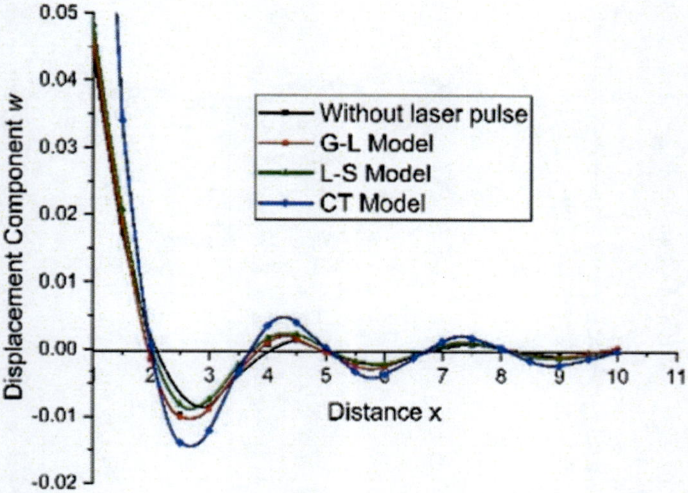

Fig. 3 Variations of displacement component w with distance x

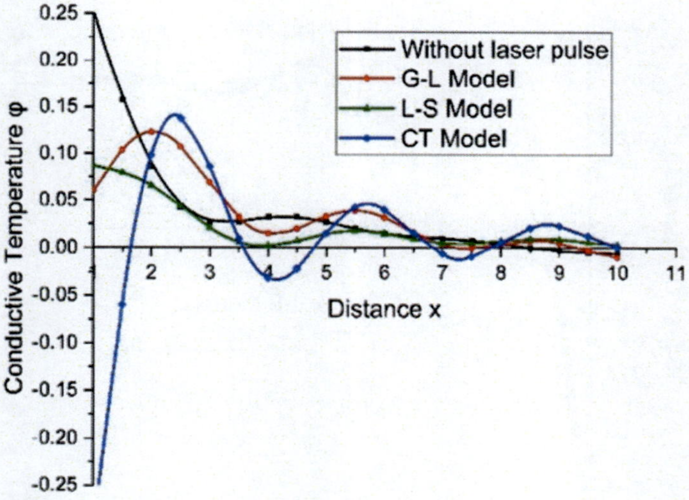

Fig. 4 Variations of conductive temperature φ with distance x

isotropic magneto-thermoelastic medium with concentrated force and with combined effects of rotation, two temperature with and without laser pulse, G-L, L-S and CT model of thermoelasticity, respectively. The displacement components w and stress components (t_{xx}, t_{xz} and t_{zz}) illustrate the same pattern but having different magnitudes for different models of thermoelasticity. The displacement component u and conductive temperature show the opposite behaviour without laser pulse and same pattern with other models of thermoelasticity. These components vary (increases or

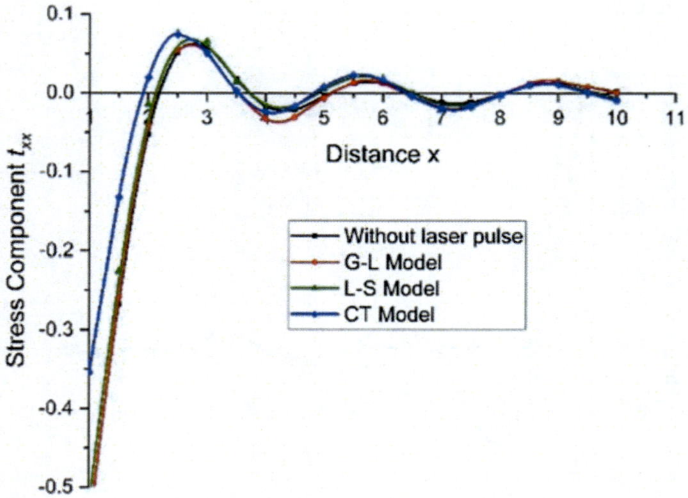

Fig. 5 Variations of stress component t_{xx} with distance x

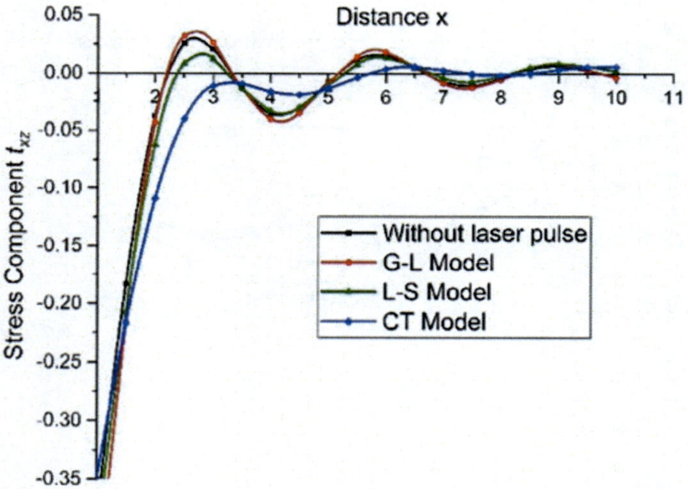

Fig. 6 Variations of the stress component t_{xz} with distance x

decreases) during the initial range of distance near the loading surface of the laser pulse and follow a small oscillatory pattern for the rest of the range of distance. CT model of thermoelasticity illustrates the more impact of laser pulse as compared to other model of thermoelasticity (Fig. 7).

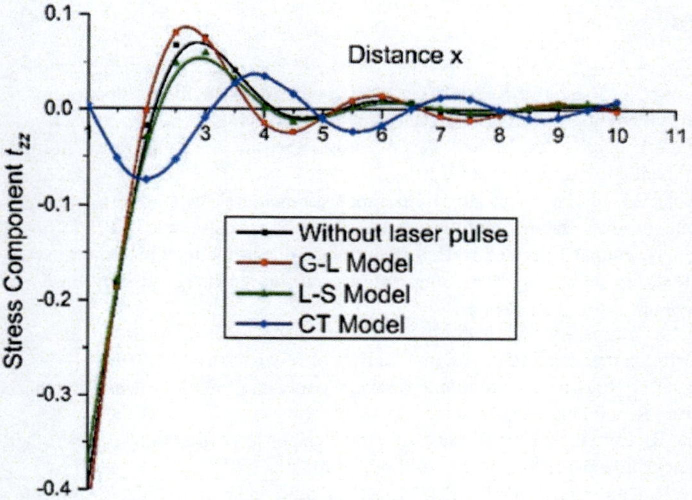

Fig. 7 Variations of the stress component t_{zz} with distance x

10 Conclusions

From the above study, it is observed as follows

- Displacement components (u and w), conductive temperature φ and stress components (t_{xx}, t_{xz} and t_{zz}) for a transversely isotropic magneto-thermoelastic medium with concentrated force and with combined effects of rotation, two temperature with and without laser pulse, G-L, L-S and CT model of thermoelasticity, respectively.
- Moreover, laser pulse, the magnetic effect of two temperature, rotation as well as the angle of inclination of the applied load play a key part in the deformation of all the physical quantities.
- The CT model shows the more oscillatory nature for the displacement components and stress components.
- The result gives the inspiration to study magneto-thermoelastic materials as an innovative domain of applicable thermoelastic solids.
- The shape of curves shows the impact of laser pulse on the body and fulfils the purpose of the study.
- The outcomes of this research are extremely helpful in the 2D problem with dynamic response of laser pulse sources in transversely isotropic magneto-thermoelastic medium with rotation and two temperature which is advantageous to discover the deformation field such as geothermal engineering, advanced aircraft structure design, electronics engineering, composite engineering and high-energy particle accelerators
- The proposed model in this research is relevant to different problems in thermoelasticity and thermodynamics.

References

1. Youssef HM, Al-Felali AS (2012) Generalized thermoelasticity problem of material subjected to thermal loading due to laser pulse. Appl Math 3(2):142–146
2. Every G, Utegulov ZN, Veres IA (2013) Laser thermoelastic generation in metals above the melt threshold. J Appl Phys 114(20)
3. Youssef HM, El-Bary AA (2014) Thermoelastic material response due to laser pulse heating in context of four theorems of thermoelasticity. J Therm Stresses 37(12):1379–1389
4. Sharma N, Kumar R, Lata P (2015) Disturbance due to inclined load in transversely isotropic thermoelastic medium with two temperatures and without energy dissipation. Mater Phys Mech 22:107–117
5. Kumar R, Kumar A, Singh D (2015) Thermomechanical interactions due to laser pulse in microstretch thermoelastic medium. Arch Appl Mech 67(6):439–456
6. Kumar A (2015) Laser pulse in micropolar thermoelastic medium with three phase lag model. Int J Eng Res Technol 4(3):1–4
7. Kumar R, Kumar A (2016) Elastodynamic response of thermal laser pulse in micropolar thermoelastic mass diffusion medium. J Thermodyn 1–9
8. Kumar R, Kumar A, Singh D (2018) Elastodynamic interactions of laser pulse in microstretch thermoelastic mass diffusion medium with dual phase lag. Microsyst Technol 24(4):1875–1884
9. Ailawalia P, Sachdeva S, Pathania D (2017) Laser pulse heating in thermo-microstretch elastic layer overlying thermoelastic halfspace. Int J Appl Phys Sci 7(4):178–192
10. Othman MIA, Marin M (2017) The effect of heat laser pulse on generalized thermoelasticity for micropolar medium. Mech Mech Eng 797–811
11. Othman MIA, Atwa SY, Elwan AW (2017) Effect of thermal loading due to laser pulse on 3-d problem of micropolar thermoelastic solid with energy dissipation. Mech Mech Eng 21(3):679–701
12. Abbas A, Marin M (2018) Analytical solutions of a two-dimensional generalized thermoelastic diffusions problem due to laser pulse. Iranian J Sci Technol Trans Mech Eng 42:57–71
13. Allam M, Tantawy R, Zenkour A (2018) Magneto-thermo-elastic response of exponentially graded piezoelectric hollow spheres. Adv Comput Des 303—318
14. Lal K, Jagtap R, Singh BN (2017) Thermo-mechanically induced finite element based nonlinear static response of elastically supported functionally graded plate with random system properties. Adv Comput Des 165–194
15. Kumar R, Sharma N, Lata P (2016) Thermomechanical interactions due to hall current in transversely isotropic thermoelastic with and without energy dissipation with two temperatures and rotation. J Solid Mech 8(4):840–858
16. Hassan M, Marin M, Ellahi R, Alamri S (2018) Exploration of convective heat transfer and flow characteristics synthesis by Cu–Ag/water hybrid-nanofluids. Heat Transf Res 49(18):1837–1848
17. Lata P, Kaur I (2019) Effect of rotation and inclined load on transversely isotropic magneto thermoelastic solid. Struct Eng Mech 70(2):245–255
18. Lata P, Kaur I (2019) Thermomechanical interactions in transversely isotropic thick circular plate with axisymmetric heat supply. Struct Eng Mech 69(6):607–614
19. Lata P, Kaur I (2019) Transversely isotropic magneto thermoelastic solid with two temperature and without energy dissipation in generalized thermoelasticity due to inclined load. SN Appl Sci 1:426
20. Lata P, Kaur I (2019) Transversely isotropic thick plate with two temperature and GN type-III in frequency domain. Coupled Syst Mech Techno Press 8(1):55–70
21. Kaur I, Lata P (2019) Effect of hall current on propagation of plane wave in transversely isotropic thermoelastic medium with two temperature and fractional order heat transfer. SN Appl Sci 1:900
22. Kaur I, Lata P (2019) Transversely isotropic thermoelastic thin circular plate with constant and periodically varying load and heat source. Int J Mech Mater Eng 14(10):1–13

23. Schoenberg M, Censor D (1973) Elastic Waves In Rotating Media. Q Appl Math 31:115–125
24. Slaughter W (2002) The linearised theory of elasticity, Birkhausar
25. Honig GH (1984) A method for the inversion of Laplace Transform. J Comput Appl Math 10:113–132
26. Honig G, Hirdes U (1984) A method for the numerical inversion of Laplace transform. J Comput Appl Math 10:113–132
27. Press WH, Teukolshy SA, Vellerling WT, Flannery BP (1986) Numerical recipes in Fortran. Cambridge University Press, Cambridge
28. Dhaliwal RS, Sherief HH (1980) Generalized thermoelasticity for anisotropic media. Q Appl Math XXXVII(1):1–8

Impact of Integration of Wind Turbine on Dynamics of a Power System

Himani Dhakla, Vijay Kumar Garg, and Sudhir Sharma

Abstract This paper investigates the impact of integration of wind turbine on the dynamic performance of system. The power system is integrated with intermittent source, i.e. wind power plant. The unpredictable nature of intermittent sources always distracts the system stability. In this study, the integrated wind turbine impacts the power grid. Integration of these intermittent sources deflects the frequency from its reference and requires a controller to keep the system frequency within limits. In this study, a comparison is performed between classical and intelligent controller by using simulation done in MATLAB simpower system toolkit.

Keywords Wind penetration · Frequency deviation · Effeciency · Settling time · Energy storage unit

1 Introduction

The combination of sustainable sources with non-renewable energy source-based power system (hybrid grid system) has expanded over the previous decades particularly after the noteworthy development of petroleum derivative expense. Specifically, the coordination of sustainable asset and diesel genset makes it the most fitting producing approach for secluded networks [1]. The principle favorable circumstances of a half and half small-scale matrix system are diminishing the fuel utilization and assure a dependable power supply. In disconnected networks with low populace densities, providing power by broadening the transmission line from national power framework is expensive, and a few areas are incomprehensible in view of topographical obstructions. Hence, those networks use their own age and conveyance system to supply power, and this plan is called smaller than usual lattice system.

H. Dhakla (✉) · V. K. Garg
Department of Electrical Engineering U.I.E.T, Kurukshetra University, Kurukshetra, India
e-mail: himani.dhakla@gmail.com

S. Sharma
Department of Electrical Engineering DAVIET, Punjab Technical University, Jalandhar, India

N. Marriwala et al. (eds.), *Mobile Radio Communications and 5G Networks*, Lecture Notes in Networks and Systems 140,
https://doi.org/10.1007/978-981-15-7130-5_44

Furthermore, because of this substitute task, the general system has drawn out life demonstrated by the release dimension of the battery is being kept ideal. The hybrid system embraces an ecological inviting innovation whereby the diesel generator is utilized irregularly and the entire task itself is a lot calmer [2–5]. This is represented by a smaller scale processor-based controller unit. The primary establishment may take up a staggering expense, yet the support cost a short time later is low, on account of the system's self-safeguarding instrument. Likewise, we can spare fuel utilization of the genset in light of the fact that the sunlight-based photovoltaic (PV) modules bolster the base power load, while the genset gives extra vitality ought to there be an unexpected crest in the vitality request.

Role of renewable energy resources (RES) (i.e. wind and solar) is increasing drastically in overall power production due to their inherent merits in the last few decades [6]. But with lots of merits, these RES have some shortcomings too. The power output of these sources is always non-uniform throughout the day [7]. Hence, again in wind power generation also, the power generated is also not uniform. Furthermore, these RES when integrated with conventional power generation unit that are already suffering from the issue of load improbability lead to more instability for overall setup [8] which further result in unintended accretion of error in frequency and tie line power interchange. Though limited number of researchers had discussed these problem overall [6–8], but no one has yet specified the effect of integration of power system with wind energy.

The paper is organized as follows: Sect. 2 discusses the problem formulation of renewable penetrated power system. Section 3 represents the model of power system integrated with wind. Section 4 discussed about the wind energy participation. In Sect. 5, the modeling of intelligent controller ANFIS is explained. The simulation results are given in Sect. 6 followed by the conclusion of the paper.

2 Problem Formulation

A unified two-area power system is used for problem formulation. Further to have supply demand ratio, wind energy sources are integrated in power system. Yet, their combination in power system turns into a reason for frequency and tie line control deviations. Hybrid energy system is a setup of at least two inexhaustible and even non-sustainable power source as fundamental wellsprings of energy usage with the goal that the limit deficiency of intensity from one source will substitute by other accessible sources to cook economic power. It is fitting intends to give power from locally accessible energy hot spots for zones where matrix augmentation is capital escalated, topographically separated spots for which power transmission from brought together utility is troublesome. Normally skilled inexhaustible sources can be outfit to create power in an economical manner to give power and make agreeable the expectation for everyday comforts of individuals. There are distinctive benefits and disadvantages of utilizing just sustainable hot spots for power age in provincial towns, merits like fuel cost slant, fuel transport cost is high, issues of a worldwide temperature alteration and environmental change is substantial. The downsides of utilizing inexhaustible sources as off-matrix/independent power systems, it has irregular nature that makes hard to direct the power yield to make do with the heap looked for. To ensure for the unwavering quality and reasonableness of the supply, consolidating ordinary diesel generator with nonconventional energy generators can take care of the issue unmistakable while working separately.

In India, more than 75% of the electricity is produced by conventional sources of energy, and about 57% of the energy is produced using only coal, and hence, it causes a lot pollution. Even after burning a lot of coal and wasting natural resources, we are short of energy. Renewable energy like solar panels or wind mills can be really helpful in tackling the ongoing situation. By interfacing wind power generation system with AC micro grid, we are using renewable energy and as well as solving the problem of shortage of power. But an abrupt change sought after will divert the frequency in the structure. This arbitrary variety in frequency causes calamitous disappointment which compiles the rotor angle to increment relentlessly [3, 4]. To maintain a strategic distance from these disappointments, a control activity ought to be connected by the essential control circle, yet it will not recoup to set point esteem.

Integration of wind energy generation is done with two-area power system model having variable generation sources. Assimilation of these sources has been achieved through simulation, and power coming from them is irregular. Figure 1 shows the behavior of power coming from renewable sources.

The basic most important unit of power generation system consists of large controllable electric generators. The amalgamation of different sources of power generation may lead to random deviations in system parameters. Such as, short-term fluctuations in wind and solar energy cause output of the power to be uncertain. These kinds of deviations make the system more challenging for the system operators in predicting how much additional power needs to be generated for future usage on hourly, monthly or yearly bases. In this paper, only wind generation is considered.

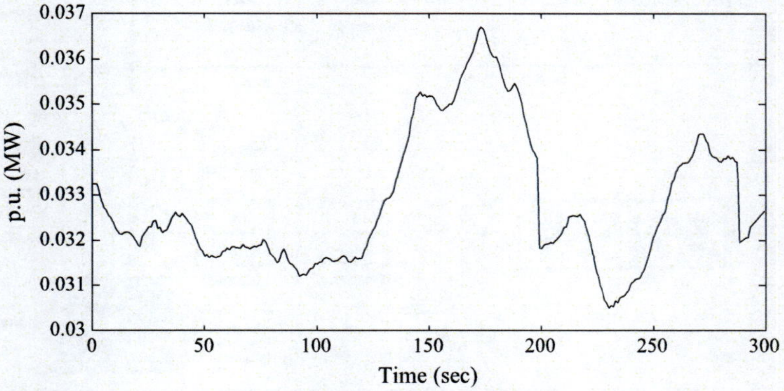

Fig. 1 Per unit power generated through wind generation unit

To fulfill the prompt power need varieties, a controller ought to be equipped for reacting to changes as quick as could reasonably be expected [5, 6]. Right now, the inclining rate of conventional generators takes numerous seconds even minutes to react to these adjustments sought after. In this way, a functioning force creating gadget is required which can give subordinate administrations. Since battery bank is fit to give dynamic power in least conceivable time, but it causes the unintentional gathering of frequency.

3 System Model

An energy generation model using wind as renewable power generation unit and energy storage unit is shown in Fig. 2.

The power generation unit consists of multi-generation sources in each area. A step change in power system always distracts the system frequency. The measurement of system unbalancing can be identified from frequency deviation and tie line power deviations. Furthermore, the above system firstly controlled through classical controller, i.e. PI controller [9–12]. Whether the classical controller does not provide satisfactory results, then moves to intelligent controller [13]. Intelligent controllers which are used in load frequency control area are artificial neural network (ANN), fuzzy logic control (FLC) and adaptive neuro fuzzy inference system (ANFIS).

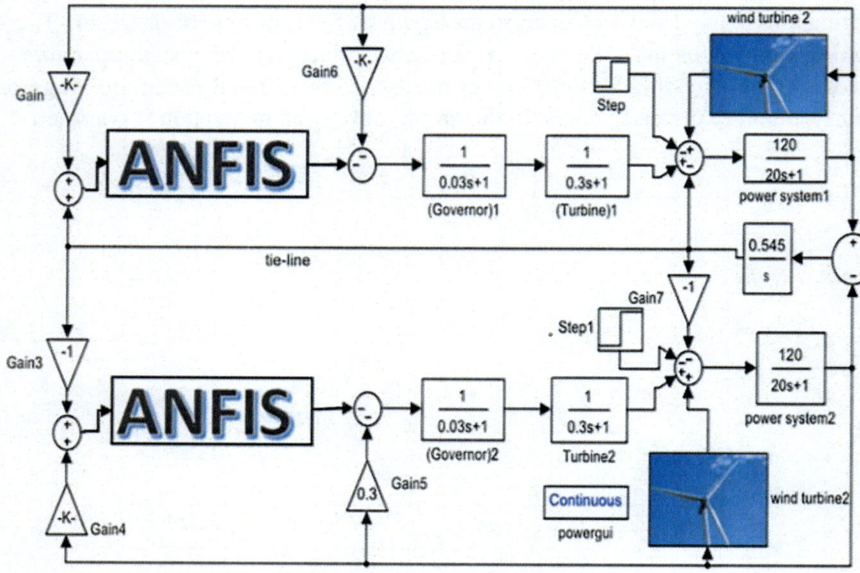

Fig. 2 Simulink model of a two-area power system integrating conventional generation unit with wind generation unit

4 Wind Energy Source

The rotor characteristics of the turbine can be obtained by the relation between the overall wind power and mechanical power absorbed by turbine blades from wind. These relations are eagerly referred initialize with wind absorbed inside rotor swept area. It is clear that the kinetic energy of air cylinder with radius drifting at speed of wind V_{wind} gives overall wind power P_{wind} within the area swept by the rotor of the wind turbine. The equation for wind power can be expressed as given in Eq. (1).

$$P_{\text{wind}} = \frac{1}{2} C_p \rho_{\text{air}} \pi R^2 V_{\text{wind}}^3 \tag{1}$$

It is practically impossible to extract all the kinetic energy from the wind. This would not allow the air to pass through the wind turbine, which is almost unachievable condition. The wind turbine reduces speed of wind, which thus extracts a fraction of the power from the wind. This fraction is expressed as the power efficiency coefficient (C_p), of the wind turbine. Therefore, the mechanical power output of the wind turbine P_{mech} considering the definition of C_p can be given by (Eqs. 2 and 3),

$$P_{\text{mech}} = C_{\text{p}} P_{\text{wind}} \tag{2}$$

$$P_{\text{mech}} = \frac{1}{2} C_{\text{p}} \rho_{\text{air}} \pi R^2 V_{\text{wind}}^3 \tag{3}$$

The maximum theoretical static limit of $C_p$is 16/27 (0.593 approx.) which is maximum possible value, i.e. 59% of power can be extracted from kinetic energy of wind. It is also called as Betz Limit. Practically, mechanical power extracted P_{mech} depends upon rotor speed, wind speed and blade angle. Hence, P_{mech} and C_p are functions of these parameters (Eq. 4)

$$P_{\text{mech}} = f(\omega_{\text{turb}}, V_{\text{turb}}, \beta) \tag{4}$$

5 ANFIS

The architecture of the ANFIS model is based on Takagi–Sugeno fuzzy interface model. The comprehensive ANFIS controller model that can be used in any plant is presented in this section. An algorithm that can describe the working of different layers of ANFIS consists of the following steps: input fuzzification, applying fuzzy operator, normalization, defuzzyification and last summation. The structure contains the same components as.fis, except for the NN block. The five interconnected network layers that describe the working of ANFIS are shown below in Fig. 3 in the form of block diagram.

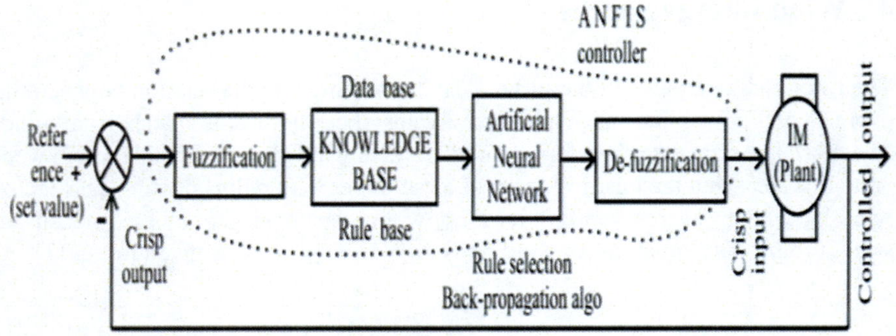

Fig. 3 Block diagram of ANFIS Controller

A controller is mandatory for modeling and feedback control of any dynamic system, which can take care of all the disturbances of the system and try to make system stable within few seconds. Mathematical model of power plant along with controller mathematical model is required to design the controller with ANFIS model which is used for simulation purpose.

6 Simulation Results

This area confirms the proposed model outcome through a point by point system recreation utilizing MATLAB simpower system tool kit. To contemplate the frequency guideline sway in a power system incorporated with or without sustainable power sources joining. The undesirable deviations from system frequency and tie line power can be expelled by picking best parameters of the PI controller with ANFIS controller [13]. This area laid out the effect of these sustainable power source assets in power system as given in the accompanying areas.

Case I: Without Wind Integration

Unexpected unsettling influences in load dependably occupy the continuous procedure strength. So also, in this power system at time 0s, an adjustment in load occupies the adjusting light emission system and happens a dunk in frequency and tie line control, which is appeared in Figs. 4, 5 and 6. After 20 s, settling happens, and past 30 s, there is zero enduring state blunder.

Case I: With Wind Integration

Presently, inexhaustible wellspring of energy for example wind is coordinated with system. Presently, its outcome is appeared in Figs. 7 and 9 showing better settling time and relentless state blunder close around zero. For this situation, system has ideal estimation of frequency deviation. In any case, this implies that with wind energy

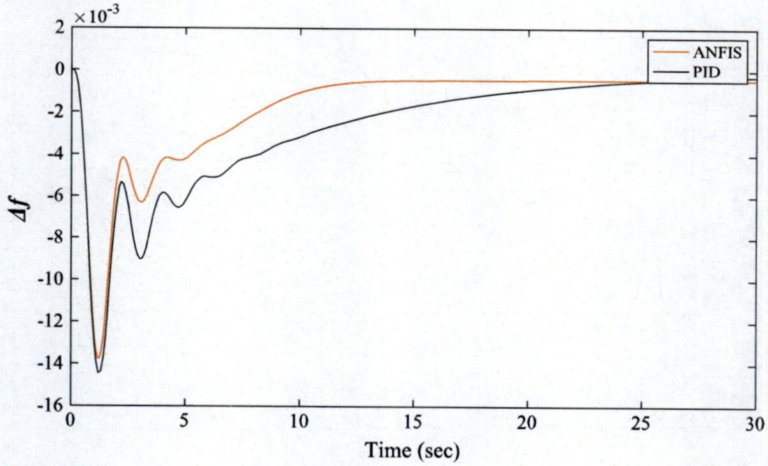

Fig. 4 Frequency response of the system in area 1

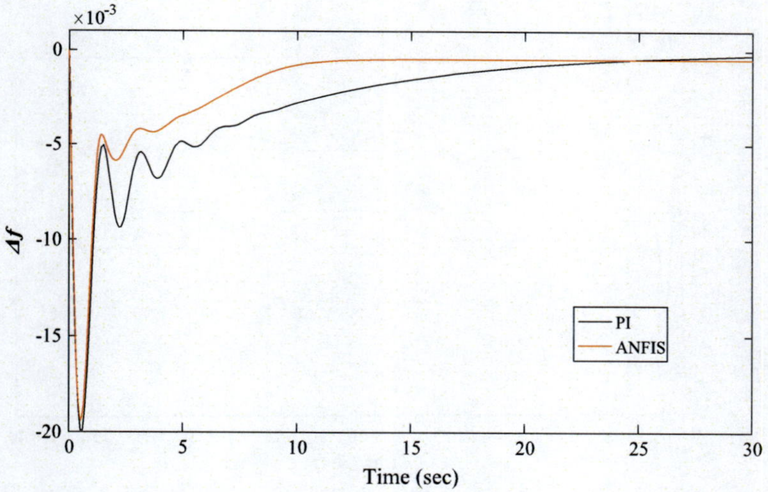

Fig. 5 Frequency response of the system in area 2

source having smooth inflow of frequency. Some climate condition will cause more variance at that point of wind sources.

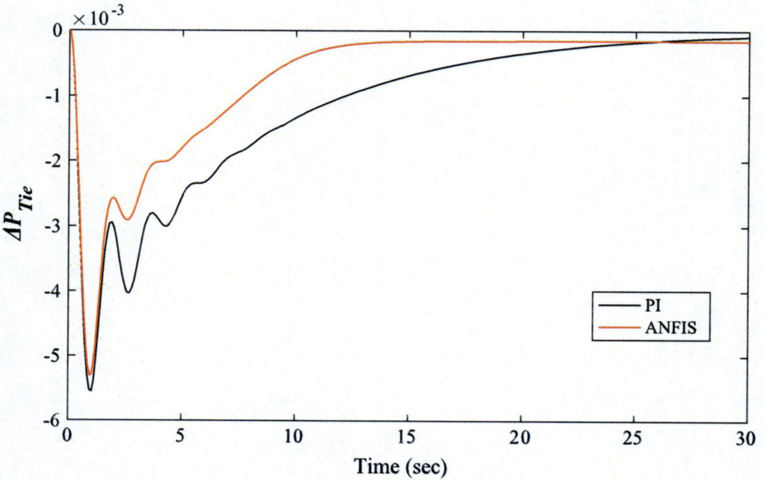

Fig. 6 Tie line power response of the system

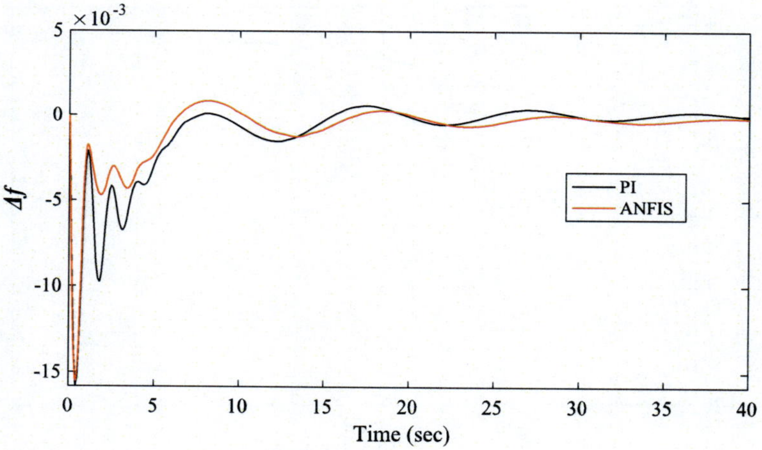

Fig. 7 Frequency response of the system in area 1

7 Conclusion

This work recommends neural network-based ANFIS controller for wind penetrated power system. The training process of ANN-based ANFIS has been described in details. The model of two-area wind integrated system is developed and employed to test robustness of ANFIS controlled system following load disturbances. Both PI and ANFIS controllers are simulated for different load changes, and results have

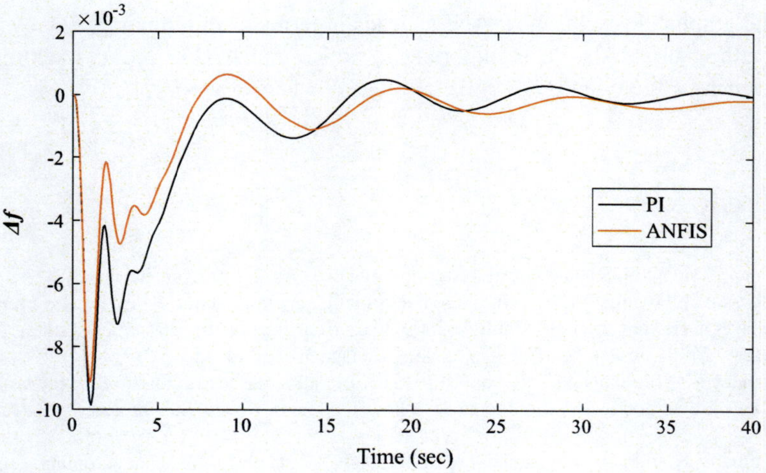

Fig. 8 Frequency response of the system in area 2

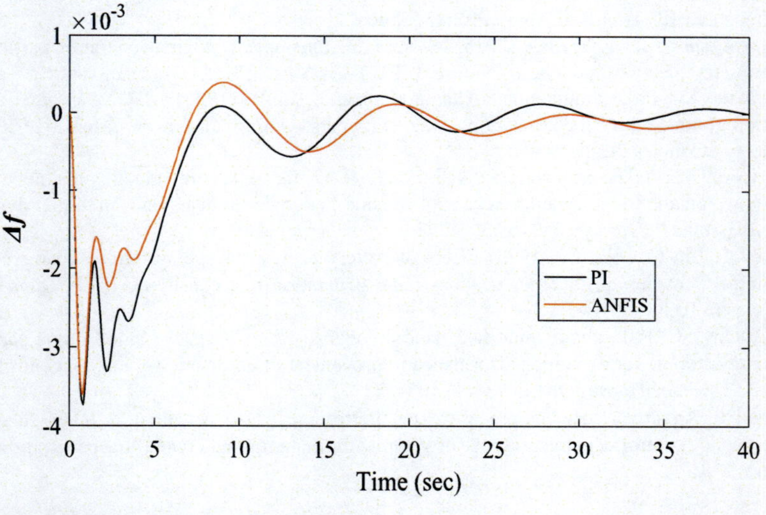

Fig. 9 Tie line power response of the system

been plotted. Two-area wind integrated transfer function model with small disturbances has been developed. The conventional PI controller tuned using Ziegler–Nichols method provides satisfactory outcomes for LFC without any nonlinearity in the system. However, wind turbines introduce nonlinearity in the system which cannot be ignored in LFC problem. Therefore, the intelligent ANN-based controller is introduced to tackle the complexity of the control area. The performances of ANFIS controller over PI controller have been compared.

The graphical results show ANFIS leads in terms of minimizing peak overshoot and settling time. Also, it is important to notice that ANFIS reduces settling time with raising system load.

References

1. Bevrani H (2009) Robust power system frequency control. Springer, Berlin
2. Bevrani H, Hiyama T (2014) Intelligent automatic generation control. Taylor and Francis
3. Shayeghi H, Shayanfar HA, Jalili A (2009) Load frequency control strategies: a state-of-the-art survey for the researcher. Energy Convers Manage 50(2):344–353
4. Pandey SK, Mohanty SR, Kishor N (2013) A literature survey on load frequency control for conventional and distribution generation power systems. Renew Sustain Energy Rev 25:318–334
5. Kumar P, Kothari DP, Kocaarslan I, Çam E (2005) Recent philosophies of automatic generation control strategies in power systems. Int J Electr Power Energy Syst 27(8):542–549
6. Mukherjee S, Teleke S, Bandaru V (2011) Frequency response and dynamic power balancing in wind and solar generation. IEEE Power Energy Soc Gen Meet 1–5
7. Liang X (2017) Emerging power quality challenges due to integration of renewable energy sources. IEEE Trans Ind Appl 53(2):853–866
8. Gevorgian V, Zhang Y, Ela E (2015) Investigating the impacts of wind generation participation in interconnection frequency response. IEEE Trans Sustain Energy 6(3):1004–1012
9. Tripathy SC, Balasubramanian R, Chandramohanan Nair PS (1992) Adaptive automatic generation control with superconducting magnetic energy storage in power systems. IEEE Trans Energy Convers 7(3):434–441
10. Tripathy SC, Balasubramanian R, Nair PSC (1992) Effect of superconducting magnetic energy storage on automatic generation control considering governor deadband and boiler dynamics. IEEE Trans Power Syst 7(3):1266–1273
11. Lu CF, Liu CC, Wu CJ (1995) Effect of battery energy storage system on load frequency control considering governor deadband and generation rate constraint. IEEE Trans Energy Convers 10(3):555–561
12. Tripathy SC, Balasubramanian R, Chandramohanan Nair PS (1991) Small rating capacitive energy storage for dynamic performance improvement of automatic generation control. IEEE Proc C-Gener Transm Distrib 138(1):103–111
13. Jood P, Aggarwal SK, Chopra V (2019) Performance assessment of a neuro-fuzzy load frequency controller in the presence of system non-linearities and renewable penetration 74(1): 362–378

Hybrid Version of Apriori Using MapReduce

Ashish Sharma and Kshitij Tripathi

Abstract The era of technology is going to be changed so frequently, data size keeps on increasing exponentially, and as the data is increasing day by day, many new things are coming up in front that has to be considered in while getting information from the dataset. One of the most popular algorithms based on frequent itemsets is the Apriori algorithm. As the data size is going to be increased every day, number of items is also increasing, in such cases, the Apriori is not able to provide the best solutions, in such cases, the solution comes like the algorithm should run parallel, but it is not feasible solution. The solution can be feasible if the right method is hybrid with the Apriori. The MapReduce is one the best approaches that can provide solution in an efficient manner. The Apriori is also working on the frequency of the data points, and MapReduce can also work on the mapping of the key value with frequent data points. Therefore, the MapReduce can provide the parallel solution to Apriori in an efficient manner. In this paper, a hybrid version of Apriori and MapReduce for the fast and efficient execution is shown. The Apriori algorithm deployed on the MapReduce platform with suitable frequent key values. The hybrid approach is executed on the dataset and provides more accurate result. Experimental results show that the algorithm scales up linearly with respect to dataset sizes.

Keywords Association rule mining (ARM) · Apriori algorithm · FP-growth · MapReduce · Parallel apriori

A. Sharma (✉) · K. Tripathi
Department of Computer Engineering and Applications, GLA University, NH#2, Delhi Mathura Highway, Post Ajhai, Mathura, UP, India
e-mail: ashishs.sharma@gla.ac.in

K. Tripathi
e-mail: kshitij.tripathi_mtcs19@gla.ac.in

N. Marriwala et al. (eds.), *Mobile Radio Communications and 5G Networks*, Lecture Notes in Networks and Systems 140,
https://doi.org/10.1007/978-981-15-7130-5_45

1 Introduction

Data mining includes a pattern discovery chain in large databases covering machine learning process methods, statistics, and database systems. It is an interdisciplinary subfield of statistical science and statistics with the overall aim of extracting information from a dataset and transforming the information from a dataset into a systematic system for future use.

Such techniques are used in various areas, and as compared to the other techniques, all the techniques have their different impact. ARM is the first method employed in the area of data mining. The Apriori algorithm is used to provide the itemsets relation. Mr. R. Agarwal and Mr. R. Srikant proposed the algorithm in 1994 [1]. The Apriori algorithm is used, and the application area-based paper is implemented by [2].

It is used in all datasets to identify the regular element sets. It is used in all datasets to identify the regular element sets, named Apriori because it is used to gain prior knowledge of the properties of frequently set objects. There are varieties of areas where Apriori is used and it is considered as a Apriori algorithm. The Apriori has also been used for the different areas, and changes are made in Apriori as needed. Yet, all of these algorithms have some positive and some negative points.

In our conventional algorithms, the positive points are as simple and easy to understand, and the results are intuitive and easy to convey to the customer. This algorithm is used to generate frequent patterns with the support and confidence specified.

These are the advantage, but there are many drawbacks to the same. The Apriori algorithm is not able to work with higher number of dimensions as well as on huge amount of dataset. A dataset has time as the main constraint or as the space and time attributes as important attributes.

These are the key parameters where it is necessary to improve the Apriori algorithm, so that it can also be used to deliver results for such application areas. The Apriori algorithm also needs to be updated for the enormous generation of commonly used patterns, multiple scanning of datasets, and consideration of only minimal support. The Apriori algorithm is unable to manage a dataset that used to be large in size.

Harikumar and Dilipkumar [3], Zaki and Zulkurnain [4] the author proposed an idea to deal with the same. Moving further as researchers know that they get a huge number of regular itemset in Apriori which will be eliminated in [5]. But nowadays, authors have seen that there is a great deal of dependence on time attributes so needed time-relevant information so [6] takes responsibility for it, and in our Apriori algorithm, we consider min (support and confidence) from [7], it can be used [8] to reduce the pruning part of our algorithm.

In today's era, the dataset used to be modified on a regular basis, so everyone need some algorithm that generates association rule according to the updated dataset in [9], and persons can improve efficiency by reducing the enormous frequency of pattern generation that can be done by [10], and nowadays, people have areas where they need to consider both time and space to get our association rule just like in traffic dataset, so it is there in [11]. But everyone still have to do several scans of our dataset, which is omitted slightly in [12, 13]. Besides that, persons have another algorithm to generate the association rule based on the FP-Growth, and the algorithm is FP-Graph-Growth which can be used to generate the association rule.

The paper further is divided into as follows, Sect. 2 provides literature review work, Sect. 3 presents technical discussion of machine learning model, the comparison model is considered in the Sect. 4, experimental result is illustrated in Sect. 5. Finally, Sect. 6 presents conclusion and future work.

2 Literature Review

In the paper of [1], authors have provided the right way to get the result in a better way for the market basket analysis. In the paper of [2], authors have taken the education dataset of an entrance exam of a college. The author has applied the Apriori algorithm, and they had to find out the relationship among the various parameters of the data. They had just implemented the Apriori algorithms.

In another paper [3], the author has given the best way to utilize the Apriori algorithm to get the relationship among the different attributes in the high-dimensional dataset. As the data is high dimensional, so it must be having a lot of dimensions, and this makes Apriori very tedious to find out the relationship. So in this, they had used the QR decomposition technique to decompose the dataset without losing its meaning, and then, they forwarded the data to the Apriori algorithm to get out the result. And in [4], the author has given a review over this paper which summarizes all the things which have been done in this paper.

The author has given in research [13] a solution for the one of the main issues of the Apriori algorithm that is the multiple times scanning of the dataset as it consumes a lot of time and it is not necessary to scan the dataset again and again, so here author gives the idea to mark the transaction a delete tag if the size of it is less than 'k' and also mark the deleting tag on those transactions who do not have any element from the candidate set Ck. And all the transaction which are marked as deleting tag will not be scanned in the next scan.

The author [14] uses two main algorithms for association rule mining that is Apriori and FP-Growth on the maintenance dataset. They have found out the symptom of failure so that they can predict the reason behind the failure and the failure parts needed for the future. This helped the maintenance team to be well aware about the same and can arrange those items in-store priory.

The author has given a new rule in [7] to find out the association rule. Basically, Apriori consider minimum (support and confidence) while calculating the association rule (AR), bur in the present scenario, data is growing rapidly, so it may loose many relevant items. So, the authors have suggested new limits for both of the parameters like support (maximum, minimum) and confidence (maximum, minimum), so it will cover all the necessary rules to get our result.

The author gave a new version of the Apriori algorithm in [6], which will be giving the result while keeping the time constraints which would be very much essential for today's world, and this makes a decrease in the size of memory occupation and the number of scans of the dataset as well. So now, the result would have frequent itemset with the time constraint intervals.

The writers give everyone the new method in [15], which will use the graph technique to mine the dataset and find out the association rule. While using the graph method they have found out that now, there is no need to have to keep the redundant items as it is kept in FP-Tree.

The author implemented a shift in the Apriori algorithm in [8]. He had just introduced the modification into the pruning part of the Apriori algorithm. In that, he has taken average support instead of the minimum support and these results with the probability and improved efficiency of the Apriori algorithm.

Teng and Chen [9] provides a full description of the algorithms that can be used to evaluate market basket. In this paper, they had detailed about three kinds of algorithms those are Apriori, partition-based algorithm, and pattern growth algorithm, and they had given the review on the working of all these algorithms while keeping the incremental approach. The main thing behind the incremental approach concept is that due to frequently updation of database, it results to association rule.

In the paper of [5], a new modification over the Apriori algorithm is done. The author has introduced the consideration of Inter Itemset Distance with support and confidence in the Apriori algorithms. It is developed while keeping in mind the generation of the huge candidate and multiple scans of the dataset. With the help of Inter Itemset Distance, they had reduced the number of frequent items and the association rule. Basically, the Inter Itemset Distance measures the number of transactions the itemset is not included between two successive occurrences of the itemsets.

The author uses Bodon's idea in [10] and implemented his algorithm on parallel computing to enhance the efficiency and get the result quickly, so the main thing is that, Bodon introduced the use of a data structure 'trie' to solve the repeated frequent itemset generation issue of the Apriori algorithm. And the author just used the partition-based approach with parallel computing and trie data structure to give it speed.

In the paper of [11], author has provided the solution for a real-life problem for traffic control and provided the solution using the Apriori.

Here in [12], the author has given five algorithms which can be used in our mining strategy namely prior MapReduce, IApriori MapReduce, SpaceAprioriMapReduce, TopAprioriMapReduce, and MaxAprioriMapReduce, and all these algorithms are developed by the author in the same manner as they are written by overcoming the disadvantage or future work of the previous one.

3 The Proposed Model

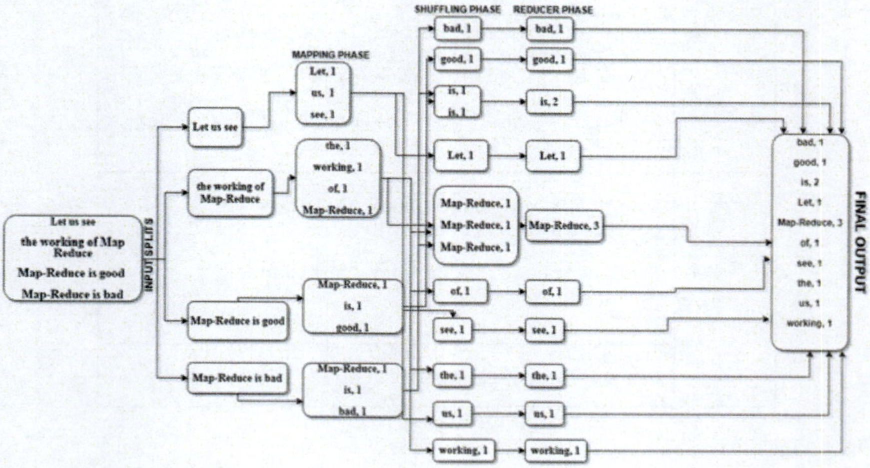

MapReduce

In the MapReduce technique, the value or frequency is measured. Thus, a map reduction technique can be used to obtain the frequency of those particular data points or keys [16].

$$\text{Map} : (\text{K1}, \text{V1}) \rightarrow (\text{K2}, \text{V2})$$

$$\text{Reduce} : (\text{K2}, \text{V2}) \rightarrow (\text{K2}, \text{V3})$$

MapReduce technique performs parallel processing on big datasets. Nonetheless, on the large amount of datasets, Apriori performs fast and parallel processing with MapReduce technique.

The proposed solution applies MapReduce technique with Apriori. To find the frequency, Apriori with a map reduction technique is used, and the value of k is then calculated using frequency (Fig. 1).

4 Result

The efficiency of the proposed solution is analyzed in this section. To get experimental results, two-dimensional dataset was chosen. Firstly, the machine assimilates data from various sources to carry out Apriori with map reduction technique. The

Step 1	Input: D: dataset
	Finding the value of k-dist plot; 1) Initialize all items in Dataset: Generate Value of k: Consider frequent value as a value of k; 2) Finding the k on the basis of frequency; Mapping item set with k;
Step2	Adopt Apriori algorithm for item set generation; Finding the Support and Confidence
	Display Most frequent items

Fig. 1 Algorithm

suggested solution evaluated on datasets of different sizes. Python is used in our proposed method and coordinates with our dataset. The Spaeth dataset is taken to apply the model. We have tested it two times, in the first instance, 100 data points are considered, and in second experiment, 200 points are considered.

Experiment: We have executed the model on the Spaeth dataset. The model is executed on two times on the different dataset, and in the first experiment, the dataset contains 100 files and calculated the purity (Tables 1 and 2).

$$\text{Purity}(\cap, C) = \frac{1}{N}\sum_{K} {}_{j}^{Max}\left|w_k \cap C_j\right|$$

Table 1 Result analysis of experiment with 100 size dataset

Dataset	Purity of apriori	Purity of apriori with MapReduce
Dataset 1	0.76	0.78
Dataset 2	0.63	0.68
Dataset 3	0.60	0.74

Table 2 Result analysis of experiment with 200 size dataset

Dataset	Purity of VDBSCAN	Purity of VDBSCAN with MapReduce
DataSet 1	0.79	0.84
DataSet 2	0.63	0.69
DataSet 3	0.62	0.89

5 Conclusion and Future Work

A complete study of Apriori algorithm is compiled in the current survey, and article possesses the various modifications in the Apriori algorithm. This makes us think before using Apriori algorithm in any field. So, to resolve all those problems, we have seen modification in the Apriori algorithm about the dataset and vulnerabilities and seen all those issue which get omitted from which algorithm. Apart from these future outcomes, readers can have a different algorithm for dimensionality reduction which they can use with Apriori to reduce the dimension and get our result and add on to that researcher can use partitioning approach with the same which will give us the good efficiency as compared to the other algorithm. Besides that, researcher can also go for the combined approach of either of the two algorithms and design a new algorithm that can solve all the problems at once. If they can have any of this approach or implementation, then that would be a new approach in the same.

MapReduce is highly lucrative for storing big data in parallel on massive commodity computer clusters. In this article, we focus primarily on the MapReduce system parallelization of the Apriori algorithm. The MapReduce computing model is well related to the Apriori algorithm measurement of regular itemsets. We reviewed numerous possible methods for parallelizing the distributed system for Apriori. MapReduce is an efficient and scalable framework for big computational problem, but on such platform, it can be hard to port some problems.

References

1. Agrawal R, Imieliński T, Swami A (1993) Mining association rules between sets of items in large databases. Acmsigmod Rec 22(2) ACM
2. Hussain S et al (2018) Classification, clustering and association rule mining in educational datasets using data mining tools: a case study. Comput Sci On-line Conf Springer, Cham
3. Harikumar S, Dilipkumar DU (2016) Apriori algorithm for association rule mining in high dimensional data. Int Conf Data Sci Eng (ICDSE). IEEE
4. Zaki FA, Zulkurnain NF (2019) Frequent itemset mining in high dimensional data: a review. Computational science technology. Springer, Singapore, pp 325–334
5. Sharma PK, Mahanta AK (2013) An apriori based algorithm to mine association rules with inter itemset distance. Int J Data Min Knowl Manage Process 3(6):73
6. Wang C, Zheng X (2019) Application of improved time series Apriori algorithm by frequent itemsets in association rule data mining based on temporal constraint. Evol Intell 1–11
7. Wang G et al (2010) Research of data mining based on Apriori algorithm in cutting database. 2010 International conference on mechanic automation and control engineering. IEEE
8. Sequeira JV, Ansari Z (2015) Analysis on improved pruning in Apriori algorithm. Int J 5.3
9. Teng WG, Chen MS (2005) Incremental mining on association rules. Foundations and advances in data mining. Springer, Berlin, Heidelberg, pp 125–162
10. Ye Y, Chiang CC (2006) A parallel Apriori algorithm for frequent itemsets mining. Fourth international conference on software engineering research, management and applications (SERA'06) IEEE
11. Xie DF, Wang MH, Zhao XM (2019) A spatiotemporal Apriori approach to capture dynamic associations of regional traffic congestion. IEEE Access

12. Luna JM et al (2017) Apriori versions based on mapreduce for mining frequent patterns on big data. IEEE Trans Cybern 48(10):2851–2865
13. Xie Y et al (2008) The optimization and improvement of the Apriori algorithm. In: Workshop on education technology and training and 2008 IEEE international workshop on geoscience and remote sensing 2
14. Rachburee N, Arunrerk J, Punlumjeak W (2018) Failure part mining using an association rules mining by FP-growth and Apriori algorithms: case of atm maintenance in Thailand. IT Convergence and Security 2017. Springer, Singapore 19–26
15. Tiwari V et al (2010) Association rule mining: a graph based approach for mining frequent itemsets. 2010 International conference on networking and information technology. IEEE
16. Sharma A, Upadhyay D (2018) VDBSCAN clustering with map-reduce technique. Recent Findings in Intelligent Computing Techniques. Springer, Singapore, 305–314

A New Time Varying Adaptive Filtering System (TVAFS) for Ambulatory ECG Signals

Reeta Devi, Hitender Kumar Tyagi, and Dinesh Kumar

Abstract Ambulatory ECG signal gets coupled with various noises. Noise and ECG signal are non-stationary in nature. Filtering system, an essential part of the ambulatory ECG system, needs to be less complex so as to minimize the overall processing cost. Present paper proposed time varying adaptive filtering system comprising of complexity reduced variable step size algorithm and a cascaded digital FIR filter. The MIT/BIH arrhythmia dataset has been used to evaluate the proposed system. Results obtained in terms of improved SNR and fast converging learning rate demonstrate that the proposed system can effectively remove noise compared with other popular adaptive filters.

Keywords Ambulatory ECG · Adaptive · LMS · NLMS · Time varying · Step size

1 Introduction

Ambulatory ECG diagnosis is conducted in routine after heart surgery. Such diagnoses are composed of various kinds of exercise testing that subject the patient to perform different motion activities. Those activities may vary from low physical works to high level of physical activities. Therefore, it is natural for the ambulatory ECG signal to get contaminated with different kinds of noises such as baseline wander, power-line interference, random noise, electrode contact displacement and motion artifacts.

These noises pose the difficulty on the filtering system as they fall in the same ECG spectrum. Ideally, a filter should be able to remove the interfering noise completely without affecting the desired signal characteristics [1]. Along with this, a desirable and very strict requirement is to keep the filtering algorithm very simple and less

R. Devi (✉) · H. K. Tyagi · D. Kumar
Kurukshetra University, Kurukshetra, Haryana, India
e-mail: reetakuk@gmail.com

D. Kumar
J C Bose University of Science and Technology, YMCA, Faridabad, Haryana, India

N. Marriwala et al. (eds.), *Mobile Radio Communications and 5G Networks*, Lecture Notes in Networks and Systems 140,
https://doi.org/10.1007/978-981-15-7130-5_46

complex so as to minimize the processing cost of the ambulatory system. The most important factors associated with the processing cost are algorithm's computational complexity producing a good quality signal and the required memory capacity to store the filter coefficients.

The conventional fixed coefficient filters are not suitable to meet above-said requirements due to non-stationary nature of the noise and the ECG signal. Thus, advanced filtering algorithms such as adaptive filters, wavelet filtering and blind source separation methods are required. The filtering performance of adaptive filters and the wavelet filters has been analysed in our previous work [2]. In that study, it was noted that DWT filter performed with a degraded performance for high noise ECG signals, whereas an adaptive recursive least squares (RLS) algorithm performed well in this environment. The results obtained in this study for RLS were also in good agreement with other previously published studies by Almahamdy et al. [3], Elias et al. [4] and Martinek et al. [5]. But this algorithm suffers from a high computational complexity with large memory requirements [6] resulting to high processing cost of the ambulatory system. Similarly, the performance of blind source separation method is good enough but at the cost of high processing cost and increased structural complexity of the system [7].

Another popular adaptive filtering method which was implemented in our previous study was the least mean square (LMS) algorithm. This method filtered high- to-low noise ECG signals with medium filtering performance and slow speed. But this method fits best for integration in real-time ambulatory ECG systems in terms of less computational complexity and less memory requirements, provided, the filtering performance and speed are modified to enhance to a level of producing acceptable quality results of noise-free ECG signal with almost no delay. Therefore, the present study was designed to improve the filtering performance and convergence rate of the LMS-based adaptive filter, so as to maximize the output SNR of a variable noise ECG signal.

Though, there are various adaptive filtering techniques proposed in literature that can be used for the noisy ECG signal [8–11]. But a dual non-stationary signal (non-stationary both in noise as well as ECG signal) requires an adaptive filtering technique that can change its properties in accordance with the changes in noise and the desired ECG signal. One such technique of cascade adaptive filtering for motion artifacts removal from the ECG signal was proposed by Kim et al. [7]. In this technique, a fixed step size fourth-order high pass filter was implemented by LMS algorithm in the first stage to remove the baseline wander. The second stage was comprised of a variable step size 16th-order normalized LMS adaptive filter to remove the motion artifacts. The overall configuration results in good quality ECG signal, besides, a high computational complexity, high memory requirements and, thus, high structural complexity and cost of the ambulatory system.

Another technique of adaptive filtering with feed-forward algorithm was proposed by Huanqian et al. [12]. This approach was also used to filter the motion artifacts from ambulatory ECG signals. Authors of this study used two adaptive filters at the input and a combining network at the output. Two adaptive filters consisting of a fixed step size normalized LMS and a variable step size normalized LMS, each of order 200 were used for the fast convergence speed and fast convergence accuracy, respectively. The combining network implements a mathematical expression to combine the two filtered outputs of the adaptive filters. In the context of the results produced by the proposed system, it can be stated that a satisfactory clean ECG signal with good morphological details was obtained. But at the same time, it is also seen that the proposed system configuration introduces a high computational complexity as well as high memory requirements which will further increase the overall structural complexity and the cost of the system.

Similarly, several other studies [13–16] have implemented variable step size adaptive filter for other applications like echo cancellation and unknown system identification, etc. However, many of these algorithms need tunable parameters that offer additional delay and complexity to the system.

This paper presents a complexity reduced variable step size normalized least mean square adaptive transversal FIR filter of order 10 with a cascaded second order fixed coefficient FIR filter configuration. The proposed method is implemented on 48 ambulatory ECG signals obtained from MIT/BIH Arrhythmia database [17]. Results obtained are compared with the existing conventional LMS filter in terms of improved signal-to-noise ratio (SNR) and the convergence of the learning rate parameter. Improved SNR is measured by the difference of the output SNR and the input SNR. It is observed from the results that the proposed system learns very fast and effectively de-noise the ambulatory ECG signal producing high values of the improved SNR compared to the conventional filter. The reduced complexity in terms of the number of components to be used in the overall structure, processing cost and the simplest algorithm for varying the step size of the proposed filtering system, will make it easy to integrate with very important medical algorithms such as early stage prediction of sudden cardiac death described in our earlier work [18].

2 Adaptive Filtering System

Figure 1 shows the proposed system block diagram with a cascaded second-order FIR filter. The adaptive filtering system consists of two components: (1) a transversal, linear, finite impulse response (FIR) filter and (2) a normalized LMS algorithm with

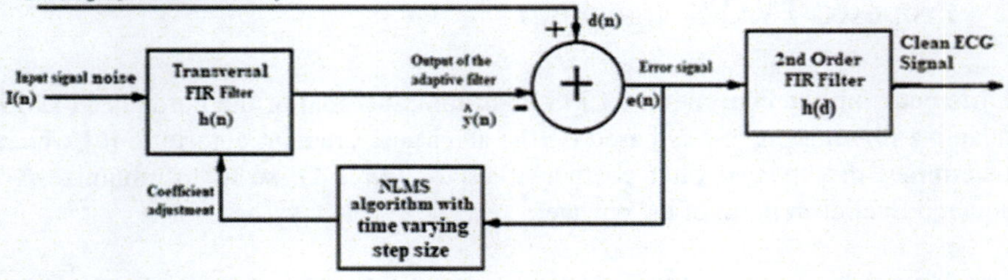

Fig. 1 Proposed system block diagram of TVAFS

time varying step size property. The transversal filter of order N convolves the input noise signal I(n), with its own impulse response function h(n) to produce an estimate of the input noise signal at a time instant n, as described by Eq. (1). The estimated noise of the filter is denoted by $\hat{y}(n)$.

$$\hat{y}(n) = \sum_{m=0}^{N-1} h(m)I(n-m) \quad (1)$$

where $h(m)$ are the coefficients of the proposed adaptive filter, and $I(n-m)$ are the noisy reference signal samples at present ($m = 0$) and past $m-1$, $(1 \le m \le N-1)$ input samples.

The estimated noise is subtracted from the desired signal $d(n)$, which is constituted by the pure ECG signal $v(n)$ added with an undesired noise signal $z(n)$ Eq. (2). The two noise signals $I(n)$ and $z(n)$ are correlated with other. Thus, the difference so obtained produces the error signal $e(n)$, an estimate of the pure ECG signal as expressed by Eq. (3) [6].

$$d(n) = v(n) + z(n). \quad (2)$$

$$e(n) = d(n) - \hat{y}(n) \quad (3)$$

The error signal $e(n)$ and the input noise signal $I(n)$ of the proposed system are designed to act on the adaptive process to adjust the coefficients of the transversal filter in order to produce an estimate of the noise mixed in the pure ECG signal. The adaptive algorithm has a time varying step size property that changes the learning rate of the filter from one iteration to the next to update the coefficients so as to minimize the mean square value of the error signal. The optimized error signal is then convolved with the impulse response function $h(d)$ of the second-order FIR filter to obtain the pure ECG signal as given in Eq. (4).

$$v(n) = \text{Pure ECG} = e(n) * h(d) \quad (4)$$

3 Proposed TVAFS Algorithm

Different from the conventional LMS algorithm, the goal of the normalized LMS adaptive filtering process is based on the stochastic gradient algorithm [6] which determines the updated filter coefficient vector $\hat{h}(n+1)$, so as to minimize the squared Euclidean norm of the change,

$$\boldsymbol{\delta\hat{h}(n+1) = \hat{h}(n+1) - \hat{h}(n)} \quad (5)$$

Subject to the constraint

$$\hat{h}^H(n+1)I(n) = d(n) \tag{6}$$

To solve the constrained optimization problem stated in Eqs. (5) and (6), the cost function $A(n)$ is designed as shown in Eq. (7) to be minimized by the method of Lagrange multipliers.

$$A(n) = \left\|\delta\hat{h}(n+1)\right\|^2 + Re\left[\lambda * \left(d(n) - \hat{h}^H(n+1)I(n)\right)\right] \tag{7}$$

By putting Eq. (5) into Eq. (7), we obtain

$$\begin{aligned} A(n) &= \left\|\left(\hat{h}(n+1) - \hat{h}(n)\right)\right\|^2 + Re\left[\lambda * \left(d(n) - \hat{h}^H(n+1)I(n)\right)\right] \\ &= \left(\hat{h}(n+1) - \hat{h}(n)\right)^H \left(\hat{h}(n+1) - \hat{h}(n)\right) \\ &\quad + Re\left[\lambda * \left(d(n) - \hat{h}^H(n+1)I(n)\right)\right] \end{aligned} \tag{8}$$

Differentiating Eq. (8) with respect to $\hat{h}(n+1)$ with the rule of differentiation of a real-valued function $(A(n))$ with respect to the complex valued coefficient vector $\hat{h}(n+1)$, we obtain

$$\frac{\partial A(n)}{\partial \hat{h}^*(n+1)} = 2\left(\hat{h}(n+1) - \hat{h}(n)\right) - \lambda^* I(n) \tag{9}$$

Setting Eq. (9) equal to zero, the optimum value of the coefficient vector may be obtained as

$$\hat{h}(n+1) = \hat{h}(n) + \frac{1}{2}\lambda^* I(n) \tag{10}$$

The value of the unknown parameter λ is obtained by solving Eqs. (6) and (10),

$$\begin{aligned} d(n) &= \left(\hat{h}(n) + \frac{1}{2}\lambda^* I(n)\right)^H I(n) \\ &= \hat{h}^H(n)I(n) + \frac{1}{2}\lambda I^H(n)I(n) \\ &= \hat{h}^H(n)I(n) + \frac{1}{2}\lambda\|I(n)\|^2 \end{aligned} \tag{11}$$

From Eq. (11), λ can be obtained as

$$\lambda = \frac{2e(n)}{\|I(n)\|^2} \tag{12}$$

where $e(n)$ the error signal defined in Eq. (3) may be reproduced as

$$e(n) = d(n) - \hat{h}^H(n)I(n) \tag{13}$$

Thus, by putting the Eqs. (10) and (12) in Eq. (5), the optimum value of the incremental change in coefficient vector is obtained as

$$\delta\hat{h}(n+1) = \frac{1}{\|I(n)\|^2} I(n)e^*(n) \tag{14}$$

In order to remove the non-stationary noise from the ECG signal, the coefficient vector needs to be updated iteratively without changing the direction vector. Thus, a positive real scaling factor μ is inserted in Eq. (14) to achieve this,

$$\delta\hat{h}(n+1) = \frac{\mu}{\|I(n)\|^2} I(n)e^*(n) \tag{15}$$

It gives the required change at iteration n to adjust the coefficients of the transversal FIR filter. Substituting Eq. (15) into Eq. (5) and reordering the terms, we obtain the updated coefficient vector at iteration $n + 1$ as

$$\hat{h}(n+1) = \hat{h}(n) + \frac{\mu}{\|I(n)\|^2} I(n)e^*(n) \tag{16}$$

Equation (16) shows the product vector $I(n)e^*(n)$ which is normalized by the squared Euclidean norm of the input noise vector $I(n)$. Hence, the algorithm is known to be normalized least mean square algorithm (NLMS).

The scale factor μ is known as the step size parameter. It is assumed a constant in the NLMS algorithm [6]. However, an ambulatory ECG signal and noise are non-stationary in nature. Thus, the NLMS algorithm may malfunction due to the large mismatch between the desired and estimated signal. For example, suppose there is an increase in the interfering noise $z(n)$; the desired signal $d(n)$ will increase, and, so will the error signal $e(n)$ if, coefficients are not adjusted in the opposite direction, which is not possible in a fixed step size NLMS algorithm. Therefore, the constant value of the step size parameter μ will appear oversized in this situation, which is required to be reduced, and vice versa.

To overcome the difficulty faced by the NLMS algorithm, a time varying step size parameter $\mu(n)$ suggested by Mader et al. [19] for echo cancellation and further improved for filtering of ambulatory ECG signal in the present study is given as

$$\mu(n) = \begin{cases} 0, & if\ \mu_c < 0.5\mu_0 \\ \mu_c, & if\ 0.5\mu_0 < \mu_c < \mu_0 \\ \mu_0, & \text{otherwise} \end{cases} \tag{17}$$

where μ_0 is the optimum step size selected by the minimum mean square error criteria for the lowest SNR value of the input signal, and μ_c is the current step size of the nth iteration which is advanced from one iteration to the next by using the following relation

$$\mu_c = \frac{S(n)E\left[I^2(n)\right]}{E\left[e^2(n)\right]} \tag{18}$$

where $S(n) = E\left[\left\|h - \hat{h}(n)\right\|^2\right]$, with h being the unknown system parameter vector. $S(n)$ is known as mean square deviation of the adaptive filter's coefficient vector. According to Eq. (17), at iteration n, the transversal filter coefficients are not updated if computed value of μ_c is less than $0.5\mu_0$. The coefficients are only updated at all iterations in accordance with the value of step size $\mu(n)$ lying in the range $0.5\mu_0 \leq \mu_c \leq \mu_0$. The final output of the TVAFS system is then obtained as per the Eq. (4). A summary of the TVAFS algorithm is presented in Table 1.

4 Experiments and Results

The performance of the designed time varying adaptive filtering system is evaluated on the standard international MIT-BIH Arrhythmia database [17]. This dataset consists of 48 ambulatory ECG recordings obtained from 25 men (age: 32–89 years) and 22 women (age: 23–89 years), with two signals collected from the same subject. Lead ML-II signal of sampling frequency 360 Hz has been used in this work. The resolution of all the collected signals is at 11 bit $\pm$ 10 mV.

The computing software MATLAB has been used to run the experiments on the Acer® laptop system embedded with an Intel core i3 processor. The results of the first two stages for removal of baseline wander by using two stage moving average filter and power-line interference by using an IIR comb notch filter have been shown in our previous work [2]. It was shown there that a pure ECG signal was obtained without any significant distortion of its desired properties after filtering the baseline wander and power-line interference at 60 Hz and its harmonics 120 and 180 Hz.

Further, this pure ECG signal has been corrupted by adding the artificial noise generated in MATLAB and proposed here to clean by a new time varying adaptive filtering system (TVAFS) with fast converging speed. The proposed algorithm has been described in Sect. 3. The artificial noise is a kind of composite noise comprising of white Gaussian random noise of varying SNR from −11 dB to 11 dB, the baseline wander, power-line interference and abrupt shift. A plot of all these noises generated artificially and added to the ECG signal is shown in Fig. 2. The performance of the TVAFS in filtering of high noise (SNR = −11 dB) and small noise (SNR = 11 dB) from the ECG signal is illustrated in Figs. 3, 4, respectively. Figures 3a and 4a show the corrupted ECG signal of known SNR value, and Figs. 3b and 4b show the

Table 1 Summary of the proposed TVAFS algorithm

Parameters
N = Filter length
$I(n)$ = Present input (noise) vector
$d(n)$ = Desired response
$\hat{h}(n)$ = Updated filter coefficient vector after adaptation at time instant n
$\hat{y}(n)$ = Output of the filter at time instant n
$e(n)$ = error signal
$h(d)$ = Filter coefficients of the second-order FIR filter
μ = Constant step size parameter of the NLMS algorithm
$\mu(n)$ = Time varying step size parameter of the TVAFS system
μ_c = Current step size parameter of the TVAFS system at iteration n given by
$\mu_c = \frac{S(n)E\left[\lvert I(n)\rvert^2\right]}{E\left[\lvert e(n)\rvert^2\right]}$
where at iteration n,
$S(n)$ = Mean square deviation
$E\left[\lvert e(n)\rvert^2\right]$ = error signal power
$E\left[\lvert I(n)\rvert^2\right]$ = input signal power
Adaptive procedure
1. Bootstrapping: Boot the filter coefficient vector to the initial values, if known a priory. Otherwise set $h(0) = 0$
2. Triggering: Trigger the TVAFS system with noise input $I(n)$ of time instant n and estimated values of the coefficients $\hat{h}(n)$
$\hat{y}(n) = \sum_{m=0}^{N-1} \hat{h}(m)I(n-m)$
3. Starting with the optimum value of step size parameter, μ_0, find out μ_c for time instant n
4. Adjustment: Adaptively adjust the coefficients of the transversal filter to next time instant $n+1$
$\hat{h}(n+1) = \hat{h}(n) + \frac{\mu}{\lVert I(n)\rVert^2} I(n)e^*(n),\ \mu = \mu_c$
5. Continuation: Increment the iteration from time instant n to $n+1$ and go back to Step 2 to obtain the optimum value of $e(n)$
6. Output: Compute Pure ECG = $e(n) * h(d)$

corresponding filtered signal by using NLMS adaptive filter with proposed variable step size property (VSS–NLMS). Figures 3c and 4c are showing the results obtained with TVAFS. The filtering performances of the conventional LMS adaptive filter for the same signals are shown in Figs. 5 and 6.

For quantitative evaluation, the performance parameters at output of the TVAFS such as mean square error (MSE), SNR, percent root deviation (PRD) and peak

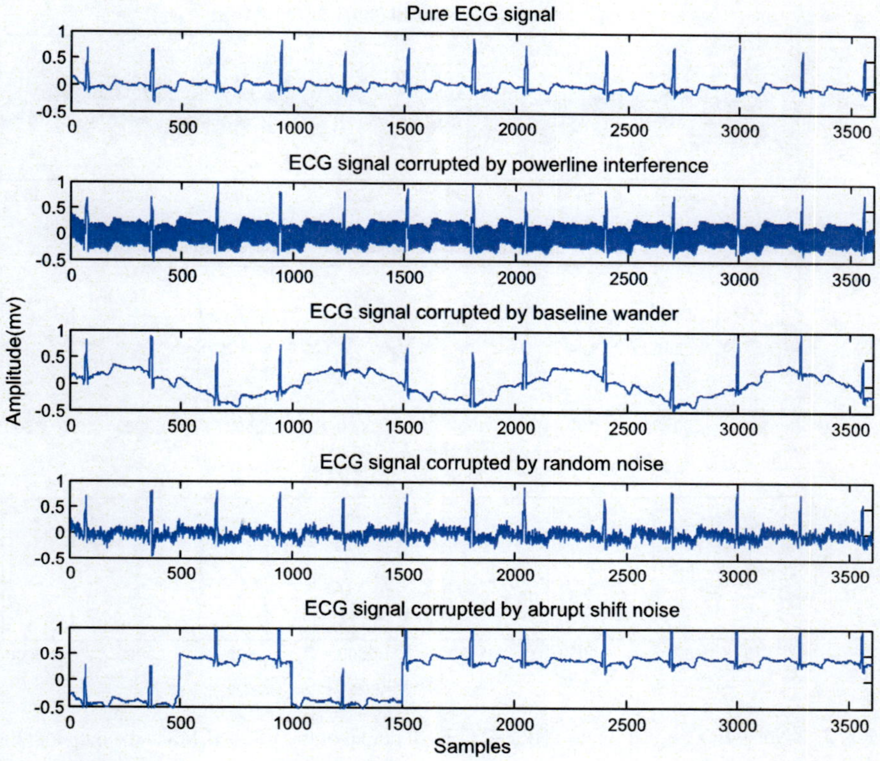

Fig. 2 Different artificial noises added to the pure ECG signal

signal-to-noise ratio (PSNR) [2] have been computed for each of 48 recordings. These results are presented in Table 2 in terms of the mean and standard deviation (Mean (STD)) values for each level of the input SNR. It shows significant high values for the output SNR of TVAFS than the other two algorithms of LMS and VSS–NLMS. For a qualitative comparison between the performances of the proposed system and the conventional adaptive LMS algorithm, the SNR improvement (dB) is plotted in Fig. 7. It can be noted from the figure that the proposed VSS–NLMS algorithm shows a significant improvement over the standard LMS algorithm. Further, the overall output of the TVAFS shows the highest SNR improvement over the others.

Figure 8 illustrates the SNR improvement obtained with proposed TVAFS versus different values of the learning rate parameter $\mu(n)$ at two different levels (-11 dB and 11 dB) of the input SNR. It can be seen in this figure that both curves achieve their maximum value at a very early stage of the learning rate parameter, i.e. 0.04 for input SNR at 11 dB and 0.09 at -11 dB, and thereafter, it starts diminishing consistently. It indicates that the designed filtering system converges very fast, due to which, clean ECG beats are provided instantly by TVAFS, right from the first beat

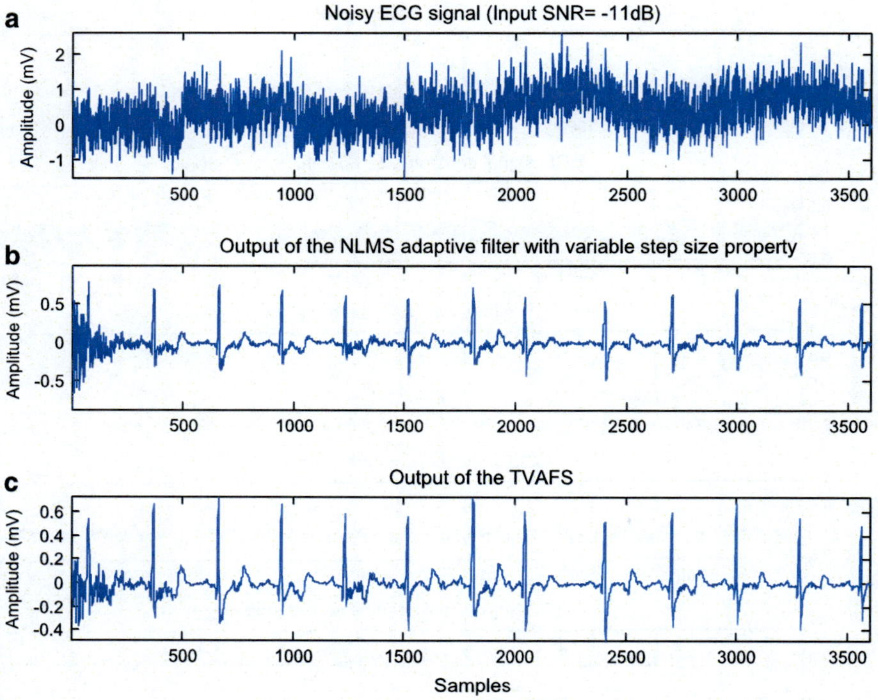

Fig. 3 **a** Noisy ECG signal (input SNR, −11 dB), **b** output of the VSS–NLMS and **c** output of the TVAFS

onwards (Figs. 3c and 4c), whereas the adaptive LMS filter suffers from the problem of gradient noise amplification, which changes the morphology of many beats in the ECG signal (Figs. 5b and 6b).

5 Conclusion

A new time varying adaptive filtering system (TVAFS) for ambulatory ECG signal is proposed in the present work. The filtering performance of the proposed system is compared in terms of SNR improvement and converging learning rate with existing conventional adaptive LMS filter. High values of the SNR improvement and fast convergence rate obtained in this work demonstrate the effective removal of the noise and improved morphological details of the filtered ECG signal as compared to the LMS filter. Also, much reduction in the complexity of the overall filtering system is achieved by using cascaded configuration of a tenth-order transversal FIR filter and a second-order fixed coefficient FIR filter. This will reduce the required memory and number of components in the final design as compared to previous

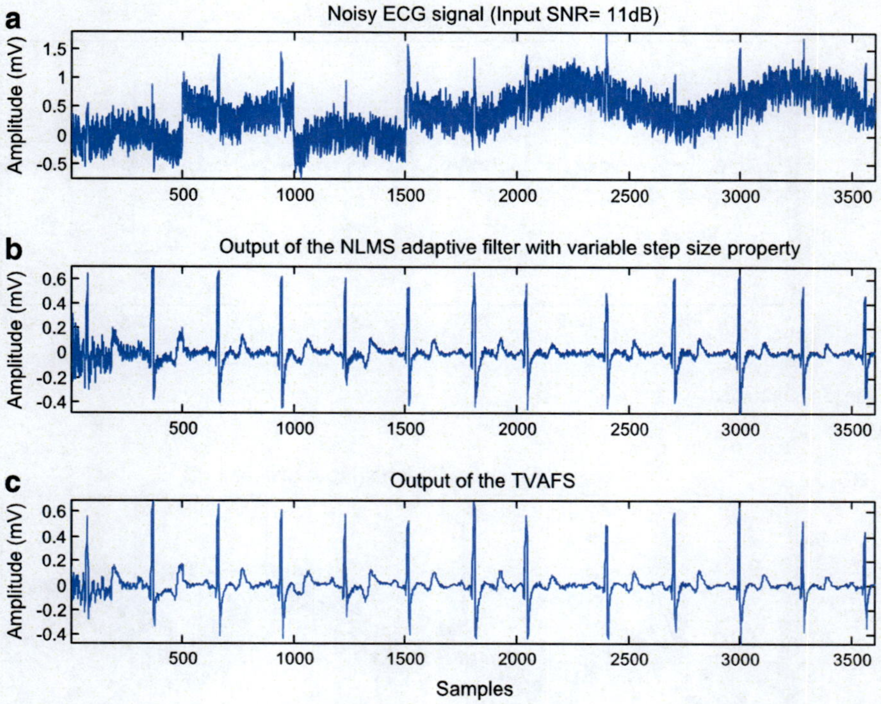

Fig. 4 **a** Noisy ECG signal (input SNR, 11 dB), **b** output of the VSS–NLMS and **c** output of the TVAFS

studies. Therefore, the lower processing cost and the simplest algorithm for varying the step size of the proposed filtering system will make it easy to integrate in low cost ambulatory systems.

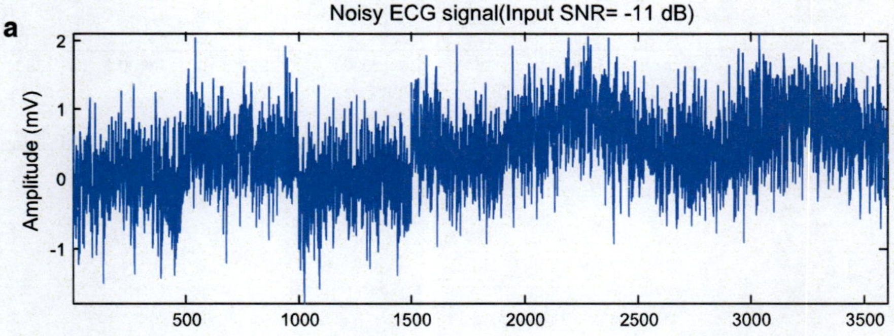

Fig. 5 **a** Noisy ECG signal (input SNR, −11 dB), **b** output of the conventional adaptive LMS filter

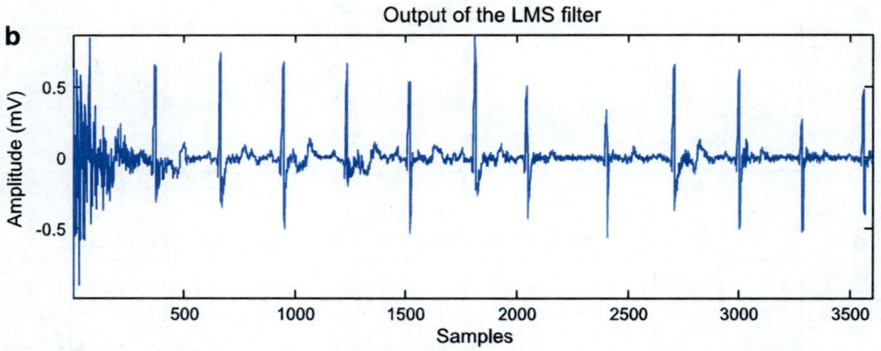

Fig. 5 (continued)

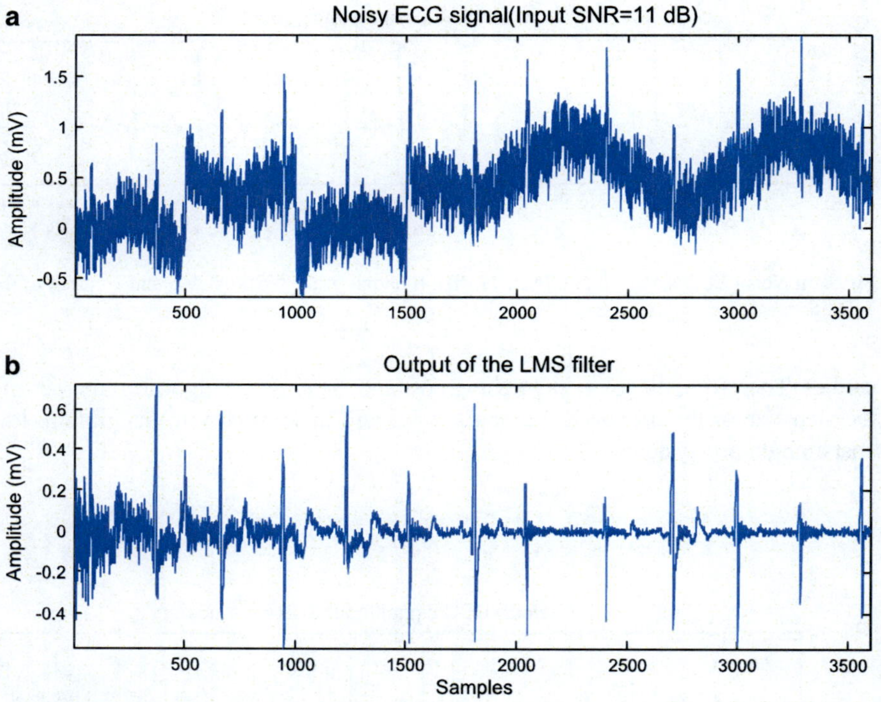

Fig. 6 **a** Noisy ECG signal (input SNR, 11 dB), **b** output of the conventional adaptive LMS filter

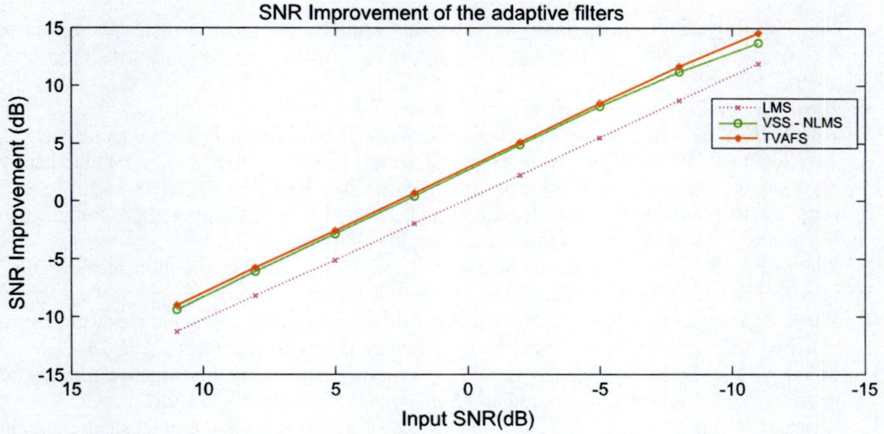

Fig. 7 SNR improvement of different adaptive filters

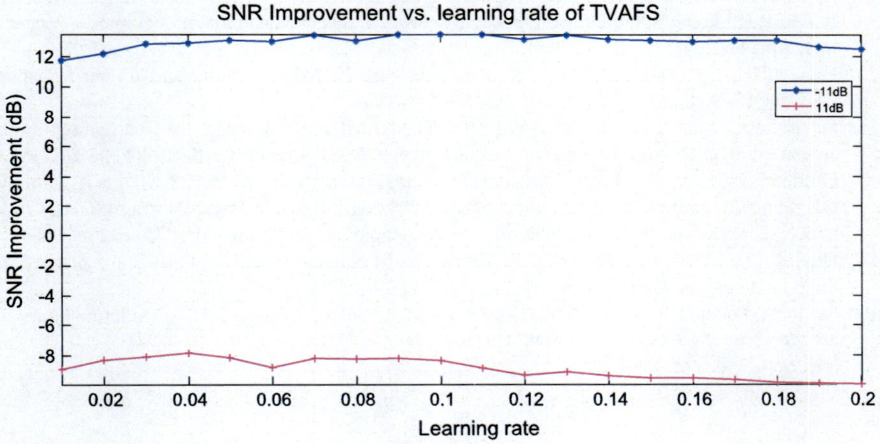

Fig. 8 SNR improvement versus learning rate of the TVAFS

References

1. Oppenheim AV, Schafer RW (1989) Discrete-time signal processing, 3rd edn
2. Devi R, Tyagi HK, Kumar D (2019) Performance comparison and applications of sparsity based techniques for denoising of ECG signal. In: 2019 6th international conference on signal processing and integrated networks (SPIN), pp 346–351
3. Almahamdy M, Riley HB (2014) Performance study of different denoising methods for ecg signals. Procedia Comput Sci 37:325–332
4. Ebrahimzadeh E, Pooyan M, Jahani S, Bijar A, Setaredan SK (2015) ECG signals noise removal: selection and optimization of the best adaptive filtering algorithm based on various algorithms comparison. Biomed Eng Appl Basis Commun 27(4):1–13

5. Martinek R, Rzidky J, Jaros R, Bilik P, Ladrova M (2019) Least mean squares and recursive least squares algorithms for total harmonic distortion reduction using shunt active power filter control. Energies 12(8):1545
6. Haykin S (2007) Adaptive Filter Theory. Pearson
7. Kim H, Kim S, Van Helleputte N, Berset T, Geng D, Romero I, Penders J, Van Hoof C, Yazicioglu RF (2012) Motion artifact removal using cascade adaptive filtering for ambulatory ecg monitoring system. IEEE Biomed Circuits Syst Conf (BioCAS) 2012:160–163
8. Hamilton PS, Curley MG, Aimi RM, Sae-Hau C (2000) Comparison of methods for adaptive removal of motion artifact. Comput Cardiol pp 383–386
9. Ziarani AK, Konrad A (2002) A nonlinear adaptive method of elimination of power line interference in ECG signals. IEEE Trans Biomed Eng 49(6):540–547
10. Wu Y, Rangayyan RM, Zhou Y, Ng S (2009) Filtering electrocardiographic signals using an unbiased and normalized adaptive noise reduction system. Med Eng Phys 31:17–26
11. Mahil J, Raja TSR, Sharmila TS (2015) Optimization algorithms for adaptive filtering of interferences in corrupted signal. Indian J Pure Appl Phys 53(April):274–281
12. Zhang H, Zhang S, Jin Q, Liu X, Li Q, Yang J, Zhao J (2016) Motion artifact suppression in ambulatory ECG with feed forward combined adaptive filter. Comput Cardiol 43(1):1–4
13. Benesty J, Rey H, Vega LR, Tressens S (2006) A nonparametric VSS NLMS algorithm. IEEE Sign Proces Lett 13(10):581–584
14. Iqbal MA, Grant SL (2008) Novel variable step size NLMS algorithms for echo cancellation. In: ICASSP, IEEE international conference on acoustics, speech and signal processing—proceedings, pp 241–244
15. Huang HC, Lee J (2012) A new variable step-size NLMS algorithm and its performance analysis. IEEE Trans Signal Proces 60(4):2055–2060
16. Huemmer C, Maas R, Kellermann W (2015) The NLMS algorithm with time-variant optimum stepsize derived from a bayesian network perspective. IEEE Sign Proces Lett 22(11):1874–1878
17. Goldberger AL, Amaral LAN, Glass L, Hausdorff JM, Ivanov PC, Mark RG, Mietus JE, Moody GB, Peng C, Stanley HE (2000) PhysioBank, PhysioToolkit, and PhysioNet components of a new research resource for complex physiologic signals. Circulation 101:e215–e220
18. Devi R, Tyagi HK, Kumar D (2019) A novel multi-class approach for early-stage prediction of sudden cardiac death. Biocybern Biomed Eng 39(3):586–598
19. Mader A, Puder H, Schmidt GU (2000) Step-size control for acoustic echo cancellation filters—an overview. Sig Proces 80(9):1697–1719

Ultrathin Compact Triple-Band Polarization-Insensitive Metamaterial Microwave Absorber

Divya and Deepak Sood

Abstract The design and characterization of a compact, ultrathin, polarization-insensitive, triple-band metamaterial microwave absorber are presented. The designed absorber consists of three metallic resonators printed on the top of a 0.8-mm-thick FR4 dielectric substrate. The structure is designed to achieve triple-band absorption at 3.92 GHz (S-band), 5.92 GHz (C-Band), and 9.2 GHz (X-Band) with 92.2%, 94.5% and 98.71% absorption, respectively. The proposed design is 0.0245λ thick, and its periodicity is 0.3068λ corresponding to its highest frequency of absorption. The design is compact and ultrathin as compared to several already reported dual- and triple-band absorbers. The absorber exhibits wide angular stability up to 60° angle of incident wave. A prototype of the designed absorber has been fabricated, and the measured results are observed in agreement with the simulated ones. The compact size and ultrathin thickness make the design fit for potential RF applications such as RCS reduction etc.

Keywords Metamaterial · Microwave absorber · Ultrathin · Compact size · RF structure

1 Introduction

Metamaterials are artificial composite materials developed from the combination of elements composed of metals/plastics. These materials are usually arranged in periodic pattern and exhibit properties not found in nature like negative electric permittivity (ε), magnetic permeability (μ), refractive index and impedance. The metamaterial showing the effect of negative refractive index was developed using the combination of metallic wire and split ring resonator (SRR) [1]. The inter-element space and size of the elements can be adjusted, to tune the effective characteristics.

Divya · D. Sood (✉)
University Institute of Engineering and Technology, Kurukshetra University, Kurukshetra, Haryana, India
e-mail: dsood2015@kuk.ac.in

N. Marriwala et al. (eds.), *Mobile Radio Communications and 5G Networks*, Lecture Notes in Networks and Systems 140,
https://doi.org/10.1007/978-981-15-7130-5_47

Due to the exotic characteristics, the metamaterials finds applications in perfect lens [2], cloaking [3], antenna [4], and absorber [5], etc. The tuning of imaginary part of the permittivity and permeability gives an idea of the development of metamaterial-based absorbers. An electromagnetic absorber is a kind of device used to efficiently absorb electromagnetic radiations. In comparison with traditional electromagnetic absorbers, the metamaterial absorbers have several advantages such as lightweight, compact size, simplicity, increased effectiveness, and wider adaptability. Despite of these unique features, the metamaterial absorbers suffer from narrow bandwidth and the sensitivity to the polarization and oblique incident angle due to their resonant structural configurations which limits its applications. Since the development of the perfect metamaterial absorber [5], several attempts have been made to improve the performance. Till date, various single-band, dual-band, multiband, and bandwidth-enhanced metamaterial absorbers were reported [6–17]. Metamaterial absorbers reported in [8, 9] are polarization insensitive, but their unit cell size and thickness are large. Similarly, multiband metamaterial absorbers reported in [10–16] are polarization insensitive and ultrathin, but their unit cell size is still large, thereby limiting their practical applications. This anticipates the requirement to develop compact and ultrathin metamaterial absorbers to fill this gap.

Therefore, in this paper, a triple-band polarization-insensitive metamaterial absorber is presented. The proposed design is compact in size and ultrathin in thickness and exhibits a wide angle stability to the incident waves. The mechanism of absorption for the design is analyzed, and the simulations are performed by using Ansys HFSS software tool. A prototype of the design is fabricated, and its characteristics are experimentally verified. A good agreement has been observed between the simulated and the measured responses. The proposed absorber is a good candidate for multiband absorption and can be used in defense and radar applications.

2 Design and Simulation

2.1 Design

The three-dimensional view of the unit cell of the proposed triple-band metamaterial absorber is shown in Fig. 1. The top layer of the structure is designed using three concentric metallic resonators (two square loops and a circular ring) printed on a 0.8 mm thick grounded FR4 ($\varepsilon_r = 4.4$) dielectric substrate. The geometrical dimensions are: $L_1 = 10$ mm, $L_2 = 9.5$ mm, $L_3 = 8.2$ mm, $W_1 = 0.25$ mm, $W_2 = 0.4$ mm, $t = 0.8$ mm, $R_1 = 3.25$ mm, $R_2 = 2.75$ mm. The top and bottom metallic layers are of copper ($\sigma = 5.8 \times 10^7$ S/m) with 0.03 mm thickness. The structure is designed with fourfold symmetry to achieve polarization insensitivity. The size of the single unit cell is (10×10) mm^2.

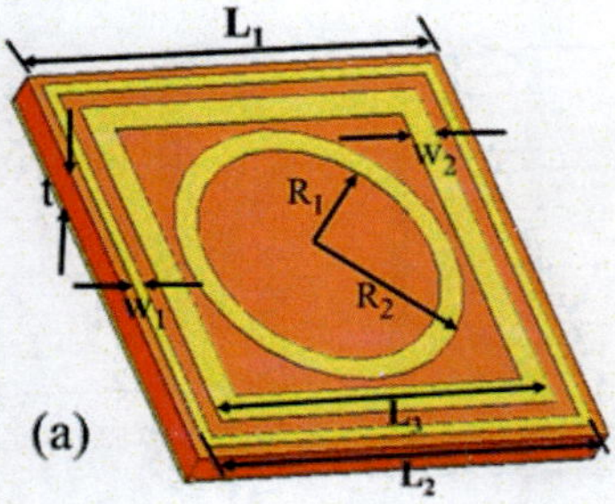

Fig. 1 **a** Unit cell of the designed triple-band absorber and its **b** representation as an array

2.2 Simulation

The simulated response of the proposed absorber under normal incidence of the electromagnetic wave is shown in Fig. 2. The structure is designed using three metallic resonators to give triple-band absorption peaks at 3.92 GHz in S-band, 5.92 GHz in C-band, and 9.2 GHz in X-band with 92.2%, 94.5% and 98.71% absorption, respectively.

The polarization insensitivity of the designed absorber structure has also been verified for different polarization angles (ϕ) from 0° to 60°. The simulated response under normal incidence for different angles of polarization (ϕ) is shown in Fig. 3. When electric field and magnetic field are at some angle 'ϕ' w.r.t X-and Y-axes, respectively, keeping the direction of propagation fixed, i.e. along Z-axis, then the polarization of incident wave changes. The designed absorber structure is fourfold symmetric, and therefore, absorption remains same for all polarization angles; thereby, the proposed design exhibits the polarization insensitivity.

Further, the performance of the absorber has also been studied for different oblique angles (θ) of incidence under TE polarization. For TE polarization, the direction of electric field vector 'E' is along 'X'-axis and magnetic field vector 'H' is at some

Fig. 2 Simulated response of the proposed triple-band metamaterial absorber for normal incidence

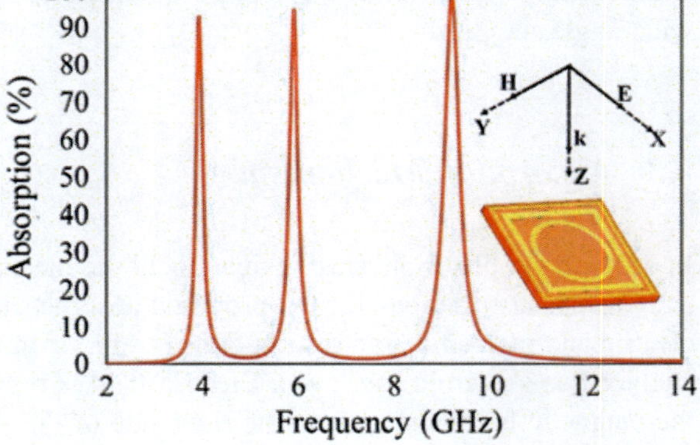

Fig. 3 Simulated absorption for different angles of polarization of the incident wave under normal incidence

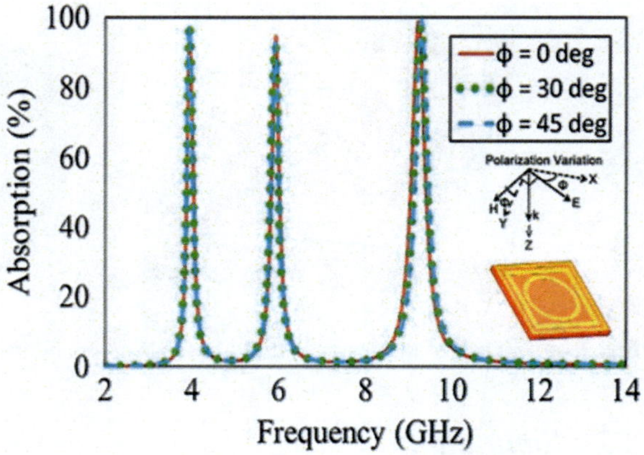

Fig. 4 Simulated absorption for different oblique angles of incident wave under TE polarization

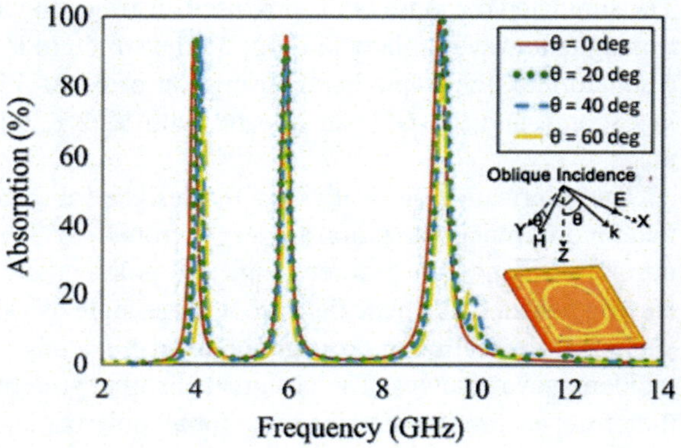

angle w.r.t *Y*-axis in such a way that the direction of incident wave is inclined by an angle 'θ' w.r.t *Z*-axis. The simulated absorption response for different oblique angles of incident wave under TE polarization is shown in Fig. 4. The absorption frequencies and amount of absorption remain same for the variation of incidence angle from 0° to 60° in the steps of 20°. Therefore, the proposed absorber exhibits wide-angle stability.

2.3 Absorption Mechanism

In order to get better physical insight about the mechanism of absorption, the field and current distributions for the proposed absorber design have been studied. The electric and magnetic field distributions for all the three absorption frequencies are analyzed as shown in Fig. 5a–c. Electric field is represented on the left side, and the magnetic field is shown on the right side of Fig. 5. It is observed that the first

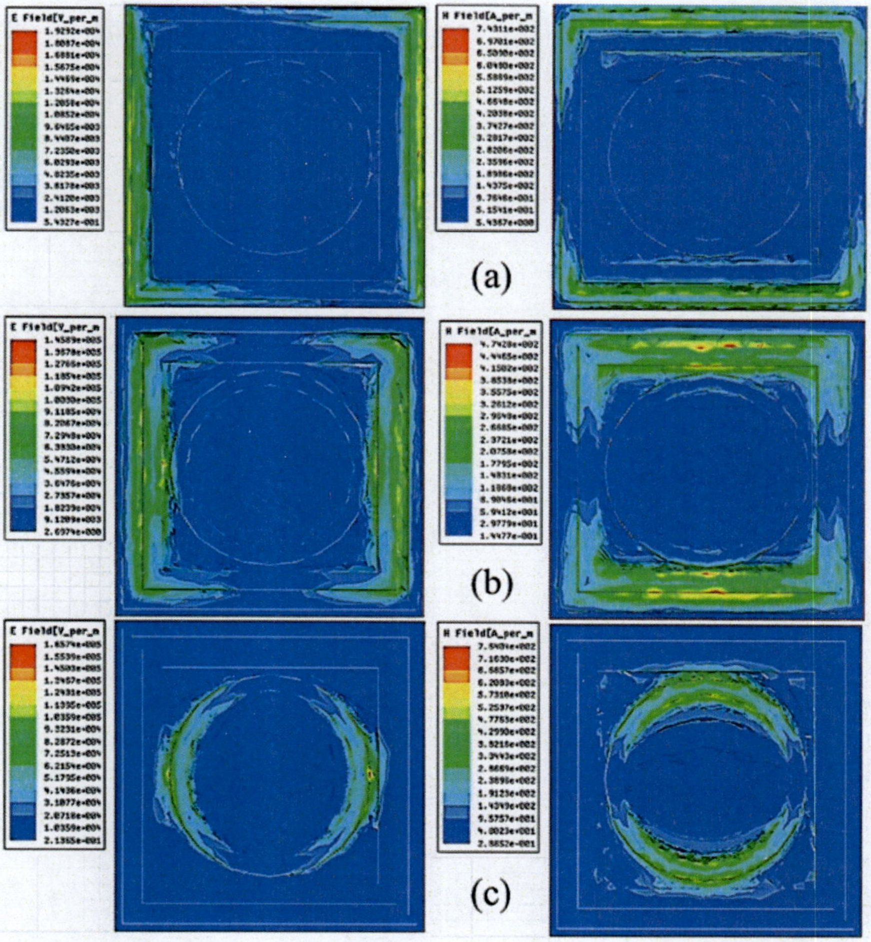

Fig. 5 Electric and magnetic field distributions for the triple-band absorber at three absorption frequencies

absorption peak is mainly contributed by the outermost square ring resonator as the electric and magnetic fields are primarily distributed around it at frequency f_1, i.e. 3.92 GHz as shown Fig. 5a. On the other hand as shown in Fig. 5b, the fields are distributed around the middle square ring resonator at f_2, i.e. 5.92 GHz which indicates that the second absorption peak is contributed by the middle resonator. Similarly, the innermost circular loop contributes the third absorption peak at f_3, i.e., 9.2 GHz, as represented by the conspicuous field distribution shown in Fig. 5c. The electric field mainly excites the top metallic resonators. Further, the current distribution for the three absorption frequency has also been examined as shown in Fig. 6a–c. In similar to the field distribution, it is observed that the first, second, and third absorption peaks are contributed by the outermost square ring, middle ring,

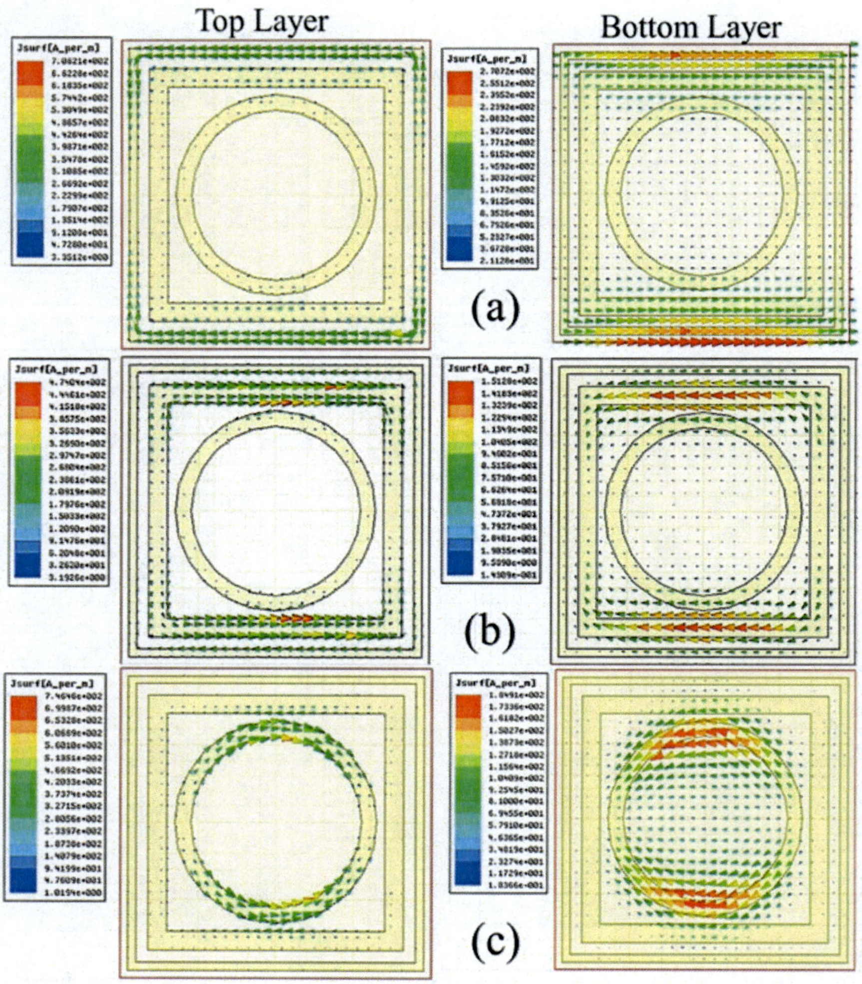

Fig. 6 Surface current distributions at the top and bottom layers for the proposed triple-band absorber at all three absorption frequencies

and innermost circular ring surface, respectively, as the surface current density is maximum around the outermost ring for f_1 and around the middle square ring for f_2 and maximum around the circular ring for f_3. The currents are antiparallel at the top and bottom surfaces which indicate the circulating flow of current around the absorber, thereby signifying that the dielectric substrate is mainly excited by the magnetic field.

Further, in order to verify the absorption contribution of each of the resonators, absorption due to individual resonators is studied as shown in Fig. 7. It is noticed that each of the metallic resonator exhibits its own absorption peak without the affecting

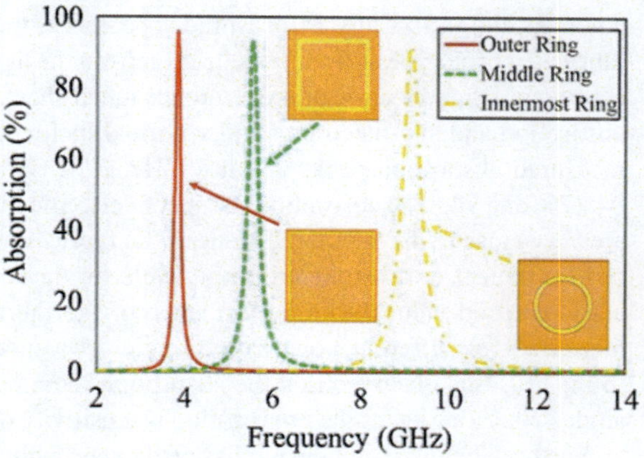

Fig. 7 Absorption contribution of individual metallic resonator rings

absorption due the others. Therefore, the absorption due each of the resonator can be controlled by adjusting the dimensional parameters of individual resonator.

3 Experimental Measurements

A prototype array of the proposed triple-band metamaterial absorber has been fabricated on a 0.8-mm-thick FR4 dielectric substrate as shown in Fig. 8. The performance of the fabricated sample is experimentally measured using Agilents vector network analyzer (VNA) model no. N5222A. Two UWB horn antennas are connected to the VNA for the measurement of reflections. For calibrating the measurement setup, a metal plate of the same size is placed at far-field distance in front of the horn

Fig. 8 Fabricated prototype of the proposed triple-band absorber

antennas and reflectivity is measured. Thereafter, the metal plate is replaced by the fabricated prototype and actual reflection from the designed structure is measured by subtracting the reflection of the reference metal sheet. A comparison of the measured absorption and simulated one under normal incidence is represented in Fig. 9. The measured absorption peaks are at 4 GHz, 5.76 GHz, and 8.90 GHz with 94.19%, 94.37% and 96.83% absorptions, respectively. The measured absorption frequencies are very close to the simulated values. The fabrication errors and nonlinear behavior of FR4 dielectric substrate are responsible for these minor differences in simulated and measured values. The performance of the fabricated prototype has also been measured for different polarization angles (ϕ) under normal incidence as shown in Fig. 10. It is observed that the absorption remains unaltered for all polarizations angles which indicates the polarization insensitivity of the proposed absorber design.

Further, the proposed design has also experimentally tested for different oblique angles under TE polarization as shown in Fig. 11. The measured and simulated

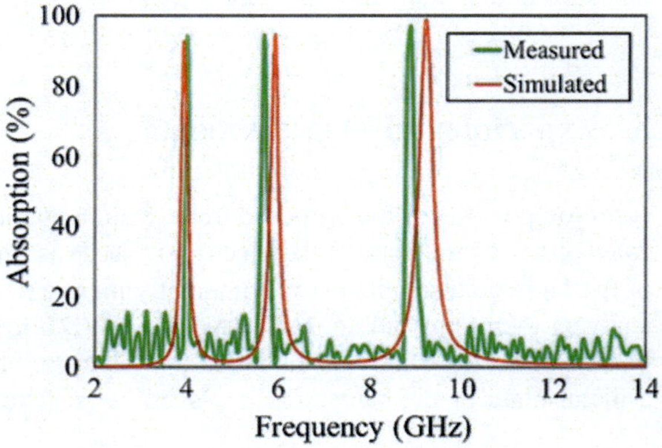

Fig. 9 Comparison of simulated and measured absorption under normal incidence

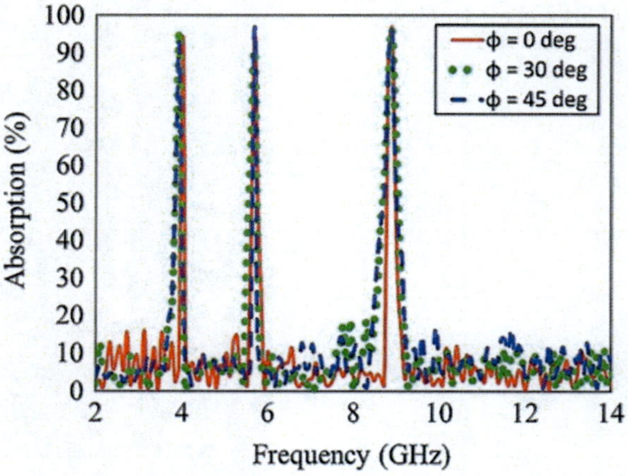

Fig. 10 Measured absorption for different polarization angles under normal incidence

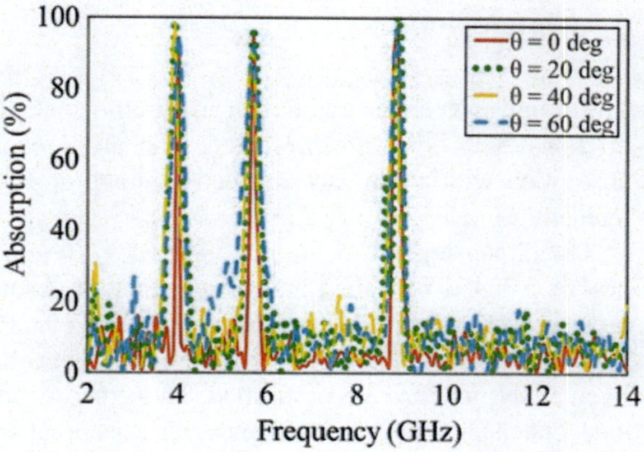

Fig. 11 Measured absorption for different oblique incidence angles under TE polarization

results are observed in agreement, and it is noticed that the design exhibit good angular stability up to 60° (Table 1).

Table 1 Comparison of the proposed absorber design with the existing one

References	Unit cell size mm^2	Thickness mm	Absorption frequency (GHz)	Absorption (%)
[8]	14 × 14 **(0.465 λ)**	0.502 (0.017 λ)	4.19, 6.64, 9.95	97.5, 96.5, 98.85
[9]	10 × 10 (0.307 λ)	0.798 (0.025 λ)	4.06, 6.73 9.22	99, 93, 95
[10]	24 × 24 **(0.58 λ)**	1 (0.024 λ)	4.3, 6.05, 7.30	96.84, 99.85, 96.99
[11]	14 × 14 **(0.43 λ)**	1 (0.031 λ)	2.90,4.18, 9.25	97, 96.45 98.20
[12]	10 × 10 **(0.45 λ)**	0.6 (0.027 λ)	5.57, 7.969,13.44	98.87, 97.99 99.28
[13]	13.8 × 13.8 **(0.64 λ)**	1 (0.046 λ)	4.4,6.05, 13.9	97
[14]	8.2 × 8.2 **(0.42 λ)**	1 (0.514 λ)	4.4,6.48, 15.44	96.6, 97.64, 85.81
[15]	6 × 12 **(0.44 λ)**	0.705 (0.026 λ)	7.7, 9.2,11.1	99.7, 98, 94
[16]	11 × 11 **(0.405 λ)**	1 (0.037 λ)	4.25,8.35,11.06	99.04, 99.62, 99.33
[17]	33.5 × 33.5 (0.29 λ)	6 (0.052 λ)	1.75, 2.17, 2.6	96.91, 96.41, 90.12
Proposed design	10 × 10 (0.3068 λ)	0.8 (0.0245 λ)	3.95, 5.92 9.21	92, 94.5 98.71

4 Conclusion

The design and characterization of an ultrathin metamaterial polarization-insensitive microwave absorber are presented. The proposed absorber provides triple-band absorption with a compact unit cell size and angular stability up to 60° angle of incident wave.

The dimensions of the unit cell are 10×10 mm^2, i.e., 0.3068λ, and the thickness is 0.0245λ w.r.t the highest frequency of absorption. The proposed structure provides absorption value of 92% at 3.95 GHz, 94.5% at 5.92 GHz, and 98.71% at 9.2 GHz. A prototype of the proposed design has been fabricated, and the experimental verifications are performed. The experimental verifications of polarization independence for the proposed microwave absorber are also accomplished by setting different orientation angles of the fabricated design. The proposed design is angular insensitive for oblique incident angles up to 60°. The experimental results are in good agreement with the simulated results. The proposed design is ultrathin and compact in comparison with the existing narrowband and multiband absorber designs.

References

1. Smith DR, Padilla WJ, Vier DC, Nemat-Nasser SC, Schultz S (2000) Composite medium with simultaneously negative permeability and permittivity. Phys Rev Lett 84:4184–4187
2. Lee H, Yi X, Fang N, Srituravanich W, Durant S, Ambati M, Sun C, Zhang X (2005) Realization of optical superlensimaging below the diffraction limit. New J Phys 7:1–16
3. Schurig D, Mock JJ, Justice BJ, Cummer SA, Pendry JB, Starr AF, Smith DR (2006) Metamaterial electromagnetic cloak at microwave frequencies. Science 314:977–980
4. Liu W, Chen ZN, Qing X (2014) Metamaterial-based low-profile broadbandmushroom antenna. IEEE Trans Antennas Propag 62:1165–1172
5. Landy NI, Sajuyigbe S, Mock JJ, Smith DR, Padilla WJ (2008) A perfect metamaterial absorber. Phys Rev Lett 100:207402
6. Ferrer JC (2019) Metamaterials and metasurfaces. Intechopen
7. Lu T, Zhang D, Qiu P, Lian J (2018) An ultra-thin dual band metamaterial absorber based on an asymmetric H shaped structure for terahertz waves. Materials 11:1–10
8. Singh AK, Abegaonkar MP, Koul SK (2019a) Dual and triple band polarization insensitive ultra-thin conformal metamaterial absorbers with wide angular stability. IEEE Trans Electromagn Compat 61:878–886
9. Shen X, Cui TJ, Zhao J (2011) Polarisation independent wide angle triple band metamaterial absorber. Opt Express 19:9401–9407
10. Chaurasiya D, Bhattacharyya S, Ghosh S, Munaga P, Srivastava KV (2015) An ultra-thin triple band polarization-insensitive metamaterial absorber for c-band applications. In: IEEE proceedings on 21st national conference on communications (NCC), Mumbai, India
11. Singh AK, Abegaonkar MP, Koul SK (2019b) A triple band polarization insensitive ultrathin metamaterial absorber for S-C-and X-bands". Prog Electromagn Res M 77:187–194
12. Dhillon AS, Mittal D, Bargota R (2019) Triple band ultrathin polarization insensitive metamaterial absorber for defense, explosive detection and airborne radar applications. Microwave Opt Technol Lett 61:89–95
13. Zeng X, Gao M, Zhang L, Wan G, Hu B (2018) Design of a triple-band metamaterial absorber using equivalent circuit model and interference theory. Microwave Opt Technol Lett 60:1676–1681

14. Sood D, Tripathi CC (2018) A Polarisation insensitive ultrathin compact triple band metamaterial absorber. Indian J Pure Appl Phys 56:149–157
15. Ramya S, Rao IS (2016) Design of new metamaterial absorber with triple band for radar cross section reduction. In: IEEE proceedings on FIFTH international conference on advances in computing and communications (ICACC), Kochi, India 303–306
16. Mishra N, Chaudhary RK (2019) Design and development of an ultrathin triple band microwave absorber using miniaturized metamaterial structure for near unity absorption characteristics. Prog Electromagn Res 8(94):89–101
17. Kaur KP, Upadhyaya T, Palandoken M, Gocen C (2019) Ultrathin dual-layer triple band flexible microwave metamaterial absorber for energy harvesting applications. Int J RF Microwave Comp Aided Eng 29:1–7

Optical Wireless Channel Characterization Based on OOK Modulation for Indoor Optical Wireless Communication System Using WT-ANN

Ankita Aggarwal and Gurmeet Kaur

Abstract In this paper, various characterizations of line-of-sight (LOS) optical wireless communication system (OWCS) based on WT-ANN are explored. To enhance the performance and to reduce the fluorescent light interference (FLI) in OWCS, the proposed system is designed and simulated using stationary wavelet transform (SWT)-based on–off keying (OOK) modulation, which has been proposed using discrete wavelet transform (DWT) in the previous study. The proposed structure enriches the performance of the OWCS, bearing in mind various parameters like reduction of noise, efficiency, data rate and overall improved performance of the scheme.

Keywords DWT · SWT · OOK · OWC

1 Introduction

In nineteenth century, the first wireless communication system was made accessible. It is a communication system, which uses light beams, propagated through the atmosphere or space to carry information. Line-of-sight (LOS), non-LOS and hybrid are the three links in which the optical wireless system works. Optical wireless system offers abundance bandwidth which means high data rate. Optical wireless communication system gives a powerful solution to resolve the forthcoming capacity crisis of radio-based wireless networks. For indoor optical communication, infrared radiation

A. Aggarwal (✉) · G. Kaur
Department of Electronics and Communication Engineering, Punjabi University, Patiala, India
e-mail: ankitahctm@gmail.com

G. Kaur
e-mail: farishta02@yahoo.co.in

A. Aggarwal
CEC, Landran, Mohali, India

N. Marriwala et al. (eds.), *Mobile Radio Communications and 5G Networks*, Lecture Notes in Networks and Systems 140,
https://doi.org/10.1007/978-981-15-7130-5_48

(IR) is a medium suitable for short range and utilized successfully in the middle of source and target. IR innovation has a huge point of interest over RF.

2 Indoor Optical Wireless Communication

Optical wireless communication (OWC) has been proposed recently to provide high-speed access for phones, PCs, printers and digital cameras in offices, shopping centres, warehouses and airplanes for indoor networks through infrared frequencies [1]. As these frequencies cannot penetrate through walls, tethering the transmission into the room thus made them inherently inoffensive, whilst allowing reuse of same optical carrier in an adjacent room. These links chiefly employ intensity modulation with direct modulation (IM/DD) resulting in refute of multipath fading [2]. However, IR system does have some severe drawback such as ambient background noise caused by intense direct sunlight, incandescent and fluorescent lighting [3]. This background noise cannot be overcome by increasing the signal power due to eye safety consideration at the wavelength of 850–950 nm [4]. Furthermore, because of multiple reflections of transmitted signal, diffuse systems sustain inter-symbol interference [5]. In spite of this, the combination of IM/DD prevents multipath fading, ambient light interference (ALI) and inter-symbol interference (ISI) [6]. A number of methods had been proposed to reduce the effect of noise including high-pass filter, angle diversity, optical filtering and linear orthogonal polarizer. In this study, the on–off keying is employed, which is widely used and most admired digital baseband modulation scheme for OWC, where binary '1' is represented by transmitting a pulse and binary '0' by an empty slot of pulse duration T_b.

In this paper, a method is discussed to reduce the FLI due to artificial light in indoor optical system. Also, to measure the performance of the system, SNR, efficiency and data rate are presented in this paper. In the previous work, DWT-based scheme was discussed to mitigate the effect of FLI. In the present work, a new technique, i.e. stationary wavelet transform (SWT) is discussed which shows improvement over the existing one. As per literature survey, this new method was never been implemented in OWCS, but shows remarkable improvement in the field of image processing [7], signal processing [8], pattern recognition, brain image classification [9] and pathological brain detection [10]. In [11], SWT was implemented for the improvement of error diffusion block truncation coding decoded image over DWT and showed favourable results. Also, Kannan in [12] described the SWT for multi-focused images, to overcome the disadvantage of DWT. After getting the idea and studying the performance of SWT in various fields, it is used to design a new IOWCS. The proposed m**odel is** using on–off keying modulation and comparison programmed with the existing one, i.e. DWT.

3 Wavelet Transform

Wavelet analysis is used to analyse the low-frequency and high-frequency components or the transient behaviour of the signal. For determining the time–frequency resolution of wavelet transform (WT) and short-term Fourier transform (STFT), the window size is considered fixed for time and frequency in STFT but not for WT. Hence, WT is very useful and effective in analysing wide varieties of signals. Different types of wavelet transform methods are marked in this paper for comparison.

3.1 *Discrete Wavelet Transform*

DWT uses the filter banks for the construction of multi-resolution time–frequency plane, having application of low-pass and high-pass filters and down sampling by two. The low-pass parts are further decomposed into bands until satisfactory level of information is obtained. The first levels of approximation y1l and detail coefficients y1h are given in [13, 14] as follows:

$$\mathrm{y}_{1\mathrm{l}} = \sum_{n} y(n)g(2k - n) \tag{1}$$

$$\mathrm{y}_{1\mathrm{h}} = \sum_{n} y(n)h(2k - n) \tag{2}$$

The high-pass and low-pass filters satisfy the condition of the quadrature mirror filter. The approximation coefficients can be further decomposed into different levels with maximum level of $\log_2 L$, with L as signal length. The original signal can be reconstructed as follows:

$$x^{'}(n) = \sum_{k} (y_{kh}(n).g(2k - n)) + (y_{kl}(n).h(2k - n)) \tag{3}$$

In DWT, the signal is convolved and decimated by accomplishing the choice of odd indices. If all the possible DWTs of the signal are performed, then there will be $2j$ decomposition level for j levels.

3.2 *Stationary Wavelet Transform*

It is designed to overcome the lack of translation invariance of the DWT. Translation invariance is achieved by removing the down samples and up samples in the DWT and up sampling the filter by a factor of $2^{(j-1)}$ in the jth level of the algorithm. The

SWT is inherently redundant scheme as the output of each level of SWT contains the same number of samples as the input, so for a decomposition of N levels there is a redundancy of N in the wavelet coefficients [15, 16]. This algorithm is more commonly known as "algorithme a trous" in French which means inserting zeros in the filters. It was introduced by Holschneider [17]. The following block diagram describes the digital implementation of SWT.

In SWT, the high-pass and low-pass filters are applied to the data at each level and those are modified at each level by padding zeroes with them (Fig. 1).

4 System Design

In this section, the designing of the proposed system based on stationary wavelet transform is described and shown in Fig. 2. The proposed system is pronounced as follows: initially, an OOK generator is used as input device, which generates an OOK signal, after which it is mixed with the channel noise. AWGN noise is considered as a standard. At the receiver side, after filtering the received data using receiver filter, denoising based on wavelet transform and artificial neural network is applied. The wavelet transform used for the proposed system is stationary wavelet transform and neural network used is back propagation neural network. Finally, the threshold is applied to the output of ANN to achieve the desired output.

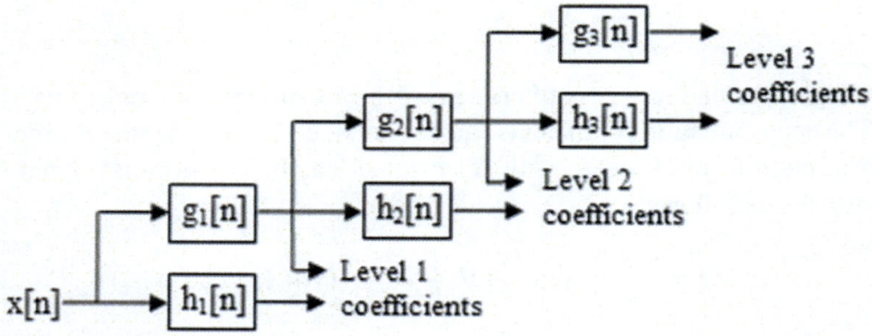

Fig. 1 Three-level SWT filter bank [15]

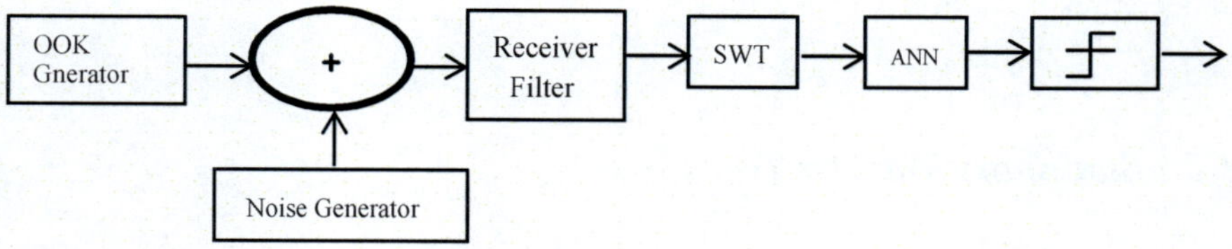

Fig. 2 Block diagram of the proposed system

Table 1 Parameters used for the proposed system

Parameter used	Value
Mother wavelet	Sym2
Window size	3
NN type	Back propagation NN
Hidden layer	1
Neurons	5(HL), 1(OL)
Transfer function	Logsigmoid
Threshold data	If last value > 0.5 then '1' else '0'

The same process as described by the block diagram is also embodied in the form of algorithm. For the proposed system, the algorithm is presented below:

1. Generate OOK Data
2. Add AWGN
3. Match filtering
4. SWT windowed data
5. Train Neural network
6. Repeat step 1–5
7. Classify using neural network
8. Threshold network output
9. Calculate BER
10. Save
11. Check BER for particular value.
    1. If yes, Train and Plot result
    2. If No, decrease AWGN and go to step 6.

5 Simulations

The simulations are done using MATLAB. The several standard parameters used for the same are given in Table 1.

6 Results and Discussions

In this section of the paper, the performance of the SWT based system is specified and verified with the existing technique. Various characterizations used to show the performance of the proposed system are the change in BER versus SNR, efficiency and data rate.

Table 2 Comparison of SNR/BER for DWT and SWT

S. NO.	SNR(dB)	BER	
		DWT	SWT
1.	5.5	0.1717	0.1567
2.	8.5	0.098	0.098
3.	10	0.0825	0.08
4.	16	0.0535	0.052
5.	19	0.0433	0.0396
6.	22	0.0418	0.0379
7.	25	0.0329	0.0319
8.	28	0.0283	0.0275

6.1 SNR Versus BER

The ratio of signal power to noise power in signal to noise ratio is used to compare to the level of desired signal to the level of background noise. And BER is the number of bit error per unit time. Both the values are calculated for the proposed system using OOK modulation with SWT-NN and DWT-NN. The comparison between the two is given in Table 2.

The above table shows that when SNR increases, BER decreases. As given in [6], there is enhancement in SNR versus BER for the proposed system which indicates the increase in performance.

6.2 Efficiency

The performance of a system is better computed by calculating its efficiency. For the IOWCS, the spectral efficiency in bits/Hz is computed for both DWT and SWT techniques for OOK modulation. Figure 3 shows the comparative efficiency or the both methods.

On viewing the graph, it can be concluded that the efficiency of the proposed system is better than the existing one.

6.3 NOPR Versus Data Rate

Comparison between two techniques is made through normalized optical power requirement which is defined as follows:

$$\text{NOPR} = \frac{\text{Optical power required to acheive a desired slot error rate}}{\text{Optical power required to achieve same slot error rate in ideal channel}}$$

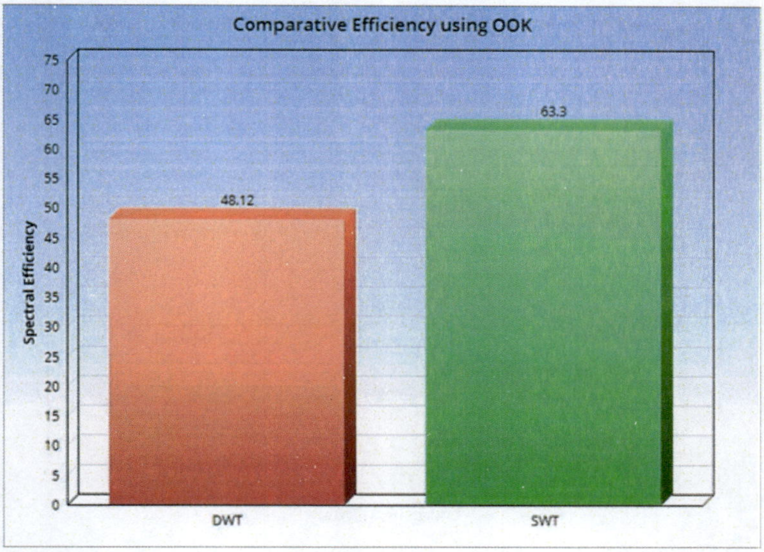

Fig. 3 Efficiency for OOK using DWT and SWT

For OOK scheme, the NOPR is plotted with respect to data rate and is given in Figs. 4 and 5. The different plots given in the figure are as follows:

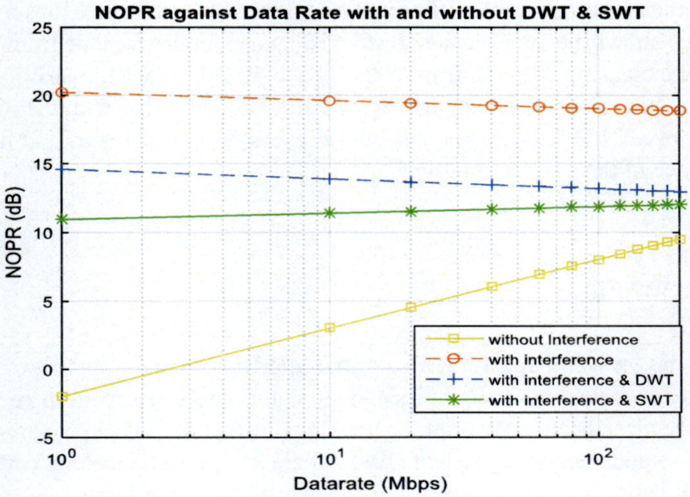

Fig. 4 NOPR against data rate

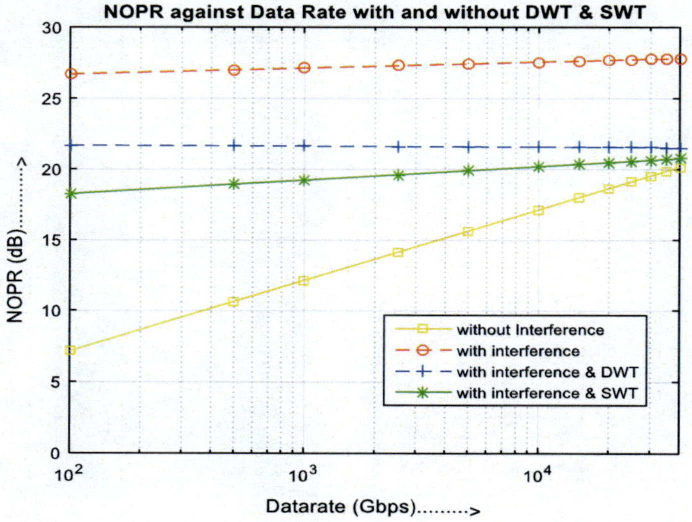

Fig. 5 NOPR against high data rate

(a) NOPR, with noise only
(b) NOPR, with FLI interference
(c) NOPR, with FLI interference and DWT denoising
(d) NOPR, with FLI interference and SWT denoising.

The same curves are plotted after increasing the date rate from Mbps to Gbps.

Figure 4 shows the improvement over existing techniques for date from 10 to 100 Mbps. On increasing date rate from 200 Mbps to 40 Gbps, the comparison is shown in Fig. 5, which also shows the increasing curve. It shows that with the increase in date rate, even NOPR increases, but interferences do not increase, and hence, the performance of the system improved.

7 Conclusion

The stationary wavelet transform has various applications and even shows improvement in many fields like image processing, signal denoising, pattern recognition, brain image classification and pathological brain detection. In the proposed system, the same is implemented in place of DWT, where it contributes a better performance in terms of better SNR/BER, spectral Efficiency and improved data rate.

References

1. Elgala H, Mesleh R, Haas H (2011) Indoor optical wireless communication: potential and state of art. IEEE Commun Mag 49(9):56–62
2. Gfeller FR, Bapst U (1979) Wireless in-home data communication via diffuse infrared radiation. Proc IEEE 67:1474–1485
3. Morairo AJC, Valadas RT, De Oliveira AM, Duarte (1995) Characterization and modeling of artificial light interference in OWC system. Proc PIMRC 1:326–331
4. Boucouvals AC (2000) Optical wireless Communication. IEEE Proc Opto Electronics 147:279–280
5. JR Barry (1994) Wireless infrared communication. Kluwer Academic Publishers, pp 79–107, Boston
6. Rajbhandari S, Ghassemlooy Z, Angelova M (2011) Wavelet Artificial Neural Network Receiver for Indoor Optical Wireless Communication. J Lightwave Technol 29(17):2651–2659
7. Mota HDO, Vasconcelos FH, Castro CLD (2016) A comparison of cycle spinning versus stationary wavelet transform for the extraction of features of a partial discharge signal. IEEE Trans DiElectric Electr Insul 23(2):1107–1118
8. Naga RA, Chandralingam S, Anjaneyulu T, Satyanarayana K (2012) Denoising EOG signal using stationary wavelet transform. Meas Sci Rev 12(2)
9. Zhang Y Feature extraction of brain MRI by stationary wavelet transform and its applications. J Biol Syst 115–132. https://doi.org/10.1142/S0218339010003652.
10. Dong Z Magnetic resonance brain image classification via stationarywavelet transform and generalized eigenvalue proximal support vector machine. J Med Imaging Health Inform 1395–1403. https://doi.org/10.1166/jmihi.2015.1542
11. H. Prasetyo, Hsia CH, Guo JM (2019) Improving EDBTC image quality using stationary and decimated wavelet transform. Int Symp Elctron Smart Devices
12. Kannan K, Perumal SA Optimal level of decomposition of SWT for region level fusion of multifocused images
13. Rajbhandari S (2009) Application of wavelets and artificial neural network for indoor optical wireless communication systems. Ph.D. Thesis, School of Computing, Engineering and Information Sciences, Northumbria University
14. https://en.wikipedia.org/wiki/discrete_wavelet_transform
15. https://en.wikipedia.org/wiki/stationary_wavelet_transform
16. Yang Y, Boling CS, Kamboh AM, Mason AJ (2015) Adaptive threshold Neural spike detector using Stationary wavelet transform. IEEE Trans Neural Syst Rehabil Eng 23(6):946–955
17. Holschneider M, Kronland-Martinet R, Morlet J, Tchamitchian P (1989) A real time algorithm for signal analysis with the help of wavelet transform. In: Wavelets, time-frequency methods and phase space, Springer-Verlag, pp 289–297

Indoor Optical Wireless Communication System Using Wavelet Transform and Neural Network Based on Pulse Position Modulation Intended for Optical Wireless Channel Characterization

Ankita Aggarwal and Gurmeet Kaur

Abstract In this paper, various characterizations of the optical wireless communication system (OWCS) based on WT-ANN for PPM modulation are explored. Stationary wavelet transform (SWT) is designed and simulated in the proposed system, to enhance the performance and to reduce the fluorescent light interference (FLI), based on pulse position modulation (PPM), in OWCS, which has been proposed using discrete wavelet transform (DWT) in the previous study. Reduction of noise, improvement in efficiency and overall improved performance are the various parameters taken into consideration to enrich the performance of the proposed structure of the OWCS.

Keywords DWT · SWT · PPM · OWC

1 Introduction

The first wireless communication system was made accessible, in the nineteenth century. The optical wireless system basically works on three links, i.e. line-of-sight (LOS), non-LOS and hybrid, where light beams are used to carry information, which propagates through atmosphere or space. High data rate is offered by wireless system due to the availability of abundance bandwidth. The impending ability disaster for radio-based networks is resolved by optical wireless communication systems. IR advancement has an enormous focal point over RF due to its suitability for short range and used effectively in the centre of source and target.

A. Aggarwal (✉)
Department of Electronics and Communication Engineering, Punjabi University, Patiala, India
e-mail: ankitahctm@gmail.com

CEC, Landran, Mohali, India

G. Kaur
Department of Electronics and Communication Engineering, Punjabi University, Patiala, India

N. Marriwala et al. (eds.), *Mobile Radio Communications and 5G Networks*, Lecture Notes in Networks and Systems 140,
https://doi.org/10.1007/978-981-15-7130-5_49

2 Indoor Optical Wireless Communication

To provide high-speed access for phones, PCs, printers and digital cameras in offices, shopping centres, warehouses and airplanes, optical wireless communication (OWC) has been proposed, for indoor networks through infrared frequencies [1]. As these frequencies cannot penetrate via walls, tethering the transmission into the room, as a consequence, made them inherently inoffensive, even as allowing reuse of identical optical carrier in an adjoining room. These links chiefly employ intensity modulation with direct modulation (IM/DD) resulting in refute of multipath fading [2]. However, IR system does have few severe disadvantage along with ambient background noise caused by excessive direct sunlight, incandescent and fluorescent lighting [3]. This background noise cannot be overcome through growing signal power owing to eye safety consideration at the wavelength of 850–950 nm [4]. Furthermore, because of multiple reflections of the transmitted signal, diffuse systems sustain inter-symbol interference [5]. In spite of this, the combination of IM/DD prevents multipath fading, ambient light interference (ALI) and inter-symbol interference (ISI) [6]. High-pass filter, angle diversity, optical filtering and linear orthogonal polarizer are few methods, which had been proposed to reduce the effect of noise. In this study, the widely used and most admired digital baseband modulation scheme, i.e. the pulse position modulation, an orthogonal baseband modulation technique is employed for OWC. For hand-held devices, PPM is preferred because of the requirement of low-power consumption. A symbol of a fixed length slots, with only one nonzero pulse of constant power and every block of $\log_2$ M information bits, mapped onto one of M conceivable symbols is occupied for PPM. Here, through the location of pulse within the symbol, information is carried. $T_s = T_b M/L$ is the slot duration for PPM, for achieving the same throughput as on–off keying (OOK), where T_b is pulse duration and M is bit resolution with $2^M = L$, and L as symbol length [6].

To reduce the FLI due to artificial light in indoor optical system, a method is discussed in this paper. Also, to measure the performance of the system SNR, efficiency and data rate are presented in this paper. In the previous work, DWT-based scheme was discussed to mitigate the effect of FLI. In the present work, a new technique, i.e. stationary wavelet transform (SWT) is discussed which shows improvement over the existing one. As per literature survey, this new method was never been implemented in OWCS, but shows remarkable improvement in the field of image processing [7], signal processing [8], pattern recognition, brain image classification [9] and pathological brain detection [10]. In [11], SWT was implemented for the improvement of error diffusion block truncation coding decoded image over DWT and showed favourable results. Also, K. Kannan in [12] described the SWT for multi-focused images, to overcome the disadvantage of DWT. After getting the idea and studying the performance of SWT in various fields, it is used to design a new IOWCS. The proposed model is designed using pulse position modulation and comparison in encoded with the existing one, i.e. DWT.

3 Discrete Wavelet Transform

Wavelet analysis is used to analyse the low-frequency and high-frequency components, or the transient behaviour of the signal. For determining the time–frequency resolution of wavelet transform (WT) and short-term Fourier transform (STFT), the window size is considered fixed for time and frequency in STFT but not for WT. Hence, WT is very useful and effective in analysing wide varieties of signals. Different types of wavelet transform methods are marked in this paper for comparison. DWT uses the filter banks for the construction of multi-resolution time–frequency plane, having application of low-pass and high-pass filters and down sampling by two. The low-pass parts are further decomposed into bands until satisfactory level of information is obtained. The first levels of approximation y1l and detail coefficients y1h are given in [13, 14] as follows:

$$\mathrm{y}_{1\mathrm{l}} = \sum_{n} y(n)g(2k-n) \tag{1}$$

$$\mathrm{y}_{1\mathrm{h}} = \sum_{n} y(n)h(2k-n) \tag{2}$$

The high-pass and low-pass filters satisfy the condition of the quadrature mirror filter. The approximation coefficients can be further decomposed into different levels with the maximum level of $\log_2 L$, with L as signal length. The original signal can be reconstructed as follows:

$$x'(n) = \sum_{k} (y_{kh}(n).g(2k-n)) + (y_{kl}(n).h(2k-n)) \tag{3}$$

In DWT, the signal is convolved and decimated by accomplishing the choice of odd indices. If all the possible DWTs of the signal are performed, then there will be $2j$ decomposition level for j levels.

4 Stationary Wavelet Transform

It is designed to overcome the lack of translation invariance of the DWT. Translation invariance is achieved by removing the down samples and up samples in the DWT and up sampling the filter by a factor of $2(j-1)$ in the jth level of the algorithm. The SWT is inherently redundant scheme as the output of each level of SWT contains the same number of samples as the input, so for a decomposition of N levels there is a redundancy of N in the wavelet coefficients [15, 16]. This algorithm is more commonly known as "algorithme a trous" in French which means inserting zeros in the filters. It was introduced by Holschneider [17]. The following block diagram describes the digital implementation of SWT.

In SWT, the high-pass and low-pass filters are applied to the data at each level, and those are modified at each level by padding zeroes with them (Fig. 1).

5 System Design

In this section, the designing of the proposed system based on the stationary wavelet transform is described and shown in Fig. 2. The proposed system is manifested as follows: initially, a PPM generator is used as input device, which generates a PPM signal, after which it is mixed with the channel noise, i.e. AWGN noise, which is considered as a standard. At the receiver side, after filtering the received data using receiver filter, denoising based on wavelet transform and artificial neural network is applied. The wavelet transform used for the proposed system is stationary wavelet transform and neural network used is back propagation neural network. Finally, the threshold is applied to the output of ANN to achieve the desired output.

The same process as described by the block diagram is also embodied in the form of algorithm. For the proposed system, the algorithm is presented below:

1. Generate PPM Data
2. Add AWGN
3. Match filtering
4. SWT windowed data
5. Train Neural network
6. Repeat step 1–5

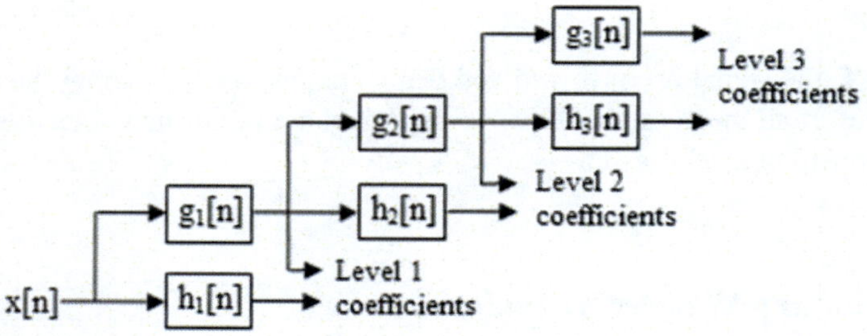

Fig. 1 Three-level SWT filter bank [15]

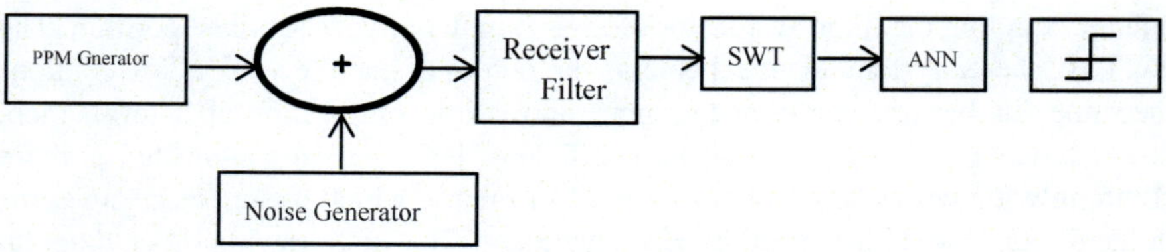

Fig. 2 Block diagram of the proposed system

7. Classify using neural network
8. Threshold network output
9. Calculate BER
10. Save
11. Check BER for particular value.
    1. If yes, Train and Plot result
    2. If No, decrease AWGN and go to step 6.

The flow chart of the proposed system is given in Fig. 3. Through the flow chart, the step by step processing of the proposed system is being described.

6 Simulations

The simulations are done using MATLAB. The several standard parameters used for the same are given in Table 1.

7 Results and Discussions

In this section of the paper, the performance of the SWT-based system is specified and verified with the existing technique. Various characterizations used to show the performance of the proposed system are the change in BER versus SNR, and efficiency.

7.1 SNR Versus BER

The ratio of signal power to noise power in signal to noise ratio is used to compare to the level of desired signal to the level of background noise.

BER is the number of bit error per unit time. Both the values are calculated for the proposed system using PPM with SWT–NN and DWT-NN. The comparison between the two is given in Table 2. The table shows that when SNR increases, BER decreases. As given in [6], there is enhancement in SNR vs BER for the proposed system which indicates the increase in performance.

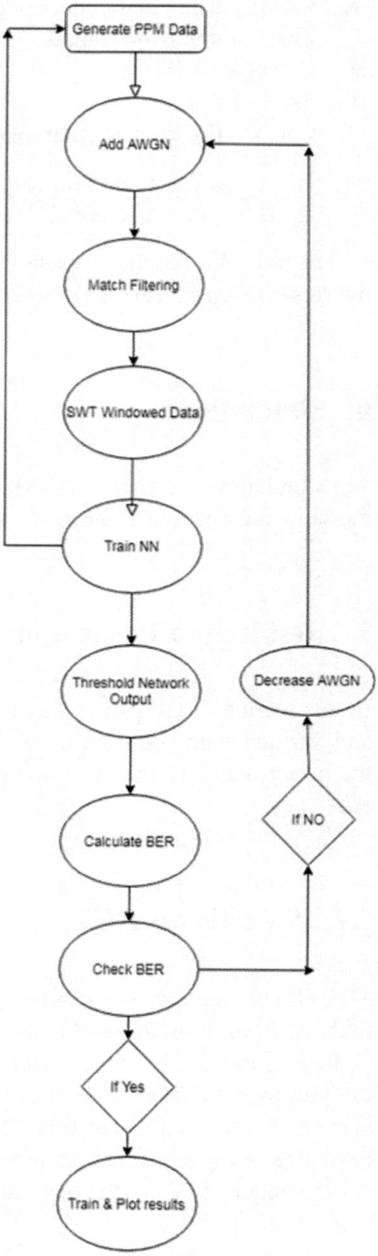

Fig. 3 Flow chart of the proposed system

Table 1 Parameters used for the proposed system

Parameter used	Value
Mother wavelet	Sym2
Window size	3
NN type	Back propagation NN
Hidden layer	1
Neurons	5(HL), 1(OL)
Transfer function	Log-sigmoid
Threshold data	Log-sigmoid

Table 2 Comparison of SNR/BER for DWT and SWT

SNR(dB)	BER					
	DWT			SWT		
	4	8	16	4	8	16
5.5	0.1767	0.1733	0.1683	0.1725	0.1683	0.1675
8.5	0.0975	0.096	0.096	0.096	0.096	0.0955
10	0.085	0.0783	0.0779	0.0838	0.0825	0.0817
16	0.0539	0.0532	0.0517	0.052	0.0518	0.0512
19	0.0466	0.0448	0.0433	0.0464	0.0454	0.0427
25	0.0328	0.0314	0.0309	0.0322	0.0303	0.0278
28	0.0291	0.029	0.0286	0.027	0.0268	0.0256

7.2 *Efficiency*

The performance of a system is better computed by calculating its efficiency. For the IOWCS, the spectral efficiency in bits/hz is computed for both DWT and SWT techniques for PPM.

Figure 4 shows the efficiency for PPM using DWT, for symbol length 4, 8 and 16. In Fig. 4, it is shown that the efficiency is more for PPM-16, i.e. 75.88%, which comes out 55.61% for PPM-8 and for 4-PPM its only 42.46%.

Figure 5 shows the efficiency for PPM using SWT, for symbol length 4, 8 and 16. It is shown that the efficiency for PPM-16 is 78.94%, which is 60.92% for PPM-8 and for 4-PPM is only 46.36%.

Figure 6 shows the efficiency for PPM using DWT and SWT. Figure shows that the efficiency is more for PPM-16 as compared to PPM 4 and 8. On viewing the graph, it can be concluded that the efficiency of the proposed system is better than the existing one.

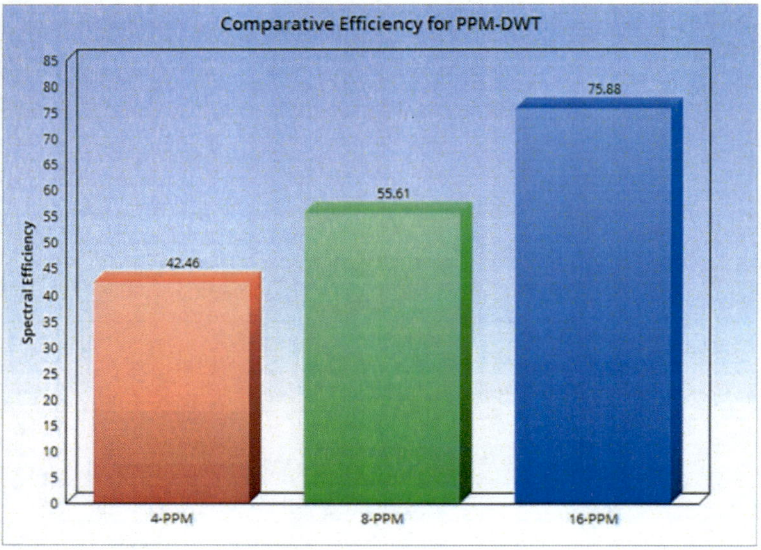

Fig. 4 Efficiency for OOK using DWT and SWT

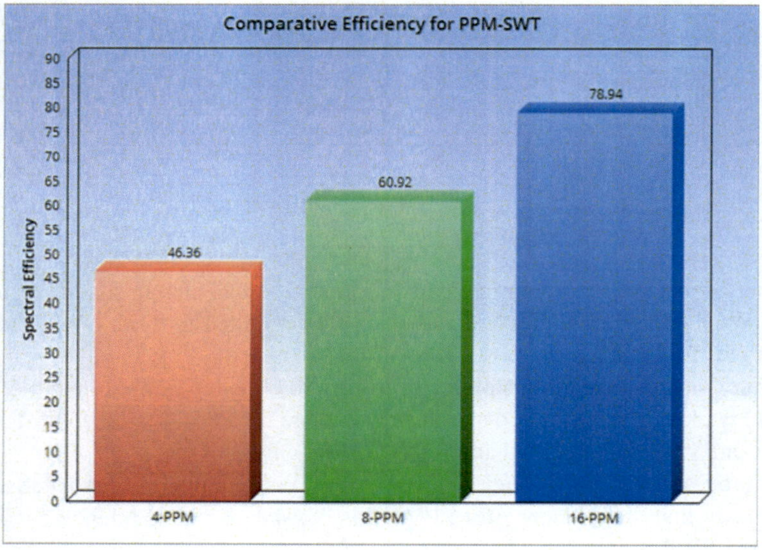

Fig. 5 Efficiency for PPM using SWT

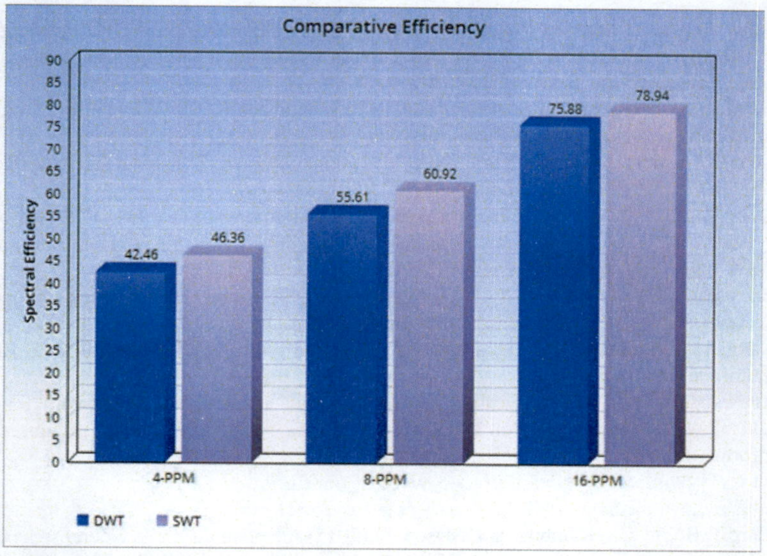

Fig. 6 Efficiency for PPM using DWT and SWT

8 Conclusion

The stationary wavelet transform has various applications and even shows improvement in many fields like image processing, signal denoising, pattern recognition, brain image classification and pathological brain detection. In the proposed system, the same is implemented in place of DWT and where it contributes a better performance in terms of better SNR/BER, spectral efficiency and improved data rate.

References

1. Elgala H, Mesleh R, Haas H (2011) Indoor optical wireless communication: potential and state of art. IEEE Commun Mag 49(9):56–62
2. Gfeller FR, Bapst U (1979) Wireless in-home data communication via diffuse infrared radiation. Proced IEEE 67:1474–1485
3. Morairo AJC, Valadas RT, De Oliveira, Duarte AM (1995). Characterization and modeling of artificial light interference in OWC System. Proceeding of PIMRC 1, pp 326–331
4. Boucouvals AC (2000) Optical wireless Communication. IEEE Proc Opto Electron 147:279–280
5. Barry JR (1994). Wireless infrared communication. Kluwer Acad Publishers, Boston, pp 79–107
6. Rajbhandari S, Ghassemlooy Z, Angelova M (2011) Wavelet artificial neural network receiver for indoor optical wireless communication. J Lightwave Technol 29(17):2651–2659

7. Mota HDO, Vasconcelos FH, Castro CLD (2016) A comparison of cycle spinning versus stationary wavelet transform for the extraction of features of a partial discharge signal. IEEE Trans DiElectric Electr Insul 23(2):1107–1118
8. Naga RA, Chandralingam S, Anjaneyulu T, Satyanarayana K (2012) Denoising EOG signal using stationary wavelet transform. Meas Sci Rev 12(2)
9. Zhang Y Feature extraction of brain MRI by stationary wavelet transform and its applications. J Biol Syst, 115–132. https://doi.org/10.1142/S0218339010003652
10. Dong Z Magnetic resonance brain image classification via stationarywavelet transform and generalized eigenvalue proximal support vector machine. J Med Imag Health Inf, pp 1395–1403. https://doi.org/10.1166/jmihi.2015.1542
11. Prasetyo H, Hsia CH, Guo JM (2019) Improving EDBTC image quality using stationary and decimated wavelet transform. Int Symp Elctron Smart Devices
12. Kannan K, Perumal SA Optimal level of decomposition of SWT for region level fusion of multifocused images
13. Rajbhandari S (2009) Application of wavelets and artificial neural network for indoor optical wireless communication systems. Ph.D. Thesis, school of computing, engineering and information sciences, northumbria university
14. https://en.wikipedia.org/wiki/discrete_wavelet_transform
15. https://en.wikipedia.org/wiki/stationary_wavelet_transform
16. Yang Y, Boling CS, Kamboh AM, Mason AJ (2015) Adaptive threshold Neural spike detector using stationary wavelet transform. IEEE Trans Neural Syst Rehabil Eng 23(6):946–955
17. Holschneider M, Kronland-Martinet R, Morlet J, Tchamitchian P (1989) A real time algorithm for signal analysis with the help of wavelet transform. In: Wavelets, time-frequency methods and phase space, 289–297. Springer-Verlag

GDH Key Exchange Protocol for Group Security Among Hypercube Deployed IoT Devices

Vimal Gaur and Rajneesh Kumar

Abstract Effective deployment of IoT devices to preserve security, with minimum computations and energy consumption will be great challenge in the current world scenarios. This research work discusses the problem of deploying IoT devices and key exchange among IoT devices. These devices are prone to physical attacks, reason is their unattended deployment. Security of IoT devices is difficult to achieve, because of the heterogeneous array of servers and other devices. In this paper, labels are assigned to each device in virtual hypercube overlay to facilitate communication. Since hypercube deployment has the least computational complexity among other private key management solutions. Finally, communication between IoT devices deployed in hypercube arrangement is done using group Diffie–Hellman key agreement protocol. The proposed protocol is more efficient in terms of assigning a new labeling scheme to IoT devices (placed at hypercube nodes) and then uses these labels to generate shared secret key. This labeling scheme does in fact allow for effective key exchanging among IoT devices.

Keywords Group diffie–hellman key exchange · Internet of things

1 Introduction

Over the past decades, IoT have been greatly developed and applied owing to its ultra-large-scale network, eco-friendliness and heterogeneous nature. A promising hybrid platform for IoT is desired. IoT when deployed in centralized framework leads to high maintenance costs, low interoperability, single point of failure (SPOF)

V. Gaur (✉)
Maharaja Surajmal Institute of Technology, New Delhi 110058, India
e-mail: vimalgaur@msit.in

R. Kumar
Maharishi Markandeshwar Deemed To Be University, Mullana, Ambala, India
e-mail: drrajneeshgujral@mmumullana.org

N. Marriwala et al. (eds.), *Mobile Radio Communications and 5G Networks*, Lecture Notes in Networks and Systems 140,
https://doi.org/10.1007/978-981-15-7130-5_50

against security and results in scalability issues when deployed in distributed platform. Distributed networks are implemented through blockchain, since blockchain can process many heterogeneous devices.

IoT devices make smarter decisions by improving efficiency in automating (real/day to day) activities like controlling washing machines, security cameras, refrigerators, digital set-top boxes, microwave oven and smart TVs remotely through IoT. Apart from personal, IoT plays a major role in serving community like health monitoring, weather monitoring, tracking automobiles and providing connections, public lightning, factories and buildings [9]. All these devices can be sensed easily and controlled remotely over the Internet. Almost all IoT devices are low power devices [9] and dependent on batteries. Since IoT devices require no human intervention to transfer information to other devices so they should be attractive, lightweight and sleek [9]. Further, for uninterrupted transfer of data over network, sources of power having prolonged lifespan independent of electrical outlets are demanding.

In order to significantly reduce the application processing time, concurrent execution of tasks is to be done. This concurrent execution emphasizes the need of interconnection networks. These networks are restricted as static networks and dynamic networks. Static networks are those in which point-to-point communication is done between devices and dynamic networks are those in which switches are used for communication. Static communication of IoT devices is unattended, which facilitates possibility of physical attacks in the system. Static communication involves networks with varying degree. Degree 1 (shared bus), Degree 2 (linear array), Degree 3 (binary tree), Degree 4 (two-dimensional mesh) and varying degree (N-Cube). In the proposed solution, we use virtual hypercube overlay, in which nodes of hypercube represent IoT devices. Also, IoT devices are typically deployed in an unattended manner. In view of securing group communication, a secure group Diffie–Hellman key exchange algorithm on IoT devices deployed in hypercube organization is presented in the paper.

2 Literature Survey

Pointcheval et al. [1] addressed the problems faced by group Diffie–Hellman and explained the assumptions under which dynamic GDH will work and proposed a generic model to be applied to many cryptographic scenarios. Further, a notion of group DH distribution and its use in computational and decisional group is presented.

Kapoor and Batra [2] signify the use of ECDHE-RSA-RC4-SHA which is elliptic curve Diffie–Hellman ephemeral, signed by an RSA key. The elliptic curve (P-256) is classified to be as strong as a 3248-bit RSA key. In this approach, communicating parties fix a finite field Fq and a base point B before starting communication. Also, both parties should agree upon EC domain parameters. This technique reduces memory requirements.

Djenouri et al. [3] presented network routing protocol attacks and their solutions. Also, MAC protocol security issues and their solutions are presented. Further, it has been analyzed that solutions proposed to network layer and MAC layer rely on cryptography. Providing key infrastructure to MANETs is difficult because of lack of any central infrastructure.

Vasundhara [4] explains how iterations are included in Diffie–Hellman and any number of participants can take part in an agreement by performing iterations of the Diffie–Hellman key agreement protocol. This technique also performs N modular exponentiations on N participants, by taking the assumption that these participants are arranged in a circle. Further, divide and conquer strategy is used and every participant perform $\log_2 (N) + 1$. After the end of every session, secret integers are discarded so as to make our session secure.

Kolagar et al. [5] explained the combinatorial framework for pre-distribution of keys in hypercube in wireless sensor networks. Different key distribution techniques like random key distribution and deterministic key distribution are discussed.

Ahmad Khan and Salah [6] explained layers of IoT and associated protocols. Further, they explained security issues at different levels. After analyzing security issues at different layers, the implications, layer affected by different levels are studied and proposed solutions are discussed. These security issues are resolved using blockchain technology.

Lavanya [7] proposed the use of distributed group key management for securing group communication. Further, she added that distributed approach for key management is robust against delay and network failure. Proposed algorithm further stated how forward and backward secrecy is maintained when a new node enters a group or withdraws a group.

Ashwini and Rojaramani [8] presented network technologies for IoT. Network technologies are compared on the basis of different parameters like compatibility, security, power consumption, range and bandwidth. Further, network technologies are explained on the basis of wired and wireless networks.

Kima et al. [9] addressed IoT-based smart grid network and its challenges and security issues. The whole city will come to halt owing to attacks on IoT.

Bhosavale and Sonavane [10] analyzed RPL-based IoT using Contiki (Cooja simulator) and proposed intrusion detection system for wormhole attack in real-life scenario. He further suggested that many IDS have already been developed using centralized and distributed approach, but Bhosavale focused on hybrid approach of IDS. Attacker and attacking nodes are identified using received signal strength indicator. Further, correlation value is analyzed for increasing number of nodes.

Sadique et al. [11] proposed a generic model with six layers of IoT and their security requirements at each layer are studied. Further, twelve security challenges for IoT paradigms are proposed and a trustworthy IoT infrastructure is required.

Chatzigiannakis et al. [12] focused on simulating GDH.3 protocol for complex message exchanges with increase in number of local computations. Further, he explained necessary properties of cryptography that must be satisfied in order to perform group key agreement. Simulation results of GDH.3 show its utility for smaller network sizes, and performance of GDH.3 is measured for sensor nodes joining a group or leaving a group.

3 Proposed Work to Preserve Group Security Among Hypercube Deployed IoT Devices

Group communication operations like one to all broadcast, all to all broadcast, prefix-sum, scatter and gather, all to all personalized are applied to different processing architectures. Here, in our proposed approach, we are applying all to all broadcast to hypercube architecture. The key points to be noted about all to all broadcast are as follows-

- Broadcast is simultaneously initiated by all p nodes.
- At each step, different dimensions of hypercube show communication.
- Different nodes may broadcast different messages at any time.
- Every node can be a source and a destination but communication is single-ported (means at any particular moment one node can either be source or destination).

$$T = t_s * \log(p) + t_w * m * (p - 1)$$

T = Time taken to communicate a message between two IoT devices in the network is the sum of t_s and t_w.

t_s= Time to prepare a message for transmission.

t_w= Time taken by the message to traverse the network to its destination (Fig. 1, Table 1).

m = Size of message.

p = Number of nodes.

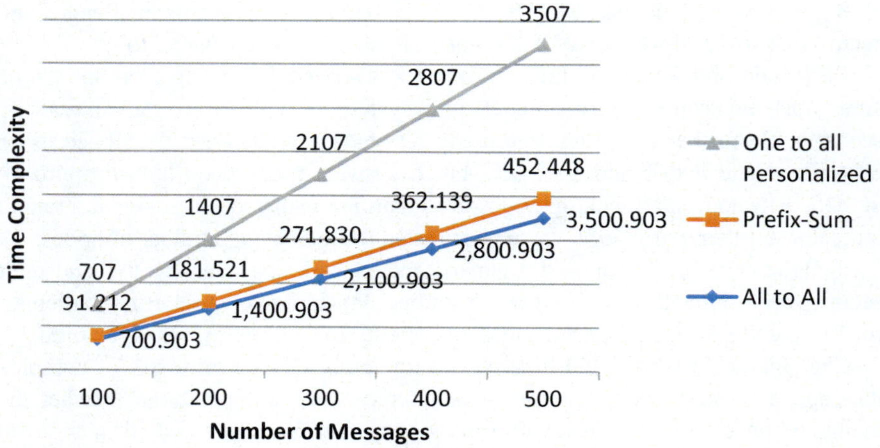

Fig. 1 Clearly depicts the time complexity of group communication operations to be used in hypercube

Table 1 Key management solutions [3]

Solutions	Computational complexity
Hypercube	2 * log(*n*)
Hypercube + Pwd authentication	4 * log(*n*)
GDH	2 * (*n*−1)
n-Party Pwd authentication	6 * (*n*−1)
Contributory n-Party Pwd authentication	4 * (*n*−1)
Octopus	2 * (log(*n*) + 1)

In [3], we have learned that computational complexity while deploying IoT devices is minimum in hypercube structures (Fig. 2).

All to all broadcast communication in hypercube

M_1, M_2, M_3.................. M_8 are IoT devices. The nodes of hypercube are labeled as IoT devices. The nodes of hypercube communicate if and only if difference in bits is 1.

M_1—000, m_2—001, m_3—010, m_4—011, m_5—100, m_6—101, m_7—110, m_8—111

As clear from above, M_1 and M_2 have difference in 1 bit, so they can communicate. In order to apply cryptography, there must be some mechanism for sharing

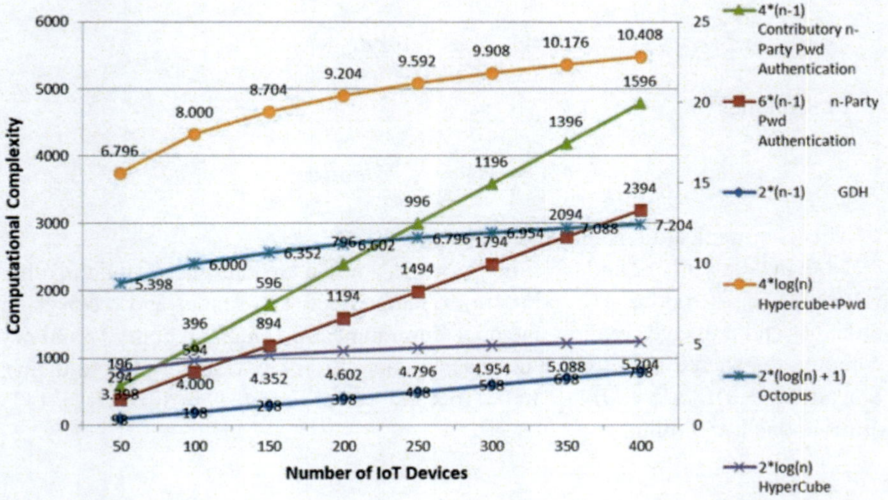

Fig. 2 Clearly depicts the computational complexity of different key management solutions with the increasing number of IoT devices

key. Each symmetric and asymmetric encryption technique has its advantages and disadvantages. So, we choose to apply Diffie–Hellman key agreement protocol to IoT devices for safely developing and exchanging keys over an insecure channel. Since Diffie–Hellman key exchange algorithm is based on multiplicative group of integers modulo q, where q is prime, and α is a primitive root modulo q. These two values are chosen such that resulting shared secret can take on any value from 1 to $q{-}1$.

- **Step 1** Since M_1 and M_2 have to communicate, they mutually agree upon two parameters α and q.

  Secret integer chosen by M_1 is $r1$, M_1 then sends $M_{2,}$

$$\boldsymbol{M_1 = \alpha^{r1} mod q}$$

  Secret integer chosen by M_2 is $r2$, M_2 then sends $M_{1,}$

$$\boldsymbol{M_2 = \alpha^{r2} mod q}$$

  Now, new value of M_1 is as follows

$$\boldsymbol{M_1 = \alpha^{r1r2} mod q}$$

  Now, new value of M_2 is as follows

$$\boldsymbol{M_2 = \alpha^{r2r1} mod q}$$

  This implies

$$\boldsymbol{M_1 = M_2 = \alpha^{r1r2} mod q}$$

  This communication is bidirectional.

Note that only $r1$, $r2$ and α^{r1r2} mod $q = \alpha^{\mathrm{r2r1}}$ mod q are kept secret. All the other values—α, q, α^{r1}mod q, α^{r2}mod q are exchanged between sender and receiver. In order to send messages in the same open communication channel, both M_1 and M_2 compute the shared secret to be used as an encryption key (known only to them). Since ($M_3 = 010$, $M_4 = 011$), ($M_5 = 100$, $M_6 = 101$) and ($M_7 = 110$, $M_8 = 111$) differ in one bit. Similarly.

$$M_3 = M_4 = \alpha^{r3r4} \, mod \, q$$

$$M_5 = M_6 = \alpha^{r5r6} \, mod \, q$$

$$M_7 = M_8 = \alpha^{r7r8} \, mod \, q$$

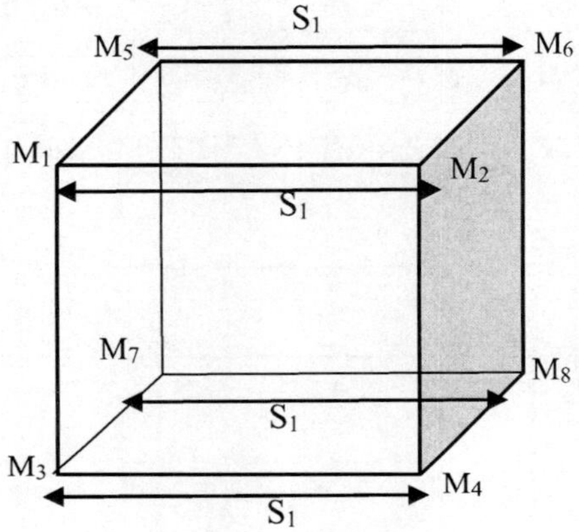

- **Step 2** M_1 communicates with M_3, M_2 with M_4, M_5 with M_7, M_6 with M_8 (as shown in the figure) The communication is bidirectional, so M_1 sends α^{r1r2} mod q to M_3 and M_3 sends α^{r3r4} mod q to M_1. We get the values of M_1 to M_8 from the following equations. Now, the new values will be as follows:

$$M_1 = M_3 = \alpha^{r1r2r3r4} \, mod \, q$$

$$M_2 = M_4 = \alpha^{r1r2r3r4} \, mod \, q$$

$$M_5 = M_7 = \alpha^{r1r2r3r4} \, mod \, q$$

$$M_6 = M_8 = \alpha^{r1r2r3r4} \, mod \, q$$

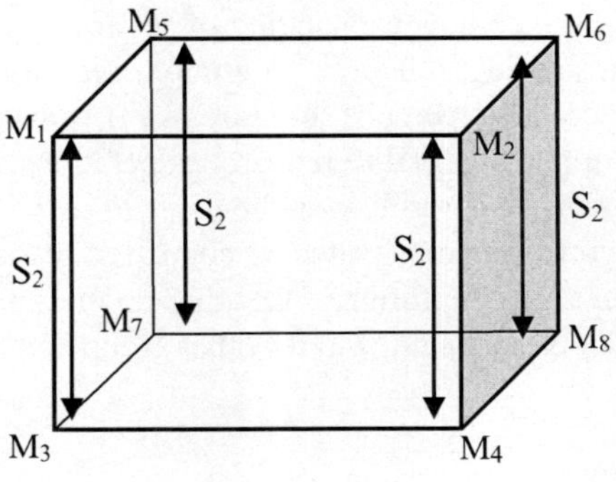

- **Step 3** M_1 communicates with M_5. So, M_1 sends $\alpha^{r1r2r3r4}$ mod q to M_5 and M_5 sends $\alpha^{r5r6r7r8}$ mod q to M_1. We get the values of M_1 to M_8 as follows:

$$M_1 = M_5 = \alpha^{r1r2r3r4r5r6r7r8} \, mod \, q$$

$$M_3 = M_7 = \alpha^{r1r2r3r4r5r6r7r8} \, mod \, q$$

$$M_2 = M_6 = \alpha^{r1r2r3r4r5r6r7r8} \, mod \, q$$

$$M_4 = M_8 = \alpha^{r1r2r3r4r5r6r7r8} \, mod \, q$$

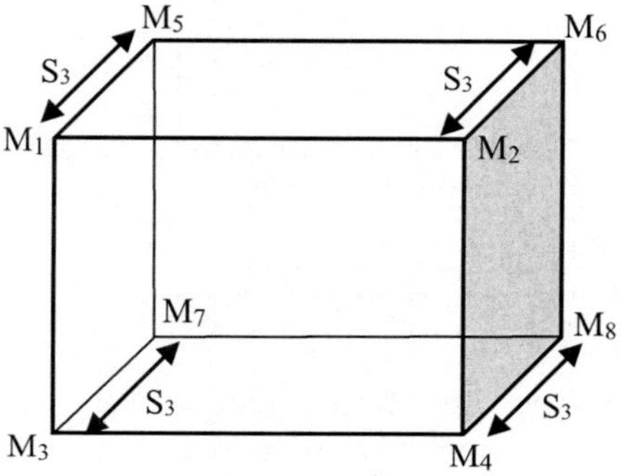

It is clear from above key exchanging mechanism that all IoT devices M_1, M_2, M_3, M_4, M_5, M_6, M_7, M_8 have key value as $\alpha^{r1r2r3r4r5r6r7r8}$ mod q. Hence, in all to all broadcast, all IoT devices have the same shared secret key; thereafter, encryption and decryption are performed using any symmetric or asymmetric key cryptography.

4 Conclusion and Future Scope

We have deployed IoT devices in hypercube structure in the above paper. We planned to further investigate how exchanging of session keys using group Diffie–Hellman is done and also dealt with different group communication operations. Our research work concludes that hypercube structures are best in terms of computational complexity and time complexity for different group operations. Further, a good choice of encryption and decryption algorithm can be made to show results effectively. As seen, hypercube networks perform better as compared to other networks. In the future, communication can be performed for higher-order dimensions of hypercube. Also, hypercube networks can be utilized for other group communication operators.

References

1. Pointcheval E. B. a. D (2007) Provably secure authenticated group diffie hellman key exchange. ACM Trans Inf Syst Secur 10(3):1–45
2. Batra MK (2016 Mar 2016) Improved Diffie-hellman key exchange using elliptic curve scheme for securing wireless sensor networks routing data. IJSRSET 239–246
3. C. C. O. R. Djamel Djenouri, A. N. B, Lyes Khelladi (2005) A survey of security issues in mobile AD HOC and sensor networks. IEEE Commun Sur Electron Mag Original Peer Rev Sur Art, 2–28
4. Vasundhara S (2017) Elliptic curve cryptography and Diffie-hellman key exchange. IOSR J Math, 56–61
5. Qasemzadeh Kolagar, H. S. J, A. M. I (2017) Hypercube bivariate-base key management for wireless sensor networks. J Sci Islamic Republic of Iran, 273–285
6. Minhaj Ahmad Khan KS (2018) IoT security: review, blockchain solutions, and open challenges. Future Gener Comput Syst, 395–411
7. Nallapaneni Manoj Kumar PKM (2018) BlockChain technology for security issues and challenges in IoT. Int Conf Comput Intell Data Sci, pp s1815–1823
8. Ashwini RP (2018) Heterogenous device communication models and data management in IoT. IJSRSET, 420–424
9. Kenneth Kima, V. O, K. L (2019) Cyber security challenges for IoT-based smart grid networks. Int J Crit Infrastruct Prot 25:36–49
10. Kazi Masum Sadique RPJ (2018) Towards security on internet of things: applications challenges in technology. Int Conf Internet Things: Appl Challenges Technol
11. S. D. B. A, Sonavane SS (2019) Real-time intrusion detection system for wormhole attack in the RPL based internet of things. Int Conf Interdisc Eng
12. "DeSensor Networkssign (2007) Analysis and performance evaluation of group key establishment in wireless. Electron Notes Theoret Comput Sci 171:17–31

Depression Detection in Cancer Communities Using Affect Analysis

Vaishali Kalra, Srishti Sharma, and Poonam Chaudhary

Abstract Despite the unceasing advancements in medical science and growing awareness, cancer continues to be a deadly disease that claims almost ten million lives every year. The alarming number of fatalities is caused due to privation of timely cancer detection, tardy medical attention or in some cases from patients losing the will to live due to a protracted and unending treatment procedure. Governments across the world are taking steps to ensure timely cancer detection and treatment. However, little attention is being paid to the seemingly unending treatment course taking a toll on the patient's mental health thus crushing the patient's spirit to continue. Through this work, we propose an approach for timely detection of depression in cancer patients by carrying out affect analysis of cancer communities on Twitter. We use the Plutchik's wheel of emotions in psychology as a tool for depression detection. We carry out the Twitter profiling of ten persons who are part of a cancer support group on Twitter and carry out our experiments on this dataset to demonstrate the suitability of the proposed technique.

Keywords Cancer · Depression · Twitter · Sentiment analysis · Affect analysis · Plutchik's wheel of emotions

1 Introduction

Cancer remains an untreatable and treacherous disease worldwide. It claims almost ten million lives every year. As per the World Health Organization (WHO), one in five

V. Kalra · S. Sharma (✉) · P. Chaudhary
Department of Computer Science and Engineering, The NorthCap University, Gurgaon, India
e-mail: srishti@ncuindia.edu

V. Kalra
e-mail: vaishalikalra@ncuindia.edu

P. Chaudhary
e-mail: poonamchaudhary@ncuindia.edu

N. Marriwala et al. (eds.), *Mobile Radio Communications and 5G Networks*, Lecture Notes in Networks and Systems 140,
https://doi.org/10.1007/978-981-15-7130-5_51

men and one in six women worldwide are detected with cancer during their lifespan, and one in eight men and one in eleven women die from the disease [1]. It leads to not just physical weakness, but also puts a significant amount of mental pressure on the patients. In most cases, it demands lifetime attention from the patients even after early and timely diagnosis. Cancer patients routinely need to make treatment-related decisions, look for self-care pathways and make daily life transformations. Researches in the domains of psychology, medicine and economics [2–4] have shown that the emotions of patients often affect their thinking process and choices. Despite taking decision based on the present medical evidences, fear, anxiety and worry affect the decisions and mental behavior of the patients [5]. Thus, tracking the mental health of cancer patients is a vital part of the treatment that often goes unnoticed. With the proliferation in online social media networks like Twitter, Facebook, etc., there are many virtual societies and communities formed to address this issue. Cancer patients can take part in the virtual communities available online. They can exchange their thoughts and get mental support from other patients across the world. Many such communities are publically accessible and garnering attention from researchers across the world as a means to extract useful information about the emotions and moods of cancer patients.

The field of Sentiment Analysis (SA), also known as opinion mining, analyzes the sentiments, attitudes, evaluations, emotions, appraisals, opinions and reviews toward entities such as services, issues, topics, organizations, individuals, events, products and their attributes. It is a large problem domain to explore. The umbrella of sentiment analysis or opinion mining comprises different tasks like opinion mining, opinion extraction, sentiment analysis, sentiment mining, subjectivity analysis, affect analysis, review mining, emotion analysis, etc. Sentiment analysis is basically a Natural Language Processing (NLP) problem and touches every not solved problems of NLP like negation handling, word sense disambiguation and co-reference resolution. So, SA enables us to explore this rich set of inter-related sub-problems and allows us to abstract a structure from the complex and intimidating unstructured natural language text. This enables researchers to design more accurate and robust systems by exploiting the correlation of these sub-problems from a practical point of view. Opinions and sentiments are subjective and allow examining a collection of opinions of many people rather than a single person's view. These views can be acquired from different resources like news articles, tweets (Twitter postings), forum discussions, blogs and Facebook postings. This makes a large collection of opinions and sentiments on the Web which inclines the research toward the summarization of these opinions [6]. This summarization is application-oriented like what is the subjectivity and emotions, what are the key concepts, etc. Sentiment analysis can be modeled as machine learning problem of classification and two sub-problems must be resolved (1) subjectivity classification problem and (2) polarity classification problem.

Emotions and opinions are not equivalent terms but inter-related in many aspects. Every opinion represents an underlying emotion. Emotion is a broader term than sentiment and encompasses a reflection of the person's mood such as whether the person was happy, sad, surprised, etc., rather than just categorizing into positive, negative or neutral. Identifying this emotional undertone in an opinion is a sub-field of SA, referred to as affect analysis.

People pass through different states of emotions depending upon the events happening in their lives. Emotion can be defined as a psychological change in their behavior that arises naturally rather than through any conscious effort and occurring due to physiological changes only [7]. Emotions signify some kind of feelings such as those of sadness, aggression, disapproval and joy. Emotions being a significant part of human behavior have been researched at length in psychology and the linked field of behavioral sciences. Surprisingly, there is very less work has been done in the area of emotion analysis, referred to as affect analysis in computational linguistics.

In this paper, we carry out affect analysis of the patients' Twitter posts to detect the mental health of the patients, identify whether the patients are depressed and need attention.

2 Related Work

Online support groups (OSGs) may provide narratives which contains the conscious or unconscious sentiments of the cancer patients. A more accurate and clearer conceptualization of the role of social media platform in diagnosing health-related support is very important. Crannell et al. [8] have studied 660 tweets of cancer patients who were describing their conditions on online social media site Twitter. They have automated the process of quantifying these self-reported diagnostic indicators by using machine learning algorithm for sentiment analysis. Sentimental comments posted by online support group (OSGs) of breast cancer patients were analyzed by Cabling et al. [9] and concluded the more likelihood of context which can provide the support for like-minded patients. Twitter as a social media platform has contributed potential support for monitoring public health trends [10–12]. Rodrigues et al. [13] have developed a sentiment analysis health tool names as "senti-health-cancer (SCH-pt)." It is tailored specifically to detect sentimental messages of patient's online, i.e., positive/negative messages over online platform. The positive or negative sentiments in conversation can help in finding the mental disorders in cancer patients specifically depression [14, 15].

Zainuddin et al. [16] have done aspect-based sentiment analysis on Twitter data and found the "hate" sentiment inside the data using machine learning algorithms. A systematic review has been provided by Giuntini et al. [17] on recognizing depression disorder in narrative provided by the patients on social network using techniques of emotion and sentiment analysis.

In [13, 15] also the researchers proposed to detect depression using social media posts. However, the researchers in these works use sentiment analysis for depression detection. We go one step ahead of these works and propose to use affect analysis for depression detection in cancer communities as affect relates to an emotion representing a person's state of mind and conveys substantially more information than sentiment which is only positive, negative or neutral. Thus, affect analysis has more applicability to depression detection than SA.

3 Proposed Work

We followed the approach of automatic analysis of emotions from tweets using knowledge taken from the field of psychology. In this work, we propose an approach for depression detection in cancer communities using affect analysis. Robert Plutchik's wheel of emotions, represented in Fig. 1, outlines the eight basic emotions as joy, trust, fear, surprise, sadness, anticipation, anger and disgust. Each primary emotion also has a polar opposite, so that: Joy is the opposite of sadness, fear is the opposite of anger [18].

For the task of depression detection from cancer communities, we identified cancer communities on Twitter. We researched and observed that most of these communities

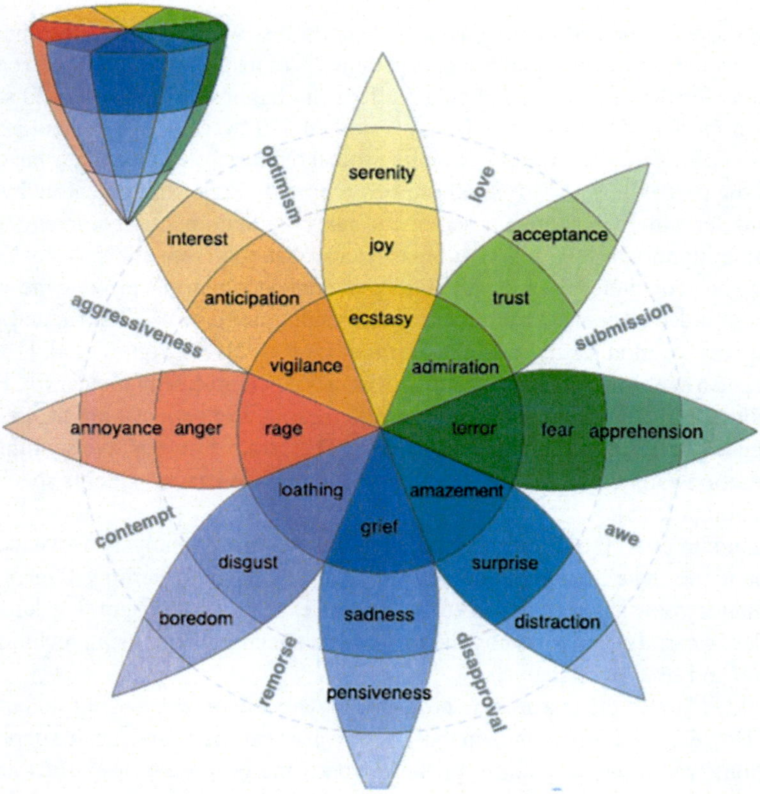

Fig. 1 Robert Plutchik's wheel of emotions

Table 1 Patients categorization based on symptoms

S. No.	Features	Categories	S. No.	Features	Categories
1	Aggression	Depression	10	Joy	Non-depressive
2	Ambiguous	Depression	11	Love	Non-depressive
3	Anger	Depression	12	Optimism	Non-depressive
4	Anticipation	Depression	13	Remorse	Non-depressive
5	Awe	Depression	14	Sadness	Depression
6	Contempt	Depression	15	Submission	Depression
7	Disapproval	Depression	16	Surprise	Non-depressive
8	Disgust	Depression	17	Trust	Non-depressive
9	Fear	Depression			

were created by cancer patients to seek help and support from each other. For the purpose of our research, we selected one such cancer community on Twitter and from this community, identified a sample of ten people. We carried out user profiling and extracted all tweets posted in the past one year by the ten users selected from Twitter using the Twitter API. We call the resultant dataset of tweets as T where $T = \sum_{i=0}^{10} t_i$ and t_i refers to the tweet dataset t pertaining to user i.

To perform affect analysis on the user tweets extracted, we use the Support Vector Machines (SVM) classifier and train it on a labeled dataset of tweets classified according to Robert Plutchik's emotion model [19]. The training dataset has tweets labeled in seventeen emotion categories derived from Plutchik's eight basic emotions. We use the SVM classifier as it has been known to outperform the other well-known classifiers for NLP tasks as shown by researchers in [20–22].

Out of the seventeen emotions, we combine the emotions such as aggression, ambiguous, anger, disgust, sadness and fear into one set and label it as D (shown in Table 1) and the rest into another set and label it as Z. This work has been carried out in consultation with doctors from the field of psychiatry.

Next, we select tweets from the user i and carry out affect analysis using SVM on all user i tweets. After affect analysis step, every user i tweet would be labeled as representative of either one of the emotions listed in set D or Z. If after labeling, the count of user i tweets labeled D exceeds those that are labeled as Z, then the user i is said to be suffering from depression. Similarly, we repeat these steps for values of $i = 1$–10, i.e., for all the users in dataset and ascertain the cancer patients suffering from depression.

4 Result and Analysis

For the analysis of the proposed model, we have consulted with the doctors of psychiatry to know the features of the depressive patients and based on their guidance we have categorized our dataset which has been labeled into 17 categories from Plutchik's wheel of emotions into two main categories of depressive and non-depressive. This has been shown in Table 1.

Based on the given 17 categories in the labeled dataset, the user's profiles tweets are categorized into labeled categories using SVM machine learning algorithm and are then summed up into two proposed categories *D* (depression) and *Z* (non-depression) as shown in Table 2, and the percentage of depressiveness and non-depressiveness is shown in Fig. 2. From the results shown in Table 2, we can conclude

Table 2 Profiling of user's tweets into depressive and non-depressive categories

User's profiles	Depression	Non-depression	Total_Tweets
p1	951	430	1381
p2	1181	855	2036
p3	134	120	254
p4	588	529	1117
p5	833	1075	1908
p6	884	721	1605
p7	863	689	1552
p8	60	61	121
p9	1104	745	1849
p10	1101	1203	2304

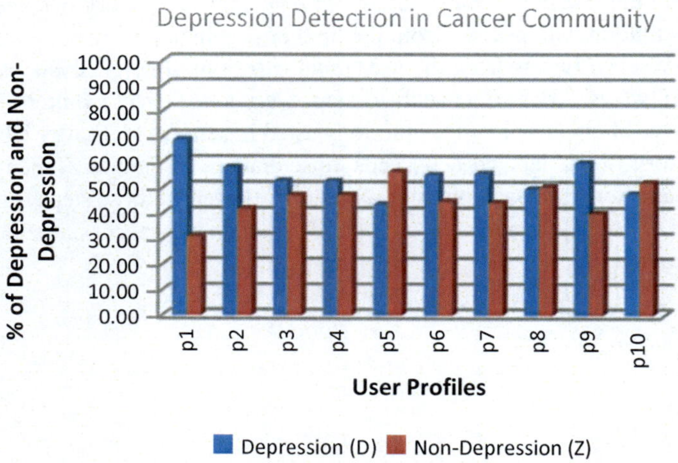

Fig. 2 Analysis graph of depression detection on user's profiles

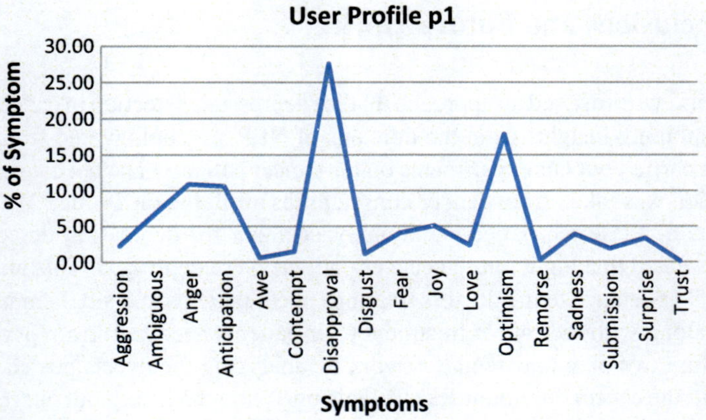

Fig. 3 Analysis graph of user profile p1

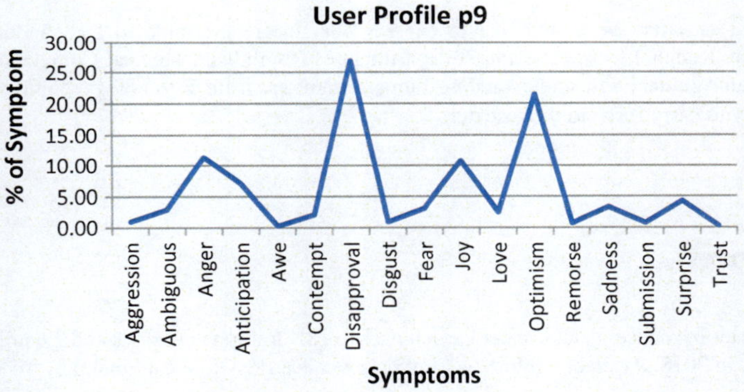

Fig. 4 Analysis graph of user profile p9

that the out of ten patients, eight patients are having symptoms of depression and this analysis can help cancer doctors to guide the patients to have consultancy from the psychiatrists for their better treatment. This analysis is carried out to help the cancer community.

From the graph shown in Fig. 2, we can clearly find out that the patient p1 and p9 are the most depressive patients, and the detailed analysis of their tweets is shown in Figs. 3 and 4, respectively.

5 Conclusions and Future Work

In this work, we proposed an approach that for depression detection in cancer communities combining insights from the domains of NLP, psychology and from psychiatrists. We carried out our experiments on ten cancer patients. The ten cancer patient's information was taken from cancer communities available on Twitter. We extracted the tweets of these ten patients and analyzed them for depression detection. Our results revealed that eight out ten cancer patients were depressed. This analysis can be of great assistance to the doctors treating cancer patients and aid them in deciding whether along with the cancer treatment their patients need help from psychiatrists. In the future, we plan to extend this work by analyzing the tweets posted by all the users in all the cancer communities available on Twitter and carry out our research on a much larger scale. Furthermore, presently, we have used the SVM classifier, but in the future, we plan to improvise our depression detection approach by incorporating deep learning.

Acknowledgements We would like to express our sincere gratitude to Dr. Viddur Arya and Dr. Abhinav Kuchhal from Psychiatrist department of Rohilkhand Medical College and Hospital for providing guidance in understanding human emotions from Psychology point of view and opportunity to carry forward the research.

References

1. Latest global cancer data: Cancer burden rises to 18.1 million new cases and 9.6 million cancer deaths in 2018. Available: https://www.who.int/cancer/PRGlobocanFinal.pdf, Accessed 2 Jan 2020
2. Finucane ML, Alhakami A, Slovic P, Johnson SM (2000) The affect heuristic in judgments of risks and benefits. J Behav Decis Making 13(1):1–17
3. Reyna VF, Nelson WL, Han PK, Pignone MP (2015) Decision making and cancer. Am Psychol 70(2):105
4. Brown SL, Whiting D, Fielden HG, Saini P, Beesley H, Holcombe C, Holcombe S, Greenhalgh L, Fairburn L, Salmon P (2017) Qualitative analysis of how patients decide that they want risk-reducing mastectomy, and the implications for surgeons in responding to emotionally-motivated patient requests. PloS one 12(5)
5. Mazzocco K, Masiero M, Carriero MC, Pravettoni G (2019) The role of emotions in cancer patients' decision-making. e-Cancer Med Sci
6. Hu M, Bing L (2004) Mining and summarizing customer reviews. In: Proceedings of ACM SIGKDD international conference on knowledge discovery and data mining (KDD-2004)
7. Emotion, Available: https://www.thefreedictionary.com/emotion. Accessed 19 July 2014
8. Crannell WC, Clark E, Jones C, James TA, Moore J (2016) A pattern-matched twitter analysis of US cancer-patient sentiments. J Surg Res 206(2):536–542
9. Cabling ML, Turner JW, Hurtado-de-Mendoza A, Zhang Y, Jiang X, Drago F, Sheppard VB (2018) Sentiment analysis of an online breast cancer support group: communicating about tamoxifen. Health Commun 33(9):1158–1165
10. Alajajian SE, Williams JR, Reagan AJ, Alajajian SC, Frank MR, Mitchell L, Lahne J, Danforth CM, Dodds PS (2017) The Lexicocalorimeter: Gauging public health through caloric input and output on social media. PLoS ONE 12:e0168893

11. Reece AG, Reagan AJ, Lix KL, Dodds PS, Danforth CM, Langer EJ (2016) Forecasting the onset and course of mental illness with Twitter data. arXiv preprint arXiv:1608.07740
12. Birjali M, Beni-Hssane A, Erritali M (2017) A method proposed for estimating depressed feeling tendencies of social media users utilizing their data. Adv Intell Syst Comput 552:413–420
13. Rodrigues RG, das Dores RM, Camilo-Junior CG, Rosa TC (2016) SentiHealth-cancer: a sentiment analysis tool to help detecting mood of patients in online social networks. Int J Med Inf 85(1):80–95
14. Zhou TH, Hu GL, Wang L (2019) Psychological disorder identifying method based on emotion perception over social networks. Int J Environ Res Public Health 16(6):953
15. Wang X, Zhang C, Ji Y (2013) A depression detection model based on sentiment analysis in micro-blog social network. In: Proceedings of the Pacific-Asia conference on knowledge discovery and data mining, Berlin, Germany, 14–17 Apr 2013
16. Zainuddin N, Selamat A, Ibrahim R (2020) Discovering hate sentiment within twitter data through aspect-based sentiment analysis. In: Journal of physics: conference series, vol. 1447, no. 1. IOP Publishing, p 012056
17. Giuntini FT, Cazzolato MT, dos Reis MDJD, Campbell AT, Traina AJ, Ueyama J (2020) A review on recognizing depression in social networks: challenges and opportunities. J Ambient Intel Human Comput 1–17
18. Plutchik's Wheel of Emotions. Available: https://www.6seconds.org/2017/04/27/plutchiks-model-of-emotions/, Accessed 2 Jan 2020
19. Primary emotions of statements by Lowri Williams, Cardiff University. Available: https://www.figure-eight.com/data-for-everyone/, Added on 25 June 2015
20. Srishti S, Shampa C (2018) Sarcasm detection in online review text. ICTACT J Soft Comput 08(03)
21. Multi-Class Text Classification Model Comparison and Selection. Available: https://towardsdatascience.com/multi-class-text-classification-model-comparison-and-selection-5eb066197568. Accessed 18 Jan 2020
22. Monalisa G, Goutam S (2018) Performance assessment of multiple classifiers based on ensemble feature selection scheme for sentiment analysis. Appl Comput Intel Soft Comput 12. https://doi.org/10.1155/2018/8909357

Triple-Band Polarization-Insensitive Wide Angle Ultrathin Hexagonal Circumscribed Metamaterial Absorber

Krishan Gopal, Deepak Sood, and Monish Gupta

Abstract An ultrathin triple-band polarization-insensitive compact metamaterial absorber for microwave frequency applications is presented. The top layer of the proposed structure is designed with a metallic hexagonal circumscribed pattern with concentric rings. It exhibits distinct absorption peaks at 7.69, 10.56, and 15.60 GHz covering C, X, and Ku bands. The proposed absorber is ultrathin with a thickness of $0.044\lambda_0$ and compact in size with unit cell dimensions of $0.44\lambda_0$ w.r.t highest frequency of absorption. An array of the proposed structure is fabricated, and experimental results are observed in agreement with the simulated responses.

Keywords Metamaterial · Absorber · Ultrathin · Triple-band · Compact

1 Introduction

Electromagnetic metamaterials (MMs) are artificial arrays of structured subwavelength elements whose electromagnetic parameters can be flexibly tailored. They have attracted much attention due to their unusual electromagnetic properties that cannot be obtained with natural materials, such as negative permittivity and permeability. Such properties lead MMs to a very wide range of potential applications like superlens [1], miniaturized antenna [2], cloaking [3], absorbers [4], negative refraction index, backward propagation, and reverse Doppler effects. Electromagnetic absorbers are used for stealth applications, radome design, EMI interference reduction, etc. The conventional absorbers are thick and bulky which limit their uses [5]. Therefore, the ultrathin thickness, compact unit cell size, and near-unity absorption property of the metamaterial-based absorbers attract much attention. They have been investigated intensively since a perfect metamaterial absorber composed of electric resonator has been reported in [4]. Thereafter, a range of narrowband [6],

K. Gopal (✉) · D. Sood · M. Gupta
Department of Electronics and Communication Engineering, University Institute of Engineering and Technology, Kurukshetra University, Kurukshetra, Haryana 136119, India
e-mail: krishanjohar@gmail.com

N. Marriwala et al. (eds.), *Mobile Radio Communications and 5G Networks*, Lecture Notes in Networks and Systems 140,
https://doi.org/10.1007/978-981-15-7130-5_52

dual-band [7, 8], multiband [9–18], and broadband [19] absorbers have been designed and analyzed till date.

In this paper, a wide angle triple-band polarization-insensitive metamaterial absorber is presented. The absorber exhibits three distinct absorption peaks at 7.69 GHz, 10.56 GHz, and 15.60 GHz with 95.27%, 98.35%, and 97.96% absorption, respectively. It exhibits high absorption for oblique incident angles up to 45° for both TE and TM polarizations. The absorption mechanism of the proposed structure is analyzed through field and current distributions. The designed structure has been fabricated and the measured results are observed in good agreement with the simulated responses. The wide angle absorption and polarization-insensitive characteristics have also been successfully accomplished through experimental measurement.

2 Unit Cell Design

Figure 1 shows the front view of the unit cell of the proposed metamaterial absorber. The top layer is designed with a metallic hexagonal circumscribed pattern concentric with circular rings on a dielectric substrate FR-4 ($\varepsilon_r = 4.4$, tan$\delta = 0.02$) with 0.8 mm thickness. The bottom metallic layer is a copper ground plane. The optimized dimensions of the proposed structure are: $a = b = 8.5$ mm, $r_1 = 3.4$ mm, $r_2 = 2.59$ mm, $W_1 = 0.6$ mm, $W_2 = 0.24$ mm, $W_3 = 0.17$ mm, $W_4 = 0.4$ mm, $L_1 = 2.92$ mm, respectively.

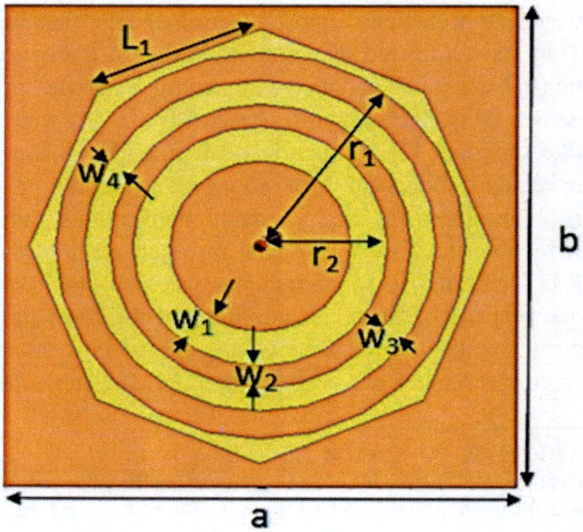

Fig. 1 Front view of the unit cell of the proposed metamaterial absorber

3 Simulation

The proposed design is simulated using ANSYS HFSS simulation software with Floquet's periodic boundary conditions. The absorptivity of the proposed structure is calculated as:

$$A(\omega) = 1 - |S_{11}(\omega)|^2 - |S_{21}(\omega)|^2 \tag{1}$$

where $A(\omega)$, $|S_{11}(\omega)|^2$, and $|S_{21}(\omega)|^2$ represent absorbed power, reflected power, and transmitted power, respectively. Due to the continuous metallic layer at the bottom of the proposed structure, there will be no power transmission through it, i.e., $|S_{21}(\omega)|^2 = 0$. Therefore, Eq. (1) becomes,

$$A(\omega) = 1 - |S_{11}(\omega)|^2 \tag{2}$$

The simulated absorptivity under normal incidence of the wave is shown in Fig. 2. It is observed that the proposed structure exhibits 95.27%, 98.35%, and 97.96% absorption at distinct frequencies of 7.69 GHz, 10.56 GHz, and 15.60 GHz, respectively covering C, X, and Ku bands. The designed structure is also examined for various polarization angles from 0° to 45° with a step size of 15° as shown in Fig. 3a. It is observed that the fourfold symmetric design configuration of the proposed structure anticipates the absorption stability at the three absorption peaks with the variations of polarization angles. Further, the performance of the proposed design is also analyzed for different oblique angles of the incident wave as shown in Fig. 3b. It is observed that the absorptivity remains same at the three frequencies with the variation of incident angle from 0° to 60° of the incident wave which proves that excellent angular stability of the proposed absorber.

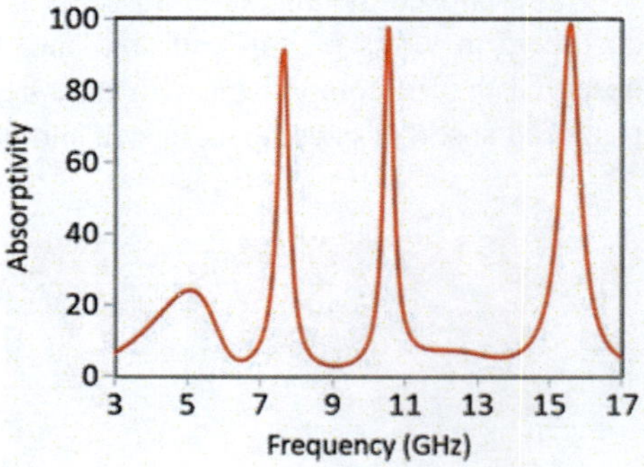

Fig. 2 Simulated absorptivity of the proposed metamaterial absorber under normal incidence

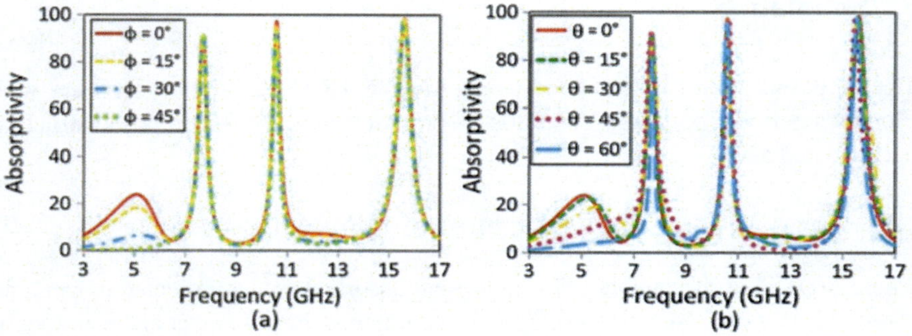

Fig. 3 **a** Simulated absorptivity of the proposed metamaterial absorber for different angles of polarization and **b** for different oblique angles of incidence wave

4 Absorption Mechanism

In order to get better physical insight of the proposed design, the absorption mechanism has been studied through electric field analysis as shown in Fig. 4. It is observed that electric field is mainly distributed at the outer hexagonal ring at the first frequency (f_1) 7.69 GHz as shown in Fig. 4a. At the second frequency (f_2) of absorption, the field is observed to be distributed around the middle circular ring as shown in Fig. 4b. In the similar way, the field is mainly concentrated on the innermost ring at the third frequency (f_3) as shown in Fig. 4c. The surface current distribution has also been analyzed for all the three absorption frequencies as shown in Fig. 5. It is observed that the current is distributed at the outer ring for the first frequency (f_1). For the second and third absorption frequencies, the currents are distributed at the middle and inner metallic ring, respectively. The currents are observed as antiparallel at the top and bottom surface.

The proposed metamaterial absorber is compared with the previously reported absorbers in terms of unit cell size and thickness as illustrated in Table 1. It is observed that in comparison with the already designed absorbers [9, 13–16], the proposed absorber is compact in size and ultrathin in thickness.

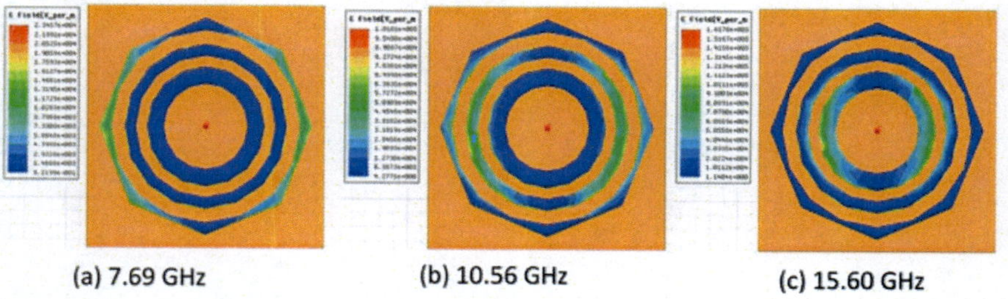

Fig. 4 Electric field distribution of the proposed absorber at the three absorption peaks

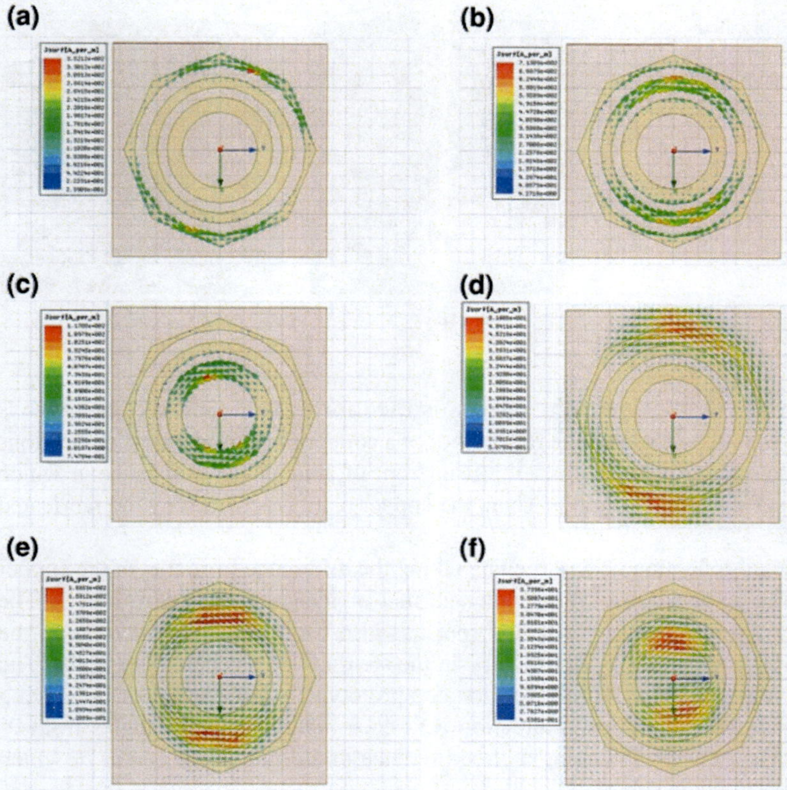

Fig. 5 Current distribution at the top and bottom surfaces

Table 1 Comparison of proposed absorber with exiting absorbers

Absorbers	Highest frequency of absorption (GHz)	Unit cell size	Thickness	Polarization insensitivity
[9]	12.19	0.82λ	0.032λ	Yes
[13]	12.54	1.25λ	0.041λ	Yes
[14]	7.30	0.58λ	0.024λ	Yes
[15]	9.86	0.49λ	0.053λ	Yes
[16]	14.2	0.44λ	0.047λ	Yes
Proposed	15.60	0.44λ	0.044λ	Yes

5 Experimental Measurements

A prototype array of the proposed structure has been fabricated as shown in Fig. 6a The performance is experimentally verified using vector network analyzer model N5222A (10 MHz to 26.5 GHz) connected to two UWB horn antennas (1–18 GHz)

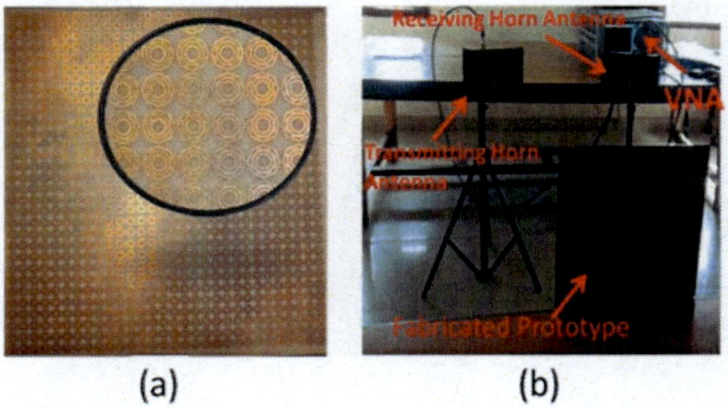

Fig. 6 **a** Fabricated prototype and **b** measurement setup

as shown in Fig. 6b. One horn antenna is used for transmission and the other is used for reception. Initially, the reflectivity of a copper plate of identical dimensions of prototype is measured by placing it in front of antennas at a distance at which near field effects are negligible. Then, the copper plate is replaced by the fabricated prototype.

The actual reflection is measured by taking the difference from that of the copper plate. The measured absorption under normal incidence is shown in Fig. 7. The polarization performance of the absorber has also been measured by rotating the structure around its axis from 0° to 45° in the steps of 15° as shown in Fig. 8a. The oblique incidence behavior of the fabricated prototype is also measured by moving the horn antennas in opposite direction w.r.t each other at oblique angles from 0° to 60° in the steps of 15° along a circumference of a circle at the center of which fabricated sample is kept fixed.

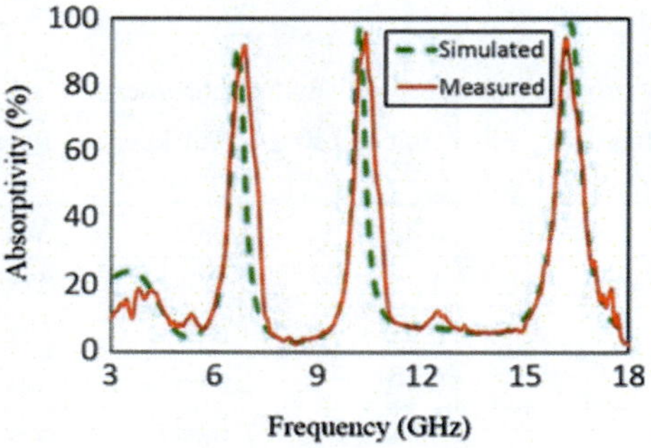

Fig. 7 Comparison of simulated and measured responses

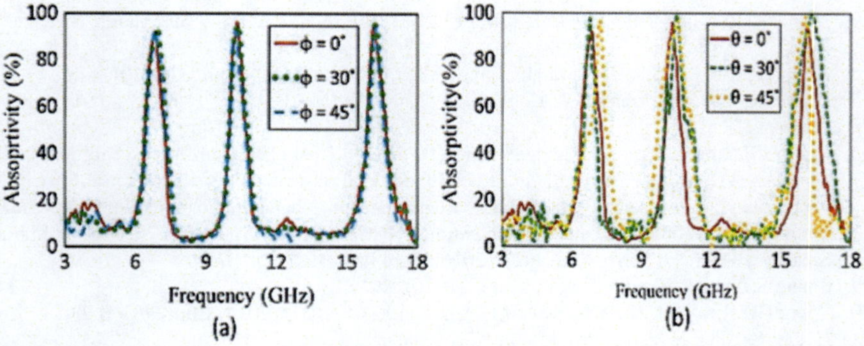

Fig. 8 Measured absorption response for **a** different polarization angles and **b** oblique incidence angles

6 Conclusion

An ultrathin, polarization insensitivity triple-band metamaterial absorber based on concentric metallic rings has been designed. The dimensions of the rings have been optimized to obtain three absorption peaks covering C, X, and Ku band. In order to understand the physical mechanism of absorption, field distributions, and surface current distribution are studied. The proposed absorber shows good absorption characteristics for wide incidence angles (up to 60°) for TE and TM polarizations. A prototype of the design is fabricated and its performance is experimentally verified. The designed absorber is ultrathin (0.044λ) corresponding to the highest absorption frequency) which makes it fit for various potential applications like air surveillance radar, thermal emitters, wireless communication, and phase shifter, etc.

References

1. Caloz C, Itoh T (2006) Electromagnetic: transmission line theory and microwave applications. Wiley, New York
2. Shelby RA, Smith DR, Schultz S (2001) Science 292:77
3. Veselago VG (12964) Soviet physics UspekhUsp. Fiz. Nauk 92:509
4. Fang N, Lee H, Sun C, Zhang X (2005) Science 308:534
5. Enoch S, Tayeb G, Vincent P (2002) Phys Rev Lett 89:3901
6. Schurig D, Mock JJ, Justice BJ, Cummer SA, Pendry JB, Starr AF, Smith DR (2006) Science 314:977
7. Landy NI, Sajuyigbe S, Mock JJ, Smith DR, Padilla WJ (2008) Phys Rev Lett 100(20):207402
8. Saville P (2005) Review of radar absorbing materials. Defence R&D, Canada-Atlantic, p 5
9. Lin B-Q, Zhao S-H, Da X-Y, Fang Y-W, Ma J-J, Li W, Zhu ZH (2015) Microw Opt Technol Lett 57:1439
10. Zhai H, Li Z, Li L, Liang C (2013) Microw Opt Technol Lett 55:1606
11. Huang YJ, Yang HL, Hou XW, Tian Y, Hou DY (2010) Prog Electrom Res 108:37
12. Guo XR, Zhang Z, Wang JH, Zhang JJ (2013) J Electrom Waves Appl 27:629

13. Bian BR, Liu SB, Zhang HF, Li BX, Ma B (2013) PIERS proceedings, Stockholm, Sweden, p 435
14. Shen XP, Cui TJ, Zhao JM, Ma HF, Jiang WX, Li H (2011) Opt Express 19(10):9401
15. Bian BR, Liu SB, Wang SY, Kong XK, Zhang HF, Ma B, Yang H (2013) J Appl Phys 114(19):194511
16. Ben M, Shaobin L, Kong X, Zhang H, Mao Z, Wang B (2014) J Electrom Waves Appl 27(5):629
17. Chaurasiya D, Bhattacharyya S, Ghosh S, Munaga P, Srivastava RV (2015) An ultra-thin triple-band polarization-insensitive metamaterial absorber for C-band applications. In: Proceedings of first national conference on communications (NCC), IIT Bombay, India, 27th Feb–1st March
18. Sharma SK, Ghosh S, Srivastava KV (2016) Appl Phys 122(12):1071
19. Shang S, Yang S, Tao L, Yang L, Cao H (2016) AIP Adv 6
20. Wang GD, Liu MH, Hu XW, Kong LH, Cheng LL, Chen ZQ (2014) Chin Phys B 23(1):175
21. Huang L, Chen H (2011) Progress In electromagnetics research 113:103
22. Wen DE, Yang HL, Ye QW, Li MH, Guo LY, Zhang LY (2013) Phys Scr 88

An Effective Fusion of a Color and Texture Descriptor for an Image Retrieval System: An Exploratory Analysis

Shikha Bhardwaj, Gitanjali Pandove, and Pawan Kumar Dahiya

Abstract Content-based image retrieval (CBIR) is one of the most prominent systems by which the desired and relevant images are recovered from a massive database by using the basic image features like color, shape, texture, spatial information, and edge. In this paper, an experimental analysis is being done to determine the most efficient combination using color and texture descriptors. Here, color moment (CM) is used for color feature extraction and is used individually in a combination with different texture descriptors, namely discrete wavelet transform (DWT), Gabor transform, Curvelet transform, graylevel co-occurrence matrix (GLCM), and local binary pattern (LBP). These color and texture features can be combined using two different levels of the system: feature-level fusion and score-level fusion. Both the feature-level and score-level combination techniques are used in this paper. The results of this analytical experimentation depict that the framework of CM, and GLCM attains the highest results among all the other combinations using feature-level fusion technique on a benchmark CBIR dataset, particularly WANG. Precision, recall, f-score are some of the evaluation parameters which are utilized in this paper to measure the effectiveness of the fused descriptors.

Keywords Color moment · Graylevel co-occurrence matrix · Local binary pattern · Feature-level fusion · Score-level fusion

S. Bhardwaj (✉)
ECE Department, DCRUST, Murthal, Sonepat, India
e-mail: shikpank@yahoo.com

ECE Department, UIET, Kurukshetra University, Kurukshetra, Haryana, India

G. Pandove · P. K. Dahiya
ECE Department, DCRUST, Murthal, Sonepat, Haryana, India

N. Marriwala et al. (eds.), *Mobile Radio Communications and 5G Networks*, Lecture Notes in Networks and Systems 140,
https://doi.org/10.1007/978-981-15-7130-5_53

1 Introduction

Due to the latest and many sophisticated advancements in the field of image processing, there is a cumulative growth in the number of available images. Technological developments in the area of mobile devices, smart cameras, other portable devices for capturing images, satellite cameras, etc., are some of the major causes for the creation of large image repositories [1]. These storehouses of images are available both offline and online. The manual searching, classification, browsing, indexing, categorization, and retrieval of images from these large datasets are very confusing and almost impossible [2]. Therefore, the retrieval efficiency and retrieval accuracy of a user become restricted, and to enhance these parameters, an effective system is required which automatically retrieves the desired and relevant images from huge datasets.

Based on the above facts, content-based image retrieval (CBIR) is a system which retrieves the desired images from massive datasets based on the visual features of an image like texture, shape, color, edge, and spatial information. The desired feature from the database is extracted using a specific feature extraction technique, and a feature vector is formed. Then, when a query image is given by the user, again the same feature is extracted and a feature vector is obtained. Afterward, both the feature vectors are compared in the similarity matching stage [3], where a distance metric is used to find the relevant images, arranging the images in increasing order of the distance. Finally, based on the requirement of the user, top N images are obtained. A basic CBIR system is shown in Fig. 1.

In contrast to CBIR, images can also be retrieved on the basis of text and annotation but this technique is prone to many errors like human errors, synonyms, homonyms, and misspellings. This technique is called text-based image retrieval (TBIR). But, CBIR has many advantages as compared to TBIR.

The most important attribute of an image is color and is indispensable to human perception. Color can be extracted by using divergent color descriptors but the most prominent are: color histogram (CH) [4]; it is the basic technique but does not include any spatial information, dominant color descriptor (DCD): It takes long computation time and includes less spatial information. Color coherence vector (CCV): It is very slow and takes more space, Color auto-correlogram (CAC): It is fast and accurate but sometimes lack in accurate results. There are many more color feature techniques but are the variants of these basic descriptors. Based on these facts about color descriptors,

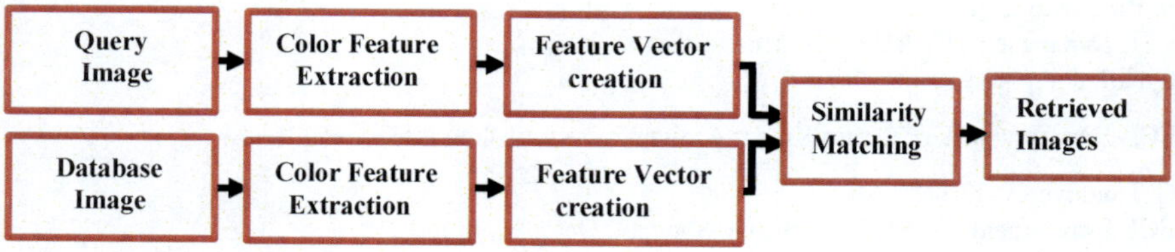

Fig. 1 A basic CBIR system

it can be concluded that color moment (CM) [5] has the maximum advantages as compared to other techniques. It includes the maximum spatial information and also is very fast in computation, more scalable and extremely robust. Moreover, the obtained moments are based on using the basic statistical measures like mean, standard deviation, skewness, and kurtosis.

Another important feature of an image is texture which describes the inborn patterns of an image like coarseness, smoothness, roughness, and circularity. Texture can be extracted by employing various texture feature extraction techniques like discrete wavelet transform (DWT) [6], local binary pattern (LBP) [7], graylevel co-occurrence matrix (GLCM), Tamura features, Gabor transform [8], and Curvelet transform.

This paper describes an analytical experimental analysis of the best color feature descriptor; namely, color moment with many texture feature extraction techniques has been done based on the two different combining levels: feature level and score level. Thus, the main benefactions of this paper are:

- To find the best color feature extraction technique based on both the theoretical and experimental analyses.
- To find the average precision of each texture descriptor individually.
- To find the best hybrid combination by using various color and texture descriptors by the inclusion of chosen color moment with different texture feature descriptors individually.
- To check the efficiency of two divergent combining techniques which have been used for the amalgamation of color and texture features. These two techniques are based on two different levels of the system: feature level and score level.
- These experiments have been carried on a prominent benchmark CBIR dataset, namely WANG.

The remaining organization of the paper is as under: Related work is given in Sect. 2; in Sect. 3, the proposed methodology has been given, experimental setup, and results are given in Sect. 4, and lastly conclusion with future trends is given in Sect. 5.

2 Related Work

In the literature, different techniques have been used to form the hybrid or fusion-based image retrieval system. A hybrid system based on the combination of color, texture, and shape features has been proposed by Pavithra et al. [9]. Color moment is used for color extraction while to extract the shape and texture information from an image, canny edge and LBP are used. Prashant et al. [10] described another hybrid system based on the combination of LBP and GLCM. LBP is utilized here in different scales.

Color and texture are indeed considered as the most important features of an image, as they give principal information regarding an image. Different techniques like block variation of local correlation coefficients (BVLC) and block difference of inverse probabilities (BDIP) [11] have also been used for the extraction of texture attributes with color histogram as a color descriptor. DCD in combination with Curvelet and wavelet features has also been used by Fadaei et al. [12].

There are many techniques which are just advanced versions of their basic counterparts. Like distance coherence vector (DCV) is another technique which is used to extract the color information from an image and is the changed version of the basic color coherence vector. Similarly, color difference histogram (CDH) [13] is also an advanced version of color histogram.

Adaptive Tetrolet transform has also been used for texture feature extraction by using a hierarchical system in the form of three levels [14]. Color and shape features are extracted by using color channel correlation histogram and edge joint histogram, respectively. Another hybrid CBIR system utilized 2D-DWT for shape extraction in combination with color and edge directivity descriptor (CEDD) [15] which is utilized for both color and texture extraction.

In the current era, many more sophisticated and intelligent techniques based on the concepts of relevance feedback, machine learning, and deep learning have also been utilized with an image retrieval system to enhance its performance. In machine learning, classifiers like support vector machines (SVM), random forest, naïve Bayes, extreme learning machine have been used successfully in this domain [16]. Convolutional neural network (CNN) [17], deep belief network (DBN), Boltzmann machine, auto-encoders, etc., are some of the techniques under deep learning [18]. But, in spite of these techniques also, there is a requirement of an effective combining technique at the ground level, which can be utilized to merge all the obtained feature vectors into a common feature vector. Generally, the obtained feature vectors can be combined using two techniques: at the feature level and at the score level.

Thus, in this paper both these combination levels are used for the creation of a hybrid model based on the extraction of color and texture features. Color moment is used for color feature extraction and is used separately in combination with many texture descriptors to identify the best group of color and texture features.

3 Proposed Methodology

Feature extraction is an important process in CBIR systems. Feature can be defined as a localized, pertinent, and perceptible attribute of an image. In this paper, different combinations of color and texture descriptors have been analyzed to determine the best combination for the formation of a hybrid CBIR system. For the extraction of color attributes, color moment has been used here and is used individually in a combination with prominent texture descriptors like LBP, GLCM, DWT, Gabor and Curvelet transform to find out the best combination.

In order to combine these color and texture descriptors, two techniques have been employed for their combination. These are used at two stages: feature-level stage and score-level stage. The detailed description of the experimental analysis is as under.

Table 1 Theoretical comparison of various color feature extraction techniques

Technique	Advantages	Disadvantages
CH	Invariant to changes in angle and rotation	Does not include any spatial information
CCV	Includes spatial information	High dimensionality and more computational time
CAC	Includes spatial information	Includes noise and is complex
DCD	Gives effective representation of colors	Produces incorrect ranking and takes more time
CM	Fast, robust, scalable and includes all spatial information	Not any specific

3.1 Color Feature Extraction

Color moment has been deployed in this paper for the purpose of color feature extraction. This color feature extraction technique has been selected based on an experimental and theoretical analysis with various other prominent color descriptors. Color moment has many advantages in contrast to many other color descriptors, and these can also be described with the help of a theoretical comparative analysis given in Table 1.

From Table 1, it can be concluded that color moment has no specific disadvantage in contrast to other color extraction techniques which suffers from one or the other pitfalls.

3.2 Texture Feature Extraction

To obtain the textural information from an image, varied types of techniques like graylevel co-occurrence matrix (GLCM), local binary pattern (LBP), discrete wavelet transform (DWT), Gabor and Curevelet have been used. GLCM gives the spatial relationship between two adjacent pixels of an image. Suppose, if an image of size 3×3 is selected then, LBP depicts the relationship of the central pixel to other remaining pixels of the image. Wavelet, Gabor, and Curvelet are transform-based feature extraction techniques which can also be used for its extraction.

3.3 Combination Techniques

In order to combine color moment with all the texture descriptors individually, two techniques have been used. These techniques can be described based on two levels of the CBIR system, namely feature level and score level.

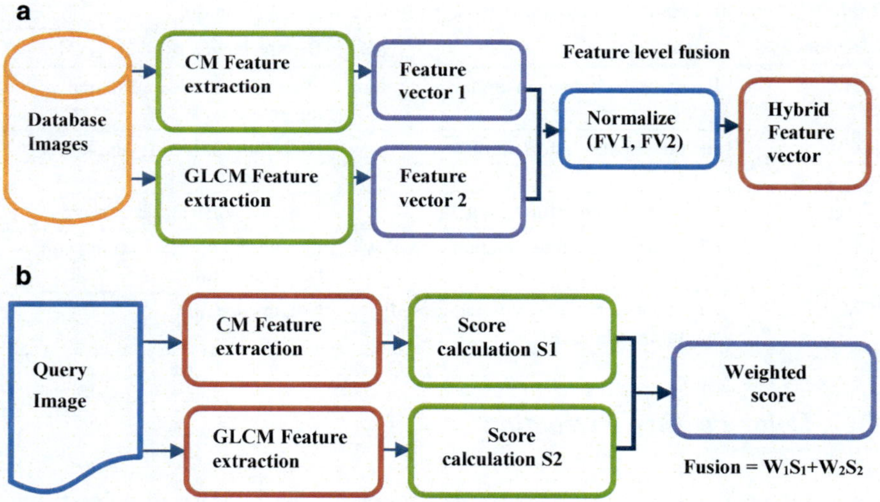

Fig. 2 **a** Architecture of a feature-level fusion using CM and GLCM. **b** Architecture of a score-level fusion using CM and GLCM

In feature-level combination, the procedure is as follows:

- Firstly, image attributes are extracted using both color and texture descriptors, and then two feature vectors are formed.
- Then, the two obtained feature vectors are normalized independently and a final or hybrid feature vector is formed. This feature-level fusion technique works in the feature extraction phase of the system, and its methodology is given in Fig. 2a
- In order to normalize the two feature vectors, minimum–maximum normalization is used as to obtain a common range for both the feature vectors are mandatory, and this technique solves all these issues and is given by:

$$\text{normalized features} = \frac{z - \text{maximum}}{\text{minimum} - \text{maximum}} \tag{1}$$

Here, $z =$ feature value

minimum = Lowest value of each feature vector

maximum = Topmost value of each feature.

In score-level combination, the procedure used is as under:

- Firstly, the extraction of color and texture features is done by using a specific technique; from the given database images, and two independent feature vectors are obtained.

- Then, during the query formulation phase, the same color and texture features are extracted from the given query image, and again two feature vectors are formed.
- Then, the similarity score of both feature vectors obtained during the query phase is calculated with respect to respective database. This fusion technique works during the query phase of the system, and the methodology for this technique is shown in Fig. 2b.
- Finally, a weighted similarity is calculated to combine all the obtained similarity scores and is given by:

$$\text{Fused vector} = G_1 * \text{similarity score } 1 + G_2 * \text{similarity score } 2$$

In the above equation $G1$ and $G2$ are the weights given to the obtained similarity scores.

4 Experimental Setup and Results

In order to carry out the experimentation, five combinations of color and texture have been evaluated. These combinations are analyzed on WANG [19] dataset. It has 1000 images with 10 categories. Each category has 100 images and includes images of mountains, beaches, dinosaur, flowers, and many more. The size of each image is 256 × 384 or 384 × 256, and sample image from each category is shown in Fig. 3.

4.1 Experimental Setup

The experiments are performed in MATLAB R2018a, 4 GB memory, 64-bit windows with core i3 processor.

Fig. 3 Sample images from WANG dataset

4.2 Results

In order to assess the performance of different fusion models, different evaluation parameters [20, 21] can be used but the main evaluation metrics which are:

$$\text{Precision} = \frac{\text{Number of relevant images retrieved}}{\text{Total number of images retrieved}} \tag{2}$$

$$\text{Recall} = \frac{\text{Number of relevant images retrieved}}{\text{Number of relevant images in database}} \tag{3}$$

$$F - \text{score} = 2\,\frac{\text{Precision} * \text{Recall}}{\text{Precision} + \text{Recall}} \tag{4}$$

In this paper, color moment has been chosen based on both the comparative theoretical and experimental analyses with other color descriptors. The average precision obtained by using various color feature extraction techniques on WANG dataset based on each category of the dataset is given in Table 2.

From Table 2, it can be clearly visualized that color moment has the highest average precision as compared to many other techniques and it has also many advantages as mentioned in Table 1. Therefore, CM has been chosen as one of the best techniques to extract the color information from an image.

Here, five prominent texture descriptors have been chosen which have been combined with CM one by one in order to detect the best combination. But, prior to each combination, average precision of each texture descriptor has also been calculated and is given in Table 3.

Table 2 Average precision (%) on WANG dataset using various color descriptors

Database	Semantic name	Avg. Precision (%)				
		DCD	CH	CM	CAC	CCV
WANG	Africa	63.00	85.00	90.00	85.00	38.75
	Beach	46.00	48.20	58.75	57.50	32.50
	Building	43.75	32.50	43.75	40.00	31.25
	Bus	60.00	73.75	68.00	65.00	25.00
	Dinosaur	92.50	92.50	100.00	100.00	42.50
	Elephant	78.12	60.62	84.37	82.00	46.25
	Flower	55.00	58.75	65.00	61.00	32.50
	Food	71.25	71.25	85.00	83.00	32.50
	Horse	86.25	86.25	95.00	91.25	22.50
	Mountain	86.25	96.25	96.25	93.00	38.75
	Average	68.21	70.50	78.61	75.77	34.25

Table 3 Average precision (%) on WANG dataset using various texture descriptors

Database	Semantic	Avg. Precision (%)				
		GLCM	DWT	Gabor	LBP	Curvelet
WANG	Africa	70	79	61	67	74
	Beaches	63	61	82	49	46
	Buildings	52	49	52	70	27
	Bus	60	43	51	90	45
	Dinosaur	99	100	57	100	98
	Elephant	65	61	67	43	54
	Flower	62	59	40	96	31
	Horse	48	40	82	65	32
	Mountain	90	88	54	36	74
	Food	98	96	63	35	84
	Average	70.7	67.6	60.9	65.1	56.5

From Table 3, it can be concluded that GLCM attains the highest results. Now, in order to validate these results, each texture descriptor has been combined with CM individually using feature-level and score-level fusion techniques. Precision, recall, and f-measure have been calculated based on each texture descriptor. These results based on feature-level and score-level fusion are given in Tables 4 and 5.

From Tables 3 and 4, it can be concluded that the combination of CM + GLCM attains the highest results in comparison with other combination techniques using

Table 4 Results (%) using feature-level fusion

Using feature-level combination			
Technique	Precision	Recall	*F*-measure
CM + GLCM	84.5	18	28.33
CM + DWT	80	17	26.33
CM + Gabor	75	14	24.33
CM + LBP	78	16	25
CM + Curvelet	70	12	22

Table 5 Results (%) using score-level fusion

Using Score-level Combination			
Technique	Precision	Recall	*F*-measure
CM + GLCM	82	16	26.34
CM + DWT	75	15	25.33
CM + Gabor	71	12	21
CM + LBP	76	14	24
CM + Curvelet	65	10	18

both feature-level fusion and score-level fusion. But, the results obtained by using feature-level fusion are higher than the score-level fusion results. Thus, it can be further added that normalization achieves better combination capabilities in contrast to combination based on similarity basis.

There are varied similarity measures [22] which can be adopted to calculate the distance between two images. The results obtained by using varied distance metrics on the feature-level fusion is shown in Fig. 4.

From Fig. 4, it can be seen that Manhattan distance metric attains the highest average precision (%) as compared to other similarity measures because it is calculated based on the absolute difference values rather than squared values.

Different color and feature extraction techniques are utilized here to carry out the analysis. The parameters of the utilized techniques are given in Table 6.

Since CM + GLCM combination attains the highest results based on feature-level fusion. The retrieval results of CM + GLCM combination based on WANG dataset, depicting top 15 retrieved images are shown in Fig. 5 with the help of a graphical user interface (GUI).

From Fig. 5, it can be visualized that with respect to the given query image, similar top 15 images are retrieved which belongs to the same category of the query image.

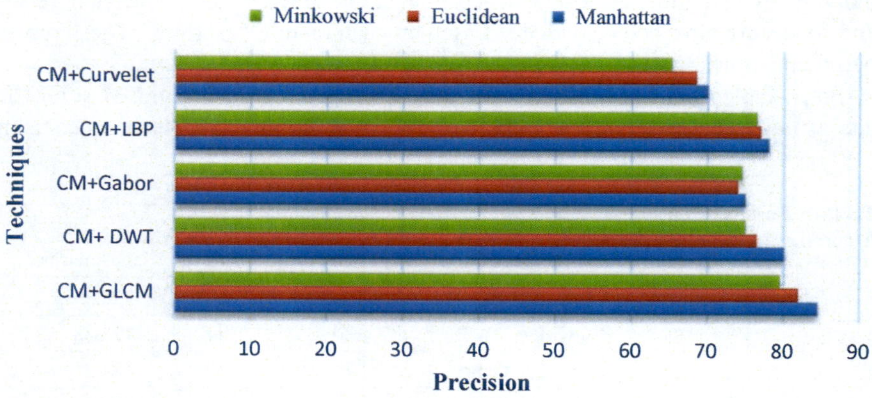

Fig. 4 Comparison of various distance metrics using each combination

Table 6 Various parameters of the utilized techniques

CM features	9
GLCM features	44
LBP features	256
DWT features	1680
Gabor features	8
Curvelet features	145
Feature-level fusion	Min–max normalization
Score-level fusion	Weighted similarity

Fig. 5 Top 15 retrieved images based on CM + GLCM combination using feature-level fusion

Thus, this CM + GLCM combination is effective in retrieving the best and most similar relevant images with respect to the query image.

5 Conclusion and Future Trends

This paper depicts an experimental analysis to determine the best combination of color and texture descriptors. For color extraction, color moment has been chosen based on its various advantages and also on the basis of a comparative analysis with various other color techniques. For texture extraction, various techniques like GLCM, LBP, DWT, Gabor, and Curvelet have been utilized. CM has been combined with each texture descriptor individually on the basis of two fusion techniques, namely feature-level fusion and score-level fusion. From this experimentation, it can be concluded that the combination of CM and GLCM attains the highest precision, recall, and f-measure of the order of 82, 16, and 26.34% by using feature-level fusion as compared to other combinations. Thus, this combination is very effective in retrieving the most relevant and similar images and can be used to form a fusion-based CBIR system. In the future, shape and edge features can also be combined with these features to retrieve more information from an image. Moreover, intelligent techniques like machine learning, deep learning, and relevance feedback can also be employed with the fusion-based model to enhance the efficiency of the system.

References

1. Alzu'bi A, Amira A, Ramzan N (2015) Semantic content-based image retrieval: a comprehensive study. J Vis Commun Image Represent 32:20–54
2. Jhanwar N, Subhasis C, Guna S, Bertrand Z (2004) Content based image retrieval using motif cooccurance matrix. Image Vis Comput 22(14):1211–1220
3. Naveena AK, Narayanan NK (2016) Image retrieval using combination of color, texture and shape descriptor. In: International proceedings on next generation intelligent systems (ICNGIS). IEEE, pp 1–5
4. Patel JM (2016) A review on feature extraction techniques in content based image retrieval. In: International proceedings on wireless communication, signal processing and networking. IEEE, pp. 2259–2263
5. Shikha B, Pandove G, Dahiya PK (2020) An extreme learning machine-relevance feedback framework for enhancing the accuracy of a hybrid image retrieval system. Int J Inter Multi Artif Intel 6(2):15–27
6. Sai NST, Patil R, Sangle S, Nemade B (2016) Truncated DCT and decomposed DWT SVD features for image retrieval. Procedia Comput Sci 79:579–588
7. Naghashi V (2018) Co-occurrence of adjacent sparse local ternary patterns: a feature descriptor for texture and face image retrieval. Optik-Int J Light Electron Opt 157:877–889
8. Singla A, Garg M (2014) CBIR approach based on combined HSV, auto correlogram, color moments and gabor wavelet. Int J Eng Comput Sci 3(10):9007–9012
9. Pavithra LK, Sharmila TS (2017) An efficient framework for image retrieval using color texture and edge features. J Comput Electr Eng 0:1–14
10. Srivastava P, Khare A (2017) Utilizing multiscale local binary pattern for content-based image retrieval. Multim Tools App 1–27. https://doi.org/10.1007/s11042-017-4894-4
11. Singh C, Preet Kaur K (2016) A fast and efficient image retrieval system based on color and texture features. J Vis Commun Image Represent 41:225–238
12. Fadaei S, Amirfattahi R, Ahmadzadeh MR (2017) New content-based image retrieval system based on optimised integration of DCD, wavelet and curvelet features. IET Image Process 11(2):89–98
13. Liu GH, Yang JY (2013) Content-based image retrieval using color difference histogram. J Pattern Recogn 46(1):188–198
14. Pradhan J, Kumar S, Pal AK, Banka H (2018) A hierarchical CBIR framework using adaptive tetrolet transform and novel histograms from color and shape features. Digit Sig Process A Rev J 82:258–281
15. Iakovidou C, Anagnostopoulos N, Kapoutsis A, Boutalis Y, Lux M, Chatzichristofis SA (2015) Localizing global descriptors for content-based image retrieval. EURASIP J Adv Sig Process 1–20. https://doi.org/10.1186/s13634-015-0262-6
16. Tomas P, Virginijus M (2017) Comparison of Naïve Bayes, random forest, decision tree, support vector machines, and logistic regression classifiers for text reviews classification. Baltic J Mod Comput 5(2):221–232
17. Arun KS, Govindan VK (2018) A hybrid deep learning architecture for latent topic-based image retrieval. Data Sci Eng 3(2):166–195
18. Weibo L, Zidong W, Xiaohui L, Nianyin Z, Yurong L, Fuad EA (2017) A survey of deep neural network architectures and their applications. Neurocomputing 234:11–26
19. Mehdi E, Aroussi E, Houssif NEl (2018) Content-based image retrieval approach using color and texture applied to two databases (Coil-100 and Wang). Springer International Publishing, pp 49–59. https://doi.org/10.1007/978-3-319-76357-6_5
20. Shikha B, Pandove G, Dahiya PK (2019) An intelligent multi-resolution and co-occuring local pattern generator for image retrieval. EAI Endors Trans Scalable Inf Syst 6(22):1–12
21. Mistry Y, Ingole DT, Ingole MD (2017) Content based image retrieval using hybrid features and various distance metric. J Electr Syst Inf Technol 2016:1–15
22. Sharma P (2017) Improved shape matching and retrieval using robust histograms of spatially distributed points and angular radial transform. Optik 145:345–364

Power Consumption Reduction in IoT Devices Through Field-Programmable Gate Array with Nanobridge Switch

Preeti Sharma, Rajit Nair, and Vidya Kant Dwivedi

Abstract In recent times, field-programmable gate array (FPGA) has done a lot of improvement in the area of IoT, a computing device that works like a microcontroller. In this segment, there is an emergence of nanobridge-FPGA that has the ability to improve the performance and minimizing the power consumption in IoT devices. It is a field-programmable gateway array that incorporates genuine nanobridge metal atom migration-type switch which is highly durable and resistive against high temperature and radiation. FPGA has already shown improvement over CPU in terms of efficiency with its group of integrated circuit that provides hardware switching using memory modules and semiconductor switches but nanobridge-FPGA enhances the power efficiency by using space-saving nanobridge switch instead of semiconductor switches and memory modules. In this work, the discussion will be on the working principle of nanobridge including the architecture of nanobridge-FPGA and its comparison with the performance of available FPGAs and CPUs.

Keywords Semiconductor · CPU · Memory · ASIC · LUT · SRAM

1 Introduction

In recent times, researchers are gaining interest in carrying out different computing research on IoT or mobile platforms, particularly for advanced computing solutions, with the progress in Internet of things (IoT) technology as well as the increasing computing power of mobile devices.

P. Sharma (✉) · V. K. Dwivedi
Bansal College of Engineering, Mandideep, India
e-mail: preetirajitnair@gmail.com

R. Nair
Jagran Lakecity University, Bhopal, India
e-mail: rajitnitbpl@gmail.com

N. Marriwala et al. (eds.), *Mobile Radio Communications and 5G Networks*, Lecture Notes in Networks and Systems 140,
https://doi.org/10.1007/978-981-15-7130-5_54

It is critical that the amount of data transmitted between IoT nodes and servers is reduced in the edge computer paradigm. The IoT nodes or mobile devices therefore require additional calculation tasks. In this way, network transfer or contact delay avoids the transmission bottleneck. The performance and energy efficiency of programs such as field-programmable gate arrays have proven superior to the use of CPUs or GPUs [1] in terms of programmable hardware devices, including FPGAs. Therefore, the framework on IoT or mobile platforms is very promising. Coarse grain reconfigurable architecture (CGRA) is another type of programmable devices. In general, CGRAs provide less complicated and more regular communication between processing elements or processor cores than FPGAs to speedup applications.

Other than these there are variety of challenges which has to be faced by society due to IoT that ranges from the social structure maintenance, establishment of safe and secure society, enhancing the efficiency of manufacturing unit, and the development of advanced transportation system for more complex society issues. To address these social issues, new administrations and advances are continually being created.

These advances include monitoring of sensor networks [2], analysis of big data [3], and image processing technology by artificial intelligence (AI) [4] although these technologies are very much dependent on cloud storage and cloud information processing [5]. In case of enhancing the real-time capabilities and reducing the amount of data, it is very much needed that processing like data analysis, data accumulation, and compression has to be done on device or edge side. The devices should have characteristics so that they can perform task or information processing on the cloud side. Characteristics required by devices are high-performance power ratio; real-time response and robustness although to develop this characteristic more integrated circuit are needed so that we can cope up with the problem of heat generation. Traditional CPUs are using conventional integrated circuit with von Neumann computing architectures, and it is becoming very difficult to improve the power consumption ratio using this old architecture.

Its not an easy task to develop a solution for the problems related to power consumption, some of the solutions are like application-specific integrated circuit can be build and which are considered to be power efficient; but at the same time, they are having constraints like limited flexibility, programmability as well as they are expensive too. There are different types of semiconductor integrated circuits which are used in IoT devices like ASIC, FPGA, and CPU or MCU [6]. They all have their own advantages and disadvantages; but in recent times, FPGA has done a great improvement in the area of Internet of things (IoT) due to its reconfiguration and low-power consumption property. So in this paper, we will study about FPGA with nanobridge technology and also discuss the SRAM-based FPGA and nanobridge-FPGA.

Figure 1, it is shown that FPGA is providing good flexibility as well as less more energy-efficient. This is the reason today most of the devices consist of FPGA, and they all are performing well in terms of performance.

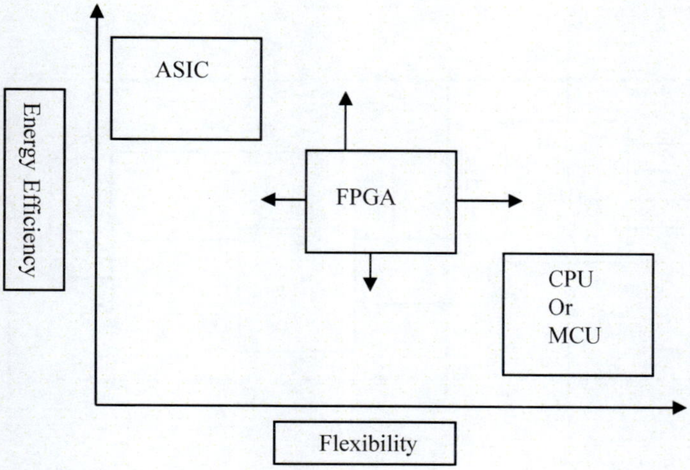

Fig. 1 Block diagram of FPGA

1.1 Field-Programmable Gate Arrays (FPGA)

Field-programmable gate array is the semiconductor devices that carry matrix of configurable logic blocks (CLBs) with wire circuitry and is used to perform different logical functions. This combinational logic blocks are connected using some hierarchy which makes FPGA to be used for different application. It is inexpensive, easy realization of logic network in hardware and simple as software. Hardware of FPGA contains PLD, logic gates, SRAM to perform digital functions and useful for storing data set. It contains a layout in matrix form, and user can configure each logic block. Along with digital functions, it has features of digital functions but is not as flexible as digital functions. Vendors like Xilinx, Actel, Altera provide FPGA which are different in programmability, interconnection, and basic functionality of logic blocks. Because of its combinational logic blocks and the interconnected matrix, FPGA becomes very powerful and flexible technology (Fig. 2).

FPGA has a benefit that it can be reprogrammable; so whenever user wants to change the design, they can easily modify, update, and download a new configuration file which can be used in new application. This functionality makes it flexible and cost-efficient, whereas other devices like ASSP and ASICs have fixed hardware. As FPGA is reprogrammable, it can be used as a prototyping purpose for ASIC. Because of the presence of large number of gates, it finished its task in a short time. FPGA design cycles are simpler as compared to CPU means designing tools check the function. FPGA performs parallel task like computing and processing because of its large resource of logic gates and wiring, which makes it fast as compared to general CPU and also multitasking (Fig. 3).

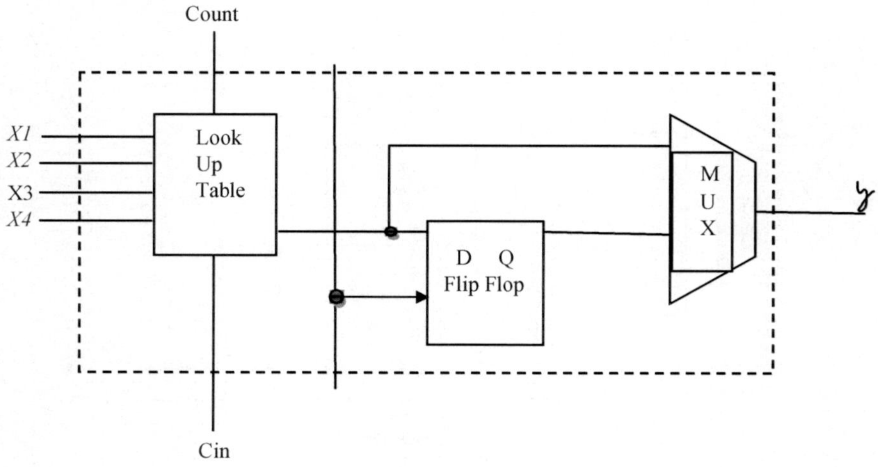

Fig. 2 Block diagram of FPGA

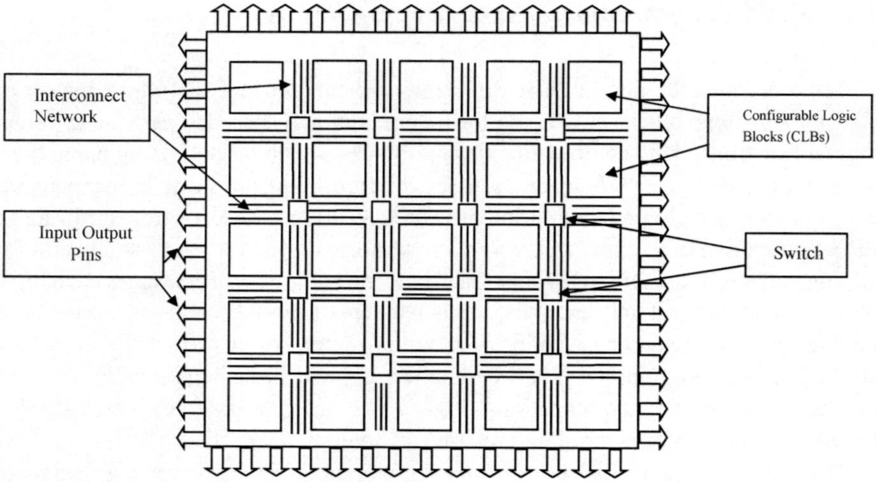

Fig. 3 FPGA architecture

Along with its advantages, there are some drawbacks of FPGA like: Its size is much greater approximately 40 times than the other device and is better for low-quantity production; if it is used for large quantity, then the cost of the product may also increase. At the time of power out, SRAM-based FPGA not able to retain its architecture; for reloading configured data, it has taken the help of external non-volatile memory at power in, and this consumes power for hundreds of second. Thus, it draws much more power nearly 12 times, the ASIC and speed are also less than ASIC. Because of more power consumption programmer does not have any control on power optimization in FPGA, and also, performance over wide range of temperature is poor. In order to reduce the problem related to power consumption and temperature fluctuation in FPGA, nanobridge is used with FPGA as a computational switch. NB-FPGA also helps in reducing the cell area and signal delay which occurs in FPGA used for large circuit.

SRAM-based FPGA consists of six transistors per cells with several leak paths, resulting in large cell leakage current and high system static current [7]. Flash-based FPGAs consist, in contrast, of only one transistor with 1000× less flow rate per cell, which results in extremely low static energy. FPGA power consumption is:

Dynamic Current—Dynamic FPGA power consumption is:

$$P = CV2F$$

where
P = Dynamic power in watts
C = Load capacitance in μF
V = Operating voltage in volts
F = Frequency in kHz.

The higher frequency of service leads to increased power consumption. The calculation of dynamic power based on resource usage and frequency is given by all FPGA suppliers. Before selecting a system unit, it is important to make the power evaluation. The cumulative power is provided by all power components as mentioned in many research papers. If a computer increases, a power-up current adds to the overall system power consumption, which in higher densities can reach a limit of several hundred mA or several amperes. While the device is operated, the power budget is controlled by static and dynamic power.

Thermal control is also a problem from a machine perspective when the average power consumption is high. If the thermal control is not taken into account in such a situation, it can cause thermal depletion.

2 Literature Survey

There has been increased in the number of portable device in the environment of wireless communication. It has been observed that nowadays normal house has almost 4–8 portable devices for their daily needs. It is predicted that in the future, it will get increased day by day, and it will become a challenge for the manufactures to make a device that consumes low power. In early stages of the development, researchers does not much bother about the power consumption but later on they feel the power must be minimized so that it can become feasible for everyone to use. In IoT, there are many papers based on power reduction techniques some are in the form of software and some are in the form of hardware.

Shambavi et al. [8] have designed BCD adder which is based on architectural low-power synthesis. There are various levels of abstraction at which the power can be minimized, but architectural low-power techniques affect more than circuits. To order to minimize energy consumption at architectural level, two different approaches have been explored, pipelines, and parallelism. VHDL and xilinx ISE 10.1 aim the Xilinx XC5VLX30-2 FPGA to evaluate and implement the proposed designs. The results show optimum power, delay, and range optimization for various designs, and a comparative analysis is provided.

Tanesh et al. [9] used the LUVCMOS IO protocol for the highest frequency of i7 processors to make ALU energy-efficient. LVCMOS12 is observed to have a 26.23, 58.37, and 75% lower IO power reductions than LVCMOS18, LVCMOS25, and LVCMOS33 with 1 GHz, respectively, as being the most energy-efficient LVCMOS12 available. They use the MOBILE DDR IO standard instead of the mainstream IO standard LVCMOS33 which was used for typical ALU.

Nehru [10] has implemented low-performance binary coded decimal (BCD), which uses a four-bit multiplexer and or gate-based look ahead adder (MOCLA) to form a basic building block; this product has reduced the silicone area of the chip and power consumption. The implemented MOCLA design uses two complete adder and PG input MUX, OR, gate diffusion input (GDI) and is used in low-power BCD adder circuits.

Cisneros et al. [11] were concerned with the design, synthesis, and execution of arithmetic logic units (ALUs) on FPGA and have a similar clock gating technique which is characterized by low-power consumption. Musavi et al. [12] designed IoT-based frame buffer to store frames and to use them in frame detecting changes in object position in a power-efficient and thermal conscious frame buffer. They insert a 128-bit IPv6 address into the frame buffer so that IoTs can be programmed. They used this configuration with various WLAN channel operating frequencies to test the reliability of this design with the wireless network. The frequency of service on WIFI networks (802.11 hp, 802.11 hp, 802.11 hp, 800 MHz, 2.4, 3.6, 5.9, and 60 GHz), respectively, is 802.11, 802.11, and 802.11 hp.

Abdullah et al. [13] aim to increase the range of Ethernet-based networks through pseudonoise generator-based optical transmitter. Structural changes are suggested to be implemented with various IEEE802.3 protocols (IEEE optical standards) in PN generator-based optical Ethernet energy-efficient transmitter (IEEE802.3az). Prashanth et al. [14] have implemented the DSP modules, such as the ALU floating point is introduced and designed. The design is based on FPGA Cyclone II's performance and implementation takes place after simulation of the functional and timing. ModelSim is the method for simulation. The Quartus II method for synthesis and implementation illustrates functional and timing analysis for all DSP modules using high-performance, Altera tools for synthesis and application.

Gaurav et al. [15] present some techniques for minimizing power that can be applied to FPGAs-centric communication designs. The literature includes various techniques, but most of them are only used at device level. This report gives an insight into reducing the power of design hierarchy architecturally, using the CAD tool XPower Analyzer. The techniques suggested are applied for analysis purposes on the arithmetical and logical circuit as ALU which are used as processing elements in portable wireless devices. On XILINX ISE to Spartan 3E XC3S250E FPGA, the circuit has been tested and realized, and the study has shown significant power consumption change.

3 Nanobridge Technology

To overcome from the problems of FPGA, there is an evolution of new technology known as nanobridge which is developed by NEC consists of memory that is non-volatile and which can replace both semiconductor switch and semiconductor-based memory. This technology is based on polymer-solid electrolyte (PSE) which is incorporated in between an inert electrode—ruthenium (Ru) and an active electrode—copper (Cu). When positive voltage is applied on the copper electrode, copper gets ionized and shift through polymer-solid electrolyte (PSE) to form a bridge between the two given electrodes. This way nanobridge switches from high resistance (off) to low resistance (on) condition. Conversely, when positive voltage is applied to the copper electrode, the copper atoms, which have formed the bridge, are retrieved by the copper electrode and transition to a high resistance condition. There is no power requirement for maintaining ON and OFF condition of switch because of its electrical properties which include lesser load capacity almost 1/10th consumption than semiconductor switch with high resistance difference of ON/OFF condition, i.e., of approximately 500 Ω and 1 GΩ. This is a reason why nanobridge can function like a semiconductor memory.

The nanobridge can be rewritten repeatedly and does not require power to maintain ON and OFF conditions (non-volatile). Its electrical characteristics include smaller load capacity than the semiconductor switch (about 1/10th) and high ON/OFF resistance difference (500 Ω and 1 GΩ). Therefore, the nanobridge can function as a semiconductor memory to switch logic signals and retain circuit data. Moreover, because the thickness of the copper bridge is estimated to be just a few nanometers, it can be incorporated in a circuit even under a design rule with advanced integration.

The voltage of transition depends on the voltage sweep rate. The pipe transition can be repeated, and every state continues when the voltage is low. This behavior is observed repeatedly for up to 3–103 cycles. The period number is of the order of 105 for devices with a hole diameter of 0.3 μm. Each state has more than a month's retention period.

Conductance switching can be clarified when the metal bridge within the Cu_2S film is formed and dissolved. When this top electrode has a negative voltage, Cu ions are electrochemically neutralized and precipitated in Cu_2S (Fig. 4). Cu ions are supplied at the Cu_2S/Cu interface through this electrochemical reaction. The Cu bridge is dissolved in the solid electrolyte by applying positive voltage, which results in an OFF state.

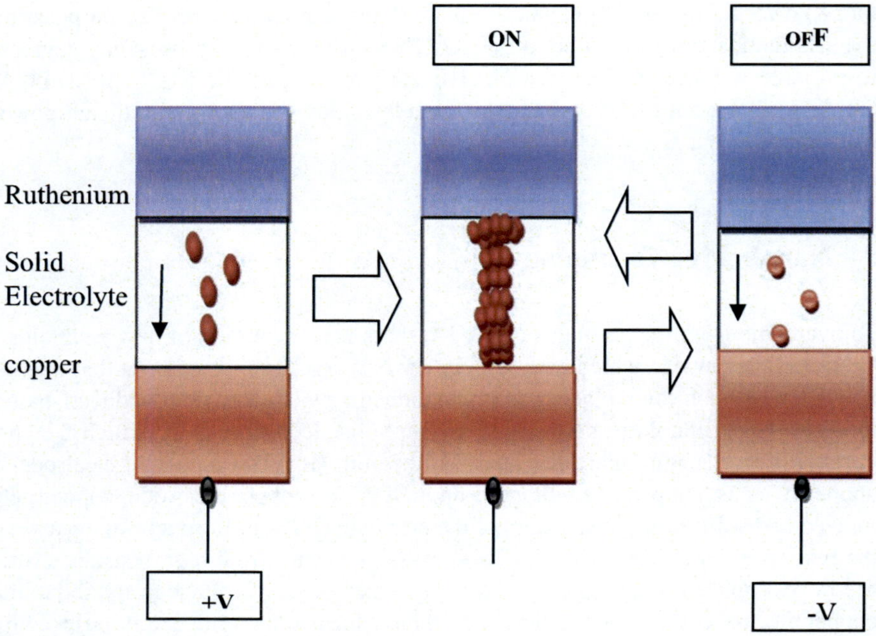

Fig.4 Nanobridge switch

3.1 *Performance Analysis of Nanobridge-FPGA*

The nanobridge-FPGA is contrasted with the commercially available SRAM-FPGA developed by a 40 nm semiconductor process in the assessment of nanobridge-FPGA's efficiency. The nanobridge-FPGA is developed by NEC, and this has high radiation tolerance which can support for the use of LSIs in space. The nanobridge-FPGA and SRAM-FPGA can be compared based on the number of the L UT as LUTs and DFFs are also used in the logic circuits of the SRAM-FPGA. In the same 40 nm semiconductor process, the density of the logical circuits provided by nanobridge-FPGA is two times the density of SRAM-FPGA. Therefore, the region occupied by the logic circuits can be minimized if a nanobridge is integrated (Table 1).

The nanobridge-FPGA device configuration is decided by the nanobridge-FPGA high and low output. NEC offers design tools for generating ON and OFF device circuit information. The application circuit is defined at the RTL level, and information from HIGH/LOW can be provided with the manufactured FPGA tools [16]. The first step is to obtain a gate-level Netlist using the RTL description as a logic synthesis tool. The Netlist is translated by a cluster package tool to 4 input LUTs. In addition, the positioning and connections of the LUTs are determined via a positioning and routing tool for the circuit-setting information.

The Verilog test bench produces the output signal pattern generated from the nanobridge-FPGA input signal pattern implemented with an application circuit. A

Table 1 Comparison of SRAM-FPGA with nanobridge-FPGA

Parameters	SRAM—FPGA	Nanobridge—FPGA
Switch	Transistor	Nanobridge
Circuit configuration memory	SRAM	Nanobridge
No. of LUTs	1320	6440
Density of LUTs (/mm^2)	1365	2577
Process node (nm)	40	40
Min. operating voltage (V)	0.92	0.68
Max. speed (@1 V) (MHz)	32	63
Operating power (μW/MHz)	40	13.5

semiconductor tester can be used to check whether the output signal pattern is compatible with nanobridge-FPGA operations and the correct circuit operation. Using the semiconductor tester, we calculated operating voltage, signal delay, and operating speed.

In this work, we have taken 40 nm process node because it provides more benefits than prior nodes like 65 nm node and latest 45 nm. One of the most important benefits is that it provides higher integration and density improvement [17]. The minimum operating voltage of a SRAM-FPGA is 0.92 V but the nanobridge-FPGA can operate at voltages of 0.68 V. SRAM-FPGA operates at a power of 40 μW/MHz while that of nanobridge-FPGA reduced to 13.5 μW/MHz. The smaller capacity of nanobridge and its short cable lengths make it possible to reduce the size of the logic circuits, which decreases the length of the cable, are largely attributable to these differences in performance.

4 Conclusion

In order to reduce the power consumption and fluctuation of temperature when FPGA is used in different IOT application, we proposed nanobridge switch. This paper focused on improving the performance in IOT devices through FPGA. The combination of FPGA with nanobridge technology provides better results for low-power consumption and temperature fluctuation and also in case of operating speed related to conventional SRAM-based FPGA architecture. In this paper, the nanobridge is used as a semiconductor memory to retain data and also as a logical signal switch.

References

1. Aluru S, Jammula N (2014) A review of hardware acceleration for computational genomics. IEEE Des Test
2. Akyildiz IF, Su W, Sankarasubramaniam Y, Cayirci E (2002) A survey on sensor networks. IEEE Commun Mag
3. Wu X, Zhu X, Wu GQ, Ding W (2014) Data mining with big data. IEEE Trans Knowl Data Eng 26(1)
4. Miller DD, Brown EW (2018) Artificial intelligence in medical practice: the question to the answer? Am J Med
5. Baliga J, Ayre RWA, Hinton K, Tucker RS (2011) Green cloud computing: balancing energy in processing, storage, and transport. In: Proceedings of IEEE
6. Nair R, Sharma P, Bhagat A, Dwivedi VK (2019) A survey on IoT (Internet of Things) emerging technologies and its application. Int J End-User Comput Dev
7. Ostler PS et al (2009) SRAM FPGA reliability analysis for harsh radiation environments. In: IEEE transactions on nuclear science
8. Mishra S, Verma G (2013) Low power and area efficient implementation of BCD adder on FPGA. In: 2013 international conference on signal processing and communication, ICSC 2013
9. Kumar T, Pandey B, Das T, Chowdhry BS (2014) Mobile DDR Io standard based high performance energy efficient portable ALU design on FPGA. Wirel Pers Commun
10. Nehru K (2018) Power analysis data set for 4-Bit MOCLA adder. Data Br
11. Ortega-Cisneros S, Raygoza-Panduro JJ, Suardíaz Muro JJ, Boemo E (2005) Rapid prototyping of a self-timed ALU with FPGAs. In: Proceedings—ReConFig 2005: 2005 international conference on reconfigurable computing and FPGAs
12. Musavi SHA, Chowdhry BS, Kumar T, Pandey B, Kumar W (2015) IoTs enable active contour modeling based energy efficient and thermal aware object tracking on FPGA. Wirel Pers Commun
13. Abdullah MFL, Das B, Mohd Shah NS, Pandey B (2015) Energy-efficient Pseudo noise generator based optical transmitter for Ethernet (IEEE 802.3az). In: I4CT 2015–2015 2nd international conference on computer, communications, and control technology, art proceeding
14. Prashanth BUV, Kumar PA, Sreenivasulu G (2012) Design and implementation of floating point ALU on a FPGA processor. In: 2012 international conference on computing, electronics and electrical technologies, ICCEET 2012
15. Verma G, Kumar M, Khare V, Pandey B (2017) Analysis of low power consumption techniques on FPGA for wireless devices. Wirel Pers Commun
16. Miyamura M et al (2015) 0.5-V highly power-efficient programmable logic using nonvolatile configuration switch in BEOL. In: FPGA 2015—2015 ACM/SIGDA international symposium on field-programmable gate arrays
17. Corporation A (2009) Leveraging the 40-nm process node to deliver the World's most advanced custom logic devices. Altera White Paper, pp 1–8

Li-Fi Technology—A Study on a Modernistic Way of Communication

Harleen Kaur

Abstract Li-Fi implies light fidelity, a wireless data transmission technique which uses illumination or light as medium of communication. Li-Fi is an innovative technology based on visible light communication (VLC) that creates light as a media of communication replacing the cable wire communication. Li-Fi is evolved to beat the speed in Wi-Fi. The paper summarizes most of the research, developments and applications achieved so far and looks at the various aspects of the strengths and weaknesses of Li-Fi's technology.

Keywords VLC · Li-Fi · Wi-Fi

1 Introduction

Li-Fi, light fidelity, is a wireless technology centered on light and not on radio waves as clear from its label. Li-Fi was proposed first time by Prof. Harald Haas, at the University of Edinburgh, and due to the practical demonstration conducted, the concept being compared with Wi-Fi, led to be appreciated and scattered quite fast. Li-Fi is an alternative of Wi-Fi [1] that transmit data using the spectrum of visible light. Li-Fi technology used light emitting diode (LED) for transmitting data wirelessly [2]. Li-Fi is fast and cheap optical adaptation of the Wi-Fi. Speed of Li-Fi with superfast broadband system is 250 times faster. Li-Fi will need a transceiver fitted LED lamps to transmit and receive information as well as can glow a room, i.e., sending data through LED bulb that varies in intensity faster than human eye can follow. The transfer of data can be in any form of light, i.e., light may be invisible, ultra violet or visible part of spectrum and the speed is extremely high. Li-Fi working technology is based on the use of visible light between red and blue. Li-Fi uses the optical spectrum but Wi-Fi uses the radio part of electromagnetic spectrum [1]. The principal of Li-Fi is simple when the LED turn ON and OFF at high speed not visible

H. Kaur (✉)
Department of Electronics, Punjab Technical University, Jalandhar, India
e-mail: harleen687@gmail.com

N. Marriwala et al. (eds.), *Mobile Radio Communications and 5G Networks*, Lecture Notes in Networks and Systems 140,
https://doi.org/10.1007/978-981-15-7130-5_55

to human eye. The amplitude of light is varied. Any LED light can set if used along with a microchip inside it can be used to detect the changes in the light flickering thus helping in converting light into data in its digital form. When LED is ON it transmits the digital signal 1, and if LED is OFF it transmit a digital signal 0. A controller is also connected at the backside of these LED bulbs to code the data to these LED.

2 Li-Fi Design and Working

It is constructed on visible light communication [3] which uses light radiations between 780 and 375 nm as optical carrier for data transmission and illumination. The main components of Li-Fi systems are: (1) high brightness white LED's which act as a transmitter, (2) silicon photodiode which act as a photoreceiver. Likewise, factors like line of sight (LOS) and presence of light are much essential.

2.1 How Does Li-Fi Work?

In practice, Li-Fi technology comprises of LED lamp as the media transmission and photodetector as a receiver of transmitted data. Lamp driver is desired to make LED working properly [2]. Although amplification and processing are capable to deal with the signal that originates from the photodetector. Fundamental perception for working norm in Li-Fi technology is directing to: Transceiver and light as a media transmission (Fig. 1).

Figure 2 is a fundamental concept block diagram for Li-Fi. This simple concept indicates as a duplex communication. The speed of Li-Fi is 14Gbps using three off-the-shelf laser diodes (red, green, and blue) and expects the rate till 100 Gbps when the entire visible spectrum is used. Rendering to [3] Li-Fi and VLC utilized a similar medium as a data communication that is light, and the difference between Li-Fi and VLC is VLC has unidirectional, point-to-point light communication at low data rates. While the Li-Fi technology is entirely organized, bidirectional, and high speed wireless communication. Others assumed Li-Fi is the fusion of Wi-Fi and VLC [1].

Transceiver. Transceiver is a unit that acts as a transmitter and receiver simultaneously. This transceiver comprises of LED to transmit the light and photodiode to receive the light. Amplifier is implanted to strengthen the power of light received from the photodiode. The modem is used to modify and demodulate the signal. The signal that emanates from the photodiode is analog and it gets converted into digital in the modem and sent by LED. The driver before the LED functions to drive the flow of current of the LED in order to get the flickering. The flickering is function of the LED for data transmission, if LED is ON then it transmits digital '1' and if OFF, it transmits digital '0' [4, 5].

Modulation. The modulation signals are used to switch LED at desired frequencies that contain information to be transmitted. According to Islim and Haas [6] there

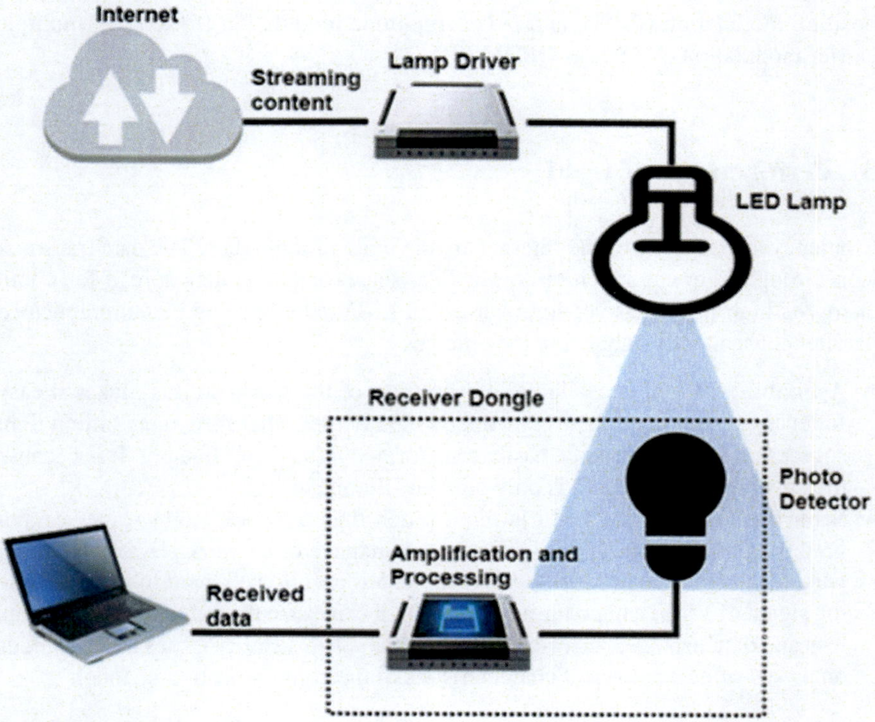

Fig. 1 Basic concept diagram of Li-Fi

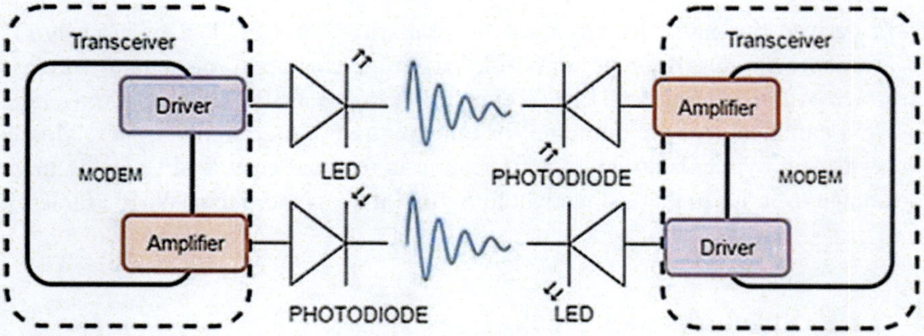

Fig. 2 Li-Fi transceiver based on VLC

are several techniques of modulation in Li-Fi. Modulation methods are needed in the communication, even the illumination is not required. Because of that, a modulation technique may support a dimmable illumination. The deviation in intensity of light links to the information in the message signal. There are many kinds of modulation in Li-Fi, i.e., single carrier modulation (SCM) [7] as On–Off keying (OOK), pulse position modulation (PPM), and pulse amplitude modulation (PAM) [8]; multiple carrier modulation (MCM) as OFDM [9].

3 Performance of Li-Fi

Efficiency and security of the Internet are the dominating issues. The data transmission through light can reach up to gigabits per second. The data spread for a unit energy is high in the case of light waves. In Li-Fi data, bits can be communicated parallel consequently enhancing the efficiency.

- **Availability**: Light is available in every part of the world so this makes it easy for every person in airplanes to work on the Internet. There are many billion light sources on earth that can be easily transformed into a Li-Fi hotspot. It is because the coverage area of Li-Fi is only on their illuminate area.
- **Security**: For security, Li-Fi is more tough than a Wi-Fi. Light waves cannot go through solid objects thus providing abundance of network privacy. No other person can split the network unless the holder has allowed them to use it. While the signal of Wi-Fi can go through the wall, it can cause the vulnerabilities in data loss and data leakage. According to Blinowski [10], security issues in VLC focus on aspect of basic physical characteristics of the communication channel.

3.1 Advantages of Li-Fi Over Wi-Fi

Li-Fi derived this name by virtue of the similarity to Wi-Fi. The Li-Fi design is to overcome the disadvantage of Wi-Fi. Wi-Fi operates good for general wireless coverage within buildings and Li-Fi is epitome for high density wireless data coverage inside a confined area or room and for relieving radio interference issues. Mostly power-driven by LEDs so its communication is fast and easy. Li-Fi has the main advantage over Wi-Fi that its bandwidth is 10,000 times more than Wi-Fi (Table 1).

4 Li-Fi Strengths and Weaknesses

Li-Fi, as a potential substitute of Wi-Fi radio waves looming congestion and wireless dead zones, has strengths and weaknesses that put it in competition with other industry specific technologies like RFID and iBeacons. First of all, we consider this new technology's speed as the most important strength. Then, security, low cost, lack of electromagnetic smog, and therefore safe for human use, more energy friendly are some other strengths that worth being considered.

Table 1 The difference between Li-Fi and Wi-Fi

Parameter	Li-Fi	Wi-Fi
Standard	IEEE 802.15.xx	IEEE 802.11xx
Frequency band	1000 times of THz	2.4 GHz
Data transmission	Bits	Radio waves
Data rate	Around 1–10 Gbps	150 Mbps–1 Gbps
Communication	Based on visible light communication	Based on radio frequency communication
Transmitter/receiver	LED	LED
Coverage range	10 m	20–100 m based on power and antenna
Power consumption	Less	More
Secure	More	Less
Efficiency	More	Less
Interference	No interference issues with RF waves	Interference with neighbor AP routers

1. **High Speed**—Data rates of 1 Gbps have been reported using single phosphor-coated white LEDs [11]; later, 3.4 Gbps has been demonstrated with an off-the-shelf red–green–blue (RGB) LED [12]. Researchers of the University of Edinburgh reported 3.5 Gbps from a single color incoherent LED [13]. Haas's group developed commercial [14, 15] speeds of up to 100 Gbps, this speed being attainable for Li-Fi when the complete visible range is exploited [16]. The University of Oxford Li-Fi scientists hit 224 Gbps in laboratory environments (at this speed, one moment is enough to download five high definition motion pictures) [17]. Latest investigations have validated data rates of 14 Gbps for Li-Fi with three off-the-shelf RGB laser diodes [18]. 42.8 Gbit/s has been accomplished for indoor OWC with two-dimensional optical beam-directing, in 2015 [19].
2. **Security**—Li-Fi optical signals are incapable to infiltrate partitions, this being a benefit concerning security issues. The similar aspect can be manipulated to exclude interferences between adjacent cells. In the course of past ten years, there have been persistent statements of improved point-to-point link data speeds using standard white LEDs under trial laboratory circumstances [12].
3. **Low cost**—In certain cases, Li-Fi's implementation can be cheaper than other communication technologies [20].
4. **Low power intake**—Taking into justification that, mostly, indoors lights are ON more often, the energy used for communication would almost be zero as a result of the piggybacking of data on illumination. Energy-efficient intensity modulation (IM) techniques authorized at a communication even if the lights are visually off [18, 21].
5. **No electromagnetic smog**—Since Li-Fi uses visible light spectrum (VLS) that it is unimpeded by RF interference and therefore generates no electromagnetic smog. This makes it an increasingly attractive option where electromagnetic

interference becomes dangerous for intrinsically hazardous environments, such as refineries, oil platforms and petrol stations, near MRI medical equipment, airplanes or electrical transformer stations.

6. **Safe for health**—The technology could even find a niche with consumers who are aware and affected by the adverse effects of electromagnetic radiation.
7. **License-free bandwidth**—The visible light spectrum includes hundreds of THz of license-free bandwidth, 10,000 times more than the entire RF spectrum. On the other side of the balance, there are still some weaknesses of the Li-Fi's technology developed so far.
8. **Distance and interferences**—One of the constraints is distance, typically not more than 10 m, without intervention of simulated light or dazzling sunlight. Li-Fi cannot be operated in direct sunbeams (or other conditions along with harsh lighting) since the photodetectors would not be able to discover the modulating light waves. This could bind the locations in which it could be used.
9. **No standard:** Li-Fi has no standard yet, so equipment from one vendor would not work with another's.

5 Applications of Li-Fi

Lots of vicinities where Li-Fi usage offers a consistent, secure, economical and extreme high speed communication. Setup has already been introduced worldwide, so we can recapitulate some of them as follows:

- **Airplanes**: Since Wi-Fi in flights with the majority of airlines is outlawed, and thus limited, Li-Fi can be an appropriate substitute for wireless communication. The use of this know-how within aircraft cubicle has benefits since hefty measures of wiring can be prevented resulting in cost saving, lowered loads and flexible layout design.
- **Hospitals**: Li-Fi can be upgraded in clinics and hospitals also where the Wi-Fi is prohibited. They can be used in several improved medical device to communicate with each other for quick data interpretation [22].
- **Nuclear power plants**: Li-Fi can be a convenient replacement of Wi-Fi in electromagnetic sensitive regions such as nuclear power plants as it does not instigate any electromagnetic intrusion.
- **In home and office purposes**: Li-Fi system can be incorporated in home appliances such as: secure systems, freeze, central heating systems, TV's and so on to lower energy consumptions for a smart energy management.
- **Smart illumination**: Public lighting can be used to provide Li-Fi hotspots and monitoring lighting and sensor arrangement [23].
- **Vehicle and traffic lights**: LED devices can be mounted as headlights and tail-lights building a rational transport system. Traffic beam scan also moves to LED with the gain of road safety and traffic control.
- **Underwater**: Light proliferate sub-aqua where radio frequencies cannot be exhausted because of saline, high conductivity and high attenuation atmosphere.

While cables build threads in transfer through marine water and can be supplanted with Li-Fi transmitters. Also, they can relay data to submarines, to surface as well as to divers alongside their head lights.

- **In well-being surveillance**: The wearable Li-Fi transmitter like LED wristlets, ear rings, wrist watches, etc., permit nonstop scrutinizing individual status of health offering instant notice to family' physician of any noteworthy changes happened in health condition by connecting to the Internet and updating the information online in real time.
- **Interior navigation**: The LED lights sources are used in like shopping malls, cinema halls, government organizations, work offices or any indoor areas, the Li-Fi technology allows an indoor routing (pinpointing an article within about 10 cm)and offers placement information (valuable for understanding, for illustration, in which direction a client is looking) [24, 25].

References

1. Ayyash M et al (2016) Coexistence of WiFi and LiFi toward 5G: concepts, opportunities, and challenges. IEEE Commun Mag 54(2):64–71
2. Haas H, Yin L, Wang Y, Chen C (2016) What is LiFi? J Light Technol 34(6):1533–44
3. Dobroslav Tsonev HH, Videv S (2014) Light fidelity (Li-Fi): towards all-optical networking. In: Proceedings of SPIE 9007, broadband access communication technologies VIII, vol 9007
4. Singh PYP, Bijnor UP (2013) A comparative and critical technical study of the Li-Fi (a future communication) v/s Wi-Fi 2(4):2011–2013
5. Miteku N (2015) Li-Fi over Wi-Fi in internet data communication. IJIREEICE 3(12):153–159
6. Islim MS, Haas H (2016) Modulation techniques for Li-Fi. Mohamed ZTE Commun 14:29–40
7. Revathi S, Aarthi G (2011) Performance analysis of wave length division and sub carrier multiplexing using different modulation techniques. Int J Eng Res Appl 1(2):317–320
8. Ghassemlooy Z, Yuan Y, Tang X, Luo P (2017) SVM detection for superposed pulse amplitude modulation in visible light communications
9. Afgani M, Haas H, Elgala H, Knipp D (2006) Visible light communication using OFDM. In: Proceedings of 2nd international Conference on Testbeds research, infrastructures networks communities, pp 129–134
10. Blinowski G (2015) Security issues in visible light communication systems. IFAC-Papers Online 28(4):234–239
11. Kalid AM, Cossu G, Corsini R, Choudhury P, Ciaramella E (2012) 1-Gb/s transmission over a phosphorescent white led by using rate-adaptive discrete multi-tone modulation. IEEE Photon J 4(5):1465–1473
12. Cossu G, Khalid AM, Choudhury P, Corsini R, Ciaramella E (2012) 3.4 Gbit/s visible optical wireless transmission based on RGB LED. Opt Exp 20(26):B501–B506
13. Tsonev D, Chun H, Rajbhandari S, McKendry JD, Videv S, Gu E (2014) A 3-Gb/s single-LED OFDM-based wireless VLC link using a gallium nitride μLED. IEEE Photon Technol Lett 26(7)
14. Kuppusamy P, Muthuraj S, Gopinath S (2016) Survey and challenges of Li-Fi with comparison of Wi-Fi. In: Proceedings of 2016 IEEE, international conference on wireless communication signal process. Networking (WiSPNET) pp 896–899
15. Virk GK, Fidelity L (2015) Li-Fi: A new communication mechanism. Int J Comput Appl 118(15):1–4

16. Tsonev D, Videv S, Haas H (2015) Towards a 100Gb/s visible light wireless access network. Opt Exp
17. Islim MS, Haas H (2016) Modulation techniques for Li-Fi. ZTE Commun 14(2):29–40
18. Gomez A, Shi K, Quintana C (2015) Beyond 100-Gb/s Indoor wide field-of-view optical wireless communications. IEEE Photon Technol Lett 27(4)
19. Oh CW, Tangdiongga E, Koonen AMJ (2015) 42.8 Gbit/s indoor optical wireless communication with 2-dimensional optical beam-steering. In: Optical fiber communications conference and exhibition (OFC)
20. Dimitrov S, Haas H (2015) Principles of LED light communications towards networked Li-Fi. Cambridge University Press, Cambridge CB2 8BS, United Kingdom
21. Dimitrov S, Haas H (2012) Optimum signal shaping in OFDM-based optical wireless communication systems. In: 64th IEEE proceedings of vehicular technology conference (VTC Fall)
22. Alao OD, Joshua JV, Franklyn AS, Komolafe O (2016) Light fidelity (Li-Fi): an emerging technology for the future. IOSR J Mob Comput Appl (IOSR-JMCA) 3(3):18–28. e-ISSN: 2394-0050, P-ISSN: 2394-0042
23. Pawar A, Anande A, Badhiye A, Khatua I (2015) 'Li-Fi': Data transmission through illumination. Int J Sci Eng Res 6(10)
24. Sharma R, Sanganal A, Pati S (2014) Implementation of a simple Li-Fi based system. IJCAT Int J Comput Technol 1(9)
25. Chatterjee S, Agarwal S, Nath A (2015) Scope and challenges in light fidelity (Li-Fi) technology in wireless data communication. Int J Innov Res Adv Eng (IJIRAE) 2(6). ISSN: 2349-2163

A 28 GHz Corporate Series-Fed Taper Antenna Array for Fifth-Generation Wireless Communication

Mohit Pant, Leeladhar Malviya, and Vineeta Choudhary

Abstract This paper proposed a 1 × 16 antenna array to cover 28 GHz bands for 5G wireless communications systems. The design approach is based on 16-element tapered antenna array with corporate (parallel)-series feed structure network. The individual element of the antenna consists of the inset feed for achieving the proper impedance matching and tapered patch for achieving the desired gain, bandwidth and miniaturization. The whole configuration is designed over a Rogers RT/duroid 5880 dielectric substrate size of $28 \times 30 \times 0.79$ mm^3. The measure peak gain is 15.85 dB at 28 GHz. The simulated result shows a return loss of −45.73 dB, bandwidth of 27.56–28.381 GHz, gain of 15.85 dBi and 93.36% of radiating efficiency.

Keywords 5G · Taper · Millimeter wave · Corporate series-fed

1 Introduction

Fifth generation of wireless application takes the benefit of high data rate, frequency spectrum saving, power saving, link improvement and multipath propagation. The compact designs of antennas made it possible for them to resonate at higher frequencies [1]. Federal Communications Commission (FCC) has suggested millimeter wave (MMW) spectrum for the deployment for fifth-generation (5G) networks which can achieve high data rates, high bandwidth, low latency and low compact size. The higher frequency spectrum for 5G band around 28 GHz covers frequencies range from 24.25–27.5 GHz, 26.5–27.5 GHz, 26.5–28.95 GHz and 27.5–28.35 GHz [2].

M. Pant (✉) · V. Choudhary
Ujjain Engineering College, Ujjain 456010, India
e-mail: ermohitpant@gmail.com

V. Choudhary
e-mail: vinita1988@rediffmail.com

L. Malviya
Shri G.S. Institute of Technology and Science, Indore, Madhya Pradesh 452003, India
e-mail: ldmalviya@gmail.com

N. Marriwala et al. (eds.), *Mobile Radio Communications and 5G Networks*, Lecture Notes in Networks and Systems 140,
https://doi.org/10.1007/978-981-15-7130-5_56

New technologies such as massive MIMO and mmWave are the main concentration for 5G wireless communication system. In view of mmWave spectrum, high gain and antenna arrays are important for reducing free space path loss, atmospheric attenuation and foliage loss at higher frequencies. Feed networks radiate very little in comparison with the patches when design on the same substrate because radiation from fringing fields on both sides of the microstrip lines cancel each other. Therefore, feed structures are frequently used in 5G patch array antenna [3].

In the recent years, several patch antenna arrays have been reported. A new arrangement of microstrip series-fed tapered array with five elements was designed for 5 GHz resonant frequency. The reported antenna has a side-lobe level of −14 dB. However, the reported antenna array is designed with only one inset feed [4]. The analysis of double-layer series-fed linear array antenna is investigated. The design antenna has a side-lobe level as low as −20 dB and measured gain of 13.5 dB across a bandwidth of 23% [5]. However, the structure complexity may lead to extra loss. A series-fed microstrip antenna array using unequal inter-element spacing has been presented in Ref. [6]. The antenna achieved the measured gain which is 14.4 dBi at 9.0 GHz operating frequency. The design of a butler matrix for a feed network combined with either patch or slot antenna array was investigated [7, 8]. The maximum gain of 15 dBi was achieved with the designs. A 28 GHz horizontally and vertically polarized 16-element array antenna was investigated that employs digital beamforming. The design had 15.65 dBi gain and 63% radiation efficiency [9]. A novel series-fed patch array antenna is realized on a duroid 5880 substrate. The bandwidth of the designed antenna array was enhanced by truncating patches with the opposite corner. The measure gain of the 4 × 16 array is around 21 dBi for frequency band from 37 to 39 GHz [10].

Microstrip patch antenna becomes a natural choice for low-cost mmW antenna array due to their low profile, lightweight and compactness. The important part to design the antenna is to choose the proper dielectric substrate because the losses such that dielectric loss, conductor loss and radiation loss are very large in the millimeter wave. Rogers RT/duroid, FR-4, polyimide, etc., are used to develop the 5G antennas [11]. A series-fed array antenna design on Rogers RT/duroid 622 substrate material of 254 mm thickness and dielectric constant 2.9 and loss tangent of 0.0015 have been reported. The maximum gain measured is of 12.39 dBi [12].

In this paper, 16-element corporate series-fed tapered antenna array is proposed, which will improve the total gain. Microstrip feed is preferred to avoid cutting of substrate and conductive layers as compared to coaxial feed structure and also helped to minimized surface waves [13]. Antenna is designed for 28 GHz resonant frequency which can be used for 5G wireless communication applications.

2 Antenna Design

In the Fig. 1, 1 × 16 elements proposed antenna is designed with each of four arms are having inset feed. In the first three arm of each series array the slot of width 0.8 mm and length of 0.5 mm is truncated. Tapering is done by varying the patch

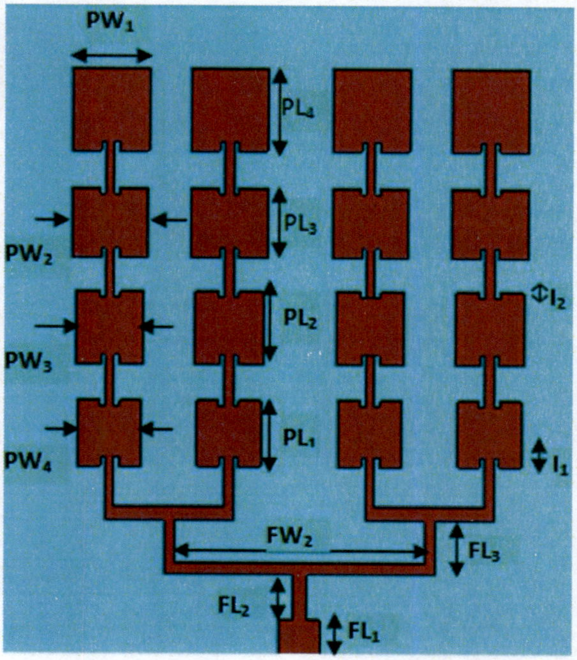

Fig. 1 Schematic front view of the proposed design

width and length. Each element of the array has a distance of λ/4 mm between adjacent elements. The proposed array antenna of 1 × 16 is designed on the Rogers RT/duroid 5880 dielectric substrate which has the relative permittivity ε_r of 2.2 with loss tangent 0.009 and thickness of 0.79 mm. The patches are fed by combination of series and parallel microstrip transmission line of the proper set of dimensions as shown in Table 1. The total size of the substrate is 36 × 33 × 0.79 mm^3.

The antenna design process consists of two steps. In the first step, 1 × 16 element antenna array is simulated without any inset feed and slot in the patches. The second step consists of inset feed in each antenna array arm, and the fixed length and width of slot in the first three arms are cut off in the antenna array as shown in Fig. 1.

Table 1 Dimensions of the proposed patch antenna array

Parameters	Values (mm)	Parameters	Values (mm)
S_L	36	S_W	33
FL_1	2.45	FL_2	2.52
FW_1	2.45	FW_2	0.80
FW_3	0.60	FW_4	0.57
PW_1	3.5	PW_2	3.7
PW_3	4.0	PW4	4.1
*I*1	0.5	*I*2	0.3

Fig. 2 Schematic back view of the proposed design

All the optimized values are shown in Table 1. The length and width arrays of each patch are varied so that each element resonates at the same frequency and maximum gain is achieved for the entire frequency band of 27.56–28.38 GHz (Fig. 2).

3 Result and Discussion

3.1 Resonant Frequency

With help of CST microwave studio (CSTMWS) 2018 simulation tool, results are calculated by using the inbuilt particle swarm optimizer (PSO) for best optimized parameters. All the design steps are compared in Fig. 3. In design step-I with 1×16 array without any inset feed and without cutting slot on patches, antennas resonate at 28.28 GHz and cover 27.86–28.63 GHz frequency band. In design step-2, 1×16 array antennas with inset feed and slot on patches, the antenna resonates exactly at 28 GHz and occupies 27.56–28.38 GHz bandwidth in 2:1 VSWR band.

The whole configuration is designed over a substrate size of $36 \times 33 \times 0.79\ mm^3$. The proposed antenna resonates at 28 GHz in 27.56–28.38 GHz frequency band. The simulated result shows a return loss of −45.73 dB and the voltage standing ratio VSWR < 2 at 28 GHz indication that the antenna has minimum reflection at this frequency. The change of inset feed as shown in Fig. 4 is investigated in the range of 0.50 mm to 1.00 mm. It was observed from the graph that as we increased the

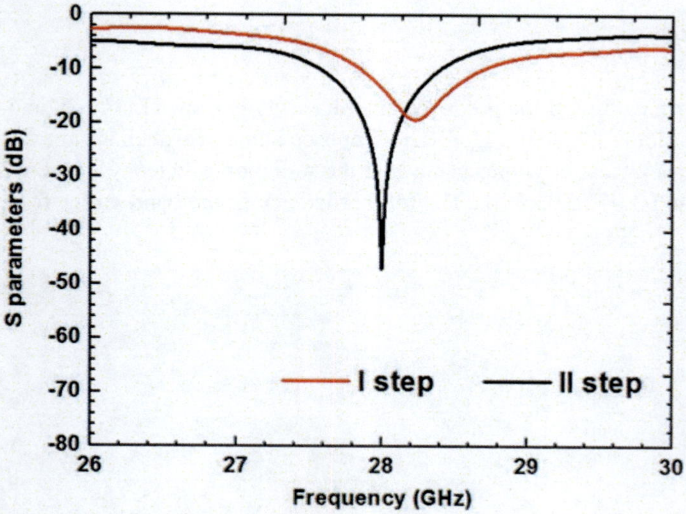

Fig. 3 S_{11} parameters for design antenna for step 1 to step 2

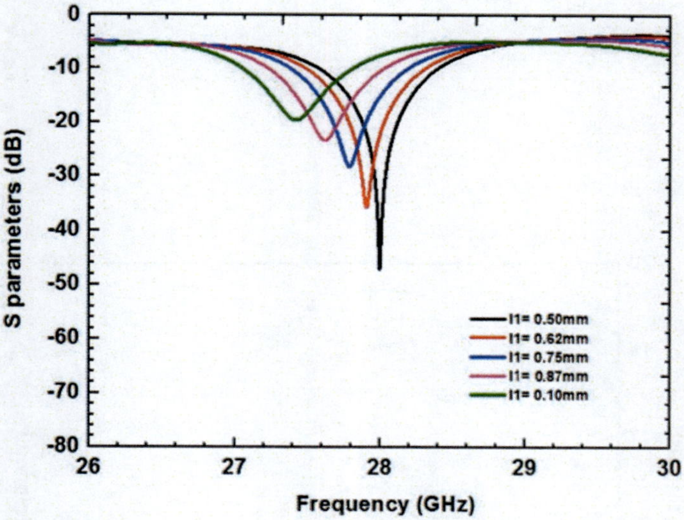

Fig. 4 Effect of different inset feed lengths on S_{11} parameters

inset feed, the resonance frequency of the design is shifted from higher frequencies to lower frequencies. It is clearly shown from Fig. 4 that the value obtained in return loss factor with the optimum resonant frequency is obtained when inset feed length is 0.5 mm which is obtained by black line.

3.2 Gain

The 3D gain pattern of the proposed antenna array is shown in Fig. 5, and the gain–efficiency plot is shown in Fig. 6. The proposed antenna array has gain varying from 15.32 to 16.42 dBi in whole band, and the radiation efficiency in the whole band varies from 93.05 to 93.78%. The total efficiency in the band varies from 79.18 to

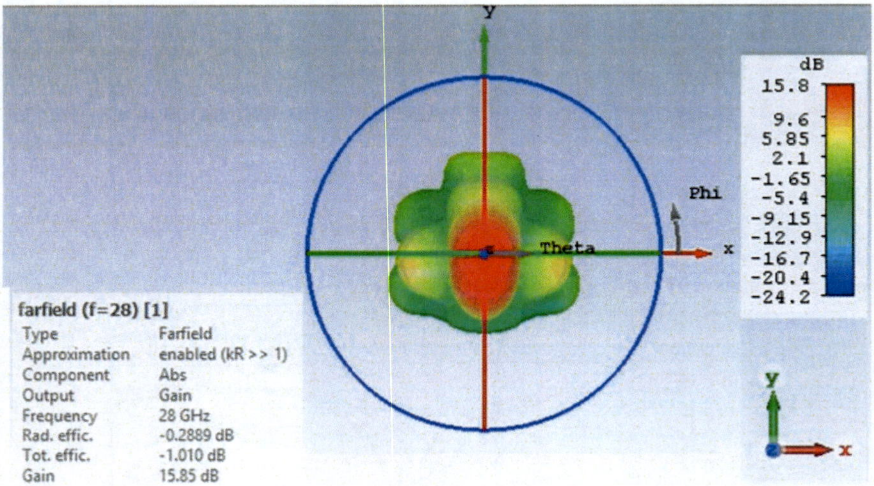

Fig. 5 3D gain pattern

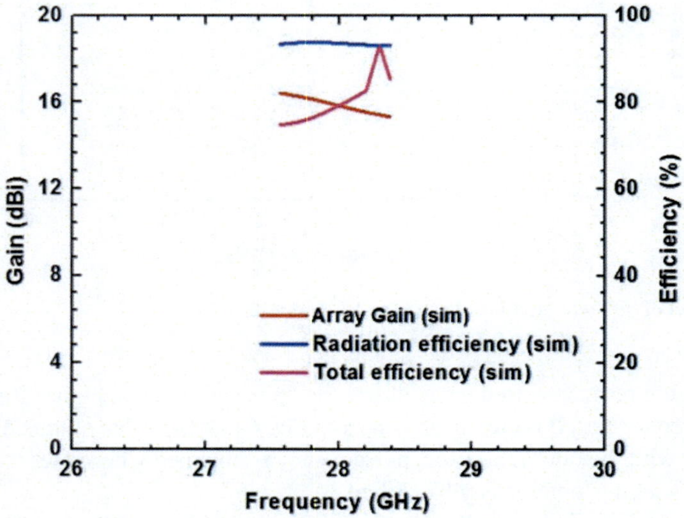

Fig. 6 Gain–efficiency

84.20% GHz. At resonant frequency, the gain measured is 15.80 dBi and the radiation efficiency is 93.56%.

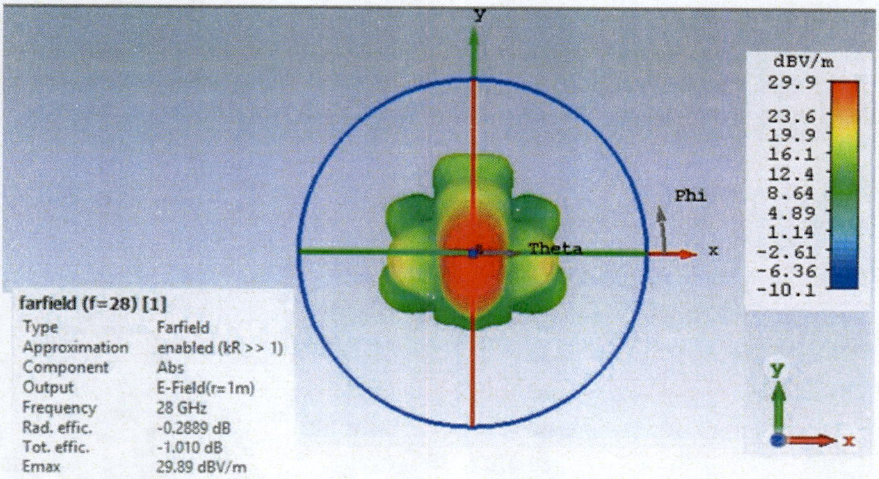

Fig. 7 3D E-field pattern

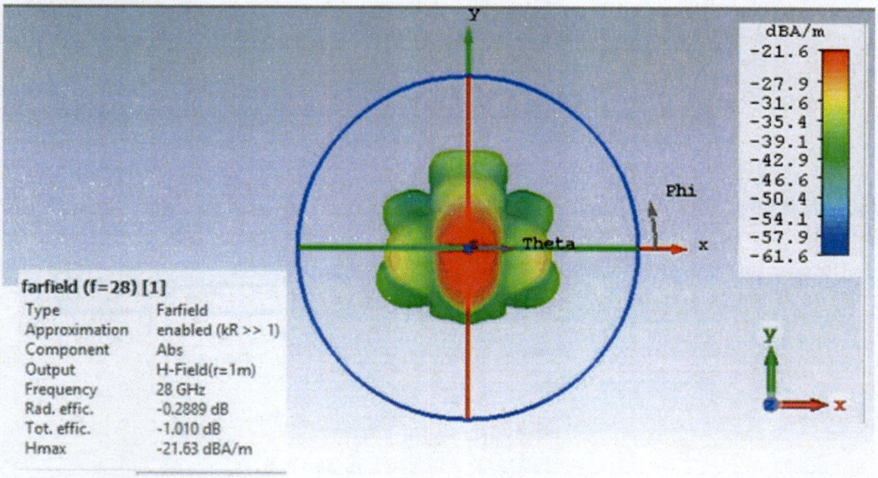

Fig. 8 3D H-field pattern

3.3 Radiation Pattern

The 3D E-field and H-field radiation patterns are shown in Figs. 7 and 8, respectively, and the normalized E-field and H-field patterns are shown in Figs. 9 and 10, respectively. The proposed antenna array has 29.9 dBV/m value of E-field at resonant frequency and −21.6 dBA/m value of H-Field. The antenna array has 31.80 beamwidth and the main lobe directions of E-field and H-field is 0°.

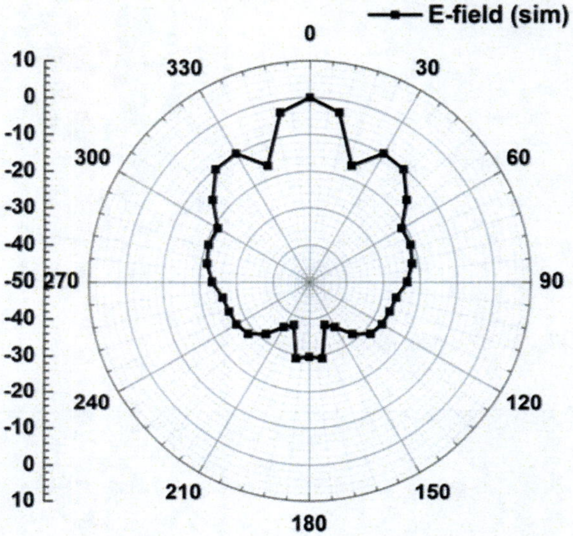

Fig. 9 Normalized E-field pattern

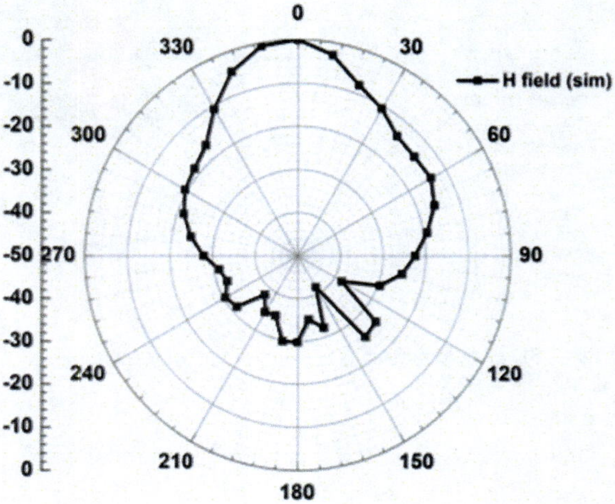

Fig. 10 Normalized H-field pattern

4 Conclusion

In this paper, an antenna array of size $36 \times 33 \times 0.79$ mm^3 is presented. The present antenna is designed for 5G communications in the 28 GHz band. The array is designed with 16 elements with measured gain which is 15.85 dBi. The antenna -10 dB impedance bandwidth varies from 27.56 to 28.381 with VSWR ≤ 2. The combined effect of adding inset feed and slot has improved the impedance matching, bandwidth and gain. The measure return loss is -45.73 dB at 28 GHz resonant frequency. The proposed antenna array suits the 5G wireless communication.

References

1. Malviya L, Panigrahi RK, Kartikeyan MV (2018) Four element planar MIMO antenna design for long-term evolution operation. IETE J Res 64:367–373
2. Gupta P, Malviya L (2019) 5G multi-element/port antenna design for wireless applications: a review. Int J Microw Wireless Technol 11:918–938
3. Milligan TA (2005) Modern antenna design, 2nd edn. Wiley Inc., Hoboken, New Jersey
4. Yuan T, Yuan N, Li L-W (2008) A novel series-fed taper antenna array design. IEEE Antennas Wireless Propag Lett 7:362–365
5. X. Chang, Huyan, P. Gao et al.: Stacked series-fed linear array antenna with reduced side lobe," Electronics Letter. vol. 50, no. 4, pp. 251- 253.(2014).
6. Yin J, Wu Q, Yu C, Wang H, Hong W (2017) Low-side lobe-level series-fed microstrip antenna array of unequal interelement spacing. IEEE Antennas Wireless Propag Lett 16:1695–1698
7. Yang QL, Ban YL, Lian JW, Yu ZF, Wu B (2016) SIW butler matrix with modified hybrid coupler for slot antenna array. IEEE Access 4:9561–9569
8. Tseng CH, Chen CJ, Chu TH (2008) A low-cost 60-GHz switched-beam patch antenna array with butler matrix network. IEEE Antennas Wireless Propag Lett 7:432–435
9. Tariq S, Psychoudakis D, Eliezer O, Khan F (2018) A new approach to antenna beamforming for millimeter-wave fifth generation (5G) systems. In: Texas symposium on wireless and microwave circuits and systems (WMCS), pp 1–5
10. Chen H-C, Chiu T, Hsu C-L (2019) Design of series-fed bandwidth-enhanced microstrip antenna array for millimetre-wave beamforming application. Int J Antennas Propag 1–10
11. Malviya L, Chouhan S (2019) Multi-cut four-port shared radiator with stepped ground and diversity effects for WLAN application. Int J Microw Wireless Technol 11(10):1044–1053
12. Moubadira M, Mchbala A, Touhamia NA, Aghoutanea M (2019) A switched beamforming network for 5G modern wireless communications applications. Sci Direct Procedia Manuf 753–761
13. Malviya L, Panigrahi R, Kartikeyan MV (2016) Circularly polarized 2×2 MIMO antenna for WLAN applications. Prog Electromagn Res C 66:97–107

Design and Analysis of Gain Enhancement THz Microstrip Curvature Patch PBG Antenna with Inset Feed

Rashmi Pant, Leeladhar Malviya, and Vineeta Choudhary

Abstract Terahertz frequencies have become significant in communication systems to execute the huge demand of the next-generation wireless communication systems for high data rate, high capacity, and low latency applications. In order to achieve high-speed wireless communication, antenna designs with high gain and high radiation efficiency are needed. The present research demonstrates the microstrip curvature patch antenna design with inset feed on polyimide inhomogeneous substrate using periodic photonic band gap crystal. Photonic crystal substrate contains several sets of air holes perforated in the polyimide substrate, where each set has its own specific radius. The proposed antenna design covers frequency range from 0.6342 THz to 0.6911 THz and resonates at 0.6588 THz with −43.47 dB return loss at 2:1 VSWR. The simulated result shows that the proposed antenna design achieves 8.956 dBi gain with 86.96% radiation efficiency, 0.6342–0.6911 THz bandwidth at the resonant frequency of 0.6588 THz which makes it suitable for the THz wireless communication system.

Keywords Terahertz · PBG · CST · Polyimide · Photonic crystal

1 Introduction

Over the last few years, a huge demand of high data rate in wireless communication attracted researchers toward Terahertz (THz) communication links. THz radiations are a part of electromagnetic spectrum that lies between microwave and infrared

R. Pant (✉) · V. Choudhary
Ujjain Engineering College, Ujjain, Madhya Pradesh 456010, India
e-mail: rashmi.dhanotia@gmail.com

V. Choudhary
e-mail: vinita1988@rediffmail.com

L. Malviya
Shri G.S. Institute of Technology and Science, Indore, Madhya Pradesh 452003, India
e-mail: ldmalviya@gmail.com

N. Marriwala et al. (eds.), *Mobile Radio Communications and 5G Networks*, Lecture Notes in Networks and Systems 140,
https://doi.org/10.1007/978-981-15-7130-5_57

regions, and it offers high bandwidth to fulfill the requirement of high-speed communication. THz technology has grown rapidly due to its capability to penetrate high with low attenuation loss, non-ionizing nature, high water absorption, high resolution imaging power, etc. [1, 2]. However, atmospheric path loss at THz band is the main challenge that affects the communication distance. The thicker dielectric substrate and large ground plane in microstrip antenna achieve high performance parameters but suffers from surface wave losses [3]. The photonic band gap (PBG) structure on substrate is utilized nowadays to suppress undesirable excited surface waves in the substrate [4]. The PBG-based substrate is able to diffract surface waves at the edges and improves the efficiency of the antenna [5–7]. There are many structures which are generalized structure and can be utilized to produce any frequency band by changing the radius of curvature of patch only [8].

The elementary structure is the rectangular patch curved on both sides. The curved-shaped rectangular patch improves the reduction in size at certain extent. The thick substrates improve the electrical performance of microstrip patch antenna. However, the surface waves introduced due to high substrate thickness ranges ($0.003\lambda \leq h \leq 0.05\lambda$) degrade the performance of antenna [9–12]. To suppress the undesired surface waves, the PBGs employed on the substrate and obtain the benefits of thick substrates, which enhance the performance of proposed design in terms of gain and efficiency. The optimized dimensions of antennas are finally calculated by using CST microwave studio tool.

2 Antenna Design

In this paper, we designed and simulated the proposed microstrip curvature patch antenna with inset feed based on inhomogeneous polyimide substrate. The substrate comprised polyimide having the dielectric constant 3.5 and loss tangent of 0.0027. The homogeneous substrate of thickness 191.29 μm is selected to excite surface wave in antenna. The substrate is having dimension of 700 μm × 800 μm × 191.29 μm, along with multiple concatenated duplicates of the air cylinder of same thickness substrate having radius of 25 μm and 87.5 μm distance apart. The PBG structure employed as substrate to improve the characteristics of antenna parameters.

The radiating element is having dimensions of 540 μm × 580 μm × 7 μm with slotted circular curvature of radius of 45 μm which is mounted over the ground plane. The microstrip feed line has feed dimension of 70 μm × 150 μm × 7 μm to achieve 50Ω impedance. The thickness of metal employed is 7 μm, which is same for patch, feeding line and ground plane. The feed line is inset to the patch with the width of 107 μm and length of 115 μm. The proposed structure is designed and simulated using CST microwave studio, and optimization is done using inbuilt particle swarm optimization.

The proposed design consists of microstrip curvature patch antenna with inset feed, and PBG structure is employed on polyimide substrate instead of homogeneous substrate. Using CST microwave studio tool, optimized dimensions are shown in

Table 1 Dimension of the proposed microstrip curvature patch PBG antenna

Parameters	Values (μm)	Parameters	Values (μm)
S_L	700	S_W	800
P_W	580	F_i	115
F_L	70	D	50
F_W	150	r_o	45
S	87.5	S_h	191.29

Table 1. The schematic front and back view of the proposed microstrip curvature patch PBG antenna is shown in Figs. 1 and 2.

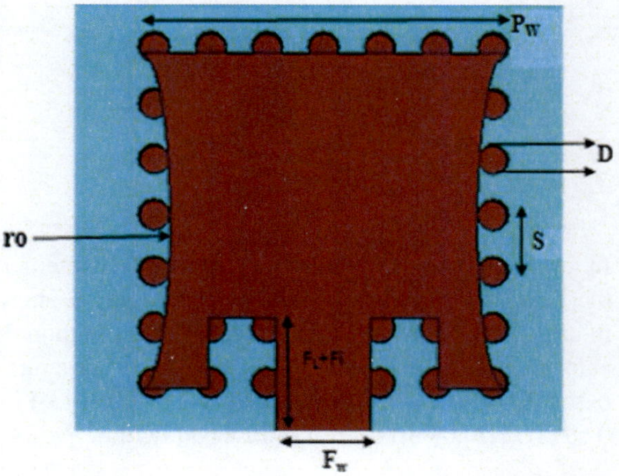

Fig. 1 Schematic front view of the proposed design

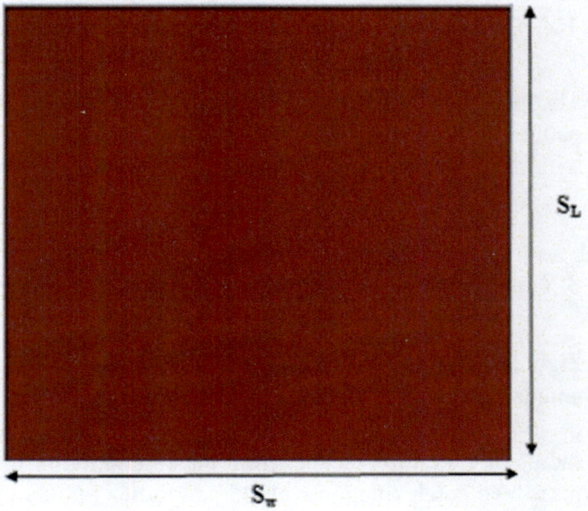

Fig. 2 Schematic back view of the proposed design

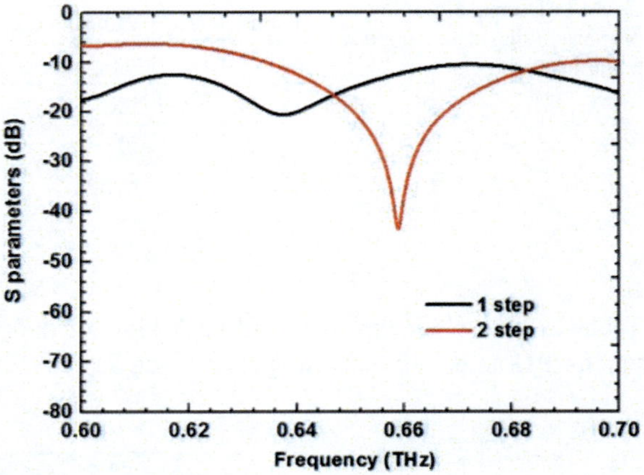

Fig. 3 S_{11} parameters for design antenna for step 1 to step 2

3 Result and Discussion

3.1 Resonant Frequency

In this work, microstrip curvature patch PBG antenna with inset feed is design with the help of computer simulation tool. Figure 3 shows return loss performance of microstrip curvature patch inset feed antenna without PBG in step 1 and the proposed antenna with PBG in step 2. The microstrip curvature patch antenna with inset feed based on PBG resonates at 0.6588 THz with −43.47 dB return loss covers 0.6342–0.6911 THz at −10 dB return loss bandwidth.

3.2 Gain

The gain and efficiency of the proposed antenna are shown in Fig. 4, and the 3D gain pattern is shown in Fig. 5. The proposed antenna has gain 8.956 dBi at 0.6588 THz with the radiation efficiencies of 86.96% in whole band.

3.3 Effect of Radius of Curvature on the Proposed Antenna

Figure 6 shows the effect of different radius of curvature (ro) of inset feed patch antenna on return loss. The change of radius of curvature is simulated for the range of 40–82 μm. It is observed that very slightly change in return loss with change in radius of curvature. Hence, from the simulation results of gain and return loss, the inset feed patch antenna with 45 μm radius of curvature has better performance at the resonant frequency 0.6588 THz.

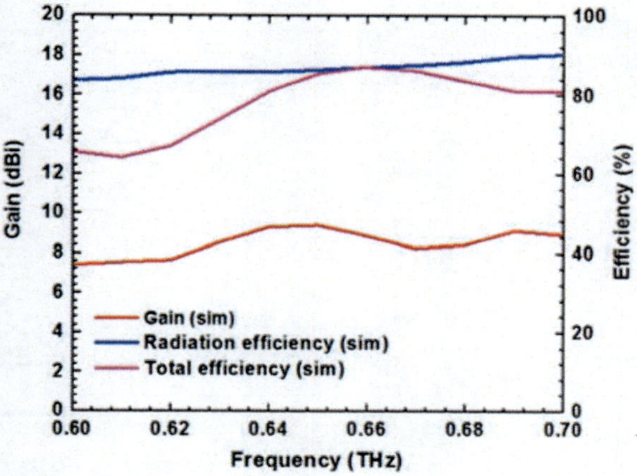

Fig. 4 Gain and efficiency of the proposed antenna

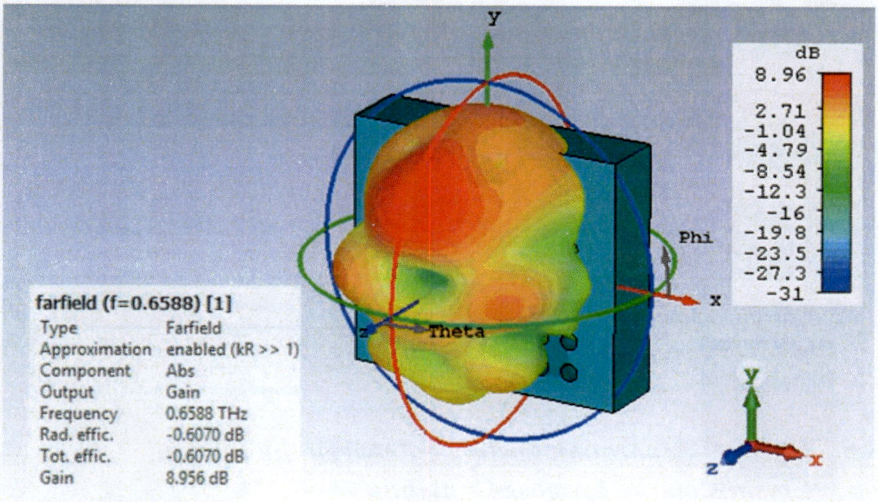

Fig. 5 3D gain pattern of the proposed antenna at 0.6588 THz

3.4 *Effect of Photonic Band Gap Hole Size on the Proposed Antenna*

Figure 7 shows the effect of PBG hole radius (r) for periodic PBG substrate structure on return loss. The change in hole radius is investigated for the range of 10–28 μm. It is observed that the resonant frequency shifted from low frequencies to higher frequencies as the hole radius increases. The most promising value of gain and return loss is achieved at hole radius 25 μm which is shown by pink line.

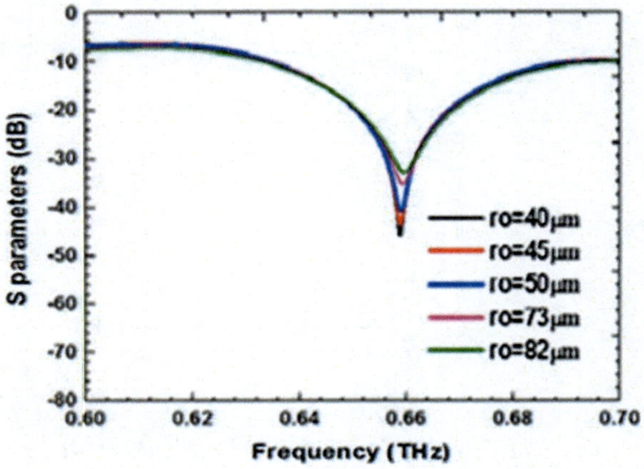

Fig. 6 Effect of different radius of curvature on return loss

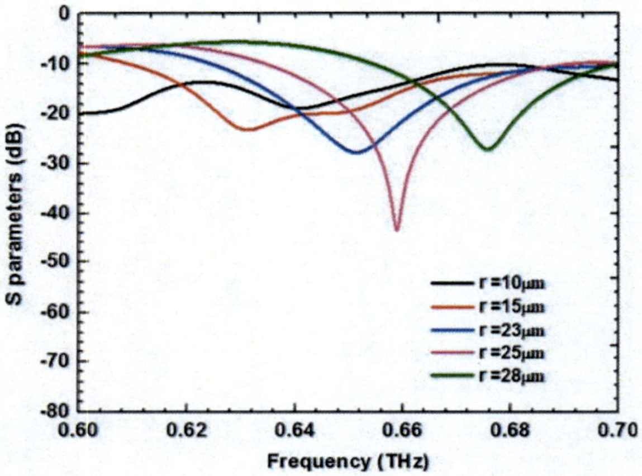

Fig. 7 Effect of different photonic band gap hole radii on return loss

3.5 Effect of Separation Distance Between the Photonic Crystals on the Proposed Antenna

Figure 8 shows the effect of separation distance (s) between photonic crystals for periodic PBG substrate structure on return loss. The change in separation distance is analyzed for the range of 55–90 μm. The variation in resonant frequency with slight change in return loss is observed with the change in separation distance between photonic crystals. The performance of gain and return loss is achieved at a distance 87.5 μm.

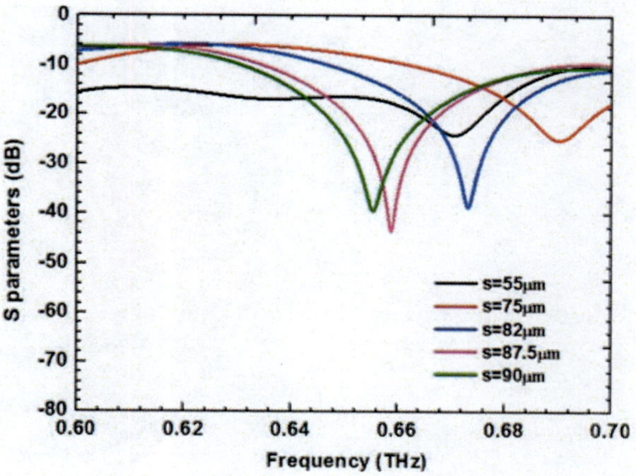

Fig. 8 Effect of separation distance (s) between photonic crystals

3.6 Radiation Pattern

The 3D E-field and H-field radiation patterns are shown in Figs. 9 and 10 at 0.6588 THz. The proposed antenna has 23.72 dBV/m value of E-field and −27.8 dBA/m value of H-field at resonant frequency at 0.6588 THz.

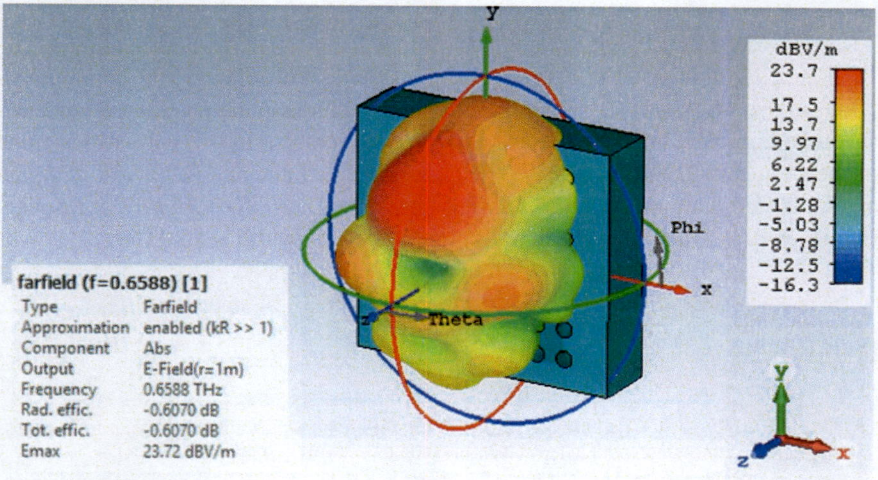

Fig. 9 3D E-field pattern

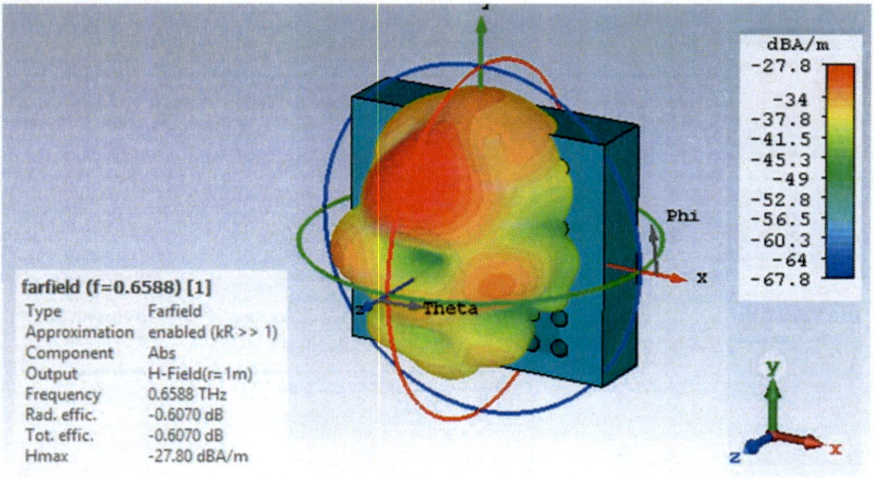

Fig. 10 3D H-field pattern

4 Conclusion

In the present work, we designed and analyzed THz microstrip curvature patch antenna based on periodic photonic band gap polyimide substrate. Antenna has been designed, based on PBG substrate in order to enhance its performance around 0.65 THz, making it more applicable in the next generation of wireless communication. Simulation results show that the inset feed with curvature patch based on PBG substrate improved the performance parameters of conventional microstrip antenna in the frequency band of 0.5–0.7 THz. The proposed antenna has 8.956 dBi gain and 86.96% radiation efficiency with minimal return loss of −43.47 dB at resonant frequency of 0.6588 THz. The overall impedance bandwidth is 56.9 GHz.

References

1. Gupta P, Malviya L, Charhate SV (2019) 5G multi-element/port antenna design for wireless applications: a review. Int J Microw Wireless Technol 11:918–938
2. Akyildiz IF, Jornet JM, Han C (2014) Terahertz band: next frontier for wireless communications. Phys Commun 12:16–32
3. Hocini A, Temmar MN, Khedrouche D, Zamani M (2019) Novel approach for the design and analysis of a terahertz microstrip patch antenna based on photonic crystals. Photon Nanostruct Fundam Appl 36:100723
4. Kushwaha RK, Karrupanan P, Malviya L (2018) Design and analysis of novel microstrip patch antenna on photonic crystal in THz. Phys B 545:107–112
5. Gonzalo R, Martınez B, Maagt P, Sorolla M (2004) Improved patch antenna performance by using photonic bandgap substrates. Microwave Opt Technol Lett 24:213–215

6. Waterhouse RB (2003) Microstrip patch antennas: a designer's guide. Springer Science Business Media, LLC
7. Temmar MN, Hocini A, Khedrouche D, Zamani M (2019) Analysis and design of terahertz microstrip antenna based on synthesized photonic band gap substrate using BPSO. J Comput Electron 18:231–240
8. Malviya L, Shakya A, Gehlod K (2018) 5.8 GHz WLAN MIMO antenna with power divider arms. In: IEEE Indian conference on antennas and propogation (InCAP)
9. Jha KR, Singh G (2011) Analysis and design of enhanced directivity microstrip antenna at terahertz frequency by using electromagnetic bandgap. Int J Numer Model Electron Netw Devices Fields 24:410–424
10. Mittala D, Sidhu E (2017) THz rectangular microstrip patch antenna employing polyimide substrate for video rate imaging and homeland defence applications. Optik Int J Light Electron 144
11. Federici J, Moeller L (2010) Review of terahertz and sub-terahertz wireless communications. J Appl Phys 107:111101–111122
12. Pozar DM (1983) Considerations for millimeter wave printed antennas. IEEE Trans Antennas Propag 31:740–747

Performance Analysis of Classification Methods in the Diagnosis of Heart Disease

Sonu Bala Garg, Priyanka Rani, and Jatinder Garg

Abstract With a mortality rate of 17.9 million per year, heart disease has emerged out to be the deadliest disease of the world. Early detection of this disease can reduce mortality. Data mining based disease diagnosis systems can aid medical professionals in the correct and timely diagnosis of the disease. In this study a Python-based data mining system, capable of diagnosing the heart disease using decision tree, KNN classifier, naive Bayes, random forest, and support vector machine (SVM) classification data mining methods, has been developed. The system was applied to four heart disease datasets obtained from the UCI machine learning repository. The relative performance of various data mining techniques was evaluated by comparing the results. The results showed that the support vector machine, with 98.7% efficiency, 98.4% precision, and 99.2% recall, has emerged out to be the best method for the diagnosis of heart disease.

Keywords Heart disease diagnosis · Data mining · Artificial intelligence

1 Introduction

In the contemporary world, heart disease has registered tremendous growth due to changing lifestyles and dietary habits. It is one of the most common diseases these days. The heart is a major organ of the human body. The ailment of the heart also affects the functioning of other organs [1]. Some of the prominent factors that trigger cardiovascular disease include high cholesterol level, hypertension, diabetes, smoking and drinking habits, and family background. Heart disease has become a major cause of death worldwide. However, timely detection and treatment can help in reducing the mortality rate significantly. The diagnosis of the disease is done

S. B. Garg · P. Rani
IK Gujral Punjab Technical University, Jalandhar, Punjab, India

J. Garg (✉)
Baba Hira Singh Bhattal Institute of Engineering and Technology, Lehragaga, Punjab, India
e-mail: jatindergarg@yahoo.com

N. Marriwala et al. (eds.), *Mobile Radio Communications and 5G Networks*, Lecture Notes in Networks and Systems 140,
https://doi.org/10.1007/978-981-15-7130-5_58

by medical professionals. However, data mining and artificial intelligence-based cardiovascular disease assessment systems have been designed to assist the doctor in making better diagnostic decisions. Such expert systems help in diagnosing the disease by matching the patient data with the disease data previously stored in the databases. Apart from increasing the effectiveness of the diagnosis, such systems also hold the potential to reduce the cost significantly [2]. However, such systems are not perfect even to date. Scientists and experts are making efforts to improve the diagnosis accuracy and efficiency of these systems to make them more powerful and reliable.

For a decision-making process, data mining is highly valuable. From a huge amount of data existing on the Internet, the valuable and only required information can be extracted through data mining. Data mining combines statistical analysis, machine learning, and database technologies to extract out hidden patterns from large databases. Researchers are using various data mining techniques to help medical professionals obtain more accuracy in dignosis decisions [3]. Healthcare industry generates a huge amount of useful data that can be used to extract out a variety of useful information [4]. With access to large amounts of patient data, healthcare organizations are now focusing on increasing the use of data mining methods to improve the quality of patient care [5]. A variety of data mining techniques have been utilized for designing heart disease diagnosis systems. These include neural networks, naive Bayes, genetic algorithm, decision tree, and support vector machine, prominently. For designing an effective system, it becomes imperative to analyses and compares these methods with the aim of finding the best one.

In the past, some researchers tried to develop disease diagnosis systems for the heart disease. A few of them also tried to evaluate the relative performance of various data mining algorithms. Thomas and Princy [6] presented a research study in which some data mining techniques were employed to provide an insight into the detection of heart diseases. Babu et al. [7] explored the usage of decision tree, MAFIA, and k-means algorithms for the diagnosis of cardiovascular diseases in various types of patients. Raju et al. [8] tried to develop an efficient treatment approach for the prediction of heart disease using various data mining classification methods. The support vector machine was found to be the best method as per the experimental outcomes. Shaji [9] developed a feature extraction-based medical expert system for the prediction of heart disease. The KNN, ANN, random forest, and SVM classifiers were implemented, on which these collected features were applied. Based on the accuracy of the results obtained, ANN classifier was found to be the best method.

In this study, a python-based heart disease diagnosis system has been developed. The developed system can carry out the disease diagnosis using decision tree, KNN classifier, naive Bayes, random forest, and support vector machine (SVM) classification data mining methods. The performance of these data mining methods has been evaluated and compared by applying them to four standard heart disease datasets obtained from the UCI repository.

2 Research Methodology

The various data mining algorithms were implemented on different datasets for the prediction of heart disease.

The methodology used for conducting the study is shown in Fig. 1.

2.1 *The Datasets*

For conducting this study, four heart disease datasets, namely Cleveland dataset, Hungarian dataset, Switzerland dataset, and VA Long Beach dataset were obtained from the UCI machine learning repository. These are the most widely used dataset for heart disease studies [10, 11]. These databases have been obtained from various healthcare systems and were donated by David W. Aha. All the data instances in these datasets have 75 attributes, but, for the sake of simplicity only 14 most relevant attribute had been used for the current study. These include age, sex, cp, trestbps, chol. Fbs, restecg, thalach, exang, oldpeak, slope, a, thal, num, and predicted attribute.

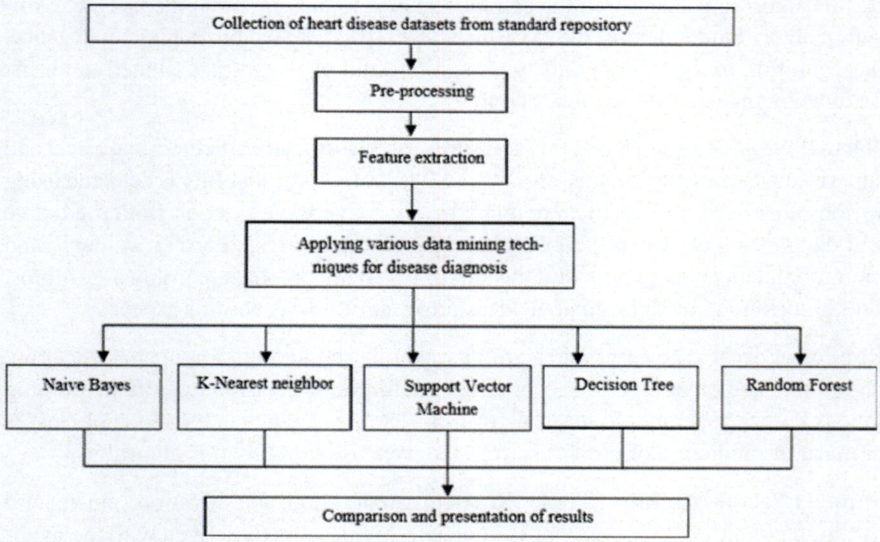

Fig. 1 Research methodology used

2.2 The Software Tool

Python is a widely used open-source high-level programming language. It works on almost all popular computer platforms like Windows, Mac, Linux, and Raspberry Pi. Python is widely used for handling big data and for performing complex mathematical tasks. The codes in Python can be executed as soon as they are written thereby providing a quick response. It is a very strong tool for statistics and offers a wide range of univariate and multivariate analysis capabilities. Python-based codes were developed for heart disease analysis using the following five data mining methods:

Decision Tree. A structure that is much similar to a flowchart is designed in several applications. A classifier that is formed to a similar structure is known as a decision tree. A test is represented on a feature value by each node available in the structure. The results of any test are represented by its respective branches. Class distribution is represented by tree leaves. This classifier is a predictive model that helps in performing classification based on the type of applications. The possible test results are labeled using the branches that extend from the nodes.

KNN Classifier. For performing learning, training samples are used by the k-nearest neighbor classifier. Each individual sample is pointed out in the n-dimensional space by this algorithm. The k-training samples nearest to unknown sample use a classification algorithm to identify the existing pattern space. Based on euclidean distance, the definition of closeness is given. A similar kind of weight is assigned to all the features by the nearest neighbor classifier.

Naive Bayes. For representing this algorithm as a classification method, statistical and supervised learning techniques are applied. The output's probability is detected using various unsure conditions that are related to the naive Bayes model. Both predictive and diagnostic issues could be handled through this process. Further, valuable and monitored data is integrated with the aim of resolving practical learning algorithms. This algorithm aims to evaluate the learning algorithms as per this aspect.

Random Forest. The ensemble learning method used for performing classification, regression, and other tasks by generating a multitude of decision trees at the training time is known as a random forest algorithm. The class which is the mode of classes or mean of prediction of individual trees is given as output by this algorithm.

Support Vector Machine (SVM). To perform regression, classification, and general pattern recognition, this classifier is known as highly effective. Even with the availability of high size input space, adding apriori knowledge is not necessary here. To distinguish members of two classes in the training data, an optimum classification function is identified here. To have an idea of the best classification, the merit can be understood geometrically. For having a linearly separable dataset, a linear classification function is equal to a separating hyperplane $f(x)$.

2.3 *The Performance Indicators*

The performance of each data mining method was evaluated in terms of the accuracy, precision, and recall.

a. **Accuracy**. The ratio of the number of instances classified correctly to the number of total instances multiply by 100 gives the value of accuracy.

$$\text{Accuracy} = \frac{\text{Number of points correctly classified}}{\text{Total Number of points}} * 100$$

b. **Precision**. The ratio of true positive instances to the total true positive and false positive instances is known as precision.

$$\text{Precision} = \frac{\text{True Positive}}{\text{True Positive} + \text{False Positive}}$$

c. **Recall**. The ratio of relevant instances retrieved to the total number of relevant instances is known as recall.

$$\text{Recall} = \frac{\text{True Positive}}{\text{True Positive} + \text{False Negative}}$$

3 Results and Discussion

The results achieved by each classifier on various datasets are given below.

Table 1 Comparison of results obtained by different algorithms on Cleveland dataset

Techniques	Accuracy (%)	Precision (%)	Recall (%)
Decision tree	72.53	79	71
KNN classifier	83.52	86	85
Naive Bayes	86.91	87	90
Random forest	87.91	87	92
SVM	86.32	86.3	91

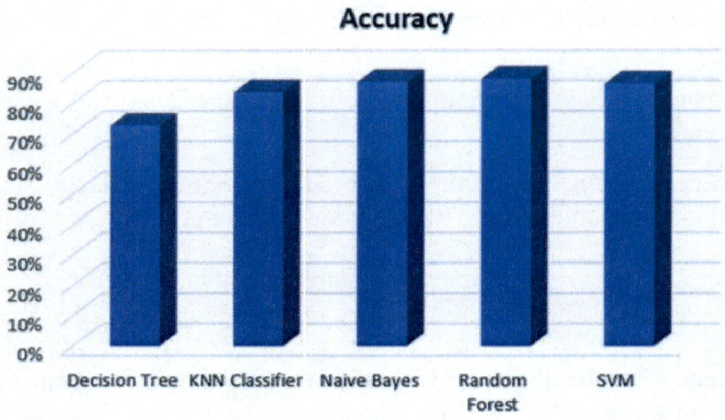

Fig. 2 Comaprison of accuracy for Cleveland dataset

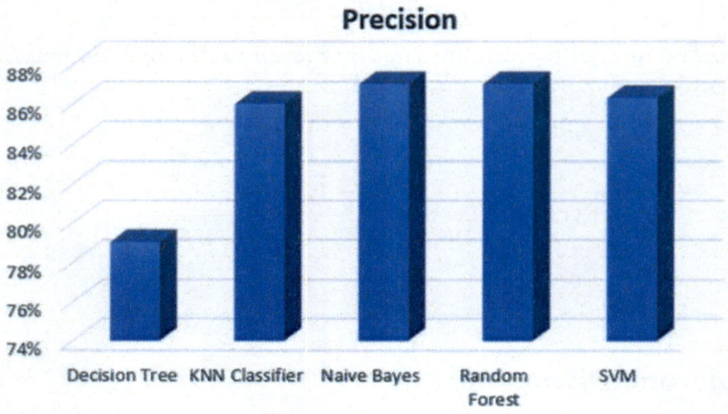

Fig. 3 Comparison of precision for Cleveland dataset

3.1 Cleveland Dataset

The results obtained for the Cleveland dataset have been presented in Table 1 and visually compared in Figs. 2, 3 and 4.

From these results, it can be clearly seen that for Cleveland dataset, the random forest method gives the best performance in terms of accuracy (87.91%), precision (87%), and recall (92%).

3.2 Hungarian Dataset

The results obtained for the Hungarian dataset have been compiled in Table 2 and compared visually from Figs. 5, 6 and 7.

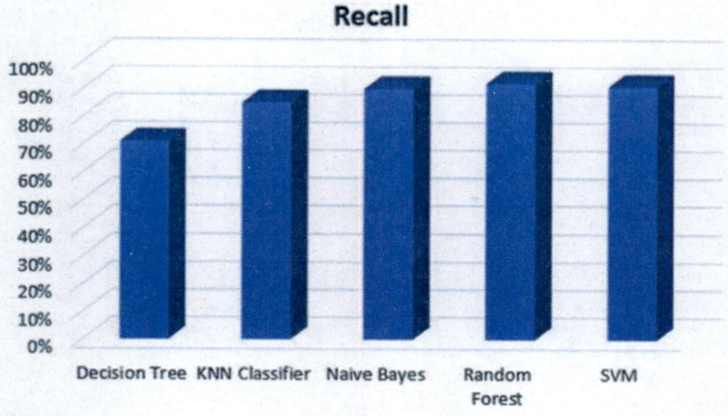

Fig. 4 Comparison of recall for Cleveland dataset

Table 2 Comparison of results obtained by different algorithms on Hungarian dataset

Techniques	Accuracy (%)	Precision (%)	Recall (%)
Decision tree	75.58	90	74
KNN classifier	69.76	76	84
Naive Bayes	80.23	92	79
Random forest	83.72	89	89
SVM	84.83	93	85

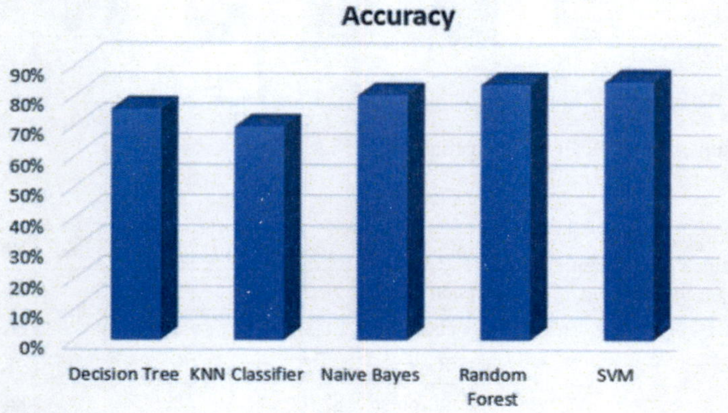

Fig. 5 Comparison of accuracy for Hungarian dataset

For the Hungarian dataset, the SVM method has emerged out to be the best performer with maximum accuracy (84.83%), precision (93%), and recall (85%).

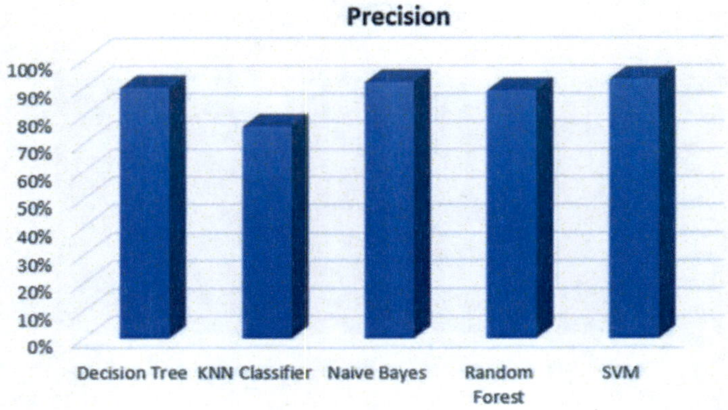

Fig. 6 Comparison of precision for Hungarian dataset

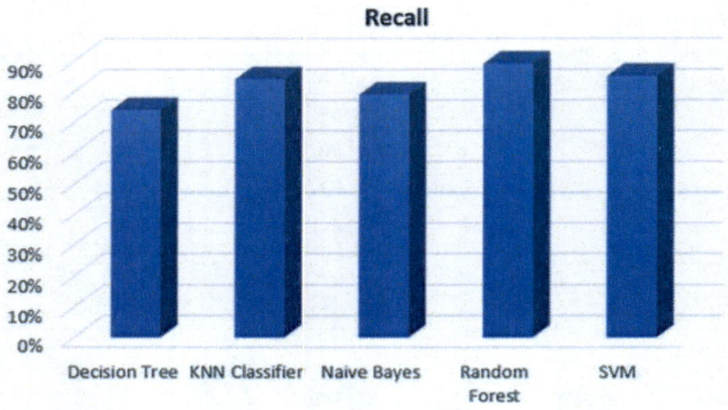

Fig. 7 Comparison of recall for Hungarian dataset

Table 3 Comparison of results obtained by different algorithms on Switzerland dataset

Techniques	Accuracy (%)	Precision (%)	Recall (%)
Decision tree	98.5	93	85
KNN classifier	36.58	50	65
Naive Bayes	98.7	97.4	98.2
Random forest	82.92	83	84
SVM	98.9	98.1	97.4

3.3 *Switzerland Dataset*

The results are shown in Table 3, Figs. 8, 9 and 10 presents the comparison of the results.

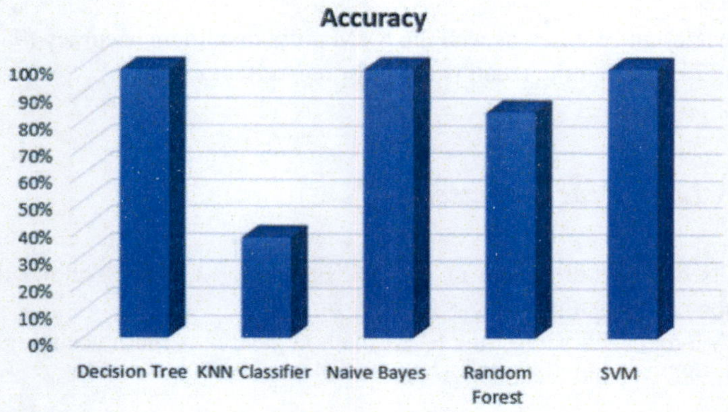

Fig. 8 Comparison of accuracy for Switzerland dataset

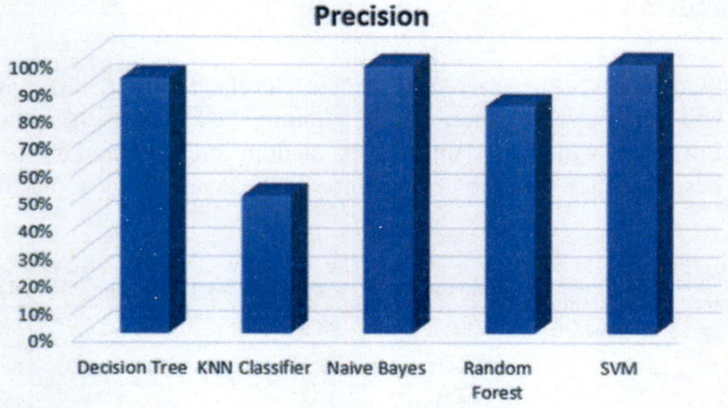

Fig. 9 Comparison of precision for Switzerland dataset

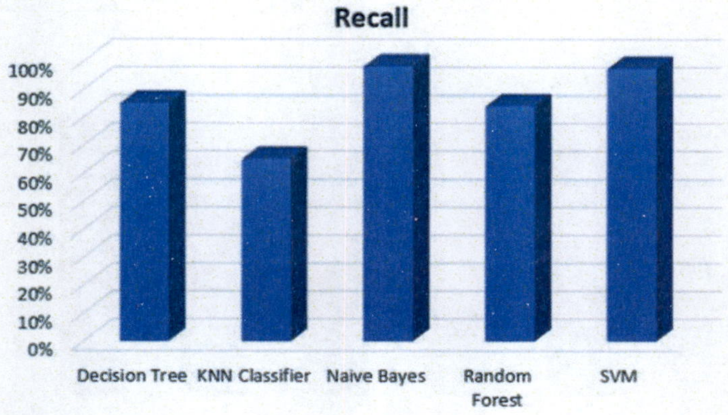

Fig. 10 Comaprison of recall for Switzerland dataset

For Switzerland dataset, as well the SVM gives maximum accuracy (98.9%) and precision (98.1%), but the recall is more for naïve Bayes (98.2%).

3.4 VA Long Beach Dataset

The results obtained are given in Table 4. Figs. 11, 12 and 13, shows the visual comparison of the results.

For VA long beach dataset the SVM method gave maximum accuracy (98.9%), precision (98.7%), and recall (99.1%).

4 Conclusion

From this study, it has been observed that except Cleveland dataset, the support vector machine was found to be the best classification method for the diagnosis of heart disease. However, for the Cleveland dataset, random forest method has been found to be the best classifier with the performance of the support vector machine being

Table 4 Comparison of results obtained by different algorithms on VA long beach dataset

Techniques	Accuracy (%)	Precision (%)	Recall (%)
Decision tree	98.3	98.2	98.5
KNN classifier	27.27	50	65
Naive Bayes	98.4	98.5	98.8
Random forest	83.83	83	84
SVM	98.9	98.7	99.1

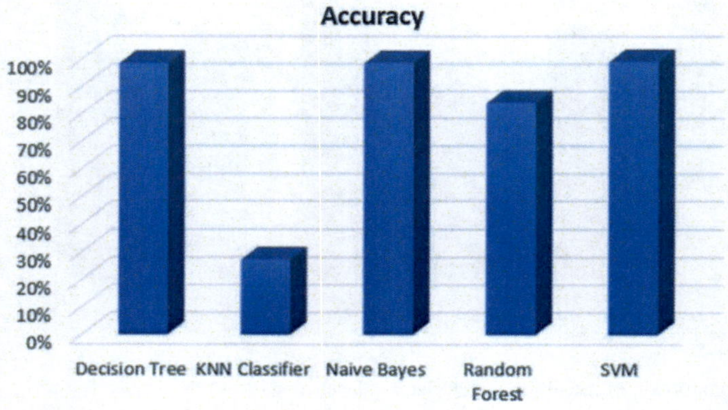

Fig. 11 Comparison of accuracy for VA long beach dataset

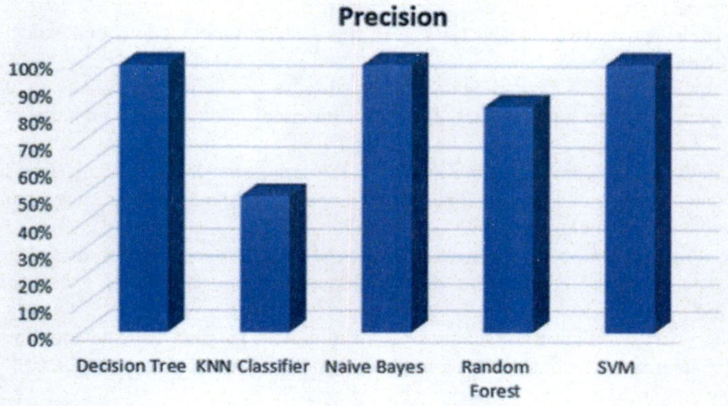

Fig. 12 Comparison of precision for VA long beach dataset

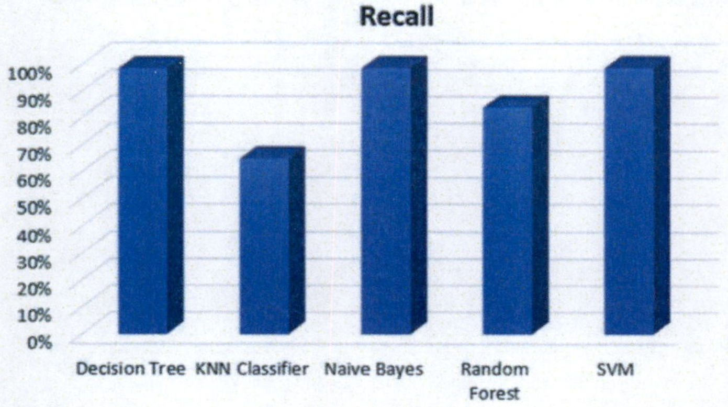

Fig. 13 Comparison of recall for VA long beach dataset

marginally lower. In general, it can be concluded that the support vector machine has been found to be the best technique for the diagnosis of heart disease.

References

1. Amin SU, Agarwal K, Beg R (2013) Genetic neural network based data mining in prediction of heart disease using risk factors.In: 2013 IEEE conference on information and communication technologies, pp 1227–1231
2. Yoo I, Alafaireet P, Marinov M, Pena-Hernandez K, Gopidi R, Chang J-F et al (2012) Data mining in healthcare and biomedicine: a survey of the literature. J Med Syst 36:2431–2448
3. Garg SB, Mahajan AK, Kamal T (2017) An investigation for detection of breast cancer using data mining classification techniques
4. Garg J, Garg SB (2018) A review of online tools for cancer genomics studies

5. Garg SB, Mahajan AK, Kamal T (2017) An approach for diabetes detection using data mining classification techniques
6. Thomas J, Princy RT (2016) Human heart disease prediction system using data mining techniques. In: 2016 international conference on circuit, power and computing technologies (ICCPCT), pp 1–5
7. Babu S, Vivek E, Famina K, Fida K, Aswathi P, Shanid M et al (2017)Heart disease diagnosis using data mining technique. In: 2017 international conference of electronics, communication and aerospace technology (ICECA), pp 750–753
8. Raju C, Philipsy E, Chacko S, Suresh LP, Rajan SD (2018)A survey on predicting heart disease using data mining techniques. In: 2018 conference on emerging devices and smart systems (ICEDSS), pp 253–255
9. Shaji SP (2019)Prediction and diagnosis of heart disease patients using data mining technique. In: 2019 international conference on communication and signal processing (ICCSP), pp 0848–0852
10. Kahramanli H, Allahverdi N (2008) Design of a hybrid system for the diabetes and heart diseases. Expert Syst Appl 35:82–89
11. Nahar J, Imam T, Tickle KS, Chen Y-PP (2013) Computational intelligence for heart disease diagnosis: a medical knowledge driven approach. Expert Syst Appl 40:96–104

A Critical Evaluation of Mathematical Modeling Approaches in the Scientific Research

Jatinder Garg, Sonu Bala Garg, and Kulwant Singh

Abstract The Scientists and engineers are engaged in finding the cause-and-effect relationships in various processes since time immemorial. It has helped them understand and optimize the interplay of various process parameters to achieve the best output. Just like any other scientific research, it is equally important to conduct experiments, collect data, and develop mathematical models for the welding processes. The increased automation and mechanization has made it further important. The mathematical modeling approaches used in welding research have evolved considerably with time. This paper provides an overview of various mathematical modeling approaches used in welding research. The statistical modeling has emerged out to be the best approach in present times.

Keywords Mathematical modeling · Welding · Manufacturing

1 Introduction

The demand for better precision and higher production rates has led to the increased use of automated welding and cladding processes in the manufacturing sector. Automation leads to better control over the heat input, which results in uniform chemistry and metallurgy of the weld metal [1]. For such processes, establishing a mathematical relation between process parameters and bead geometry is of utmost importance, to attain a closed control over the bead dimensions [2–4]. It helps in producing a weld joint as per the design requirements [5]. Similarly, it is also

J. Garg
Baba Hira Singh Bhattal Institute of Engineering and Technology, Lehragaga, Punjab, India

S. B. Garg (✉)
IK Gujral Punjab Technical University, Jalandhar, Punjab, India
e-mail: sonugarg79@yahoo.com

K. Singh
Sant Longowal Institute of Engineering and Technology, Longowal, Punjab, India

N. Marriwala et al. (eds.), *Mobile Radio Communications and 5G Networks*, Lecture Notes in Networks and Systems 140,
https://doi.org/10.1007/978-981-15-7130-5_59

important to establish mathematical models for the element transfer, the mechanical strength, and the microstructure of the welds. Over the years, scientists and researchers have tried to utilize a variety of techniques to develop such mathematical models. It has resulted in the evolution of different approaches. Apart from the trial and error approach, various techniques commonly utilized for welding research can be summarized in the following categories:

- Theoretical approach.
- Qualitative approach.
- Dimensional analysis approach.
- Specific quantitative approach.
- General quantitative approach.

2 Theoretical Approach

This approach is primarily based on the mathematical model developed by Rosenthal [6] for predicting temperature variation on infinite and semi-infinite plates due to heat transfer by a moving heat source. Though this model was capable of predicting the effects of input parameters on bead geometry parameters, yet the error in results was as much as 300% [7]. Further, being purely a conduction model, it ignored puddle motion. It also suffered from a lot of other drawbacks. This model was modified by Christensen [8] by developing dimensionless equations that could be used to relate temperature distribution with bead geometry. Tsai extended the point source model to include a Gaussian distributed heat source [9]. Another modification of Rosenthal's theory was proposed by Nunes. He added sources and sinks as per the method of images, but these were added at various locations within the weldment to simulate conditions such as fluid motion or heat sinks [10]. Some researchers attempted to relate the weld bead geometry with the welding parameters by this approach [11].

By and large, these models cannot be used to predict the bead dimensions precisely due to numerous superficial assumptions made during their development. As a result, this approach is not capable of accurately predicting the relationship between bead geometry and input parameters.

3 Qualitative Approach

The quantitative approach is also known as "one parameter at a time approach" in the common language. It is frequently used for investigating the effects of welding parameters on the bead geometry responses. In this technique, experiments are conducted by varying the values of one of the input parameters while keeping the other parameters constant. The data thus collected is used in establishing a relationship between the input and output parameters. This approach is comparatively simple as it does not require knowledge of complex mathematical techniques for conducting the experiments and interpreting the results. Historically, it has been used by many researchers to establish the mathematical relationships between welding parameters and bead geometry responses [12–15]. It had also been employed in establishing a pragmatic relationship between penetrations and welding conditions [16].

However, in this approach, as the number of input and output parameters increases, the number of experimental runs needed to establish the relationships also increases considerably. It makes the application of this technique for complex problems unreasonable. Further, by using this approach, one fails to get an idea about the interactive effects of input parameters on the responses.

But still, this technique is very useful in collecting information about a welding process over a narrow experimental range. It finds applications in conducting initial trial runs for establishing process limits and finding the workable ranges of various input variables before the application of the statistical design of experiment.

4 Dimensional Analysis Approach

To overcome the limitations of qualitative approach, dimensional analysis technique has been used by some researchers [17–19]. In this approach, all the parameters and physical properties are converted to fundamental units of mass (M), length (L), time (t), and temperature (T). Experiments are conducted as per the qualitative approach, and parameters/physical quantities are related to each other in equation form, in terms of dimensionless units with different exponents. Thereafter, these equations are converted to the final relationships that are used with the normal units of the different parameters. Then, these relationships are used in the estimation of bead geometry parameters. This technique had been used for the estimation of the relationship between the weld elements and the welding parameters [20].

The limitation of this technique lies in its inability to predict the interactive effects among the welding parameters. Further, this technique lacks the capability to optimize them.

5 Specific Quantitative Approach

In this method, the weld bead geometry is predicted on the basis of the empirical relationships developed by various researchers. Some of these relationships are checked and modified by subsequent researchers and are utilized in the prediction of bead geometry and other output parameters. Equations developed by Jackson [12] for establishing a relationship between welding parameters and the bead geometry were subsequently utilized by Gunnert [15] for the estimation of penetration, width, and reinforcement of weld beads. A mathematical model has been developed by Chandel [21] for the prediction of melting rate and penetration in submerged arc welding. The models developed by previous researchers for MIG welding were verified by Thorn et al. [22].

This approach is more practical and useful as compared to the purely quantitative approach, but is limited to the specific applications only.

6 General Quantitative Approach (Statistical Approach)

This approach is based on statistical techniques. It is being used in the field of medicine and plant research for quite a long time. Because of its inherent advantages, it has also been adopted in the field of welding research for the last few decades. The primary advantage of this approach over the classical approach is its ability to study several parameters at the same time, that too at different levels. By this approach, in addition to modeling the effect of input parameters on the responses, their interactive effect can also be studied. In general, the statistical approach can be seen as a method to extract more information from the recorded data.

In a complex system with many inputs and responses, the number of experiments required to be conducted increases rapidly. However, the statistical method of designing experiments helps in minimizing the total number of experiments to be conducted at various levels. The results obtained by this technique can be stated in quantitative terms.

In general, the statistical approach is based on a more sound logic than any other approach. It also helps in minimizing the time and the cost of experimentation, and increases the authenticity of the results. Several methods of applying the general quantitative approach to scientific research are available that are being applied by welding scientists in experimental design [1, 23–25].

7 Conclusion

Over the years, various scientific approaches for mathematical modeling of welds have got evolved. These include the theoretical approach, the qualitative approach, the dimensional analysis approach, the specific quantitative approach, and the statistical modeling approach. Due to its sound logical base, the ease of application, and the capability to develop more realistic models, the statistical modeling approach has evolved as the most popular modeling method in the welding research these days.

References

1. Senthilkumar B, Kannan T (2015) Effect of flux cored arc welding process parameters on bead geometry in super duplex stainless steel claddings. Measurement 62:127–136

2. Jeffus L (2012) Welding: principles and applications. Cengage Learning, New York
3. Palani PK, Murugan N (2006) Development of mathematical models for prediction of weld bead geometry in cladding by flux cored arc welding. Int J Adv Manuf Technol 30(7–8):669–676
4. Kumar V (2011) Modeling of weld bead geometry and shape relationships in submerged arc welding using developed fluxes. Jordan J Mech Ind Eng 5(5):461–470
5. Benyounis KY, Olabi AG (2008) Optimization of different welding processes using statistical and numerical approaches—a reference guide. Adv Eng Softw 39(6):483–496
6. Rosenthal D (1941) Mathematical theory of heat distribution during welding and cutting. Weld J 20(5):220s–234s
7. Tsai N (1983) Heat distribution and weld bead geometry in arc welding. Massachusetts Inst Technol
8. Chriestensen N, Davies V, Gjermundsen K (1965) Distribution of temperature in arc welding. Brit Weld J 12:54–74
9. Tsai CL Modeling of thermal behaviors of metals during welding. In: Trends in welding research in the United States, New Orleans, 16–18 Nov. 1981. Joining divison of American society of metal. pp 91–108
10. Nunes AC Jr (1983) An extended rosenthal weld model. Weld J 62(6):165s–170s
11. Sharapov YV (1972) Width calculation for automatic submerged arc butt welds. Weld Prod 3:45–47
12. Jackson CE, Shrubsall AE (1953) Control of penetration and melting ratio with welding technique. Weld J 32(4):172s–178s
13. Apps RL, Gourd LM, Lelson KA (1963) Effect of welding variables upon bead shape and size in submerged-arc welding. Weld Met Fab 31(11):453–457
14. Renwick BG, Patchett BM (1976) Operating characteristics of the submerged arc process. Weld J 55(3):69s–74s
15. Gunnert R (1948) Penetration and travel speed in metal arc welding. Weld J 27(7):542s–560s
16. Demyantsevich VP (1974) Dependence of penetration on automatic submerged-arc welding conditions. Weld Prod 21(7):63–66
17. Caddell RM (1967) The influence of physical properties on penetration in arc welding. J Eng Ind 89(2):328–332
18. Metzbower EA (1993) Penetration depth in laser beam welding. Weld J 72(8):403s–407s
19. Murray PE (2002) Selecting parameters for GMAW using dimensional analysis. Weld J 81(7):125s–131s
20. Zhang W, Kim CH, DebRoy T (2004) Heat and fluid flow in complex joints during gas metal arc welding—part II: application to fillet welding of mild steel. J Appl Phys 95(9):5220–5229
21. Chandel RS (1987) Mathematical modeling of melting rates for submerged arc welding. Weld J 66(5):135s–140s
22. Thorn K, Feenstra M, Young JC, Lawson WHS, Kerr HW (1982) The interaction of process variables-their influence on weld dimensions in GMA welds on steel plates. Metal Constr 14(3):128–133
23. Sreeraj P, Kannan T (2012) Modelling and prediction of stainless steel clad bead geometry deposited by GMAW using regression and artificial neural network models. Adv Mech Eng 4:237379
24. Kolahan F, Heidari M (2010) A new approach for predicting and optimizing weld bead geometry in GMAW. Int J Mech Syst Sci Eng 2(2):138–142
25. Nagesh DS, Datta GL (2002) Prediction of weld bead geometry and penetration in shielded metal-arc welding using artificial neural networks. J Mater Process Technol 123(2):303–312

Cloud Load Balancing Using Optimization Techniques

Ajay Jangra and Neeraj Mangla

Abstract Cloud computing is an integrated phenomenon that incorporates data, applications and services in a dynamic environment and enables worldwide optimization of resources. This computing technology is scalable and elastic in nature that opens door for large amount of incoming data from different venues with high velocity. Managing such data in distributive and heterogeneous environment imposes a challenge of load balancing on the service providers. They need to allocate the incoming tasks efficiently to the computing nodes to avoid imbalanced mapping and execution of the tasks. To achieve efficient load balancing, various load balancing algorithms have been proposed, and they all focus on achieving the effective distribution of data and improve the associated measurement factors. In this paper, different load balancing algorithms have been studied and analyzed with description of their techniques and focused parameters. Then, there is a brief discussion on the existing load balancing algorithms and further compares them based on parameters like throughput, scalability, resource utilization, etc., followed by the important findings thus made.

Keywords Ant colony · Agent-based algorithm · Cloud computing · Estimated finish time · Genetic algorithm · Load balancing · Honeybee foraging and throttled algorithm

A. Jangra (✉)
PhD scholar, Maharishi Markandeshwar (Deemed To Be University) MMDU, Mullana, Ambala, India
e-mail: er_jangra@yahoo.co.in

N. Mangla
CSE Department, Maharishi Markandeshwar (Deemed To Be University) MMDU, Mullana, Ambala, India
e-mail: erneerajynr@gmail.com

N. Marriwala et al. (eds.), *Mobile Radio Communications and 5G Networks*, Lecture Notes in Networks and Systems 140,
https://doi.org/10.1007/978-981-15-7130-5_60

1 Introduction

Cloud computing is a modeled technique, which equipped users with virtualized pool of resources in distributive environment and also facilitated pay as you go model for the resource utilization [1]. Cloud service providers (CSP) aim to serve request online as per the type of cloud environment (public, private, hybrid and community) and these services. In the beginning of cloud era, IaaS, PaaS and SaaS were the three accessible elastic and scalable services of cloud computing. These are Internet-based rented services made available by CSPs and assured over subscription through service-level agreement (SLAs) [2]. The client/customer does not need to own the hardware or software rather than they can use them online with the Internet facility.

With the enhancement in computing era, the cloud computing is rapidly enriching and able to offer (EaaS) ***Everything as a Service***. Besides, day to day the amount of data evolving with it is also gaining an attention to be managed efficiently for the effective and sustainable working over distributed cloud scenario [3]. It is one of the major responsibilities of data handler to ensure the reliable and secure data handling in cloud. Large amount of data in cloud increase the velocity of demands which initiates a new challenge of load balancing in cloud computing that targets the effective scheduling and resource monitoring [4–6].

This paper justifies the requirement of load balancing in cloud computing with various supporting strategies that focus on improving the process of efficient resource utilization.

1.1 Load Balancing in Cloud Computing

Load balancing is a uniform approach of scheduling jobs among the available computing nodes. Monitoring resources and their effective utilization in broad network access is the main objective of load balancing [7]. A load balancing algorithm is considered resourceful when it is fault-tolerant and scalable in nature and guarantees to produce maximum throughput. There are numerous algorithms available that are categorized as dynamic or static in nature and operate in different environment. Static algorithms possess predefined factors and states on which it runs, whereas dynamic algorithms are operable at run time depending on the current state dynamically balancing the traffic on server. Various types of load balancing algorithms are as follows [6, 8].

A. **Static Load Balancing**

Static load balancing techniques are non-preemptive in nature that has predefined strict rules to be followed based on input and does not depend upon the current state of the machine to manage the workload. It requires a prior knowledge of the system setup and resource availability. It is also known as policy-driven load balancing driven by parameters like server capacity, throughput rate, fault tolerance, response time, etc. [9]. Examples of static load balancing technique are artificial bee colony search, two phase scheduling, central manager algorithm, etc.

B. **Dynamic Load Balancing**

Dynamic load balancing techniques are preemptive in nature that do not require prior knowledge of input and depend on the current state and enhance the overall working of the system. It manages load dynamically and prevents nodes from getting overloaded as it transfers load within nodes on run time. It is also known as feedback-driven load balancing [10]. Examples of dynamic load balancing technique are artificial ant colony search, round robin algorithm, throttled load balancing algorithm, etc.

1.2 Load Balancing Optimization Algorithms

This section shows the literature review done on different load balancing optimization algorithms describing different ways to balance huge data in cloud.

Throttled Load Balancing Algorithm
Throttled algorithm is a state-based algorithm that depends on the current state of virtual machines whether it is available or busy. Load balancers [11] are the modules of operating environment that dynamically balance the load on different available virtual machines and maintain their index as well as its associated state. If the state of virtual machine is available, the request is assigned; otherwise, the action is declined and searched for safe state. Virtual machines are responsible for the execution of request after its state has been verified by the load balancer. In [12], the comparison is made by the author between round robin and throttled algorithm in terms of time and cost. It is considered superior as compared to round robin algorithm in terms of cost as it reduces the cost of virtual machine's usage per hour (Fig. 1).

Ant Colony Optimization Algorithm
Ants are small blind insects that used to find food outside their nest when they are hungry. The way they find their food by following the shortest path is associated with load balancing in cloud. These ants decompose pheromone on their way to food with same speed and same rate. This helps the other ants to follow their path to food. So, more the followers, the higher will be the concentration of the pheromone decomposed. Evaporation rate of this pheromone on the shortest path is quiet low. In [13], the author has shown that this same process can be adopted by the job scheduler of cloud computing for mapping the task to available executing nodes. Schedulers check the load on their surrounding and transfer request accordingly for effective utilization by maintaining the descriptive table containing all the necessary information about virtual machines.

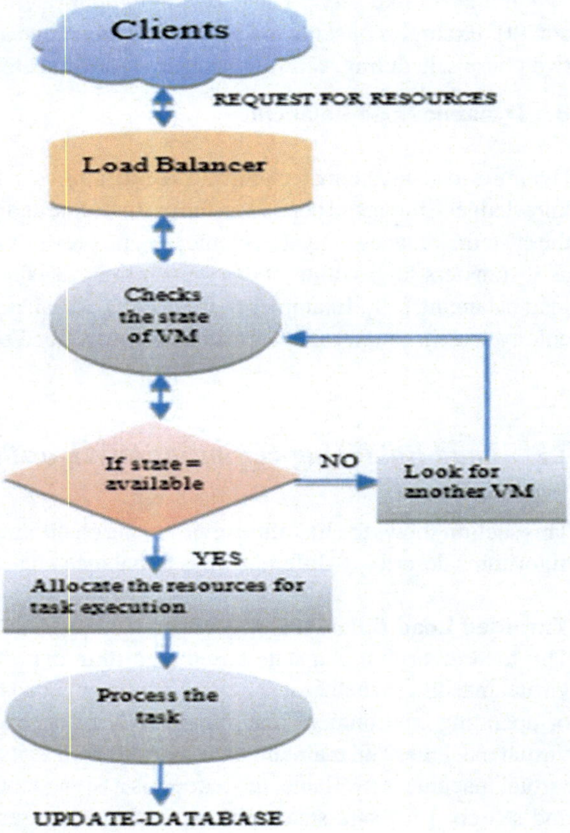

Fig. 1 Process flow of throttled algorithm

Honeybee Optimization Algorithm
It is one of the finest algorithms of load balancing in cloud computing that follows the behavior of honeybees. There are two categories of honeybees: One is detector honeybees that go out in the search of food and other one is follower honeybees that follow the path directed by leader honeybees. Leader honeybees come back and perform the well-known waggle dance in which they form numeric eight. This special dance tells the quality and quantity of the food and also the duration of the dance shows the distance of the food from beehive with its recorded profit [14]. Unemployed bees in the hive have the option to be detectors or followers. This paper [15] shows the improved artificial honeybee algorithm as the basic algorithm may create some imbalance of load among the nodes. According to this improved algorithm, threshold value is set for every server queue, and when the length of a server queue exceeds the value, the load is transferred to another server, and they are executed independently improving the throughput of the system.

Genetic Algorithm
Genetic algorithm is one of the optimal algorithms of effective search and optimization in load balancing. Simple GA follows the three-step process of selecting the population followed by genetic operations and replacement with new population. The genetic-based load balancing strategy in [16] balances the load based on process described based on genetic algorithm. It first initializes the population, which finds out the fitness factor followed by crossover and mutation. Replace the offspring with new population and do the acceptance testing. This improves the QoS requirements of the client.

Generalized Priority Task Scheduling Algorithm
Resource monitoring in cloud computing is done in three steps—discovering and filtering the resources in broad network, selecting the appropriate resource from the available ones and finally submitted the task to desired resource. In generalized priority optimal task scheduling [17], high size task is given high priority as well as the servers with high MIPS and maps the task accordingly to the virtual machine with identified id and updates the available resources. This paper shows the improved execution time as compared to round robin and first come first serve algorithms.

Agent-Based Load Balancing
In traditional load balancing, load balancers aim at scheduling the task to appropriate servers to avoid the overloading state in the system. The agent-based dynamic load balancing uses a software entity named as mobile agent which is independent software program and run on the behalf of network user. This agent covers one walk in shared pool of resources within two walks (Figs. 2, 3 and 4).

In its first walk, it gathers all the information about the status of the servers with the average calculation of the jobs, and in its second walk it analyzes the overloaded and under loaded state of the servers and transfers the load accordingly. The additional agent in dynamic load balancing improves the throughput and response time of the system [18].

A. Estimated Finish Time Task Scheduling.

There are several computing factors like throughput, processing time, finish time, response time, etc., that counts the efficient load balancing. Faster the task execution takes place more, efficient will be the strategy. In estimated finish time task scheduling [19], the characteristics of the task are judged during the allocation and processing time in order to avoid the blocking of processes in the queue.ss It estimates its finish time at earlier stage during allocation and guides it to the appropriate server that improves the performance and resource utilization as it ensures the maximum usage of virtual machines.

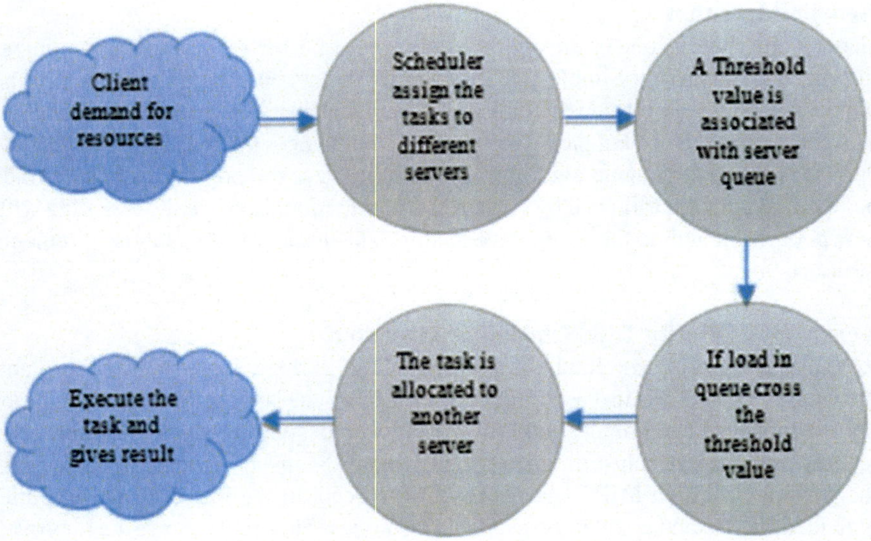

Fig. 2 Process diagram of artificial honeybee

Fig. 3 Flowchart of genetic algorithm

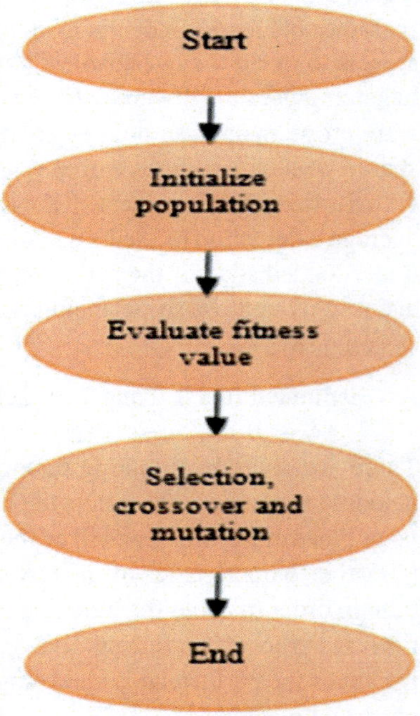

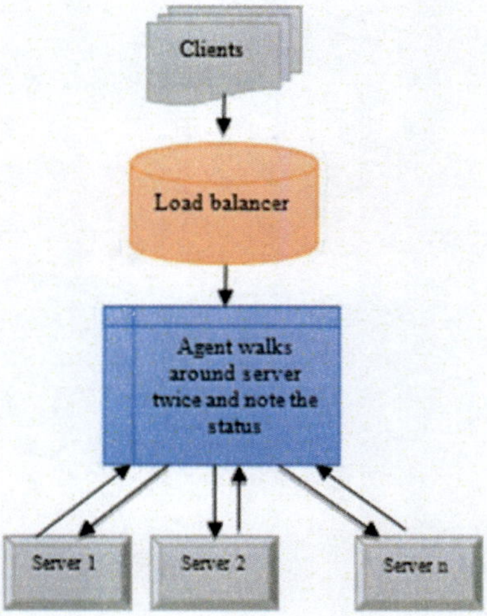

Fig. 4 Agent-based load balancing

2 Comparison and Important Findings

In this paper, comparison among several load balancing techniques has been discussed and summarized according to their performances and results. This section illustrates some important findings and performs the comparison between those existing strategies based on some measurement parameters that are listed and tabulated below:

I. Throttled and agent-based load balancing algorithm can be applied on various cost-oriented models and plays an important role in business-oriented applications and government sectors.
II. Honeybee and generalized priority focus on improving the overall execution of the system and balance the load among various nodes more efficiently as they are predictive in nature and pre-analyze the data to be allocated at different locations.
III. Resources are the most valuable assets of the computing environment and are needed to be utilized effectively so that it can contribute to the scalability and meritorious performance of the system. Ant colony estimated finish time and genetic algorithm successfully achieved the commendable resource utilization and can be modeled further with other parameters.
IV. These techniques can be integrated with other modeled load balancing techniques to achieve success in various sectors like banking, medical, forecasting, etc. (Table 1).

Table 1 Analysis of various load balancing algorithms

Metrics	Throughput	Scalability	Response time	Processing time	System stability	Cost	Performance	Resource utilization
Throttled algorithm	✗	✗	✓	✓	✗	✓	✗	✗
Ant colony optimization	✗	✗	✗	✗	✗	✗	✓	✓
Honeybee foraging	✓	✓	✗	✗	✓	✗	✗	✗
Genetic algorithm	✗	✗	✗	✗	✓	✗	✓	✓
Generalized priority	✓	✗	✗	✗	✗	✗	✗	✓
Agent-based scheduling	✓	✗	✓	✗	✗	✓	✗	✗
Estimated finish time	✗	✗	✗	✗	✗	✗	✓	✓

3 Conclusion

Cloud computing is scalable Internet-based service that aims to improve the utility of computing resources with the increase in velocity, volume and variability of incoming data. Rapid rise of data makes it difficult to handle such large amount of data and introduced a new challenge of load balancing in distributive cloud environment. Several load balancing algorithms have been proposed by intellectual researchers to efficiently direct the tasks to the computing nodes for smooth and uniform execution. Considering various proposed algorithms, this paper performs a comparative analysis among them based on different metrics. This analysis concludes that different algorithms work on different parameters, and none of them works considering all the parameters. All the proposed algorithms are efficient in one way or the other but do not claim to be the best. Therefore, these algorithms can be carried out with some new measurement parameters and can improve the quality of distribution of data with enhanced security and privacy methods.

References

1. Lakhina U, Singh N, Elamvazuthi I, Meriaudeau F, Nallagownden P, Ramasamy G, Jangra A (2018) Threshold based load handling mechanism for multi-agent micro grid using cloud computing. In: International conference on intelligent and advanced system. Kuala Lumpur
2. Linthicum DS (2017) Connecting fog and cloud computing. IEEE Cloud Comput 18–20
3. Maenhaut P-J, Moens H, Volckaert B, Ongenae V, Turck FD (2017) Resource allocation in the cloud: from simulation to experimental validation. In: IEEE 10th International conference on cloud computing (CLOUD). California
4. Geetha P, Robin CR (2017) A comparative-study of load-cloud balancing algorithms in cloud environments. In: International conference on energy, communication, data analytics and soft computing (ICECDS). Chennai
5. Jangra A, Bala R (2013) PASA: Privacy-Aware Security Algorithm for Cloud Computing. In: Abraham A, Thampi S (eds) Intelligent Informatics. Advances in intelligent systems and computing, intelligent informatics. pp 487–498, p 11, ISBN: 9783642320620
6. Singh N, Jangra A, Lakhina U An efficient load balancing algorithm for cloud computing using dynamic cluster mechanism. 3rd international conference on computing for sustainable global development, 16th–18th March 2016, IEEE, BharatiVidyapeeth's institute of computer applications and management (BVICAM), New Delhi (INDIA), Conference ID: 37465, pp 2613–2618
7. Kumar P, BundeleDM, Somwansi D (2018) An adaptive approach for load balancing in cloud computing using mtb load balancing. In 3rd International conference and workshops on recent advances and innovations in engineering (ICRAIE). Jaipur
8. Shao U, Chen J (2016) A load balancing strategy based on data correlation in cloud computing. In: 2016 IEEE/ACM 9th international conference on utility and cloud computing (UCC). Shangai.
9. Lakhina U, Singh N, Jangra A (March 2016) An efficient load balancing algorithm for cloud computing usingdynamic cluster mechanism. 3rd international conference on computing for sustainable global development. pp 2613–2618
10. Volkova VN, Chemenkaya LV, Desyatirikova EN, Hajali M, Khodar A, Osama A (2018) Load balancing in cloud computing. In: 2018 IEEE conference of russian young researchers in electrical and electronic engineering (EIConRus), pp 387–390. Moscow

11. Rahman M, Iqbal S, Gao J (April 2014) Load balancer as a service in cloud computing. In: IEEE 8th International Symposium on Service Oriented System Engineering (SOSE), 2014. pp 204–211
12. Shoja H, Nahid H, Azizi R (July 2014) A comparative survey on load balancing algorithms in cloud computing. International conference on computing, communication and networking technologies (ICCCNT), 2014. pp 1–5
13. Kumar R, Sahoo G (October 2013) Load balancing using ant colony in cloud computing. Int J InfTechnol Conver Serv (IJITCS). 3(5)
14. Dhinesh Babu LD, Krishna PV (May 2013) Honey bee behavior inspired load balancing of tasks in cloud computing environments. Appl Soft Comput 13(5):2292–2303
15. Yao J, He J-H (April 2012) Load balancing strategy of cloud computing based on artificial bee algorithm. In: 8th International conference on computing technology and information management (ICCM), 2012, vol 1. pp 185–189
16. Dasgupta K, Mandal B, Dutta P, Mandal JK, Dam S (2013) A genetic algorithm (GA) based load balancing strategy for cloud computing. First international conference on computational intelligence: modeling techniques and applications (CIMTA) 2013, vol 10. pp 340–347
17. Agarwal DA, Jain S (2014) Efficient optimal algorithm of task scheduling in cloud. Int J Comput Trends Technol (IJCTT) 9(7):344–349
18. Grover J, Katiyar S (August 2013) Agent based dynamic load balancing in cloud computing. International conference on human computer interactions (ICHCI), 2013. pp 1–6
19. Fahim Y, Lahmar EB, Labriji EH, Eddaoui A (November 2014) The load balancing based on the estimated finish time of tasks in cloud computing. Second world conference on complex systems (WCCS), 2014. pp 594–598

QoS Sensible Coalition-Based Radio Resource Management Scheme for 4G Mobile Networks

T. Ganga Prasad and MSS. Rukmini

Abstract From the last few decades, wireless communication networks (mobile) have experienced a remarkable change to attain advancement for maintaining the systems QoS with data rate for multimedia streaming. Some extensions of these networks came to picture for a transformation into speed, technology, frequency, data capacity, latency, etc. with extreme precision levels. It gives the enhanced lifetime and network connectivity of the system. This visualization of the next-generation wireless networks is of various types of radio access technologies such as Advance LTE, WiMax, and Wi-Fi. All invention has some principles, diverse capacity, and a new technique with new features which make a distinction from the previous one. These all extended forms of wireless networks are based on the heterogeneity of the network. For achieving these said objectives, a QoS with optimal confederation-aware technology, i.e. QOC-RRM method, is discussed. Hence, this predictable expose gives an idea about an LTE network for future-generation radio resource management. Our proposed technique makes use of the QOC-RRM method. In this hybrid RDNN method, i.e. recurrent deep neural network, we present differentiate operators based on multiple constraints through priority-wise. This QOC-RRM method controls the due source through the sink or in some cases base stations. The consumer not at all practiced earlier than such high-value skill that includes the entire advancement features. This proposed work is implemented by using the Network Simulator (NS2) version 2.34 and its extension NS3 tool. The performance outcome shows that the proposed work outperforms as compared to the existing work, i.e. the conventional RRM scheme. The parameters used are the radio spectrum utilization, the utmost amount of dynamic user, and the least rate of the data required.

T. G. Prasad (✉) · MSS. Rukmini
ECE, VFSTR, Vignan University, Guntur, Andhra Pradesh, India
e-mail: Prasad4_1@rediffmail.com

MSS. Rukmini
e-mail: mssrukmini@gmail.com

N. Marriwala et al. (eds.), *Mobile Radio Communications and 5G Networks*, Lecture Notes in Networks and Systems 140,
https://doi.org/10.1007/978-981-15-7130-5_61

Keywords Mobile wireless communication networks · Mobile broadband · Mobile generations · Quality of service · Wi-fi · RDNN · GSM · Radio resource management · Long-term evolution

1 Introduction

The wireless networks have constrained lifetime due to its limitations. In the previous techniques for example in the wireless mobile communication system in the first generation, it uses analog technology for communication instead of digital one [1]. The design of the structure is based on the technology used for transmission of the data packets. In comparison with the principal generation technique the 2G, i.e. the second generation will uphold text messaging. On the other hand, the increasing tool principles are inquired by the 3G networks for successive generation of the wireless network system for mobile communication. These technologies have the goals for making a standard infrastructure that should be organized in a way to sustain the obtainable as well as the prospect services [2].

These all generation styles require the transportation of the information which should be designed in an organized way that it can develop itself as the technology changes. On the other hand, all these development should be done exclusively to compromise the existing services of the current network [3]. The well-known network, i.e. wireless LAN, with predetermined Internet and maintaining the mobile Internet wirelessly since the matching QoS as predetermined Internet. Finally, the fifth-generation technique is predicated on fourth-generation technology only for making some extended uprising to the 5G technology. This transformation ill face some issues when it will process the data. These issues should be resolved that are wider exposure and improvement from one to another technology for making a difference between the technologies. For the upcoming future technologies, we can say that the sixth-generation technology designed for the wireless mobile statement network that will put together the satellites to recommend the comprehensive exposure systems [4].

The left-behind part of the proposed work is set as: Sect. 2 will give the literature review, the Sect. 3 of the work will focus on the projected clarification for the designed system, Sect. 4 of the paper derives the numerical representation of the proposed work with solution, the simulation outcomes of the proposed work is discussed in Sect. 5, and lastly, the Sect. 6 will give the conclusion of the proposed work with future scope.

2 Literature Survey

History and Background:

The statements through the mobile have to turn out to be more fashionable in most recent years due to its exceptional growth starting from various generations of first to fifth generation in mobile machinery. It happens only due to the extremely

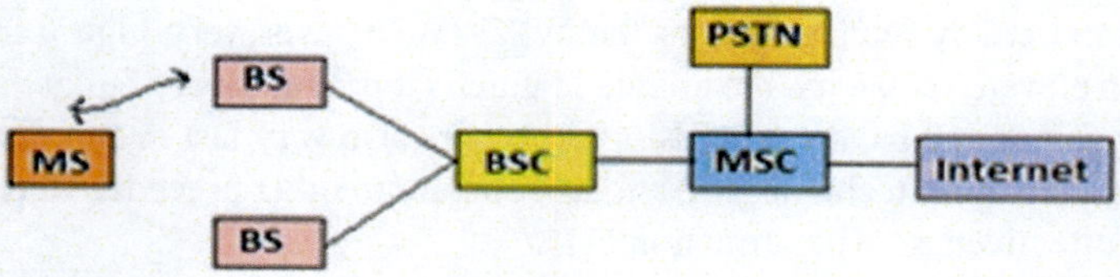

Fig. 1 Structural design of advance mobile organization

skyscraping augment in the number of users as well as the obligation of examination friendly broadcasting skills. The word generation refers to the revolutionization in the broadcasting of the information data with attuned bands of frequencies [5]. Since 1980, the mobile infrastructure has undergone significant variations from that time only this technology knowledgeable an enormous development. The first generation systems are the earliest movable phones to be used and had an introduction in 1982 till 1990. The best example of the first-generation mobile systems is the Advanced Mobile Phone System that is utilized through frequency-modulated technology. The basic features of the first-generation systems are: The 2.4 kbps is the speed of the system, it uses an analog signal, the quality of the audio is poor, the lifetime of the battery is also poor with a large size, and it also faces security issues because it provides very poor security [6].

Anyway, the main contribution of this generation is that various technologies were introduced by the first generation only. These technologies are MTS, advanced AMTS, IMTS, and very famous technology is PTT [7] (Fig. 1).

The second-generation system is started in 1980 and uses GSM technology. It uses the signals in digital form and can deliver text messages also with low speed. The key features are that its speed of information transmission is 64 kbps. The extension of this technology is called as 2.5 generation with 64–144 kbps transmission speed with camera facility [8]. Simultaneously, the third generation which comes to the picture in 2000 uses GSM technology. Its transmission speed is increased with the help of switching of packet technique. 3G gives the facility to access TVs well as MP4 files. The features of 3G technology are, the speed of the network is 2 Mbps, normally it is known as a smartphone system, it gives a faster speed of communication, buffer capacity *s* more but expensive. This generation is also called as Universal Mobile Telecommunication System and CDMA2000 due to its extended features in various counties all over the world [9]. On the other hand, the next-generation systems offer 100 Mbps speed for downloading the data. The Long-Term Evolution can be considered as other names of the fourth-generation technique due to its features. These generation systems can provide the speed of 10 M to 1 Gbps, and it can provide good security with enhanced battery lifetime; it will charge less per bit cost with some complicated design structure [10]. Finally, the fifth-generation mobile systems begin at the end of 2010. This generation of mobile systems can provide large area coverage with good connectivity. The most important thing about this technology is that it can work on Wireless World Wide Web and come as a complete wireless network system exclusive of restrictions. The fifth-generation system has, as discussed extremely manageable to the WWWW, it gives very high transmission speed with elevated capacity, the unique feature is it provides very bulky data propagation capacity of Gbps, the processing capacity s also very fast in contrast with the preceding generation technologies, and as compare to other preceding techniques it has more effectiveness with attraction [11].

LTE-QoS RRM.

The network parameters like delay, jitter, bandwidth, and packet loss play a vital role in designing of QoS-based LTE Systems. On the off chance that the system limited in terms of resources guarantying essential QoS is significant for smooth network tasks explicitly for real-time steaming and multimedia applications like online games, Voice over IP (VoIP), and IPTV. The radio resource management (RRM) holds the power consumption of every sector in each cell. These cells are calculated depending upon the output of the antenna adjustment module. And SINR also calculated depends on the channels model, interference from the base adjacent BSs, and transmission power of the BS. The limit of every user and the entire system is determined depending on the power and SINR [12].

Proposed Scheme (Fig. 2).

In parallel with the LTE radio access, packet core networks are also evolving into the flat SAE architecture. This new architecture is designed to optimize network performance, improve cost-efficiency, and facilitate the uptake of mass market IP-based services [13]. There are only two nodes in the SAE architecture user plane: the LTE base station (eNodeB) and the SAE Gateway, as shown in Fig. 1. The LTE base stations are connected to the core network using the core network–RAN interface, S1. This flat architecture reduces the number of involved nodes in the connections [14].

3 Proposed Work

QoS aware of optimal confederation radio resource management (QOC-RRM).

As per the future technology, the LTE networks should be implemented through various methods called radio resource organization. These projected designing schemes of RRM by maintaining the QoS are very important. It is an advanced technology in which advanced LTE systems were dealt with [15]. The designing of the LTE system using the QoS metric is an advanced procedure that is done with the help of an optimal confederation technique designed using the quality of services. Hence, the advanced methods which were used are the neural networks systems with recurrent deep method [16].

User classification using RDNN.

The said RDNN systems are neural systems that were used for dealing with various consecutive information. These neural systems act as a convolution neural network that is very particular for dealing with a matrix with different values of X, for example,$x^{(1)}, \ldots x^{(T)}$ (Figs. 3 and 4).

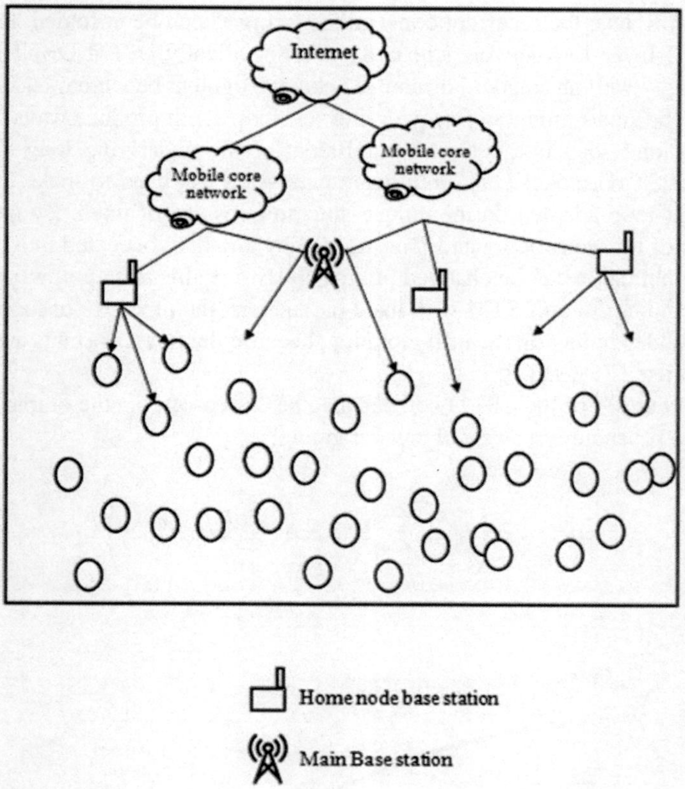

Fig. 2 A system model of the proposed QOC-RRM

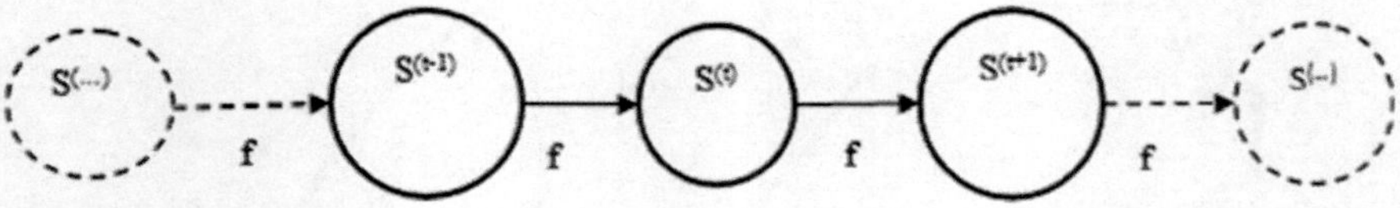

Fig. 3 Dynamical structure

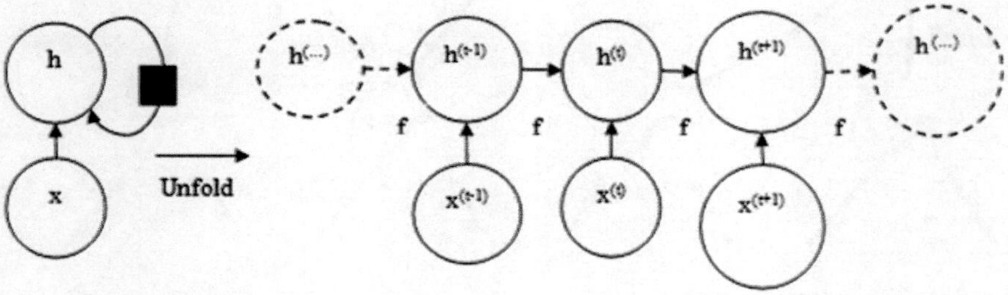

Fig. 4 User classified recurrent network

To train RNNs, the recurrent connections in Fig. 5 can be unfolded, reasonably yielding a T-layer deep network with tied loads. Specifically, LSTM-DRNNs perform outstandingly well on standard discourse acknowledgment benchmarks. The LSTM describes the smart thought of acquainting self-loops with produce ways where the slope can long span is a central commitment of the underlying long short term memory (LSTM) model [17]. Significant expansion has been to make the weight on this self-loop adapted on the unique situation, instead of fixed. By making the heaviness of this self-loop gated (constrained by another concealed unit), the time size of combination can be changed progressively. For this situation, we imply that notwithstanding for an LSTM with fixed parameters, the time size of incorporation can change depending on the info grouping, because the time constants are yield by the model itself [18, 19].

The output $h_i^{(t)}$ of the LSTM cell can also be closed-off, via the output gate $q_i^{(t)}$, which also procedures a sigmoid unit for gating:

$$q_i^{(t)} = \sigma\left(b_i^o + \sum_j U_i^o x_j^{(t)} + \sum_j W_{i,j}^o h_j^{(t-1)}\right)$$

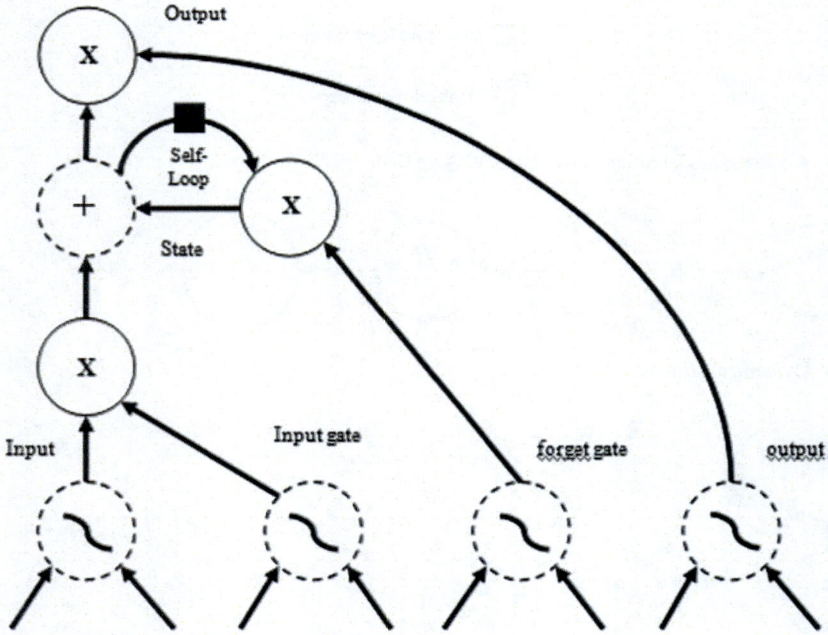

Fig. 5 LSTM system in recurrent

4 Result and Discussion

Performance evaluation and comparison.

The evaluation of the performance is done in two different ways either by using the sum rate or the guaranteed rate. These methods, i.e. QOC-RRM performance using sum rate and QOC-RRM performance using a guaranteed rate, are discussed below with their outcomes.

QOC-RRM performance using sum rate:

Figure 6 shows the basic user performance compared with MOC-based SONRRM and with some other techniques. Furthermore, due to the constrained resource for the BSs, the basic CG SON RRM gives the worst performance of interference reading of RRM in the BSs. It is not an accurate illustration of the authentic downlink interference.

QOC-RRM performance using Guaranteed rate:

Figure 7 shows that significant guaranteed rate improvement for users is given by the proposed method. Hence, we can say through the figure that the proposed technique can offer at slightest 50% QoS improvement compared to the existing works by users guaranteed rate [10].

Figure 8 shows the indoor performance of the guaranteed rate which gives the best sum rate compared with different indoor users. When building distance MBS gets increased, the macro base stations are maximum in our proposed indoor sum rate. Our proposed indoor sum rate compared with other existing systems 10% improved in our performance. Figure 9 presents the performance for the outdoor rate that gives the superlative outcome in contrast with dissimilar users for outdoor techniques. When

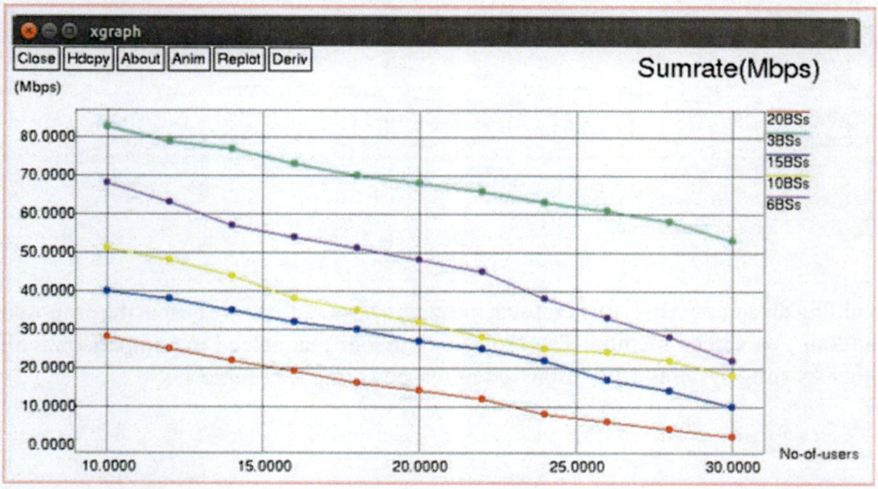

Fig. 6 User sum rate presentation

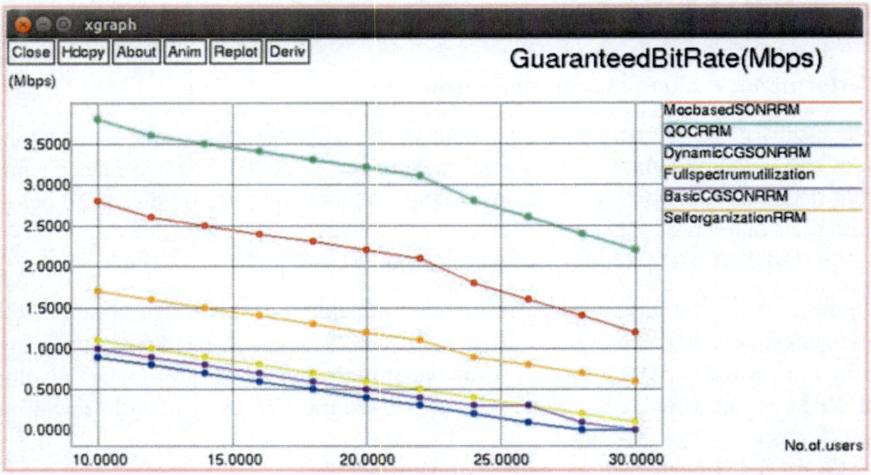

Fig. 7 Guaranteed rate performance comparison

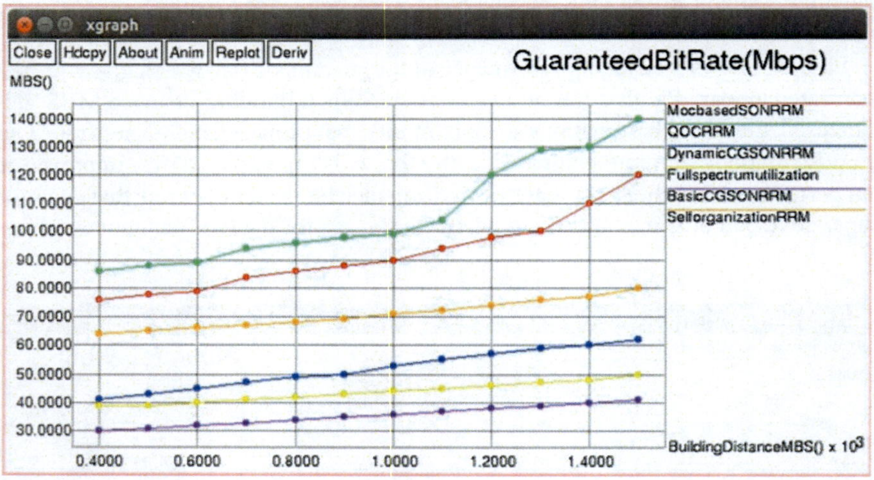

Fig. 8 Indoor guaranteed rate

building distance MBSs get increased, the macro base stations are best in the proposed outdoor guaranteed method. The projected outdoor guaranteed in comparison with other existing systems 15% improved in our performance (Table 1).

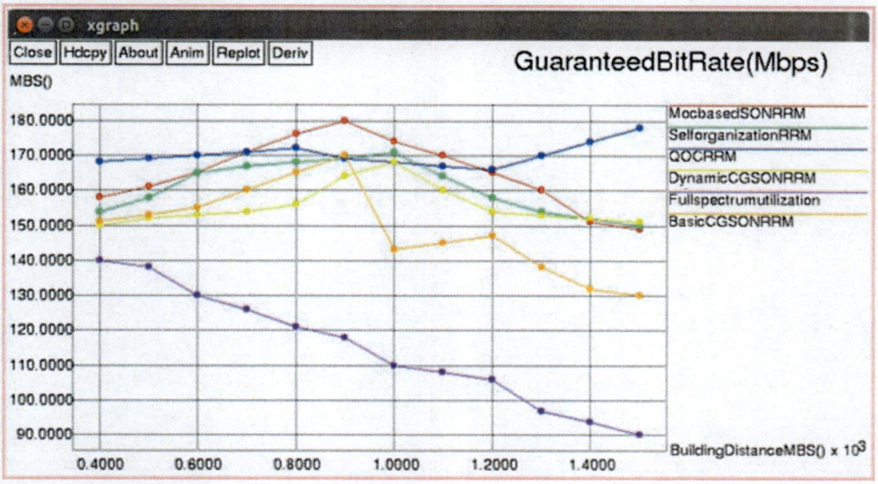

Fig. 9 Performance of outdoor guaranteed rate

5 Conclusion

The whole of humanity is annoying for turn out to be completely wireless, trying to become continuous access to the informative data anytime and wherever through improved excellence, high speed, improved bandwidth with less cost. The projected work implements an advanced LTE network with QoS and radio resource management. The globe is speedily increasing its communication through mobile wireless networks. In this case, the QOC-RRM method is used for the QoS. Additionally, the RDNN technique with allocated resource scheming through the sink for making a difference based on the numerous constraints. Hence, CWO algorithms are anticipated and analyzed with weed optimization for the queuing principle in the chaotic form with the routing process to share data. After receiving the data, the sink schedules the priority of the users for the available resources on the first come first serve bases. The outcomes show that our projected work outperforms with 10% better results in comparison with the existing exertion. The projected work is implemented on NS3, and it gives 50% improvement as compared to the obtainable method.

Table 1 Outdoors bit rate

Sr. No	Building distance	Basic CGSONRRM	Dynamic CGSONRRM ~	Full spectrum utilization	Moc-Based SONRRM	QOC-RRM	Self-Organization RRM
		Bit rate (MBS)	Bit rate (MBS)	Bit rate (MBS)	Bit rate (MBS)	Bit rate (MBS)	Bit rate (MBS)
1	400	151	150	140	158	168	154
2	500	153	152	138	161	169	158
3	600	155	153	130	165	170	165
4	700	160	154	126	171	171	167
5	800	165	156	121	176	172	168
6	900	170	164	118	180	169	169
7	1000	143	168	110	174	168	171
8	1100	145	160	108	170	167	164
9	1200	147	154	106	165	166	158
10	1300	138	153	97	160	170	154
11	1400	132	152	94	151	174	152
12	1500	130	151	90	149	178	150

References

1. Teece DJ (2018) Profiting from innovation in the digital economy: Enabling technologies, standards, and licensing models in the wireless world. Res Policy 47(8):1367–1387
2. Eluwole, Opeoluwa Tosin, Nsima Udoh, Mike Ojo, Chibuzo Okoro, and Akintayo Johnson Akinyoade. “From 1G to 5G, what next?.” IAENG International Journal of Computer Science 45, no. 3 (2018).
3. Czaja, Sara J., Walter R. Boot, Neil Charness, and Wendy A. Rogers. Designing for older adults: Principles and creative human factors approaches. CRC press, 2019.
4. Vuran MC, Salam A, Wong R, Irmak S (2018) Internet of underground things in precision agriculture: Architecture and technology aspects. Ad Hoc Netw 81:160–173
5. Lodhi, Amairullah Khan, and Syed Abdul Sattar. “Cluster Head Selection by Optimized Ability to Restrict Packet Drop in Wireless Sensor Networks.” In Soft Computing in Data Analytics, pp. 453–461. Springer, Singapore, 2019.
6. Arab, Lara Al. “Entrepreneurial Intentions of Undergraduate Students.(c2018).” PhD diss., Lebanese American University, 2018.
7. Sriram, Poorna Pravallika, Hwang-Cheng Wang, Hema Ganesh Jami, and Kathiravan Srinivasan. “5G Security: Concepts and Challenges.” In 5G Enabled Secure Wireless Networks, pp. 1–43. Springer, Cham, 2019.
8. Lodhi, Amairullah Khan, M. S. S. Rukmini, Syed Abdulsattar, and Shaikh Zeba Tabassum. “Performance improvement in wireless sensor networks by removing the packet drop from the node buffer.” Materials Today: Proceedings (2020).
9. Razzak, Md Abdur. “Design And Development Of A Multi-Channel Military Application Using Interactive Dialogue Model (Idm) Providing Enhanced Usability.” Phd Diss., Department Of Computer Science And Engineering Military Institute Of Science And Technology, 2018.
10. Lodhi, Amairullah Khan, M. S. S. Rukmini, Syed Abdulsattar,” Energy-Efficient Routing Protocol Based on Mobile Sink Node in Wireless Sensor Networks” International Journal of Innovative Technology and Exploring Engineering (IJITEE) ISSN: 2278–3075, Volume-8 Issue-7, May, 2019.
11. Porambage P, Okwuibe J, Liyanage M, Ylianttila M, Taleb T (2018) Survey on multi-access edge computing for internet of things realization. IEEE Communications Surveys & Tutorials 20(4):2961–2991
12. Ahmed F, Naeem M, Ejaz W, Iqbal M, Anpalagan A (2018) Resource management in cellular base stations powered by renewable energy sources. Journal of Network and Computer Applications 112:1–17
13. Amairullah Khan Lodhi, M. S. S. Rukmini, Syed Abdulsattar, “Energy-Efficient Routing Protocol for Network Life Enhancement in Wireless Sensor Networks” Recent Patents on Computer Science. https://doi.org/10.2174/2213275912666190619115304
14. Batista P, Araújo I, Linder N, Laraqui K, Klautau A (2019) Testbed for ICN media distribution over LTE radio access networks. Comput Netw 150:70–80
15. Ali, Khitem Ben, Faouzi Zarai, Radhia Khdhir, Mohammad S. Obaidat, and Lotfi Kamoun. “QoS aware predictive radio resource management approach based on MIH protocol.” IEEE Systems journal 12, no. 2 (2018): 1862–1873.
16. Amairullah Khan Lodhi, M.S.S Rukmini, Syed Abdulsattar, “Efficient Energy Routing Protocol based on Energy & Buffer Residual Status (EBRS) for Wireless Sensor Networks” International Journal of Engineering and Advanced Technology (IJEAT) ISSN: 2249 – 8958, Volume-9 Issue-1S5, December, 2019. https://doi.org/10.35940/ijeat.A1008.1291S52019
17. Lodhi AK, Rukmini MSS, Abdulsattar S (July 2019) Energy-efficient routing protocol for node lifetime enhancement in wireless sensor networks. Int J Adv Trends Comput Sci Eng 8(1.3):24–28. https://doi.org/10.30534/ijatcse/2019/0581.32019
18. Lodhi AK, Rukmini MSS, Abdulsattar S Energy-efficient routing protocol based on mobile sink node in wireless sensor networks
19. Puranik VV, Lodhi AK Dynamic resource management of cognitive radio networks via fuzzy logic

Reliable and Energy-Efficient Data Transfer Routing in Wireless Body Area Networks

Nikhil Marriwala

Abstract Many low-power devices mounted on a body and connected in a network form a wireless body area network (WBAN). The physiological signals generated by the body are captured by these low-power devices called as nodes and are then transmitted to the base station or sink for further processing. The sensor nodes deployed on the human body form a network and share the information by the use of different routing protocols which have a significant impact on the dissipation of energy and reliability of the networks. Sensor nodes can self arrange themselves for the configuration of cluster head in a non-centralized hierarchical routing. The nodes are unacquainted about the whole rational structure of the network during self-configuring. The base station first collects information and residual energy of each node in a planned routing technique. This paper proposes an energy-efficient reliable routing protocol for transmission of data from the nodes to the base station by cluster formation with the analysis of the residual energy of all the nodes. The comparison of the proposed protocol energy efficient and reliable routing (EERR) with that of the M-ATTEMPT protocol based on energy dissipation with time, data packet sent and the system life span of network shows that system life span is increased for the network.

Keywords Wireless body area network · Wireless sensor network (WSN) · Radio frequency (RF) · Hierarchical routing protocols · Cluster head · Specific absorption rate

N. Marriwala (✉)
Department of Electronics and Communication Engineering, University Institute of Engineering and Technology, Kurukshetra University, Kurukshetra, India
e-mail: nikhilmarriwala@gmail.com

N. Marriwala et al. (eds.), *Mobile Radio Communications and 5G Networks*, Lecture Notes in Networks and Systems 140,
https://doi.org/10.1007/978-981-15-7130-5_62

1 Introduction

Merely, a small number of sensors are used in WBAN, which are situated on the body or fixed in the body [1]. These little sensors are sited on patient's body to evaluate crucial signs like pulse rate, glucose level, blood pressure, etc. [2, 3]. These deliberate measured quantities are then sent to the doctor to estimate the state of the patient [4]. Constant monitoring of the patient at a remote position is offered by wireless sensor networks (WSN) [1, 5, 6]. Progressing in wireless technology instinctive a novel creation of this creation is appropriate for wireless networking in the human body [7]. A multi-hop topology or a point-to-point topology is used for conveying data between sensor nodes in these networks [3]. Depending on the application, we use the topology; for example, multi-hop topology is needed to calculate the posture of an athlete [8]. Among sensor nodes, the sensed data is swap, and then, it is reached to sink or base station. Sensors can be placed or fixed in the body of an athlete. The main applications of WSN technology is remote monitoring of the health of the human body [9]. Merely, a small number of sensors are used in WBAN, which are situated on the body fixed in the body [4]. These little sensors are sited on patient's body to evaluate crucial signs like pulse rate, glucose level, blood pressure, etc. [3]. These values are then forward to the doctor or the medical server to supplementary estimate the state of the patient [10]. Constant monitoring of patient at remote situate is offered by wireless sensors. Low-power devices that are placed on, in or approximately the human body formed a network known as WBAN, and these devices are used to observe activity and physiological signals for personal activity, medical and other purposes and applications Fig. 1. Because of the ever more population of the aging stage to diseases related to age and could regularly benefit from constant monitoring of signals related to physiological, healthcare applications have fascinated the researcher's notice [11]. The use of WBANs has the potential to improve the present healthcare monitoring immensely and could lead to practical and remote, analysis of many diseases at an early stage [12]. Besides an actuator controlled by a WBAN, can repeatedly liberate medicine or other agents based on its settings and capacity.

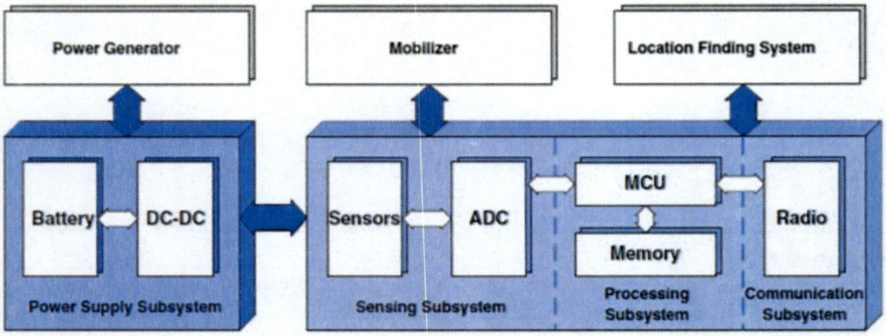

Fig. 1 Block diagram representing the architecture of the wireless sensor node

To provide insulin to a diabetic patient, actuators are used [13] without snooping the everyday activities of patient's health monitoring provided by the WBAN heal [4, 9]. To obtain information concerning the health of patient on a stable basis where the caregiver needs this, the WBANs should give, among other individuality, consistent connections that are moderately not sensitive to link or node failure [1, 14]. The mobility of the patient increases the chance of loss of packet, and it is favored that the inaccuracy rate of the packet must be kept less than 1% [6, 15]. At low power, WBAN must be transmitted to keep the patients adjacent to injurious effects of health related with the emission of radio frequency (RF) [16]. The Specific Absorption Rate (SAR) must be low down; therefore, SAR is defined as the rate of captivating the RF energy by a quantity or body [7].

2 Wireless Sensor Node

In a WSN, a sensor node is a node that can assemble sensory information, performing arts some processing, and communicating with other related nodes in the network [17, 18]. The classic architecture of the sensor node is shown in Fig. 1. The block diagram shown in Fig. 1 represents sensor nodes, microcontroller, external memory, power source, and transceiver representing the architecture of the sensor node [17].

In WSN, the deployment of sensor nodes can be done using two different ways, i.e. manual and random. In manual deployment, each sensor node is assigned a fixed particular task like temperature and moisture measurement in agriculture deployment or heat sensing in building for fire alarms, etc. In random deployment, the location of the sensor node is random, for example, dust sensors sprayed around or airdropped in war zones, etc.

3 Wireless Sensor Network

A WSN comprises many sensors (e.g. acoustic, seismic, or image sensors), which are treated as sensor nodes and used for different types of data measurements. Different sensor nodes deployed in a WSN take the desired measurements from the environment and then transmit it to the sink, i.e., from the sensor node to root node [19]. Hundreds of up to thousands of sensor nodes are used by the WSN and can be extended out as a mass or positioned out one by one. To set up a network capable of sensing, the sensor nodes work jointly with each other, i.e. a WSN [6]. Each sensor node must be low cost and small because of the potentially bulky scale of the WSNs. The accessibility of low-cost sensor nodes has resulted in the development of many application areas [4, 11, 20]. The sensor network can offer entrée to information as shown in Fig. 2. Once the data is collected by the sensor nodes, it is processed and analyzed by them. A WSN has to operate for a long time in many application areas.

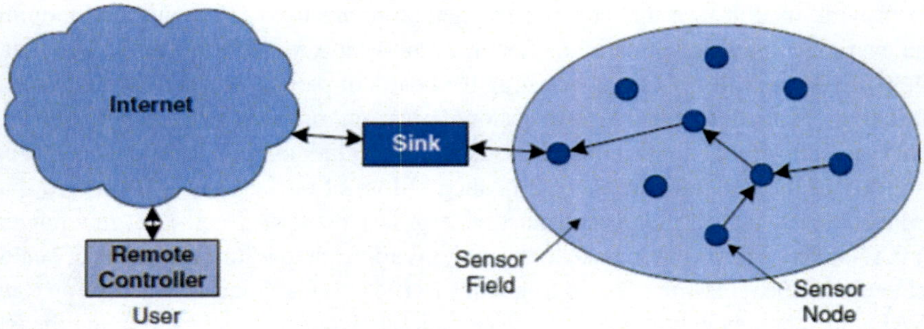

Fig. 2 Block diagram of wireless sensor network

The energy outflow of both the sensor network and individual sensor nodes together is of main significance [13]. Thus, for WSNs, energy is a very vital issue.

4 Literature Review

For reliable data communication between the nodes, Quality of Service (QOS)-aware Peering Routing Protocol (QPRR) has been proposed in [21]. QPRR is used in indoor hospital and helps in increasing the reliable delivery of critical data to the destination. In QPRR, more energy is consumed due to too much information processing. In [5], authors have projected a routing protocol using the cluster techniques, inspired by LEACH protocol. To reduce the energy utilization in node, the nodes transmit the data directly to sink using the clustering technique. CH selection is done on the basis of threshold value. The selected CH remains the CH for the next round and other rounds also if it has the energy larger than the threshold level. If its energy drops below the threshold value, a new CH with energy greater than threshold value is selected. A routing protocol for homogeneous and heterogeneous WBANs, called M-ATTEMPT, has been discussed in [22]. M-ATTEMPT protocol is a routing protocol as described which is based on thermal awareness and senses the link hot spot. This thermal-aware routing protocol routes the data away from these links. According to the data rate of the sensor nodes, they are placed in descending order around the sink node of the body.

A Reliability Enhanced-Adaptive Threshold-based Thermal-Unaware Energy-Efficient Multi-hop Protocol (RE-ATTEMPT) is proposed and discussed in [23]. The level of the energy possessed by the nodes decides the data rate and is thus placed on the body. The network uses either single-hop or multi-hop routing for data transmission. The network lifetime is increased as the data load is uniform for all the nodes; hence, the energy consumed is also even.

5 Problem Formulation

In a sensor network, tens to thousands of sensor nodes are collected which are dispersed in a broad region for communication with each other. Along with them, one or more nodes serves up as sink(s) that can communicate with the user through the offered wired networks or either directly.

An important aspect of WBNs is that their energy requirements are high and they have a limited accessibility to the energy; hence, efficient usage of energy is the only way out to enhance the life span of the network. Hierarchical routing signifies that sensor nodes self-configure themselves for the election of CHs. The paper aims at creating a routing protocol which is highly energy efficient in comparison with the other existing hierarchical routing protocols.

6 Methodology

Initially, the base station is fixed at a particular position on the human body. The other sensor nodes having equal energy of 0.5 J are then set on various body parts such as legs, hands, and heart. Initially, there is no CH as all the sensor nodes in the network possess equal energy. A hierarchical routing protocol is used for the reduction of packet loss. To improve the Quality of Service (QOS), one-hop routing is used in the proposed system. Keeping different conditions into consideration such as the residual energy and minimum distance from the base station the cluster head (CH) is selected in round 1.

The residual energy is given in Eq. 1

$$RE :(1.5 * ((\mathrm{ETX} + \mathrm{EDA}) * (B) + \mathrm{Eamp} * B * (D * D))) \tag{1}$$

where

RE	Residual energy required by the nodes to send the combined data to the sink.
D	Distance from a particular cluster head or base station.
ETX	Energy consumed by transmitter to send data.
EDA	Energy data aggregation.
Eamp	Energy consumed by transmitting amplifier.
B	Data bits need to transfer.

Once the CH is selected, all the other nodes get the information about the CH selection according to the minimum distance. Making use of the nearest distance algorithm, all the nodes in the network elect one of the nodes as their CH. Equation 2 represents the energy consumption during sending the data by the node to their respective CHs.

$$\mathrm{ETX} * (b) + \mathrm{Eamp} * b * (\text{ Min\_dis} * \text{Min\_dis}); \tag{2}$$

where

Min_dis Distance of a particular node to the cluster head.

CH now cumulates the data and sends it to a sink. Equation 3 gives the energy utilized which is determined for each node and CH.

$$1.5 * ((\mathrm{ETX} + \mathrm{EDA}) * (b) + \mathrm{Eamp} * b * (\mathrm{distance} * \mathrm{distance})) \quad (3)$$

The CH selection in round 2 is done by checking the minimum distance from the base station and threshold energy as given by Eq. 4

$$\text{Cost Function} = \text{distance/Residual Energy} \quad (4)$$

Once the CH selection is made, the nodes start sending the data to their respective CHs which are selected based on minimum distance algorithm of a particular node from CHs and energy consumed. The CH cumulates the data and sends it to the sink. The same process is repeated until all the number of rounds is finished. To measure the performance of the proposed system, different parameters such as dissipation of energy, lifetime of the network, and total no of data packets sent are taken into consideration. Figure 3 gives the flow diagram of the proposed energy-efficient protocol.

7 Implementation and Results

Owing to the verity that less energy is used by the clustering protocols, a lot of applications have given wide acceptance to the WSNs protocols. At various levels to reduce energy expenditures, several on hand WSN protocols use cluster base scheme. Based on a possibility, CH in the most cluster-based protocol is elected. A new routing protocol named energy efficient and reliable routing (EERR) has been proposed which helps to decrease packet loss as well as energy utilization. Figure 4 represents the deployment of nodes in body.

8 Radio Model

In present-day scenario, radio models of large quantity are accessible. As proposed in a first sort, radio model is used. In this, radio model among transmitter and receiver 'd' is the separation, and due to communication channel, 'd2' is the loss of energy. Table 1 signifies the two radio models used for the simulation of the network.

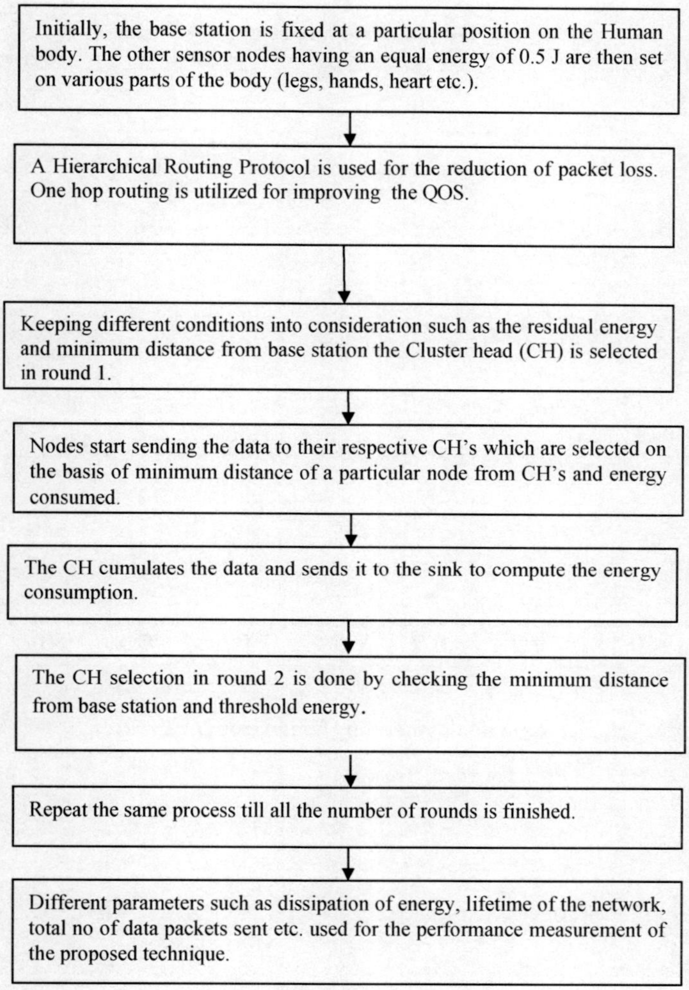

Fig. 3 Flowchart of the proposed energy-efficient protocol

9 Network Parameter

Table 2 highlights different parameters used during the simulation of the network.

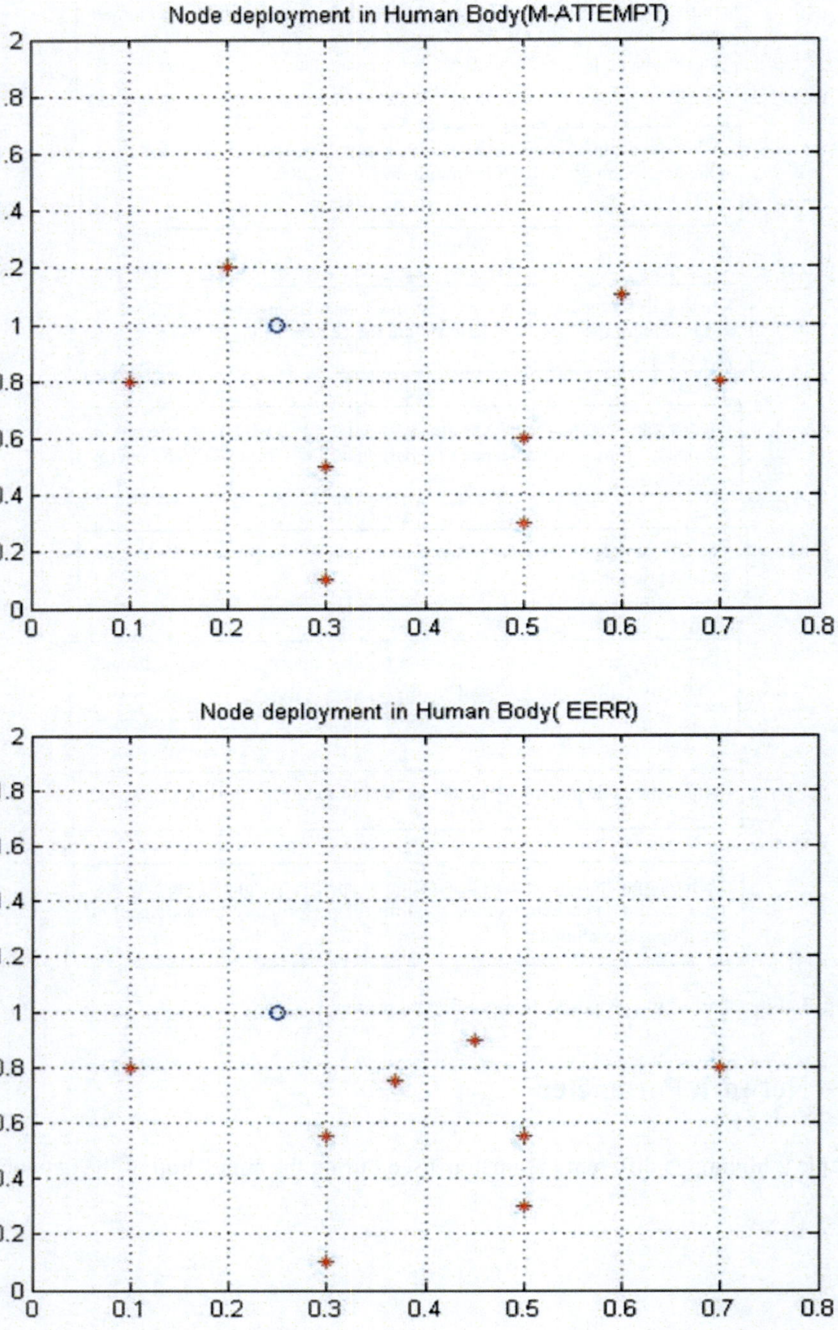

Fig. 4 Deployment of different nodes on the body

Table 1 Radio models used

Radio model	nRF 2402	NCS36510
Transmitting energy (ETX)	16.9nJ/bit	97.9 nJ/bit
Receiving energy (ERX)	36.4nJ/bit	174.8 nJ/bit
Amplifier energy (Eamp)	1.97e–9 j/b	2.71e–7 J/b

Table 2 Various parameters used during simulation

Parameters		
Initial energy	Eo	1 J
Amplifier energy	Eamp	2.71e-7 j/b
Transmitting energy	ETX	97.9nJ/bit
Receiving energy	ERX	174.8nJ/bit
Data aggregation energy	Eda	5nJ/bit
Packet size	b	5000 bits
No. of nodes	n	8

10 Discussion

The plot shown in Fig. 5 represents the number of dead nodes left after the network starts to send the data packets after each round. It can be seen that the proposed network protocol EERR remains active for much larger period as compared to M-ATTEMPT. Table 3 represents the number of dead nodes after each round for the two protocols M-ATTEMPT and EERR.

The plot shown in Fig. 6 represents the packets sent to the sink or base station after each round. It can be observed from the plot that the proposed network protocol

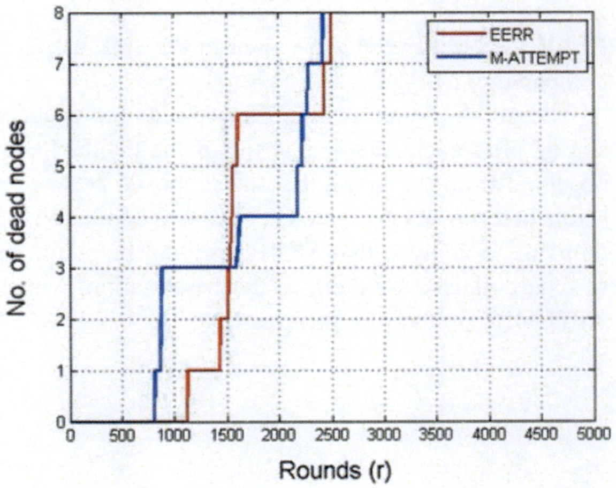

Fig. 5 Plot representing comparison between dead nodes versus rounds

Table 3 Number of nodes dead

Dead node	Round in which node dead (M-ATTEMPT)	Round in which node dead (EERR-PROPOSED)
1	800	1140
2	900	1537
3	925	1616
4	1680	1645
5	2229	1681
6	2333	1711
7	2380	2537
8	2520	2604

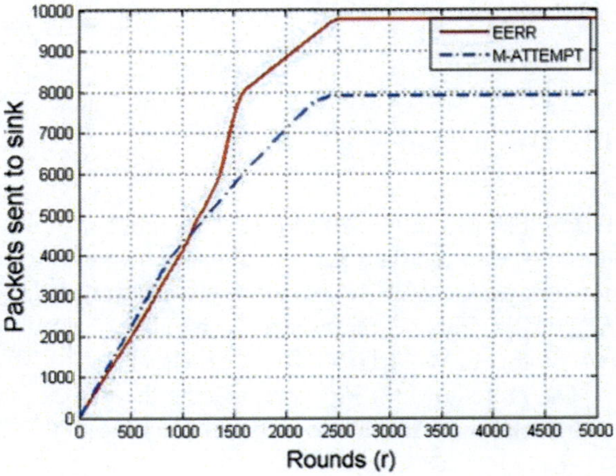

Fig. 6 Plot representing comparison between no. of rounds versus data sent to base station (SINK)

EERR is able to send more data to the sink using the same amount of energy as compared to M-ATTEMPT (Fig. 7).

The graph plotted in Fig. 8 represents the packets received by the sink or base station after each round in terms of the residual energy of the network. It can be observed from the graph that the proposed network protocol EERR is much more stable in terms of energy consumption as compared to M-ATTEMPT. The graph also helps us to analyze that EERR protocol helps the nodes to transfer large amount of data collected by them to the base station, whereas the data sent by using M-ATTEMPT protocol is quite less.

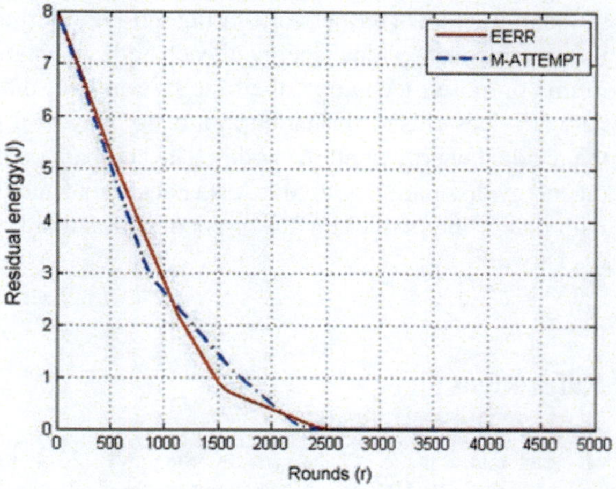

Fig. 7 Plot representing comparison between no. of rounds versus residual energy

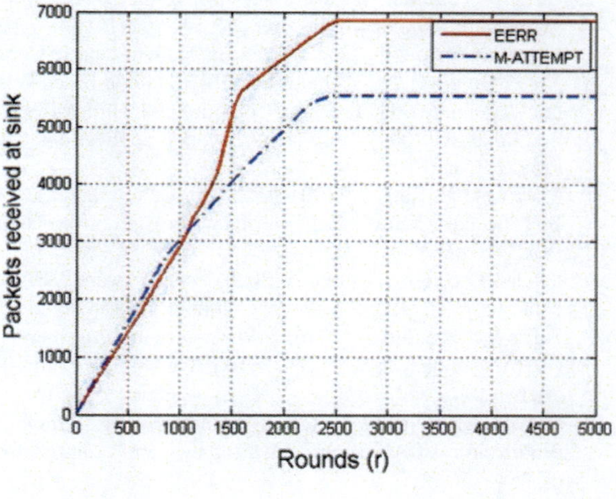

Fig. 8 Plot representing comparison between number of rounds versus data received at base station (SINK)

11 Conclusion

Routing protocols can have a vital impact on energy dissipation and the overall reliability of these networks. In several of their uniqueness, WBANs be dissimilar from traditional wireless communiqué networks. Power consciousness is one of them; due to this, the batteries of sensor nodes have a limited life span and are not easy to be replaced. To conserve the long life of the network and to reduce energy exploitation, therefore, all protocols must be intended in such a fashion. That is why in WBANs, routing protocols mean mainly to achieve power preservation they focus mainly on the Quality of Service (QoS) as in traditional networks.

Sensor nodes can self arrange themselves for the configuration of cluster head in a non-centralized hierarchical routing. The nodes are unacquainted about the whole rational structure of the network during self-configuring. The base station first collects information and residual energy of each node in a planned routing technique. Hence, in this proposed technique, the base station with the global information about the network does cluster formation in a better way as it has the information regarding the residual energy of all the nodes. The comparison of the proposed protocol based on energy dissipation with time, data packet sent and the system life span of network with that of the M-ATTEMPT protocol shows that system life span is increased for the network.

References

1. Rathee D, Rangi S, Chakarvarti SK, Singh VR (2014) Recent trends in Wireless body area network (WBAN) research and cognition based adaptive WBAN architecture for healthcare. Health Technol (Berl) 4(3):239–244
2. Ullah F, Abdullah AH, Kaiwartya O, Kumar S, Arshad MM (2017) Medium access control (MAC) for wireless body area network (WBAN): superframe structure, multiple access technique, taxonomy, and challenges. Human-centric Comput Inf Sci 7(1)
3. Xie Z, Huang G, He J, Zhang Y (2014) A clique-based WBAN scheduling for mobile wireless body area networks. Procedia Comput Sci 31(Itqm):1092–1101
4. Reichman A (2009) Body area networks: applications, architectures and challenges. IFMBE Proc. 25(5):40–43
5. Marriwala N, Rathee P (2012) An approach to increase the wireless sensor network lifetime. In: Proceedings of the 2012 world congress on information and communication technologies, WICT 2012
6. Sagar AK, Singh S, Kumar A (2020) Energy-aware WBAN for health monitoring using critical data routing (CDR). Wirel Pers Commun 123456789
7. Sarra E, Ezzedine T (2016) Performance improvement of the wireless body area network (WBAN) under interferences. 2016 IEEE 18th Int Conf e-Health Networking, Appl Serv Heal 2016
8. Sethi D , Bhattacharya PP (2016) A study on energy efficient and reliable data transfer (EERDT) protocol for WBAN. Proc—2016 2nd Int. Conf. Comput. Intell. Commun. Technol. CICT 2016. pp 254–258
9. Negra R, Jemili I, Belghith A (2016) Wireless body area networks: applications and technologies. Procedia Comput Sci 83:1274–1281
10. Ntouni GD, Lioumpas AS, Nikita KS (2014) Reliable and energy-efficient communications for wireless biomedical implant systems. IEEE J Biomed Heal Informatics 18(6):1848–1856
11. Anand J, Sethi D (2017) Comparative analysis of energy efficient routing in WBAN. 3rd IEEE Int. Conf
12. Yoo HJ (2013) Wireless body area network and its healthcare applications. Asia-Pacific Microw. Conf. Proceedings, APMC, pp 89–91
13. Dharshini S, Subashini MM (2017) An overview on wireless body area networks. 2017 Innov. Power Adv. Comput. Technol. i-PACT 2017, vol. 2017–Janua, pp 1–10
14. Xu W, Wu Q, Daneshmand M, Liu Y (2015) A data privacy protective mechanism for WBAN. Wirel Commun Mob Comput (February 2015) pp 421–430
15. Fabri D et al (Dec. 2013) A quantitative comparison of the performance of three deformable registration algorithms in radiotherapy. Z Med Phys 23(4):279–290
16. Ullah S, Alsalih W (2013) On secure and power-efficient RFID-based. pp 770–776
17. Shweta NG (2017) Review on cluster head election algorithms based on distance and energy. 4(7):767–774

18. Marriwala N, Rathee P (2012) An approach to increase the wireless sensor network lifetime. Inf Commun Technol (WICT), 2012 World Congr 495–499
19. Thanh Hiep P, Nhu Thang N, Sun G, Huy Hoang N (2019) Proposal of a hierarchical topology and spatial reuse superframe for enhancing throughput of a cluster-based WBAN. ETRI J 41(5):648–657
20. Fauzi M, Shazali K (2012) Wireless sensor network applications: a study in environment monitoring system. In: International symposium on robotics and intelligent sensors 2012 (IR IS 2012) vol. 41, pp. 1204–1210
21. Khan ZA, Sivakumar S (2013) A QoS-aware routing protocol for reliability sensitive data in hospital body a qos-aware routing protocol for reliability sensitive data in hospital body area networks. (December)
22. Javaid N, Abbas Z, Fareed MS, Khan ZA, Alrajeh N (2013) M-ATTEMPT: a new energy-e ffi cient routing protocol for wireless body area sensor networks. Procedia—Procedia Comput Sci 19(Ant):224–231
23. yAhmad A, Javaid N, Qasim U, Ishfaq M, Khan ZA, Alghamdi TA (2014) RE-ATTEMPT: a new energy-efficient routing protocol for wireless body area no. April

A Model on IoT Security Method and Protocols for IoT Security Layers

Chandra Prakash and Rakesh Kumar Saini

Abstract Internet of things (IoT) digitalized the worldwide system containing individuals, connected things, smart devices, data, and information. It is a well-known fact that as an ever-increasing number of devices interface with the Internet, the difficulties of making sure about the information that transmitted and interchanges that they start are getting progressively significant. Throughout the years, we have seen a flood in IoT devices, comprehensively in two parts: homes and manufacturing. Since these are autonomous and secure fields, the duties of making sure about the devices rest with the platform providers. In this paper, we have discussed the various application areas where IoT is applied to get an effective and reliable outcome, and majorly, we have focused on security aspects related to IoT. For that purpose, we have proposed a security model to protect the IoT network or system from unwanted threats and attacks. The proposed model is providing a choice of a suitable security method and protocols for IoT Security layers. This model is used to improve the performance of IoT system by opting the appropriate security methods for IoT layers to reduce the power and time consumption.

Keywords IoT applications · IoT security challenges · IoT communication protocols · IoT in manufacturing · IoT in healthcare

1 Introduction

IoT can be defined as a systematic setup of interrelated computing devices, individuals, connected things, advanced machines, data, and information that are given through unique identification and capable to send information over a system without direct interfering of human. A thing in the word IoT can be an individual with

C. Prakash (✉) · R. K. Saini
School of Computing, DIT University, Dehradun, Uttarakhand, India
e-mail: chandra.thukral.19@gmail.com

R. K. Saini
e-mail: rakeshcool2008@gmail.com

N. Marriwala et al. (eds.), *Mobile Radio Communications and 5G Networks*, Lecture Notes in Networks and Systems 140,
https://doi.org/10.1007/978-981-15-7130-5_63

implantable cardiac monitor, a car that has in-built sensors to caution the driver when pressure of tire is low or some other regular or man-made article that can be allotted an IP address and can move information over a system. IoT is a thought that could drastically modify our relationship with innovation. The market is as of now concentrating on the vertical spaces of IoT since it is in moderately early periods of advancement, but IoT cannot be treated as a solitary thing, or single device, or even a solitary innovation. So as to accomplish the normal quick development from IoT deals, more concentration should be put on interfaces, versatile applications, and normal/predominant guidelines [1, 2].

Web-enabled remote study classrooms will be an achievement for creating nations, making profound infiltration in regions where setting up a customary institution foundation is beyond the realm of imagination. Web-enabled industries and manufacturing units are giving separating outcomes, making them more secure and increasingly proficient through robotized process controls. Finance-related administrations are as of now utilizing the web for a large number of their administrations [3, 4]. While the potential outcomes of these new advancements are amazing, they have additionally uncovered unadorned IoT security challenges. During most recent couple of years, we have seen an increment in the number and the refinement of attacks focusing on IoT devices. The interconnectivity of individuals, objects, and groups in the present digitalized world opens up an entirely different playing field of susceptibilities passageways where the cybercriminals can get in. On the other hand, IoT technology devises a number of problems as well. Like, complexity is one of the most substantial drawbacks as IoT operations are difficult, and there is not at all flexible incorporation between nodes. There are various devices with diverse design, implementation and deployment, so any drawback in software and hardware will have severe difficulties. IoT network undergoes from validation and access control problems because smart things have diverse devices that are based on various platforms. Moreover, all devices are essential to interact with another device via different network. Therefore, security problems have the key challenge because all devices are unprotected to all kinds of attacks and threats. There are many kinds of attacks and threats that might cause severe tragedies in the network. Furthermore, all private data of users are unprotected to the most hazardous attacks. In the proposed model, that is used to create security control system for the IoT network to offer appropriate security tools for the IoT security layers. It can assist designer to reduce the time and power consumption. This paper also gives a review of the present province of IoT security challenges [5].

2 Security Challenges in IoT

There are some security challenges in Internet of things that are shown in Fig. 1.

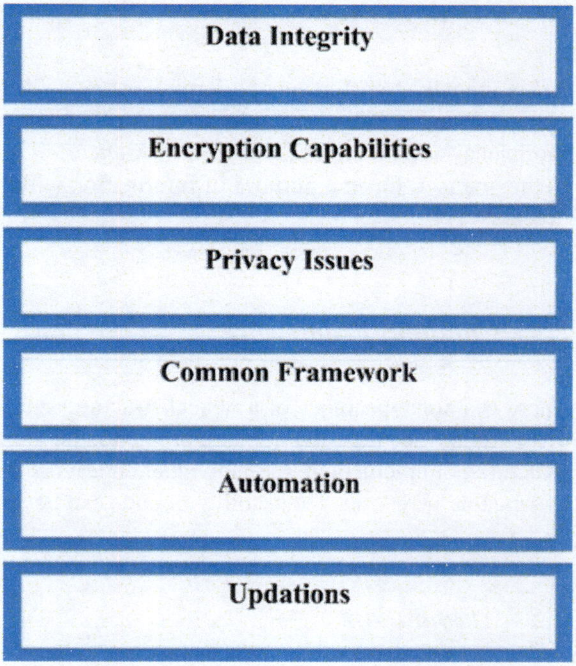

Fig. 1 Security challenges in IoT

2.1 *Data Integrity*

Data integrity is the correctness, consistency, and completeness of data. Additionally, data integrity can be defined as data security with respect to administrative consistence. It is kept up by an assortment of procedures, guidelines, and standards actualized during the structure stage. At the point when the data integrity is secure, the data will keep complete, correct, and consist in the database, regardless of what extent its put away or how regularly it is accessed. The integrity of data guarantees that your information is sheltered from any outside forces [6].

2.2 *Encryption and Decryption Capabilities*

IoT carries another set of security parameters. In contrast of VPN encryption, which protect network by an encryption and unspecified channel, IoT devices bring with their own built-in robust security and encryption protocols. VPN distributes a separated space on the system. However, any participant inside the VPN can access the devices of the network. IoT device with a VPN can have many options to create a secure network.

2.3 Privacy Issues

IoT is about the transferring of information among different devices, individuals, and platforms. IoT devices assemble information for various reasons, such as improving proficiency and experience, decision-making, offering better assistance, and so forth; in this manner, the end purpose of information will be totally made sure about and defended [7].

2.4 Common Framework

There is a nonappearance of a typical structure; thus, all the manufacturers need to deal with the security and hold the protection all alone. When a typical normalized structure is implemented, the individual endeavors will then together be used in an expandable way; thus, the usability of code can be accomplished.

2.5 Automation

In the end, industries should manage increasingly more number of IoT devices. This tremendous measure of client information can be hard to oversee. The reality cannot be denied that it requires a solitary mistake or trespassing a solitary calculation to cut down the whole foundation of the information.

2.6 Updations

Dealing with the update of a huge number of devices should be practiced individually. Sometimes, it is required to update the IoT devices manually because they did not have the support of auto-update. One should monitor the accessible updates and apply the equivalent to all the various devices. This procedure becomes tedious and convoluted and if any error occurs in the process than this will prompt escape clauses in the security later. IoT includes the utilization of a large number of information focuses and each point ought to be made sure about. For sure, the need is for the multilayer security (security at every single level). From endpoint devices, cloud, embedded applications to web and versatile, applications that influence IoT, each layer ought to be security unblemished [8].

3 Literature Review

IoT technology is a framework, which makes devices of routine work smarter and everyday conversation develops educational. Many researchers are working on IoT. In this paper, we describe some existing work proposed by many researchers that are:

Atzori et al. [9] extraordinary visions of this IoT paradigm are reported and permitting technology reviewed. What emerges is that still essential troubles shall be faced by means of the studies community.

Lee et al. [10] proposed an encryption technique based on XOR manipulation, in place of complicated encryption along with the use of the hash characteristic, for anti-counterfeiting and privacy protection. The enhancement of the safety is defined, and hardware layout method is also established.

Abomhara et al. [11] classified risk types also examine, describe IoT devices and offer a model to detect invaders and assault.

Barnaghi et al. [12] present a semantic demonstrating approach for various segments in an IoT system. It is additionally talked about how the model can be incorporated into the IoT system by utilizing mechanized affiliation components with physical elements and how the information can be found utilizing semantic pursuit and thinking instruments.

Gubbi et al. [13] present a cloud-driven vision for overall usage of IoT. The key empowering technologies and applications areas that are probably going to drive IoT to inquire about sooner rather than later are talked about. A cloud usage utilizing Aneka, which depends on interaction of private and open Clouds, is introduced. They finished IoT vision by developing the requirement for convergence of WSN, the Internet and dispersed processing coordinated at innovative research network.

Xiao et al. [14] address the troubles appearing in device finding and interplay. They develop a person interoperable system to allow managers to operate through various devices of various settings by constant semantics and syntax. In the proposed framework, a parting method is used in which a device illustration technique for actual, but not unusual, and digital devices is created. A transformable tool is offered to ensure the right transformation of device semantics and syntax.

Zorzi et al. [15] summarize what in our opinion are the primary wireless—and mobility-related technical demanding situations that lie beforehand, and description a few initial thoughts on how such challenges can be addressed in an effort to facilitate the Internet of things' development and receipt within the next few years. We also pronounce a case study on the Internet of things protocol structure.

Hongsong et al. [16] survey to protection and consider in M2M gadget is important. Different visions of the M2M security and receive as accurate with standard are specified and studied on this paper. They comprise normal generation educations development and guard creation. All these will contribute to identify safety and outdated advancing in M2M machine.

4 Proposed Model of IoT Security

The layered architecture of IoT network consists of mainly six layers named as coding, perception, network, middleware, application, and business, shown in Fig. 2. Out of these six layers of IoT-layered architecture, three of them, perception, network, and application, are used to design the architecture of IoT security concerns. Every layer of IoT security architecture has its own communication and security protocols, standards, and components. Based on this security architecture of IoT, we have proposed a security model that could help us to protect from unwanted threats and attacks, un-authentication and protect our private information.

The proposed model is consisting of three stages of development that includes concerning of the security layers, security protocols, and database servers. Figure 4 shows the proposed model of IoT security concern.

Most appropriated protocols at different layers of security architecture of the proposed model are given as:

IEEE 802.11 Protocol at Perception layer: IEEE 802.11 defines a group of determinations created by the IEEE for wireless LAN (WLAN). IEEE 802.11 indicates an over-the-air interface between a remote device and a base station or between two remote devices. This standard is utilized related to IEEE 802.2 and is intended to inter-work consistently with Ethernet and is all the time used to convey Internet Protocol traffic.

6LowPAN Protocol at Network layer: The low-power personal area networks over IPv6 (6LoPAN) have characterized embodiment and header-compression systems which permit IPv6 packet to transmit over IEEE802.15.4-based systems.

SMQTT Protocol at Application layer: SMQTT stands for secure message queue telemetry transport protocol. The SMQTT is an extended version of MQTT, which used the lightweight parameter encryption. SMQTT comprises of 4 primary

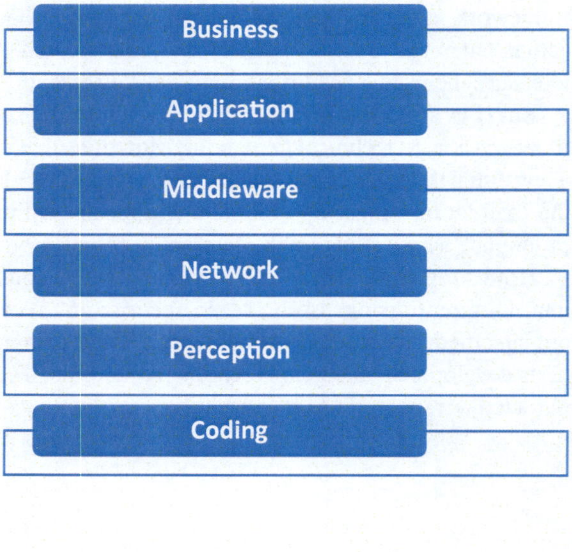

Fig. 2 IoT layered architecture

steps: setup, encryption, publishing, and decryption. In setup stage, subscribers and publishers register themselves to broker and get a master key as per the choose by key generation algorithm (KGA). At the point when the information is published, the data is encrypted and distributed by the broker which sends it to the subscribers, which is at last decrypted at the subscriber end having a similar master key.

4.1 Implementation of Proposed Model

Step 1: The first step of the proposed model consists of maintaining the security requirements of IoT security layer, like Access control, Privacy, Confidentiality, Integrity, Availability, Authorization, and Authentication, by using algorithms like Hash Algo, End-to-End authentication, Cryptography Algo, Access Control, Key Management, Intrusion Prevention System Encryption Protocol, Data privacy and integrity ACLs, Antivirus, Firewalls, and Risk Assessment from attacks like Node Capture, Fake Node, Denial of Service, Replay Attack, Node Jamming, Routing Threats, RFID Tag Spoofing and Cloning, Session Hijacking, Sybil, Flooding Attacks, Data access permission, Managing mass information and programming vulnerabilities (Fig. 3).

Step 2: In the subsequent step, i.e. step two, the security protocols or standards and security control mechanism for different layers of IoT security architecture have been described. The most appropriate communication protocols in perception layer are IEEE 802.11. The most suitable communication protocol for the network layer is 6LowPAN, which encapsulates the IPV6. SMQTT protocol is used for application layer of IoT security architecture.

Step 3: The third stage of the proposed model is database servers which store all data and parameters of security concern for every security layer, clients' profiles, security components errors, log records of the IoT framework, and access control records.

4.2 Flowchart of Proposed Model

Flowchart of the model for overall process is shown in Fig. 4 (On next page). The process is consisting of collecting the data from physical medium like sensor and converted into digital signal for further process. User can also give their instruction via user interface to control the system process. The digital signals are encrypted by using an appropriate key generation algorithm. The encrypted data is aggregated with the user data. With the interface of IoT gateway, the data is transmitted to database via web server, where the encrypted data is decrypted by using of same key and display to user. Meanwhile, the decrypted data is also stored in the database for future reference.

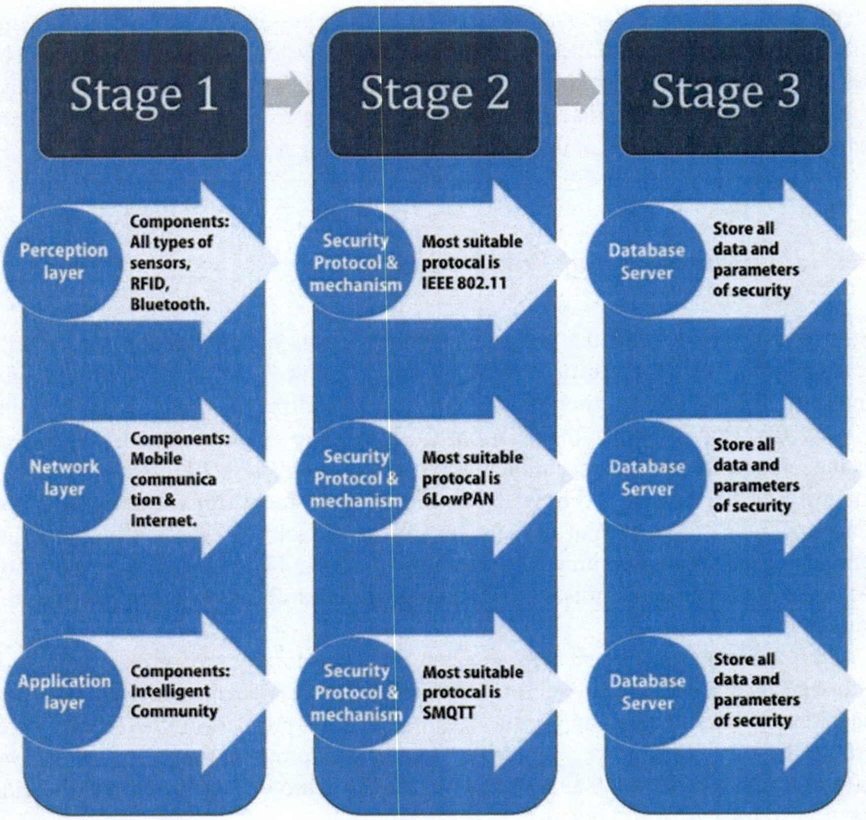

Fig. 3 Proposed Model of IoT Security Concern

5 Conclusion

The IoT framework is helpless against attacks at each layer. Subsequently, there are numerous security risks and prerequisites that should be dispatched. Current situation of research in IoT security is fundamentally focused on verification and access control conventions. However, with the quick development of innovation it is basic to merge new systems administration conventions like IPv6 and 5G to accomplish the dynamic blend of IoT technologies. The primary emphasis of this paper was to feature significant security issues of IoT. Especially, centering the security attacks and their countermeasures. The proposed model is capable enough to handle these attacks and threads to protect the sensitive data and private information. The main goal of the proposed security model is to choose and apply the appropriate security protocol and algorithm. By using these protocols and security algorithm, the system is capable enough to detect the problem and apply the suitable algorithm to protect the system from unavoidable unwanted situations. As we are aware that a lot of IoT

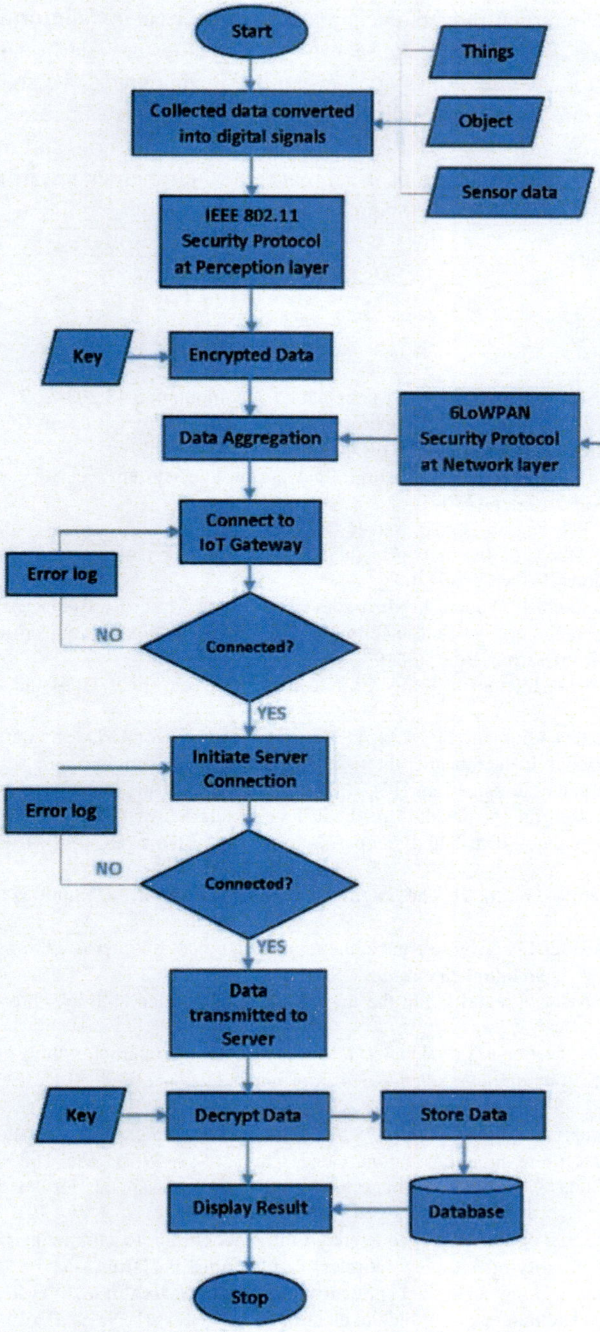

Fig. 4 Flowchart of the IoT security model

devices are come to be easy target. Indeed, even this isn't in the casualty's information on being contaminated. In this paper, the security prerequisites are additionally conferred for authentication, secrecy, integrity, and so on. In this paper, we study the existing work in this area. We have faith that this paper will be valuable for researchers in the field of security for IoT by assisting the significant issues in IoT security and giving better comprehension of the threats and their elements starting from different interlopers like intelligence agencies and organizations.

References

1. Koblitz N (1987) Elliptic curve cryptosystems. Mathematics of computation 48:203–209
2. Vignesh R, Samydurai A. 1 Student, 2 Associate professor security on internet of things (IOT) with challenges and countermeasures in 2017. IJEDR 5(1) ISSN: 2321-9939
3. Xie Y, Wang D (2014) An item-level access control framework for inter-system security in the internet of things. Appl Mech Mater 1430–1432
4. Anggorojati B, Mahalle PN, Prasad NR, Prasad R (2012) Capability-based access control delegation model on the federated IoT network. In: Int'l symposium on wireless personal multimedia communications (WPMC). 604–608
5. Castrucci M, Neri A, Caldeira F, Aubert J, Khadraoui D, Aubigny M et al (2012) Design and implementation of a mediation system enabling secure communication among critical infrastructures. Int'l J Crit Infrastruct Prot 5:86–97
6. Da Xu L, He W, Li S (2014) Internet of things in industries: a survey. IEEE Trans Ind Inf 10(4):2233–2243
7. Tarouco LMR, Bertholdo LM, Granville LZ, Arbiza LMR, Carbone F, Marotta M, de Santanna JJC (2012) Internet of things in healthcare: Interoperability and security issues. In: IEEE international conference on Communications (ICC) 2012. IEEE. pp 6121–6125
8. Mohan A (2014) Cyber security for personal medical devices internet of things. In: 2014 IEEE international conference on distributed computing in sensor systems (DCOSS). IEEE. pp 372–374
9. Atzori L, Iera A, Morabito G (2010) The internet of things: a survey. Comput Netw 54(15):2787–2805
10. Lee JY, Lin WC, Huang YH (2014) A lightweight authentication protocol for internet of things. In: Int'l symposium on next-generation electronics (ISNE). 1–2
11. Abomhara M, Køien GM Cyber security and the internet of things: vulnerabilities, threats, intruders and attacks
12. De S, Barnaghi P, Bauer M, Meissner S (2011) Service modelling for the internet of things. In 2011 Federated conference on computer science and information systems (FedCSIS). IEEE. pp 949–955
13. Gubbi J, Buyya R, Marusic S, Palaniswami M (2013) Internet of things (iot): a vision, architectural elements, and future directions. Future Gener Comput Syst 29(7):1645–1660
14. Xiao G, Guo J, Xu L, Gong Z (2014) User interoperability with heterogeneous iot devices through transformation
15. Zorzi M, Gluhak A, Lange S, Bassi A (2010) From today's intranet of things to a future internet of things: a wireless-and mobility-related view. Wireless Commun IEEE 17(6):44–51
16. Hongsong C, Zhongchuan F, Dongyan Z (2011) Security and trust research in m2m system. In: 2011 IEEE international conference on vehicular electronics and safety (ICVES). IEEE. pp 286–290

Structural Health Monitoring System for Bridges Using Internet of Things

Pravleen Kaur, Lakshya Bhardwaj, and Rohit Tanwar

Abstract Many bridges in India as well as in the world are on the edge of deteriorating, and their life span is already finished and yet they are still in use. These bridges are a risk to many lives. There are many factors which contribute in making these existing bridges dangerous which are heavy load of vehicles on a single bridge, high-level water or pressure, and heavy rains. With recent advancements in Internet of things (IOT) and other technologies integrated with wireless sensor devices, structural monitoring of bridges can be done using structural health monitoring (SHM) systems. Nowadays, wireless sensors can process real-time data and measure the parameters like displacement and hence can be useful in detecting the damage of the structure; the results are sent through a standard protocol to the servers on the Internet, i.e., cloud. In this paper, we suggest SHM damage detection technique via various sensors such as piezoelectric sensor and the usage of self -healing material Epoxy filled fiber-reinforced polymer (FRP). SHM system proposed in this paper consists of raspberry pi, analog-to-digital converter, Wi-Fi module, and various sensors to measure the various parameters of the bridge.

1 Introduction

The safety and durability of bridges are a great concern to the government. Since the structure of old bridges has deformed and hence further caused accidents in the near past, "according to the survey, 70% of bridges in India are old and need repair, and 57% are over 80 years old; in 2003, around 23% bridges in India need repair and monitoring."

Therefore, it is important to monitor the structure of bridges especially the old ones so that the further calamities can be prevented. For this purpose, we need to measure

P. Kaur · L. Bhardwaj · R. Tanwar (✉)
School of Computer Science, University of Petroleum and Energy Studies, Dehradun, India
e-mail: rohit.tanwar.cse@gmail.com

N. Marriwala et al. (eds.), *Mobile Radio Communications and 5G Networks*, Lecture Notes in Networks and Systems 140,
https://doi.org/10.1007/978-981-15-7130-5_64

physical, mechanical, and chemical aspects of the bridge; during the measurement, the structure should be refurnished, strengthen, or enlarged according to the necessity.

Structural health monitoring (SHM) technology is being used and has been recognized by many administrative authorities for monitoring the structure of existing as well as new bridges [8].

Over the past two decades, SHM has the advantage of new sensing technologies. Fiber optic sensors (FOP) and other wireless technologies are widely in use in SHM systems. The studies and applications have pointed out that FOP sensor has a long sensing range and the capability of providing strain or temperature at every spatial resolution along the entire sensing fiber, imbedded in or attached to the structures, using the fiber itself as the sensing medium, and have affirmed that FOP sensors are more accurate and reliable than other sensors [1] (i.e., strain gauge).

IOT is the network between the physical devices, vehicles, home appliances, and other various items and electronic devices, software, sensors, actuators, and connectivity which allows these items to connect over the network and exchange data [4], creating opportunities for more direct integration of the physical world to the computer-based systems.

2 Related Work

Liang et al. addressed that there is a problem of sensor deployment on large-scale sensor networks and proposed self-diagnostic and self-reconfiguration reasoning methods for SHM which was further verified by experimenting these on real-life scenarios with an aluminum plate and an actuator/sensor bonding.

As mentioned in the development of an IOT-based bridge safety monitoring system by Lian Lee et al. the system used Zigbee technology. This system consisted of many features, and some of them are mentioned below:

1. Monitoring devices installed in the bridge environment.
2. Connection between the database transmitted from the monitoring devices and the cloud server.
3. A dynamic database for storing data about the condition of the bridge of various areas.
4. A cloud-based server that calculates and analyzes.

Y. Sun worked on the sensor node deployment on the bridge in his paper "Research on the Railroad Bridge Monitoring Platform Based on the Internet of Things," in which sensor was mainly deployed for the damage detection and security state of the bridge.

Pressure sensors were deployed on the pier along with humidity sensors on the curve near the water body, deformation sensors on the surface of the bridge to detect any kind of defect or deformation in the bridge. Wireless multimedia sensor is a new kind of sensor which was deployed for collecting data of multimedia like audio and image formation [2, 3].

In the paper, "The Deployment of a Smart Monitoring System Using Wireless Sensor and Actuator Networks" implementation of smart monitoring system using wireless sensor network through the routing protocol for low-key and lossy networks (RPL) is explored. The applicability of a smart monitoring system in a typical smart grid scenario; in particular, a thorough experimental campaign has been carried out to verify the feasibility of integrating the IETF RPL protocol within such a framework [4, 7]. A paper "A Review of Accelerometry-Based Wearable Motion Detectors for Physical Activity Monitoring" by Che-Chang Yang and Yeh-Liang was based on the concept of level of mobility on different surfaces. Using accelerometry measurement, we can measure the motion on any surface, load on the surface, and detect if the pressure on certain area is more or less [5, 6].

3 Problem Formulation

The main purpose of SHM system is to check the durability sustainability and safety of bridges. Despite the advancement of SHM, there still exist some challenges in SHM such as:

1. Real-time monitoring of on-site conditions: It is observed that instant action on the damage could not be taken as there is delay in the transmission of data from the real-site to the authorities.
2. The integrity of sensor data is not maintained: sensor data is not accurately maintained as it becomes a costly system for authorities to deploy and maintain the integrity of the sensor data.
3. Data is not transmitted in real-time so actions needed for the current situation of the bridge are not taken in time and the concerned authority is not taken into picture.
4. After crack detection, no sudden action can be taken.
5. Cost-efficient.

4 Problem Statement

The aim is to develop a system that can do real-time monitoring of on-site conditions in consistent and efficient manner and timely alarming the issues if any to the concerned authorities.

5 Proposed Solution

In order to address the mentioned issues of leveraging high computation and analysis of bridges, we have used structural health monitoring system.

Figure 1 shows the block diagram of subsystems of the proposed SHM.

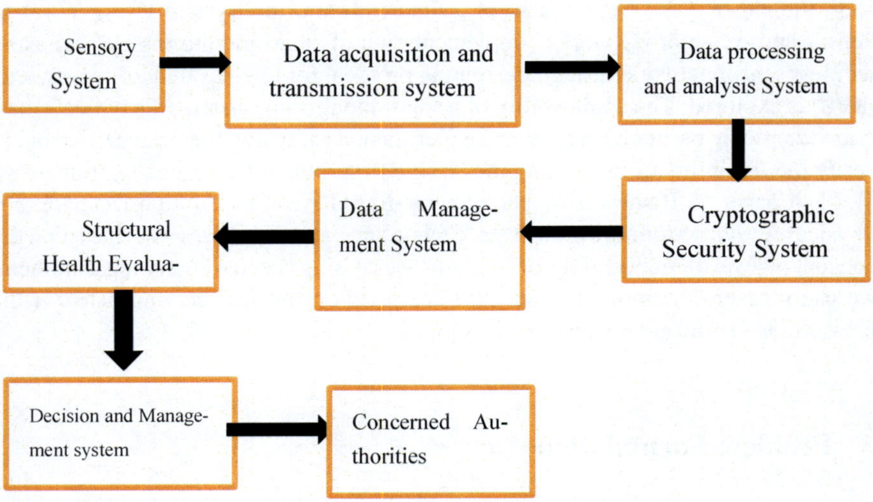

Fig. 1 Components of bridge monitoring system

Sensory subsystem–The purpose of sensory subsystem is to sense the information of the environment in which the bridge is working and safety of the bridge affected by which factors; the factors include wind speed, temperature and humidity of the environment, load vehicle, vibration, strain, cable extension and temperature of the structure.

1. **Machine sense**: It consists of accelerometer, vibration sensor, and gyroscope sensor.
2. **Temperature sensor**: It is used to detect the temperature of the material used.
3. **Anemometer sensor**: It is used to measure the wind speed and traversal distance.
4. **Magnetic float sensor**: It is used to measure the water level from the bridge.
5. **Accelerometer**: It is used to detect the bridge tilt.
6. **Ultrasonic sensor**: It is used to measure the height of sensor module from ground.
7. **Strain Gauge sensor**: It is used to measure the strain on the bridge and pillars.
8. **Humidity sensor**: It is used to monitor bridge's humidity.
9. **Fiber optic sensors**: It is used to assess the condition of existing structure in order to enhance the durability of the new bridges.
10. **Piezoelectric sensors**: It is used for damage detection as they can transmit and receive guided waves such as lamb waves in solid (lamb waves are much more cost-effective as well as reliable as they are sensitive to change in structures).

Now these sensors are interfaced with the microcontroller using signal conditioning circuit which makes the signal feasible for the microcontroller and allows an alarm and LCD; when the threshold value of sensor is exceeded, then the alarm is triggered. All the parameters are constantly transmitted through RF module, GSM module integrated with multi-hop network to a remote PC.

A driver is connecting RF module at the receiver end to the PC. GSM module is responsible for the maintenance of the bridge. Figure 2 represents the layout of sensors to be utilized in the proposed work, and Table 1 represents the variable it measures.

On the receiver side, RF module receives the data from the transmitter side. A serial is used to connect the receiver side RF module to the PC. The PC works as a control station where the data is stored sent by the transmitter side time to time.

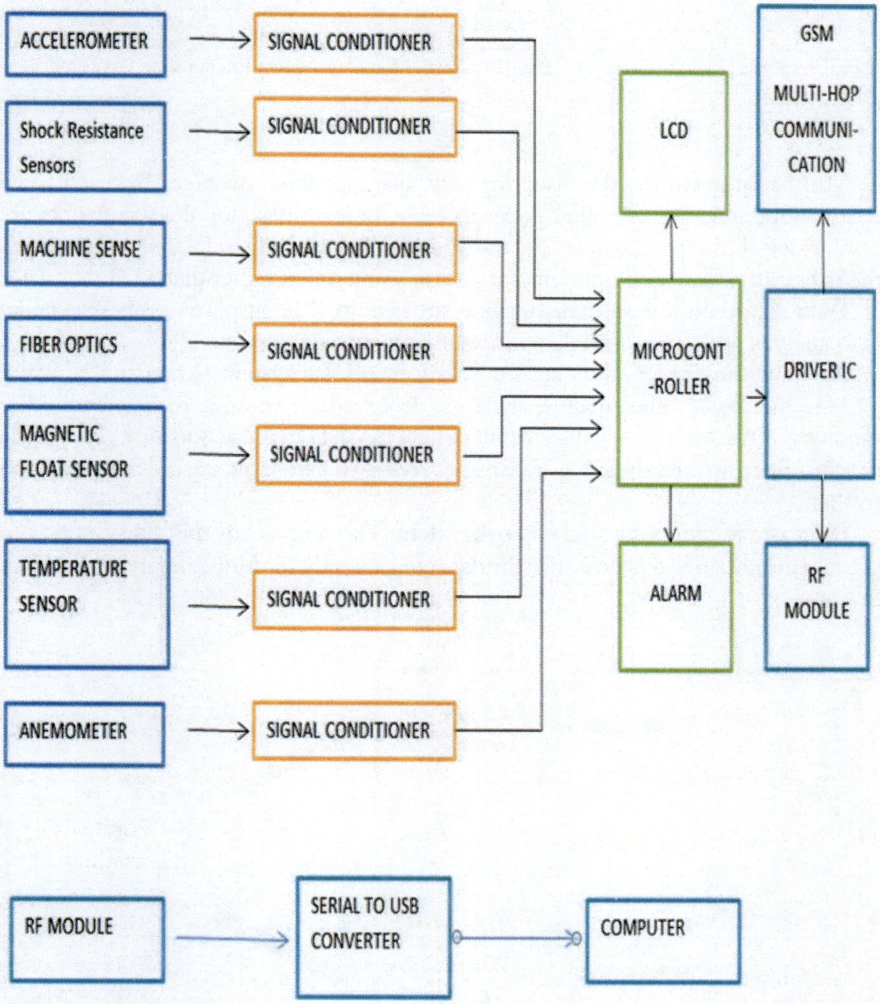

Fig. 2 Sensory subsystem

Table 1 Monitoring using sensors

Monitoring Item	Variables	Sensors
Loads and environmental actions	Vehicle Load Wind Load Earthquake ground motion Vessel collision Temperature and humidity Vibration Displacement	Weigh-in-motion Camera Ultrasonic anemometer Seismometer Accelerometer/seismometer Temp. and humidity sensor Accelerometer
Global response	Strain Bearing displacement	Optical fiber bragg grating strain sensors Magnetostrictive displacement sensors
Local response	Hanger rod/cable force	Fiber brag grating test-force rings

Multi-hop network: It is used for wide span network coverage. To reach some destination, node can use other nodes as relay. In this, mesh topology networks are used which speed the rear time data transmission from GSM model to PC; it manages the data traffic, and the important data is transmitted first rather than the other data.

Data Acquisition and transmission subsystem: The purpose of this subsystem is to analyze and to transmit data sensed by sensory subsystem.

As shown in Fig. 3., the data from the sensory subsystem is transmitted using STM32 microcontroller module, which is deployed for communicating with other modules of the sensory subsystem and collect the data in the acquisition unit before is transmitted wirelessly using transmitter so that it can be processed and analyzed further.

Data processing and analysis subsystem: The purpose of this subsystem is to process the obtained data to utilize the data conveniently for further analysis as shown in Fig. 4.

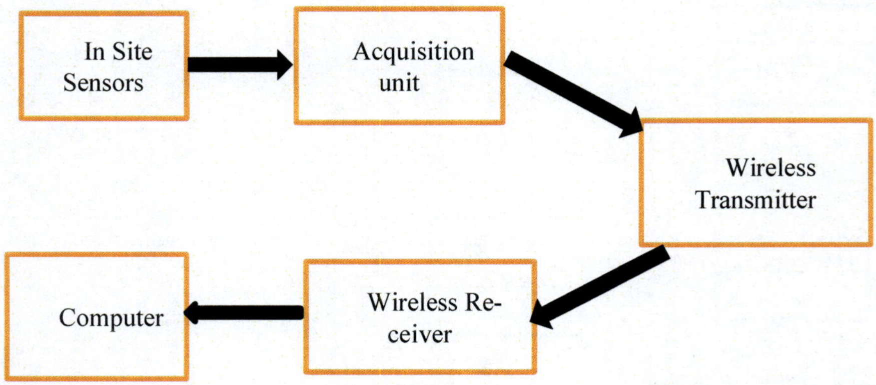

Fig. 3 Data acquisition and transmission system

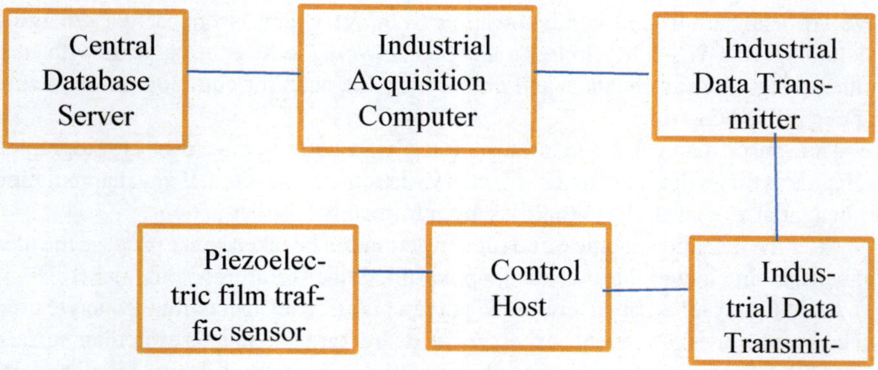

Fig. 4 Data processing and analysis subsystem

Cryptographic Security System: The sensor data can be manipulated easily so there is a need of cryptographic safeguards that will ensure that integrated violations are detected. If any manipulation is caught at any level, the authorities are informed for further actions.

Cryptographic safeguard: Using this, the encryption of sensor data is done for the security of data, and hence, it cannot be manipulated. Encryption ordinary information is converted into cypher text which is difficult to be manipulated at the center server the data is decrypted and used.

Data Management System: Data management subsystem: The purpose of this subsystem is to store and receive the analyzed data.

Structural Health Management System: We have used the SHM method for the monitoring of the bridges as it is effectively in use at the present for the structural monitoring of various infrastructures like bridges and buildings.

SHM is a process in which we can implement a damage detection and to characterize strategies for the study of structures such as bridges and buildings. For bridges, it includes monitoring of many factors such as (i) measuring the load on the bridge and (ii) effects of the loads, like traffic, wind and weather, height from the ground, stress and strain, thermal effects on bridge.

SHM DAMAGE DETECTION TECHNIQUE: Classification of damage detection in SHM systems is done in two types:

1. Local-based damage detection: Screening of structures is done in local-based damage detection.
2. Global-based damage detection: Vibrational characteristics are detected in global-based damage detection.

Various sensors like piezoelectric, optical fiber, ultrasonic, laser, image detection, and vibrational sensors are some of the sensors which are used to detect the location and the size of the damage in the SHM system.

Since the main issue with the SHM is real-time communication between sensor and the server; hence, the wireless sensor network (WSN) comes in the roleplay.

We are using multi-hop communication (routing) which is one of the emerging technologies in WSN. In wireless multi-hop network, nodes communicate with each other using wireless channels and do not have the need for common infrastructure or centralized control.

The sensor data can be manipulated easily so there is a need of cryptographic safeguards that will ensure that integrated violations are detected. If any manipulation is caught at any level, the authorities are informed for further actions.

After the detection of cracks, no sudden action can be taken so we propose the idea of self-healing material bio-cement, epoxy-filled fiber-reinforced polymer (FRP).

FRP: Epoxy-filled fiber-reinforced polymer is affected than normal concrete used in current bridges. It prevents excessive load structure and increases in their surface area. FRP composites are more durable materials which are different from the steel reinforcement for their resistance to electrochemical corrosion. FRP composites belong in anisotropic materials. The properties of anisotropic materials depend upon various factors like the matrix type, volume and alignment of fiber, construction quality.

SHM system has the main goal of providing fast services safety and sustainability.

6 Iot Platform for Shm

As shown in the figure, the proposed IOT platform consists of Wi-Fi module, Raspberry Pi, analog-to-digital converter, digital-to-analog converter, buffer, piezoelectric sensor.

There are two piezoelectric sensors that are rooted on the top of the structure and then are connected to a high-speed analog-to-digital converter. Sensors are deployed in such a way that they can catch all the possible damage. Now, analog-to-digital converter is connected to a buffer which is used for level conversion and to protect the Raspberry Pi. Then the excitation signal is generated by Raspberry Pi and sends over digital-to-analog converter to be converted into analog signal. Using the proposed SHM technique, Raspberry Pi is used for detecting any damage in the structure and the location of it. Raspberry Pi also keeps the record of the structural health of the bridge and sends it over the Internet server. The data stored on the Internet and can be monitored remotely from any mobile device. The Internet server, if any damage found sends an alert to the authority. As shown in Fig. 5, these are the various models of IOT platform for SHM:

1. Wi-Fi module: Miniature Wi-Fi (802.11b/g/n) module is a USB module that has 2.4 GHz ISM BAND. It has a data rate up to 150 Mbps. It uses IEEE 802.11n, IEEE 802.11 g, and IEEE 802.11b standards. The Wi-Fi module is used to send the data to the cloud.
2. Raspberry Pi 2: It is a single-board computer which features a full Linux operating system with diverse programming and connectivity options. The onboard

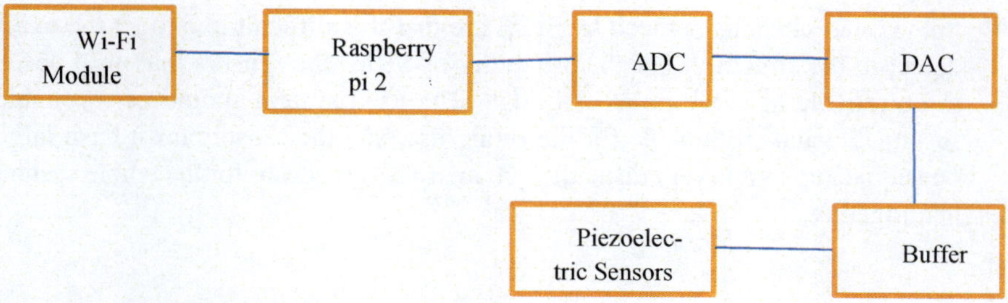

Fig. 5 IOT platform for SHM

900 MHz quad-core ARM cortex—A7 CPU is present. The Raspberry Pi is used to collect the structural health and send it to the cloud using Wi-Fi module.
3. Analog-to-digital converter (ADC): The CA 3306 is a CMOS parallel ADC designed for applications demanding both low power consumption and high-speed digitization. It is a 6-bit 15 MSPS ADC with single 5 V supply.
4. Digital-to-analog converter (DAC): The MCP 4725 is a low-power high accuracy, single channel, 12-bit buffered voltage output DAC with non-volatile memory.
5. Buffer: The 74HC4050 is a hex buffer with overvoltage tolerant inputs which is tolerant up to 15 V.
6. Piezoelectric Sensors: The PZT transducer converts mechanical energy into electrical signals or electrical to mechanical energy. When electric signals are applied to piezoelectric crystal, then it acts as an actuator which can excite an elastic lamb wave.

1. Decision and management system: After the detection of the damage in the bridge, the decision is made by the system; if it can be self-healed using FRP, it needs further help from the government. If it requires assessment, the concerned authority is notified about the damage and hence can be rectified.

7 Challenges

1. As the society progresses, the infrastructure along with technology enhancement is much important. Nowadays, a technology does not even last few weeks and a replacement of it in the market is there. Everyone wants things to be done at fast rate along with the best results.
2. Making SHM works efficiently and effectively increases the cost of the system which is not affordable for some countries with less economy as whole. So, to suggest a cost-efficient and effective idea is necessary through this project and we can make an effective SHM which already exists in the market but makes the system costly.
3. By reducing the cost of the system, we can help many countries in their development procedure.

4. Some other challenges faced by SHM are that it is difficult to extract the exact defect in the structure. For the reliability of SHM, the sensors that used needs to be reliable at every point of bridge. The used system should be a reliable sustainable and sustainable for the future use, and the sensors must have large power backup as it is very difficult to change a single sensor for the whole system in altogether.

8 Conclusion and Future Scope

This thesis presents an overview on SHM systems using IOT to measure the damage in the structures of the bridges so as to provide safety of the structure. The advantage of combining IOT TO SHM systems is that it allows systems with reduced battery consumption to continuously monitor the structure of existing as well as upcoming bridges.

In order to deploy all the advantages of IOT paradigm in the case of SHM enhancement of sensor virtualization, sensors scalability and security are required. There is a need of better investigation on how effectively we can achieve distributed real-time data collection and the synchronization of the sensors and all IOT layers.

In the future scope, the data produced by the system is expected to collect the data accurately and transmit, store, and analyze the data in a given time by the SHM system.

References

1. Abdelgawad A, Yelamarthi K. Internet of things (IoT) platform for structure health monitoring. https://www.hindawi.com/journals/wcmc/2017/6560797/,accessed date: 25th Jan 2020, Article ID 6560797
2. Bresson N, Balzac L, Bui N, Casari P, Vangelista L, Zorzi M (2010) The deployment of a smart monitoring system using wireless sensor and actuator networks, 2010 First IEEE international conference on smart grid communications, Gaithersburg, MD. pp 49–54
3. Fujino Y, Murata M, Okano S, Takeguchi M (2000). Monitoring system of the akashi kaikyo bridge and displacement measurement using GPS. Proceedings of SPIE—The international society for optical engineering, vol. 3995. p 10 1117/12.387814
4. Kekare1 A, Huddedar P, Bagde R (May–Jun 2014) Bridge health monitoring system, IOSR J Electron Commun Eng (IOSR-JECE), 9(3), Ver. IV: 08–14, ISSN: 2278-8735
5. Stephanie B (2017) Baker, lian el at, Ian atkinson, "Internet of things for smart healthcare: technologies challenges and opportunities" . Access IEEE 5:26521–26544
6. Sun Y (2014) Research on the railroad bridge monitoring platform based on the internet of things. Int J Control Autom 7(1):401–408
7. Tokognon CA, Gao B, Tian GY, Yan Y (June 2017) Structural health monitoring framework based on internet of things: a survey. IEEE Int Things J 4(3):619–635 ISSN: 2327-4662
8. Yang C-C, Hsu Y-L (2010) A review of accelerometry-based wearable motion detectors for physical activity monitoring. Sensors 10(8):7772–7788

Author Index

N. Marriwala et al. (eds.), *Mobile Radio Communications and 5G Networks*, Lecture Notes in Networks and Systems 140,
https://doi.org/10.1007/978-981-15-7130-5